The Mycota

Edited by
K. Esser

The Mycota

I   *Growth, Differentiation and Sexuality*
1st edition ed. by J.G.H.Wessels and F. Meinhardt
2nd edition ed. by U. Kües and R. Fischer

II   *Genetics and Biotechnology*
Ed. by U. Kück

III   *Biochemistry and Molecular Biology*
Ed. by R. Brambl and G. Marzluf

IV   *Environmental and Microbial Relationships*
1st edition ed. by D. Wicklow and B. Söderström
2nd edition ed. by C.P. Kubicek and I.S. Druzhinina

V   *Plant Relationships*
1st edition ed. by G. Carroll and P. Tudzynski
2nd edition ed. by H.B. Deising

VI   *Human and Animal Relationships*
1st edition ed. by D.H. Howard and J.D. Miller
2nd edition ed. by A.A. Brakhage and P.F. Zipfel

VII   *Systematics and Evolution*
Ed. by D.J. McLaughlin, E.G. McLaughlin, and P.A. Lemke[†]

VIII   *Biology of the Fungal Cell*
Ed. by R.J. Howard and N.A.R. Gow

IX   *Fungal Associations*
Ed. by B. Hock

X   *Industrial Applications*
1st edition ed. by H.D. Osiweacz
2nd edition ed. by M. Hofrichter and R. Ullrich

XI   *Agricultural Applications*
Ed. by F. Kempken

XII   *Human Fungal Pathogens*
Ed. by J.E. Domer and G.S. Kobayashi

XIII   *Fungal Genomics*
Ed. by A.J.P. Brown

XIV   *Evolution of Fungi and Fungal-like Organisms*
Ed. by J. Wöstemeyer

XV   *Physiology and Genetics: Selected Basic and Applied Aspects*
Ed. by T. Anke and D. Weber

The Mycota

A Comprehensive Treatise
on Fungi as Experimental Systems
for Basic and Applied Research

Edited by K. Esser

V

Plant Relationships
2nd Edition

Volume Editor:
H.B. Deising

 Springer

Series Editor

Professor Dr. Dr. h.c. mult. Karl Esser
Allgemeine Botanik
Ruhr-Universität
44780 Bochum, Germany

Tel.: +49 (234)32-22211
Fax.: +49 (234)32-14211
e-mail: Karl.Esser@rub.de

Volume Editor

Professor Dr. Holger B. Deising
Naturwissenschaftliche Fakultät III
Institut für Agrar- und Ernährungswissenschaften
Phytopathologie und Pflanzenschutz
Ludwig-Wucherer-Str. 2
06099 Halle (Saale), Germany

Tel.: +49 345 5522660
Fax: +49 345 5527120
e-mail: deising@landw.uni-halle.de

Library of Congress Control Number: 2008937452

ISBN 978-3-540-87406-5   e-ISBN 978-3-540-87407-2
ISBN 3-540-58006-9 (Part A)   ISBN 3-540-62018-4 (Part B) 1st ed.

springer.com
© Springer-Verlag Berlin Heidelberg 1997, 2009

Cover design: Erich Kirchner and WMXDesign GmbH, Heidelberg, Germany

Printed on acid-free paper   5 4 3 2 1 0

Karl Esser

(born 1924) is retired Professor of General Botany and Director of the Botanical Garden at the Ruhr-Universität Bochum (Germany). His scientific work focused on basic research in classical and molecular genetics in relation to practical application. His studies were carried out mostly on fungi. Together with his collaborators he was the first to detect plasmids in higher fungi. This has led to the integration of fungal genetics in biotechnology. His scientific work was distinguished by many national and international honors, especially three honorary doctoral degrees.

Holger B. Deising

(born 1956) studied agricultural sciences and botany at the University of Kiel, Germany. His PhD thesis focused on nitrate reduction by *Sphagnum* species. After graduating in 1987 he worked on the infection structures of plant-pathogenic rust fungi and he qualified as a lecturer at the University of Konstanz in 1996. In 1997 he became a Full Professor for Phytopathology and Plant Protection at Martin-Luther-University Halle-Wittenberg. His research stays include McMaster University (Hamilton, ON, Canada), the University of Georgia (Athens, GA, USA), and Purdue University (West Lafayetta, IN, USA). His scientific interest is directed to various aspects of fungus–plant interactions, with special focus on the differentiation and function of fungal infection structures and pathogenicity factors in the causal agent of maize anthracnose and stalk rot, *Colletotrichum graminicola*. Another area of research includes molecular mechanisms conferring fungicide resistance in several plant-pathogenic fungi.

Series Preface

Mycology, the study of fungi, originated as a subdiscipline of botany and was a descriptive discipline, largely neglected as an experimental science until the early years of this century. A seminal paper by Blakeslee in 1904 provided evidence for selfincompatibility, termed "heterothallism", and stimulated interest in studies related to the control of sexual reproduction in fungi by mating-type specificities. Soon to follow was the demonstration that sexually reproducing fungi exhibit Mendelian inheritance and that it was possible to conduct formal genetic analysis with fungi. The names Burgeff, Kniep and Lindegren are all associated with this early period of fungal genetics research.

These studies and the discovery of penicillin by Fleming, who shared a Nobel Prize in 1945, provided further impetus for experimental research with fungi. Thus began a period of interest in mutation induction and analysis of mutants for biochemical traits. Such fundamental research, conducted largely with Neurospora crassa, led to the one gene: one enzyme hypothesis and to a secondNobel Prize for fungal research awarded to Beadle and Tatum in 1958. Fundamental research in biochemical genetics was extended to other fungi, especially to Saccharomyces cerevisiae, and by the mid-1960s fungal systems were much favored for studies in eukaryotic molecular biology and were soon able to compete with bacterial systems in the molecular arena.

The experimental achievements in research on the genetics andmolecular biology of fungi have benefited more generally studies in the related fields of fungal biochemistry, plant pathology,medicalmycology, and systematics. Today, there ismuch interest in the geneticmanipulation of fungi for applied research. This current interest in biotechnical genetics has been augmented by the development of DNA-mediated transformation systems in fungi and by an understanding of gene expression and regulation at the molecular level. Applied research initiatives involving fungi extend broadly to areas of interest not only to industry but to agricultural and environmental sciences as well.

It is this burgeoning interest in fungi as experimental systems for applied as well as basic research that has prompted publication of this series of books under the title The Mycota. This title knowingly relegates fungi into a separate realm, distinct from that of either plants, animals, or protozoa. For consistency throughout this Series of Volumes the names adopted for major groups of fungi (representative genera in parentheses) are as follows:

Pseudomycota

Division:	Oomycota (Achlya, Phytophthora, Pythium)
Division:	Hyphochytriomycota

Eumycota

Division:	Chytridiomycota (Allomyces)
Division:	Zygomycota (Mucor, Phycomyces, Blakeslea)
Division:	Dikaryomycota

Subdivision:	Ascomycotina
	Class:	Saccharomycetes (Saccharomyces, Schizosaccharomyces)
	Class:	Ascomycetes (Neurospora, Podospora, Aspergillus)
Subdivision:	Basidiomycotina
	Class:	Heterobasidiomycetes (Ustilago, Tremella)
	Class:	Homobasidiomycetes (Schizophyllum, Coprinus)

We have made the decision to exclude from The Mycota the slime molds which, although they have traditional and strong ties to mycology, truly represent nonfungal forms insofar as they ingest nutrients by phagocytosis, lack a cell wall during the assimilative phase, and clearly show affinities with certain protozoan taxa.

The Series throughoutwill address three basic questions:what are the fungi,what do they do, andwhat is their relevance to human affairs? Such a focused and comprehensive treatment of the fungi is long overdue in the opinion of the editors.

A volume devoted to systematics would ordinarily have been the first to appear in this Series. However, the scope of such a volume, coupled with the need to give serious and sustained consideration to any reclassification of major fungal groups, has delayed early publication. We wish, however, to provide a preamble on the nature of fungi, to acquaint readers who are unfamiliar with fungi with certain characteristics that are representative of these organisms and which make them attractive subjects for experimentation.

The fungi represent a heterogeneous assemblage of eukaryotic microorganisms. Fungal metabolism is characteristically heterotrophic or assimilative for organic carbon and some nonelemental source of nitrogen. Fungal cells characteristically imbibe or absorb, rather thaningest,nutrients andtheyhave rigid cellwalls.The vastmajorityof fungi are haploid organisms reproducing either sexually or asexually through spores. The spore forms and details on theirmethod of production have been used to delineate most fungal taxa.Although there is amultitude of spore forms, fungal spores are basically only of two types: (i) asexual spores are formed followingmitosis (mitospores) and culminate vegetative growth, and (ii) sexual spores are formed following meiosis (meiospores) and are borne in or upon specialized generative structures, the latter frequently clustered in a fruit body. The vegetative forms of fungi are either unicellular, yeasts are an example, or hyphal; the latter may be branched to form an extensive mycelium.

Regardless of these details, it is the accessibility of spores, especially the direct recovery of meiospores coupled with extended vegetative haploidy, that have made fungi especially attractive as objects for experimental research.

The ability of fungi, especially the saprobic fungi, to absorb and grow on rather simple and defined substrates and to convert these substances, not only into essential metabolites but into important secondarymetabolites, is also noteworthy.Themetabolic capacities of fungi have attracted much interest in natural products chemistry and in the production of antibiotics and other bioactive compounds. Fungi, especially yeasts, are important in fermentation processes. Other fungi are important in the production of enzymes, citric acid and other organic compounds as well as in the fermentation of foods.

Fungi have invaded every conceivable ecological niche. Saprobic forms abound, especially in the decay of organic debris. Pathogenic forms exist with both plant and animal hosts. Fungi even grow on other fungi. They are found in aquatic as well as soil environments, and their spores may pollute the air. Some are edible; others are poisonous. Many are variously associated with plants as copartners in the formation of lichens and mycorrhizae, as symbiotic endophytes or as overt pathogens. Association with animal systems varies; examples include the predaceous fungi that trap nematodes, the microfungi that grow in the anaerobic environment of the rumen, the many insectassociated fungi and themedically important pathogens afflicting humans. Yes, fungi are ubiquitous and important.

There are many fungi, conservative estimates are in the order of 100,000 species, and there are many ways to study them, from descriptive accounts of organisms found in nature to laboratory experimentation at the cellular and molecular level. All such studies expand our knowledge of fungi and of fungal processes and improve our ability to utilize and to control fungi for the benefit of humankind.

We have invited leading research specialists in the field of mycology to contribute to this Series. We are especially indebted and grateful for the initiative and leadership shown by theVolumeEditors in selecting topics and assembling the experts.We have all been a bit ambitious in producing these Volumes on a timely basis and therein lies the possibility of mistakes and oversights in this first edition.We encourage the readership to draw our attention to any error, omission or inconsistency in this Series in order that improvements can be made in any subsequent edition.

Finally, we wish to acknowledge the willingness of Springer-Verlag to host this project, which is envisioned to require more than 5 years of effort and the publication of at least nine Volumes.

Bochum, Germany
Auburn, AL, USA
April 1994

KARL ESSER
PAUL A. LEMKE
Series Editors

Addendum to the Series Preface

In early 1989, encouraged by Dieter Czeschlik, Springer-Verlag, Paul A. Lemke and I began to plan The Mycota. The first volume was released in 1994, 12 volumes followed in the subsequent years, and two more volumes (Volumes XIV and XV) will be published within the next few years. Unfortunately, after a long and serious illness, Paul A. Lemke died in November 1995. Thus, it was my responsibility to proceed with the continuation of this series, which was supported by Joan W. Bennett for Volumes X–XII.

The series was evidently accepted by the scientific community, because several volumes are out of print. Therefore, Springer-Verlag has decided to publish completely revised and updated new editions of Volumes I, II, III, IV, V, VI, VIII, and X. I am glad that most of the volume editors and authors have agreed to join our project again. I would like to take this opportunity to thank Dieter Czeschlik, his colleague, Andrea Schlitzberger, and Springer-Verlag for their help in realizing this enterprise and for their excellent cooperation for many years

Bochum, Germany   KARL ESSER
May 2008

Volume Preface to the Second Edition

Joseph G.H. Wessels, in a review on fungal growth and morphogenesis, described fungi as the natural complement to plant life. He speculated that plants could probably have arisen without animals evolving, but raised doubts whether plants could have ever have evolved without the advent of fungi. The intimacy of trans-kingdom relationships between fungi and plants could not have been circumscribed any clearer.

The first edition of volume V *Plant Relationships* was been published in *The Mycota* series more than ten years ago. In their preface George Carroll and Paul Tudzynski ,the editors, commented on the large number of fungal and vascular plant species (respectively estimated to be in the order of 10^6 and $300-350\times10^3$) and emphasized the enormous number of interactions displayed by fungi and plants. The large number of fungi and plants reflects the complexity of the mechanisms of the interactions between them. Therefore, only examples can be given of fungal lifestyles and their interactions with plants, either mutualistic or pathogenic.

Significant methodological progress has been made in almost all areas of mycological research (e.g. transcriptomics, proteomics, metabolomics) since the first edition was published. For example, molecular genetics has experienced strong support from various genome sequencing efforts. To date, more than 50 fungal and several oomycete genomes have been sequenced and genome-wide gene expression profiling, functional screens for genes in yeasts or bacteria, the labelling of gene products, and the efficiency of targeted inactivation of genes have improved. These advances were accompanied by improvements in light and electron microscopy which, together with the utilization of molecular tools, allowed a proportional development in our knowledge of the cell biology of fungal interactions. Last but not least, increased sensitivies in the detection of biomolecules through analytical chemistry enabled our understanding of the chemical basis of both mutualistic and pathogenic fungus–plant interactions.

The second edition of *The Mycota*, volume V, reflects the substantial progress made in various areas of fungus–plant interactions. Organized in three parts, i.e. profiles in pathogenesis and mutualism, mechanisms of pathogenic and mutualistic interactions, and plant response to pathogen ingress, this book provides an overview of fundamental aspects of fungal lifestyles.

Chapters 1–5 focus on different fungal systems, characterize their profiles, and give examples both for pathogenic *Phytophthora* species belonging to the Oomycota and for fungi with biotrophic, necrotrophic, or mutualistic lifestyles.

Chapters 6–16 focus on mechanisms of the interactions, with detailed discussions on specific aspects, such as spore release and distribution (which are of critical importance to the success of pathogens), the basis of the specificity of fungus–plant interactions, and signal transduction. Protein secretion, the role played by cell wall-degrading enzymes and toxins, and the induction of programmed cell death represent further areas of research that significantly helped our understanding fungal pathogenicity.

The last four chapters of this section, 13–16, describe in detail mechanisms of mutualism, i.e. mycorrhizal and endophytic interactions, as well as interactions between fungi and algae in lichens. Finally, chapters 17 and 18 describe the response of plants towards attacking pathogenic fungi and focus on signal perception and transduction and the mechanisms of defense.

Forty-six internationally acknowledged scientists specialized in different areas of eukaryotic microbiology (mycology), microbial genetics and genomics, plant pathology, plant molecular biology, and plant genetics contributed to this volume. I would like to express my gratitude to all authors for their tremendous efforts, to the series editor, Karl Esser, who provided many helpful comments, and to Springer-Verlag for continuous assistance and patience.

I hope that the second edition of *Plant Relationships*, volume V in *The Mycota* series, will be appreciated by a wide variety of professional biologists, as it allows one to keep pace with the rapidly developing field of fungal research. In addition, the text and high-quality illustrations may provide a source for teaching at graduate level; and the different chapters can be used by young researchers as a helpful introduction to the relevant literature on fungus–plant interactions.

Halle (Saale), Germany   HOLGER B. DEISING
October 2008   *Volume Editor*

Volume Preface to the First Edition

The number of fungal species has been loosely estimated to be on the order of 1 million, while the number of vascular plants is known with considerably greater certainty to lie between 300000 and 350000. Clearly, any volume which purports to deal with interactions between these two vast assemblages of organisms must do so concisely and selectively. In the chapters to follow, we have made no attempt to be allinclusive, but rather have chosen examples from which general conclusions about fungus/plant interactions might be drawn. The materials presented here come from the core literature on plant pathology and from research on fungal mutualisms and on evolutionary biology. A variety of approaches are evident: biochemistry, molecular biology, cellular fine structure, genetics, epidemiology, population biology, ecology, and computer modeling. The frequent overlap of such approach es within single reviews has resulted in a rich array of insights into the factors which regulate fungus/plant interactions. In these chapters, such interactions have also been considered on a variety of scales, both geographic and temporal, from single plant cells to ecosystems, from interactions which occur within minutes of contact to mechanisms which have presumably evolved during the course of several hundred million years.

Volume V consists of two parts: Volume V, Part A, and Volume V, Part B. While section headings provide signposts, we wish to make the rationale for the organization of these volumes absolutely dear. Part A begins with a brief introduction to both volumes. A series of reviews follows (Chaps. 1-6) which deal with the temporal sequence of events from the time fungal spores make contact with a host plant until the point where fungal hyphae are either firmly ensconced within a host or the attempted infections have been repulsed. Chapters 7-12 deal with metabolic interactions between host and fungus within the host plant after infection and particularly with the roles played by low molecular weight fungal metabolites such as toxins and phytohormones in pathogenic as well as mutualistic associations.

Chapters 1-8 of Part B are grouped in a section labeled, "Profiles in Pathogenesis and Mutualism"; here, interactions between fungi and host plants are explored in a variety of important model systems. These reviews focus less on processes per se and more on the specific fungi or groups of fungi as examples of pathogens or mutualists on plants. Chapters 9-12 of Part B move from discussions of physiological interactions between individuals to considerations of interactions at an expanded geographic scale, within populations of plants. Here, Chapter 9 provides a treatment of dassical plant epidemiology, while Chapter 11 provides the same focus for mutualistic mycorrhizal associations. Chapter 10 covers the fuzzy area between population biology and microevolution in a genus of ubiquitous and pleurivorous pathogens; Chapter 12 ofters much the same approach for mutualistic endophytes of grasses.

Chapters 13-16 of Part B ofter a view of an expanded temporal scale and consider the evolution of plant/fungus interactions. Chapter 13 considers the flexibility of the fungal genome, the ultimate substrate on which evolutionary forces must act. Chapter 14 discusses the evolutionary relationships between pathogenic and mutualistic fungi in one situation which has been particularly well worked out, the clavicipitaceous endo-phytes of grasses. Chapter 15 considers the evolutionary interplay between fungi and plants as illuminated through the use of mathematical and computer-driven models. The final chapter in the volume (Chap. 16) deals with overall evolution of fungal parasitism and plant resistance and provides an appropriate coda for this series of essays.

Who is the audience for these volumes? Who might and will read them with profit? Basic literacy in mycology, in particular, and in modern biology, in general, has been assumed as a background for these chapters, and they clearly are not intended for the biological novice. However, we do expect that these volumes will be appreciated by a wide variety of professional biologists including, for example: teachers of upper division courses in general mycology engaged in the valiant (but often futile) attempt to keep their lectures up-to-date; graduate students contemplating literature reviews in connection with a thesis project; nonmycologists who wish to know what the fungi might have to offer in the way of model systems for the study of some fundamental aspect of host/parasite interactions; evolutionary biologists who have just become aware that fungi offer advantages in studying the evolutionary consequences of asexual reproduction. These, and many others, will read these chapters with pleasure. On the whole we are very pleased with the contributions presented here and believe they will prove informative and useful as entrees into the literature on fungus/plant interactions for some years to come.

Eugene, Oregon, USA   GEORGE CARROLL
Münster, Germany   PAUL TUDZYNSKI
March 1997   *Volume Editors*

Contents

Profiles in Pathogenesis and Mutualism

1 Cellular and Molecular Biology of *Phytophthora*–Plant Interactions   3
ADRIENNE R. HARDHAM, WEIXING SHAN

2 *Botrytis cinerea*: Molecular Aspects of a Necrotrophic Life Style   29
PAUL TUDZYNSKI, LEONIE KOKKELINK

3 Profiles in Pathogenesis and Mutualism: Powdery Mildews.   51
CHRISTOPHER JAMES RIDOUT

4 The Uredinales: Cytology, Biochemistry, and Molecular Biology   69
RALF T. VOEGELE, MATTHIAS HAHN, KURT MENDGEN

5 The Sebacinoid Fungus *Piriformospora indica*:
an Orchid Mycorrhiza Which May Increase Host
Plant Reproduction and Fitness   99
PATRICK SCHÄFER, KARL-HEINZ KOGEL

Mechanisms of Pathogenic and Mutualistic Interactions

6 Biomechanics of Spore Release in Phytopathogens.   115
NICHOLAS P. MONEY, MARK W.F. FISCHER

7 Gene for Gene Models and Beyond: the *Cladosporium
fulvum*–Tomato Pathosystem   135
PIERRE J.G.M. DE WIT, MATTHIEU H.A.J. JOOSTEN,
BART H.P.J. THOMMA, IOANNIS STERGIOPOULOS

8 The cAMP Signaling and MAP Kinase Pathways
in Plant Pathogenic Fungi   157
RAHIM MEHRABI, XINHUA ZHAO, YANGSEON KIM, JIN-RONG XU

9 The Secretome of Plant-Associated Fungi and Oomycetes.   173
SOPHIEN KAMOUN

10 From Tools of Survival to Weapons of Destruction:
The Role of Cell Wall-Degrading Enzymes in Plant Infection.   181
ANTONIO DI PIETRO, Mª ISABEL GONZÁLEZ RONCERO
Mª CARMEN RUIZ ROLDÁN

11 Photoactivated Perylenequinone Toxins in Plant Pathogenesis. 201
MARGARET E. DAUB, KUANG-REN CHUNG

12 Programmed Cell Death in Fungus–Plant Interactions 221
AMIR SHARON, ALIN FINKELSHTEIN

13 The Ectomycorrhizal Symbiosis: a Marriage of Convenience 237
FRANCIS MARTIN, ANDERS TUNLID

14 Establishment and Functioning of Arbuscular Mycorrhizas 259
PAOLA BONFANTE, RAFFAELLA BALESTRINI, ANDREA GENRE, LUISA LANFRANCO

15 Epichloë Endophytes: Clavicipitaceous Symbionts of Grasses. 275
CHRISTOPHER L. SCHARDL, BARRY SCOTT, SIMONA FLOREA, DONGXIU ZHANG

16 Lichen-Forming Fungi and Their Photobionts. 307
ROSMARIE HONEGGER

Plant Response

17 Signal Perception and Transduction in Plants 337
WOLFGANG KNOGGE, JUSTIN LEE, SABINE ROSAHL, DIERK SCHEEL

18 Defence Responses in Plants 363
CHIARA CONSONNI, MATT HUMPHRY, RALPH PANSTRUGA

Biosystematic Index 387

Subject Index 393

List of Contributors

RAFFAELLA BALESTRINI
(e-mail: r.balestrini@ipp.cnr.it)
Dipartimento di Biologia Vegetale, Università di Torino and Istituto Protezione Piante-CNR, Viale Mattioli 25, 10125 Torino, Italy

PAOLA BONFANTE
(e-mail: p.bonfante@ipp.cnr.it)
Dipartimento di Biologia Vegetale, Università di Torino and Istituto Protezione Piante-CNR, Viale Mattioli 25, 10125 Torino, Italy

KUANG-REN CHUNG
(e-mail: krchung@ufl.edu)
Research and Education Center, University of Florida, Lake Alfred, FL 33850, USA

CIARA CONSONNI
(Tel.: +49 221 5062 316, Fax: +49 221 5062 353)
Max-Planck Institute for Plant Breeding Research, Department of Plant–Microbe Interactions, Carl-von-Linné-Weg 10, 50829 Köln, Germany

MARGARET E. DAUB
(e-mail: margaret_daub@ncsu.edu, Tel.: +1 919 513 3807, Fax: +1 919 515 3436)
Department of Plant Biology, North Carolina State University, Raleigh, NC 27695-7612, USA

PIERRE J.G.M. DE WIT
(e-mail: pierre.dewit@wur.nl, Tel.: +31 317 483130/483410, Fax: +31 317 483412)
Laboratory of Phytopathology, Wageningen University, and Wageningen Centre for Biosystems Genomics, Binnenhaven 5, 6709 PD Wageningen, The Netherlands

ANTONIO DI PIETRO
(e-mail: ge2dipia@uco.es, Tel.: +34 957 218981, Fax: +34 957 212072)
Departamento de Genética, Universidad de Córdoba, Campus de Rabanales, Edificio Gregor Mendel, 14071 Córdoba, Spain

ALIN FINKELSHTEIN
(e-mail: alinf@ex.tau.ac.il)
Department of Plant Sciences, Tel Aviv University, Tel Aviv 69978, Israel

MARK W.F. FISCHER
Department of Chemistry and Physical Science, College of Mount St Joseph, Cincinnati, OH 45233, USA

SIMONA FLOREA
(e-mail: Simona@uky.edu, Tel.: +1 859 257 7445, Fax: +1 859 323 1961)
Department of Plant Pathology, 201F Plant Science Building, 1405 Veterans Drive, Lexington, KY 40546-0312, USA

ANDREA GENRE
(e-mail: andrea.genre@unito.it)
Dipartimento di Biologia Vegetale, Università di Torino and Istituto Protezione
Piante-CNR, Viale Mattioli 25, 10125 Torino, Italy

MA ISABEL GONZÁLEZ RONCERO
(Tel.: +34 957 218981, Fax: +34 957 212072)
Departamento de Genética, Universidad de Córdoba, Campus de Rabanales,
Edificio Gregor Mendel, 14071 Córdoba, Spain

MATTHIAS HAHN
Phytopathologie, Fachbereich Biologie, Technische Universität Kaiserslautern,
67663 Kaiserslautern, Germany

ADRIENNE R. HARDHAM
(e-mail: Adrienne.Hardham@anu.edu.au, Tel.: +61 2 6125 4168, Fax: +61 2 6125 4331)
Plant Cell Biology Group, Research School of Biological Sciences, The Australian
National University, Canberra, ACT 2601, Australia

ROSMARIE HONEGGER
(e-mail: rohonegg@botinst.unizh.ch, Tel.: +41 1 634 82 43, Fax: +41 1 634 82 04)
Institute of Plant Biology, University of Zürich, Zollikerstrasse 107, 8008 Zürich,
Switzerland

MATT HUMPHRY
(Tel.: +49 221 5062 316, Fax: +49 221 5062 353)
Max-Planck Institute for Plant Breeding Research, Department of Plant–Microbe
Interactions, Carl-von-Linné-Weg 10, 50829 Köln, Germany

MATTHIEU H.A.J. JOOSTEN
(Tel.: +31 317 483130/483410, Fax: +31 317 483412)
Laboratory of Phytopathology, Wageningen University, and Wageningen Centre
for Biosystems Genomics, Binnenhaven 5, 6709 PD Wageningen, The Netherlands

SOPHIEN KAMOUN
(e-mail: sophien.kamoun@tsl.ac.uk, Tel.: +44 1603 450410)
The Sainsbury Laboratory, Colney Lane, Norwich, NR1 3LY, United Kingdom

YANGSEON KIM
(Tel.: +1 765 496 6918, Fax: +1 765 496 6918)
Department of Botany and Plant Pathology, Purdue University, West Lafayette,
IN 47907, USA

WOLFGANG KNOGGE
(e-mail: wknogge@ipb-halle.de)
Leibniz Institute of Plant Biochemistry, Department of Stress and Developmental
Biology, Weinberg 3, 06120 Halle (Saale), Germany

KARL-HEINZ KOGEL
(e-mail: karl-heinz.kogel@agrar.uni-giessen.de, Tel.: +49 641 9937491,
Fax: +49 641 9937499)
Interdisciplinary Research Centre for BioSystems, Land Use and Nutrition, Institute
of Phytopathology and Applied Zoology, Justus Liebig University, Heinrich-Buff-Ring
26–32, 35392 Giessen, Germany

LEONIE KOKKELINK
(Tel.: +49 251 8324998, Fax: +49 251 8321601)
Institut für Botanik, Westf. Wilhelms-Universität, Schlossgarten 3, 48149 Münster,
Germany

LUISA LANFRANCO
(e-mail: luisa.lanfranco@unito.it)
Dipartimento di Biologia Vegetale, Università di Torino and Istituto Protezione
Piante-CNR, Viale Mattioli 25, 10125 Torino, Italy

JUSTIN LEE
(e-mail: jlee@ipb-halle.de)
Leibniz Institute of Plant Biochemistry, Department of Stress and Developmental
Biology, Weinberg 3, 06120 Halle (Saale), Germany

FRANCIS MARTIN
(e-mail: fmartin@nancy.inra.fr, Tel.: +33 383 39 40 80, Fax: +33 383 39 40 69)
UMR 1136, INRA-Nancy Université, Interactions Arbres/Microorganismes,
INRA-Nancy, 54280 Champenoux, France

KURT MENDGEN
(Tel.: +49 7531 88 4305, Fax: +49 7531 88 3035)
Lehrstuhl Phytopathologie, Fachbereich Biologie, Universität Konstanz,
78457 Konstanz, Germany

RAHIM MEHRABI
(Tel.: +1 765 496 6918, Fax: +1 765 496 6918)
Department of Botany and Plant Pathology, Purdue University, West Lafayette,
IN 47907, USA

NICHOLAS P. MONEY
(e-mail: moneynp@muohio.edu, Tel.: +1 513 529 2140)
Department of Botany, Miami University, Oxford, OH 45056, USA

RALPH PANSTRUGA
(e-mail: panstrug@mpiz-koeln.mpg.de, Tel.: +49 221 5062 316, Fax: +49 221 5062 353)
Max-Planck Institute for Plant Breeding Research, Department of Plant–Microbe
Interactions, Carl-von-Linné-Weg 10, 50829 Köln, Germany

CHRISTOPHER J. RIDOUT
(e-mail: ridout@bbsrc.ac.uk)
John Innes Centre, Norwich Research Park, Norwich, NR4 7UH, United Kingdom

SABINE ROSAHL
(e-mail: srosahl@ipb-halle.de)
Leibniz Institute of Plant Biochemistry, Department of Stress and Developmental
Biology, Weinberg 3, 06120 Halle (Saale), Germany

Ma CARMEN RUIZ ROLDÁN
(Tel.: +34 957 218981, Fax: +34 957 212072)
Departamento de Genética, Universidad de Córdoba, Campus de Rabanales, Edificio
Gregor Mendel, 14071 Córdoba, Spain

PATRICK SCHÄFER
(e-mail: patrick.schaefer@agrar.uni-giessen.de, Tel.: +49 641 9937494,
Fax: +49 641 9937499)
Interdisciplinary Research Centre for BioSystems, Land Use and Nutrition, Institute
of Phytopathology and Applied Zoology, Justus Liebig University, Heinrich-Buff-Ring
26–32, 35392 Giessen, Germany

CHRISTOPHER L. SCHARDL
(e-mail: schardl@uky.edu, Tel.: +1 859 257 7445, Fax: +1 859 323 1961)
Department of Plant Pathology, 201F Plant Science Building, 1405 Veterans Drive,
Lexington, KY 40546-0312, USA

DIERK SCHEEL
(e-mail: dscheel@ipb-halle.de)
Leibniz Institute of Plant Biochemistry, Department of Stress and Developmental
Biology, Weinberg 3, 06120 Halle (Saale), Germany

BARRY SCOTT
(e-mail: D.B.Scott@massey.ac.nz)
Institute of Molecular BioSciences, Massey University, Palmerston North 5321,
New Zealand

WEIXING SHAN
Plant Cell Biology Group, Research School of Biological Sciences, The Australian
National University, Canberra, ACT 2601 Australia; *current address*: Shaanxi Key
Laboratory for Molecular Biology of Agriculture and Department of Plant Pathology,
Northwest A&F University, Yangling, Shaanxi 712100, P.R. China

AMIR SHARON
(e-mail: amirsh@ex.tau.ac.il)
Department of Plant Sciences, Tel Aviv University, Tel Aviv 69978, Israel

IOANNIS STERGIOPOULOS
(Tel.: +31 317 483130/483410, Fax: +31 317 483412)
Laboratory of Phytopathology, Wageningen University, Binnenhaven 5, 6709 PD
Wageningen,
The Netherlands

BART H.P.J. THOMMA
(Tel.: +31 317 483130/483410, Fax: +31 317 483412)
Laboratory of Phytopathology, Wageningen University, and Wageningen Centre
for Biosystems Genomics, Binnenhaven 5, 6709 PD Wageningen, The Netherlands

PAUL TUDZYNSKI
(e-mail: tudzyns@uni-muenster.de, Tel.: +49 251 8324998, Fax: +49 251 8321601)
Institut für Botanik, Westf. Wilhelms-Universität, Schlossgarten 3,
48149 Münster, Germany

ANDERS TUNLID
Department of Microbial Ecology, Ecology Building, Lund University,
SE 223 62 Lund, Sweden

RALF T. VOEGELE
(e-mail: Ralf.Voegele@uni-konstanz.de, Tel.: +49 7531 88 4305, Fax: +49 7531 88 3035)
Lehrstuhl Phytopathologie, Fachbereich Biologie, Universität Konstanz, 78457
Konstanz, Germany

JIN-RONG XU
(e-mail: jinrong@purdue.edu, Tel.: +1 765 496 6918, Fax: +1 765 496 6918)
Department of Botany and Plant Pathology, Purdue University, West Lafayette,
IN 47907, USA

DONGXIU ZHANG
(e-mail: Dongxiu Zhang@email.uky.edu, Tel.: +1 859 257 7445, Fax: +1 859 323 1961)
Department of Plant Pathology, 201F Plant Science Building, 1405 Veterans Drive,
Lexington, KY 40546-0312, USA

XINHUA ZHAO
(Tel.: +1 765 496 6918, Fax: +1 765 496 6918)
Department of Botany and Plant Pathology, Purdue University, West Lafayette,
IN 47907, USA

Profiles in Pathogenesis and Mutualism

1 Cellular and Molecular Biology of *Phytophthora*–Plant Interactions

ADRIENNE R. HARDHAM[1], WEIXING SHAN[1,2]

CONTENTS

I. Introduction 3
II. Establishing Plant Infection 4
   A. Targeting Preferred Infection Sites 4
   B. Attaching to the Plant Surface 6
   C. Penetration of the Host Surface 7
   D. Nutrient Acquisition to Support
       Pathogen Growth and Reproduction...... 11
III. *Phytophthora* Effectors and Elicitors
   of the Plant Defence Response 12
   A. Extracellular *Phytophthora* Effectors...... 12
     1. Inhibitors of Plant Enzymes......... 12
     2. *Phytophthora* Elicitins 13
     3. Cellulose Binding Proteins.......... 14
   B. *Phytophthora* Effectors Translocated
       into the Host Cytoplasm 14
   C. Secreted elicitors of the plant
       defence response 15
     1. The Gp42 Cell Wall Transglutaminase... 15
     2. Necrosis and Ethylene Inducing
         Proteins 15
IV. The Plant's Response to Potential Pathogens ... 15
   A. The Products of Plant Resistance Genes
       Recognize *Phytophthora* Elicitors.......... 15
   B. Subcellular Reorganization Within
       Plant Cells During the Defence Response... 16
     1. Strengthening the Cell Wall Barrier
         to Prevent Pathogen Penetration...... 16
     2. Cytoskeletal Reorganization Plays
         a Key Role in Facilitating
         Penetration Resistance 16
     3. Actin-Based Polarized Transport
         and Secretion at the Infection Site..... 18
V. Counterdefence: Suppressing Plant Defences... 18
VI. Conclusions 19
   References................................ 20

I. Introduction

There are over 60 species of *Phytophthora* and many are aggressive plant pathogens that cause extensive losses in agricultural crops, horticulture and natural ecosystems (Erwin and Ribeiro 1996). Some species, such as *P. infestans* the causal agent of late blight of potato and *P. sojae* the cause of soybean root rot, have a limited host range. Others, such as *P. cinnamomi* and *P. nicotianae*, have extremely broad host ranges, with both of these pathogens infecting over 1000 different plant species (Erwin and Ribeiro 1996; Hardham 2005). The genus *Phytophthora* belongs to the class Oomycetes and is now grouped with a variety of other protists within the Stramenopile cluster (Adl et al. 2005; Harper et al. 2005; Yoon et al. 2002). The Stramenopiles also include the coloured algae, the diatoms and the apicomplexans (i.e. malarial parasites). One of the distinguishing structural characteristics of organisms classified within the Stramenopiles is possession of flagella adorned with tubular tripartite hairs called mastigonemes (Barr 1992; Patterson and Sogin 1992). Modern molecular analyses of gene sequences have strengthened evidence of the close phylogenetic relationships between these different groups of Stramenopile organisms (Gunderson et al. 1987; Van de Peer et al. 1996) and have provided a basis for informative comparative studies of infection strategies.

Species of *Phytophthora* produce biflagellate, asexual spores called zoospores and, in most cases, these motile zoospores are instrumental in initiating plant infection. The zoospores are formed within a multinucleate cell called a sporangium that subsequently cleaves to form and release the uninucleate zoospores (Hardham and Hyde 1997). *Phytophthora* sporangium superficially resemble fungal conidia and during vegetative growth *Phytophthora* species form hyphae whose appearance and life style are also similar to those of fungi. These similarities in morphology and mode of nutrient acquisition between *Phytophthora* and fungi are accompanied by similarities in aspects of their infection strategies (Hardham 2007). In both cases, the development of disease requires the pathogen to establish initial contact with a

[1] Plant Cell Biology Group, Research School of Biological Sciences, The Australian National University, Canberra, ACT 2601, Australia; e-mail: Adrienne.Hardham@anu.edu.au
[2] *Current address*: Shaanxi Key Laboratory for Molecular Biology of Agriculture and Department of Plant Pathology, Northwest A&F University, Yangling, Shaanxi 712100, P.R. China

Plant Relationships, 2nd Edition
The Mycota V
H. Deising (Ed.)
© Springer-Verlag Berlin Heidelberg 2009

potential host, to attach onto and then to penetrate the host surface and to obtain nutrients from the plant in order to grow and reproduce. Most *Phytophthora* species are necrotrophs that obtain the nutrients they need from dead or dying plant cells. However, while none are true biotrophs, a number of species are hemibiotrophs that initially establish a biotrophic relationship with their host plant before turning to a necrotrophic life style.

The establishment of infection through the activity of a motile *Phytophthora* spore contrasts with the situation in most fungi, however, there are clear parallels in the mechanisms employed by *Phytophthora* and fungi, in plant penetration and colonization. Both groups of organisms secrete effector molecules required for pathogenicity, including cell wall degrading enzymes and proteins that are transported into the plant cell cytoplasm (see Chap. 9). In susceptible hosts, these effectors successfully orchestrate colonization of the plant. In resistant hosts, on the other hand, these effectors or other elicitors trigger plant defence and there are strong similarities in the plant's response to attempted invasion by *Phytophthora* or fungi. In this chapter, we explore current understanding of the cellular and molecular basis of the interactions between plants and their *Phytophthora* pathogens, focusing on key aspects of *Phytophthora* pathogenicity, plant recognition of *Phytophthora* invasion and plant defence responses.

II. Establishing Plant Infection

A. Targeting Preferred Infection Sites

Species of *Phytophthora* may arrive at and initiate infection of potential host plants as hyphae, sporangia or zoospores (Fig. 1.1). These three cell types differ in the distances they may travel before reaching a host. Dissemination through hyphal growth restricts the spread of disease to the vicinity of a pre-existing infection site. In contrast, production of caducous sporangia that detach from the mycelial mass may facilitate pathogen dispersal over large distances if the sporangia are blown in the wind, as is believed to have occurred for *P. infestans* during the spread of the late blight disease through Europe in the 1840s (Aylor 2003; Erwin and Ribeiro 1996). Not all species of *Phytophthora*, however, produce caducous sporangia. Sporangia may germinate either directly through production of hyphae or indirectly though cleavage of their multinucleate cytoplasm and subsequent release

of uninucleate motile zoospores. Zoospores swim at speeds of up to about $200\,\mu m/s$ and can cover distances of several centimetres. The movement of zoospores may allow infection at nearby sites as occurs, for example, when zoospores of *P. infestans* are released from sporangia that have landed on the surface of a leaf. Zoospores may also allow the pathogen to spread over much greater distances if they get into water that is moving through the environment. The zoospores of soil-borne *Phytophthora* species, for example, may be carried downhill in streams or water-logged soils.

Phytophthora zoospores are able to swim through the action of two flagella that emerge from the centre of a groove along the ventral surface of the spore (Fig. 1.1A). The zoospore flagella are typical eukaryotic flagella based on a microtubular axoneme that consists of nine microtubule doublets surrounding a central pair of microtubules (Hardham 1987a). The axonemal microtubules are connected by protein complexes that form radial spokes and a variety of other linkages; flagellar function is achieved by the sliding of adjacent microtubule doublets relative to their neighbours, a process powered by the mechanochemical protein, dynein (Silflow and Lefebvre 2001).

Being able to swim enhances the chance that the zoospores initiate disease because they are chemotaxis and electrotactically attracted to potential infection sites on the surface of host plants (Gow 2004; Tyler 2002). In general, these tactic responses appear to be non-specific in that the zoospores move towards both host and non-host plants, being attracted by gradients of compounds such as sugars and amino acids diffusing from the plant surface (Carlile 1983). Specific recognition of a chemoattractant produced by a host plant is, however, known to occur (Tyler et al. 1996). Two isoflavones secreted by soybean roots attract zoospores of the soybean pathogen, *P. sojae*, but not zoospores of several *Phytophthora* species that do not cause disease on soybean (Morris and Ward 1992). Specificity of attraction in terms of targeting zoospore movement to particular locations on the plant is also known. For instance, zoospores of a number of soil-borne species swim to the root elongation zone rather than the root cap or root hair regions (Carlile 1983; Van West et al. 2002). Zoospores also swim towards wounds and may show auto-aggregation phenomena. There is evidence that targeting to different regions of a root surface may involve electrotaxis, a process in which the cells are able to detect and swim towards anodic or cathodic regions of the

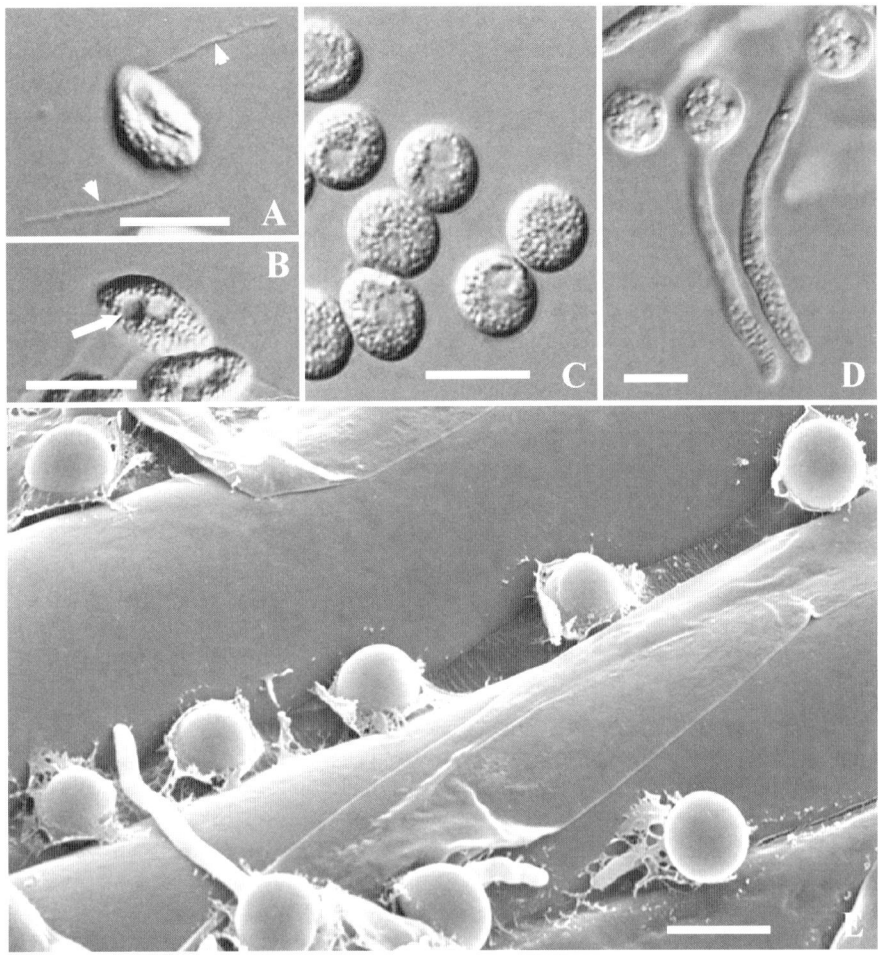

Fig. 1.1. *Phytophthora* zoospores and cysts. **A** *P. nicotianae* zoospore showing emergence of the two flagella (*arrowheads*) from the centre of the ventral groove. **B** *P. nicotianae* zoospore showing the water expulsion vacuole (*arrow*) at the anterior end of the cell. **C** *P. nicotianae* cysts. **D** *P. nico-tianae* cysts 2h after germination. **E** Scanning electron micrograph of *P. nicotianae* cysts that have targeted and settled in the grooves between the epidermal cells of a tobacco (*Nicotiana tobacum*) root. Material secreted by the spores coats the cyst and nearby plant surface. *Bars* 10 µm

root (Van West et al. 2002). At an even finer spatial scale, zoospores of foliar pathogens may be preferentially attracted to stomata and zoospores of root pathogens may target the grooves between adjacent epidermal cells (Fig. 1.1E; Gees and Hohl 1988; Hardham 2001, 2005; Judelson and Blanco 2005).

Like the zoospores of other Stramenopiles, *Phytophthora* zoospores are said to have heterokont flagella because the two flagella have different morphologies. The anteriorly directed flagellum is shorter than the posterior flagellum and possesses two rows of tubular hairs called mastigonemes about 1 µm in length (Hardham 1987a). Observations of zoospore motility suggest that the anterior flagellum pulls the cell forward while the posterior flagellum acts like a rudder, occasionally bending to change the swimming direction. Both flagella form quasi-sinusoidal waves that emanate from the base of the flagella and propagate to their tip. This form of beating of the anterior flagellum would normally propel the cell backwards but the two rows of rigid mastigonemes reverse the thrust of flagellar beat, causing the zoospore to be pulled forwards (Cahill et al. 1996; Jahn et al. 1964). Until recently, there has been little information on the nature of the components that make up the mastigonemes of *Phytophthora* or other stramenopile species. The first advances arose from immunocytochemical studies using monoclonal antibodies directed towards zoospore surface molecules that revealed that the shaft of *P. nicotianae* mastigonemes is made of a 40-kDa glycoprotein (Robold and Hardham 1998). Amino acid sequence data have now been obtained following immunoprecipitation purification of the mastigoneme protein and these data used to clone the corresponding gene (M. Arikawa, T. Suzaki, L.M. Blackman and A.R. Hardham, unpublished data). The results indicate that the *Phytophthora* mastigoneme protein Pn14B7 is related to the Sig1 and Ocm1 proteins recently cloned from two algal Stramenopile, *Scytosiphon lomentaria* and *Ochromonas danica*, respectively (Honda et al. 2007; Yamagishi et al. 2007).

As yet we have little information on the identity of zoospore proteins involved in the reception of chemotaxis or electrotaxis signals, however, recent studies of *P. infestans* genes encoding the α-subunit of a trimeric G-protein (Dong et al. 2004; Latijnhouwers et al. 2004) and a bZIP transcription factor (Blanco and Judelson 2005) indicate that both these proteins play a role in zoospore motility. Silencing of these two genes inhibits zoospore motility by causing the cells to turn more frequently or spin in tight circles. Unfortunately it has not been possible to use these mutants to assess the contribution of zoospore motility and taxis to pathogen virulence because silencing the genes also produced aberrations during infection structure development. Regulation of flagellar activity is known to involve controls of cytoplasmic Ca^{2+} concentration and two calcium-binding proteins, calmodulin and centrin, have been localized within the flagella apparatus of *P. cinnamomi* zoospores (Gubler et al. 1990; Harper et al. 1995). Genes encoding centrin, a dynein light chain protein and a radial spoke protein were recently cloned from *P. cinnamomi* and are currently being further characterized (R. Narayan, L.M. Blackman and A. R. Hardham, unpublished data).

Phytophthora zoospores are not surrounded by a cell wall and their outer surface is that of the plasma membrane (Hardham 1987b). Water from their surroundings enters the zoospores down its chemo-osmotic gradient and, in order to maintain cell volume and homeostasis, must be pumped out of the cell. Zoospores achieve this through the operation of a contractile vacuole (often called a water expulsion vacuole; Fig. 1.1B) that consists of a reticulate spongiome surrounding a central bladder (Mitchell and Hardham 1999; Patterson 1980). It is not known exactly how contractile vacuoles function in any protist but it is believed that H^+-pumping ATPases power the accumulation of solutes within the spongiome, accompanied by the passive influx of water (Stevens and Forgac 1997). Localization of vacuolar H^+-ATPase in the spongiome of *P. nicotianae* zoospores is consistent with this hypothesis (Mitchell and Hardham 1999). Water is believed then to be transferred from the spongiome to the bladder which periodically fuses with the plasma membrane and contracts to expel the accumulated water.

Phytophthora zoospores are able to swim for many hours utilizing endogenous energy stores, thought to be predominantly polysaccharides (such as mycolaminarins) and lipids (Bimpong 1975; Wang and Bartnicki-Garcia 1974). They inherit many, if not the majority, of their proteins from the sporangium and early inhibitor studies

suggested that mRNA and protein synthesis were not required for zoospore function (Penington et al. 1989). However, more recently labelling studies have shown that new proteins are synthesized in *P. infestans* zoospores (Krämer et al. 1997) and proteomic analyses have identified polypeptides that are more abundant in zoospores than in any other stage of the life cycle of *P. palmivora* (Shepherd et al. 2003). In addition, transcriptome and other studies have identified genes that are preferentially expressed in *Phytophthora* zoospores (Ambikapathy et al. 2002; Connolly et al. 2005; Judelson and Blanco 2005; Škalamera et al. 2004). Proteins synthesized in zoospores may function during this motile phase or they may be required to function in the cysts that are formed by zoospore encystment. For example, one gene that is highly expressed in *P. nicotianae* zoospores is that encoding Δ^1-pyrroline-5-carboxylate reductase, an enzyme involved in proline biosynthesis (Ambikapathy et al. 2002). High levels of proline may be required for osmoregulation in the wall-less *Phytophthora* zoospores as they are in some other protists (Steck et al. 1997). In contrast, cell wall degrading enzymes (e.g. cellulase) encoded by genes identified in the transcriptome study (Škalamera et al. 2004) may be synthesized in readiness for secretion by germinated cysts during plant invasion.

B. Attaching to the Plant Surface

Having reached the surface of a potential host plant, *Phytophthora* zoospores adjust their swimming pattern so that the ventral surface faces the plant (Hardham and Gubler 1990). While they maintain this orientation, the zoospores encyst (Fig. 1.1C, E). This is a rapid process during which the two flagella are detached, rendering the spores non-motile, and material is secreted from three different categories of spherical vesicles in the zoospore peripheral cytoplasm (Fig. 1.1E). A network of cortical cisternae also fragments, apparently fusing with the plasma membrane, possibly thereby bringing about a rapid and wholesale change in the composition of the spore plasma membrane (Hardham 1989). The material that is secreted during zoospore encystment includes adhesion proteins that firmly attach the spores to the plant surface (Hardham and Gubler 1990). Attachment of pathogen spores or other cells to the surface of their hosts is an important aspect of the infection process (Epstein and Nicholson 1997;

Tucker and Talbot 2001). Not only does it prevent the pathogen being dislodged before it penetrates the plant, but the close contact also aids the reception of signals that guide pathogen growth and that trigger the development of specialized infection structures. Strong adhesion also facilitates host penetration by hyphae or appressoria.

The secretion of adhesive and other proteins from encysting zoospores is complete within about 2 min and a cellulosic cell wall capable of withstanding cell turgor is formed within 5–10 min (Hardham and Gubler 1990). As the cell wall forms, the pulsing of the contractile vacuole slows down and become undetectable (Mitchell and Hardham 1999). Zoospore encystment is triggered by a range of physical and chemical factors and there is evidence for a role of cell surface receptors and of the phospholipase D signal transduction pathway in induction of this process (Bishop-Hurley et al. 2002; Hardham and Suzaki 1986; Latijnhouwers et al. 2002).

The regulated secretion triggered during zoospore encystment involves exocytosis of the contents of the so-called large peripheral, dorsal and ventral vesicles (Fig. 1.2; Hardham 1995, 2005; Hardham and Hyde 1997; Škalamera and Hardham 2006). Material released from the dorsal vesicles includes a high molecular weight glycoprotein that forms a mucilage-like covering that coats the cysts and the nearby plant surface (Figs. 1.1E, 1.2A, B; Gubler and Hardham 1988). This material may protect the young cysts from physical or chemical damage but evidence to support this hypothesis has not yet been obtained. Material released from the ventral vesicles includes a 220-kDa adhesive protein, named Vsv1, that attaches the cyst to the plant (Fig. 1.2C, D). Cloning of the gene encoding Vsv1 in *P. cinnamomi* has revealed that, apart from short N- and C-terminal sequences, the bulk of the PcVsv1 protein is composed of 47 copies of a domain approximately 50 amino acids in length that shows homology to thrombospondin type 1 repeats found in a number of adhesive extracellular matrix proteins in animals and secreted adhesins in apicomplexan malarial parasites (Adams and Tucker 2000; Robold and Hardham 2005; Tomley and Soldati 2001). Homologues of the PcVsv1 adhesive occur in other *Phytophthora* species and in species of *Pythium*, *Plasmopara* and *Albugo*, suggesting that the Vsv1 protein may be a spore adhesive used throughout the plant pathogenic Oomycetes.

Until recently, studies of the large peripheral vesicles in the zoospore cortex indicated that their contents were not secreted during encystment but that the vesicles moved away from the plasma membrane and became randomly distributed within the cyst cytoplasm (Gubler and Hardham 1990). However, in zoospore transcriptome studies in *P. nicotianae*, cloning of a gene encoding a complement control protein has given rise to evidence that some of the contents of large peripheral vesicles are secreted during encystment (Škalamera and Hardham 2006). Evidence for the selective secretion of PnCcp proteins from the large peripheral vesicles comes from double immunolabelling of PnCcp and Lpv proteins at both the light and electron microscope levels (Fig. 1.2E–I). In motile zoospores both proteins are localized to the large peripheral vesicles but in young cysts, PnCcp proteins are absent from the vesicles and instead coat the cyst surface. In mammals, proteins containing complement control protein modules play a number of roles in signalling and adhesion (King et al. 2003). Their role in the infection of plants by *Phytophthora* zoospores remains to be elucidated.

In addition to the adhesives secreted by zoospores during their encystment, other *Phytophthora* genes encoding putative adhesives that may function in hyphae or germinated cysts have been cloned and characterized. Hyphae and cysts of *P. nicotianae* (formerly *P. parasitica*) have been shown to synthesize and secrete a 34-kDa glycoprotein, termed CBEL, that contains two cellulose-binding domains (Séjalon-Delmas et al. 1997; Villalba Mateos et al. 1997). Silencing of the expression of the CBEL gene interferes with adhesion of the hyphae to cellophane membranes and with morphogenetic changes normally induced by contact with cellulose in vitro (Gaulin et al. 2002). Silencing of CBEL expression does not have a great effect on pathogenicity on host tobacco plants. Another family of secreted proteins that may play a role in adhesion of germinated cysts is the Car proteins (*cyst-germination-specific acid repeat) from *P. infestans* (Görnhardt et al. 2000). The Car proteins contain multiple copies of an octapeptide repeat, a motif found in mammalian mucin proteins (Guyonnet Duperat et al. 1995). *Car* genes are expressed during cyst germination and appressorium differentiation and the Car proteins are secreted onto the germling surface. By analogy with the functions of mammalian mucins, the Car proteins have been hypothesized as playing roles in protecting the germlings from desiccation or physical damage and in germling adhesion (Görnhardt et al. 2000).

C. Penetration of the Host Surface

Zoospores may carry mRNA transcripts and proteins that function in the cysts that are formed

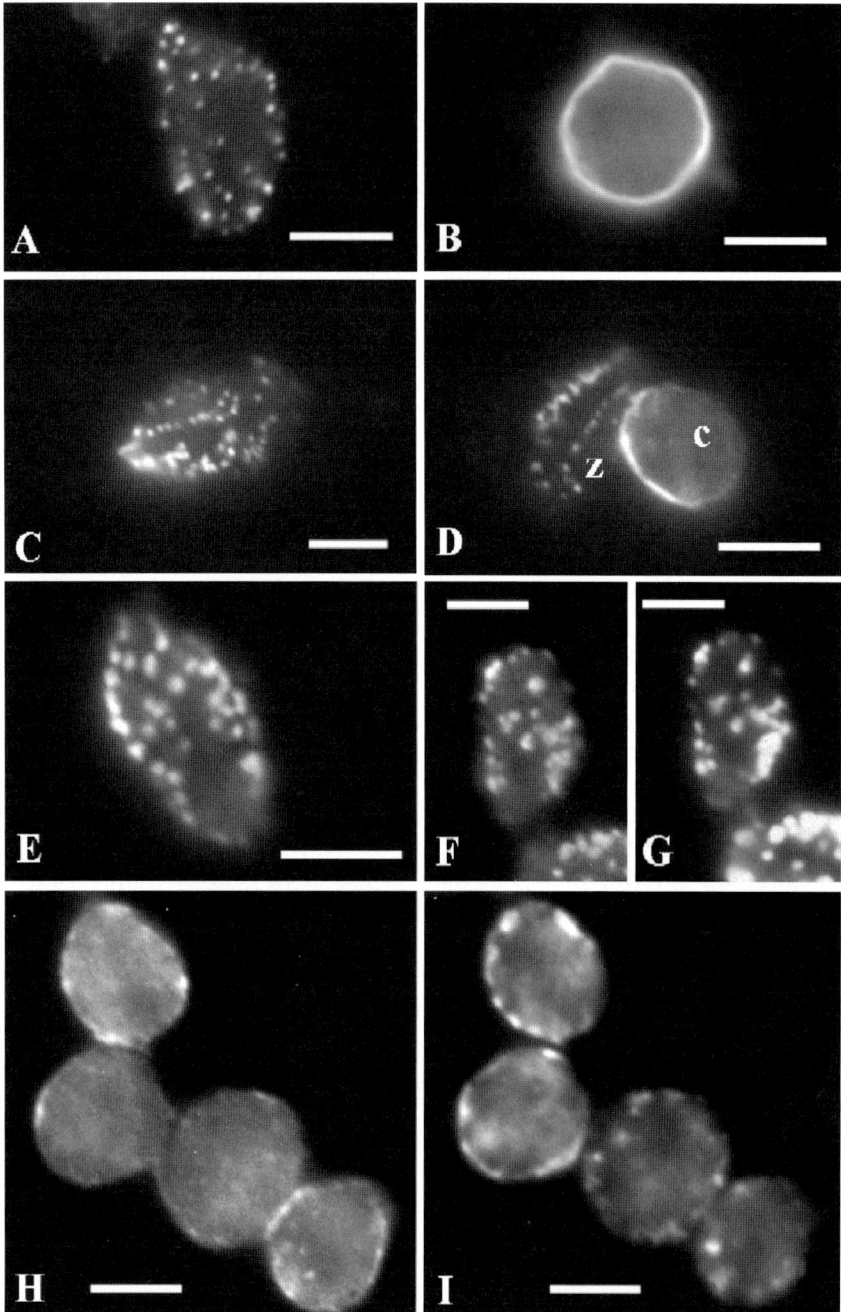

Fig. 1.2. Regulated secretion of cortical vesicles by *P. nico-tianae* zoospores. **A** Dorsal vesicles in a zoospore labelled with monoclonal antibody 8E6 (Gautam et al. 1999). **B** Cyst showing 8E6 labelling of the contents of the dorsal vesicles that have been secreted onto the cyst surface. **C** Labelling of the ventral vesicles in a zoospore with mono-clonal antibody Vsv1. The ventral vesicles preferentially line the ridges of the ventral groove. **D** Vsv1 labelling of zoospore (*z*) ventral vesicles and secreted material along one surface of a young cyst (*c*). **E** Large peripheral vesicles labelled with monoclonal antibody Lpv1 in a zoospore. **F, G** Double-labelling of large peripheral vesicles in zoospores with PnCcp polyclonal antibody (**F**) and Lpv1 (**G**). Both antibodies label the same cortical vesicles in the zoospores. **H, I** Double-labelling of cysts with PnCcp poly-clonal antibody (**H**) and Lpv1 (**I**). PnCcp antigens appear on the surface of young cysts but Lpv1 proteins remain in vesicles in the cell cortex. *Bars* 5 µm

during zoospore encystment; nevertheless, the process of encystment also triggers a new pattern of gene expression and protein synthesis, producing proteins required for cyst germination and germling growth and development (Avrova et al. 2003; Ebstrup et al. 2005; Grenville-Briggs et al. 2005; Krämer et al. 1997; Shan et al. 2004b; Shepherd et al. 2003). Proteins encoded by the genes that are up-regulated perform a range of functions, include DNA, RNA and protein synthesis, signalling, cell structure and growth. A number of studies have highlighted the increased abundance of heat shock and other proteins involved in scavenging reactive oxygen species and in protecting the pathogen against stress, functions that would be important for the pathogen's survival of the plant defence response (Avrova et al. 2003; Ebstrup et al. 2005; Shan and Hardham 2004). Future studies promise to elucidate the role of other cyst proteins identified in the gene discovery projects with exciting results for our understanding of molecular changes occurring during early infection events.

Spatial and temporal aspects of spore germination in many organisms are typically influenced by environmental factors, however, in *Phytophthora* the site of cyst germination is pre-determined and the polarity of germ tube emergence with respect to the adjacent plant is set up by the motile zoospore before it encysts (Hardham and Gubler 1990). In soil-borne *Phytophthora* species, such as *P. cinnamomi* and *P. nicotianae*, the zoospores approach the root and alter their mode of swimming so that they swim parallel to the root surface, frequently turning by 180 degrees so that they swim backwards and forwards over the same section of root surface, all the while maintaining an orientation such that their ventral surface faces the root. Just before encystment, motility decreases and, often quite suddenly, the flagella detach and the cell adopts a more spherical shape (Fig. 1.1C–E). The cysts typically germinate 20–30 min later and the germ tube emerges from the centre of what had been the ventral surface of the zoospore (Figs. 1.1D, E, 1.3A, B). Because most zoospores orient their ventral surface towards the root before encystment, the germ tube consequently forms directly opposite the plant surface and grows chemotropically towards a suitable penetration site (Hardham and Gubler 1990; Miller and Maxwell 1984). In root pathogenic species, the preferential targeting of the zoospores to grooves between epidermal cells is accompanied by subsequent preferential penetration along the anticlinal wall between the cells (Figs. 1.1E, 1.3A; Enkerli et al. 1997; Hardham 2001). In foliar pathogenic species whose zoospores target stomatal complexes, subsequent penetration of the leaf occurs via stomatal apertures (Gees and Hohl 1988; Judelson and Blanco 2005).

Phytophthora hyphae may penetrate the plant surface either along anticlinal walls or directly through the outer periclinal wall without any detectable modification of hyphal morphology (Hardham 2001). In some cases, however, penetration of the plant surface is preceded by the development of an appressorium or appressorium-like swelling of the hyphal apex (Fig. 1.3A, B; Bircher and Hohl 1997; Grenville-Briggs et al. 2005; Judelson and Blanco 2005). The appressoria of some species, such as *P. infestans*, appear to be differentiated cells that are separated from the subtending hyphae by a cross wall (Fig. 1.3B; Gees and Hohl 1988). In other cases, this degree of differentiation is not evident. In *P. infestans*, appressoria formation is induced by factors similar to those that trigger fungal appressorial differentiation, including surface topography and hydrophobicity (Bircher and Hohl 1997) and is accompanied by changes in patterns of gene expression and protein synthesis (Grenville-Briggs et al. 2005). In *P. nicotianae* and *P. cinnamomi*, the hyphal swellings that develop over an anticlinal wall tend to be disk-shaped structures oriented along the groove (Fig. 1.3A). Those that form over the periclinal wall generally assume a more globular shape (Hardham 2001). The fact that the appressorium-like swellings are flat discs rather than spherical expansion of the tip when formed over anticlinal walls suggests that swelling of the hyphal apex does not result simply from the inhibition of hyphal growth by the unyielding plant surface. Instead, their formation is indicative of apical differentiation to produces a structure better able to penetrate the underlying cell wall. Although it has been shown that Oomycete hyphae can exert a force at their apex similar to that generated by fungal hyphae (Money et al. 2004), as yet there are no data on turgor pressures within *Phytophthora* appressoria or on the mechanical pressures they exert during penetration of the plant surface.

Like fungal phytopathogens, *Phytophthora* species also use cell wall degrading enzymes to penetrate and colonize the plant (see Chap. 10). Although there is still only limited in planta evidence of the activity of wall digestion during penetration, loss of pectin from the host cell walls was recently demonstrated during infection by a species of the closely-related genus, *Pythium* (Boudjeko et al. 2006). *Phytophthora* hyphae secrete a range of enzymes that break down the polymers found in plant cell walls and genome and EST sequencing projects have catalogued the genes that encode them. In three *Phytophthora* species for which genome sequence data are available,

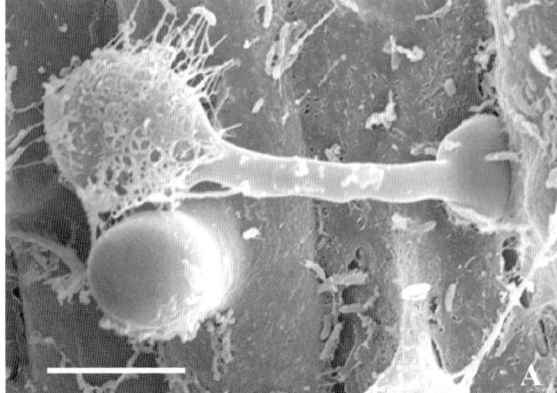

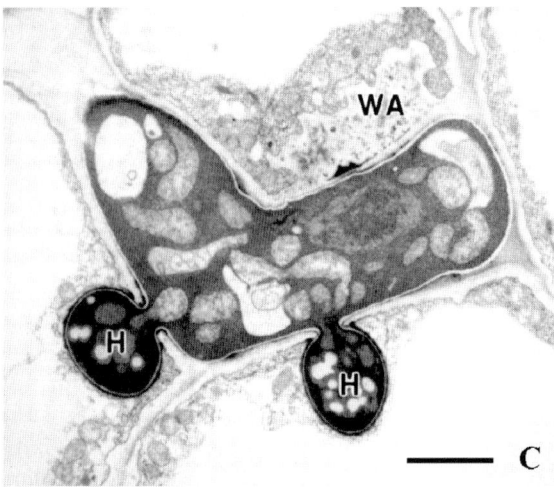

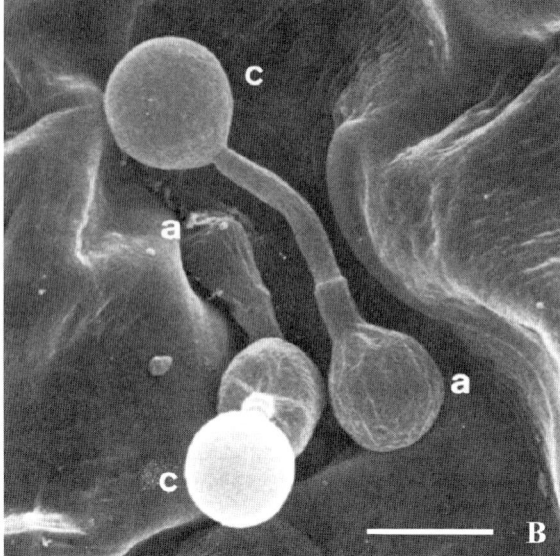

Fig. 1.3. Plant penetration and nutrient acquisition. **A** Germinated cyst of *P. cinnamomi* that has formed a disc-shaped appressorium-like structure as it grows between two epidermal cells of an alfalfa root. *Bar* 10 μm. **B** Appressoria (*a*) that have differentiated at the hyphal apices of germinated cysts (*c*) of *P. infestans. Bar* 20 μm. Micrograph reproduced with permission from Gees and Hohl (1988). **C** Two haustoria (*H*) formed by an intercellular hypha of *P. sojae* growing in a soybean (*Glycine max*) root. One root cell has deposited a cell wall apposition (*WA*) adjacent to the hypha. *Bar* 2.5 μm. Micrograph reproduced with permission from Enkerli et al. (1997)

namely *P. ramorum, P. sojae* and *P. infestans,* genes encoding glucanases, polygalacturonases, pectin esterases, pectin lyases and xylanases have been identified (http://www.genome.jgi-psf.org; http://www.broad.mit.edu/tools/data/seq.html). To date, only a small number of these genes have been characterized in any detail (Brunner et al. 2002b; Götesson et al. 2002; McLeod et al. 2003; Torto et al. 2002; Yan and Liou 2005).

The gene family that has been studied in most detail is that encoding endopolygalacturonase, an enzyme that degrades the polygalacturonan backbone of pectin molecules (Götesson et al. 2002; McLeod et al. 2003; Torto et al. 2002; Yan and Liou 2005). Polygalacturonases and other pectin degrading enzymes are secreted early in infection. Their activity exposes other wall polymers to attack and causes tissue maceration by disrupting the middle lamella that normally glues adjacent plant cells together. The gene family encoding endopolygalacturonase in *Phytophthora* contains over 20 members (Götesson et al. 2002). The reasons

for such a large gene family have not been fully elucidated but are likely to include a need for enzymes specialized to digest the diverse range of pectin molecules. During fungal infection of host plants, a cascade of expression of different members of the polygalacturonase gene family has been demonstrated. In *Botrytis cinerea,* one gene is constitutively expressed and it is thought that the products that are released by digestion of wall pectin by the enzyme encoded by this gene trigger the expression of other members of the multigene family (Ten Have et al. 2001). In *P. cinnamomi,* of the 20 or so polygalacturonase genes, only a small number have been found to be expressed during in vitro culture in both defined and undefined media or during plant infection; one of these genes is under glucose catabolic repression (E. Landgren, A. Götesson, L.M. Blackman and A.R. Hardham, unpublished data).

Cell wall degrading enzymes are just one category of molecules that are secreted by *Phytophthora* hyphae as they grow. As in fungi, material to be secreted into the external environment is thought to be transported to and released from the hyphal apex in small apical vesicles.

Although not visible in the light microscope, *Phytophthora* hyphae contain a cluster of such vesicles in the apical cytoplasm, similar to the Spitzenkörper of fungal hyphae. In addition to degradative enzymes, these apical vesicles are likely to contain adhesins, enzymes involved in cell wall synthesis and modification, hyphal wall components and molecules involved in counterdefence (Gaulin et al. 2002; Rose et al. 2002; Shapiro and Mullins 2002; Tian et al. 2005). *Phytophthora* spores and hyphae contain recognizable dictyosomes (Hardham 1987b) and proteins, glycoproteins and polysaccharides destined for secretion are synthesized and packaged in the endoplasmic reticulum and Golgi apparatus (Dearnaley and Hardham 1994). Fusion of the apical vesicles at the hyphal apex also contributes membrane to the expanding plasma membrane, including proteins that function as receptors, channels and enzymes involved in cell wall synthesis (Loprete and Hill 2002).

As in fungal hyphae, transport and distribution of apical vesicles and other cell components in the hyphal apex is dependent on the function of cytoskeletal elements, namely microtubules and actin microfilaments, which form longitudinal arrays along the hyphae (Heath 1995; Temperli et al. 1990). Near the hyphal apex, organelles are typically stratified along the hypha. Behind the cluster of apical vesicles in the very tip, the cytoplasm contains mitochondria and nuclei before becoming increasingly filled with spherical and tubular vacuoles (Ashford and Allaway 2007; Heath and Kaminskyj 1989). Experimental studies using cytoskeletal inhibitors indicate that microtubules regulate long-distance movement of hyphal organelles while actin microfilaments facilitate movement of the apical vesicles to their site of fusion at the tip (Heath 1995; Temperli et al. 1991). A cap of actin microfilaments is also sometimes seen at the apex of *Phytophthora* hyphae (Walker et al. 2006). A function in stabilizing the expanding apical dome has been suggested from studies of other Oomycetes (Jackson and Heath 1990; Kaminskyj and Heath 1995) and the lack of the actin cap has been correlated with active invasion of the surrounding medium (Walker et al. 2006). Mathematical modelling has been used to show that the rate of extension and morphology of *Phytophthora* hyphae can be predicted using parameters reflecting the rate of movement of the cluster of apical vesicles, the so-called vesicle supply centre, and the rate of vesicle fusion with the plasma membrane at the hyphal tip (Dieguez-Uribeondo et al. 2004). A decrease in the number of apical vesicles emanating from the apical cluster leads to slower hyphal growth; inhibition of movement of the vesicle cluster leads to isotropic expansion of the hyphal apex.

D. Nutrient Acquisition to Support Pathogen Growth and Reproduction

Having penetrated the plant surface, the mode of subsequent growth within the plant depends on the life style of the pathogen, that is, whether it is a necrotroph or a hemibiotroph. During necrotrophic growth, hyphae may grow intercellularly or intracellularly, acquiring the nutrients they need from dead and dying cells. However, during the initial biotrophic phase of hemibiotrophic species (*P. capsici, P. infestans, P. nicotianae, P. palmivora, P. sojae*), hyphal growth is restricted to the apoplast and disruption of host cells is minimized. During biotrophic growth, nutrients are acquired through the development of specialized haustoria that form predominantly in mesophyll cells for foliar pathogens or in cortical cells for root pathogens (Fig. 1.3C). In contrast to the situation in many biotrophic fungi, distinct haustorial mother cells do not differentiate and instead haustoria develop directly from the intercellular hyphae (Enkerli et al. 1997; Jeun and Buchenauer 2001). Formation of haustoria involves localized dissolution of the plant cell wall and invagination of the plant plasma membrane by the invading pathogen cell (Fig. 1.3C; Enkerli et al. 1997). *Phytophthora* haustoria may be globose or finger-like projections that contain the normal complement of organelles apart from nuclei (Coffey and Gees 1991). Throughout their operational lifetime, *Phytophthora* haustoria remain surrounded by the invaginated host plasma membrane, commonly termed the extrahaustorial membrane. By analogy with the situation in fungal–plant interactions, it is likely that this domain of the plant plasma membrane becomes specialized such that its properties support nutrient uptake by the haustorium. The extrahaustorial membrane is separated from the haustorial wall by an electron-dense, extrahaustorial matrix. Again, by analogy with fungal haustorial complexes, it is likely that components within the extrahaustorial matrix are of both plant and *Phytophthora* origin.

Research in recent years on biotrophic fungi has uncovered molecular evidence of specializations of both the haustorial membrane and the extrahaustorial membrane that facilitate nutrient uptake by the haustoria (Chap. 4). These features include the localization of amino acid and sugar transporters within the haustorial membrane and a concentration of H^+-ATPase to power nutrient transport (Hahn et al. 1997; Struck et al. 1996; Voegele and Mendgen 2003). Evidence for similar specializations in association with *Phytophthora* haustoria has yet to be uncovered. However, studies of *Phytophthora* proteins that are secreted from haustoria have made a major contribution to our understanding of the translocation of pathogen effector (avirulence) proteins into the host cell cytoplasm (Birch et al. 2006; Whisson et al. 2007; Chap. 9). As discussed below, an RXLR motif has been shown to direct the translocation of proteins from the apoplast into the plant cell cytoplasm from where they may orchestrate changes in host cell organization and metabolism or may be recognized by the host cell and trigger a defence response.

III. *Phytophthora* Effectors and Elicitors of the Plant Defence Response

Plants resist attack from most micro-organisms through the activation of defence reactions elicited either directly or indirectly by molecules produced by the invading microbe. Microbial elicitors may be molecules that are common to a wide group of micro-organisms. For instance, many of the elicitors that trigger basal defence responses are essential pathogen components that contain highly conserved domains, or microbe-associated molecular patterns (MAMPs), and are present across a range of organisms (Bent and Mackey 2007; Jones and Takemoto 2004). Examples of MAMP-containing elicitors include bacterial flagellin (Zipfel et al. 2004), fungal chitin (Kaku et al. 2006) and *Phytophthora* cell wall heptaglucans (Cheong et al. 1993). In many cases, the function of elicitors in this category is not specifically related to microbial pathogenicity. In contrast, other elicitors that are recognized by the plant have functions that are directly involved in the infection process. Such elicitors include not only structural components of the infection apparatus but also proteins that enable the pathogen to evade, suppress or manipulate host defences (see Chaps. 9, 18).

Because many elicitors were first isolated by genetic mapping and complementation on the basis of their induction of plant resistance, i.e. their avirulence activity, until recently, they were referred to as avirulence proteins (Martin et al. 2003). This focus on avirulence functions tended to be somewhat confusing. Why would a pathogen produce a molecule that was recognized by potential hosts and that triggered defences that subsequently thwarted infection? The explanation, as indicated above, is that these molecules are important components of the pathogen's infection machinery. The recent introduction and rapid acceptance of the term "effectors" to encompass these molecules in both their virulent and avirulent forms is a helpful development that makes their role in pathogenicity more apparent. During evolution and the on-going arms race between plants and their pathogens, the sequence of effector genes has changed in order to help pathogens avoid detection by their plant hosts. Application of modern genomic and bioinformatic approaches is greatly facilitating the identification of elicitors and effectors. A variety of such molecules has been isolated and characterized from *Phytophthora* pathogens, including pathogenicity effectors that function extracellularly or within the host cell cytoplasm as well as conserved cell wall constituents that contain MAMPs. As oomycete effectors are discussed in detail in Chap. 9, with an emphasis on the molecular level, the role(s) of these molecules is discussed here only briefly.

A. Extracellular *Phytophthora* Effectors

1. Inhibitors of Plant Enzymes

In response to pathogen attack, plants produce a variety of hydrolytic enzymes including glucanases, chitinases and proteases in defence against pathogen infection (Stintzi et al. 1993). In the case of the first two groups of enzymes, their activity also generates elicitor-active oligosaccharides that trigger further defence responses (Stintzi et al. 1993). Fungal, bacterial and oomycete pathogens have evolved mechanisms to protect themselves against these degradative enzymes by secreting effector proteins that inhibit enzymatic activity and, in the case of glucanase and chitinase inhibitors, arrest the production of potent elicitors (Abramovitch and Martin 2004).

Phytophthora cell walls are rich in β-1,3-glucans, including those that constitute the main microfibrillar component, cellulose (Bartnicki-Garcia 1970). As part of its defence response, soybean, the host of *P. sojae*, produces two basic endo-β-1,3-glucanases, designated EGaseA and EGaseB (Ham et al. 1997, Rose et al. 2002). These soybean endo-β-1,3-glucanases degrade the *P. sojae* cell wall and release elicitor-active heptaglucans (Côté et al. 2000). In a counterdefence strategy, *P. sojae* secretes at least three glucanase inhibitor proteins (Ham et al. 1997; Rose et al. 2002; York et al. 2004). One of these, designated GIP-1, has been cloned and shown to inhibit one of the two soybean endoglucanases but not either an endogenous *P. sojae* endo-β-1,3-glucanase or a tobacco endo-β-1,3-glucanase, PR-2c (Ham et al. 1997). The GIPs are similar to the trypsin class of serine proteases but are proteolytically nonfunctional. Their inhibition of plant endoglucanase activity is thus not through proteolysis of the endoglucanase but through protein–protein interactions (Rose et al. 2002). *P. sojae* GIP-1 is predicted to be a 24-kDa protein, cell wall-bound and secreted. The soybean endoglucanase that is inhibited (EGaseA) by GIP-1 is expressed constitutively in healthy plants, making it a good target for inhibition. GIP-1 reduces the ability of EGaseA to degrade the pathogen cell wall and to release elicitors. Other *Phytophthora* species also contain small GIP families (York et al. 2004).

Species of *Phytophthora* also produce proteins that inhibit the action of plant proteases synthesized in response to pathogen attack. Two families of extracellular protease inhibitors have been reported to date. The first is a group of Kazal-like protease inhibitors, designated EPI for extracellular protease inhibitor (Tian et al. 2004). There are 14 members in the EPI gene family in *P. infestans*, two of which (*EPI1*, *EPI10*) produce proteins that inhibit the tomato protease, P69B (Tian et al. 2004, 2005; Tian and Kamoun 2005). In parallel with increased expression of P69B after *P. infestans* inoculation, both *EPI1* and *EPI10* are up-regulated during tomato infection. Inhibition of P69B by both EPI1 and EPI10 suggests that suppression of the activity of this plant protease may be an important aspect of the infection of host plants by *P. infestans* (Tian et al. 2005). Evidence for expression of homologues of *EPI1* has also been found in *P. sojae* and *Plasmopara halstedii* during infection of soybean and sunflower, respectively (Tian et al. 2004), and Kazal-like proteins have also been implicated in the virulence of apicomplexan parasites (Kamoun 2006).

The second family of secreted *Phytophthora* protease inhibitors consists of cystatin-like protease inhibitors designated EPIC (Tian et al. 2007). Four members of this family have been identified in *P. infestans*, two of which (*epiC3*, *epiC4*) have orthologues in *P. sojae* and *P. ramorum*. *epiC1*

and *epiC2* are up-regulated during tomato infection and EPIC2B inhibits a tomato papain-like cysteine protease, PIP1, an enzyme that is related to the tomato protein, Rcr3. It is of interest to note that Rcr3 is also inhibited by the *Cladosporium fulvum* avirulence gene product, Avr2 (Rooney et al. 2005).

2. *Phytophthora* Elicitins

A number of fungal avirulence genes, including *Avr2*, *Avr4* and *Avr9* of *C. fulvum* and *nip1* of *Rhynchosporium secalis,* encode small (< 150 amino acid residues) secreted proteins that induce defence responses when infiltrated into plant tissues (Lauge and de Wit 1998). These proteins contain pairs of cysteine residues that form disulfide bridges that are thought to enhance protein resistance to degradation by apoplastic plant proteases and are essential for defence induction and avirulence activities (Joosten et al. 1997). *Phytophthora* species also produce small (10 kDa) cysteine-rich elicitor proteins called elicitins (Kamoun 2006).

Elicitins are encoded by large gene families in most, if not all, *Phytophthora* species and in some species of *Pythium*. *Phytophthora* elicitins were initially identified by biochemical purification and have a highly conserved 98-amino-acid elicitin domain with six conserved cysteine residues that form three disulfide bonds. Elicitins induce hypersensitive cell death and other biochemical changes associated with defence responses in *Nicotiana* species (Jiang et al. 2006; Kamoun et al. 1993; Ponchet et al. 1999; Ricci et al. 1989) and are thought to be species-specific host determinants. Strains of *P. infestans* deficient in the INF1 elicitin, for example, are pathogenic on the non-host plant *N. benthamiana*, suggesting that the INF1 protein conditions avirulence at the species level (Kamoun et al. 1998).

As well as eliciting a defence response, *Phytophthora* elicitins are pathogen effectors with known functions in lipid binding and/or processing (Osman et al. 2001a). Cryptogein, a class I elicitin produced by *P. cryptogea*, binds ergosterol and functions as a sterol-carrier protein (Boissy et al. 1999; Mikes et al. 1997, 1998; Vauthrin et al. 1999). Since *Phytophthora* cannot synthesize sterols and must assimilate them from external sources, elicitins are believed to serve an essential role in *Phytophthora* growth and development (Hendrix 1970).

Elicitin-like proteins from *P. capsici* have also been shown to have phospholipase activity (Nespoulous et al. 1999). Enhanced expression of elicitin-like genes during mating suggests that elicitins may also function in sexual reproduction (Fabritius et al. 2002). Mutational analysis of cryptogein to produce proteins with altered sterol binding capacity indicates that sterol loading is important for specific binding of the elicitor-sterol complex to a plant plasma membrane receptor and for the induction of the hypersensitive response in tobacco (Osman et al. 2001b).

3. Cellulose Binding Proteins

A 34-kDa protein isolated from cell walls of *P. nicotianae* (formerly *P. parasitica*) triggers local necrosis, defence gene expression and resistance against subsequent infections in host tobacco and in various non-host plants (Villalba-Mateos et al. 1997). As well as containing two cellulose binding domains that closely resemble the fungal type I cellulose binding domain consensus pattern, the protein exhibits elicitor and lectin activity, hence its name, CBEL (Villalba-Mateos et al. 1997). CBEL binds to crystalline cellulose and tobacco cell walls in vitro and leaf infiltration with synthetic peptides from the cellulose binding domains showed that they are essential and sufficient to trigger defence responses (Gaulin et al. 2006). Silencing *CBEL* gene expression impairs attachment of transgenic *P. nicotianae* to cellophane membranes, although the pathogen retains its ability to infect tobacco plants (Gaulin et al. 2002). CBEL-like genes are found throughout the Oomycetes, with 52 different sequences identified in five species, including the fish pathogen *Saprolegnia parasitica* (Torto-Alalibo et al. 2005).

B. *Phytophthora* Effectors Translocated into the Host Cytoplasm

A major advance in molecular plant pathology research has been the recent discovery that fungal and oomycete plant pathogens translocate effector proteins into the cytoplasm of their plant hosts (Dodds et al. 2004; Kemen et al. 2005; Rehmany et al. 2005). While the mechanisms responsible for translocation of fungal effectors are still unknown, in *Phytophthora* and *Hyaloperonospora parasitica*, another oomycete, an N-terminal RXLR motif facilitates transport into the infected host cell (Birch et al. 2006; Tyler et al. 2006; Whisson et al. 2007; Win et al. 2007).

Members of this class of effectors were first identified as avirulence factors in *P. sojae* and *P. infestans* (Armstrong et al. 2005; Shan et al. 2004a). Bioinformatic analysis of their sequence revealed the existence of the conserved RXLR motif located downstream of the secretion signal in the N-terminus. The RXLR motif is similar to the motif used by apicomplexan malarial parasites to deliver pathogen proteins into the red blood cells of mammalian hosts (Hiller et al. 2004; Marti et al. 2004). Indeed, a 30-amino-acid region flanking the RXLR motif of *P. infestans* proteins AVR3a and PH001D5 directs export of the green fluorescent protein (GFP) from the malaria parasite *Plasmodium falciparum* into host erythrocytes, indicating that the RXLR domain is functionally conserved across these stramenopile organisms (Bhattacharjee et al. 2006).

Using *P. infestans* Avr3a and *P. sojae* Avr1b avirulence proteins, it was demonstrated that RXLR-containing effectors have two major functional domains (Bos et al. 2006; Kamoun 2006; Whisson et al. 2007; B. Tyler, personal communication). The N-terminal domain containing the signal peptide and RXLR motif functions in secretion and translocation into the host cytoplasm but is not required for effector activity. The remaining C-terminal region is responsible for effector function inside the host cells.

Analysis of the fully-sequenced *Phytophthora* genomes revealed that the RXLR-containing effector secretomes consist of hundreds of proteins. Depending on the algorithm used, the genomes of *P. ramorum* and *P. sojae* contain between 350 and 672 RXLR-containing effectors (Kamoun 2006; Tyler et al. 2006; Win et al. 2007). These studies suggest that the genes encoding the RXLR-containing effectors underwent relatively rapid birth and death evolution and that, during evolution, positive selection acted mainly on the C-terminal domain (Win et al. 2007).

A second class of *Phytophthora* effector proteins that are translocated into the plant cytoplasm is exemplified by the CRN protein family. CRN1 and CRN2 were firstly identified following an in planta functional expression screen in which *P. infestans* proteins were synthesized within the plant cytoplasm following introduction of genes by a vector derived from potato virus X (Torto et al. 2003). Expression of *crn1* and *crn2* in tobacco and tomato led to a leaf-crinkling and necrosis phenotype and induction of defence response genes. Subsequent analysis revealed that the CRNs form a complex family of relatively large proteins (400–850 amino acids) in *Phytophthora* (Win et al. 2006). *P. infestans crn1* and *crn2* genes are expressed during colonization of tomato host plants and *crn8* encodes a secreted protein with a predicted kinase domain. A secreted kinase is also a major virulence factor in the apicomplexan parasite *Toxoplasma gondii* (Saeji et al. 2006; Taylor et al. 2006). The mechanism directing transport of CRN proteins into the plant cytoplasm during infection is still not known. None of the *Phytophthora* CRN proteins carries an RXLR motif; however, they do contain a conserved N-terminal LXLFLAK motif which clearly shows some similarity to the RXLR sequence (Kamoun 2006). It has been proposed that after transport into the host cytoplasm, the CRN effectors manipulate host cellular processes to cause macroscopic phenotypes such as cell death, chlorosis and tissue browning (Torto et al. 2003).

C. Secreted elicitors of the plant defence response

1. The GP42 Cell Wall Transglutaminase

GP42 is an abundant 42 kDa glycoprotein associated with the cell walls of *P. sojae* (Sacks et al. 1995). GP42 binds to a plant plasma membrane receptor in parsley cells and triggers defence gene expression and synthesis of antimicrobial phytoalexins (Nürnberger et al. 1994). GP42 is a highly conserved, Ca^{2+}-dependent transglutaminase that contains a 13-amino acid peptide (Pep-13) that is necessary and sufficient for activation of defences including cell death in parsley and potato (Brunner et al. 2002a; Halim et al. 2004). The Pep-13 motif is invariant in all examined *Phytophthora* transglutaminases and the same residues within the Pep-13 motif are important for both activation of plant defences and transglutaminase activity. While it is not yet clear whether the *Phytophthora* GP42 transglutaminases are essential for pathogen virulence or fitness, it appears that plants have evolved receptors to recognize the consensus sequence within the transglutaminase protein, which thus constitutes a MAMP-containing elicitor (Brunner et al. 2002a).

2. Necrosis and Ethylene Inducing Proteins

The isolation of a 24 kDa *necrosis and ethylene inducing protein*, Nep1, from *Fusarium* culture filtrates (Bailey et al. 1997) led to the identification of a group of proteins that had no match to any known proteins or functional domains. Nep1-like proteins (NLPs) were subsequently discovered in a wide variety of pathogens, particularly in plant-associated species (Gijzen and Nürnberger 2006; Pemberton and Salmond 2004), including other fungi (Wang et al. 2004), Oomycetes (Fellbrich et al. 2002; Qutob et al. 2002; Veit et al. 2001) and bacteria (Pemberton et al. 2005). NLPs are highly conserved and trigger defence responses in both susceptible and resistant plants (Fellbrich et al. 2002; Qutob et al. 2006). Whether this response

is due to plant recognition of an MAMP or to direct toxicity of NLPs by forming pores or interacting with crucial plant plasma membrane targets (Gijzen and Nürnberger 2006) is not known.

NLPs are also broadly distributed in *Phytophthora*, and genome analysis indicates that the NLP gene family is large and diverse, with 50–60 members in both *P. sojae* and *P. ramorum*, although in both species more than half are likely to be pseudogenes. Well-studied NLPs include *P. nicotianae* NPP1 (Fellbrich et al. 2002), *P. infestans* PiNPP1.1 (Kanneganti et al. 2006) and *P. sojae* PsojNIP (Qutob et al. 2002).
Although a role for NLPs in the virulence of necrotrophic fungal and bacterial pathogens is supported by gene disruption and over-expression experiments (Amsellem et al. 2002; Gijzen and Nürnberger 2006; Pemberton et al. 2005), as yet, the contribution of NLPs to *Phytophthora* pathogenicity is unclear. The fact that the necrosis-inducing domain is present in all *Phytophthora* NLPs (Fellbrich et al. 2002), despite its widespread recognition, suggests that the necrosis-inducing activity is functionally important. In *P. sojae*, the *NLP* genes are expressed late during host infection, consistent with their triggering host cell necrosis during the necrotrophic phase (Qutob et al. 2002). In *P. infestans*, the PiNPP1.1 protein induces host cell death in a distinct but interacting cell death pathway to that of the INF1 elicitin (Kanneganti et al. 2006).

IV. The Plant's Response to Potential Pathogens

A. The Products of Plant Resistance Genes Recognize *Phytophthora* Elicitors

Like many other plant-pathogen systems, identification and introgression of resistance genes is an effective approach for controlling *Phytophthora* diseases. Genetic analysis of plant–*Phytophthora* interaction has focused mainly on the potato–*P. infestans* and soybean–*P. sojae* interactions and has culminated in the identification of 11 resistance genes derived from wild potato, *Solanum demissum* (van der Vossen et al. 2003), and 15 resistance genes in soybean (Sandhu et al. 2004; Schmitthenner et al. 1994).

Although resistance proteins function in surveillance systems that detect elicitors from a diverse range of pathogens, they share a number of conserved domains (Martin et al. 2003). The majority of resistance proteins contain a central *nucleotide binding site* (NBS) and a carboxyl *leucine-rich repeat* (LRR) domain and are cytoplasmic. This group, commonly referred to as NBS-LRR genes, is further divided into two subclasses, those possessing an

N-terminal *c*oiled-*c*oil domain (CC-NBS-LRR proteins) and those containing an N-terminal domain resembling the cytoplasmic signalling domain of the *T*oll and *I*nterleukin-1 transmembrane *r*eceptors (TIR-NBS-LRR proteins; Martin et al. 2003). Nearly all NBS-LRR proteins are associated with resistance to biotrophic pathogens (Glazebrook 2005).

To date, all cloned plant genes conferring resistance to *Phytophthora* pathogens belong to the NBS-LRR class and encode CC-NBS-LRR proteins. Examples include the soybean *Rps1k* gene that confers resistance to *P. sojae* (Gao et al. 2005) and the *R1*, *R3a*, *Rpi-blb1* and *Rpi-blb2* genes from wild potato (*S. demissum*) that confer resistance to *P. infestans* (Ballvora et al. 2002; Huang et al. 2005; Song et al. 2003; van der Vossen et al. 2003, 2005).

B. Subcellular Reorganization Within Plant Cells During the Defence Response

Molecular recognition of invading *Phytophthora* cells triggers a cascade of responses in the plant cells under attack (see Chap. 17). The speed and nature of the response is determined by a range of factors, predominant among them the genotypes of the interacting organisms and environmental influences, including prior priming of the defence response by unsuccessful pathogens. In general, components of the defence response are activated more quickly by resistant plants than they are by susceptible hosts. Responses of plants to attack by species of *Phytophthora* include a rapid increase in cytoplasmic Ca^{2+} concentration, cytoplasmic aggregation and formation of cell wall appositions beneath invading pathogen cells, reorganization of the cytoskeleton and endomembrane system, formation of reactive oxygen species, synthesis of pathogenesis-related (PR) proteins and phytoalexins, and hypersensitive cell death (HR; Figs. 1.3C, 1.4; Able et al. 2001; Blume et al. 2000; Fellbrich et al. 2002; Takemoto et al. 2003). These defence responses fall into two main categories, namely those associated with penetration resistance and those involved in post-invasion defence. Penetration resistance involves cytoplasmic aggregation and development of wall appositions to stop ingress of the pathogen through physical and chemical barriers at the infection site. Generally, this response successfully inhibits invasion by non-adapted pathogens (i.e. non-host interactions). If it does not, in resistant plants, hypersensitive cell death usually overcomes avirulent races of pathogenic species (Enkerli et al. 1997; Huitema et al. 2003; Takemoto et al. 2003).

1. Strengthening the Cell Wall Barrier to Prevent Pathogen Penetration

Penetration resistance at the plant cell surface is the first phase of the defence response and is a key feature of basal plant resistance that operates to prevent invasion by a wide variety of potential pathogens. A major component of penetration resistance is the localized thickening of the plant cell wall underlying the invading pathogen cell (Fig. 1.3C). The wall thickening is known as a cell wall apposition or papilla and constitutes a physical and chemical barrier to penetration (Aist 1976; Schmelzer 2002). Formation of the cell wall apposition is accomplished by site-directed deposition of a variety of polysaccharides, proteins and other compounds that cross-link existing wall components, impregnate the wall with material that resists degradation and exert toxic effects on the invading pathogen. Cell wall appositions typically contain callose, phenolics, silicon, H_2O_2, peroxidase and enzyme inhibitors (Fig. 1.4H; Aist 1976; An et al. 2006; Schmelzer 2002; Zeyen et al. 2002).

Callose, a β-1,3 glucan, is likely to inhibit pathogen penetration by encasing and inhibiting degradation of other cell wall components or, perhaps, by providing a matrix in which antimicrobial compounds may be sequestered. Its rapid synthesis and deposition occurs through activation of plasma membrane-located callose synthases in the microdomain beneath the cell wall apposition (Bhat et al. 2005; Mongrand et al. 2004) and its role in penetration resistance has been demonstrated by silencing a callose synthase gene in *Arabidopsis* (Jacobs et al. 2003).

The development of cell wall appositions is preceded by the localized aggregation of cytoplasm beneath the invading pathogen cell. As in other plant–pathogen interactions, it has been shown that, during the response of plants to *Phytophthora* species, aggregation of subcellular components within the plant cell at the infection site depends upon the plant cytoskeleton.

2. Cytoskeletal Reorganization Plays a Key Role in Facilitating Penetration Resistance

Rapid translocation of cytosol and subcellular components to the infection site is one of the earliest responses of plants to attempted penetration by fungi or Oomycetes and is typically induced by compatible, incompatible and non-adapted pathogens alike (Lipka and Panstruga 2005; Takemoto and Hardham 2004). This site-directed cytoplasmic

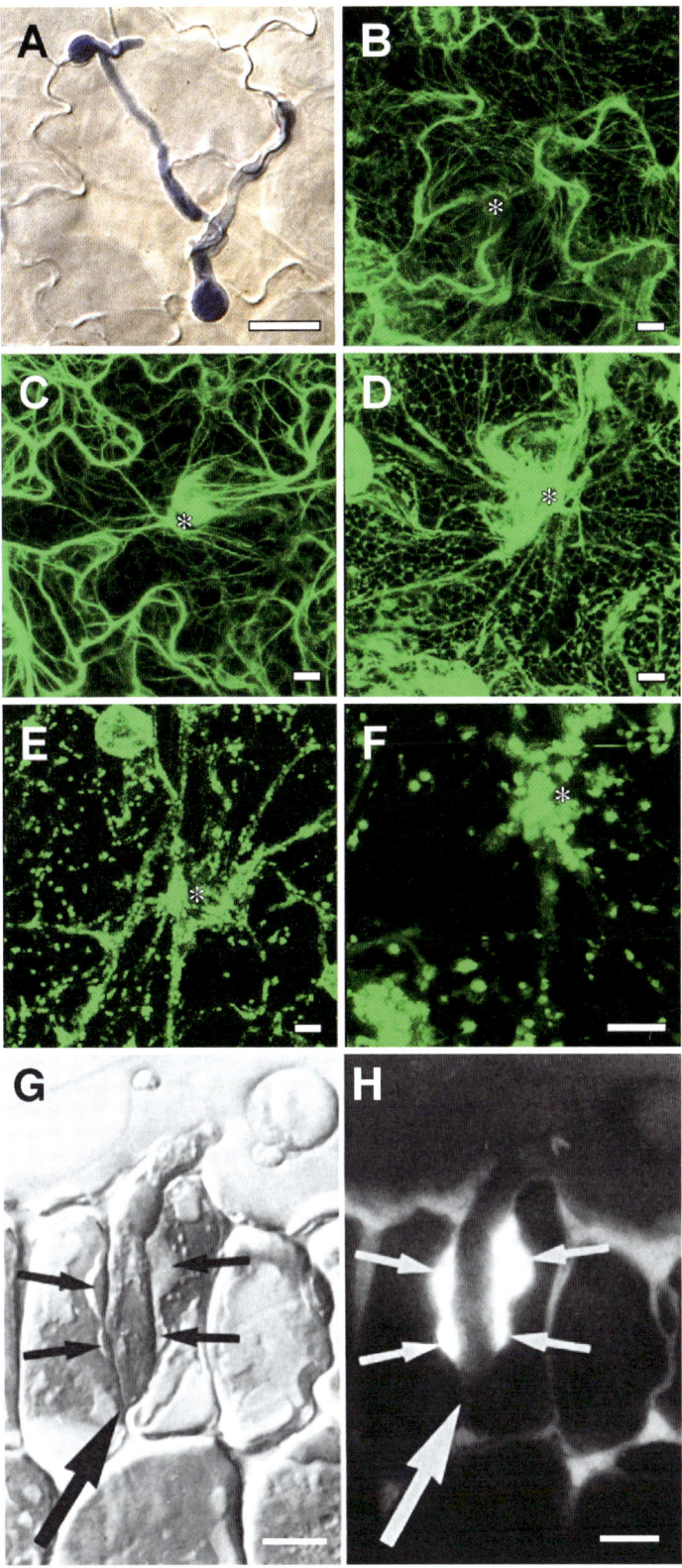

Fig. 1.4. Plant defence against *Phytophthora*. **A** Lactophenol trypan blue staining of germinated cysts of *P. sojae* growing on a cotyledon of *Arabidopsis thaliana*. *Bar* 25 μm. **B** Transgenic *A. thaliana* expressing GFP-TUA6, showing microtubule organization in cells surrounding the attempted penetration site of a *P. sojae* hypha (*asterisk*). *Bar* 10 μm. **C** Transgenic *A. thaliana* expressing GFP-hTalin, showing actin organization in cells surrounding the attempted penetration site of a *P. sojae* hypha (*asterisk*). Actin microfilaments and cables focus on the penetration site. *Bar* 10 μm. **D** Transgenic *A. thaliana* expressing GFP-tm-KKXX, showing organization of endoplasmic reticulum in cells surrounding the attempted penetration site of a *P. sojae* hypha (*asterisk*). *Bar* 10 μm. **E** Transgenic *A. thaliana* expressing GFP-STtmd, showing organization of Golgi apparatus in cells surrounding the attempted penetration site of a *P. sojae* hypha (*asterisk*). *Bar* 10 μm. **F** Transgenic *A. thaliana* expressing GFP-PTS, showing organization of peroxisomes in cells surrounding the attempted penetration site of a *P. sojae* hypha (*asterisk*). *Bar* 10 μm. For origin and details of GFP constructs, see Hardham and Takemoto (2006) and Takemoto et al. (2003). **G, H** Hypha of *P. cinnamomi* (*large arrows*) growing between two epidermal cells of a *Zea mays* root. The plant epidermal cells have deposited cell wall appositions containing callose (*small arrows*) which has been stained with aniline blue (**H**). Differential interference contrast image (**G**) and fluorescence image (**H**) of the same section. *Bar* 10 μm. Micrographs reproduced with permission from Hinch et al. (1985)

streaming is achieved by reorganization of the plant cytoskeleton.

Changes in the plant cytoskeleton during the defence response were initially studied using immunolabelling but more recently observations of transgenic plants containing GFP-tagged organelles have provided a more comprehensive picture of the dynamics and extent of the reorganization (Fig. 1.4B, C; Takemoto et al. 2003, 2006; Takemoto and Hardham 2004). Aggregation of cytoplasm beneath the invading pathogen cells occurs within minutes and is dependent on the reorganization and function of the actin cytoskeleton (Hazen and Bushnell 1983; Tomiyama et al. 1982). During their reorganization, both fine cortical actin microfilaments and the thicker transvacuolar bundles become focused on the site of attempted penetration (Fig. 1.4C; Gross et al. 1993; Takemoto et al. 2003). During the response to fungal pathogens, if this reorganization and function of actin microfilaments is prevented by treatment with pharmacological agents, the cytoplasmic aggregate fails to form and penetration resistance is lost (Kobayashi et al. 1997; Yun et al. 2003).

Microtubules also become reorganized in plant cells under attack by *Phytophthora* hyphae. The main response of the microtubule array appears to involve localized microtubule depolymerization at the infection site (Fig. 1.4B; Cahill et al. 2002; Gross et al. 1993; Takemoto et al. 2003). Immunocytochemical labelling has shown the absence of microtubules beneath the infection site in parsley and soybean cells inoculated with *P. infestans* or *P. sojae*, respectively (Cahill et al. 2002; Gross et al. 1993). However, observations of transgenic *Arabidopsis* plants expressing GFP-labelled tubulin revealed that, while microtubule depolymerization does occur, as indicated by a concentration of diffuse fluorescence at the infection site, only a subset of microtubules disappear (Fig. 1.4B). In a number of cases, the remaining microtubules form a circumferential arrangement that crosses cell boundaries around the penetration site, an arrangement that may be indicative of preferential depolymerization of microtubules that were oriented perpendicular to the wall adjacent to the invading hypha (Fig. 1.4B). Although purified elicitors may not induce the full range of defence responses, application of purified elicitin from *P. cryptogea* causes microtubule depolymerization in tobacco suspension culture cells (Binet et al. 2001; Higaki et al. 2007).

3. Actin-Based Polarized Transport and Secretion at the Infection Site

Observations of transgenic *Arabidopsis* plants expressing GFP-labelled components of the secretory pathway also highlight rapid changes in the distribution of endoplasmic reticulum, dictyosomes and peroxisomes in response to attempted penetration by *P. sojae* (Fig. 1.4D–F; Hardham and Takemoto 2006; Takemoto et al. 2003). A dense network of lamellar endoplasmic reticulum develops at the penetration site but is continuous with the reticulate network elsewhere in the cell (Fig. 1.4D). Peroxisomes and dictyosomes become aggregated at the infection site (Fig. 1.4E, F). Having aggregated, most individual peroxisomes remain relatively stationary at the infection site. Dictyosomes, however, continue to move into and out of this area, their accumulation arising from periods of immobility at the infection site. Concentrated arrays of actin and endoplasmic reticulum also form in adjacent cells along the walls adjoining those of the infected cell. This reorganization is especially evident around cells undergoing hypersensitive cell death and is indicative of signalling between infected and adjacent cells (Takemoto et al. 2003, 2006). These observations indicate that actin-dependent recruitment of the secretory apparatus and other organelles at the infection site provides the basis for cell wall apposition formation and penetration resistance.

V. Counterdefence: Suppressing Plant Defences

Studies of the *P. infestans* AVR3a effector described in Sect. III.B showed that AVR3a is capable of suppressing cell death induced by the *P. infestans* elicitin INF1 (Bos et al. 2006). The glucanase and protease inhibitors secreted by *Phytophthora* species (see Sect. III.A) also counteract the effects of the PR proteins produced as part of the plant's defence response. These observations are the latest contributions to a growing body of evidence of counterdefence measures employed by *Phytophthora* to combat the host defence response. For example, water-soluble glucans produced by *Phytophthora* zoospores or mycelia or secreted by germinated cysts inhibit hypersensitive cell death and phytoalexin production in a range of hosts (Andreu et al. 1998; Doke 1975; Doke et al. 1979; Doke et al.

1980; Sanchez et al. 1994). *Phytophthora* metabolites also suppress transient accumulation of phenylalanine ammonia lyase transcripts and generation of NADPH-dependent superoxide (Shiraishi et al. 1997). An extracellular invertase secreted by *P. sojae* mycelia also inhibits production of the soybean phytoalexin, glyceollin, triggered by a glucan elicitor from the hyphal cell walls (Zeigler and Pontzen 1982). The *P. sojae* invertase is a glycoprotein and the carbohydrate portion of the molecule is responsible for the inhibitory effect.

Plant defence against pathogen attack typically includes an oxidative burst involving elevated levels of reactive oxygen species such as hydrogen peroxide (Lamb and Dixon 1997). The reactive oxygen species are not only toxic to the pathogen but also help protect the plant cell by strengthening the cell wall barrier through enhanced cross-linking of wall components. Hydrogen peroxide is degraded by the scavenging enzyme, catalase, and a recent study of the infection of tobacco seedlings by *P. nicotianae* explored the potential role of *Phytophthora* catalases in counterdefence against hydrogen peroxide production during the oxidative burst (Blackman and Hardham 2008). The genome of *P. nicotianae* contains three catalase genes, one of which, *PnCat2*, produces a protein targeted to peroxisomes. Of the three genes, *PnCat2* is the one predominantly expressed during asexual development and it is also strongly up-regulated during the infection of susceptible tobacco plants. The activity of *P. nicotianae* catalase also increases during infection. The increased production of *Phytophthora* catalase during infection is consistent with catalase constituting a counterdefence mechanism that protects the pathogen against the oxidative burst.

Studies of oomycete plant pathogens were among the first to demonstrate that the plant defence response could include stomatal closure and that some pathogens were able to deregulate the plant's control over stomatal aperture (Farrell et al. 1969; Lindenthal et al. 2005; McDonald and Cahill 1999). Closure of stomata is believed to physically inhibit ingress of bacterial pathogens and to make it more difficult for fungal and oomycete pathogens to grow through the stomatal aperture either during initial colonization of the plant or during sporulation. On the one hand, infection of soybean with *P. sojae* causes rapid stomatal closure (McDonald and Cahill 1999) while, on the other, infection of potato leaves with *P. infestans* and of grapevine leaves with *Plasmopara viticola* causes abnormal opening of stomata in colonized tissues (Allègre et al. 2007; Farrell et al. 1969). It has been recently shown that *Pseudomonas syringae* secretes coronatine that induces stomata to re-open and allows bacteria to invade through the stomatal aperture (Melotto et al. 2006). It seems likely that some species of *Phytophthora* have also developed mechanisms to inhibit stomatal closure but the regulatory molecules remain to be identified.

VI. Conclusions

Species of *Phytophthora* constitute some of the most devastating plant pathogens known. Under the right conditions, they are able to spread at an alarming rate, threatening food production and natural ecosystems on a vast scale (Schiermeier 2001; Shearer et al. 2007). Studies of a number of key *Phytophthora* species have made an important contribution to our overall understanding of the cell and molecular biology of the interactions of plants with their eukaryotic pathogens. We have detailed knowledge of the cellular basis of the initiation of infection of host plants, including the production, structure and behaviour of the asexual motile zoospores. Recent molecular and genomic studies also provide a foundation for elucidation of the molecular basis of *Phytophthora* pathogenicity. Genes encoding spore and hyphal adhesins, cell wall degrading enzymes and pathogenicity effectors, including inhibitors of plant enzymes involved in defence, have been cloned and characterized. As well as publication of a number of transcriptome studies (Fabritius et al. 2002; Kim and Judelson 2003; Panabieres et al. 2005; Prakob and Judelson 2007; Randall et al. 2005; Shan et al. 2004b; Škalamera et al. 2004), the full genome sequences of four *Phytophthora* species are now available (http://www.genome.jgi-psf.org; http://www.broad.mit.edu/tools/data/seq.html; http://www.shake.jgi-psf.org/Phyca1/), providing an invaluable resource for researchers working not only on *Phytophthora* but on other Oomycetes and Stramenopiles. Comparisons of *Phytophthora* genomes highlight the high degree of synteny between them, a feature that is proving to be of great value in comparative studies. Bioinformatic analyses of *Phytophthora* secretomes also led to the exciting discovery of the RXLR motif responsible for translocation of *Phytophthora* effector proteins into the cytoplasm of host plant cells (Whisson et al. 2007).

Investigations of *Phytophthora*–plant inter-actions have also focused on the plant defence response. Genes encoding important resistance proteins have been cloned and their function in elicitor recognition, signalling and the induction of defence is under continued investigation. Struc-tural changes in plant cells responding to attack from *Phytophthora* pathogens have been docu-mented, with dynamic reorganization of cytoskel-etal and endomembrane systems being visualized using plants containing GFP-tagged organelles. Development of cell wall appositions through the targeted secretion of wall components and toxins at the infection site successfully inhibits ingress of non-adapted *Phytophthora* pathogens. Those adapted *Phytophthora* species or isolates that can evade or overcome the plant's basal defences may be subsequently thwarted by hypersensitive cell death. Recent advances in genomic, proteomic and bioinformatic technologies have opened up an exciting range of new opportunities in studies of *Phytophthora*–plant interactions and there is little doubt that the next few years will see an explosion in our understanding of *Phytophthora* pathogenic-ity and plant defence.

Acknowledgements. We thank Dr Daigo Takemoto for the images in Figs. 1.4A–F and Drs Hohl, Mims and Wetherbee for allowing us to reproduce their published micrographs.

References

Able AJ, Guest DI, Sutherland MW (2001) Relationship between transmembrane ion movements, production of reactive oxygen species and the hypersensitive response during the challenge of tobacco suspension cells by zoospores of *Phytophthora nicotianae*. Physiol Mol Plant Pathol 58:189–198

Abramovitch RB, Martin GB (2004) Strategies used by bac-terial pathogens to suppress plant defenses. Curr Opin Plant Biol 7:356–364

Adams JC, Tucker RP (2000) The thrombospondin type 1 repeat (TSR) superfamily: diverse proteins with related roles in neuronal development. Dev Dyn 218:280–299

Adl SM, Simpson AGB, Farmer MA, Andersen RA, Anderson OR, Barta JR, Bowser SS, Brugerolle G, Fensome RA, Fredericq S, James TY, Karpov S, Kugrens P, Krug J, Lane CE, Lewis LA, Lodge J, Lynn DH, Mann DG, McCourt RM, Mendoza L, Moestrup O, Mozley-Standridge SE, Nerad TA, Shearer CA, Smirnov AV, Spiegel FW, Tay-lor MFJR (2005) The new higher level classification of eukaryotes with emphasis on the taxonomy of protists. J Eukaryot Microbiol 52:399–451

Aist JR (1976) Cytology of penetration and infection – fungi. In: Heitefuss R, Williams PH (eds) Physiological plant pathology. Encyclopedia of plant physiology, new series, vol4. Springer, Heidelberg, pp 197–221

Allègre M, Daire X, Heloir MC, Trouvelot S, Mercier L, Adrian M, Pugin A (2007) Stomatal deregulation in *Plasmopara viticola*-infected grapevine leaves. New Phytol 173:832–840

Ambikapathy J, Marshall JS, Hocart CH, Hardham AR (2002) The role of proline in osmoregulation in *Phy-tophthora nicotianae*. Fungal Genet Biol 35:287–299

Amsellem Z, Chen BA, Gressel J (2002) Engineering hyper-virulence in a mycoherbicial fungus for efficient weed control. Nat Biotechnol 20:1035–1039

An Q, Hückelhoven R, Kogel K-H, van Bel AJE (2006) Multivesicular bodies participate in a cell wall-associ-ated defence response in barley leaves attacked by the pathogenic powdery mildew fungus. Cell Microbiol 8:1009–1019

Andreu A, Tonón C, Van Damme M, Huarte M, Daleo G (1998) Effect of glucans from different races of *Phy-tophthora infestans* on defense reactions in potato tubers. Eur J Plant Pathol 104:777–783

Armstrong MR, Whisson SC, Pritchard L, Bos JIB, Venter E, Avrova AO, Rehmany AP, Bohme U, Brooks K, Cherevach I, Hamlin N, White B, Frasers A, Lord A, Quail MA, Churcher C, Hall N, Berriman M, Huang S, Kamoun S, Beynon JL, Birch PRJ (2005) An ancestral oomyc-ete locus contains late blight avirulence gene *Avr3a*, encoding a protein that is recognized in the host cytoplasm. Proc Natl Acad Sci USA 102:7766–7771

Ashford AE, Allaway WG (2007) Motile tubular vacuole systems. In: Gow NAR, Howard RJ (eds) The Mycota. Springer, Heidelberg, pp 49–86

Avrova AO, Venter E, Birch PRJ, Whisson SC (2003) Profil-ing and quantifying differential gene transcription in *Phytophthora infestans* prior to and during the early stages of potato infection. Fungal Genet Biol 40:4–14

Aylor DE (2003) Spread of plant disease on a continental scale: role of aerial dispersal of pathogens. Ecology 84:1989–1997

Bailey BA, Jennings JC, Anderson JD (1997) The 24-kDa protein from *Fusarium oxysporum* f. *sp. erythroxyli*: occurrence in related fungi and the effect of growth medium on its production. Can J Microbiol 43:45–55

Ballvora A, Ercolano MR, Weiss J, Meksem K, Bormann CA, Oberhagemann P, Salamini F, Gebhardt C (2002) The *R1* gene for potato resistance to late blight (*Phytoph-thora infestans*) belongs to the leucine zipper/NBS/LRR class of plant resistance genes. Plant J 30:361–371

Barr DJS (1992) Evolution and kingdoms of organisms from the perspective of a mycologist. Mycologia 84:1–11

Bartnicki-Garcia S (1970) Cell wall composition and other biochemical markers in fungal phylogeny. In: Har-borne JB (ed) Phytochemical phylogeny. Academic, London, pp 81–103

Bent AF, Mackey D (2007) Elicitors, effectors, and R genes: The new paradigm and a lifetime supply of questions. Annu Rev Phytopathol 45:399–436

Bhat RA, Miklis M, Schmelzer E, Schulze-Lefert P, Pan-struga R (2005) Recruitment and interaction dynam-ics of plant penetration resistance components in a plasma membrane microdomain. Proc Natl Acad Sci USA 102:3135–3140

Bhattacharjee S, Hiller NL, Liolios K, Win J, Kanneganti TD, Young C, Kamoun S, Haldar K (2006) The malarial host-targeting signal is conserved in the Irish potato famine pathogen. PLoS Pathogens 2:0453–0465

Bimpong CE (1975) Changes in metabolic reserves and enzyme activities during zoospore motility and cyst germination in *Phytophthora palmivora*. Can J Bot 53:1411–1416

Binet M-N, Humbert C, Lecourieux D, Vantard M, Pugin A (2001) Disruption of microtubular cytoskeleton induced by cryptogein, an elicitor of hypersensitive response in tobacco cells. Plant Physiol 125:564–572

Birch PRJ, Rehmany AP, Pritchard L, Kamoun S, Beynon JL (2006) Trafficking arms: oomycete effectors enter host plant cells. Trends Microbiol 14:8–11

Bircher U, Hohl HR (1997) Environmental signalling during induction of appressorium formation in *Phytophthora*. Mycol Res 101:395–402

Bishop-Hurley SL, Mounter SA, Laskey J, Morris RO, Elder J, Roop P, Rouse C, Schmidt FJ, English JT (2002) Phage-displayed peptides as developmental agonists for *Phytophthora capsici* zoospores. Appl Environ Microbiol 68:3315–3320

Blackman LM, Hardham AR (2008) Regulation of catalase activity and gene expression during *Phytophthora nicotianae* development and infection of tobacco. Molecular Plant Pathology 9:495–510

Blanco FA, Judelson HS (2005) A bZIP transcription factor from *Phytophthora* interacts with a protein kinase and is required for zoospore motility and plant infection. Mol Microbiol 56:638–648

Blume B, Nürnberger T, Nass N, Scheel D (2000) Receptor-mediated increase in cytoplasmic free calcium required for activation of pathogen defense in parsley. Plant Cell 12:1425–1440

Boissy G, O'Donohue M, Gaudemer O, Perez V, Pernollet J-C, Brunie S (1999) The 2.1 Å structure of an elicitin–ergosterol complex: a recent addition to the sterol carrier protein family. Prot Sci 8:1191–1199

Bos JIB, Kanneganti TD, Young C, Cakir C, Huitema E, Win J, Armstrong MR, Birch PRJ, Kamoun S (2006) The C-terminal half of *Phytophthora infestans* RXLR effector AVR3a is sufficient to trigger R3a-mediated hypersensitivity and suppress INF1-induced cell death in *Nicotiana benthamiana*. Plant J 48:165–176

Boudjeko T, Andème-Onzighi C, Vicré M, Balangé A-P, Ndoumou DO, Driouich A (2006) Loss of pectin is an early event during infection of cocoyam roots by *Pythium myriotylum*. Planta 223:271–282

Brunner F, Rosahl S, Lee J, Rudd JJ, Geiler C, Kauppinen S, Rasmussen, G, Scheel D, Nürnberger T (2002a) Pep-13, a plant defense-inducing pathogen-associated pattern from *Phytophthora* transglutaminases. EMBO J 21:6681–6688

Brunner F, Wirtz W, Rose JKC, Darvill AG, Govers F, Scheel D, Nürnberger T (2002b) A beta-glucosidase/xylosidase from the phytopathogenic oomycete, *Phytophthora infestans*. Phytochem 59:689–696

Cahill D, Cope M, Hardham AR (1996) Thrust reversal by tubular mastigonemes: immunological evidence for a role of mastigonemes in forward motion of zoospores of *Phytophthora cinnamomi*. Protoplasma 194:18–28

Cahill D, Rookes J, Michalczyk A, McDonald K, Drake A (2002) Microtubule dynamics in compatible and incompatible interactions of soybean hypocotyl cells with *Phytophthora sojae*. Plant Pathol 51:629–640

Carlile MJ (1983) Motility, taxis, and tropism in *Phytophthora*. In: Erwin DC, Bartnicki-Garcia S, Tsao P (eds) Phytophthora. Its biology, taxonomy, ecology, and pathology. American Phytopathology Society, St Paul, pp 95–107

Cheong J-J, Alba R, Côté F, Enkerli J, Hahn MG (1993) Solubilization of functional plasma membrane-localized hepta-β-glucoside elicitor-binding proteins from soybean. Plant Physiol 103:1173–1182

Coffey MD, Gees R (1991) The cytology of development. In: Ingram DS, Williams PH (eds) *Phytophthora infestans*, the cause of late blight of potato. Advances in plant pathology, vol 7. Academic, London, pp 31–51

Connolly MS, Sakihama Y, Phuntumart V, Jiang Y, Warren F, Mourant L, Morris PF (2005) Heterologous expression of a pleiotropic drug resistance transporter from *Phytophthora sojae* in yeast transporter mutants. Curr Genet 48:356–365

Côté F, Roberts KA, Hahn MG (2000) Identification of high-affinity binding sites for the hepta-β-glucoside elicitor in membranes of the model legumes *Medicago truncatula* and *Lotus japonicus*. Planta 211:596–605

Dearnaley JDW, Hardham AR (1994) The Golgi apparatus of *Phytophthora cinnamomi* makes three types of secretory or storage vesicles concurrently. Protoplasma 182:75–79

Dieguez-Uribeondo J, Gierz G, Bartnicki-Garcia S (2004) Image analysis of hyphal morphogenesis in Saprolegniaceae (Oomycetes). Fungal Genet Biol 41:293–307

Dodds PN, Lawrence GJ, Catanzariti A-M, Ayliffe MA, Ellis JG (2004) The *Melampsora lini AvrL567* avirulence genes are expressed in haustoria and their products are recognized inside plant cells. Plant Cell 16:755–768

Doke N (1975) Prevention of the hypersensitive reaction of potato cells to infection with an incompatible race of *Phytophthora infestans* by constituents of the zoospores. Physiol Plant Pathol 7:1–7

Doke N, Garas NA, Kuc J (1979) Partial characterization and aspects of the mode of action of a hypersensitivity-inhibiting factor (HIF) isolated from *Phytophthora infestans*. Physiol Plant Pathol 15:127–140

Doke N, Garas NA, Kuc J (1980) Effect on host hypersensitivity of suppressors released during the germination of *Phytophthora infestans* cystospores. Phytopathology 70:35–38

Dong W, Latijnhouwers M, Jiang RHY, Meijer HJG, Govers F (2004) Downstream targets of the *Phytophthora infestans* Gα subunit PiGPA1 revealed by cDNA-AFLP. Mol Plant Pathol 5:483–494

Ebstrup T, Saalbach G, Egsgaard H (2005) A proteomics study of *in vitro* cyst germination and appressoria formation in *Phytophthora infestans*. Proteomics 5:2839–2848

Enkerli K, Hahn MG, Mims CW (1997) Ultrastructure of compatible and incompatible interactions of soybean roots infected with the plant pathogenic oomycete *Phytophthora sojae*. Can J Bot 75:1493–1508

Epstein L, Nicholson RL (1997) Adhesion of spores and hyphae to plant surfaces. In: Carroll G, Tudzynski

P (eds) Plant relationships. The Mycota, vol VII.A. Springer, Heidelberg, pp 11–25

Erwin DC, Ribeiro OK (1996) *Phytophthora* diseases worldwide. APS, St. Paul, p. 562

Fabritius AL, Cvitanich C, Judelson HS (2002) Stage-specific gene expression during sexual development in *Phytophthora infestans*. Mol Microbiol 45:1057–1066

Farrell GM, Preece TF, Wren MJ (1969) Effects of infection by *Phytophthora infestans* (Mont.) de Bary on the stomata of potato leaves. Ann Appl Biol 63:265–275

Fellbrich G, Romanski A, Varet A, Blume B, Brunner F, Engelhardt S, Felix G, Kemmerling B, Krzymowska M, Nürnberger T (2002) NPP1, a *Phytophthora*-associated trigger of plant defense in parsley and *Arabidopsis*. Plant J 32:375–390

Gao HY, Narayanan NN, Ellison L, Hattacharyya MK (2005) Two classes of highly similar coiled coil-nucleotide binding-leucine rich repeat genes isolated from the *Rps1-k* locus encode *Phytophthora* resistance in soybean. Mol Plant–Microbe Interact 18:1035–1045

Gaulin E, Jauneau A, Villalba F, Rickauer M, Esquerré-Tugayé MT, Bottin A (2002) The CBEL glycoprotein of *Phytophthora parasitica* var. *nicotianae* is involved in cell wall deposition and adhesion to cellulosic substrates. J Cell Sci 115:4565–4575

Gaulin E, Drame N, Lafitte C, Torto-Alalibo T, Martinez Y, Ameline-Torregrosa C, Khatib M, Mazarguil H, Villalba-Mateos F, Kamoun S, Mazars C, Dumas B, Bottin A, Esquerre-Tugaye MT, Rickauer M (2006) Cellulose binding domains of a *Phytophthora* cell wall protein are novel pathogen-associated molecular patterns. Plant Cell 18:1766–1777

Gautam Y, Cahill DM, Hardham AR (1999) Development of a quantitative immunodipstick assay for *Phytophthora nicotianae*. Food Agric Immunol 11:229–242

Gees R, Hohl HR (1988) Cytological comparison of specific (*R3*) and general resistance to late blight in potato leaf tissue. Phytopathology 78:350–357

Gijzen M, Nürnberger T (2006) Nep1-like proteins from plant pathogens: Recruitment and diversification of the NPP1 domain across taxa. Phytochem 67:1800–1807

Glazebrook J (2005) Contrasting mechanisms of defence against biotrophic and necrotrophic pathogens. Annu Rev Phytopathol 43:205–227

Görnhardt B, Rouhara I, Schmelzer E (2000) Cyst germination proteins of the potato pathogen *Phytophthora infestans* share homology with human mucins. Mol Plant–Microbe Interact 13:32–42

Götesson A, Marshall JS, Jones DA, Hardham AR (2002) Characterization and evolutionary analysis of a large polygalacturonase gene family in the oomycete plant pathogen *Phytophthora cinnamomi*. Mol Plant–Microbe Interact 15:907–921

Gow NAR (2004) New angles in mycology: studies in directional growth and directional motility. Mycol Res 108:5–13

Grenville-Briggs LJ, Avrova AO, Bruce CR, Williams A, Whisson SC, Birch PRJ, Van West P (2005) Elevated amino acid biosynthesis in *Phytophthora infestans* during appressorium formation and potato infection. Fungal Genet Biol 42:244–256

Gross P, Julius C, Schmelzer E, Hahlbrock K (1993) Translocation of cytoplasm and nucleus to fungal penetration sites is associated with depolymerization of microtubules and defence gene activation in infected, cultured parsley cells. EMBO J 12:1735–1744

Gubler F, Hardham AR (1988) Secretion of adhesive material during encystment of *Phytophthora cinnamomi* zoospores, characterized by immunogold labelling with monoclonal antibodies to components of peripheral vesicles. J Cell Sci 90:225–235

Gubler F, Hardham AR (1990) Protein storage in large peripheral vesicles in *Phytophthora* zoospores and its breakdown after cyst germination. Exp Mycol 14:393–404

Gubler F, Jablonsky PP, Duniec J, Hardham AR (1990) Localization of calmodulin in flagella of zoospores of *Phytophthora cinnamomi*. Protoplasma 155:233–238

Gunderson JH, Elwood H, Ingold A, Kindle K, Sogin ML (1987) Phylogenetic relationships between chlorophytes, chrysophytes, and oomycetes. Proc Natl Acad Sci USA 84:5823–5827

Guyonnet Duperat V, Audie J-P, Debailleul V, Laine A, Buisine M-P, Galiegue-Zouitina S, Pigny P, Degand P, Aubert J-P, Porchet N (1995) Characterization of the human mucin gene MUC5AC: a consensus cysteine-rich domain for 11p15 genes? Biochem J 305:211–219

Hahn M, Neef U, Struck C, Göttfert M, Mendgen K (1997) A putative amino acid transporter is specifically expressed in haustoria of the rust fungus *Uromyces fabae*. Mol Plant–Microbe Interact 10:438–445

Halim VA, Hunger A, Macioszek V, Landgraf P, Nürnberger T, Scheel D, Rosahl S (2004) The oligopeptide elicitor Pep-13 induces salicylic acid-dependent and -independent defense reactions in potato. Physiol Mol Plant Pathol 64:311–318

Ham K-S, Wu S-C, Darvill AG, Albersheim P (1997) Fungal pathogens secrete an inhibitor protein that distinguishes isoforms of plant pathogenesis-related endo-β-1,3-glucanases. Plant J 11:169–179

Hardham AR (1987a) Microtubules and the flagellar apparatus in zoospores and cysts of the fungus *Phytophthora cinnamomi*. Protoplasma 137:109–124

Hardham AR (1987b) Ultrastructure and serial section reconstruction of zoospores of the fungus *Phytophthora cinnamomi*. Exp Mycol 11:297–306

Hardham AR (1989) Lectin and antibody labelling of surface components of spores of *Phytophthora cinnamomi*. Aust J Plant Physiol 16:19–32

Hardham AR (1995) Polarity of vesicle distribution in oomycete zoospores: development of polarity and importance for infection. Can J Bot 73[Suppl 1]:S400–S407

Hardham AR (2001) The cell biology behind *Phytophthora* pathogenicity. Aust Plant Pathol 30:91–98

Hardham AR (2005) *Phytophthora cinnamomi*. Mol Plant Pathol 6:589–604

Hardham AR (2007) Cell biology of fungal and oomycete infection of plants. In: Howard RJ, Gow NAR (eds) Biology of the fungal cell, 2nd edn. The Mycota, vol VII. Springer, Heidelberg, pp 251–289

Hardham AR, Gubler F (1990) Polarity of attachment of zoospores of a root pathogen and pre-alignment of the emerging germ tube. Cell Biol Int Rep 14:947–956

Hardham AR, Hyde GJ (1997) Asexual sporulation in the oomycetes. Adv Bot Res 24:353–398

Hardham AR, Suzaki E (1986) Encystment of zoospores of the fungus, *Phytophthora cinnamomi*, is induced by specific lectin and monoclonal antibody binding to the cell surface. Protoplasma 133:165–173

Hardham AR, Takemoto D (2006) Dynamic subcellular responses in plants during interactions with fungi and oomycetes. In: Sánchez F, Quinto C, López-Lara IM, Geiger O (eds) Proceedings of the XII international congress on molecular plant–microbe interactions. ISMP-MI, St Paul, pp 70–79

Harper JDI, Gubler F, Salisbury JL, Hardham AR (1995) Centrin association with the flagellar apparatus in spores of *Phytophthora cinnamomi*. Protoplasma 188:225–235

Harper JT, Waanders E, Keeling PJ (2005) On the monophyly of chromalveolates using a six-protein phylogeny of eukaryotes. Int J Syst Evol Microbiol 55:487–496

Hazen BE, Bushnell WR (1983) Inhibition of the hypersensitive reaction in barley to powdery mildew by heat shock and cytochalasin B. Physiol Plant Pathol 23:421–438

Heath IB (1995) The cytoskeleton. In: Gow NAR, Gadd GM (eds) The growing fungus. Chapman and Hall, London, pp 99–134

Heath IB, Kaminskyj SGW (1989) The organization of tip-growth-related organelles and microtubules revealed by quantitative analysis of freeze-substituted oomycete hyphae. J Cell Sci 93:41–52

Hendrix JW (1970) Sterols in growth and reproduction of fungi. Annu Rev Phytopathol 8:111–130

Higaki T, Goh T, Hayashi T, Kutsuna N, Kadota Y, Hasezawa S, Sano T, Kuchitsu K (2007) Elicitor-induced cytoskeletal rearrangement relates to vacuolar dynamics and execution of cell death: in vivo imaging of hypersensitive cell death in tobacco BY-2 cells. Plant Cell Physiol 48:1414–1425

Hiller NL, Bhattacharjee S, van Ooij C, Liolios K, Harrison T, Lopez-Estaño C, Haldar K (2004) A host-targeting signal in virulence proteins reveals a secretome in malarial infection. Science 306:1934–1937

Hinch JM, Wetherbee R, Mallett JE, Clarke AE (1985) Response of *Zea mays* roots to infection with *Phytophthora cinnamomi* I. The epidermal layer. Protoplasma 126:178–187

Honda D, Shono T, Kimura K, Fujita S, Iseki M, Makino Y, Murakami A (2007) Homologs of the sexually induced gene 1 (*sig1*) product constitute the Stramenopile mastigonemes. Protist 158:77–88

Huang S, van der Vossen EAG, Kuang H, Vleeshouwers VGAA, Zhang N, Borm TJA, van Eck HJ, Baker B, Jacobsen E, Visser RGF (2005) Comparative genomics enabled the isolation of the *R3a* late blight resistance gene in potato. Plant J 42:251–261

Huitema E, Vleeshouwers VGAA, Francis DM, Kamoun S (2003) Active defence responses associated with non-host resistance of *Arabidopsis thaliana* to the oomycete pathogen *Phytophthora infestans*. Mol Plant Pathol 4:487–500

Jackson SL, Heath IB (1990) Visualization of actin arrays in growing hyphae of the fungus *Saprolegnia ferax*. Protoplasma 154:66–70

Jacobs AK, Lipka V, Burton RA, Panstruga R, Strizhov N, Schulze-Lefert P, Fincher GB (2003) An Arabidopsis callose synthase, GSL5, is required for wound and papillary callose formation. Plant Cell 15:2503–2513

Jahn TL, Landman MD, Fonseca JR (1964) The mechanism of locomotion of flagellates. II. function of the mastigonemes of *Ochromonas*. J Protozool 11:291–296

Jeun YCH, Buchenauer H (2001) Infection structures and localization of the pathogenesis-related protein AP24 in leaves of tomato plants exhibiting systemic acquired resistance against *Phytophthora infestans* after pre-treatment with 3-aminobutyric acid or tobacco necrosis virus. J Phytopathol 149:141–153

Jiang RHY, Tyler BM, Govers F (2006) Comparative analysis of *Phytophthora* genes encoding secreted proteins reveals conserved synteny and lineage-specific gene duplications and deletions. Mol Plant–Microbe Interact 19:1311–1321

Jones DA, Takemoto D (2004) Plant innate immunity – direct and indirect recognition of general and specific pathogen-associated molecules. Curr Opin Immunol 16:48–62

Joosten MHAJ, Vogelsang R, Cozijnsen TJ, Verberne MC, de Wit PJGM (1997) The biotrophic fungus *Cladosporium fulvum* circumvents Cf-4- mediated resistance by producing unstable AVR4 elicitors. Plant Cell 9:367–379

Judelson HS, Blanco FA (2005) The spores of *Phytophthora*: weapons of the plant destroyer. Nat Rev Microbiol 3:47–58

Kaku H, Nishizawa Y, Ishii-Minami N, Akimoto-Tomiyama C, Dohmae N, Takio K, Minami E, Shibuya N (2006) Plant cells recognize chitin fragments for defense signaling through a plasma membrane receptor. Proc Natl Acad Sci USA 103:11086–11091

Kaminskyj SGW, Heath IB (1995) Integrin and spectrin homologues, and cytoplasm-wall adhesion in tip growth. J Cell Sci 108:849–856

Kamoun S (2006) A catalogue of the effector secretome of plant pathogenic oomycetes. Annu Rev Phytopathol 44:41–60

Kamoun S, Young M, Glascock CB, Tyler BM (1993) Extracellular protein elicitors from *Phytophthora*: host-specificity and induction of resistance to bacterial and fungal phytopathogens. Mol Plant–Microbe Interact 6:15–25

Kamoun S, Van West P, Vleeshouwers VGAA, De Groot KE, Govers F (1998) Resistance of *Nicotiana benthamiana* to *Phytophthora infestans* is mediated by the recognition of the elicitor protein INF1. Plant Cell 10:1413–1425

Kanneganti TD, Huitema E, Cakir C, Kamoun S (2006) Synergistic interactions of the plant cell death pathways induced by *Phytophthora infestans* Nep1-like protein PiNPP1.1 and INF1 elicitin. Mol Plant–Microbe Interact 19:854–863

Kemen E, Kemen AC, Rafiqi M, Hempel U, Mendgen K, Hahn M, Voegele RT (2005) Identification of a protein from rust fungi transferred from haustoria into infected plant cells. Mol Plant–Microbe Interact 18:1130–1139

Kim KS, Judelson HS (2003) Sporangia-specific gene expression in the oomycete phytopathogen *Phytophthora infestans*. Eukaryot Cell 2:1376–1385

King N, Hittinger CT, Carroll SB (2003) Evolution of key cell signaling and adhesion protein families predates animal origins. Science 301:361–363

Kobayashi Y, Kobayashi I, Funaki Y, Fujimoto S, Takemoto T, Kunoh H (1997) Dynamic reorganization of microfilaments and microtubules is necessary for the expression of non-host resistance in barley coleoptile cells. Plant J 11:525–537

Krämer R, Freytag S, Schmelzer E (1997) In vitro formation of infection structures of *Phytophthora infestans* is associated with synthesis of stage specific polypeptides. Eur J Plant Pathol 103:43–53

Lamb C, Dixon RA (1997) The oxidative burst in plant disease resistance. Annu Rev Plant Physiol Plant Mol Biol 48:251–275

Latijnhouwers M, Munnik T, Govers F (2002) Phospholipase D in *Phytophthora infestans* and its role in zoospore differentiation. Mol Plant-Microbe Interact 15:939–946

Latijnhouwers M, Ligterink W, Vleeshouwers VGAA, Van West P, Govers F (2004) A G alpha subunit controls zoospore motility and virulence in the potato late blight pathogen *Phytophthora infestans*. Mol Microbiol 51:925–936

Lauge R, de Wit PJGM (1998) Fungal avirulence genes: structure and possible functions. Fungal Genet Biol 24:285–297

Lindenthal M, Steiner U, Dehne H-W, Oerke E-C (2005) Effect of downy mildew development on transcription of cucumber leaves visualized by digital infrared thermography. Plant Mol Biol 64:387–395

Lipka V, Panstruga R (2005) Dynamic cellular responses in plant-microbe interactions. Curr Opin Plant Biol 8:625–631

Loprete DM, Hill TW (2002) Isolation and characterization of an endo-(1,4)-beta-glucanase secreted by *Achlya ambisexualis*. Mycologia 94:903–911

Marti M, Good RT, Rug M, Kneupfer E, Cowman AF (2004) Targeting malaria virulence and remodeling proteins to the host erythrocyte. Science 306:1930–1933

Martin GB, Bogdanove AJ, Sessa G (2003) Understanding the functions of plant disease resistance proteins. Annu Rev Plant Biol 54:23–61

McDonald KL, Cahill DM (1999) Evidence for a transmissible factor that causes rapid stomatal closure in soybean at sites adjacent to and remote from hypersensitive cell death induced by *Phytophthora sojae*. Physiol Mol Plant Pathol 55:197–203

McLeod A, Smart CD, Fry WE (2003) Characterization of 1,3-β-glucanase and 1,3;1,4-β-glucanase genes from *Phytophthora infestans*. Fung Genet Biol 38:250–263

Melotto M, Underwood W, Koczan J, Nomura K, He SY (2006) Plant stomata function in innate immunity against bacterial invasion. Cell 126:969–980

Mikes V, Milat ML, Ponchet M, Ricci P, Blein J-P (1997) The fungal elicitor cryptogein is a sterol carrier protein. FEBS Lett 416:190–192

Mikes V, Milat ML, Ponchet M, Panabières F, Ricci P, Blein JP (1998) Elicitins, proteinaceous elicitors of plant defense, are a new class of sterol carrier proteins. Biochem Biophys Res Commun 245:133–139

Miller SA, Maxwell DP (1984) Light microscope observations of susceptible, host resistant, and nonhost resistant interactions of alfalfa with *Phytophthora megasperma*. Can J Bot 62:109–116

Mitchell HJ, Hardham AR (1999) Characterisation of the water expulsion vacuole in *Phytophthora nicotianae* zoospores. Protoplasma 206:118–130

Money NP, Davis CM, Ravishankar JP (2004) Biomechanical evidence for convergent evolution of the invasive growth process among fungi and oomycete water molds. Fungal Genet Biol 41:872–876

Mongrand S, Morel J, Laroche J, Claverol S, Carde J-P, Hartmann M-A, Bonneu M, Simon-Plas F, Lessire R, Bessoule J-J (2004) Lipid rafts in higher plant cells: purification and characterization of Triton-X-100-insoluble microdomains from tobacco plasma membrane. J Biol Chem 279:36277–36286

Morris PF, Ward EWB (1992) Chemoattraction of zoospores of the soybean pathogen, *Phytophthora sojae*, by isoflavones. Physiol Mol Plant Pathol 40:17–22

Nespoulous C, Gaudemer O, Huet JC, Pernollet JC (1999) Characterization of elicitin-like phospholipases isolated from *Phytophthora capsici* culture filtrate. FEBS Lett 452:400–406

Nürnberger T, Nennstiel D, Jabs T, Sacks WR, Hahlbrock K, Scheel D (1994) High affinity binding of a fungal oligopeptide elicitor to parsley plasma membranes triggers multiple defense responses. Cell 78:449–460

Osman H, Mikes V, Milat ML, Ponchet M, Marion D, Prangé T, Maume BF, Vauthrin S, Blein J-P (2001a) Fatty acids bind to the fungal elicitor cryptogein and compete with sterols. FEBS Lett 489:55–58

Osman H, Vauthrin S, Mikes V, Milat ML, Panabières F, Marais A, Brunie S, Maume B, Ponchet M, Blein JP (2001b) Mediation of elicitin activity on tobacco is assumed by elicitin–sterol complexes. Mol Biol Cell 12:2825–2834

Panabieres F, Amselem J, Galiana E, Le Berre JY (2005) Gene identification in the oomycete pathogen *Phytophthora parasitica* during in vitro vegetative growth through expressed sequence tags. Fungal Genet Biol 42:611–623

Patterson DJ (1980) Contractile vacuoles and associated structures: their organization and function. Biol Rev 55:1–46

Patterson DJ, Sogin ML (1992) Eukaryote origins and protistan diversity. In: Hartman H, Matsuno K (eds) The origin and evolution of the cell. World Scientific, Singapore, pp 13–46

Pemberton CL, Salmond GPC (2004) The Nep1-like proteins a growing family of microbial elicitors of plant necrosis. Mol Plant Pathol 5:353–359

Pemberton CL, Whitehead NA, Sebalhia M, Bell KS, Hyman LJ, Harris SJ, Matlin AJ, Robson ND, Birch PRJ, Carr JP, Toth IK, Salmond GPC (2005) Novel quorum-sensing-control led genes in *Erwinia carotovora* subsp *carotovora*: Identification of a fungal elicitor homologue in a soft-rotting bacterium. Mol Plant–Microbe Interact 18:343–353

Penington CJ, Iser JR, Grant BR, Gayler KR (1989) Role of RNA and protein synthesis in stimulated germination of zoospores of the pathogenic fungus *Phytophthora palmivora*. Exp Mycol 13:158–168

Ponchet M, Panabières F, Milat ML, Mikes V, Montillet J-L, Suty L, Triantaphylides C, Tirilly Y, Blein J-P (1999) Are elicitins cryptograms in plant-oomycete communications? Cell Mol Life Sci 56:1020–1047

Prakob W, Judelson HS (2007) Gene expression during oosporogenesis in heterothallic and homothallic *Phytophthora*. Fungal Genet Biol 44:726–739

Qutob D, Kamoun S, Gijzen M (2002) Expression of a *Phytophthora sojae* necrosis-inducing protein occurs during transition from biotrophy to necrotrophy. Plant J 32:361–373

Qutob D, Kemmerling B, Brunner F, Kufner I, Engelhardt S, Gust AA, Luberacki B, Seitz HU, Stahl D, Rauhut T, Glawischnig E, Schween G, Lacombe B, Watanabe N, Lam E, Schlichting R, Scheel D, Nau K, Dodt G, Hubert

D, Gijzen M, Nürnberger T (2006) Phytotoxicity and innate immune responses induced by Nep1-like proteins. Plant Cell 18:3721–3744

Randall TA, Dwyer RA, Huitema E, Beyer K, Cvitanich C, Kelkar H, Fong AMVA, Gates K, Roberts S, Yatzkan E, Gaffney T, Law M, Testa A, Torto-Alalibo T, Zhang M, Zheng L, Mueller E, Windass J, Binder A, Birch PRJ, Gisi U, Govers F, Gow NA, Mauch F, Van West P, Waugh ME, Yu J, Boller T, Kamoun S, Lam ST, Judelson HS (2005) Large-scale gene discovery in the oomycete *Phytophthora infestans* reveals likely components of phytopathogenicity shared with true fungi. Mol Plant–Microbe Interact 18:229–243

Rehmany AP, Gordon A, Rose LE, Allen RL, Armstrong MR, Whisson SC, Kamoun S, Tyler BM, Birch PRJ, Beynon JL (2005) Differential recognition of highly divergent downy mildew avirulence gene alleles by RPP1 resistance genes from two Arabidopsis lines. Plant Cell 17:1839–1850

Ricci P, Bonnet P, Huet J-C, Sallantin M, Beauvais-Cante F, Bruneteau M, Billard V, Michel G, Pernollet J-C (1989) Structure and activity of proteins from pathogenic fungi *Phytophthora* eliciting necrosis and acquired resistance in tobacco. Eur J Biochem 183:555–563

Robold AV, Hardham AR (1998) Production of species-specific monoclonal antibodies that react with surface components on zoospores and cysts of *Phytophthora nicotianae*. Can J Microbiol 44:1161–1170

Robold AV, Hardham AR (2005) During attachment *Phytophthora* spores secrete proteins containing thrombospondin type 1 repeats. Curr Genet 47:307–315

Rooney HCE, van't Klooster JW, Van der Hoorn RAL, Joosten MHAJ, Jones JDG, de Wit PJGM (2005) *Cladosporium* Avr2 inhibits tomato Rcr3 protease required for *Cf-2*-dependent disease resistance. Science 308:1783–1786

Rose JKC, Ham K-S, Darvill AG, Albersheim P (2002) Molecular cloning and characterization of glucanase inhibitor proteins: coevolution of a counter-defence mechanism by plant pathogens. Plant Cell 14:1329–1345

Sacks W, Nürnberger T, Hahlbrock K, Scheel D (1995) Molecular characterization of nucleotide sequences encoding the extracellular glycoprotein elicitor from *Phytophthora megasperma*. Mol Gen Genet 246: 45–55

Saeji JP, Boyle JP, Coller S, Taylor S, Sibley LD, Brooke-Powell ET, Ajioka JW, Boothroyd JC (2006) Polymorphic secreted kinases are key virulence factors in toxoplasmosis. Science 314:1780–1783

Sanchez LM, Doke N, Ban Y, Kawakita K (1994) Involvement of suppressor-glucans and plant epidermal cells in host-selective pathogenesis of *Phytophthora capsici*. J Phytopathol 140:153–164

Sandhu D, Gao H, Cianzio S, Bhattacharyya MK (2004) Deletion of a disease resistance nucleotide-binding-site leucine-rich-repeatlike sequence is associated with the loss of the *Phytophthora* resistance gene *Rps4* in soybean. Genetics 168:2157–2167

Schiermeier Q (2001) Russia needs help to fend off potato famine, researchers warn. Nature 410:1011

Schmelzer E (2002) Cell polarization, a crucial process in fungal defence. Trends Plant Sci 7:411–415

Schmitthenner AF, Hobe M, Bhat RG (1994) *Phytophthora sojae* races in Ohio over a 10-year interval. Plant Dis 78:269–276

Séjalon-Delmas N, Villalba Mateos F, Bottin A, Rickauer M, Dargent R, Esquerré-Tugayé MT (1997) Purification, elicitor activity, and cell wall localization of a glycoprotein from *Phytophthora parasitica* var. *nicotianae*, a fungal pathogen of tobacco. Phytopathology 87:899–909

Shan W, Hardham AR (2004) Construction of a bacterial artificial chromosome library, determination of genome size, and characterization of an *Hsp70* gene family in *Phytophthora nicotianae*. Fungal Genet Biol 41:369–380

Shan W, Cao M, Dan LU, Tyler BM (2004a) The *Avr1b* locus of *Phytophthora sojae* encodes an elicitor and a regulator required for avirulence on soybean plants carrying resistance gene *Rps1b*. Mol Plant–Microbe Interact 17:394–403

Shan W, Marshall JS, Hardham AR (2004b) Stage-specific expression of genes in germinated cysts of *Phytophthora nicotianae*. Mol Plant Pathol 5:317–330

Shapiro A, Mullins JT (2002) Hyphal tip growth in *Achlya bisexualis*. I. Distribution of 1,3-β-glucans in elongating and non-elongating regions of the wall. Mycologia 94:267–272

Shearer BL, Crane CE, Barrett S, Cochrane A (2007) *Phytophthora cinnamomi* invasion, a major threatening process to conservation of flora diversity in the South-west Botanical Province of Western Australia. Aust J Bot 55:225–238

Shepherd SJ, Van West P, Gow NAR (2003) Proteomic analysis of asexual development of *Phytophthora palmivora*. Mycol Res 107:395–400

Shiraishi T, Yamada T, Ichinose Y (1997) The role of suppressors in determining host-parasite specificites in plants cells. Int Rev Cytol 172:55–93

Silflow CD, Lefebvre PA (2001) Assembly and motility of eukaryotic cilia and flagella. Lessons from *Chlamydomonas reinhardtii*. Plant Physiol 127:1500–1507

Škalamera D, Hardham AR (2006) PnCcp, a *Phytophthora nicotianae* protein containing a single complement control protein module, is sorted into large peripheral vesicles in zoospores. Aust Plant Pathol 35:593–603

Škalamera D, Wasson AP, Hardham AR (2004) Genes expressed in zoospores of *Phytophthora nicotianae*. Mol Genet Genomics 270:549–557

Song J, Bradeen JM, Naess SK, Raasch JA, Wielgus SM, Haberlach GT, Liu J, Kuang H, Austin-Phillips S, Buell CR, Helgeson JP, Jiang J (2003) Gene *RB* cloned from *Solanum bulbocastanum* confers broad spectrum resistance to potato late blight. Proc Natl Acad Sci USA 100:9128–9133

Steck TL, Chiaraviglio L, Meredith S (1997) Osmotic homeostasis in *Dictyostelium discoideum*: excretion of amino acids and ingested solutes. J Eukaryot Microbiol 44:503–510

Stevens TH, Forgac M (1997) Structure, function and regulation of the vacuolar (H+)-ATPase. Annu Rev Cell Dev Biol 13:779–808

Stintzi A, Heitz T, Prasad V, Wiedemann-Merdinoglu S, Kauffmann S, Geoffroy P, Legrand M, Fritig B (1993)

Plant 'pathogenesis-related' proteins and their role in defense against pathogens. Biochimie 75:687–706

Struck C, Hahn M, Mendgen K (1996) Plasma membrane H+-ATPase activity in spores, germ tubes, and haustoria of the rust fungus *Uromyces viciae-fabae*. Fungal Genet Biol 20:30–35

Takemoto D, Hardham AR (2004) The cytoskeleton as a regulator and target of biotic interactions in plants. Plant Physiol 136:3864–3876

Takemoto D, Jones DA, Hardham AR (2003) GFP-tagging of cell components reveals the dynamics of subcellular re-organization in response to infection of *Arabidopsis* by oomycete pathogens. Plant J 33:775–792

Takemoto D, Jones DA, Hardham AR (2006) Re-organization of the cytoskeleton and endoplasmic reticulum in the *Arabidopsis pen1-1* mutant inoculated with the non-adapted powdery mildew pathogen, *Blumeria graminis* f. sp. *hordei*. Mol Plant Pathol 7:553–563

Taylor S, Barragan A, Su C, Fux B, Fentress SJ, Tang K, Beatty WL, Hajj HE, Jerome M, Behnke MS, White M, Wootton JC, Sibley LD (2006) A secreted serine-threonine kinase determines virulence in the eukaryotic pathogen *Toxoplasma gondii*. Science 314:1776–1780

Temperli E, Roos U-P, Hohl HR (1990) Actin and tubulin cytoskeletons in germlings of the oomycete fungus *Phytophthora infestans*. Eur J Cell Biol 53:75–88

Temperli E, Roos U-P, Hohl HR (1991) Germ tube growth and the microtubule cytoskeleton in *Phytophthora infestans*: effects of antagonists of hyphal growth, microtubule inhibitors, and ionophores. Mycol Res 95:611–617

Ten Have A, Breuil WO, Wubben JP, Visser J, van Kan JAL (2001) *Botrytis cinerea* endopolygalacturonase genes are differentially expressed in various plant tissues. Fungal Genet Biol 33:97–105

Tian M, Kamoun S (2005) A two disulfide bridge Kazal domain from *Phytophthora* exhibits stable inhibitory activity against serine proteases of the subtilisin family. Biochemistry 6:15

Tian M, Huitema E, da Cunha L, Torto-Alalibo T, Kamoun S (2004) A Kazal-like extracellular serine protease inhibitor from *Phytophthora infestans* targets the tomato pathogenesis-related protease P69B. J Biol Chem 279:26370–26377

Tian M, Benedetti B, Kamoun S (2005) A second kazal-like protease inhibitor from *Phytophthora infestans* inhibits and interacts with the apoplastic pathogenesis-related protease P69B of tomato. Plant Physiol 138:1785–1793

Tian MY, Win J, Song J, van der HR, van der KE, Kamoun S (2007) A *Phytophthora infestans* cystatin-like protein targets a novel tomato papain-like apoplastic protease. Plant Physiol 143:364–377

Tomiyama K, Sato K, Doke N (1982) Effect of cytochalasin B and colchicine on hypersensitive death of potato cells infected by incompatible race of *Phytophthora infestans*. Ann Phytopathol Soc Jpn 48:228–230

Tomley FM, Soldati DS (2001) Mix and match modules: structure and function of microneme proteins in apicomplexan parasites. Trends Parasitol 17:81–88

Torto TA, Rauser L, Kamoun S (2002) The *pipg1* gene of the oomycete *Phytophthora infestans* encodes a fungal-like endopolygalacturonase. Curr Genet 40:385–390

Torto TA, Li S, Styer A, Huitema E, Testa A, Gow NAR, Van West P, Kamoun S (2003) EST mining and functional expression assays identify extracellular effector proteins from the plant pathogen *Phytophthora*. Genome Res 13:1675–1685

Torto-Alalibo T, Tian M, Gajendran K, Waugh ME, Van West P, Kamoun S (2005) Expressed sequence tags from the oomycete fish pathogen *Saprolegnia parasitica* reveal putative virulence factors. Microbiology 5:46

Tucker SL, Talbot NJ (2001) Surface attachment and pre-penetration stage development by plant pathogenic fungi. Annu Rev Phytopathol 39:385–417

Tyler BM (2002) Molecular basis of recognition between *Phytophthora* pathogens and their hosts. Annu Rev Phytopathol 40:137–167

Tyler BM, Wu M-H, Wang J-M, Cheung W, Morris PF (1996) Chemotactic preferences and strain variation in the response of *Phytophthora sojae* zoospores to host isoflavones. Appl Environ Microbiol 62:2811–2817

Tyler BM, Tripathy S, Zhang XM, Dehal P, Jiang RHY, Aerts A, Arredondo FD, Baxter L, Bensasson D, Beynon JL, Chapman J, Damasceno CMB, Dorrance AE, Dou DL, Dickerman AW, Dubchak IL, Garbelotto M, Gijzen M, Gordon SG, Govers F, Grunwald NJ, Huang W, Ivors KL, Jones RW, Kamoun S (2006) *Phytophthora* genome sequences uncover evolutionary origins and mechanisms of pathogenesis. Science 313:1261–1266

Van de Peer Y, Van der Auwera G, De Wachter R (1996) The evolution of stramenopiles and alveolates as derived by "substitution rate calibration" of small ribosomal subunit RNA. J Mol Evol 42:201–210

van der Vossen EAG, Sikkema A, Hekkert BL, Gros J, Stevens P, Muskens M, Wouters D, Pereira A, Stiekema W, Allefs S (2003) An ancient *R* gene from the wild potato species *Solanum bulbocastanum* confers broad-spectrum resistance to *Phytophthora infestans* in cultivated potato and tomato. Plant J 36:867–882

van der Vossen EAG, Gros J, Sikkema A, Muskens M, Wouters D, Wolters P, Pereira A, Allefs S (2005) The *Rpi-blb2* gene from *Solanum bulbocastanum* is an *Mi-1* gene homolog conferring broad-spectrum late blight resistance in potato. Plant J 44:208–222

Van West P, Morris BM, Reid B, Appiah AA, Osborne MC, Campbell TA, Shepherd SJ, Gow NAR (2002) Oomycete plant pathogens use electric fields to target roots. Mol Plant–Microbe Interact 15:790–798

Vauthrin S, Mikes V, Milat ML, Ponchet M, Maume B, Osman H, Blein JP (1999) Elicitins trap and transfer sterols from micelles, liposomes and plant plasma membranes. Biochim Biophys Acta Bio-Membr 1419:335–342

Veit S, Worle JM, Nürnberger T, Koch W, Seitz HU (2001) A novel protein elicitor (PaNie) from *Pythium aphanidermatum* induces multiple defense responses in carrot, *Arabidopsis*, and tobacco. Plant Physiol 127:832–841

Villalba Mateos F, Rickauer M, Esquerré-Tugayé MT (1997) Cloning and characterization of a cDNA encoding an elicitor of *Phytophthora parasitica* var. *nicotianae* that shows cellulose-binding and lectin-like activities. Mol Plant–Microbe Interact 10:1045–1053

Voegele RT, Mendgen K (2003) Rust haustoria: nutrient uptake and beyond. New Phytol 159:93–100

Walker SK, Chitcholtan K, Yu YP, Christenhusz GM, Garrill A (2006) Invasive hyphal growth: An F-actin depleted zone is associated with invasive hyphae of the oomycetes *Achlya bisexualis* and *Phytophthora cinnamomi*. Fungal Genet Biol 43:357–365

Wang JY, Cai Y, Gou JY, Mao YB, Xu YH, Jiang WH, Chen XY (2004) VdNEP, an elicitor from *Verticillium dahliae*, induces cotton plant wilting. Appl Environ Microbiol 70:4989–4995

Wang MC, Bartnicki-Garcia S (1974) Mycolaminarans: Storage (1,3)-β-D-glucans from the cytoplasm of the fungus *Phytophthora palmivora*. Carbohydr Res 37:331–338

Whisson SC, Boevink P, Moleleki L, Avrova AO, Morales JG, Gilroy AM, Armstrong MR, Grouffaud S, Van West P, Chapman S, Hein I, Toth IK, Pritchard L, Birch PRJ (2007) A translocation signal for delivery of oomycete effector proteins into host plant cells. Nature 450:115–119

Win J, Kanneganti T-D, Torto-Alalibo T, Kamoun S (2006) Computational and comparative analyses of 150 full-length cDNA sequences from the oomycete plant pathogen *Phytophthora infestans*. Fungal Genet Biol 43:20–33

Win J, Morgan W, Bos J, Krasileva KV, Cano LM, Chaparro-Garcia A, Ammar R, Staskawicz BJ, Kamoun S (2007) Adaptive evolution has targeted the C-terminal domain of the RXLR effectors of plant pathogenic Oomycetes. Plant Cell 19:2349–2369

Yamagishi T, Motomura T, Nagasato C, Kato A, Kawai H (2007) A tubular mastigoneme-related protein, Ocm1, isolated from the flagellum of a Chromophyte alga, *Ochromonas danica*. J Phycol 43:519–527

Yan H-Z, Liou R-F (2005) Cloning and analysis of *pppg1*, an inducible endopolygalacturonase gene from the oomycete plant pathogen *Phytophthora parasitica*. Fungal Genet Biol 42:339–350

Yoon HS, Hackett JD, Pinto G, Bhattacharya D (2002) The single, ancient origin of chromist plastids. Proc Natl Acad Sci USA 99:15507–15512

York WS, Qin Q, Rose JKC (2004) Proteinaceous inhibitors of *endo*-β-glucanases. Biochem Biophys Acta 1696:223–233

Yun B-W, Atkinson HA, Gaborit C, Greenland A, Read ND, Pallas JA, Loake GJ (2003) Loss of actin cytoskeletal function and EDS1 activity, in combination, severely compromises non-host resistance in *Arabidopsis* against wheat powdery mildew. Plant J 34:768–777

Zeigler E, Pontzen R (1982) Specific inhibition of glucan-elicited glyceollin accumulation in soybeans by an extracellular mannan-glycoprotein of *Phytophthora megasperma* f. sp. *glycinea*. Physiol Plant Pathol 20:321–331

Zeyen RJ, Carver TLW, Lyngkjær MF (2002) Epidermal cell papillae. In: Bélanger RR, Bushnell WR, Dik AJ, Carver TLW (eds) The powdery mildews: a comprehensive treatise. American Phytopathological Society, St Paul, pp 107–125

Zipfel C, Robatzek S, Navarro L, Oakeley EJ, Jones JDG, Felix G, Boller T (2004) Bacterial disease resistance in Arabidopsis through flagellin perception. Nature 428:764–767

2 *Botrytis cinerea*: Molecular Aspects of a Necrotrophic Life Style

Paul Tudzynski[1], Leonie Kokkelink[1]

CONTENTS

I. Introduction........................... 29
II. *Botrytis cinerea*, the Grey Mould Fungus 29
III. Molecular Tools 31
IV. Pathogenicity Determinants 38
   A. Cell Wall Degrading Enzymes 38
   B. Phytotoxic Compounds 39
   C. Organic Acids....................... 40
   D. Phytohormones 40
   E. Reactive Oxygen Species............... 40
V. Signalling 42
   A. The cAMP-Dependent Signalling
      Pathway........................... 42
   B. MAP Kinase-Controlled Signalling
      Pathways 43
   C. Small G-Proteins 43
   D. The Ca^{2+}/Calmodulin-Dependent
      Signalling Pathway................... 44
   E. Cell Surface Receptors................. 44
VI. Special Aspects of Host Defence 45
VII. Perspectives 46
   References............................. 46

I. Introduction

Pathogenic fungi have developed a wide range of strategies to infect and colonize plants. Traditionally they are grouped into the major classes necrotroph/hemi-biotroph/biotroph, according to major criteria like their source of nutrition (living vs dead cells), their ability to infect young and healthy tissue or a preference to older or senescent ones, the formation of specialized infection structures (haustoria), and more recently the type of plant defence reaction they provoke (e.g. jasmonate vs salicylic acid pathway). In many cases these criteria do not allow unequivocal decisions for the grouping of pathogens, as pointed out in the recent review by Oliver and Ipcho (2004). As an example, they list references grouping *Phythophthora infestans* in all three classes, and other references naming

Magnaporthe grisea a necrotroph or a hemi-biotroph. Detailed cytological analyses only recently brought unequivocal evidence about the true biotrophic nature of the early infection stage of *M. grisea* (Kankanala et al. 2007), but still the exact mode of the switch to necrotrophic growth is unclear. Thus it needs much more detailed structural and physiological studies to fully understand the nature of a specific fungus–host interaction; and the question remains whether the old classification system is still helpful, because of the large degree of variation observed in nature.

One of the few pathogens considered a "genuine" necrotroph is the grey mould fungus *Botrytis cinerea*. It meets all the classic criteria of being a necrotroph: it has a broad host range, it secretes a broad set of cell wall degrading enzymes (CWDE) and phytotoxic low molecular weight compounds, it rapidly kills the host tissue, and it is able to draw nutrients exclusively from dead tissue. From the view of classic phytopathology, this strategy requires no close interaction with the host and no long adaptation phases, i.e. no real co-evolution (hence no specialization). Recent research data raise doubts whether this adequately describes the mode of interaction between *B. cinerea* and its host plants.

Since the recently published monograph on *Botrytis* species and their pathogenicity profiles (Elad et al. 2004) covers the classic literature and was recently updated by the review of Williamson et al. (2007), we focus here more on the results of molecular genetic research which emerged in the past few years, trying to substantiate and update the excellent focused review of van Kan (2006).

II. *Botrytis cinerea*, the Grey Mould Fungus

Botrytis cinerea Pers. Fr. [teleomorph *Botryotinia fuckeliana* (de Bary) Whetzel] is the causative

[1] Institut für Botanik, Westf. Wilhelms-Universität, Schlossgarten 3, 48149 Münster, Germany; e-mail: tudzyns@uni-muenster.de

Plant Relationships, 2nd Edition
The Mycota V
H. Deising (Ed.)
© Springer-Verlag Berlin Heidelberg 2009

agent of grey mould disease in more than 200 crop species. It causes severe economic losses especially in greenhouse crops and ornamentals, but it also is one of the major reasons for post-harvest losses, because it often enters plant tissues at early stages of development and only causes symptoms after ripening of the fruits. Since it is highly variable, *B. cinerea* rapidly develops resistance against fungicides, and to date only a very few stable resistance genes against this pathogen are known in plants. Therefore development of alternative control strategies is highly important; however, these require a much better understanding of the exact life style and pathogenicity strategy of the fungus on its various host plants (Elad et al. 2004).

Figure 2.1 illustrates the major steps in the life- and pathogenicity cycle of the fungus, including enzymes and signalling components shown to be involved in the different steps. Primary infection of plant tissue normally occurs by air-borne multi-nuclear conidia, which attach to the plant surface and germinate (Holz et al. 2004). This process is facilitated by water droplets (rain), but also

dry inoculation (in high humidity) is possible. The exact mode of penetration is not known yet; in many cases (not all!) the germ tubes form apical swellings before they penetrate the cuticle and grow into the host tissue. These swellings are generally called appressoria, but – as outlined by Tenberge (2004) – they lack a sealing septum and have no melanin incorporated in the cell wall, i.e. they are distinct from the genuine appressoria formed by hemi-biotrophs like *M. grisea* and *Colletotrichum* species. Penetration of a leaf surface is normally accompanied by formation of a necrotic spot or primary lesion, a few millimetres wide, which rapidly expands after a few days, forming so-called secondary or spreading lesions. Finally the leaf is fully destroyed and covered by conidiating fungal mycelium. However, in many cases *Botrytis* infects through flowers, and it often stays in a longer quiescent stage (causing no symptoms), which resembles a kind of endophytic growth, and it only attacks and destroys the plant tissue after the fruits ripen (Williamson et al. 2007). In this case early infection occurs mostly through the stigma, the fungus following the route

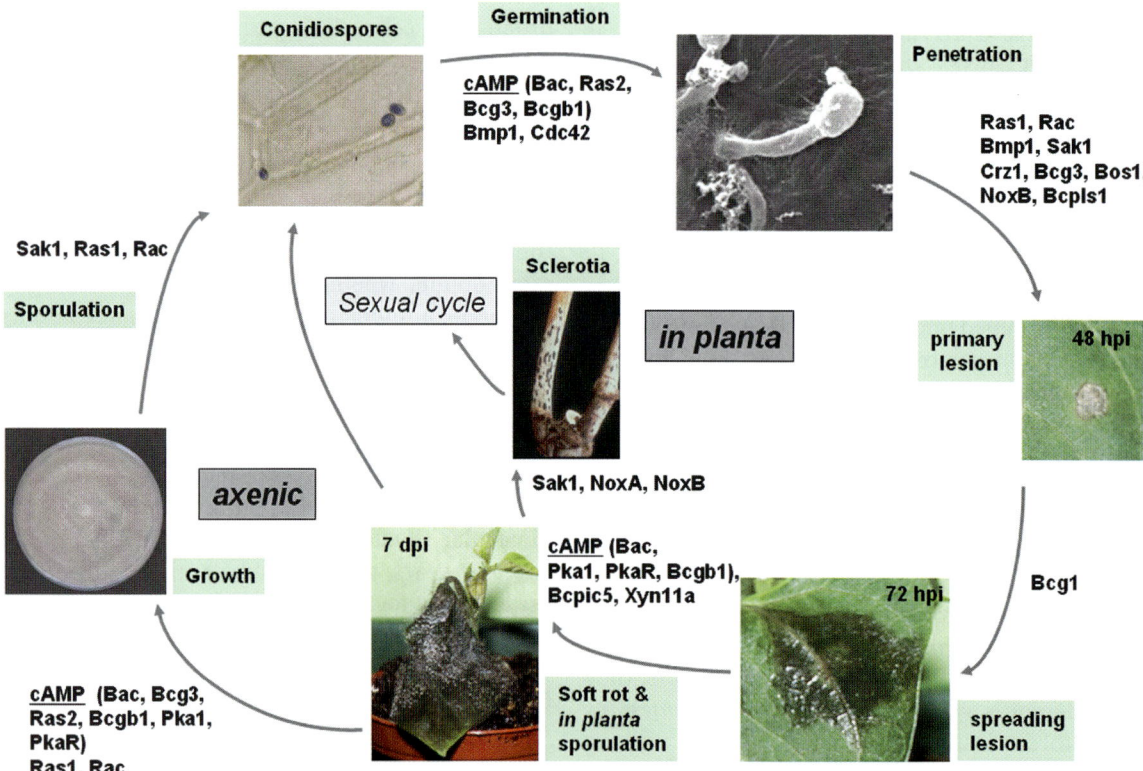

Fig. 2.1. Life cycle of *Botrytis cinerea*. Signalling components/enzymes involved in specific steps are indicated (for gene symbols, see Table 2.1)

of the pollen tube, like e.g. the biotrophic, highly specialized ergot fungus *Claviceps purpurea* (Tudzynski and Scheffer 2004). The fungal conidia in the plant debris can serve as source for infection in the next vegetation period. In axenic culture the fungus regularly produces sclerotia, which are rarely observed in nature, but they could serve as long-term survival structures. Sclerotia are also the tissue in which mating occurs and which germinate to produce the sexual fruitbodies, called apothecia. The sexual cycle can be performed under controlled conditions in the laboratory (Faretra and Pollastro 1996) and thus genetic analyses are possible, though it takes several months to complete the cycle.

Thus, *B. cinerea* is a highly versatile pathogen: it can penetrate directly into any above-soil plant tissue or organ at any stage of development, it causes immediate killing of cells or stays undetected for a long period, it can survive long periods outside living plant material, and it can use the highly efficient sexual recombination system.

III. Molecular Tools

Molecular genetics of *Botrytis* began only in 1989, when Huang and co-workers published the first successful transformation of *B. squamosa*. Today, the *Botrytis* community includes more than a dozen teams using molecular techniques to study host–parasite interactions. Since the relevant molecular tools are available, such as transformation protocols, vectors, mutants, genomic, and cDNA libraries, *B. cinerea* has become one of best studied necrotrophic pathogens, also triggered by its great economic importance and the resulting strong industrial interest. Molecular research was especially stimulated when the first genome sequence was released (strain B05.10, 4× coverage, Broad Institute; http://www.broad.mit.edu/annotation/genome/botrytis_cinerea/Home.html). Since this sequence is of limited quality and coverage, a non-industrial genome sequencing initiative guided by an international consortium was initiated; the sequence will soon be publicly available (strain T4, 10× coverage, INRA/Genoscope; http://www.urgi.versailles.inra.fr/projects/Botrytis/). The T4 sequence is currently annotated by the international consortium. Strains T4 and B05.10 differ significantly in virulence (and candidate

genes have revealed considerably genetic variability, see below) and the availability of genome sequences from these two different *B. cinerea* strains is a great opportunity to use comparative genomics tools to study variability in this species. Especially valuable is the release of the genomic sequence of a close relative, *Sclerotinia sclerotiorum*, the white mould fungus (Bolton et al. 2005), which has an even broader host range than *B. cinerea*, but differs in several important parameters (e.g. it produces no conidia). A comparison of the genome structure and coding capacity of these two major necrotrophic pathogens will be highly interesting. Automatic gene annotation led to 14 000–16 000 gene calls for both species, revealing a common set of 8400 genes, of which 1200 are absent from other species (Fillinger et al. 2007). A detailed evaluation of the genetic equipment will increase our understanding of the genetic potential of these pathogens considerably. Apart from these great perspectives for comparative genomics, there are several valuable spin-offs of the genome project (transcriptomics, proteomics, evolutionary analyses). The availability of the genome sequence has already significantly stimulated basic research in *B. cinerea*.

An important aspect of molecular genetics of *B. cinerea* (in contrast to *Sclerotinia*) is the high efficiency of targeted gene inactivation, which allows a rapid functional analysis of putative pathogenicity-related genes. Gene inactivation by single cross-over is difficult, whereas gene-replacement approaches using linear transformation cassettes yield high knock-out rates (70–100%). The standard technique of targeted gene inactivation has become gene replacement using long flanks (500–1000 bp); in most laboratories these constructs are designed by PCR. Good transformation rates can be also obtained with the *Agrobacterium*-T-DNA transfer system (Rolland et al. 2003; N. Segmüller and P. Tudzynski, unpublished data), though this system has so far been used in *B. cinerea* only for insertional mutagenesis, not for targeted gene inactivation. As an alternative for knock-out approaches, RNAi has been used in other phytopathogens for the silencing of genes (e.g. Fitzgerald et al. 2004) and it was applied successfully also in *B. cinerea* (Schumacher et al. 2008a; R. Patel, G.D. Foster and J. van Kan, unpublished data). This knock-down approach will be especially valuable for the analysis of gene families and essential genes.

Table 2.1. Deletion mutants and resulting phenotypes of *B. cinerea* generated by targeted gene inactivation

Protein/function	Symbol	Recipient strain	Effect on pathogenicity	Other phenotype	Reference(s)
Hydrolytic enzyme encoding genes					
Cutinase	*cutA*	B05.10	None		van Kan et al. (1997)
Lipase	*lip1*	B05.10	None	No lipolytic activity	Reis et al. (2005)
Lipase/cutinase	*lip1/cutA*	B05.10	None	Extracellular esterases largely eliminated	Reis et al. (2005)
Endopolygalacturonase	*bcpg1*	B05.10	Slower lesion outgrowth on different host plants		Ten Have et al. (1998)
Endopolygalacturonase	*bcpg2*	B05.10	Reduced virulence		Kars et al. (2005a)
Endopolygalacturonase	*bcpg3*	B05.10	None		Kars and van Kan, personal communication
Endopolygalacturonase	*bcpg4*	B05.10	None		Kars and van Kan, personal communication
Endopolygalacturonase	*bcpg5*	B05.10	None		Kars and van Kan, personal communication
Endopolygalacturonase	*bcpg6*	B05.10	None		Kars and van Kan, personal communication
Pectin methylesterase	*bcpme1*	Bd90	Reduced virulence on different host plants	Growth on rich media like wild type	Vallette-Collet et al. (2003)
Pectin methylesterase	*bcpme1*	B05.10	None	Wild-type growth rate on pectin-containing minimal media	Kars et al. (2005b)
Pectin methylesterase	*bcpme2*	B05.10	None	Wild-type growth rate on pectin-containing minimal media	Kars et al. (2005b)
Pectin methylesterase	*bcpme1/b cpme2*	B05.10	None	Wild-type growth rate on pectin-containing minimal media	Kars et al. (2005b)
Endopolygalacturonase + pectin methylesterase	*bcpg1/bcpme1*	B05.10	Reduced virulence, similar to Δ*bcpg1* strains		Kars et al. (2005b)
Aspartic protease	*bcap1*	B05.10	None		ten Have and van Kan, personal communication
Aspartic protease	*bcap2*	B05.10	None		ten Have and van Kan, personal communication
Aspartic protease	*bcap3*	B05.10	None		ten Have and van Kan, personal communication
Aspartic protease	*bcap4*	B05.10	None		ten Have and van Kan, personal communication
Aspartic protease	*bcap5*	B05.10	None		ten Have and van Kan, personal communication
Aspartic protease	*bcap1/bcap3*	B05.10	None		ten Have and van Kan, personal communication

Function	Gene	Strain	Phenotype (in planta)	Phenotype (in vitro)	Reference
Aspartic protease	*bcap1/bcap4*	B05.10	None		ten Have and van Kan, personal communication
Aspartic protease	*bcap1/bcap5*	B05.10	None		ten Have and van Kan, personal communication
Aspartic protease	*bcap3/bcap4*	B05.10	None		ten Have and van Kan, personal communication
Aspartic protease	*bcap3/ bcap5*	B05.10	None		ten Have and van Kan, personal communication
Endo-β-1,4-xylanase	*xyn11a*	B05.10	Delay in secondary lesion formation, reduced lesion size	Decrease in xylanase activity and growth rate on xylan	Brito et al. (2006)
Endo-β-1,4-glucanase	*cel5a*	B05.10	Infection of tomato leaves or gerbera petals like wild type	No significant reduction in beta-1,4-glucanase activity	Espino et al. (2005)
Endo-β-1,4-Xylanase	*bcxyl2*	B05.10	None		J. Schumacher et al., personal communication
Transporter encoding genes					
Fructose symporter	*frt1*	B05.10	None	Normal growth, delayed germination	Döhlemann et al. (2005)
ABC transporter	*bcatrA*	B05.10	None		del Sorbo et al. (2008)
ABC transporter	*bcatrB*	B05.10	Reduced virulence on grapevine leaves	Increased sensitivity to resveratrol	Schoonbeek et al. (2001)
ABC transporter	*bcatrD*	B05.10	None	Increased sensitivity to DMI fungicides	Hayashi et al. (2002a)
MFS transporter	*bcmfs1*	B05.10	None	Increased sensitivity to camptothecin and cercosporin	Hayashi et al. (2002b)
ABC + MFS transporter	*bcatrD/bcmfs1*	B05.10	Like ΔbcatrB	Higher sensitivity to DMI fungicides than ΔbcatrB single mutants	Hayashi et al. (2002b)
ABC + MFS transporter	*bcatrB/bcmfs1*	B05.10	None	No additional effects to single mutants	Hayashi et al. (2002b)
Signalling factor encoding genes					
Tetraspanin	*bcpls1*	T4	Non-pathogenic on bean and tomato leaves	No penetration of onion epidermis	Gourgues et al. (2004)
Adenylate cyclase	*bac*	B05.10	Slower infection process, stops in secondary lesion	Compact growth, strong sporulation	Klimpel et al. (2002)
G protein α subunit	*bcg1*	B05.10	Infection stops after primary lesion formation on bean leaves	Compact growth, no synthesis of botrydial/botcinines	Schulze Gronover et al. (2001)
G protein α subunit	*bcg2*	B05.10	Slightly reduced infection progress	Growth like wild type	Schulze Gronover et al. (2001)
G protein α subunit	*bcg3*	B05.10	Delay in penetration	Impaired in sugar-induced germination	Döhlemann et al. (2006a)
G protein β subunit	*bcgb1*	B05.10	Delay in penetration, stops in secondary lesion	Compact colonies, increased formation of aerial hyphae, no sclerotia	Schumacher et al., unpublished data

continued

Table 2.1 continued

Protein/function	Symbol	Recipient strain	Effect on pathogenicity	Other phenotype	Reference(s)
PKA regularory subunit	*bcpkaR*	B05.10	Slower infection process, stops in secondary lesion	Compact growth, strong sporulation	Schumacher et al. (2008c)
PKA catalytic subunit	*bcpka1*	B05.10	Slower infection process, stops in secondary lesion	Compact growth, strong sporulation	Schumacher et al. (2008c)
PKA catalytic subunit	*bcpka2*	B05.10	None	Not impaired during in vitro growth	Schumacher et al. (2008c)
CyclophilinA	*bcp1*	T4	Reduced infection on bean and tomato leaves	Resistant to cyclosporin A	Viaud et al. (2003)
Phospholipase C	*bcplc1*	B05.10	Infection from mycelia delayed, no infection by conidia	Very poor growth, reduced conidiation, almost no conidial germination	Schumacher et al. (2008a)
Transcription factor	*bccrz1*	B05.10	Mycelia only infect wounded tissue, conidial infection like wildtype	Very poor growth on standard media	Schumacher et al. (2008b)
MAP kinase (Fus3-homologue)	*bmp1*	A-1-3/B05.10	Non-pathogenic on carnation flowers and tomato leaves	Impaired in hydrophobicity induced germination	Zheng et al. (2000); Döhlemann et al. (2006a)
MAP kinase (HOG-homologue)	*bcsak1*	B05.10	Mycelia unable to penetrate unwounded tissue	No sporulation	Segmüller et al. (2007)
MAP kinase (Slt2-homologue)	*bmp3*	B05.10	Retarded development of necrotic lesions	Reduced vegetative growth, no sclerotia, reduced sporulation	Rui and Hahn (2007)
Protein kinase (SCH9 homologue)	*bcpk2*	B05.10	None		Schulze Gronover and B. Tudzynski, personal communication
Protein kinase	*bcsnf1*	B05.10	Reduced lesion size on bean leaves	Altered colony morphology on all tested carbon sources	Schumacher et al., unpublished data
Histidine kinase	*bos1*	UWS111/L	Severely reduced in planta growth	No conidiation, higher resistance to menadione	Viaud et al. (2006)
Histidine kinase	*bos1*	B05.10	Impaired in penetration		Liu et al. (2008)
Histidine kinase	*bchk5*	B05.10	None		Cuesta Arenas and van Kan, personal communication
Histidine kinase	*bchk1*	B05.10	None	Impaired in growth on 5 mM H_2O_2, very slow growth, aberrant hyphal morphology	N. Temme et al., unpublished data
Ras homologue	*bcras1*	B05.10	Non-pathogenic	Very slow growth, aberrant hyphal morphology, no conidiation	Kokkelink et al., unpublished data
Ras homologue	*bcras2*	B05.10	Delayed infection process on bean leaves	Compact growth, delay in germination	Schumacher et al. (2008c)

Protein	Gene	Strain	Phenotype (pathogenicity)	Phenotype (other)	Reference
Rho-type GTPase	*bcrac*	B05.10	Non-pathogenic	Very slow growth, aberrant hyphal morphology, no conidiation	Kokkelink et al., unpublished data
Rho-type GTPase	*bccdc42*	B05.10	Delayed infection process on bean leaves	Reduced growth and conidiation	Kokkelink et al., unpublished data
Rab-type GTPase	*bcrab2*	B05.10	No formation of spreading lesions		Hantsch and B. Tudzynski, unpublished data
FKBP12 homologue	*bcpic5*	T4	Reduced lesion formation, infection stops	Not impaired during in vitro growth	Gioti et al. (2006)
FKBP12 homologue		B05.10	None		Mey et al., personal communication
Transcription factor	*bap1*	B05.10	None	Not impaired during in vitro growth	Temme et al., unpublished data
Transcription factor	*bcidi4*	B05.10	None	Wild type	Segmüller et al., unpublished data
Transmembrane protein	*btp1*	B05.10	Slightly impaired in pathogenicity		Schulze Gronover et al. (2005)
Rhodopsin	*bop1*	B05.10	None		Heller et al., unpublished data
Primary metabolism					
Trehalose-6-phosphate synthase	*tps1*	B05.10	None	No trehalose synthesis	Döhlemann et al. (2006b)
Neutral trehalase	*tre1*	B05.10	None	Elevated trehalose levels	Döhlemann et al. (2006b)
Glucokinase	*glk1*	B05.10	None	Wild-type phenotype	Rui and Hahn (2007)
Hexokinase	*hxk1*	B05.10	Slightly retarded lesion formation on leaves; small lesions on apples, strawberries, tomatoes	Pleiotropic growth defect	Rui and Hahn (2007)
Oxaloacetate hydrolase	*bcoahA*	B05.10	Lesion size like wild type but with altered appearance (tomato leaf)		van Kan, personal communication
Glutamin synthetase	*bcglnA*	B05.10	NT	Retarded growth, glutamine auxotrophy	B. Tudzynski, personal communication
Secondary metabolism					
Ent-copalyl diphosphate synthase/ent-kaurene synthase	*bccps/ks1*	B05.10	None		Siewers and Tudzynski, unpublished data
FPP cyclase	*bctri5*	T4/K1	Reduced lesion development on bean leaves		Viaud et al., personal communication
Sesquiterpene cyclase	*cnd15*	B05.10		No botrydial production	Viaud et al., personal communication

continued

Table 2.1 continued

Protein/function	Symbol	Recipient strain	Effect on pathogenicity	Other phenotype	Reference(s)
Cytochrome P450 reductase	bccpr1	ATCC58025	None	Reduced growth rate	Siewers et al. (2004)
P450 monooxygenase	bcaba1	ATCC58025	None	Blocked abscisic acid synthesis	Siewers et al. (2004)
P450 monooxygenase	bcaba1	B05.10	None	Phenotype like wild type	Siewers et al., unpublished data
P450 monooxygenase	bcaba2	ATCC58025	None	Blocked abscisic acid synthesis	Siewers et al. (2006)
P450 monooxygenase	bcaba2	B05.10	None	Phenotype like wild type	Siewers et al., unpublished data
ABA biosynthesis	bcaba3	ATCC58025	None	blocked abscisic acid synthesis	Siewers et al. (2006)
ABA biosynthesis	bcaba3	B05.10	not tested	Phenotype like wild type	Siewers et al., unpublished data
Short-chain dehydrogenase/reductase	bcaba4	ATCC58025	None	Altered abscisic acid synthesis	Siewers et al. (2006)
Short-chain dehydrogenase/reductase	bcaba4	B05.10	none	Phenotype like wild type	Siewers et al., unpublished data
P450 monooxygenase	bcbot1	T4/SAS56	T4: reduced virulence; B05.10: none	Impaired botrydial synthesis	Siewers et al. (2005)
Polyketide synthase	bcpks1	B05.10	None		Kars und van Kan, personal communication
Stress-reated genes					
Catalase	bccat2	B05.10	None	Increased sensitivity to hydrogen peroxide	Schouten et al. (2002a)
Glucose oxidase	bcgod1	B05.10	None		Rolke et al. (2004)
Glutathione-S-transferase	bcgst1	B05.10	None		Prins et al. (2000)
Superoxide dismutase	bcsod1	B05.10	Smaller lesions on bean leaves		Rolke et al. (2004)
Saponinase	sap1	B05.10	None	Increased sensitivity to paraquat	Quidde et al. (1999)
Laccase	bclcc1	B05.10	None	Increased sensitivity to avenacin	Schouten et al. (2002b)
Laccase	bclcc2	B05.10	None	No inhibition by resveratrol	Schouten et al. (2002b)
NADPH oxidase	bcnoxA	B05.10	Slower infection progress	No formation of sclerotia	Segmüller et al. (2008)
NADPH oxidase	bcnoxB	B05.10	Delayed penetration, then normal lesion development	Less formation of sclerotia	Segmüller et al. (2008)
NADPH oxidases	bcnoxA/bcnoxB	B05.10	Delayed penetration and lesion development	No formation of sclerotia	Segmüller et al. (2008)
NADPH oxidase regulatory subunit	bcnoxR	B05.10	Delayed penetration and lesion development	No formation of sclerotia	Segmüller et al. (2008)

Other

Function	Gene	Strain	Phenotype		Reference
Chitin synthase	*bcchs1*	Bd90	About 32% reduced lesion diameter on vine leaves	Cell wall sensitivity to enzymatic degradation	Soulie et al. (2003)
Chitin synthase	*bcchs3a*	Bd90	Reduced virulence on *Vitis vinifera* and *Arabidopsis thaliana* leaves	Reduction in chitin content, impaired growth on solid media	Soulie et al. (2006)
Chitin synthase	*bcchs4*	Bd90	Slight delay in infection (*A. thaliana, V. vinifera*)		Soulie et al., personal communication
Necrosis/ethylene inducing protein	*bcnep1*	B05.10	Virulence not affected on tomato and *Nicotiana benthamiana*		Cuesta Arenas and van Kan, personal communication
Necrosis/ethylene inducing protein	*bcnep2*	B05.10	Virulence not affected on tomato and *N. benthamiana*		Cuesta Arenas and van Kan, personal communication
Emoparmil-binding protein	*bcpie3*	T4/B05.10	Reduced colonization on tomato and bean leaves		Gioti et al., personal communication

The number of functionally analysed genes is rapidly expanding, as shown in Table 2.1 (see also Tudzynski and Siewers 2004).

The majority of the knock-out mutants reported so far were derived from strain B05.10 (for a description, see Tudzynski and Siewers 2004). This strain is highly virulent on several host plants and is genetically stable. Since it consistently yields high transformation rates, it is now one of the standard recipient strains in most laboratories; and it was also used for the first genome sequencing project (see above). As shown in Table 2.1 (and discussed in detail in the next paragraph), the role of specific genes can differ between *B. cinerea* strains: e.g. targeted mutation of the pectin methylesterase *Bcpme1* led to reduced virulence in strain Bd90 (Valette-Collet et al. 2003), but not in strain B05.10 (Kars et al. 2005b). These data emphasize the importance of the choice of strain to be used in such experiments and the urgent need to standardize these parameters in the *Botrytis* research community. However, these data show the high degree of flexibility of the pathogen. The first whole-genome comparisons of different field isolates of *B. cinerea* using high-speed sequencing technology substantiate these data (Kliebenstein and Rowe 2007). From this point of view it is interesting to have access to genome sequences of moderate coverage from two strains (T4, B05.10) that differ considerably in virulence.

As the large number of genes investigated by the targeted inactivation approach shows (Table 2.1), the classic "candidate genes" have more or less been functionally analysed in *Botrytis*; therefore unbiased cloning approaches are more interesting because they offer the perspective of identifying genes with novel functions. The genome analyses show that about 30% of the bona fide ORFs encode proteins of unknown function; therefore this approach is highly valid for a better understanding of the complex system. Two general strategies are applied. Firstly, screening can be performed for genes that are differentially expressed, e.g. specifically in planta or during certain developmental stages, and these genes can subsequently be analysed for their contribution to virulence by targeted inactivation. Techniques applied in *B. cinerea* include differential screening of cDNA libraries, differential display (DD) RT-PCR (Benito et al. 1996), suppression subtractive hybridization (Schulze Gronover et al. 2004), and macroarrays (Viaud et al. 2003; Chagué et al. 2006; Schumacher et al. 2008a). In the near future, microarrays will become available and allow whole-genome screening approaches.

A second strategy is based on random insertional mutagenesis, which has the great advantage of starting from a known phenotype. While the classic REMI technique was not successful in *B. cinerea* (see Tudzynski and Siewers 2004), *Agrobacterium*-mediated transformation has been established for *B. cinerea* in two laboratories (Rolland et al. 2003; N. Segmüller and P. Tudzynski, unpublished data). It could be shown that this system fulfils two major criteria that are essential for use in insertional mutagenesis: most transformants carry single-copy integrations, and the integration sites appear to be random. The availability of *B. cinerea* genomic sequences considerably facilitates the identification of the tagged genes. An insertional mutant library has been established and, of the 2800 mutants derived so far, more than 30 are significantly impaired in virulence (S. Giesbert, I. Haeuser-Hahn, P. Schreier, P. Tudzynski, unpublished data). This represents a larger collection of virulence mutants than any obtained so far with the targeted inactivation approach!

So far these un-biased approaches – promising though they are – have not yielded "new" virulence determinants. The next section provides a more detailed discussion of the results from the "classic" candidate gene approach, which have already significantly broadened our understanding of the pathobiology of *Botrytis*.

IV. Pathogenicity Determinants

Table 2.1 lists altogether 102 knock-out mutants of *B. cinerea*, of which only a limited number showed a pronounced defect in virulence. In the following a few of these data will be discussed in the context of classic pathogenicity determinants.

A. Cell Wall Degrading Enzymes

As genome analyses have confirmed, *B. cinerea* has a large set of genes encoding cell wall degrading enzymes (CWDEs). Therefore it must be expected that – like in the pioneering work of John Walton's group in the necrotroph *Cochliobolus carbonum* (e.g. Scott-Craig et al. 1998) – disruption of single or even multiple genes would not yield a significant reduction in virulence. At least with respect to the penetration process, the role of cutinolytic enzymes should be important, because the (pseudo-)appressoria are not equipped to generate high pressure. Nevertheless, knock-outs of a cutinase and a lipase gene showed no effect (see Table 2.1). However, the genomic sequence revealed that there are more putative cutinase/lipase encoding genes, so that the role of these enzymes in pathogenicity is open. TEM analyses indicate that there is no strong pressure involved in penetration (see Fig. 2.2). Interestingly, *B. cinerea* needs a protein for penetration (Bcpls1; Gourgues et al. 2004) which is a homologue of a tetraspanin-like

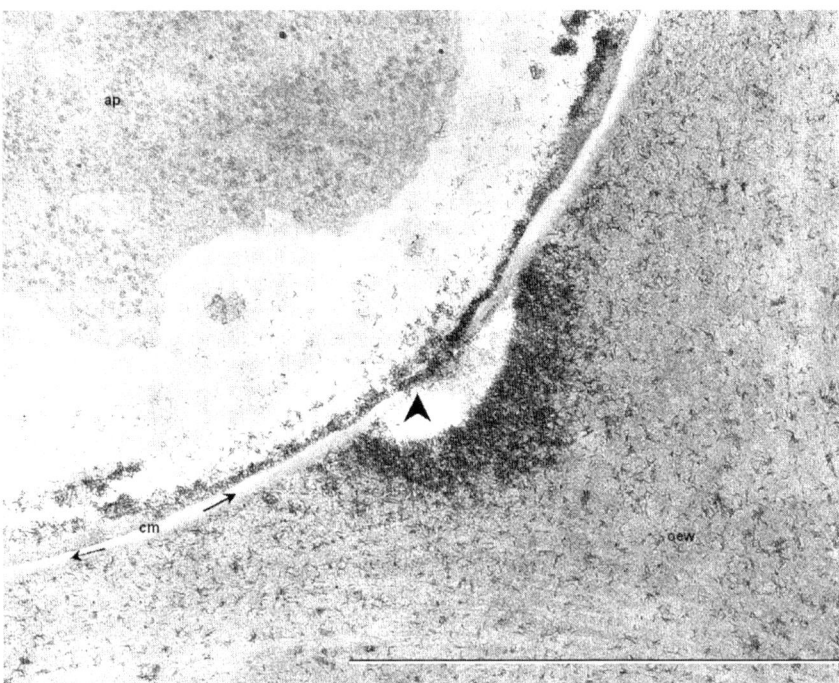

Fig. 2.2. In situ localization of H_2O_2 by the cerium chloride technique at the interface of a *B. cinerea* appressorium (*ap*) and the outer epidermal wall (*oew*) of a tomato leaf at 12 h post inoculation. In the host leaf, a small pore (*arrow head*) is present in the cuticle (*cm*) and in the cell wall layer underneath. *Bar* 1 μm (courtesy of Dr. K.B.Tenberge)

protein required for appressoria function in *M. grisea*, suggesting that there are common mechanisms in penetration (but probably beyond turgor pressor generation).

The van Kan group performed a detailed analysis of the role of pectinases in this interaction. *B. cinerea* possesses at least six endo-polygalacturonase (endo-PG) genes, which show different expression profiles during the interaction (ten Have et al. 2001), indicating specific functions. This is substantiated by knock-out analyses: only two of the six enzymes seem to be relevant for the infection process and cannot be complemented by the isoenzymes (Kars et al. 2005a; Williamson et al. 2007). Since the native pectin usually is highly methylated, a demethylating enzyme would be required for endo-PG activity. However, deletion of two pectinmethylesterase genes (*bcpme1, -2*) in strain B05.10 had no effect on virulence (Kars et al. 2005b). Surprisingly, deletion of *bcpme1* in another strain had an impact on virulence (Valette-Colett et al. 2003), suggesting considerable genetic divergency of different *B. cinerea* strains; this was substantiated by other analyses (see below). Not many functional analyses of other CWDE genes have been

performed so far; knock-out of a cellulase gene had no effect (as expected, since there are several in the genome; Espino et al. 2005). However, deletion of one of the several xylanase genes led to significantly reduced virulence (Brito et al. 2006). The reason for the unexpected crucial role of this single enzyme is open; it could be involved in the initiation of plant defense (parts of which are actually needed by *Botrytis*, see below) or could specifically weaken a cell wall component. In *C. carbonum*, deletion of a gene encoding a serine threonin protein kinase (*snf1*), which controls expression of most secreted enzymes, leads to avirulence (Tonukari et al. 2000). Deletion of an orthologue of this gene, *bcsnf1*, reduced virulence in *B. cinerea*, but there is so far no evidence that this effect is due to reduced expression of genes encoding extracellular enzymes (J. Schumacher and B. Tudzynski, unpublished data).

B. Phytotoxic Compounds

As a classic necrotroph, *B. cinerea* secretes a whole set of phytotoxic compounds. The best studied toxin is botrydial, a sesquiterpene causing necrotic lesions on many host plants; its toxic effect is light-dependent – as is infection by *B. cinerea* – and it could be detected both in submersed culture and in planta in all wild-type strains analysed so far. Therefore it was considered a bona fide pathogenicity factor (Deigton et al. 2001; Colmenares et al. 2002). Since it was supposed that a cytochrome P450 monooxygenase was involved in the biosynthesis of botrydial, Siewers et al. (2005) performed a systematic functional analysis of P450 genes and identified the first gene involved in botrydial biosynthesis, *bcbot1*. Interestingly, Δbcbot1 mutants, which lack the toxin, are fully virulent in strain B05.10, while they show severely reduced virulence in strain T4, substantiating the earlier findings of genetic variability between different *B. cinerea* isolates. In this case the different importance of botrydial for infection could be due to the presence/lack of a second class of toxins, botcinines (Reino et al. 2004). Since their biosynthetic pathway has yet to be established, functional analyses are not yet possible. An interesting recently detected group of small phytotoxic proteins (NEP1-like proteins; Staats et al. 2007) seems to be involved in the induction of host cell death; their role in the infection process is still open.

C. Organic Acids

Finally, *B. cinerea* – like many fungi – can secrete organic acids to acidify its surroundings to optimize the pH for its secreted enzymes. However, secretion of oxalic acid serves additional purposes: it is able to form complexes with Ca^{2+} ions bound by pectin, thus helping in making the major component of the middle lamella accessible for the fungus. Oxalate is synthesized by the oxaloacetate hydrolase (BcOAH1). Δbcoah1 mutants do not produce oxalate in vitro (Han et al. 2007). However, their impact on virulence is not yet clear (J. van Kan, personal communication).

D. Phytohormones

B. cinerea has also been shown to produce several phytohormones (Sharon et al. 2004). The best investigated system so far is abscisic acid; its biosynthetic pathway has been shown to be different from that of plants and functional analyses have identified a cluster of genes involved in

this pathway (Siewers et al. 2004, 2006). Mutants of genes involved in late pathway steps are not impaired in virulence; however, the analysis of early pathway mutants is needed before coming to final conclusions on the role of ABA in virulence (V. Siewers, L. Kokkelink, P. Tudzynski, unpublished data). Chague et al. (2002) showed that ethylene biosynthesis in *B. cinerea* does not follow the common pathways via 1-aminocyclopropane-1-carboxylic acid (ACC) or 2-oxoglutarate, but through α-keto γ-methylthiobutyric acid (KMBA). Genes encoding enzymes of this pathway are not yet available; however, recent studies indicate that ethylene might have significant impact on both partners during the interaction, though the impact of fungal ethylene is not yet clear (Chague et al. 2006).

E. Reactive Oxygen Species

Reactive oxygen species play an important role in host–pathogen interaction in *Botrytis*, though the details still have to be elucidated. *Botrytis* causes an oxidative burst during cuticle penetration and lesion formation; cytological analyses indicate that the fungus contributes to the elevated ROS levels during the interaction (Fig. 2.3; Tenberge 2004); the TEM analyses (Fig. 2.2) also clearly show that there is no high pressure involved in the penetration, as no deformation of the plant's cell wall structure is evident. Williamson and collaborators presented evidence that in the infected tissue free radicals accumulate and there is a depletion of antioxidants (for a review, see Lyon et al. 2004). Biotrophic and hemi-biotrophic fungi tend to avoid the initiation of an oxidative burst; the aggressive mode of ROS provocation in *B. cinerea* thus seems to be a special trait of the necrotrophic life style. There are two interesting aspects of these findings: how does the fungus cope with the detrimental effect of ROS on its own cellular components, and how is the fungally secreted ROS produced (Fig. 2.3). Analyses of the *B. cinerea* oxidative stress response system have so far yielded no specificities: a comparative genome analysis showed that *B. cinerea* has more or less the same set of genes encoding antioxidants as other pathogens and even saprophytes (B. Oeser, M. Dickman, M. Levy, P. Tudzynski, unpublished data). Deletion of a gene encoding a transcription factor involved in oxidative stress response, Bap1, led to increased sensitivity against H_2O_2, but

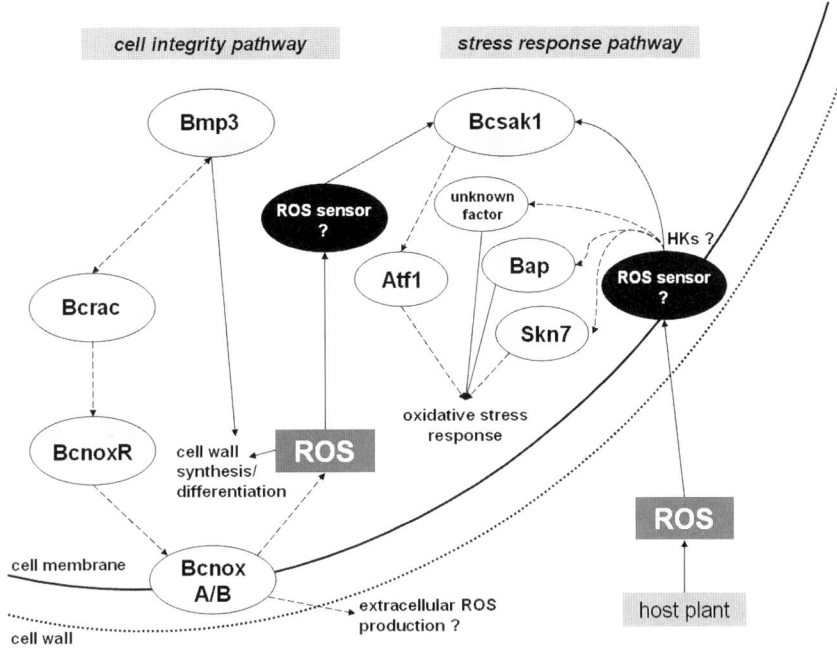

Fig. 2.3. Scheme of the ROS homeostasis regulatory pathways in *B. cinerea*. *Atf1* bZIP transcription factor (for details, see text); *Bcp67phox* regulatory subunit of the nox complex; for protein designations, see Table 2.1

had no impact on virulence; and the same holds true for a histidine kinase, Bchk1, which is obviously involved in ROS sensing, but is not essential for virulence (N. Temme, P. Tudzynski, unpublished data). The only signalling factor involved in oxidative stress response for which an impact on virulence has been shown so far is the stress-inactivated MAPKinase Bcsak1 (see below). The exact mode of generation of extracellular ROS by *Botrytis* is not yet clear; deletion of a gene encoding a glucose oxidase had no impact on virulence (Rolke et al. 2004). A secreted SOD (yielding H_2O_2 from O_2-) is a virulence factor (Rolke et al. 2004), though its exact role is open (either detoxification of plant O_2- or generation of H_2O_2). Recently new candidates for O_2- sources in *B. cinerea* have been analysed: genes encoding putative NADPH oxidases (*bcnoxA, -B*) have been deleted. The single mutants are indeed impaired in virulence. A double mutant is almost avirulent, as is a ΔbcnoxR mutant lacking the regulatory subunit of the Nox complex (Segmüller et al. 2008), and their impact on the ROS status is under investigation. The mutants are also affected in formation of sclerotia. Since these enzymes are not exclusive for *Botrytis* (or even pathogens) and are involved in differentia-

tion in various fungi (Aguirre et al. 2005; Takemoto et al. 2007), their role in *Botrytis* could also primarily be in differentiation of infection structures. Recently Egen et al. (2007) presented evidence that NADPH oxidases are involved in differentiation of appressoria in *M. grisea*; B. Scott and collaborators showed that, in the endophyte *Epichloe festucae*, Nox is involved in maintaining the subtle balance between fungus and host (Tanaka et al 2006). It is possible that also in *B. cinerea* ROS homeostasis is essential for differentiation steps, e.g. involving cell wall biogenesis/re-organization. Of course the cell must make sure that elevated ROS level triggering/accompanying such differentiation steps are locally and timely restricted, by induction of the oxidative stress system, this requires an internal ROS sensing system (see Fig. 2.3). Unexpectedly, BcnoxA and B obviously do not contribute significantly to the extracellular pool of ROS observed during infection (Segmüller et al. 2008); therefore the source for the fungal part of the ROS produced in planta is still unkown. This external ROS accumulation (partly fungal, partly from the plant's oxidative burst) requires an external ROS sensor to trigger the protective oxidative stress response system.

To the fungus a fine-tuning of all these systems is essential; how this is achieved is an interesting question for further research, as is the source of external ROS and its role during the interaction.

Taken together, the candidate gene approach yields valuable information on the contribution of specific genes to virulence and e.g. indicates a high degree of variability, making the classic definition of virulence or pathogenicity genes questionable. However, many – if not most – of the single knock-outs have no effect on virulence, because of the complexity of the system. Therefore an alternative approach addressing groups of genes activated under certain conditions is more promising, i.e. the functional analysis of transcription factors and signal cascades in general.

V. Signalling

Conserved signal transduction pathways, such as the cAMP-dependent and several MAP kinase pathways, have been shown to be important for morphogenesis, differentiation and virulence in many phytopathogenic fungi (Tudzynski and Sharon 2003; Xu 2000; Xu et al. 2006). In recent years, significant progress has been made in characterization of the major signalling cascades in *B. cinerea* and their impact on development and pathogenicity (for recent reviews, see Tudzynski and Schulze Gronover 2004; Williamson et al. 2007). Figure 2.1 shows the roles of the so far analysed signalling compounds/cascades in the different steps of the life cycle.

A. The cAMP-Dependent Signalling Pathway

The cAMP-dependent signalling pathway is involved in multiple processes in plant-pathogenic fungi, including growth, conidiation and spore germination, nutrient sensing, and virulence (Kronstad 1997). In *B. cinerea*, most components of this pathway are now fully characterized. They include three Gα subunits of heterotrimeric G-proteins (Bcg1, Bcg2, Bcg3; Schulze Gronover et al. 2001; Döhlemann et al. 2006a), the adenylate cyclase (Bac; Klimpel et al. 2002), two catalytic subunits (BcPka1, BcPka2) and the regulatory subunit (BcPkaR) of the cAMP-dependent protein kinase (PKA; Schumacher et al. 2008c).

Apart from bcpka2, deletion of each of the corresponding genes individually results in impaired virulence, but never in a total loss of pathogenicity. A very pronounced effect is observed in the Δbcg1 mutant, which is still able to produce conidia and penetrate plant tissue, but stops the infection process in primary lesions; i.e. spreading lesions and soft rot development by Δbcg1 mutants have never been observed (Schulze Gronover et al. 2001). Both Δbcg2 and Δbcg3 mutants have a less severe phenotype; they are able to invade the plant and cause spreading lesion but delayed compared to the wild type. The Δbcg3 mutant showed reduced conidial production, prolonged germ tubes and a reduced penetration efficiency, which may account for the delayed infection process (Döhlemann et al. 2006a). This phenotype can be restored by supplementation with cAMP, i.e. it is exclusively governed by the cAMP pathway. In contrast, several functions of Bcg1 seem to be cAMP-independent: addition of cAMP to the Δbcg1 mutant restores the wild-type colony morphology but not e.g. the loss of protease secretion and production of the phytotoxin botrydial, suggesting that Bcg1 controls at least one additional signalling pathway. The adenylate cyclase Bac is activated by two Gα subunits, Bcg1 and Bcg3, as indicated by the similar and partly overlapping phenotypes of the Δbac mutant and the Δbcg1 and the Δbcg3 mutants: Δbcg1 and Δbac form compact colonies on high sucrose-containing medium (Klimpel et al. 2002), whereas both, Δbcg3 and Δbac are impaired in spore germination (Döhlemann et al. 2006a). In contrast to the Δbcg3 mutant, the Δbac mutant was unable to sporulate in planta and in vitro conidiation was significantly reduced (Klimpel et al. 2002; Schumacher et al 2008c).

Recently, the genes encoding the two catalytic subunits (*bcpka1/2*) and the regulatory subunit (*bcpkaR*) of the PKA were also functionally characterized (Schumacher et al. 2008c). Whereas Δbcpka2 mutants showed wild-type growth, conidiation, germination, and infection, the Δbcpka1 mutants showed a strong phenotype in vitro and in planta: they grow slowly on different complete and synthetic media, and – similar to the *Δbac* mutants – the development of spreading lesions is delayed and soft rot of whole leaves does not occur. In contrast to the Δbac mutant, the Δbcpka1 mutants are able to sporulate in planta.

Interestingly, mutants affected in the regulatory subunit of PKA (BcPkaR) show the same phenotype as Δbcpka1 mutants, in contrast to other systems where lack of the regulatory subunit induces constitutive PKA activity (Schumacher et al. 2008c).

Recently, the gene for the Gβ-subunit (Bcgb1) of the heterotrimeric G-protein complex was cloned and deleted. The Δbcgb1 mutants showed increased formation of aerial hyphae, delayed and reduced conidiation, lack of sclerotia formation, and delayed penetration of plant tissue. Infection was arrested at the stage of secondary lesion formation, preventing soft rot development (J. Schumacher and B. Tudzynski, unpublished data). Preliminary array analyses revealed little overlap in the *B. cinerea* genes that are regulated by Bcg1 (Gα) and Bcgb1 (Gβ). In contrast, in the chestnut pathogen *Cryphonectria parasitica* the transcripts of about 100 genes showed altered expression levels in both the Gα mutant Δcpg1 and the Gβ mutant Δcpgb1. In most cases, these transcripts appeared to be co-regulated, suggesting a considerable redundancy in pathway control or extensive cross-talk between the Gα and Gβ subunit-controlled pathways (Dawe et al. 2004). The explanation for the overlapping functions is probably the degradation of the Gα subunit Gcpg1 in Δcpgb1 mutants and vice versa (Parsley et al. 2003). These differences between *B. cinerea* and *C. parasitica* (and the unexpected phenotype of the ΔbcpkaR mutants) illustrate that highly conserved signalling components in fungal pathogens may have quite different functions, confirming the need to analyse not only a few model systems!

B. MAP Kinase-Controlled Signalling Pathway

MAP kinase-controlled signalling pathways have been shown in several plant pathogens to be essential for the early phases of infection, specifically for the penetration of plant surfaces (Xu and Hamer 1996; Xu 2000; Mey et al. 2002; Jenczmionka and Schäfer 2005; Solomon et al. 2005). In filamentous fungi, three of the five yeast MAPK cascades are conserved (Xu 2000): homologues of Fus3 (*M. grisea*: Pmk1), Slt2 (*M. grisea*: Mps1), and Hog (*M. grisea*: Osm1). In *B. cinerea*, deletion of the *pmk1*-homologous gene, *bmp1*, resulted in altered growth rate, reduced conidiation, and total inability to penetrate host tissue (Zheng et al. 2000). The same mutant in the B05.10 background showed fewer pleiotropic growth defects in vitro; detailed

analysis revealed a role of Bmp1 in carbon source-induced germination of conidia (Döhlemann et al. 2006a).

Deletion of the *bmp3* gene (encoding a Slt2 homologue) led to reduced vegetative growth on media with low osmolarity, impaired conidiation, and failure to form sclerotia (Rui and Hahn 2007). Although some defects in this mutant are similar to those in other pathogenic fungi (e.g. the impaired ability to invade plant tissue), Bmp3 has features unique for a Slt2-type MAP kinase, such as low osmolarity-induced growth inhibition.

Also the *B. cinerea* Hog1 homologue, Bcsak1, has been functionally characterized. It shows unique features: it is phosphorylated when *B. cinerea* is exposed to certain fungicides, osmotic stress, and oxidative stress. The Δbcsak1 mutants are significantly impaired in vegetative and pathogenic development: they fail to produce conidia, show increased sclerotial development, and are unable to penetrate unwounded plant tissues (Segmüller et al. 2007). This is by far the strongest phenotype associated with a stress-activated MAP cascade in phytopathogenic fungi. To study the impact of stress on pathogenic development in detail, homologues of yeast transcription factors involved in stress response are currently being characterized. A homologue of the *Saccharomyces cerevisiae yap1* gene, *bap1*, was functionally analysed. The Δbap1 mutants were more sensitive to oxidative stress in vitro, but showed normal virulence. Northern analysis showed that Bap1 controls several typical oxidative stress response genes, but these genes differ from the ones regulated by Bcsak1, indicating that Bap1 acts independently from the Bcsak1 cascade (N. Temme and P. Tudzynski, unpublished data). The *B. cinerea* genes encoding homologues of the yeast response regulator Skn7 and the bZIP factor Atf1 (which acts in yeast downstream of the HOG cascade) are currently being functionally analysed (N. Temme and P. Tudzynski, unpublished data).

C. Small G-Proteins

Small G-proteins of the RAS superfamily have been shown to play important roles in cell polarization, differentiation, and development in a wide range of eukaryotes. Like other filamentous fungi, *B. cinerea* contains two genes encoding proteins of the Ras subfamily, Bcras1 and Bcras2. Δbcras1 mutants are viable but show strong growth defects, including an irregular hyphal morphology and loss of polarized growth; Δbcras1 mutants do not sporulate and are totally apathogenic (L. Kokkelink et al., unpublished data). The phenotype of Δbcras2 mutants was not as strong, including slightly impaired spore formation and pathogenicity, and decreased growth rates on solid media. Interestingly this phenotype can be partially restored by addition of cAMP, indicating an impact of Bcras2 on Bac (Schumacher et al. 2008c).

Unlike yeast, filamentous fungi have Rac-like proteins from the Rho subfamily in addition to Cdc42 (Boyce et al. 2005). In *B. cinerea*, Δbcrac mutants displayed a similarly strong phenotype as Δbcras1 mutants (loss of polarized growth, spore formation, pathogenicity), suggesting that Bcras1 and Bcrac act in the same signalling pathway. In contrast, growth and pathogenic development in Δbccdc42 mutants are only slightly affected (L. Kokkelink and P. Tudzynski, unpublished data).

In addition, several members of the Rab subfamily of small GTPases have been identified in *B. cinerea*. In eukaryotic cells, Rab-like GTPases are major regulators of vesicular trafficking and are involved in essential processes including exocytosis, endocytosis and cellular differentiation (Siriputthaiwan et al. 2005). So far, *bcrab2* in *B. cinerea* has been characterized in more detail. Δbcrab2 mutants can penetrate plant tissue and develop small lesions but spreading lesions have never been observed. These results indicate that the Rab/GTPase Bcrab2 is essential for invading host cells, probably by regulating the intracellular transport of secretory vesicles involved in the delivery of proteins to the extracellular medium (P. Hantsch and B. Tudzynski, unpublished data).

D. The Ca²⁺/Calmodulin-Dependent Signalling Pathway

The Ca²⁺/calmodulin-dependent signalling pathway, which has so far not been in the focus of research in plant pathogen signalling, was recently shown to be important for pathogenicity of *B. cinerea*. The roles of the PP2B phosphatase calcineurin and cyclophilin A, highly conserved components of the Ca²⁺/calmodulin-dependent signalling pathway, were investigated using drug effectors. Immunosuppressive drugs, such as cyclosporin A (CsA) and FK506, inhibit calcineurin activity by binding to the peptidyl-prolyl isomerases cyclophilin A

and FKBP12, respectively. The protein–drug complexes bind to the hydrophobic interface between both subunits, thus inhibiting the calcineurin phosphatase (for a review, see Kraus and Heitman 2003). In *B. cinerea*, the genes encoding the cellular targets of both drugs, cyclophilin A and FKBP12 (*bcp1* and *bcpic5*), have been deleted, yielding drug-resistant mutants affected in virulence on bean and tomato leaves (Viaud et al. 2003; Gioti et al. 2006). Targeted disruption of the calcineurin gene was unsuccessful in a first approach (Viaud et al. 2003); however, using a specific medium composition, a calcineurin knock-out mutant was recently generated (J. Schumacher and B. Tudzynski, personal communication). Viaud and collaborators studied the role of calcineurin using the inhibitor cyclosporin A (CsA) in a macroarray analysis. Altogether 18 calcineurin-dependent (*CND*) genes (among 2839 *B. cinerea* genes) were down-regulated by CsA. Among the co-regulated *CND* genes, three were organized as a physical cluster that could be involved in secondary metabolism (Viaud et al. 2003), which later appeared to be required for botrydial biosynthesis (Siewers et al. 2005).

As mentioned before, Δbcg1 deletion mutants lost the ability to produce the phytotoxic secondary metabolites botrydial and botcinines (Schulze Gronover et al. 2001, 2004). An extended macroarray analyses with the wild-type and the *bcg1* mutant, treated or not treated with CsA, revealed a new gene cluster responsible for a yet unknown PKS-derived compound to be under control of Bcg1 in a calcineurin-dependent manner, confirming the interconnection between these signalling pathways (Schumacher et al. 2008a). In addition, it could be shown that phospholipase C (Bcplc1) is a component of the Bcg1/Calcineurin-dependent signalling pathway (Schumacher et al. 2008a). Also, a homologue of the calcineurin-responsive transcription factor Crz1 was functionally analysed. Δbccrz1 mutants show impaired sporulation and vegetative growth; interestingly, spore-mediated penetration is normal, while mycelium cannot penetrate the host tissue; development of spreading lesions is retarded (Schumacher et al. 2008b).

E. Cell Surface Receptors

Cell surface receptors are almost a black box in the molecular biology of plant pathogens. These receptors perceive environmental signals and relay them

A 7468 ?

//Kprivdon\A7636\ - HP Laserjet 4065

Botrytis cinerea: Molecular Aspects of a Necrotrophic life style

Tudzynski, P.; Kokkelink, L.

Springer 2009

in Plant Relationships: The Mycota

Volume 5 pp 29-50

and-services)

Our network evolution
(http://explore.ee.co.uk
/network-evolution)

Newsroom
(https://explore.ee.co.uk

company

sroom)

onsibility

://explore.ee.co.uk

company

gresponsible)

Clone Phone
(http://explore.ee.co.uk
/clone-phone)

Recycle and Reward
(https://recycle.ee.co.uk/)

e-safety and e-skills
(http://explore.ee.co.uk

/digital-living)

(http://explore.ee.co.uk
/regulatory)

Accessibility
(http://explore.ee.co.uk
/digital-living/digital-
for-all)

HOW TO FIND EE

Coverage checker
(https://explore.ee.co.uk
/coverage-checker)

Store Finder
(http://storefinder.ee.co.ul
/ee/stores.html)

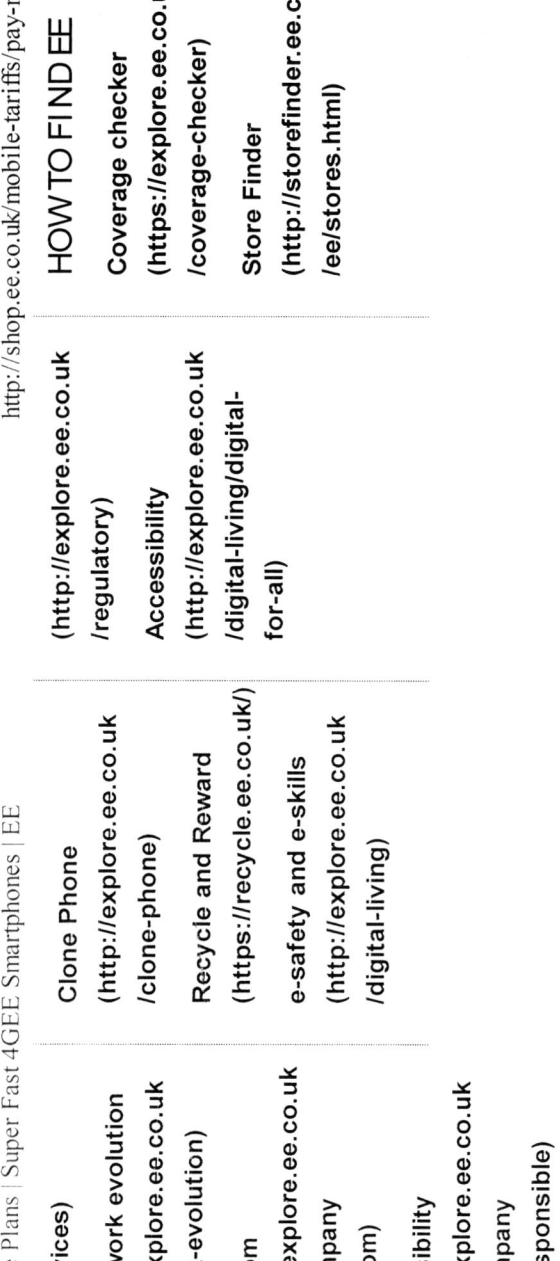

©2013 EE Limited

(htt (htt brought to you by EE

to intracellular signalling pathways. The growing number of sequenced fungal genomes only recently allowed identification of classes of proteins that could be involved in signal perception. Especially families of G-protein-coupled receptors (GPCRs) have attracted major attention in plant pathogenic fungi (DeZwaan et al. 1999; Kulkarni et al. 2005). Two GPCR subfamilies have been defined in fungi on the basis of the presence or absence of an amino-terminal extracellular cysteine-rich domain (cystein-rich fungal extracellular membrane; CFEM domain), which is characteristic for human GPCRs. A prototype representative of a fungal GPCR with such a domain is the *M. grisea* PTH11. In *M. grisea*, 61 PTH11-related proteins were identified, which constitute the largest number of GPCR-like proteins reported in fungi to date (Kulkarni et al. 2005). In *B. cinerea*, *btp*1, encoding a protein with seven transmembrane domains and significant homology to PTH11 (but without a CFEM domain), has been functionally characterized. The Δbtp1 mutants were only slightly impaired in virulence, and Btp1 thus probably does not interact with Bcg1 during pathogenesis (Schulze Gronover et al. 2005).

Another group of eukaryotic sensor proteins are the two-component histidine kinases (HKs). In response to a specific signal, HK autophosphorylates a conserved histidine residue. The phosphate is then transferred to a conserved aspartic acid residue in a response regulator (RR) protein, resulting directly in changed transcription or in regulation of down-stream components, e.g. a MAP kinase cascade (Wolanin et al. 2002). The *B. cinerea* genome sequence contains at least 20 HKs in 11 classes (Catlett et al. 2003). So far only members of class III have been characterized in fungi; they are mainly involved in osmoregulation and fungicide resistance, e.g. Os1 in *M. grisea* (Motoyama et al. 2005). Also in *B. cinerea* Bos1, a class III enzyme, was shown to be involved in osmoregulation, resistance to dicarboximide and phenylpyrrole fungicides, and virulence. Interestingly, Δbos1 mutants display a phenotype similar in several aspects to that of Δbcsak1 mutants (loss of conidiation, osmosensitivity). However, Δbos1 strains differ in their resistance to fungicides, and they are able to penetrate, but are significantly reduced in the ability to invade host cells (Viaud et al. 2006). The difference in penetration capacity is probably due to the use of different strains as recipients, since deletion of the same gene in B05.10 led to the same penetration defect as in Δbcsak1. These data indicate that Bos1 is the major upstream component of the Bcsak1 cascade (Liu et al. 2008).

Recently, Bchk1, a member of class X HKs, homologous to *Schizosaccharomyces pombe* Mak2/3 which is involved in oxidative stress responses (Buck et al. 2001), was functionally analysed in *B. cinerea*. The mutant is impaired in growth on low concentrations of H_2O_2, suggesting that this phospho-relay system is involved in oxidative stress response caused by low doses, whereas the Bcsak1 cascade is required for responses to high doses of H_2O_2 (N. Temme and P. Tudzynski, unpublished data). Deletion of the *bchk*5 gene, encoding a homologue of the single HK in *Saccharomyces cerevisiae*, Sln1, showed no obvious phenotype. (Y. Cuesta Arenas and J. van Kan, unpublished data). This result shows that homologues of yeast signalling components may have totally different functions in filamentous fungi, limiting the model character of the yeast system for the understanding of signalling networks in phytopathogens.

VI. Special Aspects of Host Defence

Host defence reactions against *B. cinerea* have been studied in recent years, mainly in the model plant *Arabidopsis* and in tomato. The data were reviewed by Williamson et al. (2007), therefore only a few special aspects should be mentioned here. In general it turns out that *B. cinerea* – like other necrotrophs – induces mainly the jasmonate and ethylene defence systems (Thomma et al. 2001; Mengiste 2008). A very interesting aspect of the interaction between *B. cinerea* and its host plant was brought up by the pioneering work of A. Levine and collaborators: they showed that *B. cinerea* triggers (and needs) the plant's defence-related cell death. A hypersensitive response (HR)-defective mutant of *A. thaliana* was less susceptible to *Botrytis* (Govrin and Levine 2000). Also, in tomato, infection by *Botrytis* caused up-regulation of genes involved in apoptosis (Hoeberichts et al. 2003). These data indicate that *Botrytis* is not a simple "killer", but uses part of the plant's response system for its own purposes (see van Baarlen et al. 2007). This sophisticated strategy indicates that– in contrast to the above-mentioned idea of necrotrophy – there has to be a close encounter between fungus and plant; and it suggests some kind of co-evolution, which was previously considered specific for biotrophs.

VII. Perspectives

The data compiled here show that *Botrytis cinerea* has become an important model system for the study of necrotrophic plant pathogens. Though our understanding of the pathogenicity mechanisms of *B. cinerea* has significantly increased, there are still no real breakthroughs in the development of new defence strategies. Apart from new developments in generating resistant tomato varieties (e.g. Finkers et al. 2007), the generation of stable resistant plants is still in its infancy. Since the infection strategy of *B. cinerea* includes the triggering of programmed cell death, which may hamper the design of plants with increased resistance without a concomitant reduction of resistance against biotrophic pathogens might be difficult. Much basic research, especially e.g. more transcriptome analyses (AbuQamar et al. 2006), is necessary to better understand the role of different plant resistance pathways and the way by which *B. cinerea* triggers them for its own purposes (van Baarlen et al. 2007). Data obtained from the model systems *Arabidopsis*, a non-natural host, will have to be substantiated by data from major host crop plants, such as tomato or bean. Also the alternative approach, the "intelligent" design of more specific fungicides has not yet been successful: currently there is no fungicide on the market that was developed on the basis of targeted molecular research. Still, agribusiness companies obviously appreciate the great potential resource arising in the rapidly accumulating knowledge on this model pathogen. Of course the great variability and broad host range of this pathogen present a special technical challenge. The availability of the genome sequence facilitates corresponding research drastically: it will allow genome-wide transcript analysis, and it is already used for detailed proteome studies (e.g. Fernandez-Acero et al. 2007; Marra et al. 2007). A very interesting approach is the study of genetic heterogeneity in *B. cinerea* populations, focusing on specific genes like the highly variable polygalacturonase-encoding gene *bcpg1*, or even studying genome-wide variation using high-speed sequence technology (Kliebenstein and Rowe 2007). Comparative genomics will help to study evolution of pathogenicity, e.g. comparison of the two closely related necrotrophs, *B. cinerea* and *Sclerotinia sclerotiorum*, both with related hemi-biotrophs and biotrophs and with more distantly related necrotrophs such as *Alternaria* spp.

There are still several aspects of the life cycle and pathogenitiy of *B. cinerea* which need further analysis: the exact mechanisms of adhesion and penetration, the details of signalling (especially sensing mechanisms and cross-talking between different cascades), the role of low molecular weight effectors (secondary metabolites, toxic polypeptides), and especially the scarcely analysed phenomenon of quiescent infection stages, which could give a valuable insight into the communication between plant and fungus. In a way, we are just beginning to understand the many interesting facets of this highly variable fungus.

Acknowledgements. We thank S. Fillinger, M. Viaud (Versailles), J. van Kan (Wageningen), B. Tudzynski, J. Schumacher, and N. Temme for providing us with unpublished data, K.B. Tenberge for electronmicroscopic photographs, and B. Tudzynski for critical reading of the manuscript. Our own experimental work was supported by the Deutsche Forschungsgemeinschaft.

References

AbuQamar S, Chen X, Dhawan R, Bluhm B, Salmeron J, Lam S, Dietrich RA, Mengiste T (2006) Expression profiling and mutant analysis reveals complex regulatory networks involved in *Arabidopsis* response to *Botrytis* infection. Plant J 48:28–44

Aguirre J, Ríos-Momberg M, Hewitt D, Hansberg W (2005) Reactive oxygen species and development in microbial eukaryotes. Trends Microbiol 13:111–118

Benito EP, Prins TW, van Kan JAL (1996) Application of differential display RT-PCR to the analysis of gene expression in a plant–fungus interaction. Plant Mol Biol 32:947–957

Bolton MD, Thomma BPHJ, Nelson BD (2006) *Sclerotinia sclerotiorum* (Lib.) de Bary: biology and molecular traits of a cosmopolitan pathogen. Mol Plant Pathol 7:1–16

Boyce KJ, Hynes MJ, Andrianopoulos A (2005) The Ras and Rho GTPases genetically interact to co-ordinately regulate cell polarity during development in *Penicillium marneffei*. Mol Microbiol 55:1487–1501

Brito N, Espinoso JJ, Gonzalez C (2006) The endo-β-1,4-xylanase Xyn11A is required for virulence in *Botrytis cinerea*. Mol Plant-Microbe Interact 19:25–32

Buck V, Quinn J, Soto Pino T, Martin H, Saldanha J, Makino M, Morgan BA, Millar JBA (2001) Peroxide sensors for the fission yeast stress-activated mitogen-activated protein kinase pathway. Mol Biol Cell 12:407–419

Catlett NL, Yoder OC, Turgeon BG (2003) Whole genome analysis of two-component signal transduction genes in fungal pathogens. Eukaryot Cell 2:1151–1161

Chagué V, Elad Y, Barakat R, Tudzynski P, Sharon A (2002) Ethylene biosynthesis in *Botrytis cinerea*. FEMS Microbiol Ecol 40:143–149

Chagué V, Danit LV, Siewers V, Schulze Gronover C, Tudzynski P, Tudzynski B, Sharon A (2006) Ethylene sensing and gene activation in *Botrytis cinerea*: a missing link in ethylene regulation of fungus-plant interactions? Mol Plant–Microbe Interact 19:33–42

Colmenares AJ, Aleu J, Duran-Patron R, Collado IG, Hernandez-Galan R (2002) The putative role of botrydial and related metabolites in the infection mechanism of *Botrytis cinerea*. J Chem Ecol 28:997–1005

Deighton N, Muckenschnabel I, Colmenares AJ, Collado IG, Williamson B (2001) Botrydial is produced in plant tissues infected by *Botrytis cinerea*. Phytochem 57:689–692

del Sorbo G, Ruocco M, Schoonbeek HJ, Scala F, Pane C, Vinale F, de Waard MA (2008) Cloning and functional charakterization of BcatrA, a gene encoding an ABC tansporter of the plant pathogenic fungus *Botryotinia fuckeliana* (*Botrytis cinerea*). Mycol Res 112:737–746

DeZwaan TM, Carroll AM, Valent B, Sweigard JA (1999) *Magnaporthe grisea* Pth11p is a novel plasma membrane protein that mediates appressorium differentiation in response to inductive substrate cues. Plant Cell 11:2013–2030

Doehlemann G, Molitor F, Hahn M (2005) Molecular and functional characterization of a fructose specific transporter from the gray mold fungus *Botrytis cinerea*. Fungal Gen Biol 42:601–610

Doehlemann G, Berndt P, Hahn M (2006a) Different signaling pathways involving a G-alpha protein, cAMP and a MAP kinase control germination of *Botrytis cinerea* conidia. Mol Microbiol 59:821–835

Doehlemann G, Berndt P, Hahn M (2006b) Trehalose metabolism is important for heat stress tolerance and spore germination of *Botrytis cinerea*. Microbiology 152:2625–2634

Egen MJ, Wang Z-Y, Jones MA, Smimoff N, Talbot NJ (2007) Generation of reactive oxygen species by fungal NADPH oxidases is required for rice blast disease. Proc Natl Acad Sci USA 104:11772–11777

Elad Y, Williamson B, Tudzynski P, Delen N (eds) (2004) *Botrytis* spp.: biology, pathology and control. Kluwer, Dordrecht

Espino JJ, Brito N, Noda J, Gonzalez C (2005) *Botrytis cinerea* endo-β-1,4-glucanase Cel5A is expressed during infection but is not required for pathogenesis. Physiol Mol Plant Pathol 66:213–221

Faretra F, Pollastro S (1996) Genetic studies of the phytopathogenic fungus *Botryotinia fuckeliana* (*Botrytis cinerea*) by analysis of ordered tetrads. Mycol Res 100:620–624

Fernandez-Acero FJ, Jorge I, Calvo E, Vallejo I, Carbu M, Camafeita E, Garrido C, Lopez JA, Jorrin J, Cantoral JM (2007) Proteomic analysis of phytopathogenic fungus *Botrytis cinerea* as a potential tool for identifying pathogenicity factors, therapeutic targets and for basic research. Arch Microbiol 187:207–215

Fillinger S, Amselem J, Artiguenave F, Billaut A, Choquer M, Couloux A, Cuomo C, Dickman MB, Fournier E, Gioti A, Giraud C, Kodira C, Kohn L, Legeai F, Levis C, Mauceli E, Pommier C, Pradier JM, Quevillon E, Rollins J, Ségurens B, Simon A, Viaud M, Weissenbach J, Wincker P, Lebrun M-H (2007) The genome projects of the plant pathogenic fungi *Botrytis cinerea* and *Sclerotinia sclerotiorum*. In: Jeandet P, Clément C, Conreux A (eds) Macromolecules of grape and wines. Lavoisier, Paris, pp 125–133

Finkers R, van den Berg P, van Berloo R, ten Have A, van Heusden AW, van Kan JAL, Lindhout P, (2007) Three QTLs for *Botrytis cinerea* resisitance in tomato. Theoret Appl Gen 114:585–593

Fitzgerald A, van Kan JAL, Plummer KM (2004) Simultaneous silencing of multiple genes in the apple scab fungus, *Venturia inaequalis*, by expression of RNA with chimeric inverted repeats. Fungal Genet Biol 41:963–971

Gioti A, Simon A, Le Pecheur P, Giraud C, Pradier JM, Viaud M, Levis C (2006) Expression profiling of *Botrytis cinerea* genes identifies three patterns of up-regulation in planta and an FKBP12 protein affecting pathogenicity. J Mol Biol 358:372–386. Erratum: J Mol Biol 364:550

Gourgues M, Brunet-Simon A, Lebrun M-H, Levis C (2004) The tetraspanin BcPls1p is required for appressorium-mediated penetration of *Botrytis cinerea* into host plant leaves. Mol Microbiol 51:619–629

Govrin EM, Levine A (2000) The hypersensitive response facilitates plant infection by the necrotrophic pathogen *Botrytis cinerea*. Curr Biol 10:751–757

Han Y, Joosten HJ, Niu W, Zhao Z, Mariano PS, McCalman MT, van Kan JAL, Schaap PJ, Dunaway-Mariano D (2007) Oxaloacetate hydrolase: the C–C bond lyase of oxalate secreting fungi. J Biol Chem 282:9581–9590

Hayashi K, Schoonbeek HJ, De Waard MA (2002a) Expression of the ABC transporter BcatrD from *Botrytis cinerea* reduces sensitivity to sterol demethylation inhibitor fungicides. Pesticide Biochem Physiol 73:110–121

Hayashi K, Schoonbeek HJ, De Waard MA (2002b) Bcmfs1, a novel major facilitator superfamily transporter from *Botrytis cinerea*, provides tolerance towards the natural toxic compounds camptothecin and cercosporin and towards fungicides. Appl Environ Microbiol 68:4996–5004

Hoeberichts FA, ten Have A, Woltering EJ (2003) A tomato metacaspase gene is upregulated during programmed cell death in *Botrytis cinerea*-infected leaves. Planta 217:517–522

Holz G, Coertze S, Wiiliamson B (2004) The ecology of *Botrytis* on plant surfaces. In: Elad Y, Williamson B, Tudzynski P, Delen N (eds) *Botrytis*, biology, pathology and control. Kluwer, Dordrecht, pp 9–27

Huang D, Bhairi S, Staples RC (1989) A transformation procedure for *Botrytis squamosa*. Current Genet 15:411–414

Jenczmionka NJ, Schäfer W (2005) The Gpmk1 MAP kinase of *Fusarium graminearum* regulates the induction of specific secreted enzymes. Curr Genet 47:29–36

Kankanala P, Czymmek K, Valent B (2007) Roles for rice membrane dynamics and plasmodesmata during biotrophic invasion by the blast fungus. Plant Cell 19:796–724

Kars I, Krooshof G, Wagemakers CAM, Joosten R, Benen JAE, van Kan JAL (2005a) Necrotising activity of five *Botrytis cinerea* endopolygalacturonases produced in *Pichia pastoris*. Plant J 43:213–225

Kars I, Wagemakers CAM, McCalman M, van Kan JAL (2005b) Functional analysis of *Botrytis cinerea* pectin methylesterase genes by PCR-based targeted mutagenesis. *Bcpme*1 and *Bcpme*2 are dispensable for virulence of strain B05.10. Mol Plant Pathol 6:641–652

Kliebenstein D, Rowe H (2007) Genomics of natural variation for signal production, perception and transduction in both plant host and fungal pathogen. In: Lorito M, Woo S, Scala F (eds) Biology of molecular plant–microbe interactions vol 6 (in press)

Klimpel A, Schulze Gronover C, Williamson B, Stewart JA, Tudzynski B (2002) The adenylate cyclase (BAC) in *Botrytis cinerea* is required for full pathogenicity. Mol Plant Pathol 3:439–450

Kraus PR, Heitman J (2003) Coping with stress: calmodulin and calcineurin in model and pathogenic fungi. Biochem Biophys Res Commun 311:1151–1157

Kronstad JW (1997) Virulence and cAMP in smuts, blasts and blights. Trends Plant Sci 2:193–199

Kulkarni RD, Thon MR, Pan H, Dean R (2005) Novel G-protein-coupled receptor-like proteins in the plant pathogenic fungus *Magnaporthe grisea*. Genome Biol 6:R24

Kunz C, Vandelle E, Rolland S. Poinssoit B, Bruel C, Cimerman A, Zotti C, Moreau E, Vedel R, Pugin A, Boccara M (2006) Characterization of a new, nonpathogenic mutant of *Botrytis cinerea* with impaired plant colonization capacity. New Phytol 170:537–550

Liu W, Leroux P, Fillinger S (2008) The HOG1-like MAP kinase Sak1 of *Botrytis cinerea* is negatively regulated by the upstream histidine kinase Bos1 and is not involved in dicarboximide- and phenylpyrrole-resistance. Fungal Genet Biol 45:1062–1074

Lyon GD, Goodman BA, Williamson B (2004) *Botrytis cinerea* pertubs redox processes as an attack strategy in plants. In: Elad Y, Williamson B, Tudzynski P, Delen, N (eds) *Botrytis*: biology, pathology and control. Kluwer, Dordrecht, pp 119–141

Marra R, Ambrosino P, Carbone V, Vinale F, Woo SL, Ruocco M, Ciliento R, Lanzuise S, Ferraioli S, Soriente I, Gigante S. Turrà D, Fogliano V, Scala F, Lorito M (2007) Study of the three-way interaction between *Trichoderma atroviride*, plant and fungal pathogens by using a proteomic approach. Curr Genet (in press)

Mengiste T (2008) Regulatory networks in plant responses to necrotrophic infections. In: Lorito M, Woo S, Scala F (eds) Biology of molecular plant–microbe interactions, vol 6 (in press)

Mey G, Held K, Scheffer J, Tenberge KB, Tudzynski P (2002) CPMK2, an SLT2-homologous mitogen-activated protein (MAP) kinase, is essential for pathogenesis of *Claviceps purpurea* on rye: evidence for a second conserved pathogenesis-related MAP kinase cascade in phytopathogenic fungi. Mol Microbiol 46:305–318

Motoyama T, Ohira T, Kadokura K, Ichiishi A, Fujimura M, Yamaguchi I, Kudo T (2005) An Os1 family histidine kinase from a filamentous fungus confers fungicide sensitivity to yeast. Curr Genet 47:298–306

Oliver RP, Ipcho SVS (2004) Arabidopsis pathology breathes new life into the necrotrophs-vs.biotrophs classification of fungal pathogens. Mol Plant Pathol 5:347–352

Parsley TB, Segers GC, Nuss DL, Dawe AL (2003) Analysis of altered G-protein subunit accumulation in *Cryphonectria parasitica* reveals a third Galpha homologue. Curr Genet 43:24–33

Prins TW, Wagemakers L, Schouten A, van Kan JAL (2000) Cloning and characterization of a glutathione S-transferase homologue from the plant-pathogenic fungus *Botrytis cinerea*. Mol Plant Pathol 105:273–283

Quidde T, Büttner P, Tudzynski P (1999) Evidence for three different specific saponin-detoxifying activities in Botrytis cinerea and cloning and functional analysis of a gene coding for a putative avenacinase. Eur J Plant Pathol 1:169–178

Reino JL, Hernández-Galán R, Durán-Patrón R, Collado IG (2004) Virulence-toxin production relationship in isolates of the plant pathogenic fungus *Botrytis cinerea*. J Phytopathol 152:563–566

Reis H, Pfiffi S, Hahn M (2005) Molecular and functional characterization of a secreted lipase from *Botrytis cinerea*. Mol Plant Pathol 6:257–267

Rolke Y, Liu S, Quidde T, Williamson B, Schouten A, Weltring K-M, Siewers V, Tenberge KB, Tudzynski B, Tudzynski P (2004) Functional analysis of H_2O_2-generating systems in *Botrytis cinerea*: the major Cu-Zn-superoxide dismutase (BCSOD1) contributes to virulence on French bean, whereas a glucose oxidase (BCGOD1) is dispensable. Mol Plant Pathol 5:17–23

Rolland S, Jobic C, Fevre M, Bruel C (2003) *Agrobacterium*-mediated transformation of *Botrytis cinerea*, simple purification of monokaryotic transformants and rapid conidia-based identification of the transfer-DNA host genomic DNA flanking sequences. Curr Genet 44:164–171

Rui O, Hahn M (2007) The Slt2-type Map kinase Bmp3 of *Botrytis cinerea* is required for normal saprotrophic growth, conidiation, plant surface sensing, and host tissue colonization. Mol Plant Pathol 8:173–184

Schoonbeek H, Del Sorbo G, De Waard MA (2001) The ABC transporter BcatrB affects the sensitivity of *Botrytis cinerea* to the phytoalexin resveratrol and the fungicide fenpiclonil. Mol Plant–Microbe Interact 14:562–571

Schouten A, Tenberge KB, Vermeer J, Stewart J, Wagemakers L, Williamson B, van Kan JAL (2002a) Functional analysis of an extracellular catalase of *Botrytis cinerea*. Mol Plant Pathol 3:227–238

Schouten A, Wagemakers L, Stefanato FL, van der Kaaij RM, van Kan JA. (2002b) Resveratrol acts as a natural profungicide and induces self-intoxication by a specific laccase. Mol Microbiol 43:883–894

Schulze Gronover C, Kasulke D, Tudzynski P, Tudzynski B (2001) The role of G protein alpha subunits in the infection process of the gray mold fungus *Botrytis cinerea*. Mol Plant–Microbe Interact 14:1293–1302

Schulze Gronover C, Schorn C, Tudzynski, B (2004). Identification of *Botrytis cinerea* genes up-regulated during infection and controlled by the Gα subunit BCG1 using suppression subtractive hybridization (SSH). Mol Plant–Microbe Interact 17:537–546

Schulze Gronover C, Schumacher J, Hantsch P, Tudzynski B (2005) A novel seven-helix transmembrane protein BTP1 of *Botrytis cinerea* controls expression of GST-encoding genes, but is not essential for pathogenicity. Mol Plant Pathol 6:243–256

Schumacher J, Viaud M, Simon A, Tudzynski B (2008a) The Galpha subunit BCG1, the phospholipase C (BcPLC1) and the calcineurin phosphatase co-ordinately regulate gene expression in the grey mould fungus *Botrytis cinerea*. Mol Microbiol 67:1027–1050

Schumacher J, de Larrinoa IF, Tudzynski B (2008b) Calcineurin-responsive zinc finger transcription factor CRZ1 of *Botrytis cinerea* is required for growth, development, and full virulanece on bean plants. Eukaryot Cell 7:584–601

Schumacher J, Kokkelink L, Huesmann C, Jimenez-Teja D, Collado I, Barakat R, Tudzynski P, Tudzynski B (2008c) The cAMP-dependent signaling pathway and its role in conidial germination, growth and virulence of the grey mould *B. cinerea*. Mol Plant–Microbe Interact (in press)

Scott-Craig JS, Cheng Y-Q, Cervone F, de Lorenzo G, Pitkin JW, Walton JW (1998) Targeted mutants of *Cochliobolus carbonum* lacking the two major extracellular polygalacturonases. Appl Environ Microbiol 64:497–1503

Segmüller N, Ellendorf U, Tudzynski B, Tudzynski P (2007) BcSAK1, a stress-activated MAPkinase is involved in vegetative differentiation and pathogenicity in *Botrytis cinerea*. Eukaryot Cell 6:211–221

Segmüller, Kokkelink L, Giesbert S, Odinius D, van Kan JAL, Tudzynski P (2008) NADPH oxidases are involved in differentiation and pathogenicity in *Botrytis cinerea*. Mol Plant–Microbe Interact 21:808–819

Sharon A, Elad Y, Barakat R, Tudzynski P (2004) Phytohormones in *Botrytis*–plant interactions. In: Elad Y, Williamson B, Tudzynski P, Delen, N (eds) *Botrytis*: biology, pathology and control. Kluwer, Dordrecht, pp 163–179

Siewers V, Smedsgaard J, Tudzynski P (2004) The P450 monooxygenase BcABA1 is essential for abscisic acid biosynthesis in *Botrytis cinerea*. Appl Environ Microbiol 70:3868–3876

Siewers V, Viaud M, Jimez-Teja D, Collado IG, Schulze Gronover C, Pradier J-M, Tudzynski B, Tudzynski P (2005) Functional analysis of the cytochrome P450 monooxygenase gene *bcbot1* of *Botrytis cinerea* indicates that botrydial is a strain-specific viruence factor. Mol Plant–Microbe Interact 18:602–612

Siewers V, Kokkelink L, Smedsgaard J, Tudzynski P (2006) Identification of an abscisic acid gene cluster in the grey mould *Botrytis cinerea*. Appl Env Microbiol 72:4619–4626

Siriputthaiwan P, Jauneau A, Herbert C, Garcin D, Dumas B (2005) Functional analysis of CLPT1, a Rab/GTPase required for protein secretion and pathogenesis in the plant fungal pathogen *Colletotrichum lindemuthianum*. J Cell Sci 118:323–329

Solomon PS, Waters OD, Simmonds J, Cooper RM, Oliver R (2005) The Mak2 MAP kinase signal transduction pathway is required for pathogenicity in *Stagonospora nodorum*. Curr Genet 48:60–68

Soulie MC, Piffeteau A, Choquer M, Boccara M, Vidal-Cros A (2003) Disruption of *Botrytis cinerea* class I chitin synthase gene Bcchs1 results in cell wall weakening and reduced virulence. Fungal Genet Biol 40:38–46

Soulie MC, Perino C, Piffeteau A, Choquer M, Malfatti P, Cimerman A, Kunz C, Boccara M, Vidal-Cros A (2006) *Botrytis cinerea* virulence is drastically reduced after disruption of chitin synthase class III gene (Bcchs3a). Cell Microbiol 8:1310–1321

Staats M, van Baarlen P, Schouten A, van Kan JAL, Bakker FT (2007) Positive selection in phytotoxic protein-encoding genes of *Botrytis* species. Fungal Genet Biol 44:52–63

Takemoto P, Tanaka A, Scott B (2007) NADPH oxidases in fungi: diverse roles of reactive oxygen species in fungal cellular differentiation. Fungal Genet Biol 44:1065–1076

Tanaka A, Christensen MJ, Takemoto D, Park P, Scott B (2006) Reactive oxigen species play a role in regulating a fungus–perennial ryegrass mutualistic interaction. Plant Cell 18:1052–1066

ten Have A, Mulder W, Visser J, van Kan JA (1998) The endopolygalacturonase gene Bcpg1 is required for full virulence of *Botrytis cinerea*. Mol Plant–Microbe Interact 11:1009–1016

ten Have A, Oude-Breuil W, Wubben JP, Visser J, van Kan JAL (2001) *Botrytis cinerea* endopolygalacturonase genes are differentially expressed in various plant tissues. Fungal Genet Biol 33:97–105

Tenberge KB (2004) Morphology and cellular organization in *Botrytis* interaction with plants. In: Elad Y, Williamson B, Tudzynski P, Delen N (eds) *Botrytis*: biology, pathology and control. Kluwer, Dordrecht, pp 67–84

Thomma BP, Penninckx IA, Cammue BP, Broekaert WF (2001) The complexity of diesease signaling in *Arabidopsis*. Curr Opin Immunol 13:63–68

Tonukari NJ, Scott-Craig JS, Walton JD (2000) The *Cochliobolus carbonum snf1* gene is required for cell wall-degrading enzyme expression and virulence on maize. Plant Cell 12:237–247

Tudzynski B, Schulze Gronover C (2004) Signaling in *Botrytis cinerea*. In: Elad Y, Williamson B, Tudzynski P, Delen N (eds), *Botrytis* spp.: biology, pathology and control. Kluwer, Dordrecht

Tudzynski P, Scheffer J (2004) *Claviceps purpurea*: Molecular aspects of a unique pathogenic lifestyle. Mol Plant Pathol 5:377–388

Tudzynski P, Sharon A (2003) Fungal pathogenicity genes. In: Arora DK, Khachatourians GG (eds) Fungal genomics. Applied mycology and biotechnology, vol 3. Elsevier, Amsterdam, pp 187–212

Tudzynski P, Siewers V (2004) Approaches to molecular genetics and genomics of *Botrytis*. In: Elad Y, Williamson B, Tudzynski P, Delen N (eds) *Botrytis*: biology, pathology and control. Kluwer, Dordrecht, pp 53–66

Valette-Collet O, Cimerman A, Reignault P, Levis C, Boccara M (2003) Disruption of *Botrytis cinerea* pectin methylesterase gene Bcpme1 reduces virulence on several host plants. Mol Plant–Microbe Interact 16:360–367

van Baarlen P, Woltering EJ, Staats M, van Kan JAL (2007) Histochemical and genetic analysis of host and non-host interactions of *Arabidopsis* with three *Botrytis* species: an important role for cell death control. Mol Plant Pathol 8:41–54

van Kan JA (2006) Licensed to kill: the lifestyle of a necrotrophic plant pathogen. Trends Plant 11:247–253

van Kan JA, van't Klooster JW, Wagemakers CA, Dees DC, van der Vlugt-Bergmans CJ. (1997) Cutinase A of *Botrytis cinerea* is expressed, but not essential, during penetration of gerbera and tomato. Mol Plant–Microbe Interact 10:30–38

Viaud M, Brunet-Simon A, Brygoo Y, Pradier J-M and Levis C (2003) Cyclophilin A and calcineurin functions investigated by gene inactivation, cyclosporin A inhibition and cDNA arrays approaches in the phytopathogenic fungus *Botrytis cinerea*. Mol Microbiol 50:1451–1465

Viaud M, Fillinger S, Liu W, Polepalli JS, Kunduru AR, Laroux P, Legendre L (2006) A class III kinase Acts as a novel virulance factor in *Botrytis cinerea*. Mol Plant–Microbe Interact 19:1042–1050

Williamson B, Tudzynski B, Tudzynski P, van Kan JAL (2007) *Botrytis cinerea*: the cause of grey mould disease. Mol Plant Pathol (in press)

Wolanin PM, Thomason PA, Stock JB (2002) Histidine protein kinases: key signal transducers outside the animal kingdom. Genome Biol 3:3013.7

Xu JR (2000) MAP kinases in fungal pathogens. Fungal Genet Biol 31:137–152

Xu J-R, Hamer JE (1996) MAP kinase and cAMP signaling regulate infection structure formation and pathogenic growth in the rice blast fungus *Magnaporthe grisea*. Genes Dev 10:2696–2706

Xu J-R, Peng YL, Dickman MB, SharonA (2006) The dawn of fungal pathogen genomics. Annu Rev Phytopathol 44:337–366

Zheng L, Campbell M, Murphy J, Lam S, Xu JR (2000) The BMP1 gene is essential for pathogenicity in the gray mold fungus *Botrytis cinerea*. Mol Plant–Microbe Interact 13:724–732

3 Profiles in Pathogenesis and Mutualism: Powdery Mildews

Christopher James Ridout[1]

CONTENTS

I. The Importance of Powdery Mildews. 51
II. Life Cycle of Powdery Mildews 53
   A. Asexual Reproduction. 53
   B. Sexual Reproduction. 53
III. Interactions with the Host Plant 55
   A. Influence on Host Metabolism. 56
   B. Induced Susceptibility and Resistance 56
IV. Establishment of Haustoria in Host Cells and
   Structure of the Haustorium/
   Plant Interface 57
V. Susceptibility and Resistance in Powdery
   Mildew Interactions 59
   A. Gene for Gene Interactions 59
   B. Genetic Analysis of Avirulence Genes. 59
   C. Identification of Powdery Mildew
      Avirulence Genes. 60
   D. Race-Specific Resistance (R) Genes 60
   E. Race-Non-Specific Resistance Genes 61
   F. Cell Entry Control and Non-Host
      Resistance 62
VI. Significance in Agriculture: Breakdown of
   Resistance in the Field 63
   References. 64

I. The Importance of Powdery Mildews

Powdery mildews are parasitic fungi which infect and cause substantial economic losses on a wide range of agricultural and ornamental plants. There are nearly 700 species of powdery mildews, occurring on about 7600 species of angiosperms, including cereals, fruit crops, cucurbits and ornamentals (Braun et al. 2002). Infection can damage tissue or whole plants, causing defoliation, cosmetic damage and reducing yields and quality. Illustrations of the ranges of symptoms are shown in Fig. 3.1A–E. Further descriptions and suggested control measures are available from the extension service of Oregon State University (http://plant-disease.ippc.orst.edu/index.cfm)

and in chapters 14–18 of the book 'The powdery mildews: a comprehensive treatise' (for details, see Braun et al. 2002).

All powdery mildews are obligate parasites, meaning that they require a live host to grow and reproduce. They obtain their nutrients through specialized feeding structures known as haustoria which grow inside living host cells. In early work, a wide range of species were studied (Harper 1905) and detailed cytology was performed on *Phyllactinia corylea*, the powdery mildew pathogen of hazel (*Corylus avellana*; Colson 1938). Barley powdery mildew (*Blumeria graminis* f. sp *hordei*, *Bgh*) is the most intensively studied of all powdery mildew fungi due to its economic importance and easily observed developmental biology (Both et al. 2005a, b; Zhang et al. 2005). Additional information on the biology of *Bgh* is available on the '*Blumeria graminis* Sequencing Project' website (http://www.blugen.org/). Powdery mildew of *Arabidopsis thaliana* caused by *Erysiphe cichoracearum* (*Ec*) has also received considerable attention because of the genetic tractability of the model host plant. In this chapter I cite mostly work from *Bgh* and *Ec*, whilst referring to work on other powdery mildews where appropriate.

Powdery mildews belong to the order Erysiphales within the Ascomycetes (Sac Fungi). There are about 18 genera currently described, which include the important plant pathogens: *Blumeria*, *Erysiphe*, *Sphaerotheca*, *Uncinula*, *Microsphaera*, *Phyllactinia*, *Podosphaera* and *Leveillula*. Characteristics of value for the taxonomy of powdery mildews include primary and secondary mycelia, the shape of appressoria and haustoria and the shape and fine structure of conidia. For those genera with teleomorphic states, the shape and structure of cleistothecia, asci and ascospores are important. The history of powdery mildew taxonomy and details of the important characteristics for their classification are provided by Braun et al. (2002). Illustrations of cleistothecia, asci and ascospores are shown in Fig. 3.1G–I.

[1] John Innes Centre, Norwich Research Park, Norwich, NR4 7UH, United Kingdom; e-mail: ridout@bbsrc.ac.uk

Plant Relationships, 2nd Edition
The Mycota V
H. Deising (Ed.)
© Springer-Verlag Berlin Heidelberg 2009

II. Life Cycle of Powdery Mildews

A. Asexual Reproduction

The predominant mode of reproduction in powdery mildew fungi is the dispersal of asexual conidiospores, produced abundantly on leaf surfaces and other aerial parts of plants. These oval-shaped spores are produced in chains from conidiophores and are dispersed to plants by wind. Conidiospores are susceptible to desiccation and remain viable for only a few days at ambient temperatures although they can survive for 3–4 weeks below −4 °C (Cherewick 1944). They can be maintained in the laboratory by regularly transferring spores onto fresh, living host plant tissue, although a method of freezing spores at −80 °C for long-term storage has been reported (Pérez-Garcia et al. 2006). I describe here the infection process of *Bgh* as an example of that which occurs in other powdery mildew fungi and which is summarized in Fig. 3.2. The conidiospores of *Bgh* land on the leaf surface and within minutes release enzymes onto the leaf surface; these may prepare the infection court and facilitate subsequent fungal development (Carver et al. 1999). Within 1 h, a short primary germ tube (PGT) forms to only breach the epidermal cuticle. *Bgh* is unique in the formation of a PGT, but otherwise the infection process is similar in other powdery mildews. The PGT may contribute to the absorption of water and solutes present in the host cell wall for subsequent use by the conidiospore (Edwards 2002). Extracellular material is secreted beneath the PGT, which adheres to the leaf. A second appressorial germ tube (AGT) is then produced by 3–4 h after infection. The AGT develops into an appressorium and an infection peg forms beneath this to penetrate the host cell cuticle. In a successful infection, haustoria develop within the cell and are visible by 12–14 h. Nutrients are absorbed through the haustoria, enabling the fungus to grow and develop. Within 2 days, conidiophores bearing conidiospores start to be produced and within 4 days micro-colonies are visible on the leaf surface. The developing fungus eventually forms new conidia in chains by 4–6 days after inoculation.

B. Sexual Reproduction

Many powdery mildews have a sexual phase resulting in the formation of ascospores within a cleistothecium, a closed spherical ascocarp (Fig. 3.1G–I). Formation of the pseudoantheridium and pseudoascogonium and events leading to ascus formation were described for some Erysiphales in the first half of the twentieth century using light microscopy (Harper 1905; Colson 1938). Wheat powdery mildew (*B. graminis* f. sp. *tritici*, *Bgt*) is heterothallic and hermaphroditic, so either of two compatible isolates can act as the maternal parent (Robinson et al. 2002); and this may indeed be the case for other powdery mildews. Details of the sexual cycle of most powdery mildews are, however, poorly understood. Cleistothecia remain dormant through periods of drought or temperature extremes when host plants are not present. When conditions become favourable, the cleitothecia absorb water and break open enabling ascospores to be ejected from the asci. For some powdery mildews, ascospores are the primary inoculum. For example the primary inoculum of

Fig. 3.1. Examples of powdery mildew diseases and illustrations of haustoria and cleistothecia. **A** Courgette (zucchini) powdery mildew (*Podosphaera fusca*). **B** Grape powdery mildew (*Erysiphe necator*). **C** Illustration of the range of infection phenotypes of barley powdery mildew (*Blumeria graminis* f. sp *hordei*) isolate CC148 on barley cultivars: *left to right* P08B (containing the resistance gene *Mla9*), Hordeum 1063 (*Mlk1*), Julia (*Ml/cp*), Hordeum 1036 (*Mla3*), W37/36 (*Mlh*). These are described, respectively, as infection types 0 (fully avirulent), 1, 2, 3 and 4 (fully virulent). **D, E** Powdery mildew disease on *Rhododendron* spp. (*Erysiphe azaleae*; **D**) and gooseberry (*Podosphaera mors-uvae*; **E**). **F** Green islands underneath colonies of *Blumeria graminis* colonies growing on leaves of couch grass. **G** Cleistothecium of *Microsphaeria azaleae*. **H, I** Cleistothecial squash preparations to reveal asci and ascospores from *Erysiphe necator* (**H**) and *Podosphaera fusca* (**I**). **J** Haustoria of *B. graminis* f. sp. *hordei* growing in barley. **K, L** Morphology of *Arabidopsis thaliana* powdery mildew (*Erysiphe cichoracearum*) haustorial complexes. Methods used for the visualization of haustoria are described by Koh et al. (2005). **K** 3D volume-rendered isolated haustorial complex. Elaborated lobe-like structures are visible around the mature haustorium (*Ha*) within the extrahaustorial membrane (*EHM*). **L** 3D volume-rendered in vivo haustorial complex. The intact EHM (*arrow*) was also visible around the haustorium (*Ha*), but distinctive from the GFP-labelled plasma membrane (*PM*) of the *Arabidopsis* epidermal cell. The appressorium (*Ap*) is also shown on the leaf surface. Figures reproduced by permission of Alejandro Pérez, (**A**), David Gadoury (**B**), Jay Pscheidt (**D, E, G**), Melodie Putnam (**H**), Juan Antonio Torés (**I**), Pietro Spanu (**J**) and Serry Koh (**K, L**)

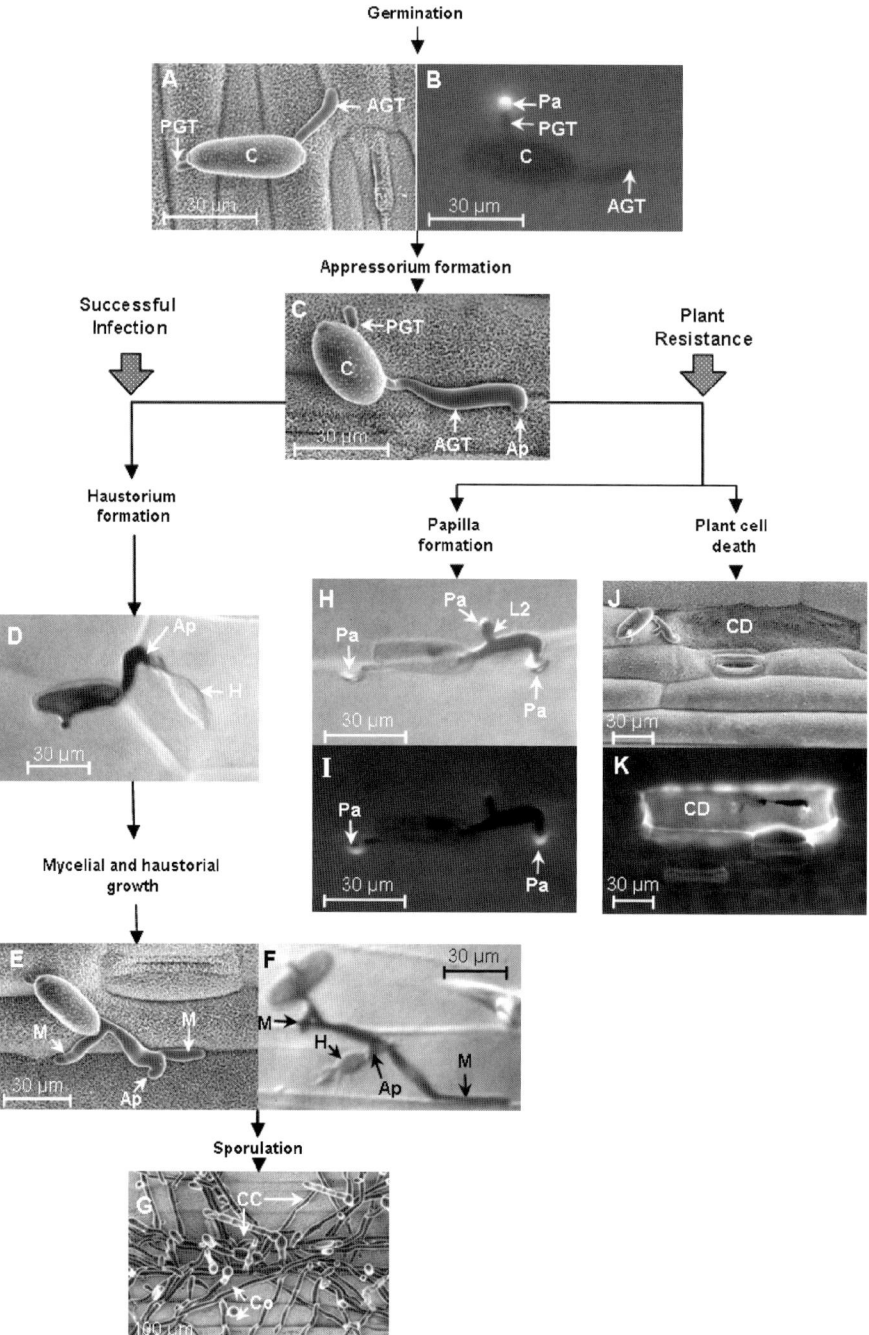

Fig. 3.2. Micrographs by cryo-SEM, transmitted light and fluorescence (blue light excitation) of key stages in *Blumeria gramins* f. sp. *hordei* development and barley cell responses. **A–C** Up to 10 h after inoculation (h.a.i.), fungal development is indistinguishable on susceptible and resistant plant genotypes such as Pallas, P01 and P22. **A** By 5 h.a.i., the primary germ tube (*PGT*) has emerged from the conidium (*C*) and adhered to the leaf surface. The appressorial germ tube (*AGT*) has elongated partially but not yet differentiated an apical lobe. **B** Fluorescence microscopy reveals autofluorogenic material accumulated within a small papilla (*Pa*) deposited as a plant epidermal cell response to the PGT. **C** By 10 h.a.i., the appressorium (*Ap*) has differentiated a hooked apical lobe. **D–G** Successful infection of a susceptible host. **D** By 15–18 h.a.i., a penetration peg emerging beneath the Ap has penetrated the plant cell and its tip has swollen to form a rudimentary haustorium (*H*) visible by light microscopy. Images by scanning electron (**E**) and light microscope (**F**) show developing colonies at 30 h.a.i. By 30 h.a.i., ectophytic mycelium (*M*) has grown from the AGT as the haustorium develops a digitate process from each end of its central body. **G** By 96 h.a.i., mycelial growth is extensive. Repeated penetration from hyphal appressoria results in the formation of further haustoria (not shown) and bulbous conidiophores

grape powdery mildew (*Erysiphe necator*, formerly *Uncinula necator*) can come from cleistothecia that overwinter primarily in bark crevices on the grapevine (Pearson and Gadoury 1987). Ascospores, released in the spring, are carried by wind and germinate on the leaf surface of developing vines. For other powdery mildews, the importance of the sexual phase is not so obvious or may depend on the local climatic conditions. For example, ascospores of *Bgt* are considered to be an important source of inoculum in the highlands of China, but less so in the lowland regions. This is because, in the lowlands, autumn wheat seedlings can become infected with conidiospores which subsequently function as the primary inoculum (Yu 2000). The occurrence of a sexual stage permits genetic analysis to be performed in some powdery mildew fungi (see Sect. V.B). Many powdery mildews, including *Ec*, do not have a sexual stage.

III. Interactions with the Host Plant

Early in the infection process, visible changes can be detected in the physiology and structure of the host cell. Upon attempted penetration, polarized cytoplasmic streaming occurs in the plant cell directed towards the site of attempted penetration (Kobayashi et al. 1997) and papillae develop at the point of first contact (Fig. 3.2B, I). Papillae function as physical and chemical barriers to resist infection and are comprised of inorganic and organic constituents, including callose and autofluorogenic phenolics. Their deposition involves: generation of nitric oxide (NO; Prats et al. 2005) and hydrogen peroxide (H_2O_2; Vanacker et al. 2000), cytoskeletal re-arrangement (Opalski et al. 2005) and redirected cytoplasmic streaming and aggregation (Zeyen et al. 2002). In barley/*Bgh* interactions, NO is generated at sites of papilla deposition commencing at around 10 h after infection (a.i.) and persisting until around 14–16 h a.i. NO may also be involved

in the complex second step that involves synthesis, marshalling and assembly of papilla components (Prats et al. 2005). Although the layers of the papillae can appear heterogeneous, they may develop in a certain chronological order with compounds created at and/or delivered to the site of fungal penetration at specific times (Celio et al. 2004). Using GFP-tagged plasma membrane marker proteins, rings of fluorescence were observed around *Ec* penetration sites on *A. thaliana* plants, which extended across cell wall boundaries and into neighbouring cells. These rings, however, seem to be localized to those infection sites where papillae were deposited (Koh et al. 1995). Even in susceptible hosts, papillae formed in the plant cell may succeed in preventing infection.

Staining by 3,3-diaminobenzidine (DAB) reveals that H_2O_2 is produced by plant cells at infection sites. H_2O_2 can be detected at developing papillae in the epidermal cell subjacent to the primary germ tube from 6 h after inoculation, and underneath the appressorium after 15 h. The presence of H_2O_2 could catalyse the cross-linking of proteins in the papillae, resulting in a stronger physical barrier to invasion (Thordal-Christensen et al. 1997). During development of the fungal haustorium, multivesicular bodies, intravacuolar vesicle aggregates and paramural bodies develop in the penetrated epidermal cell. These structures also form at the periphery of intact cells adjacent to cells undergoing hypersensitive cell death (An et al. 2006). The implication from these investigations is that the vesicular compartments participate in secretion of chemical components required for cell wall appositions. They may also contribute to the internalization of damaged membranes, deleterious materials, nutrients, elicitors and elicitor receptors.

Although *Bgh* is an obligate parasite, it appears to have all the main metabolic pathways of a filamentous fungus; it has not lost metabolic capacity nor the ability to modulate its metabolism (Thomas et al. 2001, 2002; Soanes et al. 2002; Giles et al. 2003; Soanes and Talbot 2006). The distinct lifestyle of

---

(*Co*) have started generating chains of conidia (*CC*) for wind dispersal. **H–K** Resistance responses viewed 30 h.a.i. Images by transmitted light (**H**) and fluorescence (**I**) microscopy show a germling that failed to penetrate a living plant cell from its first appressorial lobe and therefore formed a second lobe (*L2*). Refractive, autofluorescent papillae subtend both appressorial lobes and the PGT, although fluorescence is weak in the smaller papilla subtending L2. **J, K** Epidermal cell death (*CD*) as a result of single gene-controlled

hypersensitivity that prevents further pathogen growth. By SEM (**J**), dead epidermal cells are obviously collapsed while they show whole-cell autofluorescence viewed by fluorescence microscopy (**K**). In lines carrying *Mla1* attacked by an avirulent fungal isolate, most cells that do not form an effective papilla, collapse and become autofluorescent by 24 h.a.i. This figure and legend are reproduced from Prats et al. (2006) by permission of Dr. Elena Prats and Oxford University Press

B. graminis is reflected in its gene inventory based on EST analysis, having significantly fewer sequences in common with the genomes of filamentous ascomycetes than any of the other plant pathogenic fungi. Nearly half of the *B. graminis* unisequences currently available at NCBI have no homologues in the NCBI database of sequenced proteins or in the sequenced fungal genomes (Soanes and Talbot 2006). These observations suggest that the biotrophic life cycle of *B. graminis* necessitates a large number of gene products not found in necrotrophic and hemibiotrophic phytopathogens. Transcript profiling using cDNA microarrays indicates that wholesale changes in fungal gene expression occur during the switch from pre-infection development to biotrophic growth, including the co-ordinate regulation of entire suites of genes encoding enzymes in similar pathways of primary metabolism (Both et al. 2005a).

A. Influence on Host Metabolism

Powdery mildews influence host cell metabolism in many ways. One striking manifestion of powdery mildew infection is the appearance of 'green islands' underneath and around developing colonies (Fig. 3.1F). Green islands become apparent during the later stages of infection when the remainder of the leaf has senesced. Cytokinins are known to delay senescence and play a role in the synthesis and maintenance of chlorophyll and are known to influence chloroplast development and metabolism. Indeed, cytokinins have also been shown to promote re-greening of senescent leaf tissue (Zavatleta-Manchera et al. 1999). Therefore it is not surprising that cytokinins are implicated in green island formation (Brian 1967; Dekhuijzen 1976). It is not clear whether the cytokinins are of fungal origin, or whether they are formed by the plant as a direct result of the infection or by manipulation of host metabolism by the fungus (Walters and Mc Roberts 2006).

Leaves infected with obligately biotrophic fungal pathogens, including powdery mildews, often exhibit reduced rates of net photosynthesis. It is not known whether the fungus directly causes this or whether it is a response of the plant to infection (Walters and Mc Roberts 2006). Powdery mildews can, however, influence the expression of some genes of their host plants, which may help to establish successful infection. Single epidermal cells containing haustoria show reduced accumulation of transcripts from several pathogenesis-related genes, including peroxidase and an oxalate oxidase-like gene. These genes are normally up-regulated in whole-leaf tissue in both compatible and incompatible interactions, so these results indicate that the fungus can suppress defence gene activation in physical proximity to haustoria (Gregersen et al. 1997). Suppression of basal defence-related transcripts including those genes involved in the shikimate pathway, signal transduction and defence can be measured 16 h after infection in compatible interactions (Caldo et al. 2006).

B. Induced Susceptibility and Resistance

The formation of a haustorium during a compatible interaction influences the ability of the host cell to resist subsequent attack, further illustrating that powdery mildews can influence host metabolism to assist infection (Lyngkjær and Carver 1999). Even in compatible interactions, not all infection attempts are successful. For example, 67% of infections of compatible *Bgh* isolate GE3 on barley line Risø 5678S went on to produce a haustorium. Using a double-inoculation procedure, cells were first attacked by an 'inducer', followed by a second inoculation with a 'challenger'. Successful infections by the challenger increased to over 90% where cells already had an inducer haustorium in them. Conversely, if the inducer attack failed, attacked cells were rendered highly inaccessible to subsequent attack. Inaccessibility was also induced in cells immediately adjacent to the attacked cell, but not in more distant cells. Induced accessibility appeared to be associated with suppression of localized autofluorescence characteristic of host cell death. Inaccessibility was, however, associated with increased frequency and intensity of cell death responses. The results from these experiments indicate that induced changes may relate to modification in a host cell's ability to synthesize phenolic compounds. These observations are also consistent with the transcript changes during infection described earlier.

The phenomenon of induced susceptibility extends to powdery mildew interactions that are normally incompatible. When *Bgh* isolates virulent on barley varieties carrying the resistance gene *Mla1* formed haustoria, the cells became highly susceptible to isolates that are normally avirulent (i.e. cannot grow) on these barley lines (Lyngkjær et al. 2001). The conclusion from this work is that factors released by the fungus are able to suppress defence responses. However, the suppressive effect was confined to the epidermis and defence responses were observed as normal in underlying mesophyll cells. Thus, the suppressive effect only occurs in the immediate vicinity of the infected cell. *Blumeria graminis* exists as a number of formae speciales (ff. spp.), each adapted to different grass hosts. For example, *B. graminis* f. sp. *hordei* normally grows on barley, but cannot infect oats. If, however, grasses are first infected with compatible ff. spp., infection by non-compatible ff. spp. can occur (Moseman et al. 1964). In subsequent work, double-inoculation experiments, as described above, showed that infection by the inducer f. sp. enhanced infection by an incompatible f. sp. (Olesen et al. 2003). Enhanced infection correlated with suppression of defence responses within epidermal cells containing the inducer haustorium. The suppressive effect extended to adjacent cells, but did not occur at two cells distance. Suppression of penetration resistance allowed most challenger attacks by inappropriate ff. spp. to form haustoria, so enabling the fungus to survive and develop a colony. Induced susceptibility even extends beyond related powdery mildew, since barley coleoptile cells penetrated by *Bgh* and containing haustoria are more susceptibile to subsequent attack by the non-host pathogen, *Erysiphe pisi* (Kunoh et al. 1985, 1986).

IV. Establishment of Haustoria in Host Cells and Structure of the Haustorium/Plant Interface

Haustoria are specialized structures that are formed inside living host cells and which are likely to be involved in the exchange of substances between host and fungus. In fungal and oomycete diseases of higher plants, haustoria are formed by downy mildews, white rusts (peronosporales), powdery mildews and rusts. They are also formed by other fungi, including lichens, endotropic mycorrhizae, some specialized filamentous fungi in the order Mucorales and aquatic phycomycetes in the orders Chytridiales, Hyphochytridiales and Saprolegniales (Bushnell 1972). Considerable attention has been focused on investigations into the formation, structure and function of haustoria in rusts (see Chap. 4) and powdery mildews which are reviewed in this chapter.

Haustoria are considered to be the location for exchange of nutrients, and there is evidence for this in other biotrophic plant pathogens. cDNA libraries prepared from haustoria of bean rust (*Uromyces fabae*) reveal the presence of many in planta induced genes, including hexose transporters, invertase, amino acid transporters and H^+ ATPase, all consistent with the increased nutrient uptake (Hahn and Mendgen 1997; Struck et al. 2002; Voegele et al. 2001, 2006; Jakubovic 2006). In both *Bgh* and *U. fabae*, transcript profiling using cDNA microarrays indicates that wholesale changes in fungal gene expression occur during the switch from pre-infection development to biotrophic growth (Both et al. 2005a; Jakupovic et al. 2006). The failure to detect plant wall components in powdery mildew extrahaustorial matrix (EHMAT) led to the suggestion that either the pathogen suppresses their synthesis and secretion at the interface or that they become degraded by fungal hydrolytic enzymes after secretion into the matrix, perhaps providing a source of nutrition to the pathogen (Green et al. 2002). Effectors which suppress defence responses may also be released from haustoria, although so far there is no direct evidence for this.

The primary haustorium of *Bgh* starts to form in an epidermal cell during the first day after infection and continues to develop for five of six days growing at a rate of about 250 μm/day (Hirata 1967; for a review, see Bushnell 1972). Haustoria of *B. graminis* ff. spp. are unique among powdery mildew fungi in that they characteristically have finger-like branches (Fig. 3.1J), other species typically having a more spherical structure. The extrahaustorial membrane (EHM) encases the haustorium, separating it from the host cytoplasm. The EHM generally appears liquid or gel-like in consistency (Manners and Gay 1983) and is considered to be a specialized membrane derived from the host although the pathogen may also contribute components (Koh et al. 2005). The EHMAT (Green et al. 2002) lies between the EHM and the fungal haustorial wall and is a gel-like layer enriched in carbohydrates of both fungal and host origin (Koh et al. 2005). To move from plant cytoplasm to haustorial cytoplasm, substances must pass sequentially through the EHM, EHMAT, the haustorial wall and the haustorial plasma membrane. The EHM, as well as the EHMAT, is thought to act as a molecular sieve and there is evidence for pore-like structures rendering the EHM permeable to molecules up to 40 kDa (Gil and Gay 1977). Despite its apparent continuity, the EHM differs from the host cell plasma membrane both physically and chemically. The use of high-pressure frozen and freeze-substituted electron microscopy provides exceptional preservation of the host pathogen interface to reveal the fine detail of haustoria EHM and papillae (Fig. 3.3).

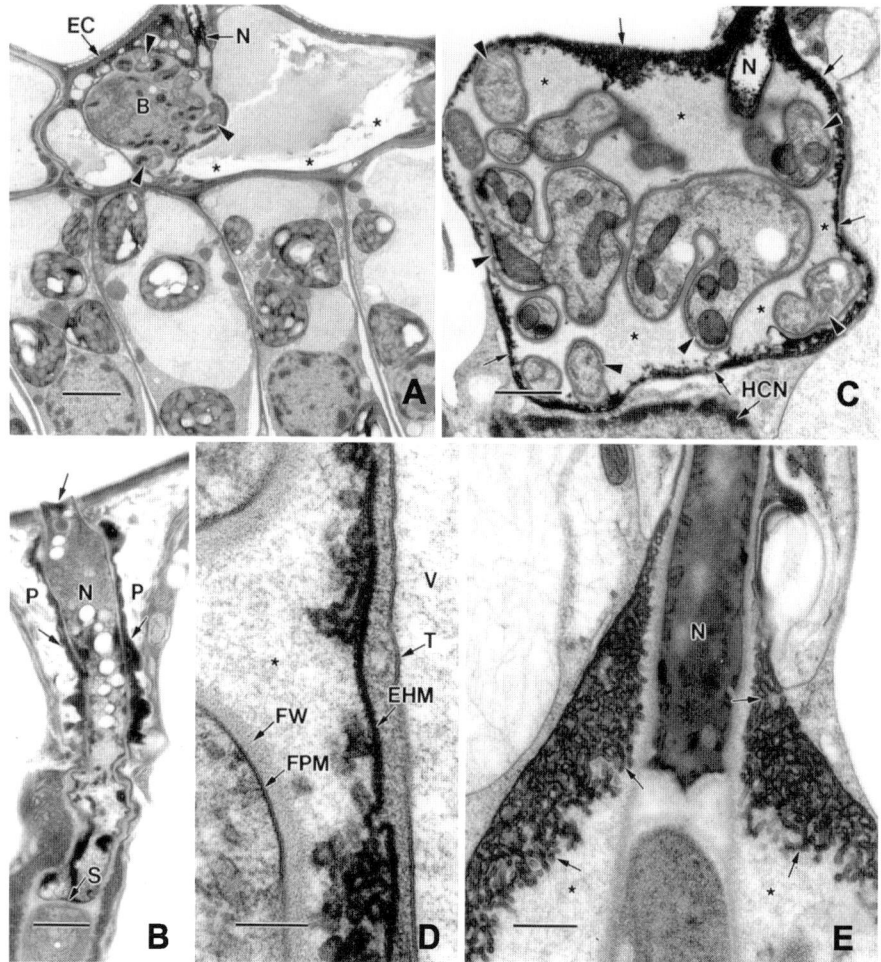

Fig. 3.3. TEMs of high-pressure frozen and freeze-substituted samples of poinsettia leaves infected with *Oidium* sp. **A** Portion of a leaf showing an epidermal cell (*EC*) containing a haustorium. Both the neck (*N*) and body (*B*) of the haustorium are visible as well as numerous slender, coiled lobes (*arrowheads*) surrounding the body. The large vacuole of the epidermal cell shows considerable mechanical damage (*asterisks*). Note the well preserved nature of the palisade cells visible below the epidermal cells. *Bar* 2.5 μm. **B** Example of the neck (*N*) of a haustorium surrounded by a papilla (*P*). Note the layer of electron-dense material (*arrows*) coating the neck of the haustorium. A septum (*S*) is visible at the base of the neck. The appressorium that formed this haustorium was lost during freezing. The sheared end of the neck of the haustorium that was continuous with the appressorium is shown at the arrow. *Bar* 0.5 μm. (**C** Section showing the highly convoluted extrahaustorial membrane (*arrows*) surrounding a haustorium. A portion of the haustorium neck is visible at *N*. Although most of the body of the haustorium is out of the plane of the section, numerous lobes of the haustorium are visible at the arrowheads. Note the extrahaustorial matrix material (*asterisks*) surrounding the lobes. A portion of the host cell nucleus is visible at HCN. *Bar* 0.5 μm.(**D** Highly magnified view of the host–pathogen interface. Visible are a portion of a host cell vacuole (*V*) and its membrane or tonoplast (*T*), the thickened and highly convoluted extrahaustorial membrane (*EHM*) and parts of the fungal plasma membrane (*FPM*) and wall (*FW*) of a lobe of the haustorium. Extrahaustorial matrix material is visible at the *asterisk*. *Bar* 0.15 μm. **E** Section showing the highly convoluted nature of the extrahaustorial membrane (*arrows*) near the base of a haustorium neck (*N*). Extrahaustorial matrix material is visible at the *asterisks*. *Bar* 0.3 μm. This figure is reproduced from Figs. 1–5 of Celio et al. (2004) with permission of Dr. Gail Celio and the National Research Council, Canada

Microscopic investigation of the barley/*Bgh* interaction reveals that multivesicular bodies (MVBs) of host plant origin proliferate near to haustoria in the penetrated epidermal cell (An et al. 2006). MVBs probably participate in the secretion of building blocks for cell wall appositions to arrest fungal penetration. They may also participate in the internalization of damaged membranes, deleterious materials, nutrients, elicitors and elicitor receptors. They may also be involved in plasma membrane extension during haustorium development if they fuse with the plasma membrane. Large bubble-like structures resembling MBVs have also been observed adjacent to the plasma membrane near to young haustoria of *Ec* in *Arabidopsis thaliana* (Koh et al. 2005).

V. Susceptibility and Resistance in Powdery Mildew Interactions

Investigations into the basis of susceptibility and resistance in powdery mildew interactions have preoccupied scientists for decades. The practical basis for these studies is to understand the nature of resistance for the development and breeding of plant varieties for agriculture and horticulture. The establishment of powdery mildew fungi on the host plant is particularly suited to microscopic investigations, and investigations of this kind have advanced our understanding of infection processes in both mildews and in other parasitic micro-organisms. The ability to perform genetic analysis in powdery mildews has advanced our understanding of the molecular basis of resistance, especially gene for gene interactions.

A. Gene for Gene Interactions

The gene for gene (GFG) relationship between resistance in a host plant and avirulence in a pathogen was first described in the interaction between flax rust (*Melampsora lini*) and flax (*Linum usitatissimum*; Flor 1955). Following Flor's work, the powdery mildews of wheat and barley were among the next diseases for which the GFG relationship was demonstrated (Moseman 1959; Powers and Sando 1960). Since then, the GFG relationship has been demonstrated or inferred in many other plant diseases, including those caused by fungi, oomycetes, bacteria, viruses, nematodes, insects and parasitic plants (Thompson and Burdon 1992; Crute et al. 1997). In GFG interactions, avirulence (AVR) molecules of pathogens and pests are recognized in plants by race-specific resistance (R) proteins. Specificity is a crucial feature of GFG relationships, with the plant mounting an effective defence against infection only if it has a resistance allele matching a specific avirulence allele in the pathogen.

GFG resistance is usually associated with a hypersensitive response (HR) in the attacked host cell. In this, recognition of the AVR gene product by the R protein causes the attacked cell to die quickly. It is often assumed that this rapid destruction of host cells is directly responsible for arrest of pathogen growth, and this is probably true in many cases. Cell death can result in a visible manifestation of the defence response that is relatively easy to observe and measure. There are also examples where HR is not associated with pathogen arrest. It has been proposed that HR could function in signaling to neighbouring cells rather than a direct defence mechanism (Heath 2000). Two examples of this type of response which occur in powdery mildew interactions are described later in this section.

B. Genetic Analysis of Avirulence Genes

Crosses between *Bgh* isolates can be made, enabling the inheritance and linkage of over 30 *AVR* loci to be investigated (Jørgensen 1988; Brown and Simpson 1994; Brown and Jessop 1995; Jensen et al. 1995; Brown et al. 1996; Caffier et al. 1996; Pedersen et al. 2002). Most *Bgh AVR* genes map as single loci as expected for the standard GFG pattern. There is also evidence for more than one avirulence gene matching a single barley resistance gene, indicating that there may be modifier and inhibitor genes which affect AVR recognition or expression (for a review, see Brown 2002). Two regions of the *Bgh* genome are known to contain clusters of linked *AVR* genes. The AVR_{a10} cluster comprises AVR_{a10}, AVR_{k1}, AVR_{a22}, AVR_{a9}, AVR_{a13}-1 (one of two *AVR* genes recognized by *Mla13*), AVR_g and possibly AVR_{a6} and AVR_{a7} (Jørgensen 1988; Christiansen and Giese 1990; Brown and Simpson 1994; Jensen et al. 1995; Caffier et al. 1996; Pedersen et al. 2002; Ridout et al. 2006). Another possible cluster occurs at the AVR_{a12} locus, and was identified in a cross between the *Bgh* isolates CC146 and DH14 (Brown et al. 1996). In this, AVR_{a12} was linked to AVR_{P17} and AVR_{a6} by 21.2 cM and 22.0 cM, respectively, and AVR_{La} was linked to AVR_{P17} at a distance of 21.2 cM. However, in all other crosses studied, no linkage of AVR_{a12} to any other *AVR* gene was established (Brown and Simpson 1994; Jensen et al. 1995; Caffier et al. 1996).

In addition to cereal powdery mildews, there is evidence of GFG resistance in the interaction between other powdery mildews and their hosts. However, these investigations were observed with naturally occurring isolates and were not subjected to rigorous genetic analysis. The most comprehensive study of the genetics of GFG interactions in powdery mildews, apart from that of cereals, was with *Golovinomyces cichoracearum* var. *fischeri* (syn. *Erysiphe cichoracearum* var. *fischeri*) on groundsel (*Senecio vulgaris*; for a review, see Clarke 1997). The patterns of interaction were more complex than those between *B. graminis* and barley or wheat varieties (Harry and Clarke 1986; Bevan et al. 1993a, b). Some interactions between particular pairs of *G. cichoracearum* var. *fischeri* isolates and groundsel lines could be classified as compatible or incompatible, but in others a clear classification was not possible. Difficulties in ascribing the nature of the interaction were due to partial resistance in the host or because the pathogen was not very aggressive. Six races of *Podosphaeria fuliginea* were defined on the basis of their specific virulence on varieties of melon (Bardin et al. 1997; Hosoya et al. 1999). In addition, pathotypes of the fungus have been defined by their ability to infect different cucurbit species. There is some evidence for specific interactions between isolates of *Erysiphe pisi* lines of the host plant, pea. (Tiwari et al. 1997).

C. Identification of Powdery Mildew Avirulence Genes

Two *AVR* genes have been isolated from *Bgh*, AVR_{k1} and AVR_{a10}, recognized by barley varieties containing Ml_{k1} and *Mla10*, respectively (Ridout et al. 2006). AVR_{k1} and AVR_{a10} belong to a large family with >30 paralogues in the genome of *Bgh* and homologous sequences are present in other ff. spp. of the fungus which infect other grasses (Ridout et al. 2006). The central core of the AVR_{k1} protein is highly conserved, whereas the amino acid sequences towards the N- and C-terminal regions are divergent. No signal peptide is predicted in the proteins, suggesting that they are not delivered to the host cell in a conventional secretion pathway. The proteins are rich in the amino acid residues lysine and arginine and are very basic. There are no homologues of AVR_{k1} or the related paralogues in any sequence database. AVR_{k1} and AVR_{a10} were demonstrated to have a dual function, both as: (a) elicitors of the resistance response and (b) as effectors which enhance infection on susceptible hosts. These findings suggest that *Bgh* and other ff. spp. of *B. graminis* might have a repertoire of related effectors, some of which may encode AVR proteins. This would explain why individual *AVR* genes can be lost without apparent loss of fitness, so enabling the fungus to overcome plant *R* genes (Ridout et al. 2006). Whether other members of the gene family may there encode AVR proteins remains to be determined.

D. Race-Specific Resistance (*R*) Genes

R genes encode receptors which recognize specific AVR elicitors in certain races of an adapted pathogen. More than 85 barley *R* genes, each conferring resistance to specific *Bgh* AVR elicitors, have been described, including over 28 alleles at the *Mla* locus on barley chromosome 5 (Jørgensen 1994). The *Mla* locus encodes allelic receptors containing an N-terminal coiled-coil (CC) structure, a central nucleotide-binding (NB) site and a leucine-rich repeat (LRR) region. The six isolated *Mla* alleles (*Mla1, Mla6, Mla7, Mla10, Mla12, Mla13*) are predicted to encode proteins that share >90% amino acid sequence identity (Zhou et al. 2001; Halterman et al. 2003; Shen et al. 2003; Halterman and Wise 2004). Analysis of *Mla1/Mla6* chimeras revealed that recognition specificity is determined by different but overlapping LRRs and a C-terminal non-LRR region (CT; Shen et al. 2003). Mla protein steady-state levels are critical for effective resistance and are subject to control by cytosolic heat-shock protein 90 (Hsp90) and the co-chaperone-like proteins RAR1 and SGT1 (Bieri et al. 2004; Hein et al. 2005). Details of the molecular functioning of chaperones in Mla resistance protein function are reviewed by Shirasu and Schulze-Lefert (2000). Co-expression of the *Bgh* AVR_{a10} effector induces nuclear associations between the Mla10 receptor and a WRKY-2 transcription factor, which could initiate downstream defence responses (Shen et al. 2007). It is proposed that the WRKY proteins repress defence responses and that activation of Mla proteins by pathogens blocks this repressor activity allowing for rapid and efficient disease resistance.

Resistance to *Bgh* controlled by a GFG interaction is generally associated with a strong HR. However, a halt in cytoplasmic streaming is the first visible sign of incompatibility in Mla/AVR interactions and preceeds hypersensitive cell death by 1–3 h (Bushnell 1981). Resistance phenotypes from AVR/Mla interactions on barley can vary considerably ranging from complete sensitivity to complete resistance (Fig. 3.1C). The extent of necrosis and sporulation varies for resistant reactions on varieties with the same *Mla* gene but with different genetic backgrounds. There is even variation between individual colonies on the same leaf and the phenotypes can be influenced by spore density and environmental conditions. Thus, consistency in inoculation and growth conditions is required for accurate scoring of phenotypes. Resistance responses for each *Mla* allele can be defined as slow or fast acting, on a continuous scale between the two extremes. For example, on *Mla1* and *Mla6* plants, a rapid HR was associated with a higher percentage of germlings arrested in their development of the haustorial stage. There was also a significant papilla response corresponding to a larger number of germlings arrested in *Mla6* plants (Boyd et al. 1995). On *Mla3* and *Mla7* plants, a slower hypersensitive response was associated with more haustoria and elongating secondary hyphae were formed, indicating that some fungal colonies were starting to form. However, in the latter cases there was a more extensive combined epidermal and mesophyll HR. It was proposed that the differences in speed on *Mla* reaction could result from different stage specific delivery of AVR molecules (Shen et al. 2003). The varying responses could also result from different amounts of each type of Mla protein, which are present in only small amounts in the cell. Evidence for this was the alteration of the reaction kinetics to produce a more rapid response by over-expressing the *Mla12* resistance gene (Shen et al. 2003).

Wheat powdery mildew (*Pm*) resistance genes follow the GFG model and the resistance reaction is associated with rapid host cell death. Major host resistance genes have been identified at 33 loci in wheat (Huang and Röder 2004; Zhu et al. 2005). Five of these loci (*Pm1, Pm3, Pm4, Pm5, Pm8*) have more than one allele conferring resistance, making a total of 49 named *Pm* resistance alleles. *Pm3* was one of the first described loci among the *Pm* genes. *Pm3* is

a single, dominant locus on the short arm of wheat chromosome 1A, and carries ten different resistance specificities (*Pm3a–Pm3j*). *Pm3* resistance alleles were generated in agricultural ecosystems after domestication of wheat 10 000 years ago (Yahiaoui et al. 2006). Since that time, *Pm3* alleles have been widely and successfully employed in breeding programmes. Some of these alleles remain effective in conferring resistance. Phylogenetic analysis of the Pm3B protein indicates that it is more similar to rice disease resistance-like proteins rather than Mla (Yahiaoui et al. 2004).

GFG resistance not associated with HR has been identified in powdery mildew interactions. Three lines of evidence strongly suggest that HR is not required for the *Mlg*-associated resistance to *Bgh*:

1. The inhibition of the HR by addition of the transcriptional inhibitor, cordycepin did not result in the release of fungal growth arrest.
2. The growth arrest of a *Bgh* isolate occurred in the absence of a detectable HR in two barley genetic backgrounds.
3. Gene dosage experiments showed that heterozygous *Mlg/mlg* plants, in comparison to the *Mlg/Mlg* genotype, show a drastic reduction of single-cell HR frequency without a proportional increase of haustorium formation (Görg et al. 1993; Schiffer et al. 1997).

Again in *Bgh*, the AVR_{Ab} gene segregates as a single Mendelian locus and follows classic GFG interaction. In this case, however, resistance is manifest as a reduction in the number of colonies formed by approximately 85% (Brown and Jessop 1995). No microscopic investigations were made in this study, so it is not known whether a reduction in colony number correlates with increased HR. In wheat, the powdery mildew resistance gene *Pm2* also governs penetration success, but not HR. Since effective papillae and HR did not occur in the same cells, papilla deposition may be independent of the HR response despite the fact that both defence mechanisms were associated with high H_2O_2 accumulation (Li et al. 2005). HR may act as a second line of defence to contain infection when the papilla defence fails.

E. Race-Non-Specific Resistance Genes

Genes conferring resistance to all powdery mildew races have been identified. *RPW8.1* and *RPW8.2* in *A. thaliana* are naturally occurring dominant alleles closely linked to each other and are required for defences associated with the hypersensitive response. Unlike *R* genes, however, they confer resistance to a range of powdery mildew pathogens, apparently not through a GFG interaction. The *RPW8* resistance locus is unusual because it mediates dominant resistance to diverse powdery mildew species, including 15 tested isolates of *Erysiphe cichoracearum*, *E. cruciferearum*, *E. orontii* and *Oidium lycopersici* (Xiao et al. 1997, 2001).

A transmembrane protein in barley known as Mlo is a pre-requisite for successful colonization by *Bgh* (Büschges et al. 1997). Plants carrying loss-of-function alleles (*mlo*) of the *Mlo* locus are resistant against all known isolates of *Bgh* (Piffanelli et al. 2006). The *mlo* mutation does not, however, affect a range of other foliar pathogens. In the absence of Mlo protein function (such as in barley *mlo* mutants), barley plants are resistant because germinated fungal spores fail to enter epidermal host cells. A characteristic feature of *mlo* resistance to *Bgh* is an early cessation of penetration through the epidermal cell wall that is not accompanied by the HR, a typical response of most *R* gene-triggered resistance (Wolter et al. 1993; Shirasu and Schulze-Lefert 2000). Since the first identification of *mlo* mutant barley, *mlo* resistance has been widely used in barley cultivation (Jørgensen 1992). Resistance to powdery mildew equivalent to *mlo* is not known in other crop species. However, non-host resistance and mlo-based immunity in *Arabidopsis thaliana* and barley respectively share similar features (see Sect. V.F). No naturally occurring broad-spectrum resistance against powdery mildew attack has been demonstrated in wheat (*Triticum aestivum*) against *B. graminis* f. sp. *tritici* and no *mlo* mutants have been detected. This may be because of the hexaploid nature of bread wheat and the likelihood that mutations may have to occur in all six copies of presumptive *Mlo* orthologues. Homologues of barley *Mlo* are, however, found in syntenic positions in all three genomes of bread wheat and also present in rice, *Oryza sativa* (Elliott et al. 2002). The Mlo protein also has homologues in *Arabidopsis*, indicating that a common host cell entry mechanism of powdery mildew fungi evolved once and at least 200 million years ago, suggesting that, within the powdery mildews, the ability to cause disease has been a stable trait throughout phylogenesis (Consonni et al. 2006).

The Mlo protein resides in the plasma membrane and has seven transmembrane domains reminiscent of the transmembrane receptors in fungi and animals. In animals, these are known as G-protein-coupled receptors and exist in three main families, lacking sequence similarity. A domain in Mlo mediates a calcium-dependent interaction with calmodulin in vitro. Loss of calmodulin binding reduces the ability of Mlo to negatively regulate defence against powdery mildew in vivo. Based on these investigations, a sensor role for Mlo in the modulation of defence reactions was proposed (Kim et al. 2002). Using noninvasive fluorescence-based imaging techniques, the Mlo protein was shown to be redistributed in the plasma membrane and accumulate beneath fungal appressoria coincident with pathogen entry into host cells (Bhat et al. 2005). Polarized Mlo accumulation occurs upon fungal attack and is independent of actin cytoskeleton function. Since *mlo* resistance is effective only against barley powdery mildew, the fungus may be targeting Mlo to achieve defence suppression (Kim et al. 2002). There are also reports that *mlo* resistance to powdery mildew increases susceptibility to other barley pathogens, including *Magnaporthe grisea* (Jarosch et al. 1999).

F. Cell Entry Control and Non-Host Resistance

Attack by *E. cichoracearum* usually results in successful penetration and rapid proliferation of the fungus on *A. thaliana*. By contrast, the non-host pathogen *Bgh* typically fails to penetrate *A. thaliana* epidermal cells (this is defined as non-host resistance). Genetic screens for mutations that result in increased penetration of *Bgh* on *A. thaliana* enabled the identification of penetration (pen) mutants. PEN1 is a syntaxin and has a close homologue SYP122 (Assaad et al. 2004). Both of these proteins are members of a large family of SNAREs (soluble N-ethylmaleimide-sensitive factor adaptorprotein receptors), present in the *A. thaliana* genome. Host proteins exhibit focal accumulation (local aggregation) at powdery mildew entry sites. Localization and genetic studies suggest that PEN1 plays an active role in the polarized secretion events that give rise to the formation of papillae during fungal attack. The *pen1* phenotype can therefore be described as the converse of the *mlo* phenotype;

mlo mutants have an increased penetration resistance whereas *pen1* mutants have a decreased penetration resistance (Collins et al. 2003). The barley orthologue of *pen1*, *ror2*, was identified as a locus 'required for ml*o* resistance' in screens for suppressors of *mlo* (Freialdenhoven et al. 1996; Collins et al. 2003). During the establishment of a compatible interaction, the timing of papilla formation is potentially critical and may affect the frequency of fungal penetration. SYP122 may have a general function in secretion, including a role in cell wall deposition. These investigations illustrate that there are multiple layers of resistance in the context of the non-host resistance and provide evidence for the existence of a vesicle-associated resistance mechanism preventing powdery mildew infection.

Pen2 encoding a glycosyl hydrolase was identified and characterized in *A. thaliana* and shown to act as a component of an inducible pre-invasion resistance mechanism (Lipka et al. 2005). Impairment of pre- and post-invasion resistance results in *A. thaliana* becoming a host for non-adapted fungi. *Pen3* encodes a putative ATP binding cassette transporter and *pen3* mutant plants permitted both increased invasion into epidermal cells and initiation of hyphae by non-host *Bgh* (Stein et al. 2006). The Pen3 protein is concentrated at infection sites and probably contributes to defences at the cell wall by exporting toxic materials to attempted invasion sites. Although *pen* mutants identified in such screens enable efficient entry of non-host powdery mildews, post-invasive fungal growth invariably ceases, coincident with a cell death response of epidermal cells containing haustoria. Thus, other factors are responsible for maintaining a compatible interaction. Inoculating *Bgt* onto a defence-related mutant *eds1* (*enhanced disease susceptibility*) of non-host *A. thaliana* resulted in partial development of the fungus (Yun et al. 2003). Bilateral haustoria were also observed, which resembled those typically produced in the compatible host, wheat. A similar decrease in non-host resistance was observed following treatment with cytochalasin E which inhibits microfilament polymerization. In *eds1* mutants, inhibition of actin polymerization severely compromised non-host resistance in *A. thaliana* against *Bgt*. Results from these investigations reveal that cytoskeletal function and *eds1* activity contribute to non-host resistance in *A. thaliana*.

VI. Significance in Agriculture: Breakdown of Resistance in the Field

When left uncontrolled, powdery mildew infections can have a significant effect on crop yield and quality. Understanding the physiology of powdery mildews and their interaction with the host plant can help to improve disease management strategies, assisting in the deployment of resistance genes and establishing when chemical control measures can be used most successfully. For example, the duration of surface wetness is important for infection by several powdery mildew species, and a model for control of glasshouse rose powdery mildew based on this has been developed. However, excessive wetness is generally detrimental to powdery mildew infection. Soil conditions and crop nutrition also significantly affect the development of powdery mildew infections. In particular, nitrogen fertilizer can be directly correlated with the amount of infection in *Bgh*. A comprehensive summary of the effects of temperature, moisture, soil conditions and light on powdery mildew epidemiology is given by Jarvis et al. (2002).

Varietal resistance is an important control measure that can be used in the management of powdery mildew diseases. However, powdery mildews can rapidly evolve to overcome resistance. When a new resistance gene is introduced, the population of the pathogen may respond by rapid growth of a few virulent clones, which spread quickly to become predominant leading to field breakdown of resistance. This has been well described in the GFG interactions of *Bgh* with barley, where barley varieties lost their *Mla* resistance within a few years of being introduced (Brown 1994; Hovmøller et al. 2000). Breakdown of resistance has been reported in other powdery mildew interactions. Pl2 is a major resistance gene used in apple breeding programmes. Virulent isolates appeared within six years after planting apple P12 genotypes resulting in a breakdown of the resistance. Ten years after planting, the percentage of genotypes that were still resistant to powdery mildew varied between 2% and 56% (Caffier and Laurens 2005). Long-term cultivation of varieties with widely used resistance genes results in significant shifts in virulence frequencies of *Bgt* on wheat (for a review, see Hsam and Zeller 2002). To delay the development of virulence, the pyramiding of several resistance genes into a single cultivar has been proposed, since the pathogen would need to undergo multiple simultaneous changes to become virulent (McIntosh and Brown 1997). The availability of molecular markers linked to resistance genes in wheat breeding could assist in this process.

There have been reports of breakdown of *mlo* resistance, but *mlo* virulence has not developed in field populations and does not seem to present a problem for barley growing. The breakdown of *mlo* resistance is associated with the relief of water stress following a period of drought (Baker et al. 2000). In these situations, *Bgh* infection increases on both *Mlo*-susceptible and *mlo*-resistant spring barley cultivars. The breakdown of *mlo* resistance is temporary and is determined by the genetic background of the host barley plant rather than the specific resistance allele.

Investigations into the molecular mechanism of resistance described earlier could eventually provide leads for the development of more durable disease control. Powdery mildews are highly adapted pathogens with a limited host range, and infection does not succeed on non-host plants. Some of the molecular components of non-host resistance are starting to be identified. Selective breeding for natural variants in non-host resistance components could provide a basis for the development of broad-spectrum resistance. The apparent inverse relationship between *pen1* non-host resistance and *mlo* illustrates that more needs to be learnt about control of pathogen entry before such resistance can be exploited in agriculture. Barley *Mlo* homologues are present in other cereals and in *A. thaliana*, illustrating that this molecule, associated with mildew resistance, could potentially be used in a range of crops. It has been proposed that R proteins guard essential virulence targets in host plant cells, which pathogens attack to establish infection (Mackey et al. 2002). If powdery mildew effectors and their host targets can be identified, this principle can be exploited in agriculture to develop durable disease control. Selecting crop varieties with natural polymorphisms in such targets could prevent attack by adapted powdery mildews, so bypassing the reliance on R proteins which are easily defeated.

Acknowledgements. I would like to acknowledge the help of all those who provided figures for this review (detailed in the figure legends) and the financial support of the Biotechnology and Biological Sciences Research Council and EU framework VI (BIOEXPLOIT)

References

An QL, Ehlers K Kogel KH, van Bel AJE, Huckelhoven R (2006) Multivesicular compartments proliferate in susceptible and resistant MLA12-barley leaves in response to infection by the biotrophic powdery mildew fungus. New Phytol 172:563–576

Assaad F F, Qiu J L, Youngs H, Ehrhardt D, Zimmerli L, Kalde M, Wanner G, Peck SC, Edwards H, Ramonell K et al (2004) The PEN1 syntaxin defines a novel cellular compartment upon fungal attack and is required for the timely assembly of papillae Mol. Biol. Cell 15:5118–5129

Baker SJ, Newton AC, Gurr SJ (2000) Cellular characteristics of temporary partial breakdown of mlo-resistance in barley to powdery mildew. Physiol Mol. Plant Pathol. 56:1–11

Bardin M, Nicot PC, Normand P, Lemaire JM (1997) Virulence variation and DNA polymorphism in *Sphaerotheca fuliginea*, causal agent of powdery mildew of cucurbits. Eur J Plant Pathol 103:545–554

Bevan JR, Clarke DD, Crute IR (1993a) Resistance to *Erysiphe fischeri* in two populations of *Senecio vulgaris*. Plant Pathol 42:636–646

Bevan JR, Crute IR, Clarke DD (1993b) Variation for virulence in *Erysiphe fischeri* from *Senecio vulgaris*. Plant Pathol 42:622–635

Bhat RA, Miklis M, Schmelzer E, Schulze-Lefert P, Panstruga R (2005) Recruitment and interaction dynamics of plant penetration resistance components in a plasma membrane microdomain. Proc Natl Acad Sci USA 102:3135–3140

Bieri S, Mauch S, Shen QH, Peart J, Devoto A, Casais C, Ceron F, Schulze S, Steinbiss HH, Shirasu K, Schulze-Lefert P (2004) RAR1 positively controls steady state levels of barley MLA resistance proteins and enables sufficient MLA6 accumulation for effective resistance. Plant Cell 16:3480–3495

Both M, Csukai M, Stumpf MPH, Spanu PD (2005a) Gene expression profiles of *Blumeria graminis* indicate dynamic changes to primary metabolism during development of an obligate biotrophic pathogen. Plant Cell 17:2107–2122

Both M, Eckert SE, Csukai M, Müller E, Dimopoulos G, Spanu PD (2005b) Transcript profiles of *Blumeria graminis* development during infection reveal a cluster of genes that are potential virulence determinants. Mol Plant–Microbe Interact 18:125–133

Boyd LA, Smith, PH, Foster EM, Brown JKM (1995) The effects of allelic variation at the MLA resistance locus in barley on the early development of *Erysiphe-graminis* f. sp. *hordei* and host responses. Plant J 7:959–968

Braun U, Cook RTA, Inman AJ, Shin, HD (2002) The taxonomy of the powdery mildew fungi. In: Belanger RR, Bushnell WR, Dik AJ, Carver TLW (eds) The powdery mildews: a comprehensive treatise. APS, St Paul, pp 13–55

Brian PW (1967) Obligate parasitism in fungi. Proc R Soc Lond B Biol Sci 168:101–118

Brown JKM (1994) Chance and selection in the evolution of barley mildew. Trends Microbiol 2:470–475

Brown JKM (2002) Comparative genetics of avirulence and fungicide resistance in the powdery mildew fungi. In: Belanger RR, Bushnell WR, Dik AJ, Carver TLW (eds) The powdery mildews: a comprehensive treatise. APS, St Paul, pp 56–65

Brown JKM, Jessop AC (1995). Genetics of avirulences in *Erysiphe graminis* f.sp. *hordei*. Plant Pathol 44:1039–1049

Brown JKM, Simpson CG (1994) Genetic analysis of DNA fingerprints and virulences in *Erysiphe graminis* f.sp. *hordei*. Curr Genet 26:172–178.

Brown JKM, LeBoulaire S, Evans N (1996) Genetics of responses to morpholine-type fungicides and of avirulences in *Erysiphe graminis* f.sp. *hordei*. Eur J Plant Pathol 102:479–490

Büschges R, Hollricher K, Panstruga R, Simons G, Wolter M, Frijters A, vanDaelen R, vanderLee T, Diergaarde P, Groenendijk J, Töpsch S, Vos P, Salamini F, SchulzeLefert P (1997) The barley *Mlo* gene: a novel control element of plant pathogen resistance. Cell 88:695–705

Bushnell WR (1972) Physiology of fungal haustoria. Annu Rev Phytopathol 10:151–176

Bushnell WR (1981) Incompatibility conditioned by the *Mla* gene in powdery mildew of barley: the halt in cytoplasmic streaming. Phytopathology 71:1062–1066

Caffier V, Laurens F (2005) Breakdown of Pl2, a major gene of resistance to apple powdery mildew, in a French experimental orchard. Plant Pathol 54:116–124

Caffier V, de Vallavieille-Pope C, Brown JKM (1996) Segregation of avirulences and genetic basis of infection types in *Erysiphe graminis* f.sp. *hordei*. Phytopathology 86:1112–1121

Caldo RA, Nettleton D, Peng JQ, Wise RP (2006) Stage-specific suppression of basal defense discriminates barley plants containing fast- and delayed-acting *Mla* powdery mildew resistance alleles. Mol Plant–Microbe Interact 19:939–947

Carver TLW, Kunoh H, Thomas BJ, Nicholson RL (1999) Release and visualisation of the extracellular matrix of conidia of *Blumeria graminis*. Mycol Res 103:547–560

Celio GJ, Mims CW, Richardson EA (2004) Ultrastructure and immunocytochemistry of the host–pathogen interface in poinsettia leaves infected with powdery mildew. Can J Bot 82:421–429

Cherewick WJ (1944) Studies on the biology of *Erysiphe graminis* DC. Can J Res 22:52–86

Christiansen SK, Giese H (1990) Genetic analysis of the obligate parasitic barley powdery mildew fungus based on RFLP and virulence loci. Theor Appl Genet 79:705–712

Clarke DD (1997) The genetic structure of natural pathosystems. In: Crute IR, Holub EB, Burdon JJ (eds) The gene-for-gene relationship in plant–parasite interactions. CAB International, Wallingford, pp 231–243

Collins NC, Thordal-Christensen H, Lipka V, Bau S, Kombrink E, Qiu J, Hückelhoven R, Stein M, Freialdenhoven A, Somerville SC, Schulze-Lefert P (2003) SNARE-protein-mediated disease resistance at the plant cell wall. Nature 425:973–977

Colson B (1938) The cytology and development of *Phyllactinia corylea*. Lev. Ann Bot 2:381–402

Consonni C, Humphry ME, Hartmann HA, Livaja M, Durner J, Westphal L, Vogel J, Lipka V, Kemmerling B, Schulze-Lefert P, Somerville SC, Panstruga R (2006) Conserved requirement for a plant host cell protein in powdery mildew pathogenesis. Nat Genet 38:616–720

Crute IR, Holub EB, Burdon JJ (eds) (1997) The gene-for-gene relationship in plant–parasite interactions. CAB International, Wallingford

Dekhuijzen HM (1976) Endogenous cytokinins in healthy and diseased plants. In: Heitefuss R, Williams PH (eds) Physiological plant pathology, encyclopedia of plant physiology, new series, vol 4. Springer, Heidelberg, pp 526–559

Edwards HH (2002) Development of primary germ tubes by conidia of *Blumeria graminis* f.sp. *hordei* on leaf epidermal cells of Hordeum vulgare. Can J Bot 80:1121–1125

Elliott C, Zhou F, Spielmeyer W, Panstruga R, Schulze-Lefert P (2002) Functional conservation of wheat and rice Mlo orthologs in plant defense modulation to powdery mildew. Mol Plant–Microbe Interact 15:1069–1077

Flor HH (1955) Host–parasite interaction in flax rust – its genetics and other implications. Phytopathology 45:680–685

Freialdenhoven A, Peterhansel C, Kurth J, Kreuzaler F, Schulze-Lefert P (1996) Identification of genes required for the function of non-race-specific mlo resistance to powdery mildew in barley. Plant Cell 8:5–14

Gil F, Gay JL (1977) Ultrastructural and physiological properties of the host interfacial components of haustoria of *Erysiphe pisi* in vivo and in vitro. Physiol Plant Pathol 10:1–12

Giles PF, Soanes DM, Talbot NJ (2003) A relational database for the discovery of genes encoding amino acid biosynthetic enzymes in pathogenic fungi. Compar Funct Genomics 4:4–15

Görg R, Hollricher K, Schulze-Lefert P (1993) Functional analysis and RFLP-mediated mapping of the *Mlg* resistance locus in barley. Plant J 3:857–866

Green JR, Carver TLW, Gurr SJ (2002) The formation and function of infection and feeding structures. In: Belanger RR, Bushnell WR, Dik AJ, Carver TLW (eds) In powdery mildews: a comprehensive treatise. APS, St Paul, pp 66–82

Gregersen PL, Thordal-Christensen H, Forster H, Collinge DB (1997) Differential gene transcript accumulation in barley leaf epidermis and mesophyll in response to attack by *Blumeria graminis* f.sp. *hordei* (syn. *Erysiphe graminis* f.sp. *hordei*). Physiol Mol Plant Pathol 51:85–97

Hahn M, Mendgen K (1997) Characterization of in planta induced rust genes isolated from a haustorium-specific cDNA library. Mol Plant––Microbe Interact 10:427–437

Halterman DA, Wise RP (2004) A single-amino acid substitution in the sixth leucine-rich repeat of barley MLA6 and MLA13 alleviates dependence on RAR1 for disease resistance signaling. Plant J 38:215–226

Halterman DA, Wei FS, Wise RP (2003) Powdery mildew induced *Mla* mRNAs are alternatively spliced and contain multiple upstream open reading frames. Plant Physiol 131:558–567

Harper RA (1905) Sexual reproduction and the organization of the nucleus in certain mildews. Carnegie Inst Wash 37:1–105

Harry IB, Clarke DD (1986) Race-specific resistance in groundsel (*Senecio vulgaris*) to the powdery mildew *Erysiphe fischeri*. New Phytolol 103:167–175

Hein I, Pacak MB, Hrubikova K, Williamson S, Dinesen M, Soenderby IE, Sundar S, Jarmolowski A, Shirasu K, Lacomme C (2005)Virus-induced gene silencing-based functional characterization of genes associated with powdery mildew resistance in barley Plant Physiol 138:2155–2164

Hirata K (1967) Notes on the haustoria, hyphae and conidia of the powdery mildew fungus on barley *Erysiphe graminis* f.sp. *hordei*. Mem Fac Agric Niigata Univ 6:207–259

Hosoya K, Narisawa K, Pitrat M, Ezura H (1999) Race identification in powdery mildew (*Sphaerotheca fuliginea*) on melon (*Cucumis melo*) in Japan. Plant Breed 118:259–262

Hovmøller MS, Caffier V, Jalli M, Andersen O, Besenhofer G, Czembor JH, Dreiseitl A, Felsenstein F, Fleck A, Heinrics F, Jonsson R, Limpert E, Mercer P, Plesnik S, Rashal I, Skinnes H, Slater S, Vronska O (2000) The European barley powdery mildew virulence survey and disease nursery 1993–1999. Agronomie 20:729–743

Hsam SLK, Zeller FJ (2002) Breeding for powdery mildew resistance in common wheat (*Triticum aestivum* L.). In: Belanger RR, Bushnell WR, Dik AJ, Carver TLW (eds) In powdery mildews: a comprehensive treatise. APS, St Paul, pp 219–238

Huang XQ, Roder MS (2004) Molecular mapping of powdery mildew resistance genes in wheat: a review. Euphytica 137:203–222

Jakupovic M, Heintz M, Reichmann P, Mendgen K, Hahn M (2006) Microarray analysis of expressed sequence tags from haustoria of the rust fungus *Uromyces fabae*. Fungal Genet Biol 43:8–19

Jarosch B, Kogel KH, Schaffrath U (1999) The ambivalence of the barley *Mlo* locus: mutations conferring resistance against powdery mildew (*Blumeria graminis* f.sp. *hordei*) enhance susceptibility to the rice blast fungus *Magnaporthe grisea*. Mol Plant–Microbe Interact 12:508–514

Jarvis WR, Gubler WD, Grove GG (2002) Epidemiology of powdery mildews in agricultural systems. In: Belanger RR, Bushnell WR, Dik AJ, Carver TLW (eds) The powdery mildews: a comprehensive treatise. APS, St Paul, pp 169–199

Jensen J, Jensen HP, Jørgensen JH (1995) Linkage studies of barley powdery mildew virulence loci. Hereditas 122:197–209

Jørgensen JH (1988) *Erysiphe graminis*, powdery mildew of cereals and grasses. Adv Plant Pathol 6:135–157

Jørgensen JH (1992) Discovery, characterization and exploitation of *Mlo* powdery mildew resistance in barley. Euphytica 63:141–152

Jørgensen JH (1994) Genetics of powdery mildew resistance in barley. Crit Rev Plant Sci 13:97–119

Kim MC, Panstruga R, Elliott C, Müller J, Devoto A, Yoon HW, Park H, Cho MJ, Schulze-Lefert P (2002) Calmodulin interacts with MLO to regulate defence against mildew in barley. Nature 416:447–450

Kobayashi J, Kobayashi I, Funaki Y, Fujimoto S, Takemoto T, Kunoh H (1997) Dynamic reorganization of microfilaments and microtubules is necessary for the expression of non-host resistance in barley coleoptile cells. Plant J 11:525–537

Koh S, André A, Edwards H, Ehrhardt D, Somerville S (2005) *Arabidopsis thaliana* subcellular responses to compatible *Erysiphe cichoracearum* infections Plant J 44:516–529

Kunoh H, Hayashimoto A, Harui M, Ishizaki H (1985) Induced susceptibility and enhanced resistance at the cellular level in barley coleoptiles. I. The significance of timing of fungal invasion. Physiol Plant Pathol 27:43–54

Kunoh H, Kuroda K, Hayashimoto A, Ishizaki H (1986) Induced susceptibility and enhanced resistance at the cellular level in barley coleoptiles. II. The timing and localization of induced susceptibility in a single coleoptile cell and its transfer to an adjacent cell. Can J Bot 64:889–895

Li AL, Wang ML, Zhou RH, Kong XY, Huo NX, Wang WS, Jia JZ (2005) Comparative analysis of early H_2O_2 accumulation in compatible and incompatible wheat–powdery mildew interactions. Plant Pathol 54:308–316

Lipka V, Dittgen J, Bednarek P, Bhat RA, Stein M, Landtag J, Brandt W, Scheel D, Llorente F, Molina A, Wiermer M, Parker J, Somerville SC, Schulze-Lefert P (2005) Pre- and post-invasion defenses both contribute to non-host resistance in *Arabidopsis*. Science 310:1180–1183

Lyngkjær MT, Carver TLW (1999) Induced accessibility and inaccessibility to *Blumeria graminis* f.sp. *hordei* in barley epidermal cells attacked by a compatible isolate. Physiol Mol Plant Pathol 55:151–162

Lyngkjær MT, Carver TLW, Zeyen RJ (2001) Virulent *Blumeria graminis* infection induces penetration susceptibility and suppresses race-specifc hypersensitive resistance against avirulent attack in *Mla1*-barley. Physiol Mol Plant Pathol 59:243–256

Mackey D, Holt BF III, Wiig A, Dangl JL (2002) RIN4 interacts with *Pseudomonas syringae* type III effector molecules and is required for RPM1-mediated resistance in *Arabidopsis*. Cell 108:753–754

Manners JM, Gay JL (1983) The host–parasite interface and nutrient transfer in biotrophic parasitism. In: Callow JA (ed) Biochemical plant pathology. Wiley, Chichester, pp 163–195

McIntosh RA, Brown GN (1997) Anticipatory breeding for resistance to rust diseases in wheat. Annu Rev Phytopathol 35:311–326

Moseman JG (1959) Host–pathogen interaction of the genes for resistance in *Hordeum vulgare* and for pathogenicity in *Erysiphe graminis* f.sp. *hordei*. Phytopathology 49:469–472

Moseman JG, Scharen AL, Greely LW (1964) Propagation of *Erysiphe graminis* f.sp. *tritici* on barley and *Erysiphe graminis* f. sp *hordei* on wheat. Phytopathology 55:92–96

Olesen KL, Carver TLW, Lyngkjaer MF (2003) Fungal suppression of resistance against inappropriate *Blumeria graminis formae speciales* in barley, oat and wheat. Physiol Mol Plant Pathol 62:37–50

Opalski KS, Schultheiss H, Kogel KH, Huckelhoven R (2005) The receptor-like MLO protein and the RAC/ROP family G-protein RACB modulate actin reorganization in barley attacked by the biotrophic powdery mildew fungus *Blumeria graminis* f.sp. *hordei*. Plant J 41:291–303

Pearson RC, Gadoury DM (1987) Cleistothecia, the source of primary inoculum for grape powdery mildew in New York. Phytopathology 77:1509–1579

Pedersen C, Rasmussen SW, Giese H (2002). A genetic map of *Blumeria graminis* based on functional genes, avirulence genes, and molecular markers. Fungal Genet Biol 35:235–246

Pérez-Garcia A, Mignorance E, Rivera ME, Del Pin D, Romero D, Torés JA, De Vicente A (2006) Long-term preservation of *Podosphaera fusca* using silica gel. J Phytopathol 154:190–192

Piffanelli P, Ramsay L, Waugh R, Benabdelmouna A, D'Hont A, Hollricher K, Jorgensen JH, Schulze-Lefert P, Panstruga R (2006) A barley cultivation-associated polymorphism conveys resistance to powdery mildew. Nature 430:887–891

Powers HR, Sando WJ (1960) Genetic control of the host–parasite relationship in wheat powdery mildew. Phytopathology 50:454–457

Prats E, Mur LAJ, Sanderson R, Carver TLW (2005) Nitric oxide contributes both to papilla-based resistance and the hypersensitive response in barley attacked by *Blumeria graminis* f.sp. *hordei*. Mol Plant Pathol 6:65–78

Prats E, Gay A, Mur L, Thomas B, Carver T (2006) Stomatal lock-open, a consequence of epidermal cell death, follows transient suppression of stomatal opening in barley attacked by *Blumeria graminis*. J Exp Bot 57:2211–2226

Ridout CJ, Skamnioti P, Porritt O, Sacristan S, Jones JDG, Brown JKM (2006) Multiple avirulence paralogues in cereal powdery mildew fungi may contribute to parasite fitness and defeat of plant resistance. Plant Cell 18:2402–2414

Robinson HL, Ridout CJ, Sierotzki H, Gisi U, Brown JKM (2002) Isogamous, hermaphroditic inheritance of mitochondrion-encoded resistance to Qo inhibitor fungicides in *Blumeria graminis* f. sp. *tritici*. Fungal Genet Biol 36:98–106

Schiffer R, Görg R, Jarosch B, Beckhove U, Bahrenberg G, Kogel KH, Schulze-Lefert P (1997) Tissue dependence and differential cordycepin sensitivity of race-specific resistance responses in the barley powdery mildew interaction. Mol Plant–Microbe Interact 10:830–839

Shen QH, Zhou FS, Bieri S, Haizel T, Shirasu K, Schulze-Lefert P (2003) Recognition specificity and RAR1/SGT1 dependence in barley *Mla* disease resistance genes to the powdery mildew fungus. Plant Cell 15:732–744

Shen QH, Saijo Y, Mauch S, Biskup C, Bieri S, Keller B, Seki H, Ulker B, Somssich IE, Schulze-Lefert P (2007) Nuclear activity of MLA immune receptors links isolate-specific and basal disease-resistance responses. Science 315:1098–1103

Shirasu K, Schulze-Lefert P (2000) Regulators of cell death in disease resistance. Plant Mol Biol 44:371–385

Soanes DM, Talbot NJ (2006) Comparative genomic analysis of phytopathogenic fungi using expressed sequence tag (EST) collections. Mol Plant Pathol 7:61–70

Soanes DM, Skinner W, Keon J, Hargreaves J, Talbot NJ (2002) Genomics of phytopathogenic fungi and the development of bioinformatic resources. Mol Plant–Microbe Interact 15:421–427

Stein M, Dittgen J, Sanchez-Rodriguez C, Hou BH, Molina A, Schulze-Lefert P, Lipka V, Somerville S (2006) *Arabidopsis* PEN3/PDR8, an ATP binding cassette transporter, contributes to nonhost resistance to inappropriate pathogens that enter by direct penetration. Plant Cell 18:731–746

Struck C, Ernst M, Hahn M (2002) Characterization of a developmentally regulated amino acid transporter (AAT1p) of the rust fungus *Uromyces fabae*. Mol Plant Pathol 3:23–30

Thomas SW, Rasmussen SW, Glaring MA, Rouster JA, Christiansen SK, Oliver RP (2001) Gene identification in the obligate fungal pathogen *Blumeria graminis* by expressed sequence tag analysis. Fungal Genet Biol 33:195–211

Thomas SW, Glaring MA, Rasmussen SW, Kinane JT, Oliver RP (2002) Transcript profiling in the barley mildew pathogen *Blumeria graminis* by serial analysis of gene expression (SAGE). Mol Plant–Microbe Interact 15:847–856

Thompson JN, Burdon JJ (1992) Gene-for-gene coevolution between plants and parasites. Nature 360:121–125

Thordal-Christensen H, Zhang ZG, Wei YD, Collinge DB (1997) Subcellular localization of H_2O_2 in plants. H_2O_2 accumulation in papillae and hypersensitive response during the barley–powdery mildew interaction. Plant J 11:1187–1194

Tiwari KR, Penner GA, Warkentin TD, Rashid KY (1997) Pathogenic variation in *Erysiphe pisi*, the causal organism of powdery mildew of pea. Can J Plant Pathol 19:267–271

Vanacker H, Carver TLW, Foyer CH (2000) Early H_2O_2 accumulation in mesophyll cells leads to induction of glutathione during the hypersensitive response in the barley–powdery mildew interaction. Plant Physiol 123:1289–1300

Voegele RT, Struck C, Hahn M, Mendgen K (2001) The role of haustoria in sugar supply during infection of broad bean by the rust fungus *Uromyces fabae*. Proc Natl Acad Sci USA 98:8133–8138

Voegele RT, Wirsel S, Möll U, Lechner M, Mendgen K (2006) Cloning and Characterization of a novel invertase from the obligate biotroph *Uromyces fabae* and analysis of expression patterns of host and pathogen invertases in the course of infection. Mol Plant–Microbe Interact 19:625–634

Walters DR, McRoberts N (2006) Plants and biotrophs: a pivotal role for cytokinins. Trends Plant Sci 11:581–586

Wolter M, Hollricher K, Salamini F, Schulze-Lefert P (1993) The *Mlo* resistance alleles to powdery mildew infection in barley trigger a developmentally controlled defense mimic phenotype. Mol Gen Genet 239:122–128

Xiao S, Ellwood S, Findlay K, Oliver RP, Turner JG (1997) Characterization of three loci controlling resistance of *Arabidopsis thaliana* accession Ms-0 to two powdery mildew diseases. Plant J 12:757–768

Xiao S, Ellwood S, Calis O, Patrick E, Li T, Coleman M, Turner JG (2001) Broad-spectrum mildew resistance in *Arabidopsis thaliana* mediated by RPW8. Science 291:118–120

Yahiaoui N, Srichumpa P, Dudler R, Keller B (2004) Genome analysis at different ploidy levels allows cloning of the powdery mildew resistance gene Pm3b from hexaploid wheat. Plant J 37:528–538

Yahiaoui N, Brunner S, Keller B (2006) Rapid generation of new powdery mildew resistance genes after wheat domestication. Plant J 47:85–98

Yu D (2000) Wheat powdery mildew in central China. Dissertation, Univerity of Wageningen

Yun BW, Atkinson HA, Gaborit C, Greenland A, Read ND, Pallas JA, Loake GJ (2003) Loss of actin cytoskeletal function and EDS1 activity, in combination, severely compromises non-host resistance in *Arabidopsis* against wheat powdery mildew Plant J 34:768–777

Zavaleta-Mancera H, Franklin K, Ougham H, Thomas H, Scott I (1999) Regreening of senescent *Nicotiana* leaves. I. Reappearance ofNADPH-protochlorophyllide oxidoreductase and light harvesting chlorophyll a/b-binding protein. J Exp Bot 50:1677–1682

Zeyen RJ, Carver TLW, Lyngkjaer MF (2002) Epidermal cell papillae. In: Belanger RR, Bushnell WR, Dik AJ, Carver TLW (eds) The powdery mildews: a comprehensive treatise. APS, St Paul, pp 107–125

Zhang Z, Henderson C, Perfect E, Carver TLW, Thomas BJ, Skamnioti P, Gurr SJ (2005) Of genes and genomes, needles and haystacks: *Blumeria graminis* and functionality. Mol Plant Pathol 6:561–575

Zhou FS, Kurth JC, Wei FS, Elliott C, Vale G, Yahiaoui N, Keller B, Somerville S, Wise R, Schulze-Lefert P (2001) Cell-autonomous expression of barley Mla1 confers race-specific resistance to the powdery mildew fungus via a Rar1-independent signalling pathway. Plant Cell 13:337–350

Zhu ZD, Zhou RH, Kong XY, Dong YC, Jia JZ (2005) Microsatellite markers linked to 2 powdery mildew resistance genes introgressed from *Triticum carthlicum* accession PS5 into common wheat. Genome 48:585–590

4 The Uredinales: Cytology, Biochemistry, and Molecular Biology

Ralf T. Voegele[1], Matthias Hahn[2], Kurt Mendgen[1]

CONTENTS

I.	Introduction	69
II.	A Brief History of Rust Fungi and Rust Research	70
III.	Phylogeny and Taxonomy	71
IV.	Life Cycle	73
V.	Epidemiology	74
VI.	Spore Germination and the Formation of Infection Structures	75
VII.	Features of Urediospore Infection	76
VIII.	Structural Aspects of the Dikaryotic Haustorium	78
IX.	Biochemical and Molecular Analyses of Rust Fungi	80
X.	Suppression of Host Defenses	83
XI.	Host Responses to Rust Infection	85
XII.	Control of Rust Disease	86
XIII.	Genetics and Molecular Biology of Rust Resistance	87
XIV.	Imminent Threats: The Cases of *P. pachyrhizi* and *P. graminis* Ug99	88
XV.	Conclusions and Perspectives	89
	References	90

I. Introduction

Fungi belonging to the order *Uredinales* are commonly referred to as rust fungi. All members of the *Uredinales* are parasitic on plants, often causing dramatic losses in various important crop plants (Alexopoulos et al. 1996). Together with the powdery mildew fungi and the downy mildew-causing oomycetes, rust fungi form an extremely successful group of parasites, the obligate biotrophs. The term **obligate biotrophic** characterizes a specific lifestyle in which the pathogen is absolutely dependent on a living host to complete its life cycle. In turn, the host plant as a whole usually suffers only limited damage over an extended period of time (Staples 2000). By contrast, **necrotrophic parasites** kill their hosts quickly after infection and subsequently thrive on the dead plant material (Staples 2001). **Hemibiotrophic fungi**, such as *Colletotrichum* spp., are characterized by a more or less extended biotrophic phase before switching to necrotrophic growth and killing their host (Perfect and Green 2001). In order to separate the true obligate biotrophic pathogens from hemibiotrophs and necrotrophs we suggest the following six criteria:

1. Obligate biotrophs are not culturable in vitro (at least not to a point representing the parasitic phase)
2. They form highly differentiated infection structures (variations of the normally tubular cell shape, which are necessary for pathogenesis)
3. They have limited secretory activity
4. They establish a narrow contact zone separating fungal and plant plasma membranes
5. They engage in a long-term suppression of host defense responses
6. They form haustoria (specialized hyphae that penetrate host cells).

The peculiarities of the lifestyle of obligate biotrophs, paired with their huge economic impact, make rust fungi a versatile field of study at both the fundamental and the applied level. This chapter on *Uredinales* can by no means cover the complete literature on rust fungi. It is intended to summarize key references, review articles, and books to provide the interested reader with a gateway to more specialized literature on most aspects of research involving rust fungi. Readers new to the field are encouraged to consult the excellent textbooks by Alexopoulos et al. (1996) and Webster and Weber (2007) to gain easier access into the exciting field of mycology in general and obligate biotrophic plant parasites like the rust fungi in particular.

[1] Lehrstuhl Phytopathologie, Fachbereich Biologie, Universität Konstanz, 78457 Konstanz, Germany;
e-mail: Ralf.Voegele@uni-konstanz.de
[2] Phytopathologie, Fachbereich Biologie, Technische Universität Kaiserslautern, 67663 Kaiserslautern, Germany

Plant Relationships, 2nd Edition
The Mycota V
H. Deising (Ed.)
© Springer-Verlag Berlin Heidelberg 2009

II. A Brief History of Rust Fungi and Rust Research

There is evidence for a deep-rooted association of rust fungi with food and forage crops. For example, wheat leaf fragments infected with *Puccinia graminis*, the causative agent of stem rust of wheat, have been found in a storage jar from the Late Bronze Age (Kislev 1982). During the reign of the second Roman king Numa Pompilius, the festival of Robigalia was reported by Pliny the Elder to be introduced around 700 BC to appease the fertility god Robigus, god of rusts and mildews (Pliny 69). Thus it appears that rusts have plagued farmers around the globe throughout history. Many cereals and legumes, the two plant families most important for humans (Graham and Vance 2003), suffer from rust infection. Cereal rusts have been a recurring problem in many parts of the world, occasionally causing yield losses of sometimes more than 75% in some areas (Rapilly 1979; Eversmeyer and Kramer 2000; Long 2003). Cereal rusts have been under reasonable control for the past decades mainly through crop management and breeding of resistant wheat lines (see Sects. XII, XIII). However, a new hypervirulent strain of *P. graminis*, Ug99 or TTKS, which seems able to infect about 90% of 12 000 wheat lines tested, was recently found to spread from its original point of discovery in Africa, threatening the world's wheat production yet again (Stokstad 2007; see Sect. XIV). Legume rusts have so far prevailed in Africa, Asia, and Oceania. For example, yield losses of up to 50% have been reported due to infection of fava beans (*Vicia faba*) with *Uromyces fabae* (Tissera and Ayres 1986). Another legume rust, *Phakopsora pachyrhizi*, the causative agent of Asian soybean rust (ASR), has lately spread into the continental United States threatening soybean production there (Schneider et al. 2005; see Sect. XIV). This fact has made the United States Department of Agriculture (USDA) and soybean farmers go on high alert. The possible consequences of such a global spread of a pathogen are exemplified by another rust fungus, *Hemileia vastatrix*, the causative agent of coffee rust. After a first report of the fungus in Ceylon (formerly Sri Lanka) in 1869, it took less than three decades to annihilate the entire coffee production of the island, leaving the British society only tea as a social drink (Staples 2000).

Fontana (1767) was the first to link rust disease to a parasitic fungus. The first comprehensive description of rust fungi, comprising some 120 species, was published by Unger (1833). He found rust fungi on most plant families and correlated the extent of infection with humidity. He also studied cross-sections through infected leaves and noted the degradation of chlorophyll in diseased areas. De Bary (1853) was the first to notice the importance of the germ pore in urediospore walls for production of the germ tube. In addition, he discussed the significance of tip growth for the ability of a fungus to penetrate through stomatal openings. It was also de Bary (1863) who introduced the term haustorium to describe the only hyphae of obligate biotrophic parasites that invade plant host cells (see Sect. VIII). These structures were first described by Zanardini a decade earlier (von Mohl 1853). A few years later de Bary (1865) elucidated the life cycle of *P. graminis*, coined the term teleutospores as the final spore form in the life cycle of macrocyclic rusts, and defined the terms autoecious and heteroecious (see Sect. IV). Eriksson (1894) described the specialization of the rust fungi on cereals, and Stakman and Piemeisal (1917) identified different races of wheat stem rust (see Sect. III). This result was the basis for an effective breeding program for resistance in cereals and other plants (Kolmer 1996). In 1927, Craigie (1927) discovered heterothallism of *P. graminis* and revealed the function of the pycnia as sexual organs. This was the final step in the elucidation of the rust life cycle (see Sect. IV). Rusts, in particular *P. graminis*, gained notoriety through the attention paid to them by biological warfare researchers of both superpowers during the Cold War (Line and Griffith 2001). While biological warfare programs involving rust fungi were discontinued in the early 1970s, *P. graminis* today is considered one of the most important potential bio-terrorism threats to agriculture in the United States (Madden and Wheelis 2003). The "gene for gene" concept, describing the interaction between pathogenic microorganisms and their host plants, introduced by Flor (1955, 1956), resulted from experiments with the flax rust, *Melampsora lini*, and its host *Linum usitatissimum* (see Sect. XIII). Up to the middle of the past century, research involving rust fungi was mainly based on infection studies and cytological analyses using the light microscope. In the early 1960s, cytological analysis of the host–parasite interface was raised to a new level with the introduction of electron microscopy to the field (Moore and McAlear 1961; Keen 2000). Another significant event of this decade was the report of the first axenic culture of a rust fungus (Williams et al. 1966; Keen 2000). However, in retrospect this method did not quite meet the expectations originally put into it. The 1960s were also characterized by a number of studies analyzing the physiology of host and parasite (for a review, see Bushnell 1972). These cytological and physiological studies continued through the 1970s and 1980s. Axenic cultures and the generation of infection structures by germinating spores on artificial surfaces such as collodion membranes (Dickinson 1949), polystyrene replicas of leaf surfaces (Wynn 1976), or structured polyethylene sheets (Staples et al. 1983) made biochemical analyses of proteins possible during the 1980s and 1990s (Mendgen et al. 1996). A significant event during that time was the finding that appressorium formation could be induced in vitro by simple topographic signals (Hoch et al. 1987; see Sect. VII). Another milestone in rust research was the introduction of a method to isolate rust haustoria from infected plant tissue (Hahn and Mendgen 1992). This work paved the way for more than a decade of molecular work mainly on *U. fabae* as a model organism (for a review, see Voegele 2006). The same period coincides with the molecular reconstruction of the gene for gene hypothesis fueled by the isolation and characterization of several rust resistance genes from flax and the corresponding avirulence genes from *M. lini* (Ellis et al. 2007a, b). Presently, new vistas are being opened to rust research, with the first rust genomes that are currently sequenced: (a) *Melampsora larici-populina*, the causative agent of poplar rust, and (b) *P. graminis* f. sp. *tritici*. The choice for *P. graminis* f. sp. *tritici* was based on its huge economic impact, whereas that for *M. larici-populina* was based on the fact that the host (*Populus trichocarpa*) genome has also been sequenced (Tuskan et al. 2006). In addition, the genomes of *P. trichocarpa* symbiotic fungal associates *Laccaria bicolor* and *Glomus intraradices* are also at or near completion. With the sequencing of the soybean pathogen *Puccinia pachyrhizi* in progress, three rust genomes will be available shortly. However, considering the phylogenic analysis by Maier and coworkers (2003) it would also be highly desirable to obtain genomic sequence information from a member of the genus *Uromyces*, the second largest genus among the rust fungi (see Sect. III).

Cytological, biochemical, and molecular work during the past five decades have mainly focused on five species of rust fungi: *P. graminis, P. triticina* (formerly *P. recondita* f. sp. *tritici*), *U. appendiculatus, U. fabae,* and *M. lini*. As already mentioned, *M. lini* and its host flax were used by Flor (1956) to demonstrate the gene for gene hypothesis. *U. appendiculatus* and *P. graminis* have been used in a number of cytological and physiological studies (Zhou et al. 1991; Leonard and Szabo 2005). Today molecular analyses of rust fungi mainly focus on *P. triticina* (Thara et al. 2003), *M. lini* (Catanzariti et al. 2006), and *U. fabae* (Jakupovic et al. 2006). Consequently, this chapter primarily focuses on work done using these organisms.

III. Phylogeny and Taxonomy

Like other man-made concepts of categorization, taxonomic placement and phylogenetic classification change over time as established methods improve and new methods are introduced. Traditionally, rust fungi are grouped together with smut fungi in the class *Teliomycetes* (Jülich 1981). However, recent molecular and ultrastructural data showed that rusts and smut fungi are only distantly related. Currently, a separation of three classes, namely *Urediniomycetes* (including the rust fungi), *Ustilaginomycetes*, and *Hymenomycetes* under the phylum *Basidiomycota* seems to be the best established classification (Swann et al. 1981; Cummins and Hiratsuka 2003). An important feature that distinguishes *Urediniomycetes* from other members of the *Basidiomycota* is the absence of the formation of clamp connections. Within this class of fungi nuclei in the growing hyphal tip divide conjugately, and as the daughter nuclei separate, a septum is formed to delimit two binucleate compartments, with the apical one continuing to elongate (Alexopoulos et al. 1996). Septum morphology is another characteristic to identify members of the *Urediniomycetes*. Septa are simple with a single open or plugged pore; a dolipore arrangement typical for other *Basidiomycota* is missing (Webster and Weber 2007). A third distinctive characteristic for members of the *Urediniomycetes* is the formation of transversely septated metabasidia from which basidiospores are formed laterally (Gäumann 1959).

Today the *Uredinales* are thought to comprise more than 100 genera and around 7000 species (Maier et al. 2003). These numbers correspond to about 75% of the genera and even 95% of the species of the class *Urediniomycetes*. Based on recent data the order *Uredinales* can be considered to be monophyletic (Swann et al. 1981). The order is divided into 13 families, each consisting of between three and 30 genera (Cummins and Hiratsuka 2003). Classic taxonomic classification is mainly based on spore and fruiting structure morphology, with a strong emphasis on teliospores morphology (usually two-celled for *Puccinia*, one-celled for *Uromyces* species; Cummins and Hiratsuka 2003). Some of these classifications, however, are controversial because the morphology of different spore types may leave some ambiguity with respect to final classification.

Rust fungi and their host plants are excellent examples of coevolution. Rusts infecting members of such old plant divisions as ferns or conifers are almost exclusively heteroecious and macrocyclic (see Sect. IV). Since the order *Uredinales* seems to be monophyletic (Swann et al. 1981), it can be inferred that the extraordinarily complex heteroecious macrocyclic life cycle evolved only once. Reductions seem to have occurred at many different stages of evolution (Laundon 1973). Wahl et al. (1984) discussed host–parasite coevolution for cereal rusts. In centers of coevolution, genes responsible for plant defense and genes for fungal virulence have accumulated. Redistribution of a host subsequently gave rise to independent evolution (Anikster 1984). This diversification may at least in part be responsible for some of the complications associated with rust taxonomy. It would therefore be highly desirable to scrutinize the classic taxonomical system based on morphological and physiological characters and amend/correct it using more DNA sequence data as they become available, in order to better define the phylogeny and taxonomy of rust fungi (Aime 2006).

In terms of number of species the *Pucciniaceae* are by far the largest among all rust families (Maier et al. 2007). Within this family, the genus *Puccinia*, with about 4000 species, and the genus *Uromyces*, with about 600 species, together represent almost two-thirds of all known rust species (Cummins and Hiratsuka 2003). While these two genera form a strongly supported group together with two more genera, the analysis by Maier and coworkers (2003) also suggests that these two genera are polyphyletic. This pioneering work on molecular phylogeny of rust fungi has lately been substantiated by two further molecular studies (Maier et al. 2007; van der Merwe et al. 2007). Findings from these studies indicate that some of the morphological characteristics, i.e. the number of cells per teliospore, may have arisen many times during evolution. Since certain of these characteristics were used in classic taxonomy, some species

may have been mislabeled. However, more work on this topic is needed before any taxonomic and nomenclatural changes should be considered.

Using physiological characters for taxonomical classification often does not allow unambiguous resolution down to the species level. Such taxa consisting of clusters of closely related, but reproductively isolated individuals are usually referred to as a species complex. Species complexes are a common phenomenon among rust fungi (Gäumann 1959). Great caution has therefore to be taken, whether a rust fungus is named sensu strictu (in a strict sense) according to the classical concept of a species, or sensu lato (in a broader sense) describing a species complex.

Most rusts can attack more than one host. *P. graminis* for example can infect at least 365 species of cereals and grasses (Anikster 1984). Such rust species are sometimes subdivided into more specialized categories, each designated a *forma specialis* (f. sp.; variety, specialized form). There are virtually no distinctive morphological characteristics for the *formae speciales*, and they are identified by determination of the host species. This type of specialization was first described in the 1890s by Eriksson and Henning, working on cereal rusts (Eriksson 1894; Eriksson and Henning 1896). *P. graminis* f. sp. *tritici* for example exhibits a host preference for wheat and barley, while *P. graminis* f. sp. *avenae* shows a preference for oat.

Within rust species or *formae speciales*, a further specialization is commonly observed. It is known that certain genotypes of a pathogen are able to attack only certain host cultivars. Such races of the pathogen are typically assigned a number in the order of their identification. The race concept is tightly linked to the virulence/avirulence pattern of rust fungi and the susceptibility/resistance pattern of their respective host plants according to the gene for gene hypothesis introduced by Flor (1955). It was shown that broadly virulent pathogens occur more frequently in highly resistant host populations, whereas avirulent pathogens dominate susceptible populations (Thrall and Burdon 2003). The non-random spatial distribution maintained despite high pathogen mobility implies that selection favors virulent races in resistant hosts and avirulent races in susceptible hosts. Physiological races were first described by Stakman and Piemeisal (1917), who established a first set of wheat cultivar differentials which allowed the identification of different *P. graminis* f. sp. *tritici* races. Now extended and refined, this system still provides the basis for modern plant breeding (Kolmer 1996).

IV. Life Cycle

Rusts have one of the most complex life cycles of all fungi (Littlefield 1981). In its complete form the cycle includes five different spore forms. The already complex cycle also exhibits a high degree of plasticity, generating many different variations (see below). To make things even more complicated, there is also some ambiguity in the literature about the designation of the different spore types and fruiting structures (sori). Table 4.1 provides an overview of the terminology used for the different spore types and fruiting structures, with the most commonly used terms printed in bold. In addition to the morphological classification system, Table 4.1 also lists the Roman numerals assigned to the different developmental stages used in the ontogenic classification system (Littlefield 1981; Alexopoulos et al. 1996; Webster and Weber 2007).

Figure 4.1 depicts the life cycle of *U. fabae* and also indicates the nuclear condition during the different stages. After overwintering on residual plant material, diploid teliospores germinate in the spring with a metabasidium. After meiosis, the latter produces four haploid basidiospores with two different mating types. These are ejected from the metabasidium by the aid of a drop of liquid (Buller's drop; Webster et al. 1995), and after landing on the leaf surface of a host plant, they germinate and produce monokaryotic infection structures. Pycnia are produced on the upper surface of the

Table 4.1. Terminology (and synonyms) of spores and fruiting structures of rust fungi; morphologial terminology and developmental classification according to ontogenic terminology. The most commonly used terms are given in *bold*

Spore	Fruiting structure	Developmental stage
Pycniospore	**Pycnium**	0
Pycnospore	Spermogonium	
Spermatium		
Aeciospore	**Aecium**	I
Aecidiospore	Aecidium	
	Aecidiosorus	
Urediospore	**Uredium**	II
Urediniospore	Uredinium	
Uredospore	Uredosorus	
Teliospore	**Telium**	III
Teleutospore	Teleutosorus	
Basidiospore	**Metabasidium**	IV
Sporidium	Basidium	
	Promycelium	

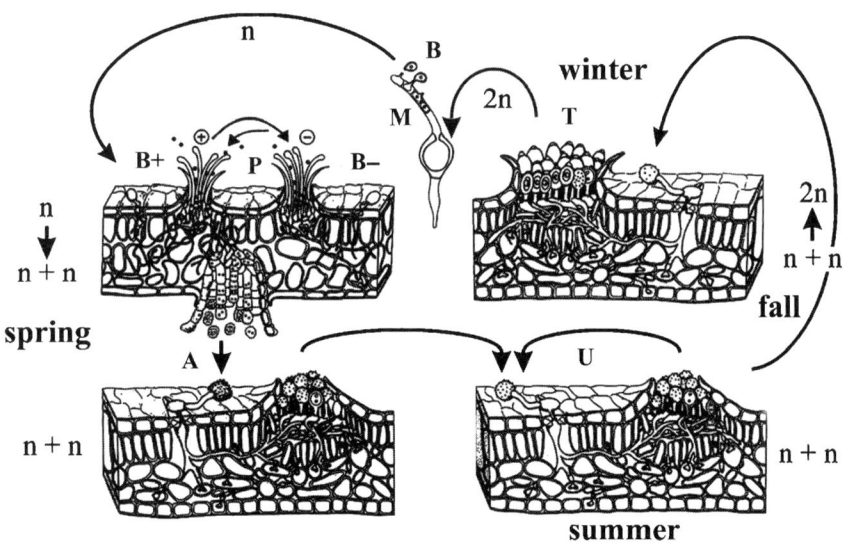

Fig. 4.1. Life cycle of *Uromyces fabae*. Overwintering diploid (*2n*) teliospores (*T*) germinate in the spring with a metabasidium (*M*) from which four haploid (*n*) basidiospores (*B*) of two mating types (+,−) are formed. Haploid pycniospores (*P*) are exchanged between pycnia of different mating types on the upper surface of a leaf. After spermatization dikaryotic (*n* + *n*) aeciospores (*A*) are formed in aecia at the lower surface of the leaf. Infecting aeciospores produce uredia from which dikaryotic urediospores (*U*) are formed. At the end of summer uredia differentiate into telia from which teliospores are formed and the cycle closes. Drawing taken from Voegele (2006)

leaf, which contain pycniospores and receptive hyphae. Pycniospores are exchanged between pycnia of different mating types (heterothallism), and after spermatization, dikaryotization occurs in aecial primordia. An aecium differentiates at the lower side of the leaf and dikaryotic aeciospores are produced. After landing on a leaf surface, these aeciospores germinate and form infection structures from which uredia which produce urediospores are formed. Urediospores are the major asexual spore form of rust fungi produced in massive amounts through repeated infection of host plants during the summer. In the fall, uredia differentiate into telia, the nuclei fuse during sporogenesis and single-celled, diploid teliospores develop for the winter, which closes the rust infection cycle.

Rusts capable of completing their entire life cycle on a single host species are called autoecious (de Bary 1865). Examples for such species are *U. fabae* on broad bean and *M. lini* on flax. Rust fungi requiring two host species in order to complete their life cycle are termed heteroecious (de Bary 1865). The two host species are typically well separated taxonomically. The classic example for an heteroecious rust is *P. graminis* which switches between cereals as main host and barberry as alternate host (Arthur 1962). Host alteration takes place after the aecial and telial stages. *P. graminis* occurs as pycnia and aecia on barberry and as uredia and telia on cereals. However, the term alternate host is used to denote either the pycnial/aecial host or the uredia/telia host and is usually applied to the host of lesser economic importance.

Not all rust fungi go through all five known spore forms. Rusts exhibiting all five spore forms are called macrocyclic. In so-called demicyclic rusts, the uredia stage (and sometimes the pycnial stage) is missing, and in microcyclic rust fungi usually only the pycnial and the telial stage are present (sometimes even only the telial stage). Since the aecial stage is missing in the latter case, all microcyclic rusts are necessarily autoecious. Macrocyclic and demicyclic rusts may be either autoecious or heteroecious. A more extensive description of the many variations of this topic can be found in the review by Petersen (1974) and the book by Cummins and Hiratsuka (2003).

V. Epidemiology

The different spore forms of rust fungi have different modes of dispersal. Pycniospores are released from their supporting cells into a viscous liquid and locally allocated by insects, splashing water, and contact among host plant organs (Littlefield 1981). Aeciospores are produced in tightly packed chains, released by the dissolution of intercalary cells, and aerially disseminated (Littlefield and Heath 1979). Teliospores may remain attached to the host organ they were produced on (Littlefield 1981). Alternatively, the pedicels on which teliospores are produced may break, and teliospores and attached pedicels are dispersed by the wind (Littlefield 1981). In any case, teliospores germinate to produce basidiospores, which are forcefully ejected from the metabasidium (involving Buller's drop; Webster et al. 1995) and then aerially dispersed (Littlefield 1981). Basidiospores are only suited for local dispersal since they desiccate rapidly.

Urediospores are the most important asexual spore form of most rust fungi. They are produced in enormous numbers through repeated infection of host plants for short- and long-range distribution during the vegetation period. Rust fungi are therefore typical "*r*-strategists" (Deising et al. 2002). For *P. graminis* f. sp. *tritici* for example, it was determined that a single uredium can produce about 600 urediospores day^{-1} (Eversmeyer and Kramer 2000). Even moderate infection can thus easily result in the production of 10^{12}–10^{13} urediospores day^{-1} ha^{-1} (Deising et al. 2002). While it is generally accepted that the spores are dry-dispersed by wind (Littlefield 1981) or carried by vectors (Wandeler and Bacher 2006), rain also seems to have a dramatic effect at least for local dissemination (Geagea et al. 1999). Spores can also be carried over distances of several hundreds or even thousands of kilometers by winds causing dissemination across or even between continents (Nagarajan and Singh 1990; Eversmeyer and Kramer 2000; Brown and Hovmøller 2002; Kolmer 2005). Annual long-distance transport of *P. graminis* occurs across the Great Plains in North America along the "Puccinia Path". Meteorological data support the idea that urediospores of *H. vastatrix* produced during a coffee rust epidemic in Angola in 1966 were carried across the Atlantic at an altitude of 1500–2000 m and deposited 5–7 days later over the coffee estates of Bahia, Brazil (Bowden et al. 1971; n.b. urediospores show a significant loss of viability only after five days; Nagarajan and Singh 1990). Similarly, urediospores could easily spread from South Africa to Australia in less then 5 days traveling at an altitude of 12 000 m (Nagarajan and Singh 1990). Deposition of spores may occur simply by sedimentation caused by gravity, or spores may be washed from the air during rainfall. Recently it was shown that, at least in the case of *P. pachyrhizi*, a rain event seems to be necessary to wash the spores from the air and cause infection (Barnes et al. 2006a; Krupa et al. 2006).

Upon successful infection the fungus colonizes the host tissue inter- and intracellularly (Fig. 4.2). However, symptoms do not become visible until several days thereafter (*sporulation phase* in Fig. 4.2). Typical symptoms include the spore-releasing fruiting structures breaking through the epidermis of the host plant. These structures and the released spores usually have a yellow, orange, red, or brownish coloration, eponymous for the disease. However, there are a number of other symptoms, like dwarfing and various tissue and organ malformations related to rust disease (Littlefield 1981). These are observed mainly when rust fungi spread systemically through the plant, which usually occurs after infection with monokaryotic basidiospores (Larous and Lösel 1993). Noteworthy examples of tissue malformation are the pseudoflower-inducing rust fungi (Roy 1993). Here the plant is inhibited from flowering. Instead, the host is induced by the fungus to form pseudoflowers which resemble true flowers in color and shape (Pfunder et al. 2001). Pseudoflower-inducing rust fungi are also a good example for the role insects play, at least in short-distance allocation (Pfunder and Roy 2000). Quantification of the pathogen within the infected host plant during the *parasitic phase* and *sporulation phase* (Fig. 4.2) proves quite difficult. Traditional methods mostly rely on visual methods, either scoring symptoms according to a macroscopically visible phenotype, or by using microscopical methods in order to estimate the fungal contribution to the total biomass of an infected plant (Winton et al. 2003; Mendgen, unpublished data). However, these methods cannot give an accurate measure of the fungal fraction at a given point of infection. Biochemical methods are mostly based on the quantification of the fungus-specific sterol ergosterol (Winton et al. 2003). Yet, ergosterol determination

cannot discriminate between different fungi and its content may vary between different species and even between the different developmental stages of a single organism (Zhao et al. 2005). Another biochemical method based on the quantification of chitin also has its limitations (Mayama et al. 1975). A versatile tool which allows species specific quantification is real time PCR (Higuchi et al. 1992). Using one of several modifications of the original method, Boyle et al. (2005) were able to quantify poplar rust caused by *Melampsora medusae* f. sp. *deltoidae* and *M. larici-populina*. Their estimates based on DNA as a template were in the order of 20% fungal contribution to the total DNA sample. However, estimates by Jakupovic et al. (2006) and our own quantifications (Voegele and Schmid, unpublished data) based on mRNA provided a considerably higher value, between 40% and 50% fungal contribution.

VI. Spore Germination and the Formation of Infection Structures

Telio- and pycniospores do not infect plants, whereas basidio-, aecio-, and urediospores do. Teliospores represent the final spore form of rust fungi and provide the main basis for their nomenclature (Mendgen 1984). They have mostly been studied using cytological techniques (Gold and Littlefield 1979; Mims and Thurston 1979; Mims 1981a; Anikster 1986). Pycniospores have also been studied primarily on an ultrastructural basis (Gold and Littlefield 1979; Gold et al. 1979). However, noteworthy are the measurements of nuclear DNA content in the early 1990s (Eilam et al. 1994). Pycniospores are important for the sexual reproduction of rust fungi (Craigie 1927). Aeciospores were also the subject of ultrastructural analysis (Mims 1981b). However, what is more important is the finding that aeciospores behave similar to urediospores, at least with respect to germination and response to topographical stimuli (Stark-Urnau and Mendgen 1993). Some studies have been performed regarding the nuclear DNA content of basidiospores (Eilam et al. 1992), their ultrastructure (Mims 1981a), and their derived infection structures (Kapooria 1971; Freytag et al. 1988; Gold and Mendgen 1991). Yet, the best studied rust spore form is the urediospore (Staples and Macko 1984, Deising et al. 1992). Almost all biochemical and all recent molecular studies are based on infection structures derived from urediospores (Mendgen et al. 1996, 2000; Hahn et al. 1997a; Hahn 2000; Voegele and Mendgen 2003; Struck et al. 2004a; Voegele 2006). The fact that infection structures from both basidio- and urediospores of *U. fabae*

penetration phase   **sporulation phase**
          parasitic phase

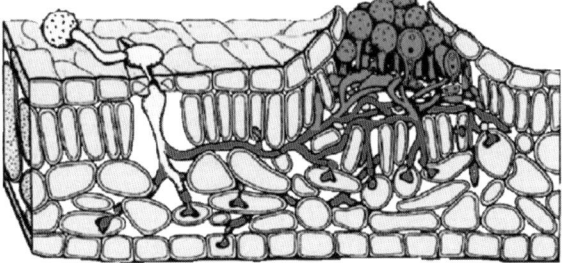

Fig. 4.2. Developmental phases of urediospore infection. Early infection structures of the *penetration phase*, structures of the *parasitic phase*, and structures of the *sporulation phase*. Drawing taken from Voegele (2006)

have been analyzed morphologically allows a comparison between mono- and dikaryotic infection structures on the same host plant (Fig. 4.3; Mendgen et al. 1996). Thick-walled, darkly pigmented and ornamented urediospores (Fig. 4.3A) germinate with a germ tube which differentiates into a well defined appressorium upon contact of the germ tube with a topographic signal of the correct magnitude (Hoch and Staples 1987; Hoch et al. 1987). A penetration hypha is formed at the base of the appressorium, which enters the leaf through the stomatal opening. A vesicle is formed within the stomatal cavity from which an infection hypha emerges. Upon contact with a mesophyll cell a haustorial mother cell is differentiated from which a haustorium is formed. Basidiospores (Fig. 4.3B) by contrast are smooth and thin-walled. There is no evidence for topographical signals involved in surface recognition, and infection structures like appressorium, vesicle, and haustorium are noticeably less differentiated. Moreover, the penetration mechanism seems to be completely different, since in this developmental stage the fungus enters the plant by direct penetration into epidermal cells. Further studies at the molecular level are needed

to determine the causes and consequences of these differences. It is noteworthy to mention that, in contrast to other rusts, urediospores of some *Phakopsora* species produce infection structures which penetrate the leaf surface directly (Bonde et al. 1976; Hoppe and Koch 1989).

VII. Features of Urediospore Infection

Urediospores are single-celled, thick-walled, hydrophobic, usually darkly pigmented, and carry spines on their surface (Woods and Beckett 1987). An important morphological feature used to distinguish different rust species is the number and position of germ pores on the surface (Gäumann 1959). Premature germination of spores, for example within uredia, is prevented by the presence of germination inhibitors (Wolf 1982). While methyl *cis*-3,4-dimethoxycinnamate has been identified in some rust species (Macko et al. 1970, 1971), there is also a large number of rusts where self-inhibitors were reported but the compounds could not be identified (Marte 1971; Macko et al. 1976). The time-frame within which these inhibitors are effective is restricted to the first 30 min after the initiation of hydration of the spore (Wolf 1982). There also seem to be endogenous germination stimulators. One of the first stimulators to be identified and one of the most widely distributed is pelargonaldehyde (*n*-nonanal; French and Weintraub 1957). However, many more chemically unrelated compounds were also shown to have stimulatory effects (French 1992). There is also evidence for exogenous stimulators and inhibitors produced by the host plant which might contribute to the regulation of germination (Gold and Mendgen 1983; Staples and Hoch 1997). Mendgen et al. (2006) were able to show that *U. fabae* stimulates the emission of specific volatiles by its host which in turn control the differentiation of infection structures (Fig. 4.4).

Fully developed urediospores are almost completely dehydrated upon release from uredia, which gives them an irregular shape (Clement et al. 1998). Only upon hydration do spores adopt their typically round to ellipsoid form. Although dry urediospores hydrate rapidly, their surface is non-wettable (Clement et al. 1994). It is this hydrophobicity which seems to be responsible for the initial adhesion of spores to the host surface (Clement et al. 1993b). This interaction also seems to involve the spines. The initial contact is quickly

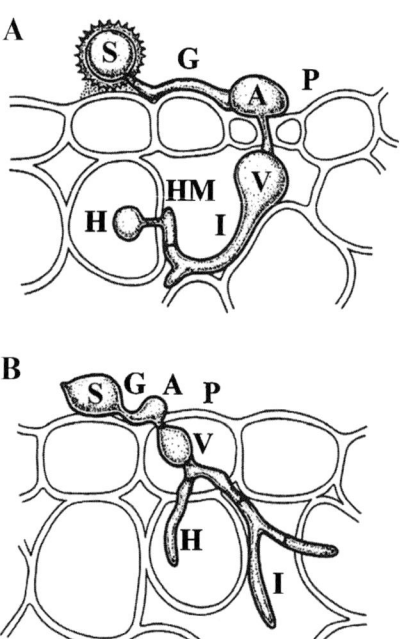

Fig. 4.3. Infection structures derived from: urediospores (**A**) and basidiospores (**B**). *A* appressorium, *G* germ tube, *H* haustorium, *HM* haustorial mother cell, *I* infection hypha, *P* penetration hypha, *S* spore, *V* vesicle. Drawing taken from Voegele (2006)

Fig. 4.4. Control of rust disease by leaf fragrances. Leaf fragrances can induce or suppress haustorium differentiation and may be used to control rust disease. Here, soybean was inoculated with *P. pachyrhizi*. The right half of the leaf was treated with 10 μl farnesyl acetate according to Mendgen et al. (2006)

followed by the production of an extracellular matrix consisting of low-molecular-weight carbohydrates and glycosylated polypeptides (Clement et al. 1993a). This matrix seems to originate from solubilization of surface components and lysis of the germ pore plug. The next step is the formation of an adhesion pad underneath the attached spore. Both seem to be exclusively of fungal origin, since they are also formed on artificial surfaces (Deising et al. 1992). Cutinases and esterases seem to be involved in the adhesion process, since autoclaved spores or spores treated with esterase inhibitors form an adhesion pad but fail to adhere (Deising et al. 1992).

Aside from liquid water or high humidity (Clement et al. 1997), light and temperature are also important for germination. For *U. fabae* a period of at least 40 min of darkness is required to induce germination (Joseph and Hering 1997). Especially harmful to germination seem to be wavelengths in the far red. Urediospores of *U. fabae* germinate in a range between 5 °C and 26 °C, with the optimal germination temperature being 20 °C (Joseph and Hering 1997). Given the correct physical parameters the spore germinates on almost any surface, indicating that no additional signals are needed to induce germination. Spores even germinate on a water surface or submerged in water if proper aeration is provided (Struck et al. 1996). However, no infection structures are formed in the absence of a structured surface. The cytoplasm of the spore moves into the growing germ tube as the developing germ tube

meanders across the surface attached to it via matrix-like material (Hoch et al. 1987; Clement et al. 1994). Germ tubes are tube-like structures with a hemispherical or hemi-ellipsoidal apical region at which growth occurs (Wessels 1993). Vesicles originating from the Golgi migrate to the apex and accumulate in the so-called Spitzenkörper (Mendgen et al. 1996). Within the apex, the Spitzenkörper is shifted towards the substrate, which results in a sort of "nose down" growth of the hypha, which might aid in the recognition of physical stimuli (Hoffmann and Mendgen 1998).

In order to produce infection structures further signals are needed for differentiation. It was shown for a number of rust species that a topographical signal is needed for the differentiation of an appressorium (Wynn 1976; Hoch et al. 1987; Allen et al. 1991; Read et al. 1997). *U. appendiculatus* and *Uromyces vignae* were found to form appressoria if a ridge of 0.4–0.8 μm in height is provided (Allen et al. 1991). The values determined roughly correspond to the height of the stomatal guard cell lips of the respective host plants. Other studies involving *Puccinia hordei* and different accessions of the host *Hordeum chinense* showed that a failure to differentiate appressoria was due to the presence of a prominent layer of wax over the guard cells of particular lines probably obscuring the relevant topographic signals (Vaz Patto and Niks 2001). Studies involving *Uromyces striatus* on artificial surfaces indicated a range of 0.1–1.2 μm ridge height as capable of inducing appressorium formation (Kemen et al. 2005a). This wider range might be correlated with the broad host range of *U. striatus*, which comprises at least 141 species and subspecies from the tribes Trifolieae, Cicereae, and Vicieae. In the membrane fraction derived from *U. appendiculatus* germ tubes a mechanosensitive channel was identified which may be involved in the transduction of the topographic signal into a differentiation response (Zhou et al. 1991). There is also some data available about the involvement of the cytoskeleton in thigmotropic signaling and appressorium associated differentiation processes (Bourett et al. 1987; Kwon et al. 1991). The cytoplasm is transferred to the growing appressorium and the vacuolated germ tube is separated from the newly developed structure by a septum. The differentiation of the appressorium coincides with the release of a number of lytic enzymes (Fig. 4.5; Deising et al. 1995b). At the base of the appressorium a penetration hypha is formed (Terhune et al. 1993). For *U. appendiculatus* a turgor pressure of 0.35 MPa has been reported. This is considerably

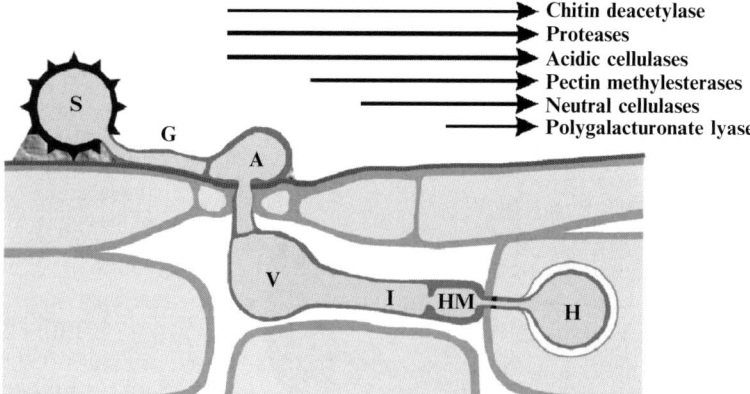

Fig. 4.5. Lytic enzymes in early dikaryotic infection structures. *A* appressorium, *G* germ tube, *H* haustorium, *HM* haustorial mother cell, *I* infection hypha, *P* penetration hypha, *S* spore, *V* vesicle. Drawing modified from Mendgen and Deising (1993)

less than the turgor pressure reported for example for appressoria of *Magnaporthe grisea*, but still enough to distort artificial surfaces or stomatal guard cell lips (Terhune et al. 1993). Within the stomatal cavity a substomatal vesicle is formed which is separated from the appressorium and the penetration hypha by a septum (Kapooria and Mendgen 1985). Only upon contact with a leaf mesophyll cell a haustorial mother cell is differentiated, which is again separated from the infection hypha by a septum. Again most of the cytoplasm moves into the differentiating haustorial mother cell and earlier structures become more or less vacuolated. Similar to appressoria, haustorial mother cells have a thick, multilayered wall that attaches firmly to the host cell wall and forms a penetration hypha to invade the host cell (Heath 1997). The haustorial mother cell therefore functionally resembles an appressorium. However, it remains to be elucidated whether the functional similarity extends to the molecular level. Results from research on the penetration process support the idea that pressure and the controlled secretion of lytic enzymes act together to prepare successful penetration of the host cell wall (Hahn et al. 1997a).

During the penetration phase (Fig. 4.2), infection structures up to the haustorial mother cell can be induced in vitro by germinating spores on artificial surfaces such as collodion membranes (Dickinson 1949), or on structured polyethylene sheets (Staples et al. 1983). While high humidity and the correct topographical signal seem to be sufficient for legume rusts to efficiently produce infection structures in vitro, the situation seems to be more complex for rust fungi infecting monocotyledonous host plants (Wiethölter et al. 2003). A number of physical and chemical stimuli such as a mild heat shock (Maheshwari et al.

1967), organic compounds (Macko et al. 1978), host epicuticular waxes (Grambow 1977), leaf volatiles (Grambow 1977), or combinations thereof (Collins et al. 2001; Wiethölter et al. 2003) have been reported to trigger the sequential in vitro development of appressoria, substomatal vesicles, infection hyphae, and haustorial mother cells of *P. graminis*. Although there are some reports about the formation of haustoria in vitro (Heath 1989, 1990a, b; Mendgen et al. 2006), true functional haustoria and structures of the parasitic phase and sporulation phase (Fig. 4.2) are only formed in planta.

The fact that haustoria and structures of the parasitic phase and sporulation phase (Fig. 4.2) are only formed in planta makes it extremely difficult to analyze processes involving these structures at a molecular level. Although conditions for axenic cultures have been established for some biotrophic fungi (Maclean 1982; Fasters et al. 1993), most of the economically important biotrophic parasites remain non-culturable, at least not to a point equivalent to the biotrophic phase (Mendgen and Hahn 2002). Studies by Heath (1990b) indicated that it is not a mere lack of specific nutrients that prevent haustoria to be formed in vitro, but rather a lack of appropriate signals from the host plant. The pathogen and the host together seem to form a new entity, the aegricorpus (disease body; Loeghering 1984). A study of the pathogen (axenic culture) or the host alone can therefore not explain the physiology of the diseased plant.

VIII. Structural Aspects of the Dikaryotic Haustorium

The haustorium represents one of the hallmarks of obligate biotrophic parasites. These structures have generated the interest of plant pathologists ever since their first description by Zanardini about 150 years ago (von Mohl 1853). When naming these structures [Latin: *haurire* (*haurio, hausi, haustum*), to drink, to draw] de Bary (1863) proposed one of the possible functions for haustoria – the uptake of nutrients from the host. However, until recently there was evidence for an involvement of haustoria in nutrient uptake for powdery mildew fungi (Ascomycota) only (for a review, see Hall and Williams 2000).

The dikaryotic rust haustorium develops from the haustorial mother cell with a slender neck and a haustorial body that forms distally to the neck (Heath and Skalamera 1997). During formation of the haustorium the cell wall of the host cell is breeched. The expanding haustorium invaginates the host plasma membrane and new membrane is probably synthesized. There is some evidence that the membrane of the host enclosing the haustorial body, the so-called extrahaustorial membrane, is modified and therefore no longer resembles a conventional plant plasma membrane. Harder and Chong (1991) summarized results obtained by freeze fracture electron microscopy with bean rust and oat crown rust. In both interactions the extrahaustorial membrane lacks intramembranous particles and exhibits a dramatic reduction of sterols (Harder and Mendgen 1982). Cytochemical studies on powdery mildew haustoria (Gay et al. 1987; Manners 1989) and later work on rust haustoria (Baka et al. 1995) suggested that the extrahaustorial membrane lacks ATPase activity. This implies that there is no control over solute fluxes from the host cell. The neck region of the haustorium is characterized by electron-dense material apparently joining the two plasma membranes of host and parasite (Harder and Chong 1984). This "neckband" (Fig. 4.6) seals the extrahaustorial matrix against the bulk apoplast, not unlike the Casparian strip in the endodermis (Heath 1976). The haustorium is therefore not truly intracellular, it remains outside the physiological barrier of the host cell (Fig. 4.6). With the development of the haustorial body, a zone of separation between the plasma membranes of parasite and host is formed. It is composed of the fungal cell wall and the extrahaustorial matrix (Hahn et al. 1997a). It seems noteworthy to mention that, while normally a cell wall is formed from the plant cell cytoplasmic membrane, no such structure is formed from the extrahaustorial membrane.

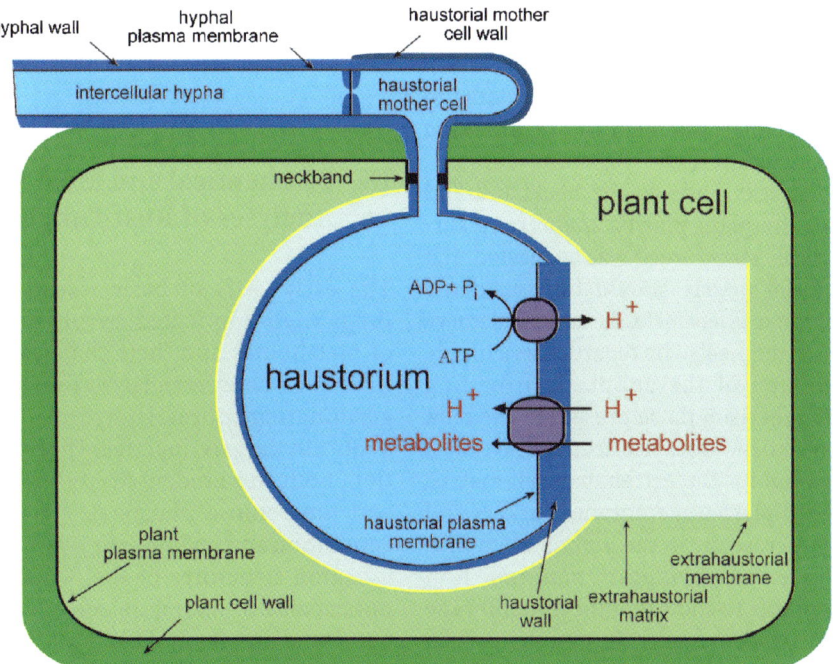

Fig. 4.6. Schematic representation of a dikaryotic rust haustorium. Structures derived from the fungus are depicted in *blue*, structures contributed by the plant are shown in *green*. The extrahaustorial matrix is shown in *light blue* and the extrahaustorial membrane in *yellow*. Drawing taken from Voegele (2006)

The extrahaustorial matrix resembles an amorphous mixture of components, mainly carbohydrates and proteins, partly of fungal but primarily of plant origin (Harder and Chong 1991), and provides the most intimate contact between host and parasite. This view is supported by the cytological analysis of hemibiotrophic parasites. The initial biotrophic phase of some hemibiotrophs, like for example *Colletotrichum* spp., is also characterized by the presence of a narrow contact zone between host and parasite (Perfect and Green 2001; Mendgen and Hahn 2002). Upon switching to necrotrophic growth the host plasma membrane surrounding the hyphae disintegrates and parasitic growth continues with narrower unsheathed hyphae. It therefore seems likely that this zone of separation plays an important role in the maintenance of the biotrophic lifestyle. Undoubtedly, the extrahaustorial matrix represents a formidable trading place for the exchange of nutrients and information between the host and the fungus (Heath and Skalamera 1997). In a recent study on *Puccinia hemerocallidinis* Mims and coworkers (2002) showed long tubular extensions of the extrahaustorial membrane contiguous with the extrahaustorial matrix. Similar structures were already described by Stark-Urnau and Mendgen (1995) for monokaryotic haustoria (haustoria derived from basidiospore infection) of *U. vignae*. These structures reach far into the host cytoplasm and exhibit coated vesicles at their tip. However, it remains to be shown whether there is any kind of trafficking linked to these structures.

Based on the seal made by the neckband and the presence of the plant plasma membrane surrounding the whole structure, it was suggested that the extrahaustorial matrix should be considered a symplastic compartment (Heath and Skalamera 1997). However, it might also be regarded as a highly specialized portion of the apoplast, providing conditions different from those present in the bulk apoplast. The neckband does not seem to be only a line of demarcation for the extrahaustorial matrix. Using GFP-tagged plasma membrane markers and laser scanning microscopy in the pathosystem *Erysiphe cichoracearum/Arabidopsis thaliana*, Koh et al. (2005) were able to show that these membrane proteins were excluded from the extrahaustorial membrane and accumulated in rings around the neckband. Although done on a different system, this work corroborates earlier findings that the composition of the extrahaustorial membrane

seems to be different from that of a conventional plasma membrane.

A further special feature of the haustorium is a highly dilated ER, which exhibits a shift from mostly parallel sheets to a predominantly tubular-vesicular network (Welter et al. 1988; Mims et al. 2002). Based on the distribution of ER markers such as BIP- and HDEL-containing proteins, this network appears to be a functional sub-compartment of the ER (Bachem and Mendgen 1995). The highly increased ER may indicate an enhanced synthesis of secreted proteins. In addition, haustoria may also contain more than two nuclei (Chong et al. 1992).

Analysis of the potential role(s) of rust haustoria has been hampered by the fact that haustoria are exclusively formed in planta and that their isolation encountered numerous problems (Bushnell 1972). As a result, haustoria have been mostly studied using cytological techniques (Harder and Chong 1991). The introduction of biochemistry and molecular biology into the field of phytopathology opened up a new dimension to investigate the role(s) of haustoria (see below). A picture is beginning to emerge indicating that haustoria do not serve only in nutrient uptake – the task postulated for these structures ever since their discovery. In fact, they seem to perform enormous biosynthetic duties and are thought to be engaged in the suppression of host defense responses and in redirecting and/or reprogramming the host's metabolic flow.

IX. Biochemical and Molecular Analyses of Rust Fungi

The early years of basic research involving rust fungi were dominated by physiological analyses of metabolites and their changes in the course of infection. One aspect analyzed was the metabolism of germinating spores. Unhydrated spores are metabolically largely inactive. However, upon hydration, both the Emden–Mayerhof–Parnas pathway and the pentose phosphate pathway seem to be active (Shu and Ledingham 1956). Apparently, early infection structures of the penetration phase are not capable of taking up considerable amounts of nutrients (Staples and Macko 1984). Therefore, the fungus relies largely on metabolites stored within the urediospore until the first haustorium is formed and contact is made to the rich resources of the host plant. This manifests itself in the fact that the cytoplasm and all its content is always kept close to the growing hyphal tip, while older structures are more or less vacuolated and separated by septum formation. Some studies indicate that it is mainly lipids which are utilized as nutrients during the germination process (Langenbach and Knoche 1971a,b). Re-synthesis of these and further compounds occurs after a period of degradation. By contrast, using *P. graminis* as a model, Daly and coworkers (1967) found that lipids and carbohydrate, mainly in the form of polyols, were the metabolites primarily used during germination. Accumulation of acyclic polyols is a common phenomenon among fungi (Lewis and Smith 1967; Jennings 1984), and there are a number of studies indicating that mannitol, D-arabitol, and/or D-sorbitol are present in urediospores of rusts (Lewis and Smith 1967; Maclean and Scott 1976; Maclean 1982; Manners et al. 1982, 1984). Recent research on this topic sheds a new light on the roles these polyols may play in pathogenesis (Link et al. 2005; Voegele et al. 2005; Voegele 2006; see below). Another aspect of this early research was the analyses of nutrient fluxes between host and parasite. These approaches were based on feeding experiments involving radioactive tracer substances. Mendgen (1979, 1981) for example employed ^{3}H-labeled amino acids using *Uromyces* spp. These experiments gave indirect evidence for a role of haustoria in nutrient uptake without providing conclusive proof.

Biochemical analyses initially focused on structures that could be generated in vitro and almost exclusively specialized on *U. fabae* as a model (Deising et al. 1991, 1995b). Acidic and neutral cellulase (Heiler et al. 1993), extracellular protease (Rauscher et al. 1995), chitin deacetylase (Deising and Siegrist 1995), pectin esterase (Frittrang et al. 1992), pectin methylesterase (Deising et al. 1995a), neutral cellulase (Heiler et al. 1993), and polygalacturonate lyase (Deising et al. 1995a) activities were found (Fig. 4.5). No significant protein secretion could be found before the onset of appressorium formation, protein secretion continuously increased thereafter, and reached a maximum upon the differentiation of infection hyphae and haustorial mother cells (Hahn et al. 1997a). Based on the biochemical characteristics of the enzymes identified, a model was suggested for obligate biotrophs explaining the highly coordinate action of these enzymes in a localized breakdown of the plant cell wall (Deising et al. 1995b). In addition to host cell wall-degrading enzymes, two proteinaceous elicitors of plant

defense responses were purified and characterized from *U. vignae* (D'Silva and Heath 1997). Molecular approaches analyzing differentiation regulated gene expression started as early as 1989 with the isolation and characterization of *INF24* from *U. appendiculatus* (Bhairi et al. 1989). This work was soon followed by the characterization of another differentiation-specific gene, *INF56* (Xuei et al. 1992, 1993), and the analysis of a larger set of stage specifically regulated genes from *P. graminis* (Liu et al. 1993).

Much of the recent biochemical and molecular work on rust fungi involves haustoria. However, biochemical and molecular work on haustoria is greatly hindered by the fact that haustoria are only formed in planta. During the early 1990s several methods were introduced to isolate haustoria from infected plant tissue for further analysis (Hahn and Mendgen 1992; Tiburzy et al. 1992; Cantrill and Deverall 1993). While some of these methods were too laborious and inefficient, the chromatographic method developed by Hahn and Mendgen (1992) proved a milestone in the research involving rust haustoria. The method is based on a selective binding of oligosaccharides present in the haustorial wall to the lectin concanavalin A immobilized on a Sepharose 6MB backbone. Repeated cycling of cell extracts of infected leaves yielded considerable quantities of highly enriched haustoria. This method provided the basis for biochemical and molecular analyses of rust haustoria.

Using this method, a number of genes preferentially or exclusively expressed in haustoria, so-called in planta induced genes (*PIG*s), were identified (Hahn and Mendgen 1997). Two of the most abundant genes in a haustorial cDNA library encode enzymes involved in vitamin B1 synthesis (Hahn and Mendgen 1997). *THI1* and *THI2* together make up about 5% of the total transcripts in haustoria. Vitamin B1 is a co-factor required for the activity of several enzymes of the central carbon metabolism (Sohn et al. 2000). Therefore, haustoria can be considered as power plants providing essential nutrients through de novo synthesis.

Other work on *U. fabae* revealed an increased plasma membrane H$^+$-ATPase activity for haustorial membranes compared to membranes isolated from spores and germlings (Struck et al. 1996, 1998). The proton gradient generated by this ATPase was suggested to drive secondary active transport systems engaged in nutrient uptake by the parasite (Hahn et al. 1997a). The finding that some of the *PIG*s encoded putative secondary transporters for amino acids was further evidence for a special role of haustoria in nutrient uptake (Hahn and Mendgen 1997; Hahn et al. 1997a, b). However, while an exclusive localization of AAT2p in haustoria could be shown, no transport activity could be detected (Mendgen et al. 2000). AAT1p was characterized as a broad-specificity amino acid secondary active transporter with a main specificity for L-histidine and L-lysine, but immunolocalization data could not be provided (Struck et al. 2002). AAT3p was shown to exhibit substrate preference for L-leucine and the sulfur-containing amino acids L-methionine and L-cysteine (Struck et al. 2004b). AAT1p and AAT3p are clearly energized by the proton-motive force and show a preference for amino acids present in low abundance in infected leaves (Struck et al. 2004b). Taken together, it seems that amino acid uptake in *U. fabae* may not be limited to haustoria. By contrast, hexose uptake seems to proceed exclusively via haustoria (Voegele et al. 2001). HXT1p was localized preferentially at the tip of monokaryotic haustoria (Voegele and Mendgen 2003) and in the periphery of the body of dikaryotic haustoria (Voegele et al. 2001). No specific labeling was found in intercellular hyphae. Neither nested PCR, nor genomic Southern blot analyses under low stringency conditions yielded evidence for additional hexose transporters present in *U. fabae* in any of the developmental stages tested (Voegele et al. 2001). Heterologous expression in yeast and *Xenopus* oocytes revealed that HXT1p is a proton-motive force-driven monosaccharide transporter. The transporter exhibits specificity for D-glucose, 2-deoxy-D-glucose, D-fructose and D-mannose, with increasing K_m values in this order (Voegele et al. 2001). This work provided the first conclusive evidence that rust haustoria are indeed nutrient uptake organs. Taken together, these data indicate that *U. fabae* makes use of several strategies to cover its nutritional demands. While amino acids seem to be taken up via both haustoria and intercellular hyphae, uptake of carbohydrates seems to be limited to haustoria. Substrate translocation is executed by secondary active transport systems which allow direct coupling of transport to the proton gradient established by the H^+-ATPase (Fig. 4.6).

Elucidating the mechanism and specificity of carbohydrate uptake in *U. fabae* provided an important advance in understanding the biotrophic relationship, but at the same time put forward a series of new challenging questions (Szabo and Bushnell 2001). Focusing on carbohydrate

metabolism, we identified a β-glucosidase (EC 3.2.1.21; Haerter and Voegele 2004) and an invertase (EC 3.2.1.26; Voegele et al. 2006) in *U. fabae*. Both enzymes could contribute substrates for the hexose transporter; however, other roles are also possible (see below). In the lumen of haustoria we identified two alcohol dehydrogenases. One NADP-dependent mannitol dehydrogenase (MAD1p; EC 1.1.1.138; Voegele et al. 2005), and a novel enzyme, an NADP-dependent D-arabitol dehydrogenase (ARD1p; EC 1.1.1.287; Link et al. 2005). MAD1p seems to be responsible for the formation of mannitol from D-fructose in haustoria. Although apparently not made in urediospores, MAD1p seems to be deposited there together with large amounts of mannitol. Assuming spores have a water content of 20%, the concentration of mannitol found in spores is close to its solubility level. The polyol disappeared rapidly from spores during germination indicating a role of this polyol in carbon storage. While there is evidence from other systems that lipids and proteins constitute the major substrates during spore germination (Shu et al. 1954; Solomon et al. 2003), utilizing the pool of mannitol first would enable a quick start of glycolysis, since the conversion of mannitol to D-fructose is a single enzyme step. At the same time, oxidation of mannitol to D-fructose would provide reducing power for anabolic processes. D-arabitol is most likely produced in haustoria by the action of ARD1p from D-ribulose and D-xylulose in an NADP-dependent reaction (Link et al. 2005). The coupling of NADP reduction to D-arabitol oxidation constitutes a novel enzymatic mechanism. Although D-arabitol is also deposited in spores and rapidly consumed during germination, no ARD1p could be detected in spores. Most likely utilization of D-arabitol in spores occurs via another enzymatic pathway. Aside from serving as carbohydrate storage compounds, there is evidence that both polyols have a role in the suppression of host defenses (see below).

The original analysis of *U. fabae* PIGs by Hahn and Mendgen (1997) was considerably extended by further haustorial EST – and microarray analysis (Jakupovic et al. 2006). The authors found very strong in planta expression for two PIGs encoding putative metallothioneins. Furthermore, several genes involved in ribosome biogenesis and translation, glycolysis, amino acid metabolism, stress response, and detoxification showed an increased expression in the parasitic mycelium. These data indicate a strong shift in gene expression in *U. fabae* between the penetration phase and parasitic phase (Fig. 4.2) and provide the basis for future analyses of the metabolism of *U. fabae*. Similar analyses involving both host and parasite genes were performed using barley plants infected and *P. triticina*. While Zhang and coworkers (2003) used the AFLP technique and Northern blot and RT-PCR, Thara and coworkers (2003) used a suppression subtractive hybridization approach and differential expression analysis. Analysis of the

M. larici-populina/poplar interaction also included a comparison of compatible and incompatible host–parasite interactions (Rinaldi et al. 2007). A different, yet interesting approach is the EST analysis of germinating urediospores from *P. pachyrhizi* (Posada-Buitrago and Frederick 2005). The molecular response of soybean to *P. pachyrhizi* infection was recently analyzed by the groups of Whitham and Baum (van de Mortel et al. 2007). These data will add considerably to the information which will become available upon completion of the *P. pachyrhizi* genome sequence. Genomic sequencing projects for *M. larici-populina* and *P. graminis* f. sp. *tritici* have already been completed. The sequence data and the possibility of comparing the different genomes will add a new impetus to rust research.

Recently, proteomics approaches were also introduced to rust fungi. Cooper et al. (2006) analyzed proteins from urediospores of *U. appendiculatus* and indicated similar upcoming analyses for the germ tube stage and infection structures. Rampitsch et al. (2006) have started to analyze the proteome of *P. triticina* during its interaction with its host. Integration of results from such approaches with array, EST, and genomic data will provide useful information to understand the molecular details of obligate biotrophy.

A serious drawback in research involving rust fungi remains the lack of a system for the stable transformation of obligate biotrophs. There are a number of reports on the transient transformation of rust fungi involving either microinjection of anti-sense oligonucleotides (Barja et al. 1998) or particle bombardment using the β-glucoronidase (GUS) gene as a color marker (Bhairi and Staples 1992; Li et al. 1993; Schillberg et al. 2000). However, all these studies used infection structures grown in vitro, and although in the latter cases stable transformation may have been achieved, no propagules could be recovered due to the failure of rust fungi to reach the sporulation phase in vitro. Currently, there is a promising approach using insertional mutagenesis into avirulence genes and selection for infection of *P. triticina* using wheat lines resistant to the wild type (Webb et al. 2006). Yet, a convincing demonstration of the stability of transformation through repeated spore cycling and a recovery of the plasmid marker remains to be provided. Based on the initial work by Wirsel et al. (2004), we recently started a different approach using the introduction of single point mutations in the genes encoding β-tubulin (*TBB1*) and the Fe-S-subunit of succinate dehydro-

genase (*SucDH1*) and in planta selection of potential transformants using the fungicides benomyl and carboxin (Voegele et al, unpublished data). In addition, a variety of different color markers were introduced into the plasmids used for transformation in order to visualize successful transformation events. Expression of the transgenes is driven by regulatory elements derived from *U. fabae* and DNA delivery is either accomplished using biolistics or *Agrobacterium tumefaciens*-mediated transformation.

The increasing availability of sequence data paired with the prospects of a stable transformation system will once again open new vistas to research focusing on the molecular principles underlying the obligate biotrophic lifestyle.

X. Suppression of Host Defenses

The establishment of biotrophy requires the evasion or suppression of host defense reactions. Rust fungi seem to have evolved a number of mechanisms to avoid recognition through host surveillance systems. Analyses of cell wall components of early infection structures, for example, indicated the most obvious differences between germlings and appressoria, which are outside the plant tissue and stained by the chitin-specific lectin WGA, and infection structures produced inside, which are not or only weakly stained by WGA (Kapooria and Mendgen 1985; Freytag and Mendgen 1991a). This observation can be explained either by the masking of chitin (Freytag and Mendgen 1991b) or by the conversion of chitin to chitosan via chitin deacetylase (Deising and Siegrist 1995; El Gueddari et al. 2002).

The β-glucosidase identified might also play a role in the suppression of host defenses. The protein shows high homology to other fungal β-glucosidases involved in the detoxification of saponins (Haerter and Voegele 2004). It is therefore possible that BGL1p has additional or alternative functions other than providing substrate for HXT1p. There is also evidence that mannitol and D-arabitol are released from the fungal mycelium into the apoplast in significant amounts (Link et al. 2005; Voegele et al. 2005). Results from mammalian (Chaturvedi et al. 1996) and plant pathosystems (Jennings et al. 2002) indicate that polyols, especially mannitol, can effectively suppress host defense responses involving reactive oxygen species. The concentrations of mannitol and D-arabitol found in infected *V. faba* tissue have been shown to be sufficient to effectively quench reactive oxygen species (Link et al. 2005; Voegele et al. 2005).

Differences in the morphology of extrahaustorial membranes produced by *P. graminis* or *Puccinia coronata* on oat suggest that formation of the fine structure of the haustorial host–parasite interface is under the control of species-specific signals from the fungus (Harder and Chong 1991). Such signals may include suppressors which have been implicated in maintaining basic compatibility between the parasite and its host plants (Bushnell and Rowell 1981). Evidence for such suppressors comes from a phenomenon called induced susceptibility. French bean tissue already infected by *U. vignae* supported additional infections by several non-host pathogens (Fernandez and Heath 1991). Suppressors for plant defense responses have been described, but they are either poorly characterized or non-proteinaceous (Basse et al. 1992; Knogge 1997; Moerschbacher et al. 1999).

A very active field of research today is the analysis of proteins secreted by a pathogen. These proteins include virulence factors, toxins, or avirulence gene products, which are now combined under the term effectors (Kamoun 2006). Such effector molecules are thought to manipulate host cell structure and/or function thereby facilitating infection and/or triggering defense responses. They might be released by the parasite into the apoplast, or alternatively they might be transferred into the host cell. Transfer of effectors into the host cell is especially interesting in the view of taking direct influence on host metabolism. Because of the intimate contact between host and parasite around haustoria (Fig. 4.7), these structures resemble the ideal location for such a transfer. Recent work on

the *M. lini/L. usitatissimum* pathosystem by the groups of Jeff Ellis and Peter Dodds identified a number of haustorium-specific secreted proteins for some of which a direct interaction with the corresponding host resistance gene products could be shown (for a review, see Ellis et al. 2007a; see Sect. XIII). This work nicely confirms the gene for gene hypothesis put forward by Flor (1955, 1956) more than 60 years ago at the molecular level. However, it has to be kept in mind that the interaction of avirulence gene products and resistance gene products results in an incompatible interaction, in other words a failure of the pathogen to establish an infection. While this is certainly an interesting aspect with respect to the basic understanding of resistance reactions and the identification of new avirulence gene/resistance gene combinations and therefore also advantageous for breeders, this situation does not reflect the true obligate biotrophic lifestyle, which is based on a long-lasting interaction of host and parasite. Recently, Kemen and coworkers (2005b) were able to show that one of the *PIG*s identified by Hahn and Mendgen (1997) is not only secreted into the extrahaustorial matrix as expected from its targeting sequences, but is further transferred to the host cell cytoplasm. However, it remains to be shown what the functions and targets of rust transferred protein 1 (RTP1p) are and how this and other effectors are translocated into the host cell.

It now seems well established that in eukaryotes, like in their bacterial counterparts, effectors can be directly transferred into the targeted host cell (Catanzariti et al. 2006, 2007). However, the mechanism by which this transfer is achieved still remains enigmatic (Ellis et al. 2006). For bacterial pathogens, specialized type III secretion systems (T3SS) have been shown to be implicated in the transfer of effectors directly into the target cells (Ghosh 2004; Mota et al. 2005). Yet, in eukaryotic pathogens no such "molecular syringes" could be identified. A clue as to how such a transfer could be accomplished comes from the identification of specific targeting signals in oomycete plant pathogens (Rehmany et al. 2005). This RXLR motif is conserved in all known avirulence proteins of oomycetes (Kamoun 2006) and is reminiscent of a host-targeting signal in malaria parasites (*Plasmodium* sp.) that is required for translocation of proteins into the cytoplasm of host cells (Hiller et al. 2004). However, such a consensus translocation signal has so far not been identified in rust effectors. Furthermore,

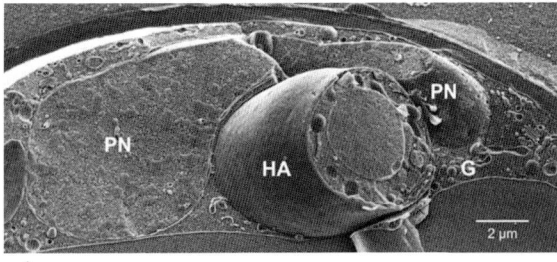

Fig. 4.7. Tight association of haustorium and plant cell nucleus. Fracture through a high-pressure frozen broad bean cell (*V. faba*) revealing the haustorium (*HA*) of *U. fabae* as seen with a cryo-scanning electron microscope. The haustorium is closely surrounded by the plant nucleus (*PN*) and Golgi bodies (*G*) of the host cell (Kemen and Mendgen, unpublished data).

the mode of translocation of any of the known effectors is still enigmatic. There are a couple of possible routes an effector molecule may take when secreted from the haustorium. These include: *I* direct membrane transfer, *II* protein-mediated translocation,*III* endosomal transfer (possibly involving the tubular vesicular structures), and *IV* retrograde transport through the host Golgi system (Fig. 4.8). Thus not only will it be interesting to determine the function and potential targets of these effectors, but it will also be motivating to clarify the transfer mechanisms for these proteins.

XI. Host Responses to Rust Infection

Successful rust infection leads to profound changes in host plant metabolism. Alterations in host physiology after rust infection have been described in detail in a number of studies (for summaries, see Farrar and Lewis 1987; Hahn 2000).

Rust-infected plant tissue shows a general decrease in chlorophyll content and photosynthesis, in a complex spatial and temporal fashion (Scholes and Farrar 1985). Marked differences are observed between colonized and non-colonized tissue of the same leaf. Photosynthesis and chlorophyll are often kept at higher levels for longer times in infected tissue regions than in surrounding non-infected tissue (Scholes and Farrar 1985 1986). This fact became generally known as the "green island" phenomenon (Scott 1972; Bushnell 1984; Walters and McRoberts 2006). While chlorophyll and photosynthetic activity are retained in infected regions, the surrounding tissue shows premature senescence and chlorosis. Video-based quantitative imaging of chlorophyll fluorescence was used for sensitive, high-resolution measurement of photosynthesis in living leaf tissue. By using this technique, changes in photosynthesis of rust-infected bean and crown rust-infected oat plants were found to follow a complex spatial and temporal pattern during disease development (Peterson and Aylor 1995; Scholes and Rolfe 1996).

Infection-related changes in carbon metabolism are considered to be of central importance during rust infection. Rust infection leads to a massive relocation of carbohydrates within the plant. In rust infection sites, starch accumulation is observed, and carbohydrate export from the leaf is

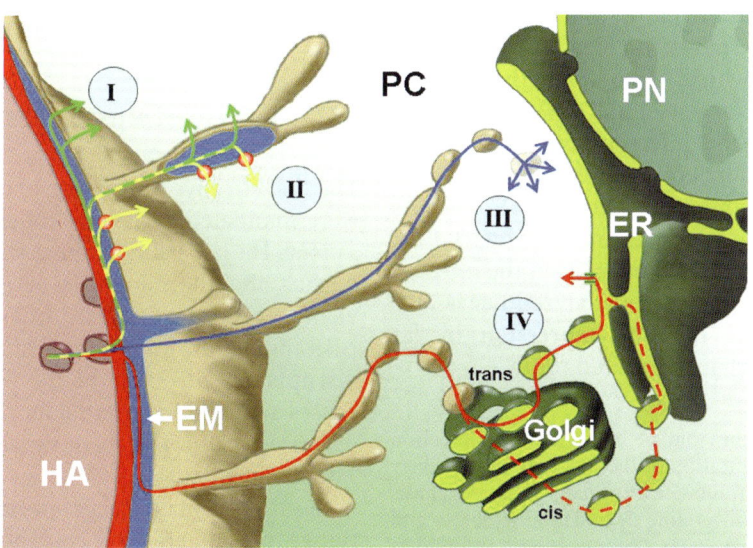

Fig. 4.8. Possible routes for effector proteins transferred from the haustorium into the host cell. *I* Direct transfer of the effector across the extrahaustorial membrane due to special characteristics of the protein. *II* Transfer mediated through specific transporters. *III* Participation of the tubular extensions in endosomal transfer. Budding vesicles fuse with lytic vesicles and become early endosomes. *IV* Retrograde transfer of effectors with participation of the complete Golgi apparatus (*dotted line*) or of the trans-Golgi network only (*solid line*). *EM* extrahaustorial matrix, *ER* endoplasmic reticulum, *HA* haustorium, *PC* plant cytoplasm, *PN* plant nucleus (Kemen and Mendgen, unpublished data)

progressively reduced (Bushnell 1984; Scholes and Farrar 1987). At later stages of infection, systemic effects become increasingly evident. Rust-infected leaves can develop into particularly effective sinks for carbohydrates. This again can be related to the well known phenomenon of "green island" formation (see above). Increased activities of apoplastic invertases at the sites of infection have been measured in a variety of pathosystems with biotrophic fungi. They are likely to be involved in the accumulation of soluble hexoses and formation of a sink tissue. Evidence for the plant or fungal origin of these activities was provided for the interactions between *A. thaliana* with white blister rust and powdery mildew, respectively (Chou et al. 2000; Fotopoulos et al. 2003). Increased expression of a host cell wall invertase gene was also observed in rust-infected *V. faba* leaves, in addition to the INV1p invertase that is secreted by the rust fungus (Voegele et al. 2006). Interestingly, induction of the host cell wall invertase gene was also observed in root tissue, which confirms the systemic effects of rust infection on the host plant. The expression of the fungal invertase INV1p in early infection structures, in which no uptake system for the produced monosaccharides is detectable, could play a role in increasing the sink strength of the invaded tissue. Apoplastic hydrolysis of sucrose would limit export of carbohydrates from the infected tissue via the phloem and therefore would condition the infected organ for conversion from a source tissue to a sink tissue which competes with naturally occurring sinks (Voegele et al. 2006). Wirsel et al. (2001) have shown that infection with a rust fungus can have far-reaching effects on host metabolism, exceeding the boundary of the infected cell. As expected, several of the genes analyzed by RT-PCR showed altered expression patterns in the infected organ. However, some of the analyzed genes also showed alterations in gene expression in far remote organs, such as stems and roots.

The search for flax genes that are induced by rust infection resulted in the identification of the *fis1* gene that is likely to encode Δ-1-pyrroline-5-carboxylate dehydrogenase, an enzyme involved in proline degradation (Roberts and Pryor 1995; Ayliffe et al. 2002). The induction was found to occur only after rust infection but not after infection with other pathogens or after wounding. Disruption of *fis1*, however, did not alter the response of the plant to rust infection (Mitchell et al. 2006). Thus, its role during the interaction remains unclear. In *Festuca rubra* leaves infected with *Puccinia* spp., several plant genes were found to be up-regulated, including a MAP kinase gene, a putative resistance gene,

and a gene encoding a Hsp70 protein (Ergen et al. 2007). In the cowpea/*U. vignae* interaction, infection related changes of gene expression were studied by extracting RNA isolated from individual epidermal cells in which the cell walls but not yet the cell lumen were penetrated by the fungal hypha (Mould et al. 2003). More than 24 genes were found to be up-regulated both in susceptible and resistant cells. Few genes were found to be specifically induced in resistant cells during rust penetration; mentionable are two encoding a PR10 pathogenesis-related protein and a phenylalanine ammonia lyase (Mould et al. 2003). A comprehensive genome-wide study of rust-induced changes in gene expression during several stages of infection was performed with resistant and susceptible soybean varieties (van de Mortel et al. 2007). In both the compatible and incompatible interaction, a biphasic pattern of differential gene expression response to rust infection was observed. Significant changes in gene expression were detected within the first 12 h after inoculation, but thereafter most of the differentially expressed genes returned to normal levels. A second wave of rust-induced up-regulation of several hundred genes was observed earlier in the incompatible (starting at 48 h) than in the compatible (72 h) interaction. The early changes in host gene expression might reflect the unusual direct mode of penetration of the soybean rust fungus. The subsequent down-regulation of early-induced genes might be interpreted as the ability of the rust fungus to suppress host defense gene induction, even when this has already been turned on (van de Mortel et al. 2007).

Taken together, changes in plant gene expression induced by rust infection seem to be both pathogen- and host-specific and follow a complex kinetic pattern, which makes them difficult to interpret in terms of molecular signals and responses.

XII. Control of Rust Disease

The first effective procedures to control rust disease were mandated by law in Rouen, France, as early as 1660. The law called for the destruction of barberry bushes, the alternate host of *P. graminis*, in the vicinity of grain fields in order to control cereal rust epidemics. Elimination of the alternate host disrupts the life cycle of the pathogen and thus causes a reduction of initial inoculum, along with a decrease in pathogen genetic variability (Roelfs 1982). The Connecticut barberry law of 1726, together with subsequent barberry laws in other states, continues to be an effective control mechanism today (Peterson et al. 2005). However, federal funding for the Barberry Eradication Program was discontinued in 1980. While no effects were detectable until 2002, it remains to be seen what consequences the discontinuation of the program will bring in the future (Peterson et al. 2005).

Another important means of controlling rust disease is the use of fungicides. The latest rust specific compilation of fungicides dates back to the early 1980s (Buchenauer 1982). However, azole- and dithiocarbamate-based formulations are still among the most effective fungicides since they exhibit high biological activity at low application rates. Moreover, the resistance risks associated with these fungicides used on cereal rusts is considered medium or even low (Brent 1995).

Biological control of rusts with hyperparasitic fungi has been described (Kranz 1981; Buchenauer and Leinhos 1982; Sharma and Sankaran 1988). However, no real breakthrough has been achieved to date. One possible explanation might be the hyperparasites' requirement for high humidity. These control methods can therefore not be very effective in temperate or arid regions (Grabski and Mendgen 1986). However, better results may be obtained in the tropics (Saksirirat and Hoppe 1990). A different, more successful story is the use of rust fungi as a biological control agent of invasive plant species (Wandeler and Bacher 2006; Fisher et al. 2007; Wood and Morris 2007).

Genetic resistance remains the most economic and environmentally friendly method to minimize yield losses due to rust fungi (Deising et al. 2002; Webb and Fellers 2006). Most commercial cereal cultivars remain resistant to rust infection for less than a decade, which is about the life span of an active breeding program (Roelfs et al. 1992). Others remain resistant for many years. The *Rpg1* gene, for example, provided North American barley cultivars with resistance to the stem rust fungus *P. graminis* f. sp. *tritici* for more than six decades (Staples 2003). A more detailed description of genetic resistance to rust fungi is provided in Sect. XIII.

XIII. Genetics and Molecular Biology of Rust Resistance

Because of the obligate biotrophic mode of rust infection, it is not surprising that successful defense of resistant plants against rust fungi is usually coupled to programmed cell death of the infected cells, the so-called hypersensitive response (HR). In most cases, HR is the consequence of genetic interactions between resistance (R) genes in the host plants and so-called avirulence (Avr) genes in certain races of the pathogen. Remarkably, the gene for gene hypothesis that appropriately describes such interactions in a variety of different plant-

pathosystems was developed by detailed studies on the flax/flax rust interaction (Flor 1955, 1956). A flax plant is resistant to rust infection when it carries at least one R gene that corresponds to or matches a specific avirulence gene present in the attacking rust strain. The simplest interpretation in molecular terms for this situation is that the R genes encode specific receptors for the Avr gene products. Molecular studies on a variety of gene for gene systems revealed that direct interactions between R proteins and Avr proteins can occur, but also showed that R proteins can recognize Avr proteins indirectly, e.g. by monitoring the integrity of host cellular targets of effector action (Jones and Dangl 2006). This "guard hypothesis" implies that R proteins recognize pathogen effectors only indirectly (see Chaps. 17, 18).

Many R genes against rust fungi were identified a long time ago by classic genetics; and their introgression into commercial cultivars of wheat, barley, oat and other crops remains a major task of agricultural resistance breeding today. A catalogue of resistance genes for different crop plants is available from the USDA (http://www.ars.usda.gov). This catalogue lists, for example, more than 45 wheat R genes against *P. graminis* f. sp. *tritici*. For the same host plant, more than 56 R genes against *P. triticina* are described. At least 19 R genes are identified in barley against *P. hordei*, and in oat more than 96 R genes against *P. coronata*.

During the past decade, a number of R genes have been cloned. One of the best characterized plants is flax, in which several members and alleles from all of the five R gene loci were sequenced (Dodds et al. 2001; Catanzariti et al. 2006, 2007). They encode typical R proteins of the Toll/interleukin resistance (TIR)–nucleotide binding site (NBS)–leucine-rich repeats (LRR) type (Ellis et al. 2007a). Comparative sequence analyses of the flax R gene alleles and artificially created hybrid genes reveal that the LRR domains are the major determinants of Avr protein recognition specificity (Ellis et al. 1999, 2007b; Dodds et al. 2001). In cereal crops, the cloning of several agronomically important R genes has been achieved. A prominent example is the *Rpg1* gene from barley which provides resistance to most pathotypes of *P. graminis* f. sp. *tritici*. The gene is incorporated into all major North American barley cultivars and has protected them from significant stem rust losses for more than 60 years (Staples 2003). As in other plants, cloning of *Rpg1* was achieved by a map-based approach (Brueggeman et al. 2002). The Rpg1 protein contains two tandem serine/threonine kinase motifs which are both required for Rpg1-mediated resistance (Nirmala et al. 2006). The known rust-specific R genes from grasses encode either NBS-LRR-type R proteins without an N-terminal TIR domain (barley *Lr1*, *Lr10*, *Lr21*, maize *Rp1*), or in the case of barley *Rpg1*, a serine/threonine protein kinase (Feuillet et al. 2003; Huang et al. 2003; Nirmala et al. 2006; Cloutier et al. 2007). Transfer of *Rpg1* into a susceptible commercial barley variety by transformation rendered it highly resistant against stem rust infection, illustrating the potential for transformation-assisted breeding (Horvath et al. 2003).

The sequencing of R alleles from resistant and susceptible plants provided insights into the evolution of rust resistance genes. Evidence for diversifying selection was obtained for different alleles of R genes in the L and P locus of flax (Ellis et al. 2007a). In the complex *Rp1* locus of maize, more than 30 similar R-like genes exist in the vicinity of the active *Rp1* gene. Unequal intragenic recombination between members of this gene cluster was shown to be an important source for the generation of novel resistance specificities towards different races of the maize rust *Puccinia sorghi* (Smith and Hulbert 2005).

The cloning of both R genes and their corresponding Avr genes, in the flax/flax rust pathosystem provided the opportunity to study the molecular interaction between R proteins and effector/avirulence proteins. The predicted amino acid sequences of the flax R proteins suggest that they are located in the cytoplasm, the L and M proteins possibly including N terminal membrane anchors (Ellis et al. 2007a). Transient expression of the *M. lini* Avr proteins in flax cells, using *A. tumefaciens*-mediated transformation, led to an HR in plants carrying the corresponding R genes (Dodds et al. 2004). For the L6 protein, a direct binding to the cognate AvrL567 protein was shown using yeast-two-hybrid interaction studies (Dodds et al. 2006).

Genes involved in non-host resistance against rust fungi are of interest for breeders because, despite their usually lower efficiency against pathogens compared to R genes, they are thought to confer greater durability under field conditions. Compared to R-gene-dependent resistance, non-host resistance is less frequently correlated with an HR. Instead, common observations include failure of germ tubes to invade non-host plants via stomata (possibly due to the absence of the correct topographical cues), abortive growth and premature death of the infection hypha before formation of the first haustorium, or rapid encasement of the haustorium (Heath and Skalamera 1997). A cross between a barley cultivar that is hypersusceptible to several non-host rust fungi and a normal barley cultivar revealed a number of quantitative trait-like loci that condition the defense against different

rust fungi (Jafary et al. 2006). In a similar study, the progeny of a cross between two barley lines with different degrees of resistance to wheat leaf rust revealed that both pre- and post-haustorial mechanisms are involved in non-host rust resistance (Neu et al. 2003). In both cases, R-type genes were found to be possibly involved in certain types of non-host resistance. The model plant *A. thaliana* was described to be infected by the rust fungus *Puccinia thlaspeos* (Gäumann 1959). However, host invasion occurs by monokaryotic basidiospores in roots and gives rise to systemic infections, which is hardly reproducible in the laboratory (K. Mendgen, unpublished data). Nevertheless, *A. thaliana* with its large collection of mutants is a useful model for the analysis of non-host resistance against rust fungi. *A. thaliana* mutants defective in salicylic acid-dependent defense signaling were found to allow an increased development of heterologous rust fungi, in particular *U. vignae*, up to the establishment of functional biotrophic relationships (Mellersh and Heath 2003).

Resistance to rust fungi can be induced in susceptible plants by chemical treatments, leading to systemic acquired resistance (SAR). In *V. faba*, SAR induced by treatment with salicylic acid or 2,6-dichloro-isonicotinic acid leads to significant inhibition of the invading rust fungus *U. fabae* (Rauscher et al. 1999). The inhibition correlated with the inhibitory activity of apoplastic fluids obtained from induced resistant plants and was proposed to be due to the increased expression of the pathogenesis-related protein PR-1.

XIV. Imminent Threats: The Cases of *P. pachyrhizi* and *P. graminis* Ug99

The effectiveness of fungicide treatment and genetic and molecular breeding programs has successfully protected farmers from rust infection for more than half a century. However, this success may lead to a false sense of security. A possible scenario is that relaxed efforts to generate new resistant cultivars, the discontinuation of eradication programs, and a lack of funding for basic research engaged to clarify the molecular nature of biotrophic infections may lead to dramatic consequences in the future. The recent emergence of a new race of *P. graminis* f. sp. *tritici*, UG99 or TTKS, and the spread of Asian soybean rust (ASR) into the continental United States illustrate the danger that obligate biotrophs still pose to agriculture.

Apart from the introduction of new rust pathogens, the boom and bust cycle poses a special threat to the farmers. In the boom, a resistant cultivar with single, major resistance is introduced. As a result of its good agronomic qualities, it may be widely accepted and planted over large areas. If the pathogen population is exposed to this resistance gene, races with a mutation from avirulence to virulence have a better chance to occur. In the bust part of this cycle, the virulent pathotypes spread, infect fields with the resistant cultivar, and cause an epidemic. The cycle begins again with the introduction of a new resistant cultivar (McDonald 2004). Typical examples are leaf rust of wheat caused by *P. triticina* and stem rust of wheat and barley caused by *P. graminis* f. sp. *tritici*. Some 40–50 races of leaf rust are identified annually in the United States. This high degree of variability has allowed the fungus to adapt to new resistant wheat cultivars, very often within a short period of time (Kolmer et al. 2007). By contrast, only three to five races of stem rust are found annually in the United States. This is most likely a result of the eradication of the alternate host barberry (Roelfs et al. 1992). However, many of the currently used wheat cultivars, albeit not all, are susceptible to race UG99 detected in Uganda in 1999 (Pretorius et al. 2000). This new race is spreading rapidly in the Eastern African highlands and poses a new threat to wheat production wherever the *Sr31* resistance gene is used (Wanyera et al. 2006). However, several lines of resistance genes appear to be effective against Ug99, both at the seedling and adult plant stages (Jin et al. 2007). A rapid introduction of such genes into cultivars used throughout Eastern Africa and Asia is now in progress. Wheat and barley breeders in other areas where wheat and barley are major crops should be prepared to prevent a major outbreak of Ug99.

The spread of ASR into the United States is a typical example for the introduction of an old foe into a previously pathogen-free region. In the past few years, the disease has spread from Asia to Hawaii in 1994, Africa in 1996, and South America in 2001 (Pivonia and Yang 2004). Aerial transport seems crucial for the recent transport of the disease to Florida. Model simulations suggest a transport of soybean rust spores from South America north of the equator into the United States with a tropical cyclone in August 2003 (Isard et al. 2005). Recently, soybean rust reached the eastern and central United States (Barnes et al. 2006b; Krupa et al. 2006). Sensitive PCR assays have been developed in order to detect and differentiate both *P. pachyrhizi* and the less aggressive *Phakopsora meibomiae* (Frederick et al. 2002). This technology is now widely used to trace the pathogen in rain samples (Barnes et al. 2006b; Krupa et al. 2006). The USDA has introduced an online Pest Information Platform for Extension and Education (PIPE), which allows farmers to continuously update on disease progress (http://www.sbrusa.net/). The pathogen overwinters on Kudzu (*Pueraria lobata*) in parts of Florida, Georgia, and Alabama, where leaves stay green during the winter and urediospores can survive and start a new epidemic the next year. Although teliospores are produced and may develop basidiospores, no alternate host has yet been found (Saksiriat and Hoppe 1991). Therefore, it seems unlikely that teliospores are responsible for spread of the pathogen. However, the broad host range, which comprises 31 species in 17 genera of legumes, including common bean cultivars and the widespread Kudzu, favor spread of the pathogen and makes it difficult to breed for resistance (Bromfield 1984; Miles 2007). At the moment it is therefore only the climatic conditions that prevent a massive ASR outbreak in the soybean growing areas of the United States.

XV. Conclusions and Perspectives

During the past decade much progress has been made in determining some of the aspects of obligate biotrophic growth. The increasing availability of sequence data, especially through genomic sequencing projects, paired with expression analysis in macro- and microarray format will further our understanding of the molecular details underlying this intricate plant–parasite interaction. While we are far from establishing culture conditions to produce "parasitic phase" infection structures in vitro, functional, stable transformation has drawn a step closer (Wirsel et al. 2004; Webb et al. 2006). Combined with gene expression and protein localization studies, the availability of transgenic rust fungi would greatly facilitate future molecular and biochemical work on rust fungi. For instance, gene silencing methodology paired with the identification of stage-specific promoter sequences could help to identify elements crucial for the establishment and maintenance of the obligate biotrophic lifestyle.

Acknowledgements. We would like to apologize to all researchers whose work could not be cited in this book chapter due to spatial restrictions. We would also like to extend our gratitude to all colleagues who contributed to our current understanding of rust fungi in general and the molecular biology of rust fungi in particular.

References

Aime MC (2006) Toward resolving family-level relationships in rust fungi (Uredinales). Mycoscience 47:112–122

Alexopoulos CJ, Mims CW, Blackwell M (1996) Introductory mycology. Wiley, New York

Allen EA, Hazen BE, Hoch HC, Kwon Y, Leinhos GME, Staples RC, Stumpf MA, Terhune BT (1991) Appressorium formation in response to topographical signals by 27 rust species. Phytopathology 81:323–331

Anikster Y (1984) The formae specialis. In: Bushnell WR, Roelfs AP (eds) Origins, specificity, structure, and physiology. The cereal rusts, vol 1. Academic, Orlando, pp 115–129

Anikster Y (1986) Teliospore germination in some rust fungi. Phytopathology 76:1026–1030

Arthur JC (1962) Manual of the rusts in the United States and Canada. Hafner, New York

Ayliffe MA, Roberts JK, Mitchell HJ, Zhang R, Lawrence GJ, Ellis JG, Pryor TJ (2002) A plant gene up-regulated at rust infection sites. Plant Physiol 129:169–180

Bachem U, Mendgen K (1995) Endoplasmic reticulum subcompartments in a plant parasitic fungus and in baker's yeast: differential distribution of lumenal proteins. Exp Mycol 19:137–152

Baka ZA, Larous L, Losel DM (1995) Distribution of ATPase activity at the host-pathogen interfaces of rust infections. Physiol Mol Plant Pathol 47:67–82

Barja F, Correa A Jr, Staples RC, Hoch HC (1998) Microinjected antisense Inf24 oligonucleotides inhibit appressorium development in Uromyces. Mycol Res 102:1513–1518

Barnes C, Szabo L, Johnson J, K-PN, Floyd C, Kurle J (2006a) Detection of Phakopsora pachyrhizi DNA in rain using qPCR and a portable rain collector. APS CPS MSA Joint Meeting. American Phytopathological Society, Saint Paul

Barnes CW, Szabo L, Johnson JL, Bowersox VC, Harlin KS (2006b) Detection of Phakopsora pachyrhizi spores using a real-time PCR assay. Phytopathology 96:S9

Basse CW, Bock K, Boller T (1992) Elicitors and suppressors of the defense response in tomato cells. Purification and characterization of glycopeptide elicitors and glycan suppressors generated by enzymatic cleavage of yeast invertase. J Biol Chem 267:10258–10265

Bhairi SM, Staples RC (1992) Transient expression of the β-glucuronidase gene introduced into Uromyces appendiculatus uredospores by particle bombardment. Phytopathology 82:986–989

Bhairi SM, Staples RC, Freve P, Yoder OC (1989) Characterization of an infection structure-specific gene from the rust fungus Uromyces appendiculatus. Gene 81:237–243

Bonde MR, Melching JS, Bromfield KR (1976) Histology of the suscept pathogen relationship between Glycine max and Phakopsora pachyrhizi, the cause of soybean rust. Phytopathology 66:1290–1294

Bourett T, Hoch HC, Staples RC (1987) Association of the microtubule cytoskeleton with the thigmotropic signal for appressorium formation in Uromyces. Mycologia 79:540–545

Bowden J, Gregory PH, Johnson CG (1971) Possible wind transport of coffee leaf rust across the Atlantic Ocean. Nature 229:500–501

Boyle B, Hamelin RC, Seguin A (2005) In vivo monitoring of obligate biotrophic pathogen growth by kinetic PCR. Appl Environ Microbiol 71:1546–1552

Brent KJ (1995) Fungicide resistance in crop pathogens: how can it be managed? Global Crop Protection Federation, Brussels

Bromfield KR (1984) Soybean rust. Monograph 11. American Phytopathological Society, St Paul

Brown JK, Hovmøller MS (2002) Aerial dispersal of pathogens on the global and continental scales and its impact on plant disease. Science 297:537–541

Brueggeman R, Rostoks N, Kudrna D, Kilian A, Han F, Chen J, Druka A, Steffenson B, Kleinhofs A (2002) The barley stem rust-resistance gene Rpg1 is a novel disease-resistance gene with homology to receptor kinases. Proc Natl Acad Sci USA 99:9328–9333

Buchenauer H (1982) Chemical and biological control of cereal rusts. In: Scott KJ, Chakravorty AK (eds) The rust fungi. Academic, London, pp 247–279

Buchenauer H, Leinhos G (1982) Einfluss der Mycoparasiten Verticillium psalliotae und Aphanocladium spectabile auf Getreideroste. Med Fac Landbouw Rijksuniv 47:819–830

Bushnell WR (1972) Physiology of fungal haustoria. Annu Rev Phytopathol 10:151–176

Bushnell WR (1984) Structural and physiological alterations in susceptible host tissue. In: Bushnell WR, Roelfs AP (eds) Origins, specificity, structure, and physiology. The cereal rusts, vol 1. Academic, Orlando, pp 477–507

Bushnell WR, Rowell JB (1981) Suppressors of defense reactions: a model for roles in specificity. Phytopathology 71:1012–1014

Cantrill LC, Deverall BJ (1993) Isolation of haustoria from wheat leaves infected by the leaf rust fungus. Physiol Mol Plant Pathol 42:337–341

Catanzariti AM, Dodds PN, Lawrence GJ, Ayliffe MA, Ellis JG (2006) Haustorially expressed secreted proteins from flax rust are highly enriched for avirulence elicitors. Plant Cell 18:243–256

Catanzariti AM, Dodds PN, Ellis JG (2007) Avirulence proteins from haustoria-forming pathogens. FEMS Microbiol Lett 269:181–188

Chaturvedi V, Wong B, Newman SL (1996) Oxidative killing of Cryptococcus neoformans by human neutrophils. Evidence that fungal mannitol protects by scavenging reactive oxygen intermediates. J Immunol 156:3836–3840

Chong J, Kang Z, Kim WM, Rohringer R (1992) Multinucleate condition of Puccinia striiformis in colonies isolated from infected wheat leaves with macerating enzymes. Can J Bot 70:222–224

Chou HM, Bundock N, Rolfe SA, Scholes JD (2000) Infection of Arabidopsis thaliana leaves with Albugo candida (white blister rust) causes a reprogramming of host metabolism. Mol Plant Pathol 1:99–113

Clement JA, Butt TM, Beckett A (1993a) Characterization of the extracellular matrix produced in vitro by urediniospores and sporelings of *Uromyces viciae-fabae*. Mycol Res 97:594–602

Clement JA, Martin SG, Porter R, Butt TM, Beckett A (1993b) Germination and the role of extracellular matrix in adhesion of urediniospores of *Uromyces viciae-fabae* to synthetic surfaces. Mycol Res 97:585–593

Clement JA, Porter R, Butt TM, Beckett A (1994) The role of hydrophobicity in attachment of urediniospores and sporelings of *Uromyces viciae-fabae*. Mycol Res 98:1217–1228

Clement JA, Porter R, Butt TM, Beckett A (1997) Characteristics of adhesion pads formed during imbibition and germination of urediniospores of *Uromyces viciae-fabae*. Mycol Res 101:1445–1458

Clement JA, Porter R, Beckett A (1998) The orientation of urediniospores of *Uromyces viciae-fabae* during fall and after landing. Mycol Res 102:907–913

Cloutier S, McCallum BD, Loutre C, Banks TW, Wicker T, Feuillet C, Keller B, Jordan MC (2007) Leaf rust resistance gene *Lr1*, isolated from bread wheat (*Triticum aestivum* L.) is a member of the large psr567 gene family. Plant Mol Biol 65:93–106

Collins TJ, Moerschbacher BM, Read ND (2001) Synergistic induction of wheat stem rust appressoria by chemical and topographical signals. Physiol Mol Plant Pathol 58:259–266

Cooper B, Garrett WM, Campbell KB (2006) Shotgun identification of proteins from uredospores of the bean rust *Uromyces appendiculatus*. Proteomics 6:2477–2484

Craigie JH (1927) Experiments on sex in rust fungi. Nature 120:116–117

Cummins GB, Hiratsuka Y (2003) Illustrated genera of rust fungi. American Phytopathological Society, St Paul

D'Silva I, Heath MC (1997) Purification and characterization of two novel hypersensitive response-inducing specific elicitors produced by the cowpea rust fungus. J Biol Chem 272:3924–3927

Daly JM, Knoche HW, Wiese MV (1967) Carbohydrate and lipid metabolism during germination of uredospores of *Puccinia graminis tritici*. Plant Physiol 42:1633–1642

de Bary HA (1853) Untersuchungen über Brandpilze und die durch sie verursachten Krankheiten der Pflanzen. Müller, Berlin

de Bary HA (1863) Recherches sur le developpement de quelques champignons parasites. Ann Sci Nat Part Bot 20:5–148

de Bary HA (1865) Neue Untersuchungen über die Uredineen, insbesondere die Entwicklung der *Puccinia graminis* und den Zusammenhang derselben mit *Aecidium berberidis*. Monatsber Königl Preuss Akad Wiss Berlin 1865:15–50

Deising H, Siegrist J (1995) Chitin deacetylase activity of the rust *Uromyces viciae-fabae* is controlled by fungal morphogenesis. FEMS Microbiol Lett 127:207211

Deising H, Jungblut PR, Mendgen K (1991) Differentiation-related proteins of the broad bean rust fungus *Uromyces viciae-fabae*, as revealed by high resolution two-dimensional polyacrylamide gel electrophoresis. Arch Microbiol 155:191–198

Deising H, Nicholson RL, Haug M, Howard RJ, Mendgen K (1992) Adhesion pad formation and the involvement of cutinase and esterases in the attachment of uredospores to the host cuticle. Plant Cell 4:1101–1111

Deising H, Frittrang AK, Kunz S, Mendgen K (1995a) Regulation of pectin methylesterase and polygalacturonate lyase activity during differentiation of infection structures in *Uromyces viciae-fabae*. Microbiology 141:561–571

Deising H, Rauscher M, Haug M, Heiler S (1995b) Differentiation and cell wall degrading enzymes in the obligately biotrophic rust fungus *Uromyces viciae-fabae*. Can J Bot 73:S624–S631

Deising H, Reimann S, Peil A, Weber WE (2002) Disease managment of rusts and powdery mildews. In: Kempken F (ed) Agricultural applications. The Mycota XI. Springer, Berlin, pp 243–269

Dickinson S (1949) Studies in the physiology of obligate parasitism: II. The behaviour of the germ-tubes of certain rusts in contact with various membranes. Ann Bot 13:219–236

Dodds P, Lawrence G, Ellis J (2001) Six amino acid changes confined to the leucine-rich repeat β-strand/β-turn motif determine the difference between the *P* and *P2* rust resistance specificities in flax. Plant Cell 13:163–178

Dodds PN, Lawrence GJ, Catanzariti AM, Ayliffe MA, Ellis JG (2004) The *Melampsora lini AvrL567* avirulence genes are expressed in haustoria and their products are recognized inside plant cells. Plant Cell 16:755–768

Dodds PN, Lawrence GJ, Catanzariti AM, Teh T, Wang CI, Ayliffe MA, Kobe B, Ellis JG (2006) Direct protein interaction underlies gene-for-gene specificity and coevolution of the flax resistance genes and flax rust avirulence genes. Proc Natl Acad Sci USA 103:8888–8893

Eilam T, Bushnell WR, Anikster Y, McLaughlin DJ (1992) Nuclear DNA content of basidiospores of selected rust fungi as estimated from fluorescence of propidium iodide-stained nuclei. Phytopathology 82:705–712

Eilam T, Bushnell WR, Anikster Y (1994) Relative nuclear DNA content of rust fungi estimated by flow cytometry of propidium iodide-stained pycniospores. Phytopathology 84:728–735

El Gueddari NE, Rauchhaus U, Moerschbacher BM, Deising HB (2002) Developmentally regulated conversion of surface-exposed chitin to chitosan in cell walls of plant pathogenic fungi. New Phytol 156:103–112

Ellis JG, Lawrence GJ, Luck JE, Dodds PN (1999) Identification of regions in alleles of the flax rust resistance gene *L* that determine differences in gene-for-gene specificity. Plant Cell 11:495–506

Ellis J, Catanzariti AM, Dodds P (2006) The problem of how fungal and oomycete avirulence proteins enter plant cells. Trends Plant Sci 11:61–63

Ellis JG, Dodds PN, Lawrence GJ (2007a) Flax rust resistance gene specificity is based on direct resistance-avirulence protein interactions. Annu Rev Phytopathol 45:289–306

Ellis JG, Lawrence GJ, Dodds PN (2007b) Further analysis of gene-for-gene disease resistance specificity in flax. Mol Plant Pathol 8:103–109

Ergen NZ, Dinler G, Shearman RC, Budak H (2007) Identifying, cloning and structural analysis of differentially expressed genes upon *Puccinia* infection of *Festuca rubra* var. *rubra*. Gene 393:145–152

Eriksson J (1894) Über die Spezialisierung des Parasitismus bei den Getreiderostpilzen. Ber Dtsch Bot Ges 12:292–331

Eriksson J, Henning E (1896) Die Getreideroste. Ihre Geschichte und Natur sowie Massregeln gegen dieselben. Norstedt and Soner, Stockholm

Eversmeyer MG, Kramer CL (2000) Epidemiology of wheat leaf and stem rust in the Central Great Plains of the USA. Annu Rev Phytopathol 38:491–513

Farrar JF, Lewis DH (1987) Nutrient relations in biotrophic infections. In: Pegg GF, Ayres PG (eds) Fungal Infection of Plants, vol 13. Cambridge University Press, Cambridge, pp 92–132

Fasters MK, Daniels U, Moerschbacher BM (1993) A simple and reliable method for growing the wheat stem rust fungus, *Puccinia graminis* f.sp. *tritici*, in liquid culture. Physiol Mol Plant Pathol 42:259–265

Fernandez MR, Heath MC (1991) Interactions of the non-host French bean plant (*Phaseolus vulgaris*) parasitic and saprophytic fungi. IV. Effect of preinoculation with the bean rust fungus on growth of parasitic fungi nonpathogenic on beans. Can J Bot 69:1642–1646

Feuillet C, Travella S, Stein N, Albar L, Nublat A, Keller B (2003) Map-based isolation of the leaf rust disease resistance gene *Lr10* from the hexaploid wheat (*Triticum aestivum* L.) genome. Proc Natl Acad Sci USA 100:15253–15258

Fisher AJ, Woods DM, Smith L, Bruckart WL, III (2007) Developing an optimal release strategy for the rust fungus *Puccinia jaceae* var. *solstitialis* for biological control of *Centaurea solstitialis* (yellow starthistle). Biol Control 42:161–171

Flor HH (1955) Host-parasite interaction in flax rust – its genetics and other implications. Phytopathology 45:680–685

Flor HH (1956) The complementary genetic systems in flax and flax rust. Adv Genet 8:29–54

Fontana F (1767) Observations on the rust of grain (translation by PP Pirrone, 1932). American Phytopathological Society, Washington, D.C.

Fotopoulos V, Gilbert MJ, Pittman JK, Marvier AC, Buchanan AJ, Sauer N, Hall JL, Williams LE (2003) The monosaccharide transporter gene, *AtSTP4*, and the cell-wall invertase, *At fruct1*, are induced in *Arabidopsis* during infection with the fungal biotroph *Erysiphe cichoracearum*. Plant Physiol 132:821–829

Frederick RD, Snyder CL, Peterson GL, Bonde MR (2002) Polymerase chain reaction assays for the detection and discrimination of the soybean rust pathogens *Phakopsora pachyrhizi* and *P. meibomiae*. Phytopathology 92:217–227

French RC (1992) Volatile chemical germination stimulators of rust and other fungal spores. Mycologia 84:277–288

French RC, Weintraub RL (1957) Pelargonaldehyde as an endogenous germination stimulator of wheat rust spores. Arch Biochem Biophys 72:235–237

Freytag S, Mendgen K (1991a) Surface carbohydrates and cell wall structure of in vitro-induced uredospore infection structures of *Uromyces viciae-fabae* before and after treatment with enzymes and alkali. Protoplasma 161:94–103

Freytag S, Mendgen K (1991b) Carbohydrates on the surface of urediniospore- and basidiospore-derived infection structures of heteroecious and autoecious rust fungi. New Phytol 119:527–534

Freytag S, Bruscaglioni L, Gold RE, Mendgen K (1988) Basidiospores of rust fungi (*Uromyces* species) differentiate infection structures in vitro. Exp Mycol 12:275–283

Frittrang AK, Deising H, Mendgen K (1992) Characterization and partial purification of pectinesterase, a differentiation-specific enzyme of *Uromyces viciae-fabae*. J Gen Microbiol 138:2213–2218

Gäumann E (1959) Die Rostpilze Mitteleuropas. Bücheler, Bern

Gay JL, Salzberg A, Woods AM (1987) Dynamic experimental evidence for the plasma membrane ATPase domain hypothesis of haustorial transport and for ionic coupling of the haustorium of *Erysiphe graminis* to the host cell (*Hordeum vulgare*). New Phytol 107:541–548

Geagea L, Huber L, Sache I (1999) Dry-dispersal and rain-splash of brown (*Puccinia recondita* f.sp. *tritci*) and yellow (*P. striiformis*) rust spores from infected wheat leaves exposed to simulated raindrops. Plant Pathol 48:472–482

Ghosh P (2004) Process of protein transport by the type III secretion system. Microbiol Mol Biol Rev 68:771–795

Gold RE, Littlefield LJ (1979) Light and scanning electron microscopy of the telial, pycnial, and aecial stages of *Melampsora lini*. Can J Bot 57:629–638

Gold RE, Mendgen K (1983) Activation of teliospore germination in *Uromyces appendiculatus* var. *appendiculatus*. II. Light and host volatiles. J Phytopathol 108:281–293

Gold RE, Mendgen K (1991) Rust basidiospore germlings and disease initiation. In: Cole GT, Hoch HC (eds) The fungal spore and disease initiation in plants and animals, Plenum, New York, pp 67–99

Gold RE, Littlefield LJ, Statler GD (1979) Ultrastructure of the pycnial and aecial stages of *Puccinia recondita*. Can J Bot 57:74–86

Grabski GC, Mendgen K (1986) Die Parasitierung des Bohnenrostes *Uromyces appendiculatus* var. *appendiculatus* durch den Hyperparasiten *Verticillium lecanii*: Untersuchungen zur Wirt-Erkennung, Penetration und Abbau der Rostpilzsporen. J Phytopathol 115:116–123

Graham PH, Vance CP (2003) Legumes: importance and constraints to greater use. Plant Physiol 131:872–877

Grambow HJ (1977) The influence of volatile leaf constituents on the in vitro differentiation and growth of *Puccinia graminis* f.sp. *tritici*. Z Pflanzenphysiol 85:361–372

Haerter AC, Voegele RT (2004) A novel β-glucosidase in *Uromyces fabae*: Feast or fight? Curr Genet 45:96–103

Hahn M (2000) The rust fungi: cytology, physiology and molecular biology of infection. In: Kronstad J (ed) Fungal pathology, Kluwer, Dordrecht, pp 267–306

Hahn M, Mendgen K (1992) Isolation of ConA binding haustoria from different rust fungi and comparison of their surface qualities. Protoplasma 170:95–103

Hahn M, Mendgen K (1997) Characterization of in planta-induced rust genes isolated from a haustorium-specific cDNA library. Mol Plant–Microbe Interact 10:427–437

Hahn M, Deising H, Struck C, Mendgen K (1997a) Fungal morphogenesis and enzyme secretion during pathogenesis. In: Hartleb H, Heitefuss R, Hoppe H-H (eds) Resistance of crop plants against fungi. Fischer, Jena, pp 33–57

Hahn M, Neef U, Struck C, Göttfert M, Mendgen K (1997b) A putative amino acid transporter is specifically expressed in haustoria of the rust fungus *Uromyces fabae*. Mol Plant–Microbe Interact 10:438–445

Hall JL, Williams LE (2000) Assimilate transport and partitioning in fungal biotrophic interactions. Aust J Plant Physiol 27:549–560

Harder DE, Chong J (1984) Structure and physiology of haustoria. In: Bushnell WR, Roelfs AP (eds) Origins, specificity, structure, and physiology. The cereal rusts, vol 1. Academic, Orlando, pp 431–476

Harder DE, Chong J (1991) Rust haustoria. In: Mendgen K, Lesemann D-E (eds) Electron microscopy of plant pathogens. Springer, Berlin, pp 235–250

Harder DE, Mendgen K (1982) Filipin-sterol complexes in bean rust- and oat crown rust-fungal/plant interactions: freeze-etch electron microscopy *Uromyces appendiculatus*. Protoplasma 112:46–54

Heath MC (1976) Ultrastructural and functional similarity of the haustorial neckband of rust fungi and the Casparian strip of vascular plants. Can J Bot 54:2484–2489

Heath MC (1989) In vitro formation of haustoria of the cowpea rust fungus, *Uromyces vignae*, in the absence of a living plant cell. I. Light microscopy. Physiol Mol Plant Pathol 35:357–366

Heath MC (1990a) In vitro formation of haustoria of the cowpea rust fungus *Uromyces vignae* in the absence of a living plant cell. II. Electron microscopy. Can J Bot 68:278–287

Heath MC (1990b) Influence of carbohydrates on the induction of haustoria of the cowpea rust fungus in vitro. Exp Mycol 14:84–88

Heath MC (1997) Signalling between pathogenic rust fungi and resistant or susceptible host plants. Ann Bot 80:713–720

Heath MC, Skalamera D (1997) Cellular interactions between plants and biotrophic fungal parasites. Adv Bot Res 24:195–225

Heiler S, Mendgen K, Deising H (1993) Cellulolytic enzymes of the obligated biotrophic rust fungus *Uromyces viciae-fabae* are regulated differentiation-specifically. Mycol Res 97:77–85

Higuchi R, Dollinger G, Walsh PS, Griffith R (1992) Simultaneous amplification and detection of specific DNA sequences. Biotechnology 10:413–417

Hiller NL, Bhattacharjee S, van Ooij C, Liolios K, Harrison T, Lopez-Estrano C, Haldar K (2004) A host-targeting signal in virulence proteins reveals a secretome in malarial infection. Science 306:1934–1937

Hoch HC, Staples RC (1987) Structural and chemical changes among the rust fungi during appressorium development. Annu Rev Phytopathol 25:231–247

Hoch HC, Staples RC, Whitehead B, Comeau J, Wolf ED (1987) Signaling for growth orientation and cell differentiation by surface topography in *Uromyces*. Science 235:1659–1662

Hoffmann J, Mendgen K (1998) Endocytosis and membrane turnover in the germ tube of *Uromyces fabae*. Fungal Genet Biol 24:77–85

Hoppe HH, Koch E (1989) Defense reactions in host and nonhost plants against the soybean rust fungus (*Phakopsora pachyrhizi* Syd.). J Phytopathol 125:77–88

Horvath H, Rostoks N, Brueggeman R, Steffenson B, von Wettstein D, Kleinhofs A (2003) Genetically engineered stem rust resistance in barley using the *Rpg1* gene. Proc Natl Acad Sci USA 100:364–369

Huang L, Brooks SA, Li W, Fellers JP, Trick HN, Gill BS (2003) Map-based cloning of leaf rust resistance gene *Lr21* from the large and polyploid genome of bread wheat. Genetics 164:655–664

Isard SA, Gage SH, Comtois P, Russo JM (2005) Principles of the atmospheric pathway for invasive species applied to soybean rust. Bioscience 55:851–861

Jafary H, Szabo LJ, Niks RE (2006) Innate nonhost immunity in barley to different heterologous rust fungi is controlled by sets of resistance genes with different and overlapping specificities. Mol Plant–Microbe Interact 19:1270–1279

Jakupovic M, Heintz M, Reichmann P, Mendgen K, Hahn M (2006) Microarray analysis of expressed sequence tags from haustoria of the rust fungus *Uromyces fabae*. Fungal Genet Biol 43:8–19

Jennings DB, Daub ME, Pharr DM, Williamson JD (2002) Constitutive expression of a celery mannitol dehydrogenase in tobacco enhances resistance to the mannitol-secreting fungal pathogen *Alternaria alternata*. Plant J 32:41–49

Jennings DH (1984) Polyol metabolism in fungi. Adv Microb Physiol 25:149–193

Jin Y, Singh RP, Ward RW, Wanyera R, Kinyua MG, Njau P, Fetch T, Pretorius ZA, Yahyaoui A (2007) Characterization of seedling infection types and adult plant infection responses of monogenic *Sr* gene lines to race TTKS of *Puccinia graminis* f.sp. *tritici*. Plant Dis 91:1096–1099

Jones JD, Dangl JL (2006) The plant immune system. Nature 444:323–329

Joseph ME, Hering TF (1997) Effects of environment on spore germination and infection by broad bean rust (*Uromyces viciae-fabae*). J Agric Sci 128:73–78

Jülich W (1981) Higher taxa of Basidiomycetes. Bibliotheca Mycologica 85:1–485

Kamoun S (2006) A catalogue of the effector secretome of plant pathogenic oomycetes. Annu Rev Phytopathol 44:41–60

Kapooria RG (1971) A cytological study of promycelia and basidiospores and the chromosome number in *Uromyces fabae*. Neth J Plant Pathol 77:91–96

Kapooria RG, Mendgen K (1985) Infection structures and their surface changes during differentiation in *Uromyces fabae*. J Phytopathol 113:317–323

Keen NT (2000) A century of plant pathology: a retrospective view on understanding host–parasite interactions. Annu Rev Phytopathol 38:31–48

Kemen E, Hahn M, Mendgen K, Struck C (2005a) Different resistance mechanisms of *Medicago truncatula* ecotypes against the rust fungus *Uromyces striatus*. Phytopathology 95:153–157

Kemen E, Kemen AC, Rafiqi M, Hempel U, Mendgen K, Hahn M, Voegele RT (2005b) Identification of a protein from rust fungi transferred from haustoria into infected plant cells. Mol Plant–Microbe Interact 18:1130–1139

Kislev ME (1982) Stem rust of wheat 3300 years old found in Israel. Science 216:993–994

Knogge W (1997) Elicitors and suppressors of the resistance response. In: Hartleb H, Heitefuss R, Hoppe H-H (eds) Resistance of crop plants against fungi. Fischer, Jena, pp 159–182

Koh S, Andre A, Edwards H, Ehrhardt D, Somerville S
    (2005) *Arabidopsis thaliana* subcellular responses to
    compatible *Erysiphe cichoracearum* infections. Plant J
    44:516–529
Kolmer J (1996) Genetics of resistance to wheat leaf rust.
    Annu Rev Phytopathol 34:435–455
Kolmer J (2005) Tracking wheat rust on a continental scale.
    Curr Opin Plant Biol 8:441–449
Kolmer J, Jin Y, Long D (2007) Leaf and stem rust of wheat
    in the United States. Aust J Agric Res 58:631–638
Kranz J (1981) Hyperparasitism of biotrophic fungi. In:
    Blakeman JP (ed.) Microbial ecology of the phylo-
    plane. Academic, London, pp 327–352
Krupa S, Bowersox V, Claybrooke R, Barnes CW, Szabo L,
    Harlin K, Kurle J (2006) Introduction of asian soy-
    bean rust urediniospores into the midwestern United
    States: a case study. Plant Dis 90:1254–1259
Kwon YH, Hoch HC, Staples RC (1991) Cytoskeletal organi-
    zation in *Uromyces* urediospore germling apices dur-
    ing appressorium formation. Protoplasma 165:37–50
Langenbach RJ, Knoche HW (1971a) Phospholipids in the
    uredospores of *Uromyces phaseoli*: I. identification
    and localization. Plant Physiol 48:728–734
Langenbach RJ, Knoche HW (1971b) Phospholipids in the
    uredospores of *Uromyces phaseoli*: II. metabolism
    during germination. Plant Physiol 48:735–739
Larous L, Lösel DM (1993) Strategies of pathogenicity in
    monokaryotic and dikaryotic phases of rust fungi,
    with special reference to vascular infection. Mycol Res
    97:415–420
Laundon GF (1973) Uredinales. In: Ainsworth GC, Sparrow
    FK, Sussman AS (eds) The fungi: an advanced treatise,
    vol IV B. Academic, London, pp 247–279
Leonard KJ, Szabo LJ (2005) Stem rust of small grains and
    grasses caused by *Puccinia graminis*. Mol Plant Pathol
    6:99–111
Lewis DH, Smith DC (1967) Sugar alcohols (polyols) in
    fungi and green plants. I. Distribution, physiology and
    metabolism. New Phytol 66:143–184
Li A, Altosaar I, Heath MC, Horgen PA (1993) Transient
    expression of the beta-glucuronidase gene delivered
    into urediniospores of *Uromyces appendiculatus* by
    particle bombardment. Can J Plant Pathol 15:1–6
Line RF, Griffith CS (2001) Research on the epidemiology
    of stem rust of wheat during the Cold War. In: Peter-
    son PD (ed.) Stem rust of wheat. From ancient enemy
    to modern foe. American Phytopathological Society,
    St Paul, pp 83–118
Link T, Lohaus G, Heiser I, Mendgen K, Hahn M, Voegele RT
    (2005) Characterization of a novel NADP$^+$-dependent
    D-arabitol dehydrogenase from the plant pathogen
    Uromyces fabae. Biochem J 389:289–295
Littlefield LJ (1981) Biology of the plant rusts. An introduction.
    Iowa State University Press, Ames
Littlefield LJ, Heath MC (1979) Ultrastructure of rust fungi.
    Academic, New York
Liu Z, Szabo LJ, Bushnell WR (1993) Molecular cloning
    and analysis of abundant and stage-specific mRNAs
    from *Puccinia graminis*. Mol Plant–Microbe Interact
    6:84–91
Loeghering WQ (1984) Genetics of the pathogen–host
    association. In: Bushnell WR, Roelfs AP (eds) Origins,

specificity, structure, and physiology. The cereal rusts,
    vol 1. Academic, Orlando, pp 165–192
Long DL (2003) Cereal rust bulletin: final report. CDL, USDA.
    http://www.cdl.umn.edu/crb/2003crb/03crbfin.html
Macko V, Staples RC, Allen PJ, Renwick JAA (1970)
    Self-inhibitor of bean rust uredospores: methyl
    3,4-dimethoxycinnamate. Science 170:539
Macko V, Staples RC, Allen PJ, Renwick JAA (1971) Identi-
    fication of the germination self-inhibitor from wheat
    stem rust uredospores. Science 173:835–836
Macko V, Staples RC, Yaniv Z, Granados RR (1976) Self-
    inhibitors of fungal spore germination. In: Weber
    DJ, Hess WM (eds) Fungal spore: form and function.
    Wiley, New York, pp 73–100
Macko V, Renwick JAA, Rissler JF (1978) Acrolein induces
    differentiation of infection structures in the wheat
    stem rust fungus. Science 199:442–443
Maclean DJ (1982) Axenic culture and metabolism of rust
    fungi. In: Scott KJ, Chakravorty AK (eds) The rust
    fungi. Academic, London, pp 37–120
Maclean DJ, Scott KJ (1976) Identification of glucitol (sorb-
    itol) and ribitol in a rust fungus, *Puccinia graminis*
    f.sp. *tritici*. J Gen Microbiol 97:83–89
Madden LV, Wheelis M (2003) The threat of plant pathogens
    as weapons against US crops. Annu Rev Phytopathol
    41:155–176
Maheshwari R, Allen PJ, Hildebrandt AC (1967) Physical
    and chemical factors controlling the development of
    infection structures from urediospore germ tubes of
    rust fungi. Phytopathology 57:855–862
Maier W, Begerow D, Weiß M, Oberwinkler F (2003) Phy-
    logeny of the rust fungi: An approach using nuclear
    large subunit ribosomal DNA sequences. Can J Bot
    81:12–23
Maier W, Wingfield BD, Mennicken M, Wingfield MJ (2007)
    Polyphyly and two emerging lineages in the rust gen-
    era *Puccinia* and *Uromyces*. Mycol Res 111:176–185
Manners JM (1989) The host-haustorium interface in pow-
    dery mildews. Aust J Plant Physiol 16:45–52
Manners JM, Maclean DJ, Scott KJ (1982) Pathways of glu-
    cose assimilation in *Puccinia graminis*. J Gen Micro-
    biol 128:2621–2630
Manners JM, Maclean DJ, Scott KJ (1984) Hexitols as major
    intermediates of glucose assimilation by mycelium of
    Puccinia graminis. Arch Microbiol 139:158–161
Marte M (1971) Studies on self-inhibition of *Uromyces
    fabae* (Pers.) De Bary. J Phytopathol 72:335–343
Mayama S, Rehfeld DW, Daly JM (1975) A comparison of the
    development of *Puccinia graminis tritici* in resistant
    and susceptible wheat based on glucosamine content.
    Physiol Plant Pathol 7:243–257
McDonald BA (2004) Population genetics of plant pathogens.
    Interaction between mutation and selection. APSnet.
    doi:10.1094/PHI-A-2004-0524-01
Mellersh DG, Heath MC (2003) An investigation into the
    involvement of defense signaling pathways in compo-
    nents of the nonhost resistance of *Arabidopsis thaliana*
    to rust fungi also reveals a model system for studying
    rust fungal compatibility. Mol Plant–Microbe Interact
    16:398–404
Mendgen K (1979) Microautoradiographic studies on host-
    parasite interactions. II. The exchange of ^{3}H-lysine

between *Uromyces phaseoli* and *Phaseolus vulgaris*. Arch Microbiol 123:129–135

Mendgen K (1981) Nutrient uptake in rust fungi. Phytopathology 71:983–989

Mendgen K (1984) Development and physiology of teliospores. In: Bushnell WR, Roelfs AP (eds) Origins, specificity, structure, and physiology. The cereal rusts, vol 1. Academic, Orlando, pp 375–398

Mendgen K, Deising H (1993) Infection structures of fungal plant pathogens – A cytological and physiological evaluation. New Phytol 124:193–213

Mendgen K, Hahn M (2002) Plant infection and the establishment of fungal biotrophy. Trends Plant Sci 7:352–356

Mendgen K, Hahn M, Deising H (1996) Morphogenesis and mechanism of penetration by plant pathogenic fungi. Annu Rev Phytopathol 34:367–386

Mendgen K, Struck C, Voegele RT, Hahn M (2000) Biotrophy and rust haustoria. Physiol Mol Plant Pathol 56:141–145

Mendgen K, Wirsel SG, Jux A, Hoffmann J, Boland W (2006) Volatiles modulate the development of plant pathogenic rust fungi. Planta 224:1353–1361

Miles MR (2007) Differential response of common bean cultivars to *Phakopsora pachyrhizi*. Plant Dis 91:698–704

Mims CW (1981a) Ultrastructure of teliospore germination and basidiospore formation in the rust fungus *Gymnosporangium clavipes*. Can J Bot 59:1041–1049

Mims CW (1981b) SEM of aeciospore formation in *Puccinia bolleyana*. Scanning Electron Microsc 4:299–303

Mims CW, Thurston EL (1979) Ultrastructure of teliospore in the rust fungus *Puccinia podophylli*. Can J Bot 57:2533–2538

Mims CW, Rodriguez-Lother C, Richardson EA (2002) Ultrastructure of the host-pathogen interface in daylily leaves infected by the rust fungus *Puccinia hemerocallidis*. Protoplasma 219:221–226

Mitchell HJ, Ayliffe MA, Rashid KY, Pryor AJ (2006) A rust-inducible gene from flax (*fis1*) is involved in proline catabolism. Planta 223:213–222

Moerschbacher BM, Mierau M, Graessner B, Noll U, Mort AJ (1999) Small oligomers of galacturonic acid are endogenous suppressors of disease resistance reactions in wheat leaves. J Exp Bot 50:605–612

Moore RT, McAlear JH (1961) Fine structure of the mycota. 8. On the aecidial stage of *Uromyces caladii*. J Phytopathol 42:297–304

Mota LJ, Sorg I, Cornelis GR (2005) Type III secretion: the bacteria–eukaryotic cell express. FEMS Microbiol Lett 252:1–10

Mould MJ, Xu T, Barbara M, Iscove NN, Heath MC (2003) cDNAs generated from individual epidermal cells reveal that differential gene expression predicting subsequent resistance or susceptibility to rust fungal infection occurs prior to the fungus entering the cell lumen. Mol Plant–Microbe Interact 16:835–845

Nagarajan S, Singh DV (1990) Long-distance dispersion of rust pathogens. Annu Rev Phytopathol 28:139–153

Neu C, Keller B, Feuillet C (2003) Cytological and molecular analysis of the *Hordeum vulgare–Puccinia triticina* nonhost interaction. Mol Plant–Microbe Interact 16:626–633

Nirmala J, Brueggeman R, Maier C, Clay C, Rostoks N, Kannangara CG, von Wettstein D, Steffenson BJ, Kleinhofs A (2006) Subcellular localization and functions of the barley stem rust resistance receptor-like serine/threonine-specific protein kinase Rpg1. Proc Natl Acad Sci USA 103:7518–7523

Perfect SE, Green JR (2001) Infection structures of biotrophic and hemibiotrophic fungal plant pathogens. Mol Plant Pathol 2:101–108

Petersen RH (1974) The rust fungus life cycle. Bot Rev 40:453–513

Peterson PD, Leonard KJ, Roelfs AP, Sutton TB (2005) Effect of barberry eradication on changes in populations of *Puccinia graminis* in Minnesota. Plant Dis 89:935–940

Peterson RB, Aylor DE (1995) Chlorophyll fluorescence induction in leaves of *Phaseolus vulgaris* infected with bean rust (*Uromyces appendiculatus*). Plant Physiol 108:163–171

Pfunder M, Roy BA (2000) Pollinator-mediated interactions between a pathogenic fungus, *Uromyces pisi* (Pucciniaceae), and its host plant, *Euphorbia cyparissias* (Euphorbiaceae). Am J Bot 87:48–55

Pfunder M, Schürch S, Roy BA (2001) Sequence variation and geographic distribution of pseudoflower-forming rust fungi (*Uromyces pisi* s. lat.) on *Euphorbia cyparissias*. Mycol Res 105:57–66

Pivonia S, Yang XB (2004) Assessment of the potential year-round establishment of soybean rust throughout the world. Plant Dis 88:523–529

Pliny (69) Historia naturalis, book XVIII

Posada-Buitrago ML, Frederick RD (2005) Expressed sequence tag analysis of the soybean rust pathogen *Phakopsora pachyrhizi*. Fungal Genet Biol 42:949–962

Pretorius ZA, Singh RP, Wagoire WW, Payne TS (2000) Detection of virulence to wheat stem rust resistance Gene *Sr31* in *Puccinia graminis* f.sp. *tritici* in Uganda. Plant Dis 84:203

Rampitsch C, Bykova NV, McCallum B, Beimcik E, Ens W (2006) Analysis of the wheat and *Puccinia triticina* (leaf rust) proteomes during a susceptible host-pathogen interaction. Proteomics 6:1897–1907

Rapilly F (1979) Yellow rust epidemiology. Annu Rev Phytopathol 17:59–73

Rauscher M, Mendgen K, Deising H (1995) Extracellular proteases of the rust fungus *Uromyces viciae-fabae*. Exp Mycol 19:26–34

Rauscher M, Adam AL, Wirtz S, Guggenheim R, Mendgen K, Deising HB (1999) PR-1 protein inhibits the differentiation of rust infection hyphae in leaves of acquired resistant broad bean. Plant J 19:625–633

Read ND, Kellock LJ, Collins TJ, Gundlach AM (1997) Role of topography sensing for infection-structure differentiation in cereal rust fungi. Planta 202:163–170

Rehmany AP, Gordon A, Rose LE, Allen RL, Armstrong MR, Whisson SC, Kamoun S, Tylcr BM, Birch PR, Beynon JL (2005) Differential recognition of highly divergent downy mildew avirulence gene alleles by *RPP1* resistance genes from two *Arabidopsis* lines. Plant Cell 17:1839–1850

Rinaldi C, Kohler A, Frey P, Duchaussoy F, Ningre N, Couloux A, Wincker P, Le Thiec D, Fluch S, Martin F, Duplessis S (2007) Transcript profiling of poplar leaves upon infection with compatible and incompatible strains of the foliar rust *Melampsora larici-populina*. Plant Physiol 144:347–366

Roberts JK, Pryor A (1995) Isolation of a flax (*Linum usitatissimum*) gene induced during susceptible infection by flax rust (*Melampsora lini*). Plant J 8:1–8

Roelfs AP (1982) Effects of barberry eradication on stem rust in the United States. Plant Dis 66:177–181

Roelfs AP, Singh RP, Saari EE (1992) Rust diseases of wheat: concepts and methods of disease managment. CIMMYT, Mexico, D.F.

Roy BA (1993) Floral mimicry by a plant pathogen. Nature 362:56–58

Saksirirat W, Hoppe HH (1990) *Verticillium psalliotae*, an effective mycoparasite of the soybean rust fungus *Phakopsora pachyrhizi* Syd. J Plant Dis Prot 97:622–633

Saksiriat W, Hoppe HH (1991) Teliospore germination of soybean rust fungus (*Phakopsora pachyrhizi* Syd.). J Phytopathol 132:339–342

Schillberg S, Tiburzy R, Fischer R (2000) Transient transformation of the rust fungus *Puccinia graminis* f.sp. *tritici*. Mol Gen Genet 262:911–915

Schneider RW, Hollier CA, Whitam HK, Palm ME, McKemy JM, Hernandez JR, Levy L, DeVries Paterson R (2005) First report of soybean rust caused by *Phakopsora pachyrhizi* in the continental United States. Plant Dis 89:774

Scholes J, Farrar J (1985) Photosynthesis and chloroplast functioning within individual pustules of *Uromyces muscari* on bluebell leaves. Physiol Plant Pathol 27:387–400

Scholes J, Farrar J (1986) Increased rates of photosynthesis in localized regions of a barley leaf infected with brown rust. New Phytol 104:601–612

Scholes JD, Farrar JF (1987) Development of symptoms of brown rust of barley in relation to the distribution of fungal mycelium, starch accumulation and localized changes in the concentration of chlorophyll. New Phytol 107:103–117

Scholes JD, Rolfe SA (1996) Photosynthesis in localised regions of oat leaves infected with crown rust (*Puccinia coronata*): quantitative imaging of chlorophyll fluorescence. Planta 199:573–582

Scott KJ (1972) Obligate parasitism by phytopathogenic fungi. Biol Rev 47:537–572

Sharma JK, Sankaran KV (1988) Biocontrol of rust and leaf spot diseases. In: Mukerji KJ, Gary KL (eds) Biocontrol of plant diseases. CRC, Boca Raton, pp 1–23

Shu P, Ledingham GA (1956) Enzymes related to carbohydrate metabolism in uredospores of wheat stem rust. Can J Microbiol 2:489–495

Shu P, Tanner KG, Ledingham GA (1954) Studies on the respiration of resting and germinating uredospores of wheat stem rust. Can J Bot 32:16–23

Smith SM, Hulbert SH (2005) Recombination events generating a novel *Rp1* race specificity. Mol Plant–Microbe Interact 18:220–228

Sohn J, Voegele RT, Mendgen K, Hahn M (2000) High level activation of vitamin B1 biosynthesis genes in haustoria of the rust fungus *Uromyces fabae*. Mol Plant–Microbe Interact 13:629–636

Solomon PS, Tan K-C, Oliver RP (2003) The nutrient supply of pathogenic fungi; a fertile field for study. Mol Plant Pathol 4:203–210

Stakman EC, Piemeisal FJ (1917) Biological forms of *Puccinia graminis* on cereals and grasses. J Agric Res 10:429–495

Staples RC (2000) Research on the rust fungi during the twentieth century. Annu Rev Phytopathol 38:49–69

Staples RC (2001) Nutrients for a rust fungus: The role of haustoria. Trends Plant Sci 6:496–498

Staples RC (2003) A novel gene for rust resistance. Trends Plant Sci 8:149–151

Staples RC, Macko V (1984) Germination of urediospores and differentiation of infection structures. In: Roelfs AP, Bushnell WR (eds) Origins, specificity, structure, and physiology. The cereal rusts, vol 1. Academic, Orlando, pp 255–289

Staples RC, Hoch HC (1997) Physical and chemical cues for spore germination and appressorium formation by fungal pathogens. In: Carroll GC, Tudzynski P (eds) Plant relationships, part A. The Mycota, vol V. Springer, Berlin, pp 27–40

Staples RC, Grambow HJ, Hoch HC, Wynn WK (1983) Contact with membrane grooves induces wheat stem rust uredospore germlings to differentiate appressoria but not vesicles. Phytopathology 73:1436–1439

Stark-Urnau M, Mendgen K (1993) Differentiation of aeciiospore- and uredospore-derived infection structures on cowpea leaves and on artificial surfaces by *Uromyces vignae*. Can J Bot 71:1236–1242

Stark-Urnau M, Mendgen K (1995) Sequential deposition of plant glycoproteins and polysaccharides at the host–parasite interface of *Uromyces vignae* and *Vigna sinensis*. Protoplasma 186:1–11

Stokstad E (2007) Deadly wheat fungus threatens world's breadbaskets. Science 315:1786–1787

Struck C, Hahn M, Mendgen K (1996) Plasma membrane H+-ATPase activity in spores, germ tubes, and haustoria of the rust fungus *Uromyces viciae-fabae*. Fungal Genet Biol 20:30–35

Struck C, Ernst M, Hahn M (2002) Characterization of a developmentally regulated amino acid transporter (AAT1p) of the rust fungus *Uromyces fabae*. Mol Plant Pathol 3:23–30

Struck C, Voegele RT, Hahn M, Mendgen K (2004a) Rust haustoria as sink in plant tissues or – how to survive in leaves. In: Tikhonovich I, Lugtenberg B, Provorov N (eds) Biology of plant-microbe interactions, vol 4. International Society for Molecular Plant-Microbe Interactions, St Paul, pp 177–179

Struck C, Müller E, Martin H, Lohaus G (2004b) The *Uromyces fabae UfAAT3* gene encodes a general amino acid permease that prefers uptake of *in planta* scarce amino acids. Mol Plant Pathol 5:183–189

Struck C, Siebels C, Rommel O, Wernitz M, Hahn M (1998) The plasma membrane H+-ATPase from the biotrophic rust fungus *Uromyces fabae*: Molecular characterization of the gene (*PMA1*) and functional expression of the enzyme in yeast. Mol Plant–Microbe Interact 11:458–465

Swann EC, Frieders EM, McLaughlin DJ (1981) Urediniomycetes. In: McLaughlin DJ, McLaughlin EG, Lemke PA (eds) Systematics and evolution, part B. The Mycota VII. Springer, Berlin, pp 37–56

Szabo LJ, Bushnell WR (2001) Hidden robbers: the role of fungal haustoria in parasitism of plants. Proc Natl Acad Sci USA 98:7654–7655

Terhune BT, Bojko RJ, Hoch HC (1993) Deformation of stomatal guard cell lips and microfabricated artificial topographies during appressorium formation. Exp Mycol 17:70–78

Thara VK, Fellers JP, Zhou JM (2003) *In planta* induced genes of *Puccinia triticina*. Mol Plant Pathol 4:51–56

Thrall PH, Burdon JJ (2003) Evolution of virulence in a plant host–pathogen metapopulation. Science 299:1735–1737

Tiburzy R, Martins EMF, Reisener HJ (1992) Isolation of haustoria of *Puccinia graminis* f.sp. *tritici* from wheat leaves. Exp Mycol 16:324–328

Tissera P, Ayres PG (1986) Transpiration and the water relations of faba bean (*Vicia faba*) infected by rust (*Uromyces viciae-fabae*). New Phytol 102:385–395

Tuskan GA, Difazio S, Jansson S, Bohlmann J, Grigoriev I, Hellsten U, Putnam N, Ralph S, Rombauts S, Salamov A, Schein J, Sterck L, Aerts A, Bhalerao RR, Bhalerao RP, Blaudez D, Boerjan W, Brun A, Brunner A, Busov V, Campbell M, Carlson J, Chalot M, Chapman J, Chen GL, Cooper D, Coutinho PM, Couturier J, Covert S, Cronk Q, Cunningham R, Davis J, Degroeve S, Dejardin A, Depamphilis C, Detter J, Dirks B, Dubchak I, Duplessis S, Ehlting J, Ellis B, Gendler K, Goodstein D, Gribskov M, Grimwood J, Groover A, Gunter L, Hamberger B, Heinze B, Helariutta Y, Henrissat B, Holligan D, Holt R, Huang W, Islam-Faridi N, Jones S, Jones-Rhoades M, Jorgensen R, Joshi C, Kangasjarvi J, Karlsson J, Kelleher C, Kirkpatrick R, Kirst M, Kohler A, Kalluri U, Larimer F, Leebens-Mack J, Leple JC, Locascio P, Lou Y, Lucas S, Martin F, Montanini B, Napoli C, Nelson DR, Nelson C, Nieminen K, Nilsson O, Pereda V, Peter G, Philippe R, Pilate G, Poliakov A, Razumovskaya J, Richardson P, Rinaldi C, Ritland K, Rouze P, Ryaboy D, Schmutz J, Schrader J, Segerman B, Shin H, Siddiqui A, Sterky F, Terry A, Tsai CJ, Uberbacher E, Unneberg P, Vahala J, Wall K, Wessler S, Yang G, Yin T, Douglas C, Marra M, Sandberg G, Van de Peer Y, Rokhsar D (2006) The genome of black cottonwood, *Populus trichocarpa* (Torr. & Gray). Science 313:1596–1604

Unger F (1833) Die Exantheme der Pflanzen und einige mit diesen verwandten Krankheiten dieser Gewächse, pathogenetisch und nosographisch dargestellt. Gerold, Vienna

van de Mortel M, Recknor JC, Graham MA, Nettleton D, Dittman JD, Nelson RT, Godoy CV, Abdelnoor RV, Almeida ÁMR, Baum TJ, Whitham SA (2007) Distinct biphasic mRNA changes in response to Asian soybean rust infection. Mol Plant–Microbe Interact 20:887–899

van der Merwe M, Ericson L, Walker J, Thrall PH, Burdon JJ (2007) Evolutionary relationships among species of *Puccinia* and *Uromyces* (*Pucciniaceae, Uredinales*) inferred from partial protein coding gene phylogenies. Mycol Res 111:163–175

Vaz Patto MC, Niks RE (2001) Leaf wax layer may prevent appressorium differentiation but does not influence orientation of the leaf rust fungus *Puccinia hordei* on *Hordeum chilense* leaves. Eur J Plant Pathol 107:795–803

Voegele RT (2006) *Uromyces fabae*: development, metabolism, and interactions with its host *Vicia faba*. FEMS Microbiol Lett 259:165–173

Voegele RT, Mendgen K (2003) Rust haustoria: Nutrient uptake and beyond. New Phytol 159:93–100

Voegele RT, Struck C, Hahn M, Mendgen K (2001) The role of haustoria in sugar supply during infection of broad bean by the rust fungus *Uromyces fabae*. Proc Natl Acad Sci USA 98:8133–8138

Voegele RT, Wirsel S, Möll U, Lechner M, Mendgen K (2006) Cloning and characterization of a novel invertase from the obligate biotroph *Uromyces fabae* and analysis of expression patterns of host and pathogen invertases in the course of infection. Mol Plant–Microbe Interact 19:625–634

Voegele RT, Hahn M, Lohaus G, Link T, Heiser I, Mendgen K (2005) Possible roles for mannitol and mannitol dehydrogenase in the biotrophic plant pathogen *Uromyces fabae*. Plant Physiol 137:190–198

von Mohl H (1853) Ueber die Traubenkrankheit. Bot Z 11:585–590

Wahl I, Anikster Y, Manisterski J, Segal A (1984) Evolution at the center of origin. In: Bushnell WR, Roelfs AP (eds) Origins, specificity, structure, and physiology. The cereal rusts, vol 1. Academic, Orlando, pp 39–77

Walters DR, McRoberts N (2006) Plants and biotrophs: A pivotal role for cytokinins? Trends Plant Sci 11:581–586

Wandeler H, Bacher S (2006) Insect-transmitted urediniospores of the rust *Puccinia punctiformis* cause systemic infections in established *Cirsium arvense* plants. Phytopathology 96:813–818

Wanyera R, Kinyua MG, Jin Y, Singh RP (2006) The spread of stem rust caused by *Puccinia graminis* f.sp. *tritici*, with virulence on *Sr31* in wheat in Eastern Africa. Plant Dis 90:113

Webb CA, Fellers JP (2006) Cereal rust fungi genomics and the pursuit of virulence and avirulence factors. FEMS Microbiol Lett 264:1–7

Webb CA, Szabo LJ, Bakkeren G, Garry C, Staples RC, Eversmeyer M, Fellers JP (2006) Transient expression and insertional mutagenesis of *Puccinia triticina* using biolistics. Funct Integr Genomics 6:250–260

Webster J, Weber RWS (2007) Introduction to fungi. Cambridge University Press, Cambridge

Webster J, Davey RA, Smirnoff N, Fricke W, Hinde P, Tomos D, Turner JCR (1995) Mannitol and hexoses are components of Buller's drop. Mycol Res 99:833–838

Welter K, Müller M, Mendgen K (1988) The hyphae of *Uromyces appendiculatus* within the leaf tissue after high pressure freezing and freeze substitution. Protoplasma 147:91–99

Wessels JGH (1993) Wall growth, protein excretion and morphogenesis in fungi. New Phytol 123:397–413

Wiethölter N, Horn S, Reisige K, Beike U, Moerschbacher BM (2003) *In vitro* differentiation of haustorial mother cells of the wheat stem rust fungus, *Puccinia graminis* f.sp. *tritici*, triggered by the synergistic action of chemical and physical signals. Fungal Genet Biol 38:320–326

Williams PG, Scott KJ, Kuhl JL (1966) Vegetative growth of *Puccinia graminis* f.sp. *tritici* in vitro. Phytopathology 56:1418–1419

Winton LM, Manter DK, Stone JK, Hansen EM (2003) Comparison of biochemical, molecular, and visual methods to quantify *Phaeocryptopus gaeumannii* in Douglas-Fir foliage. Phytopathology 93:121–126

Wirsel SG, Voegele RT, Mendgen KW (2001) Differential regulation of gene expression in the obligate biotrophic interaction of *Uromyces fabae* with its host *Vicia faba*. Mol Plant–Microbe Interact 14:1319–1326

Wirsel SGR, Voegele RT, Bänninger R, Mendgen KW (2004) Cloning of β-tubulin and succinate dehydrogenase genes from *Uromyces fabae* and establishing selection conditions for their use in transformation. Eur J Plant Pathol 110:767–777

Wolf G (1982) Physiology and Biochemistry of spore germination. In: Scott KJ, Chakravorty AK (eds) The rust fungi. Academic, London, pp 151–178

Wood AR, Morris MJ (2007) Impact of the gall-forming rust fungus *Uromycladium tepperianum* on the invasive tree *Acacia saligna* in South Africa: 15 years of monitoring. Biol Control 41:68–77

Woods AM, Beckett A (1987) Wall structure and ornamentation of the urediniospores of *Uromyces viciae-fabae*. Can J Bot 65:2007–2016

Wynn WK (1976) Appressorium formation over stomates by the bean rust fungus: Response to a surface contact stimulus. Phytopathology 66:136–146

Xuei X, Bhairi S, Staples RC, Yoder OC (1992) Characterization of *INF56*, a gene expressed during infection structure development of *Uromyces appendiculatus*. Gene 110:49–55

Xuei X, Bhairi S, Staples RC, Yoder OC (1993) *INF56* represents a family of differentiation-specific genes from *Uromyces appendiculatus*. Curr Genet 24:84–88

Zhang L, Meakin H, Dickinson M (2003) Isolation of genes expressed during compatible interactions between leaf rust (*Puccinia triticina*) and wheat using cDNA-AFLP. Mol Plant Pathol 4:469–477

Zhao XR, Lin Q, Brookes PC (2005) Does soil ergosterol concentration provide a reliable estimate of soil fungal biomass. Soil Biol Biochem 37:311–317

Zhou XL, Stumpf MA, Hoch HC, Kung C (1991) A mechanosensitive channel in whole cells and in membrane patches of the fungus *Uromyces*. Science 253:1415–1417

5 The Sebacinoid Fungus *Piriformospora indica:* an Orchid Mycorrhiza Which May Increase Host Plant Reproduction and Fitness

Patrick Schäfer[1], Karl-Heinz Kogel[1]

CONTENTS

I.   Introduction 99
II. The Mycorrhizal Order *Sebacinales* 100
III. *Piriformospora indica* – an Orchid
     Mycorrhizal Fungus? 101
IV. Benefits of *P. indica* Symbiosis
     for Host Plants.......................... 103
V.   Cell Death Makes a Difference 104
VI. Parasitic Associations of Plants with *P. indica*... 105
VII. Factors Involved in Plant Colonisation
     by *P. indica*............................. 106
VIII. Impact of Various Plant Mutations
     on *P. indica*-Induced Resistance............. 108
IX. Bacterial Endosymbiotic Associations
     Within *Sebacinales*........................ 108
X.   Conclusions 110
     References................................ 110

I. Introduction

Plants are potential targets (hosts) for a broad spectrum of microbial organisms. The outcome of these associations can be roughly categorised into mutualistic, commensalistic or pathogenic relationships. Interactions with certain mutualistic fungal microbes can benefit plants, resulting for example in an improved plant development even under unfavourable environmental conditions (Chap. 15). Simultaneously, the microbial partners acquire nutrients from the host and can be protected from environmental stress or competitors (Schulz and Boyle 2005). In other cases it is the microbes that primarily profit from the association, with the host fitness being either apparently unaffected (commensalism) or thoroughly impaired (pathogenesis; Redman et al. 2001).

[1] Interdisciplinary Research Centre for BioSystems,
Land Use and Nutrition. Institute of Phytopathology and Applied Zoology, Justus Liebig University, Heinrich-Buff-Ring 26–32, 35392 Giessen, Germany.
 e-mail: patrick.schaefer@agrar.uni-giessen.de,
 e-mail: karl-heinz.kogel@agrar.uni-giessen.de

Prokaryotic or eukaryotic organisms with the capability of colonising plants are generally called endophytes. An endophytic lifestyle was reported among fungi, bacteria, algae, plants and even insects (Schulz and Boyle 2005). This broad defintion of endophytism was later specified to more strongly emphasise infection strategies or the physiological character of interaction types. However, due to the broad spectrum of endophytes and their flexibility (phenotypic plasticity) in host colonisation, along with their ability to adapt to environmental factors and the host's physiological status, a more restrictive general definition does not exist. Focusing on fungal microbes, endophytes were defined as organisms that grow in living plant tissue during their entire life cycle (or a significant part of it) without causing disease symptoms (Petrini 1991; Saikkonen et al. 1998; Brundrett 2004). Schulz and Boyle (2005) broadened this definition by describing endophytes as plant inhabitants that have not yet triggered disease symptoms in plants at the time of detection. This definition excludes the impact of endophytes on host fitness at later interaction stages; depending on their lifestyle in plants or impact on host fitness such fungal endophytes range under this definition from mutualistic to pathogenic microbes (Redman et al. 2001; Schulz and Boyle 2005). In order to simplify this heterogeneity, we follow a rather restricted definition of endophytes encompassing microbes with an asymptomatic lifestyle throughout their interaction with plants. The intention of this definition is to address those fungi whose association and reproduction in plants cause neutral or beneficial rather than detrimental effects in their hosts.

Described in a broad sense, mycorrhizas are highly specialised beneficial associations between plant roots and fungi based on the bilateral exchange of nutrients, defence against pathogens and abiotic stress or an improved water balance. Variations in the benefits for each symbiotic partner gave rise to the terms balanced and exploitive mycorrhizas. Whereas in the former both partners

Plant Relationships, 2nd Edition
The Mycota V
H. Deising (Ed.)
© Springer-Verlag Berlin Heidelberg 2009

benefit equally from each other, the latter type of interaction favours the plant partner. Due to their beneficial potential for plants, mycorrhizal fungi are among the best-characterised fungal symbionts (Chaps. 13, 14). According to the above definition, mycorrhizal fungi would be considered as endophytes displaying mutualistic interactions with plants. However, in order to distinguish mycorrhizas from endophytes, a more precise definition was conceived: Endophytic plant–microbe associations lack a synchronised plant–fungus development, specialised microbial structures serving as localised plant–microbe interfaces and nutrient transfer to the plant (Brundrett 2004). Irrespective of these characteristics and as mentioned above, host plants are well known to benefit from nonmycorrhizal endophytes to their hosts. A common example is the release of toxic or antimicrobial compounds distracting herbivoric and microbial competitors (Schulz et al. 2002; Chap. 15). In other cases plant fitness is enhanced by improved water use efficiency, drought tolerance and enhanced germination rates (Saikkonen et al. 1998; Brundrett 2004). In addition, several endophytes promote plant growth and confer local and systemic induced resistance to plant pathogens (Varma et al. 1999; Schulz and Boyle 2005; Waller et al. 2005).

The fungal basidiomycete *Piriformospora indica* has drawn attention since its discovery in India during the final decade of the past century – not least due to its versatile beneficial effects conferred to a broad variety of host plant species, e.g. barley, maize, parsley, poplar, tobacco and wheat (Sahay and Varma 1999; Varma et al. 1999; Waller et al. 2005; Serfling et al. 2007). This broad host range, combined with its easy handling, makes the fungus a potential agent for protecting plants against abiotic and biotic stresses under greenhouse or field conditions. Hence, *P. indica* could support sustainability in horticulture and agriculture. Because of the reported beneficial effects, it was rather unexpected that colonisation of barley roots was found to be associated with cell death (Deshmukh et al. 2006). In agreement with other endophytic plant–fungus interactions, colonised plants were observed to lack visible disease symptoms (e.g. stunted root and shoot development, or root necrosis). Due to its colonising behaviour, the lack of distinctive colonisation structures and the as yet missing evidence for nutrient transfer to its host plants, *P. indica* was suggested to be a

fungal endophyte rather than a representative of the mycorrhizal fungi. In this chapter we discuss current results showing beneficial associations of *P. indica* with plants, especially emphasising its life strategies in host plants. Intriguingly, it has been shown that root colonisation by *P. indica* and its lifestyle in planta may vary depending on environmental factors, the genetic predisposition and the developmental stage of host plants and plant organs, respectively. These findings are discussed in the context of the phylogenetic classification of *P. indica* within the newly defined mycorrhizal order *Sebacinales*.

II. The Mycorrhizal Order *Sebacinales*

Based on morphological and ultrastructural characteristics, members of the order *Sebacinales* were originally classified as wood-decaying basidiomycetes of the order *Auriculariales* (Bandoni 1984; Weiss et al. 2004). However, recent phylogenetic studies using the nuclear DNA sequence of the large ribosomal subunit resulted in the definition of the fungal order *Sebacinales*, occupying a central position within the *Hymenomycetidae*. The order *Sebacinales* exclusively harbours beneficial fungi; however these show an extraordinary diversity, encompassing ectomycorrhizas, orchid mycorrhizas, ericoid mycorrhizas, cavendishioid mycorrhizas and jungermannioid mycorrhizas in liverworts (McKendrick et al. 2002; Selosse et al. 2002, 2007; Kottke et al. 2003; Urban et al. 2003; Weiss et al. 2004; Setaro et al. 2006). Hence, the *Sebacinales* might possess remarkable significance in natural ecosystems (Weiss et al. 2004).

Phylogenetic analysis divided the *Sebacinales* into two subgroups. Subgroup A harbours ectomycorrhizas and orchid mycorrhizas that usually form hyphal sheaths and occasionally intracellular hyphae. Fungi of this group are associated with achlorophyllous or rather heterotrophic orchids (Weiss et al. 2004). Recently, ectendomycorrhizal sebacinoids were isolated from *Ericaceae*. In addition to hyphal sheaths, colonised roots showed intercellular networks as well as intracellular structures (Selosse et al. 2007). Since some members of subgroup A are thought to form tripartite symbioses connecting trees with orchids, it is speculated that most of these fungi are able to form both ecto- and orchid mycorrhizal interactions. Subgroup B is more heterogenic with respect to the types of mycorrhizal associations. It mainly consists of *Sebacina vermifera* isolates from autotrophic orchids, ericoid mycorrhizas associated with *Gaultheria shallon*, cavendishioid mycorrhizas and liverwort-associated jungermannioid mycorrhizas (Weiss et al. 2004; Selosse et al. 2007). Within this group, isolates of *S. vermifera* represent a particularly interesting complex. These fungi can be axenically cultivated, which distinguishes them from sebacinoid mycobionts of group A. Interestingly, Warcup (1988) isolated several orchid symbionts of the *S. vermifera* complex that were shown to form hyphal coils in orchids. However, only those isolates that were isolated from ectomycorrhizal hosts were able to establish ectomycorrhizal interactions. Furthermore it was confirmed that the symbionts can only colonise a limited number of orchid hosts (Warcup 1988). In conclusion, *S. vermifera* isolates were proposed to represent a conglomerate of species rather than one diverse species (Warcup 1988; Weiss et al. 2004). It is even speculated that all members of subgroup B belong to the *S. vermifera* complex. However, this open question can only be answered when more knowledge on teleomorph stages of jungermannioid and ericoid mycorrhizas becomes available (Weiss et al. 2004).

Although exhaustive fungal sampling has not been performed, *Sebacinales* have been identified worldwide (Verma et al. 1998; Weiss et al. 2004; Setaro et al. 2006; Selosse et al. 2007) with specific branches isolated in Australia, Europe and North America (Weiss et al. 2004; Selosse et al. 2007). To date it is not known whether all *Sebacinales* are beneficial for their hosts. However, those members of the *Sebacinales* (*S. vermifera* isolates, *P. indica*, multinucleate *Rhizoctonia*) that have been examined for their mutualistic activity were able to promote growth and/or enhance disease resistance in monocotyledonous and dicotyledonous plants (Waller et al. 2005; Deshmukh et al. 2006), or support seed germination in orchids (Warcup 1988). These studies revealed that the fungi exhibit broad host specificity, although the majority were isolated from orchids, where they exhibit a rather narrow host range (Warcup 1988).

The recently described fungus *P. indica* was shown to be embedded within this group of mutualistic fungi, with the closest relationship to *S. vermifera* and multinucleate *Rhizoctonia* (Weiss et al. 2004). Although the latter was originally designated as *Rhizoctonia* sp., due to its morphological traits, recent phylogenetic studies clearly identified this fungus as a member of the *Sebacinales*. Hence, this isolate is not closely related with the pathogenic *Rhizoctonia solani* spp. (teleomorphs = *Thanatephorus*) and binucleate *Rhizoctonia* spp. (teleomorphs = *Ceratobasidium*), which are grouped within the *Ceratobasidiales* (Ogoshi 1987; Weiss et al. 2004; Gonzalez et al. 2006). Considering the beneficial effects caused by *P. indica* and the related *Sebacina* spp. or the multinucleate *Rhizoctonia*, *P. indica* might be regarded as a representative member of a huge group of microorganisms with considerable biological activities, significant agronomical potential and high ecological relevance.

III. *Piriformospora indica* – an Orchid Mycorrhizal Fungus?

P. indica was isolated for the first time from an association with a spore of *Glomus mosseae* in the rhizosphere of two shrubs of the Indian Thar desert, northwest Rajasthan (Verma et al. 1998). The fungus shows morphological traits common to members of the *Sebacinales*. In particular it possesses dolipores with imperforated parenthosomes (Verma et al. 1998) and does not have clamp connections. The structure of the basidia is unknown, since teleomorphs have not yet been isolated. However, these ultrastructural characteristics are in accordance with the phylogenetic analyses classifying *P. indica* as a member of the *Sebacinales* (Weiss et al. 2004). Whether *P. indica* coexists with *Glomus* spp. under natural conditions, or if its isolation from *Glomus mosseae* reflects a coincidence of circumstances, has not yet been investigated. It is known that *Sebacinales* often live in association with ascomycetes in their hosts and even colonise the same cells (Selosse et al. 2002; Urban et al. 2003; Setaro et al. 2006); but the reason for this coexistence is not known.

As mentioned above, the order *Sebacinales* harbours almost all mycorrhizal types other than vesicular arbuscular mycorrhizal (AM) fungi, which belong to the phylum *Glomeromycota*. Within the *Sebacinales*, *P. indica* exhibits the closest relationships to *S. vermifera* and multinucleate *Rhizoctonia*. The various *S. vermifera* isolates were sampled from diverse orchid plants and shown to support orchid seed germination (Warcup 1988; Weiss et al. 2004). The natural mycorrhizal plant partner(s) of multinucleate *Rhizoctonia* has not been definitively determined. Interestingly, the endophyte was isolated from vesicles of *Glomus fasciculatum* in pot cultures of *Trifolium subterraneum* L. (Williams 1985). In analogy to *P. indica*, the interfungal relationship to this AM fungus in nature is unknown. Both *S. vermifera* isolates and multinucleate *Rhizoctonia* exhibit a pronounced host specificity among orchids regarding their beneficial impact, e.g. by supporting seed germination. These fungi were determined to form intracellular hyphal coils (Milligan and Williams 1988; Warcup 1988), which represent characteristic traits of orchid mycorrhizas (Peterson and Massicotte 2004). In contrast to its closest neighbours, *P. indica* was reported to be isolated from the rhizosphere of the shrubs *Zizyphus nummularia* and *Prosopis juliflora* which belong, respectively, to the Rhamnaceae and Fabaceae (Verma et al. 1998).

Specific colonisation types classify each member of the *Sebacinales* to defined mycorrhizal categories. Sebacinoid fungi develop hyphal sheaths, Hartig nets and intracellular coils (ericoid and cavendishioid mycorrhiza, arbutoid ectendomycorrhiza), solely build intracellular coils (orchid mycorrhiza), or even colonise roots intercellularly (ectomycorrhiza ;Brundrett 2004; Peterson and Massicotte 2004; Selosse et al. 2007). Compared to these mycorrhizal types, epifluorescence microscopy revealed a divergent colonisation type for *P. indica* in barley and *Arabidopsis thaliana* roots (Deshmukh et al. 2006; Schäfer, unpublished data). Upon root contact, the fungus starts forming extracellular hyphal mats, which progressively develop. In parallel, it initiates intercellular root colonisation and frequently penetrates rhizodermal and cortical cells. As colonisation proceeds, the root is densely covered with extracellular hyphae and harbours thorough inter- and intracellular networks. However, the fungus never enters vascular tissue. Eventually, fungal colonisation leads to extracellular and intracellular sporulation (formation of chlamydospores; Deshmukh et al. 2006).

Some of these colonisation traits bear similarities to mycorrhizal symbioses. For instance, although the mycelium of *P. indica* is less densely packed and never covers the whole root surface, the extracellular colonisation pattern of *P. indica* is reminiscent of hyphal sheaths (Deshmukh et al. 2006). External hyphal growth was regarded not to be a characteristic of endophytes and rather treated as a mycorrhizal trait

(Saikkonen et al. 1998). Some dark septate endophytes (DSE) exhibit an asymptomatic colonisation pattern intringuingly similar to that of *P. indica* (Jumpponen and Trappe 1998). *Phialocephala fortinii*, a representative member of DSE, forms an extensive extracellular hyphal net prior to inter- and intracellular colonisation of rhizodermal, cortical, or root hair cells. Moreover, the fungus often builds intracellular coiled structures in ericaceous plants and even forms Hartig nets or labyrinthine hyphae when associated with ectomycorrhizal hosts. Similarly, *P. indica* was shown to occasionally produce intracellular coils in the monocotyledonous hosts maize and barley (Varma et al. 1999; Deshmukh et al. 2006), reminiscent of hyphal pelotons seen in the cortical cells of orchid mycorrhizas. Similar structures have occasionally been observed in *A. thaliana* (Fig. 5.1). Illustratively, Blechert et al. (1999) ana-

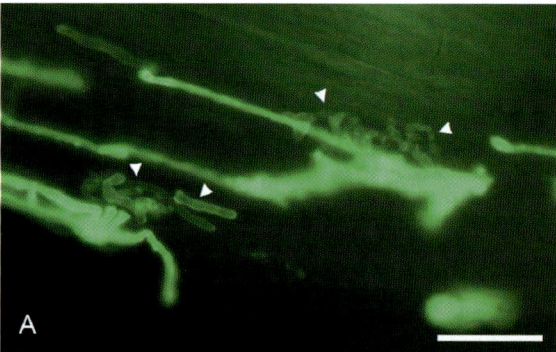

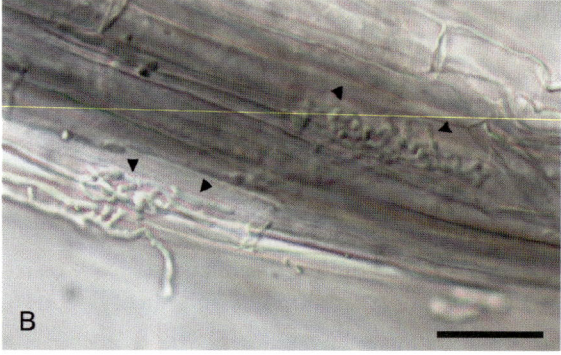

Fig. 5.1. Formation of intracellular coil-like structure of *Piriformospora indica*. Rhizodermal cell of the *Arabidopsis thaliana* root differentiation zone showing intracellular colonisation by *P. indica* at 7 days after inoculation. The fungus has begun to form coil-like hyphal structures (*arrowheads*) that eventually fills the entire plant cell (**A**). *P. indica* was stained with chitin-specific WGA-AF488. **B** Bright-field interference contrast image from the same colonised plant cells. Images were taken using an Axioplan 2 microscope. Root segments were excited at 470/20 nm and detected at 505–530 nm for WGA-AF 488. *Bar* 20 μm

lysed the colonisation of protocorms and roots of autotrophic *Dactylorhiza* spp. (Orchidaceae) by *P. indica* and found hyphal coils (pelotons) to be the typical intracellular structure. In these experiments, *P. indica* was shown to support the development of *D. maculata*. Moreover, comparison of the intracellular pelotons formed in protocorms of two *Dactylorhiza* spp. by *P. indica* were similar in morphology to pelotons formed in naturally grown *D. majalis* by an unknown orchid mycorrhiza. In orchid mycorrhizas, these pelotons are surrounded by perifungal membranes and interfacial matrices separating them from the host cytoplasm. These complexes represent plant–fungus interfaces and function specifically in nutrient exchange (Peterson and Massicotte 2004). In analogy to plant–*P. indica* associations, orchid mycorrhizas do not build the Hartig nets or arbuscules commonly observed, respectively, in ecto-mycorrhizas or arbuscular mycorrhizas (Peterson and Massicotte 2004). Recapitulating, following the definition of Brundrett (2004) it might be tempting to classify *P. indica* as an orchid mycorrhizal fungus. However, it remains of principal importance to determine whether the coiled or non-coiled intracellular hyphae possess perifungal membranes as well as interfacial matrices enabling these organs to exchange nutrients, as reported for orchid mycorrhizas. Interestingly, most members of the *Sebacinales* exhibit some host flexibility, enabling them to form ectomycorrhizas or orchid mycorrhizas (Warcup 1988; Weiss et al. 2004). It should be emphasised that all of the above-mentioned mycorrhizal traits are variable and depend on environmental factors as well as the colonised host. As a consequence, AM and ectomycorrhizal fungi colonise non-host plants or older root regions of hosts in an endophytic manner, presumably in order to guarantee survival (Brundrett 2004, Johnson et al. 1997). Thus, based on the above presumptions, *P. indica* might be regarded as a mycorrhizal fungus in associations with certain hosts (e.g. orchids), while its endophytic non-mycorrhizal activity might be predominant in alternative hosts such as barley and *A. thaliana*.

IV. Benefits of *P. indica* Symbiosis for Host Plants

The beneficial effects conveyed by *P. indica* and related *Sebacina* spp. to the plant companion have been extensively studied in barley (Waller et al. 2005; Deshmukh et al. 2006). Colonised plant seedlings show up to 30% increase in shoot biomass under greenhouse conditions. Importantly, this positive growth effect is also verifiable under field test conditions: When the spring barley elite cultivar Annabel grown in Mitcherlich test pots is colonised by *P. indica*, both the plant biomass compared to non-colonised plants and the grain yield are increased by about 10%. Unlike arbuscular mycorrhiza, growth promotion governed by *P. indica* has been demonstrated to be unaffected by P or N fertilisation (Achatz and Waller, unpublished data).

P. indica-colonised plants also acquire improved disease resistance towards the necrotrophic root pathogens *Fusarium culmorum* (Waller et al. 2005) and *F. graminearum* (Deshmukh and Kogel 2007). The molecular mechanism of this antifungal activity is not clear, because most of the defence-related *PR* genes in barley roots are only moderately and transiently induced by *P. indica* at early penetration stages (Schäfer et al., unpublished data), and evidence for antimicrobial compounds was not found. Significantly, barley leaves are very efficiently protected from infections by the powdery mildew fungus *Blumeria graminis* f.sp. *hordei* (up to 70% reduction in pustule frequencies), suggesting that a systemic resistance response is elicited by root colonisation. Systemic activation of the plants' defence machinery is corroborated by the detection of *P. indica*-mediated elevation of subcellular plant defence responses, such as cell wall apposition (papillae) and hypersensitive response, in association with attempted infection by *B. graminis*. Likewise, seven *S. vermifera* isolates originating from Australian and European sources confirmed a systemic protection activity in barley seedlings ranging from 10% to 80% reduction of powdery mildew colonies. It remains to be shown to what extent this activity spectrum may reflect a variable constitutive biological potential of single *Sebacina* strains or, alternatively, host cultivar-specific associations and thus varying degrees of specialisation of the mutualistic symbiosis.

Growth promotion as well as enhanced resistance conferred by *P. indica* against pathogens colonising roots (*Fusarium culmorum*), stem bases (*Pseudocercosporella herpotrichoides* (teleomorph: *Tapesia yallundae*) and leaves (*B. graminis* f.sp. *tritici*) were also observed in wheat under greenhouse conditions. Interestingly, the effects were mainly recorded when plants were grown on sand. However, similar effects could not be observed under field conditions, with the exception of a reduced disease development of *P. herpotrichoides* and a higher straw production in a field with poor soil quality (Serfling et al. 2007). It is noteworthy that the defence potential of *Piriformospora indica* against *Pseudocercosporella herpotrichoides* might rely on systemic effects, since both fungi colonise different plant organs.

Unexpectedly, *A. thaliana* is also among the wide range of host plants of *Piriformospora indica* (Pham et al. 2004). Fungal colonisation influences expression of specific genes in roots of *A. thaliana*, both before and after root contact with the fungal mycelium and promotes plant growth (Shahollari et al. 2005; Sherameti et al. 2005). In addition, colonised plants exhibit better growth performance and are more resistant against *Golovinomyces orontii,* the causal agent of powdery mildew on *A. thaliana* leaves. This systemic character of the induction of disease resistance becomes apparent in the reduced potential of the pathogen to propagate, due to reduced numbers of conidiophores per area unit mycelium and reduced numbers of conidia produced per leaf fresh weight (Stein and Molitor, unpublished data).

V. Cell Death Makes a Difference

Despite extensive colonisation by *P. indica*, barley and *A. thaliana* roots do not display any macroscopic evidence for impairment or even necrotisation (Fig. 5.2). Importantly, the colonisation patterns of the various root regions harbour some quantitative as well as qualitative differences, which additionally distinguish *P. indica* on barley (and *Arabidopsis*) from endomycorrhizal fungi. Fungal root colonisation increases with root maturation and the highest fungal biomass has been found in the differentiation and particularly the root hair zones. Cytological studies revealed the various interaction types of *P. indica* with different barley root regions and showed that the root hair zone (as oldest root zone) was mostly severely colonised by intracellular hyphae. In contrast, cells of the differentiation zone were often filled with fungal hyphae reminiscent of hyphal coils (Deshmukh et al. 2006), while the meristematic zone was barely and solely extracellularly colonised. Root colonisation by *P. indica* differs from that of AM fungi, which are known to preferentially colonise younger root parts, since the physiological activity of host cells is a prerequisite for efficient nutrient exchange between the symbiotic partners.

One of the main qualitative differences between *P. indica* and other mycorrhizas is the requirement of cell death for root colonisation (Deshmukh et al. 2006). Recent transmission electron microscopic studies revealed that cells are not dead at penetration stages, but show ultrastructural changes

Fig. 5.2. *A. thaliana* root responses towards *P. indica* colonisation. *A. thaliana* plants at 7 days after inoculation with *P. indica* (**A**) or mock-treatment (**B**). Plants were grown for 3 weeks on 0.5 MS medium (mod. 4; Duchefa, The Netherlands) in Petri dishes before inoculation of plants with spore suspension (500 000 spores ml^{-1}) or mock-treatment with 0.02% Tween water. At this stage, roots show intensive inter- and intracellular colonisation without causing visible colonisation symptoms in host roots

as cell colonisation becomes established (Schäfer and Zechmann, unpublished data). These findings suggest that the fungal colonisation strategy is not simply focused on the perception and subsequent colonisation of dead cells. In other words, penetrated host cells obviously die at one defined point of cell colonisation. The fact that this colonisation strategy crucially depends on host cell death at a certain interaction stage was shown in barley plants constitutively overexpressing the negative cell death regulator *Bax Inhibitor-1*. As a result of the genetically increased cell viability, fungal root colonisation was significantly reduced in these transgenic plants (Deshmukh et al. 2006). Conspicuously, in roots of wild-type barley inoculated with *P. indica*, *Bax Inhibitor-1* was found to be suppressed 5 days after inoculation and thereafter. The question arises whether this cell death-associated host response reflects a general colonisation strategy of the endophyte to benefit from plants. Alternatively, it may reflect some kind of imbalanced interaction with unfavourable host plants. As mentioned above, other mycorrhizal fungi are capable of colonising non-host roots and older root regions of host plants in an endophytic manner (Brundrett 2004). Nevertheless, AM fungi are incapable of initiating reproduction in these situations and nutrient supply is apparently not sufficient to guarantee long-term survival. In other words, the AM fungus is changing its life strategy in order to survive hostile conditions (Brundrett 2004). In contrast, *P. indica* is able to sporulate in barley and *A. thaliana* roots and, during the establishment of an initial biotrophic phase, the fungus does not induce apparent molecular and structural defence mechanisms, implying a certain degree of adaptation to these plants (Schäfer and Zechmann, unpublished data).

VI. Parasitic Associations of Plants with *P. indica*

Despite the flexibility of endophytes in colonising plants, the environmental factors, developmental stages and genetic predispositions of the interacting organisms can turn an asymptomatic association into parasitic or incompatible interactions, in which the endophyte either exhibits detrimental growth in plants traceable by disease development (and yield decrease), or has lost the capability to enter the plant tissue. Schulz and Boyle (2005) found that endophytes tended to exhibit a parasitic

lifestyle on host plants in laboratory or greenhouse studies, most probably due to unfavourable environmental conditions, whereas symptomless associations were observed in field experiments. This view is supported by investigations on dark septate endophytes in which experimental conditions resulted in a switch to a parasitic lifestyle (Jumpponen and Trappe 1998). Another example is given by *Lophodermium*. This endophyte was shown to asymptomatically colonise young needles of white pines but to switch to a more extensive and parasitic colonisation pattern during needle senescence (Deckert et al. 2001). Experiments with the root endophyte *Epichloë festucae* showed that such switches in endophytic life strategies do not necessarily depend on polygenetic traits. In contrast to the symptomless colonisation of the wild type in the host *Lolium perenne*, endophytic mutants defective in a *NADPH oxidase* (*noxA*) and a proposed regulator (*noxR*) displayed pathogenic colonisation (Takemoto et al. 2006; Tanaka et al. 2006; Chap. 15). Interestingly, the cucurbit pathogen *Colletotrichum magna* was also converted into a fungal mutalist by disruption of a single gene, although the respective gene has not yet been identified (Freeman and Rodriguez 1993; Redman et al. 1999). Hence, endophytism and even parasitism is a matter of harbouring or lacking certain genes or sets of genes. Such rather simplified genetic switches might represent a significant advantage. For example they might support a physiological flexibility under various environmental conditions and, thus, promote fungal reproduction. These genes might represent determinants of the life strategy of these microbes and help them to occupy ecological niches.

Schulz and Boyle (2005) hypothesised a "balanced antagonism" of endophyte–plant interactions, meaning an equilibrium between endophytic virulence factors and host defence responses that enable restricted non-pathogenic tissue colonisation. As soon as external or internal factors are misbalanced the asymptomatic interaction can turn into a parasitic one.

Conditions that are unfavourable for *P. indica*-plant associations, for example an antagonistic genetic background of the host plant or environmental factors, have been reported to impair or even change the outcome of the symbiosis (Kaldorf et al. 2005). However, stunted root development, as recently observed in sterile culture (Sirrenberg et al. 2007), might neither display unfavourable conditions nor be misinterpreted as a parasitic trait of *P. indica*. It rather indicates the auxin-producing capacities of the fungus that are pronounced under certain inoculation conditions. Under natural conditions, similar plant reactions might not be triggered by the fungus. For example, the use of chlamydospores of *P. indica* for plant root inoculation under comparable sterile conditions does not provoke stunted root growth (Fig. 5.2), despite intensive root colonisation. Plant root inhabitation by *P. indica* appears to depend on the developmental stage of the root tissue. In healthy roots, younger root tissue of the meristematic region is barely colonised in *A. thaliana* and barley; and in those rare cases only extracellular colonisation occurs. Deshmukh et al. (2006) comparatively quantified colonisation of root tip regions and the rest of the root and reported on significantly reduced fungal biomass in the former tissue. At the interaction sites of the meristematic zone, where *P. indica* started occupying rhizodermal cells, the plant showed a hypersensitive response-like defence reaction (Fig. 5.3). These defence responses were not detected in equally invaded cells of older root parts (Fig. 5.1; Schäfer, unpublished data). The assumption that mature root zones represent an accumulation of dead or inactive cells, and thus unprotected entry points, can be

excluded. By investigating diverse *A. thaliana* plants in which both the structural components of root cells (e.g. actin, tubulin) and the cellular organelles (e.g. nucleus, endoplasmic reticulum, plasma membrane) were tagged with green fluorescing protein, it became obvious that even mature cells were alive at the time of fungal penetration (Schäfer, unpublished data). Taken together, these studies demonstrate a clear preference of the fungus for mature root tissue. Second, the fungus is obviously recognised by its host. The colonisation pattern might be due either to a less active host immunity surveillance system in mature root parts, or to a facilitated access due to the elimination of adverse host activity by the fungus. In contrast, due to the exceptional importance of the meristematic zone for plant survival, this zone might be particularly guarded by the innate immunity system. Obviously the host is capable of restricting fungal colonisation; and this control appears to gradually decrease as root tissue matures.

As discussed above for other plant–endophyte associations, environmental conditions can provoke a pathogenic lifestyle of *P. indica* in host plants, as described by Kaldorf et al. (2005), whose study showed that the beneficial effects of *P. indica* on populus seedlings were redirected into reduced root growth and leaf necrosis when ammonium instead of nitrate was provided as single nitrogen source during plant–fungus co-cultivation. Under these experimental conditions, the fungus exhibited an unrestricted invasion of all plant organs including aerial parts. By adopting the same experimental setup, we reproduced these detrimental effects of *P. indica* in *A. thaliana* and barley (Schäfer and Kogel, unpublished data).

VII. Factors Involved in Plant Colonisation by *P. indica*

As mentioned in the previous sections, the host range of endophytes can be restricted by their genetic predisposition as well as by plant factors. Under natural conditions some endophytes display a certain degree of host specificity, so that not all plant taxa are equally infested. Failed colonisation may be accompanied by the development of disease symptoms (Schulz and Boyle 2005).

So far, *P. indica* has not been shown to possess a distinct host specificity, nor have non-host plants been detected. The fungus colonises monocotyledonous and dicotyledonous plants equally well.

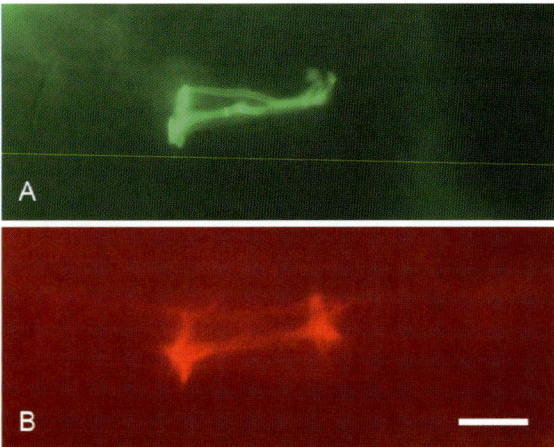

Fig. 5.3. Defence response in a cell colonised by *P. indica*. A rhizodermal cell of the meristematic *A. thaliana* root zone reacts with a hypersensitive-like response after penetration by *P. indica* at 3 days after inoculation. **A** *P. indica* was stained with WGA-AF488. **B** Under UV light the colonised cell shows autofluorescence. Images were taken using an Axioplan 2 microscope. Root segments were either excited at 470/20 nm and detected at 505–530 nm for WGA-AF 488, or excited at 546/12 nm and detected at 590 nm for detection of autofluorescence. *Bar* 20 μm

Hosts include orchids (*Dactylorhiza* sp.) and members of the Poaceae (e.g. barley, maize, rice, wheat) and Brassicaceae (e.g. *A. thaliana*; Verma et al. 1998; Blechert et al. 1999: Varma et al. 1999; Waller et al. 2005); and colonisation is asymptomatic, although these plants are both inter- and intracellularly colonised. The question arises to what extent the plant innate immunity is activated by *P. indica*. In barley, defence genes are moderately and transiently induced, as indicated by a marker gene (*PR-1, PR-2, PR-5*) expression study (Deshmukh and Kogel 2007; Waller et al. 2008) and microarray-based investigations (Schäfer et al., unpublished data). This is reminiscent of findings reported for plants colonised by AM fungi (Harrison 2005). Some common defence reactions were found in plant–endophyte interactions, e.g. papillae formation, cell wall lignification, H_2O_2 accumulation, enhanced peroxidase activity, or accumulation of phenolic compounds (Schulz and Boyle 2005). Whether these responses significantly contribute to the restriction of endophytic colonisation is unknown. Studies with tobacco and *Nicotiana sylvestris* constitutively expressing different plant chitinases demonstrated that defence-related proteins do not per se exhibit antimicrobial activity against the AM fungus *Glomus mosseae* (Vierheilig et al. 1993, 1995). Certain chitinases are even reported to support mycorrhizal root colonisation by hydrolysing chitin (Salzer et al. 1997), which would otherwise be recognised by the plant innate immunity system and induce pathogen-associated molecular pattern (PAMP)-triggered immunity (Jones and Dangl 2006; Kaku et al. 2006; Miya et al. 2007). Analogously, greenhouse experiments revealed that barley plants constitutively overexpressing an endochitinase of the soilborne fungus *Trichoderma harzianum* were equally well colonised by *G. mosseae* as control plants. Since these plants were shown to synthesise and secrete a highly active recombinant protein, antimicrobial activities of chitinases might not impair the mycorrhizal fungus (Kogel, von Wettstein and Schäfer, unpublished data).

However, recent studies identified some host genes of *A. thaliana* which restrict or support the colonisation of plants by *P. indica*. As reported above, root cell death regulation might be of importance since overexpression of the negative cell death regulator *Bax Inhibitor-1* reduces fungal colonisation in barley and in *A. thaliana* roots (Deshmukh et al. 2006; Schäfer and Kogel, unpublished data). In a genetic screen for host factors

regulating fungal colonisation of plant roots, the ethylene-insensitive *A. thaliana* mutant line *etr1-3* was identified (Khatabi and Schäfer, unpublished data). QPCR-based quantification of fungal biomass revealed a lower colonisation of this mutant, which was defective in an ethylene receptor and thus impaired in ethylene-mediated signalling responses (Bleecker et al.1988; Benavente and Alonso 2005). Despite the lower colonisation rate of *etr1-3*, induction of *P. indica*-mediated resistance to powdery mildew was not impaired. Similarly, the *A. thaliana* mutant line *ctr1*, which is defective in a serine/threonine protein kinase and acts as a negative regulator of the ethylene response pathway (Kieber et al. 1993), displayed a constitutive expression of ethylene responsive genes (Zhong and Burns 2003) and showed 3- to 4-fold higher colonisation. The extent to which ethylene production is affected in roots of *A. thaliana* interacting with *P. indica* is currently under investigation.

Recent studies on the *Nicotiana attenuata–S. vermifera* interaction indicated that infested seedlings showed reduced sensitivity to the ethylene precursor 1-aminocyclopropane-1-carboxylic acid (ACC). After applying ACC to dark-grown seedlings, morphological effects known as the triple response (a shortened and thickened hypocotyl, the inhibition of root elongation growth, a pronounced apical hook) were no longer detected in *S. vermifera*-colonised tobacco plants. Moreover, silencing of 1-aminocyclopropane-1-carboxylic acid oxidase (ACO) in *N. attenuata*, which is involved in ethylene synthesis, led to taller plants under non-inoculated conditions. Furthermore, the respective mutants no longer showed growth promoting effects after inoculation with *S. vermifera* (Barazani et al. 2007). Hence, it was postulated that endophyte-triggered growth performance might be the result of impaired ethylene synthesis and/or signalling in colonised plants. Interestingly, Barazani et al. (2005) detected a reduced herbivore resistance in *N. attenuata* leaves after inoculation with *S. vermifera*. After application of an oral secretion of a herbivore to *N. attenuata* leaves, *S. vermifera*-inoculated plants displayed a reduced ethylene burst and suppressed transcript accumulation of ethylene synthesis genes (*NaACS3, NaACO1, NaACO3*). In these experiments, neither the accumulation of jasmonic acid (JA) and JA-isoleucine nor JA signalling was affected (Barazani et al. 2007). Taken together, the inhibition of ethylene production by *S. vermifera* has positive and negative effects on the plant. While growth is promoted, herbivore resistance is impaired. The studies of Barazani et al. (2007) and our results with *A. thaliana* may indicate that the beneficial systemic effects, growth promotion of plants and resistance induction against *Golovinomyces orontii*, mediated by sebacinoid mycobionts, are not mediated by the same pathways.

VIII. Impact of Various Plant Mutations on *P. indica*-Induced Resistance

P. indica-induced systemic resistance to the powdery mildew fungus *G. orontii* is largely compromised in *A. thaliana* mutants defective in components of the JA/ethylene (ET) defence pathway. However, this resistance is independent of the salicylate (SA) pathway. *A. thaliana* genotypes showing enhanced resistance to *G. orontii* after *P. indica* colonisation can be clearly distinguished from those with no response to *P. indica*: While induced resistance still occurs in NahG plants not accumulating SA and in the SAR regulatory mutant non-expressor of PR genes1-3 (*npr1-3*), it is abolished in jasmonate response1-1 mutants (*jar1-1*, insensitive to jasmonate; Stein et al., unpublished data). Unlike *npr1-3*, the *npr1-1* null mutant (which exhibits compromised pathways for both salicylate and jasmonate) is also non-responsive to *P. indica* and thus shows a higher susceptibility to *G. orontii*. In contrast to *npr1-1*, the mutant *npr1-3* still supports a cytoplasmic function of NPR1, in spite of the fact that nuclear localisation of this protein is impaired in both mutants. Hence, a compromised defence response in *npr1-1* demonstrates a requirement for the cytoplasmic function of NPR1 for *P. indica*-induced resistance. Since root colonisation with *P. indica* is not compromised in the non-responding mutants, the mutant analyses suggest that JAR1 and NPR1 are genes required for *P. indica*-mediated resistance to powdery mildew. Interestingly, the *jar1-1* mutant is characterised by reduced JA sensitivity, leading to an impaired induced systemic resistance (ISR) reaction and reduced resistance to the opportunistic soil fungus *Pythium irregulare* (Staswick et al. 1992, 1998; Pieterse et al. 1998). JAR1 is able to adenylate JA, an enzymatic step initiating covalent modifications such as coupling to amino acids (Staswick et al. 2002). JA-isoleucine was recently shown to promote the binding of COI1 and JAZ1, crucial elements in JA signalling and possible JA receptor candidates (Chini et al. 2007; Thines et al. 2007). The requirement for JAR1 thus suggests that *P. indica*-mediated resistance requires the formation of JA conjugates. These are active in transmitting several, but not all, JA-mediated responses.

IX. Bacterial Endosymbiotic Associations Within *Sebacinales*

Recent molecular analyses have shown that both *P. indica* and *S. vermifera* are intimately associated with bacteria (Sharma et al. 2008). Based on PCR analyses and sequencing of the 16S ribosomal RNA, an association of *P. indica* with *Rhizobium radiobacter*, a gram-negative α-proteobacterium, was traced back to the original *P. indica* isolate deposited in the culture collection of the German Resource Centre for Biological Material, Braunschweig. This isolate had been deposited immediately after its discovery in the mid-1990s. While bacterial cells are not present in culture filtrates of *P. indica*, they are released after crushing the fungal mycelium, suggesting that *R. radiobacter* is closely associated with the hyphal walls or even lives endosymbiotically. Isolated bacteria show biological activities on barley similar to those mediated by *P. indica*, including systemic resistance induction against powdery mildew and growth promotion. Since *R. radiobacter* has not been successfully eliminated from *P. indica*, it remains an open question to what extent fungus and bacterium contribute to the biological effects on their host plants.

A PCR-based screen of various *Sebacina vermifera* cultures for the presence of bacteria clearly revealed fungal isolates from various original sources to be stably associated with single bacterial species. For instance, *Sebacina vermifera* strain MAFF305838 lives associated with *Paenibacillus* spp. Using fluorescence in situ hybridisation (FISH) with eubacterial fluorescent primers, bacterial cells were localised inside fungal hyphae and chlamydospores (Fig. 5.4). In contrast to *R. radiobacter*, *Paenibacillus* could not be cultivated in axenic cultures. Thus the biological activity of this bacterium and its contribution to a more complex tripartite symbiosis has not been resolved. In essence, the above findings show that the *Sebacinales* undergo complex symbioses involving host plants and bacteria.

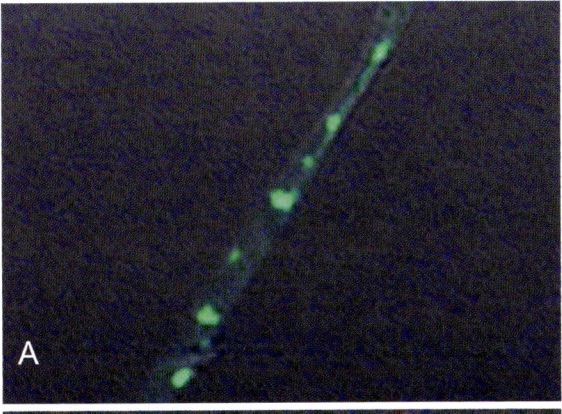

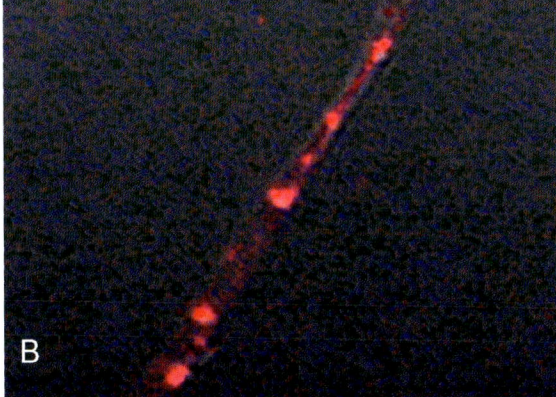

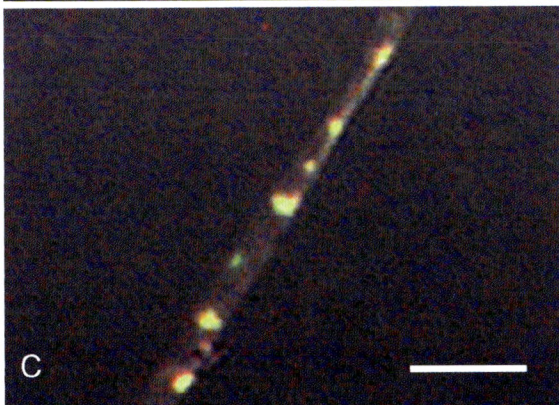

Fig. 5.4. *Sebacina vermifera* strain MAFF305838 harbours endosymbiotic bacteria. Bacteria (*Paenibacillus* sp.) associated with *S. vermifera* strain MAFF305838 were localised within hyphal cells by using fluorescent in situ hybridisation (FISH) and confocal laser-scanning microscopy. Bacteria in association with the fungus were labeled with: (**A**) a EUB-338 FITC-labeled probe mix (excitation/emission: 488 nm/517 nm) for detection of eubacteria and (**B**) a LGC-354 Cy3-labeled probe mix (543 nm/562 nm) for specific detection of Firmicutes, to which *Paenibacillus* sp. belongs. Merged images (**C**) show the congruence of both labels, indicating the endosymbiotic localisation of the bacteria. A EUK-516 Cy5-labeled probe for detecting eucaryotes (633 nm/664 nm) did not bind to endosymbiotic bacteria (data not shown). *Bar* 10 µm

There are several reports showing a mutualistic association of mycorrhizal fungi with bacteria in which, for instance, bacteria improve spore germination and the formation of mycorrhizal interactions. In addition, plant growth-promoting bacteria (PGPR) have been shown to interact physically with fungal hyphae. *Rhizobium* and *Pseudomonas* species attach to germinated AM fungal spores and hyphae (Bianciotto et al. 1996a), but no specificity for either fungal or inorganic surfaces could be detected among the bacteria tested.

True endosymbiotic bacteria have been reported in only a few fungi, including members of the *Glomeromycota* (e.g. *Gigaspora* sp., *Geosiphon pyriforme*) and the ectomycorrhizal basidiomycete *Laccaria bicolor*. For example, endobacteria have been detected in five species of *Gigasporaceae* and various fungal cells, including spores, germtubes and extra- and intraradical hyphae, but not in arbuscules (Bianciotto et al. 1996b). Endosymbiotic bacteria were first identified in the AM fungus *Glomus margarita*, and this association is the best studied interaction of AM fungi and endobacteria (Bianciotto et al. 1996b). Recent studies estimated an average of about 20 000 bacteria per *G. margarita* spore (Bianciotto et al. 2004; Jargeat et al. 2004). Although initially assigned to the genus *Burkholderia*, recent phylogenetic analyses based on 16S ribosomal RNA gene sequences proposed the introduction of a new taxon termed *Candidatus Glomeribacter gigasporarum* (Bianciotto et al. 2003). The small bacterial genomes (about 1.4 Mb) consist of a single chromosome and a single plasmid (Jargeat et al. 2004). Recently, Lumini et al. (2007) published a procedure for dilution of the bacteria by using successive in vitro single-spore inocula. The absence of bacteria severely affected presymbiotic fungal growth with deficiencies in spore shape and hyphal elongation, delays in growth onset of germinating mycelium and in branching after root exudate treatment. These results suggest that endobacteria contribute to regular development of its fungal host.

In the plant pathogen *Rhizopus microsporus* endosymbiotic bacteria play a crucial role in fungal infection strategies. Until recently, *R. microsporus* was thought to produce a toxin that kills plant root cells. However, Partida-Martinez and Hartweck (2005) demonstrated that the toxin was not produced by the fungus but by endogenous bacteria. On the basis of the 16S ribosomal RNA gene sequence, they found that the bacteria belong to the genus *Burkholderia*, a member of the beta subdivision of proteobacteria. The bacteria and bacteria-free fungus were each isolated in pure culture. There was a strong correlation between the presence of bacteria and the toxin-producing capability of *Rhizopus*. In the absence of endobacteria, *Rhizopus macrosporus* was not capable of vegetative reproduction (Partida-Martinez et al. 2007). Formation of sporangia and spores was restored only upon reintroduction of endobacteria. The motile rod-shaped bacteria appeared to be prone to chemotaxis, since they migrated toward the tips of the hyphae, the region best supplied with nutrients and where sporangia were formed.

X. Conclusions

Present knowledge characterises *P. indica* as a potential orchid mycorrhiza fungus that can be clearly distinguished from ectomycorrhizas or arbuscular mycorrhizas. However, its endophytic life style might be predominant in associations with certain plants. Even during these types of evolutionarily inappropriate interactions, the fungus is able to confer beneficial effects to its hosts; this phenomenon distinguishes *P. indica* from ectomycorrhizal and AM fungi.

    P. indica is a model organism of the newly defined order *Sebacinales* within the phylum Basidiomycota, comprising a group of mycorrhizal fungi that form mutualistic symbioses with an as yet widely unrevealed function in natural ecosystems as well as cropping systems. In contrast to AM fungi, *P. indica* and related species of the *S. vermifera* complex confer systemic resistance against root and leaf pathogens to a wide range of monocotyledonous and dicotyledonous plants. Moreover, these fungi bear a significant agronomical potential, since they increase grain yield. Their application in horticulture or agriculture is economically and practically feasible through the facilitated propragation of fungal inoculum using liquid or axenic cultures. The huge prospective biodiversity in the *Sebacinales* provides the perspective that appropriate sebacinalean mutualists might be discovered for many crop plants. Research on *Sebacinales*, however, may not only enable new crop production strategies but additionally may eminently expand our basic knowledge on host–microbe interactions. Recent discovery of fungus-associated endobacteria demonstrated that

Sebacinales can participate in a more complex symbiosis. Although the exact contributions of the partners are not fully elucidated, it is clear that the bacteria perform activities that were formerly ascribed to the fungal partner. In addition, it is obvious that *P. indica* shows properties that clearly contrast with those ascribed to AM fungi:

1. In comparison to known endophytic strategies, *P. indica* requires host cell death for successful plant colonisation, implying that fungal effector molecules interfere with the host cell death machinery.
2. *P. indica* conveys systemic disease resistance to fungal leaf pathogens, which has rarely been observed in monocotyledonous plants.
3. *P. indica* is the sole fungal mutualist identified to date that colonises *A. thaliana* and mediates a type of systemic resistance to powdery mildew which depends on jasmonate signal pathways.

The power of available *A. thaliana* signal transduction mutants and reverse genetics will further accelerate disclosure of the molecular basis of the symbiosis and its beneficial effects on the plant.

References

Bandoni RJ (1984) The Tremellales and Auriculariales: An alternative classification. Trans Mycol Soc Jpn 25:489–530

Barazani O, Benderoth M, Groten K, Kuhlemeier C, Baldwin IT (2005) *Piriformospora indica* and *Sebacina vermifera* increase growth performance at the expense of herbivore resistance in Nicotiana attenuata. Oecologia 146:234–243

Barazani O, von Dahl CC, Baldwin IT (2007) *Sebacina vermifera* promotes the growth and fitness of *Nicotiana attenuata* by inhibiting ethylene signaling. Plant Physiol 144:1223–1232

Benavente LM, Alonso JM (2005) Molecular mechanisms of ethylene signaling in *Arabidopsis*. Mol BioSyst 2:165–173

Bianciotto V, Minerdi D, Perotto S, Bonfante P (1996a) Cellular interactions between arbuscular mycorrhizal fungi and rhizosphere bacteria. Protoplasma 193:123–131

Bianciotto V, Bandi C, Minerdi D, Sironi M, Tichy HV, Bonfante P (1996b) An obligately endosymbiotic mycorrhizal fungus itself harbors obligately intracellular bacteria. Appl Environ Microbiol 62:3005–3010

Bianciotto V, Lumini E, Bonfante P, Vandamme P. (2003) 'Candidatus glomeribacter gigasporarum' gen. nov., sp. nov., an endosymbiont of arbuscular mycorrhizal fungi. Int J Syst Evol Microbiol 53:121–124

Bianciotto V, Genre A, Jargeat P, Lumini E, Becard G, Bonfante P (2004) Vertical transmission of endobacteria in the arbuscular mycorrhizal fungus *Gigaspora margarita* through generation of vegetative spores. Appl Environ Microbiol 70:3600–3608

Blechert O, Kost G, Hassel A, Rexer KH, Varma A (1999) First remarks on the symbiotic interaction between *Piriformospora indica* and terrestrial orchids. In: Varma A, Hock B (eds) Mycorrhiza, 2nd edn. Springer Heidelberg, pp 683–688

Bleecker AB, Estelle MA, Somerville C, Kende H (1988) Insensitivity to ethylene conferred by a dominant mutation in *Arabidopsis thaliana*. Science 241:1086–1089

Brundrett MC (2004) Diversity and classification of mycorrhizal associations. Biol Rev 79:473–495

Chini A, Fonseca S, Fernandez G, Adie B, Chico JM, Lorenzo O, Garcia-Casado G, Lopez-Vidriero I, Lozano FM, Ponce MR, Micol JL, Solano R (2007) The JAZ family of repressors is the missing link in jasmonate signalling. Nature 448:666–671

Deckert RJ, Melville L, Peterson RL (2001) Structural features of a *Lophodermium* endophyte during the cryptic lifecycle in the foliage of *Pinus strobus*. Mycol Res 105:991–997

Deshmukh S, Kogel KH (2007) *Piriformospora indica* protects barley from root rot caused by *Fusarium graminearum*. J Plant Dis Prot 114:263–268

Deshmukh S, Hueckelhoven R, Schäfer P, Imani J, Sharma M, Weiss M, Waller F, Kogel KH (2006) The root endophytic fungus *Piriformospora indica* requires host cell death for proliferation during mutualistic symbiosis with barley. Proc Natl Acad Sci USA 103:18450–18457

Freeman S, Rodriguez RJ (1993) Genetic conversion of a fungal plant pathogen to a nonpathogenic, endophytic mutualist. Science 260:75–78

Gonzalez D, Cubeta MA, Vilgalys R (2006) Phylogenetic utility of indels within ribosomal DNA and β-tubulin sequences from fungi in the *Rhizoctonia solani* species complex. Mol Phylogenet Evol 40:459–470

Harrison MJ (2005) Signaling in the arbuscular mycorrhizal symbiosis. Annu Rev Microbiol 59:19–42

Jargeat P, Cosseau C, Ola'h B, Jauneau A, Bonfante P, Batut J, Becard G (2004) Isolation, free-living capacities, and genome structure of "*Candidatus Glomeribacter gigasporarum*", the endocellular bacterium of the mycorrhizal fungus Gigaspora margarita. J Bacteriol 186:6876–6884

Johnson NC, Graham JH, Smith FA (1997) Functioning of mycorrhizal associations along the mutualism-parasitism continuum. New Phytol 135:575–585

Jones JDG, Dangl JL (2006) The plant innate immunity. Nature 444:323–329

Jumpponen A, Trappe JM (1998) Dark septate endophytes: a review of facultative biotrophic root-colonizing fungi. New Phytol 140:295–310

Kaku H, Nishizawa Y, Ishii-Minami N, Akimoto-Tomiyama C, Dohmae N, Takio K, Minami E, Shibuya N (2006) Plant cells recognize chitin fragments for defense signaling through a plasma membrane receptor. Proc Natl Acad Sci USA 103:11086–11091

Kaldorf M, Koch B, Rexer KH, Kost G, Varma A (2005) Patterns of interaction between *Populus* Esch5 and *Piriformospora indica*: a transition from mutualism to antagonism. Plant Biol 7:210–218

Kieber JJ, Rothenberg M, Roman G, Feldman KA, Ecker JR (1993) *CTR1*, a negative regulator of the ethylene response pathway in *Arabidopsis*, encodes a member of the Raf family of protein kinases. Cell 72:427–441

Kottke I, Beiter A, Weiß M, Haug I, Oberwinkler F, Nebel M (2003) Heterobasidiomycetes form symbiotic associations with hepatics: Jungermanniales have sebacinoid mycobionts while *Aneura pinguis* (Metzgeriales) is associated with a *Tulasnella* species. Mycol Res 107:957–968

Lumini E, Bianciotto V, Jargeat P, Novero M, Salvioli A, Faccio A, Becard G, Bonfante P (2007) Presymbiotic growth and sporal morphology are affected in the arbuscular mycorrhizal fungus *Gigaspora margarita* cured of its endobacteria. Cell Microbiol 9:1716–1729

McKendrick SL, Leake JR, Taylor DL, Read DJ (2002) Symbiotic germination and development of the myco-heterotrophic orchid *Neottia nidus-avis* in nature and its requirement for locally distributed *Sebacina* spp. New Phytol 154:233–247

Milligan MJ, Williams PG (1988) The mycorrhizal relationship of multinucleate *Rhizoctonias* from nonorchids with *Microtis* (Orchidaceae). New Phytol 108:205–209

Miya A, Albert P, Shinya T, Desaki Y, Ichimura K, Shirasu K, Narusaka Y, Kawakami N, Kaku H, Shibuya N (2007) CERK1, a LysM receptor kinase, is essential for chitin elicitor signaling in *Arabidopsis*. Proc Natl Acad Sci USA 104:19613–19618

Ogoshi A (1987) Ecology and pathogenicity of *Anastomosis* and intraspecific groups of *Rhizoctonia solani* Kühn. Annu Rev Phytopathol 25:125–143

Partida-Martinez LP, Hartweck C (2005) Pathogenic fungus harbours endosymbiotic bacteria for toxin production. Nature 437:848–888

Partida-Martinez LP, Monajembashi S, Greulich KO, Hertweck C (2007) Endosymbiont-dependent host reproduction maintains bacterial-fungal mutualism. Curr Biol 17:773–777

Peterson RL, Massicotte HB (2004) Exploring structural definitions of mycorrhizas, with emphasis on nutrient-exchange interfaces. Can J Bot 82:1074–1088

Petrini O (1991) Fungal endophytes of tree leaves. In: Andrews J, Hirano S (eds) Microbial ecology of leaves, Springer, New York, pp 179–197

Pham GH, Singh A, Malla R, Kumari M, Prasad R, Sachdev M, Rexer KH, Kost G, Luis P, Kaldorf M, Buscot F, Herrmann S, Peskan T, Oelmüller R, Saxena AK, Declerck S, Mittag M, Stabentheiner E, Hehl S, Varma A (2004) Interaction of *Piriformospora indica* with diverse microorganisms and plants. In: Varma A, Abbot L, Werner D, Hampp R (eds) Plant surface microbiology, Springer, Berlin, pp 237–264

Pieterse CM, van Wees SC, van Pelt JA, Knoester M, Laan R, Gerrits H, Weisbeek PJ, van Loon LC (1998) A novel signaling pathway controlling induced systemic resistance in *Arabidopsis*. Plant Cell 10:1571–1580

Redman RS, Ranson JC, Rodriguez RJ (1999) Conversion of the pathogenic fungus *Colletotrichum magna* to a nonpathogenic, endophytic mutualist by gene disruption. Mol Plant–Microbe Interact 12:969–975

Redman RS, Dunigan DD, Rodriguez RJ (2001) Fungal symbiosis from mutualism to parasitism: who controls the outcome, host or invader? New Phytol 151:705–716

Sahay NS, Varma A (1999) *Piriformospora indica*: a new biological hardening tool for micropropagated plants. FEMS Microbiol Lett 181:297–302

Saikkonen K, Faeth SH, Helander M, Sullivan TJ (1998) Fungal endophytes: a continuum of interactions with host plant. Annu Rev Ecol Syst 29:319–343

Salzer P, Hebe G, Hager A (1997) Cleavage of chitinous elicitors from the ectomycorrhizal fungus *Hebeloma crustuliniforme* by host chitinases prevents induction of K$^+$ and Cl$^-$ release, extracellular alkalinisation and H$_2$O$_2$ synthesis of *Picea abies* cells. Planta 203:470–479

Schulze B, Boyle C (2005) The endophytic continuum. Mycol Res 109:661–686

Schulz B, Boyle C, Draeger S, Römmert AK, Krohn K (2002) Endophytic fungi: a source of biologically active secondary metabolites. Mycol Res 106:996–1004

Selosse MA, Bauer R, Moyersoen B (2002) Basal hymenomycetes belonging to the *Sebacinaceae* are ectomycorrhizal on temperate deciduous trees. New Phytol 155:183–195

Selosse MA, Setaro S, Glatard F, Richard F, Urcelay C, Weiss M (2007) *Sebacinales* are common mycorrhizal associates of Ericaceae. New Phytol 174:864–878

Serfling A, Wirsel SGR, Lind V, Deising HB (2007) Performance of the biocontrol fungus *Piriformospora indica* on wheat under greenhouse and field conditions. Phytopathology 97:523–531

Setaro S, Weiß M, Oberwinkler F, Kottke I (2006) *Sebacinales* form ectendomycorrhizas with *Cavendishia nobilis*, a member of the Andean clade of Ericaceae, in the mountain rain forest of southern Ecuador. New Phytol 169:355–365

Shahollari B, Varma A, Oelmüller R (2005) Expression of a receptor kinase in *Arabidopsis* roots is stimulated by the basidiomycete *Piriformospora indica* and the protein accumulates in Triton X-100 insoluble plasma membrane microdomains. J Plant Physiol 162:945–958

Sharma M, Schmid M, Rothballer M, Hause G, Zuccaro A, Imani J, Kämpfer P, Domann E, Schäfer P, Hartmann A, Kogel KH (2008) Detection and identification of mycorrhiza helper bacteria intimately associated with representatives of the order *Sebacinales*. Cell Microbiol (in press)

Sherameti I, Shahollari B, Venus Y, Altschmied L, Varma A, Oelmüller R (2005) The endophytic fungus *Piriformospora indica* stimulates the expression of nitrate reductase and the starch-degrading enzyme glucan-water dikinase in tobacco and *Arabidopsis* roots through a homeodomain transcription factor that binds to a conserved motif in their promoters. J Biol Chem 280:26241–26247

Sirrenberg A, Göbel C, Grond S, Czempinski N, Ratzinger A, Karlovsky P, Santos P, Feussner I, Pawlowski K (2007) *Piriformospora indica* affects plant growth by auxin production. Physiol Plant 131:581–589

Staswick PE, Su W, Howell SH (1992) Methyl jasmonate inhibition of root growth and induction of leaf protein are decreased in an *Arabidopsis thaliana* mutant. Proc Natl Acad Sci USA 89:6837–6840

Staswick PE, Yuen GY, Lehman CC (1998) Jasmonate signaling mutants of *Arabidopsis* are susceptible to the soil fungus *Pythium irregulare*. Plant J 16:747–754

Staswick PE, Tiryaki I, Rowe ML (2002) Jasmonate response locus JAR1 and several related *Arabidopsis* genes encode enzymes of the firefly luciferase superfamily that show activity on jasmonic, salicylic, and indole-3-acetic acids in an assay for adenylation. Plant Cell 14:1405–1415

Takemoto D, Tanaka A, Scott B (2006) A p67Phox-like regulator is recruited to control hyphal branching in a fungal-grass mutualistic symbiosis. Plant Cell 18:2807–2821

Tanaka A, Christensen MJ, Takemoto D, Park P, Scott B (2006) Reactive oxygen species play a role in regulating a fungus-perennial ryegrass mutualistic interaction. Plant Cell 18:1052–1066

Thines B, Katsir L, Melotto M, Niu Y, Mandaokar A, Liu G, Nomura K, He SY, Howe GA, Browse J (2007) JAZ repressor proteins are targets of the SCF(COI1) complex during jasmonate signalling. Nature 448:661–665

Urban A, Weiß M, Bauer R, (2003) Ectomycorrhizae involving sebacinoid mycobionts. Mycol Res 107:3–14

Varma A, Verma S, Sudha, Sahay N, Butehorn B, Franken P (1999) *Piriformospora indica*, a cultivable plant-growth-promoting root endophyte. Appl Environ Microbiol 65:2741–2744

Verma S, Varma A, Rexer K, Hassel A, Kost G, Sarbhoy A, Bisen P, Bütehorn B, Franken P (1998) *Piriformospora indica*, gen. et sp. nov., a new root-colonizing fungus. Mycologia 90:896–903

Vierheilig H, Alt M, Neuhaus JM, Boller T, Wiemken A (1993) Colonization of transgenic *Nicotiana sylvestris* plants, expressing different forms of *Nicotiana tabacum* chitinase, by the root pathogen *Rhizoctonia solani* and by the mycorrhizal symbiont *Glomus mosseae*. Mol Plant–Microbe Interact 6:261–264

Vierheilig H, Alt M, Lange J, Gut-Rella M, Wiemken A, Boller T (1995) Colonization of transgenic tobacco constitutively expressing pathogenesis-related proteins by the vesicular-arbuscular mycorrhizal fungus *Glomus mosseae*. Appl Environ Microbiol 61:3031–3034

Waller F, Achatz B, Baltruschat H, Fodor J, Becker K, Fischer M, Heier T, Hückelhoven R, Neumann C, von Wettstein D, Franken P, Kogel KH (2005) The endophytic fungus *Piriformospora indica* reprograms barley to salt-stress tolerance, disease resistance, and higher yield. Proc Natl Acad Sci USA 102:13386–13391

Waller F, Mukherjee K, Achatz B, Deshmukh S, Sharma S, Schäfer P, Kogel KH (2008) Local and systemic modulation of plant responses by *Piriformospora indica* and related *Sebacinales* species. J Plant Physiol 165:60–70

Warcup JH (1988) Mycorrhizal associations of isolates of *Sebacina vermifera*. New Phytol 110:227–231

Weiss M, Selosse MA, Rexer KH, Urban A, Oberwinkler F (2004) *Sebacinales*: a hitherto overlooked cosm of heterobasidiomycetes with a broad mycorrhizal potential. Mycol Res 108:1003–1010

Williams PG (1985) Orchidaceous rhizoctonias in pot cultures of vesicular-arbuscular mycorrhizal fungi. Can J Bot 63:1329–1333

Zhong GV, Burns JK (2003) Profiling ethylene-regulated gene expression in *Arabidopsis thaliana* by microarray analysis. Plant Mol Biol 53:117–131

Mechanisms of Pathogenic and Mutualistic Interactions

6 Biomechanics of Spore Release in Phytopathogens

Nicholas P. Money[1], Mark W.F. Fischer[2]

CONTENTS

I. Introduction............................ 115
II. Categories of Discharge Mechanism 115
III. Air Viscosity and Boundary Layers 116
IV. Airflow and Drying, Electrostatics,
    and Cavitation 116
    A. Airflow and Drying 116
    B. Electrostatics 118
    C. Cavitation 118
V. Raindrops and Vibration 119
VI. Animals............................... 120
VII. Turgor Pressure 121
    A. Ascospore Discharge................... 121
    B. Active Conidial Discharge 123
    C. Zoospore Discharge................... 124
    D. *Pilobolus*.......................... 127
    E. *Sphaerobolus* 128
VIII. Surface Tension 128
IX. Conclusions and Future Research.......... 131
    References............................ 131

I. Introduction

Movement is one of the defining characteristics of living organisms. Contrary to common perceptions, fungi show a remarkable range of motion. Motion inside fungal cells, including mass flow of cytoplasm, was first observed by Antonie van Leeuwenhoek and influenced the eighteenth-century view of fungi as an eccentric branch of the animal kingdom (Ainsworth 1976). This flow of cytoplasm accompanies the extension of hyphae, and there are a number of similarities between this growth process and amoeboid locomotion (Heath and Steinberg 1999). Faster movements include invertebrate capture by constricting rings and microscopic harpoons (Müller 1958; Beakes and Glocking 1998) and a series of spectacular mechanisms that launch fungal spores into air (Ingold 1971). Spore discharge and dispersal are

[1] Department of Botany, Miami University, Oxford,
OH 45056, USA; e-mail: moneynp@muohio.edu
[2] Department of Chemistry and Physical Science,
College of Mount St Joseph, Cincinnati, OH 45233, USA

related and it is important to distinguish between them. Discharge refers to the mechanical process that separates the spore, or sporangium, from its parent mycelium; dispersal follows discharge. Both processes are vital to the activities of phytopathogens. This chapter emphasizes spore discharge in pathogens, but mechanisms among saprobes are also discussed to provide an overview of the diversity of launch processes among the fungi.

II. Categories of Discharge Mechanism

The spores of many fungal pathogens move only when they are shaken, splashed, or blown from their mycelia, or are dislodged by invertebrates. The majority of conidia, for example, loosen their connection to their conidiophores and are dispersed when subjected to the appropriate physical disturbance (Fig. 6.1). The term passive discharge is applied to these processes. Other mechanisms are categorized as active: osmotically generated pressures propel spores into air or water via the squirting of asci and zoosporangia, and a catapult energized by surface tension launches ballistosporic basidiospores (Ingold 1971). Although the terms active and passive are useful, they do not allow an unambiguous classification of discharge mechanisms. Some spores, for example, are launched in response to the twisting and cavitation of their conidiophores as they desiccate in dry air. Are these active or passive mechanisms? The process of getting spores airborne is always active in terms of energy consumption, although the greatest energy expenditure may occur when the colony constructs conidiophores that poke into the air above a leaf surface. Similarly, when an insect is coopted for dispersal, the fungus invests energy in luring the vector with chemoattractants and/or nectar. In search of clarity, the processes considered in this chapter are categorized in terms of the mechanism that separates the spore from its parent mycelium.

Plant Relationships, 2nd Edition
The Mycota V
H. Deising (Ed.)
© Springer-Verlag Berlin Heidelberg 2009

Fig. 6.1. Examples of wet- or sticky-spored and dry-spored ascomycete anamorphs. **A** Conidiophore of *Stachybotrys chartarum* supporting cluster of sticky conidia. Collapsed phialides are visible beneath the spore cluster. *Bar* 10 μm. **B** Conidiophore of *Penicillum* sp. producing chains of dry conidia. *Bar* 5 μm. Courtesy of Matthew Duley and Richard Edelmann, (Miami University, Oxford, Ohio)

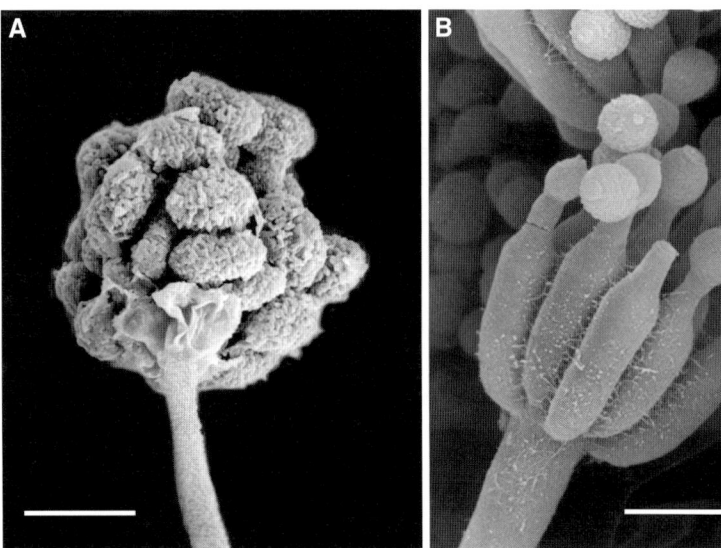

III. Air Viscosity and Boundary Layers

The physical challenges encountered by fungi and other microorganisms are very different from those experienced by large animals (Vogel 1988). Gazelles and exceptional humans achieve horizontal leaps of 9 m. The arc of their flight paths is dominated by inertial forces and neither animal is slowed by the viscosity of the air through which it moves. Things are very different for pathogenic ascomycetes that shoot their spores from foliar lesions and for basidiomycete yeasts that propel their ballistospores from blotched petals. This is because air represents a viscous obstacle to fungal movement and remarkable launch speeds are necessary to propel spores even for short distances (Skotheim and Mahadevan 2005; Vogel 2005a, b). The ratio of inertial to viscous forces is described by the non-dimensional term Reynolds number (*Re*): big animals experience high *Re*; spore movement is a low *Re* process. The same scaling principles make greater intuitive sense in the more viscous medium of water. A dolphin can glide many meters through the ocean after a single stroke of its tail fluke (high *Re*, inertia trumps viscosity), whereas a chytrid zoospore stops dead the instant it stops lashing its posterior flagellum (low *Re*, viscosity rules).

The horizontal movement of actively discharged fungal spores ceases within a fraction of a second after launch and the spore falls toward the ground. This is the case with coprophilous fungi whose spores (or spore clusters or sporangia) are shot onto vegetation surrounding their growth

substrates. For the majority of fungi, however, discharge mechanisms get spores airborne and longer-distance dispersal is driven by wind. But wind dispersal cannot occur, obviously, unless the fungus reaches moving air currents. This poses a problem for a microscopic particle that develops within the boundary layer of slow-moving air close to the growth substrate. When air flows around any solid object, drag creates a mantle of slow-moving air close to the object's surface; at the interface itself, the air is stationary. There is an inverse relationship between airspeed and boundary layer thickness and this is another situation in which the *Re* term is useful. At low airspeed, air movement is dominated by viscosity, the boundary layer is thickest, and *Re* is low. Inertia dominates airflow patterns at higher airspeeds, and under these conditions of elevated *Re*, eddies and vortices develop. The boundary layer surrounding leaves of terrestrial plants varies from approximately 0.1 mm to 9.0 mm in thickness depending upon leaf size and wind speed (Nobel 1991). The same considerations apply to the motion of water around solids, but boundary layers are much thicker, and *Re* values lower, in the aquatic environment.

IV. Airflow and Drying, Electrostatics, and Cavitation

A. Airflow and Drying

The physical disturbance of fungal colonies by airflow is thought to serve as the primary stimulus that separates the conidia of most terrestrial anamorphs from their conidiophores. Airflow also causes the release of other spore types, including sporangiospores in the Zygomycota, uredospores and teliospores of rusts, and sporangia in pathogenic Peronosporomycetes (Oomycota). Despite its importance the process of passive spore release in response to airflow has not been studied carefully in many fungi. This is significant because assumptions about the primacy of wind-driven spore release may not be valid for some fungi, including important pathogens.

Many of the most informative papers on this topic were published in the 1960s. Using miniature wind tunnels, Smith (1966) showed that airspeeds in excess of $1.3\,m\,s^{-1}$ ($4.7\,km\,h^{-1}$) were necessary for the release of uredospores of *Puccinia graminis* f. sp. *tritici* from uredosori on wheat leaves. This study also showed that the stalked uredospores of this rust were often released in groups. The release of clumps of spores seems to be a widespread phenomenon among diverse taxa. Zoberi (1961) experimented on the effects of airflow on spore release by driving air over colonies in culture tubes at controlled speeds. Spore release was monitored by collecting spores from the outflow of the culture tubes on adhesive-coated microscope slides. Working with a variety of ascomycete anamorphs and Mucorales (Zygomycota), he showed that airflow over colonies led to the ejection of a dense cloud of spores, followed by a sharp decline in the number of released spores that was sustained during continuous airflow. The airspeed in these experiments was set at $5\,m\,s^{-1}$ ($18\,km\,h^{-1}$), which is considered a gentle breeze. Relative humidity affected the number of spores that were released, with dry air enhancing discharge, but the trend of initial burst/sharp decline was the same under all conditions. The same pattern of spore release was documented in many other early studies on spore liberation (Ingold 1971), and more recently, Górny et al. (2001) established this behavior in *Aspergillus versicolor*, *Cladosporium cladosporioides*, and *Penicillium melinii* colonies subjected to airflow and vibration. Detailed biomechanical studies in the author's laboratory are pertinent to this topic (Tucker et al. 2007). This study used a miniature wind tunnel that allowed continuous microscopic observation of conidiophores during airflow. The maximum rate of dispersal in *Aspergillus niger*, *Cladosporium sphaerospermum*, and *Penicillium chrysogenum* occurred during the first 5 min of airflow of $1.6\,m\,s^{-1}$, followed by a dramatic reduction in dispersal that left more than 98% of the conidia attached to their conidiophores (Fig. 6.2). Sluggish spore release was even more pronounced in *Stachybotrys chartarum*, whose conidia are produced in sticky heads at the tips of short conidi-

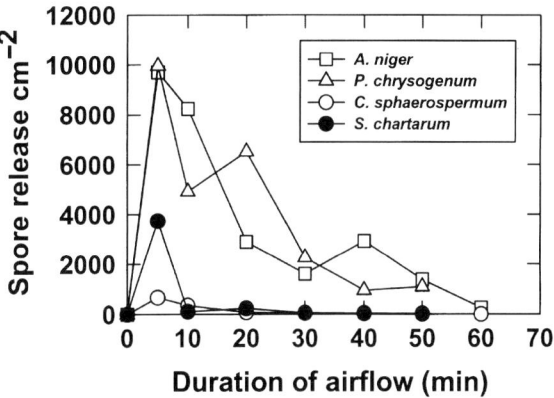

Fig. 6.2. Patterns of conidial release in *Aspergillus niger*, *Penicillium chrysogenum*, *Cladosporium sphaerospermum*, and *Stachybotrys chartarum* at low airspeed ($1.6\,m\,s^{-1}$). From Tucker et al. (2007) with permission

ophores (Fig. 6.1A). In this species, studied for its significance as an indoor air contaminant, the initial burst of release left more than 99% of the conidia attached to their conidiophores. Micromanipulation of the colonies of this species showed that micronewton (μN) forces were needed to dislodge spore clusters from their supporting conidiophores. Calculations show that low airspeeds ($0.3–1.6\,m\,s^{-1}$) disturb colonies with forces that are 1000-fold lower, in the nanonewton (nN) range. Low-velocity airflow does not, therefore, cause sufficient disturbance to disperse a large proportion of the conidia of *S. chartarum*. Although the dry-spored species showed greater mobility, a surprisingly large proportion of these conidia are also retained on the colony surface at low airspeeds. Interestingly, Aylor (1975) calculated that a force of about $0.1\,\mu N$ was necessary to detach conidia of *Helminthosporium maydis*, corresponding to a wind speed of $10\,m\,s^{-1}$.

The phenomenon of the initial burst of spore release is somewhat counterintuitive. Once a particular fungus ceases sporulation, one might anticipate continuous release of spores as long as the appropriate airflow is sustained. Instead, experiments show that the majority of spores remain on the colony surface even at relatively high airspeeds. The observed pattern of spore release may be caused by a succession of drying and disarticulation events: a cloud of highly mobile spores is released in response to the initiation of airflow, followed by a period of drying that mobilizes more spores for release. This may help explain the importance of intermittent wind in pathogen dispersal (Aylor 1990; Aylor et al. 1993). Such a mechanism could enhance pathogen survival by maintaining a continuous trickle of relatively small numbers of spores into the airstream so that the organism will increase its probability of encountering susceptible hosts and environmental conditions that are conducive for growth.

The ubiquity of the fungi, conidial or otherwise, testifies to the success of passive dispersal mechanisms on a windy planet, but there is another way of considering the apparent inefficiency of this mode of spore release. The fact that a high proportion of spores is retained on the colony surface may simply reflect the difficulty with which microorganisms become separated from surfaces. Once formed, spores must sever their connections to the parent colony. This is probably achieved by enzymatic action to weaken the interface at the adjoining cell walls and/or by drying. Even then, surface tension associated with any fluid on the spore surface, residual connections between spores and conidiophores, or electrostatic effects may impede dispersal. The limited mobility of spores may be one reason for the astonishing numbers of propagules formed by most fungi.

In most conidial fungi, spore release is maximized at low humidity. Jarvis (1962) studied the release of Botrytis cinerea conidia under field conditions. In these experiments, spore release peaked as temperature and windspeed increased, and humidity decreased. Similar patterns of spore release are seen in other pathogenic fungi that show maximum dispersal following periods of high humidity (e.g., Pady et al. 1969). The logic here is that conidial production is maximized when humidity is high and then release occurs in response to exposure to a drying airstream. The way that dry air stimulates release is not obvious, but probably relates to: (a) drying of any extracellular matrix materials on the spore surface that otherwise act as adhesives and impede release, and (b) cracking of connections between spores as shared wall layers dehydrate. Digestion of contiguous wall layers may be part of the program of conidiogenesis, but dessication of this material may also be essential. Studies on conidial mutants demonstrate that normal cell wall assembly during conidiogenesis is a prerequisite for subsequent disarticulation and dispersal of spores (e.g., Chabane et al. 2006). Contrary to this desiccation-induced pattern of spore release, conidial dispersal is stimulated by high humidity in some pathogens. For example, the rice blast fungus, Magnaporthe grisea, shows maximal release during nighttime and early morning humidity peaks (Barksdale and Asai 1961; Kim et al. 1990). This behavior is related to its unusual discharge mechanism (see Sect. VII.B).

It has been suggested that dry airflow can also effect spore release in some species by drying conidiophores (or sporangiophores) that flick spores (or sporangia) into the air when they recoil. The strongest case for this was made for Peronospora tabacina by Pinckard (1942), but this should be confirmed by modern video microscopy under conditions of controlled airflow and humidity. The forces required for separation of spores (see above) demand violent flexing of conidiophores for an effective mechanism of discharge. Jarvis (1962) concluded that hygroscopic twisting of the conidiophores of Botrytis cinerea could loosen the spores but was not capable of launching them into the air. Spore release by hygroscopic movements may occur in the non-ballistosporic basidiomycetes Podaxis and Battarraea that form capillitial strands with spiral thickenings. These cells appear similar in structure to the elaters of liverworts and slime molds whose twisting movements liberate masses of spores (Ingold 1971), but verification of this putative discharge mechanism in fungi is necessary.

B. Electrostatics

The surface of fungal spores becomes electrically charged in dry air and these charges have significant effects upon their motion. Moriura et al. (2006) studied conidiogenesis in Blumeria graminis f. sp. hordei (Erysiphales) by collecting conidia as they matured on the surface of barley leaves. In these experiments, individual spores were harvested on the tip of a negatively charged ebonite probe that was positioned 40–60 µm from the conidiophores. Blumeria produces chains of conidia that mature in a basipetal fashion and apical conidia secede from the chain through the formation of septa (Fig. 6.3). By collecting each conidium as it matured and was abstricted from the chain, Moriura and colleagues showed that the average lifespan of a conidiophore was 107 h, during which it generated 33 conidia.

Complementary experiments on conidial separation from short chains of spores in the powdery mildew Oidium neolycopersici are described by Oichi et al. (2006). Leach (1976) proposed that electrostatic charges were involved in the release of conidia of Drechslera turcica, and electrostatic charges have also been implicated in the discharge of zoospores from oomycete sporangia (Borkowski 1972) and from ballistospores of Basidiomycota (Gregory 1957; Saville 1965). Although spores, like other microscopic particles, respond to electrostatic charge, there is no evidence that charges play a significant role in spore release mechanisms. Aylor and Paw U (1980) argued that naturally-occurring electrostatic forces are too small to detach conidia from their conidiophores. Other studies also show that the charges carried on the surface of fungal spores are insufficient to effect discharge or to affect spore deposition on surfaces (McCartney et al. 1982; Webster et al. 1988).

C. Cavitation

Cavitation occurs in cells whose contents are subjected to negative pressures that are sufficient to exceed the tensile strength of the cytoplasm. The

Fig. 6.3. Diagram of conidium formation and secession in *Blumeria graminis* f. sp. *hordei*. The time intervals (*a–f*; hours) between the release of four spores (*C1-1* through *C1-4*) is shown along the bottom of the figure. Note the birth of conidial initials via the growth and septation of the basal generative cell. From Moriura et al. (2006) with permission

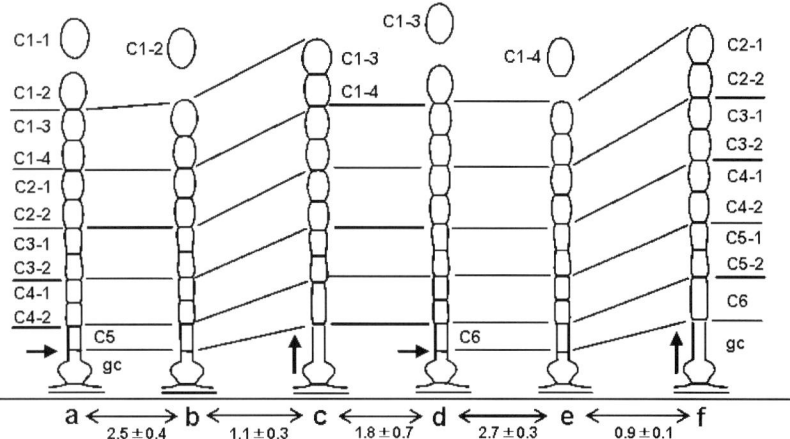

gas bubbles seen inside air-dried ascospores of Sordariomycetes are a familiar example of the cavitation process for mycologists (Ingold 1956). As water evaporates from the cytoplasm of any spore, its walls are placed under increasing tension. If the cell wall is pliable it crumples, allowing the dried cytoplasm to occupy a reduced volume; if the cell wall is rigid it resists compression, and if drying continues, the water in the cytoplasm will fracture to create a vapor-filled bubble. Cavitation can also be induced by placing spores in solutions of osmolytes whose molecular radius exceeds the exclusion threshold of the wall. This trick was used by Milburn (1970) to study the osmotic relations of the thick-walled spores of *Sordaria fimicola*; and it was also adopted by Money et al. (1998) to estimate the turgor pressure of the melanized hyphopodia of *Gaeumannomyces graminis* var. *graminis*. In principle, cavitation can occur in any cell whose wall is strong enough to resist collapse under negative pressure. Mechanisms of spore discharge powered by cavitation within conidiophores have evolved among numerous conidial fungi. Almost all of the work on these mechanisms was done by Donald S. Meredith, who discovered the process in pathogens of banana when he worked at the Banana Board Research Department in Kingstown, Jamaica, in the 1960s. Drying of the conidiophore shrinks its cytoplasm and places the surrounding cell wall under tension; spore discharge is driven when this tension is relieved by cavitation, causing the wall of the conidiophore to flex outward. The conidiophores of most species collapse and twist when they are dried (Sect. IV.A.); cavitation requires a thickened cell wall, and for cavitation to power spore discharge, this thickening must be

arranged in a specific pattern. Cavitation-based spore discharge was first observed in *Deightoniella torulosa*, which causes a variety of disease symptoms in banana (Meredith 1961). The apical cell of the conidiophore of this fungus has a bulbous tip whose side walls are thickened. Cavitation within this structure causes the thin-walled tip of the bulb to flex outward and this is responsible for conidial discharge (Fig. 6.4). Similar mechanisms are responsible for spore discharge in other fungi that grow on banana (Meredith 1962, 1963), but its distribution among fungi that develop on other host plants is unknown. Meredith (1965) also reported cavitation-based discharge in *Helminthosporium turcicum*, raising the possibility that this may play a significant role in the dispersal of many other fungi. The analogous cavitation-powered sporangium of ferns is well known among botanists (King 1944).

V. Raindrops and Vibration

There is a vast phytopathological literature on the dispersal of spores by raindrops. In mechanistic terms, the process is straightforward: dry spores and wet spores impacted by raindrops are, respectively, puffed and splashed from lesions on plant surfaces (Fitt and McCartney 1986). After discharge from the plant surface, the spores can land on adjacent plants or be carried over longer distances by wind (either as free spores or associated with water droplets). The evolution of raindrop-driven dispersal mechanisms makes a lot of sense when we consider the size of raindrops relative

Fig. 6.4. Spore discharge in *Deightoniella torulosa*. **A** Large septate conidium attached to bulbous apex of conidiophore. **B** Cavitation and resulting expansion of the bulb discharges the conidium. Images captured with conventional digital video camera with time interval of 43.5 ms between **A** and **B**. *Bar* = 50 μm

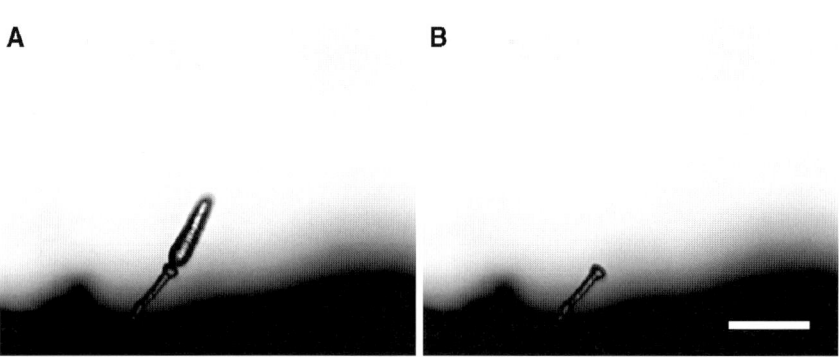

to spores and the energy released when they hit a colony. The terminal velocity of a raindrop with a radius of 1.0 mm (most that reach the ground are 0.2–5.0 mm) is about 4 m s^{-1} (Gunn and Kinzer 1949) and it will hit a colony with a force of 5 μN. The effects of micronewton forces are very interesting when we consider the nanonewton forces available from low speed airflow and the apparent need for much larger forces for conidial discharge (Sect. IV.A). A 0.5 mm raindrop is 10^6 times bigger and heavier than a spore; scaling up to human dimensions, a comparably small "raindrop" would weigh 100 kt and splatter the unfortunate target with the force of 200 kg of high explosive! Experimental data on the kinetic energy of impacting droplets is presented by Lovell et al. (2002), and Saint-Jean et al. (2006) offer a recent treatment of the effects of inertial forces and surface tension in splash dispersal. Raindrops can discharge spores when they hit lesions at terminal velocity after an unimpeded free-fall from a height of a few meters or as lower-speed secondary droplets dripping from vegetation (Herwitz 2006). Raindrops can also cause spore discharge indirectly as a source of vibration of infected tissues. Finally, drops of water moving in mist can capture and disperse spores from plant lesions. Gregory (1973) remains the classic work on fungal dispersal that explains many of the principles of splash dispersal.

Pathogens whose spores are splashed into the air by raindrops include species of *Fusarium* (Horberg 2002; Paul et al. 2004), *Colletotrichum* (Peres et al. 2005), and *Botryosphaeria* (Ahimera et al. 2004), and the uredospores of rusts (Sache 2000). This list serves to highlight some of the more recent studies on splash dispersal, but the process has widespread significance in plant disease. When a raindrop hits a dry leaf, it spreads outward, forming a thin disc before rising to form a thin-walled crater; the crater continues to widen, its rim thickens to form beads, and these beads are stretched into rays that fragment into droplets. Spores within the crater are carried on and within those

droplets. Raindrops impacting dry spores can create small clouds of spores that are dispersed independently of the water droplets ejected from the splash crater. The effect of each rain drop is of course maximized when it hits a concentrated lesion containing millions of spores. Weston and Taylor (1948) showed that a single raindrop falling on foliar lesions of *Botrytis* could contaminate an area of 2.5 m^2. The iconic "*milk drop coronet*" photograph taken by Harold Edgerton in 1957 helps us to visualize the events associated with raindrop-driven discharge of spores.

Other fungi make spectacular use of raindrops for spore discharge. These include the puffballs (Agaricales) and earth-stars (Geastrales), whose basidiomes act like bellows, crushing at impact and spritzing clouds of spores through a pore (the ostiole) on upper surface of peridium, and bird's nest fungi (Agaricales), whose packets of spores called peridioles are ejected for distances of many meters from their fluted basidiomes (Gregory 1949; Brodie 1975). These fungi are saprobes and are not known to have any deleterious effects upon plant growth.

VI. Animals

Animals act as important vectors for the dispersal of many fungal diseases of plants (Agrios 2006). Well known examples include the transmission of chestnut blight by insects and birds, and Dutch elm disease by beetles. Mechanically, there is little of interest in these interactions: the forces exerted on fungal colonies by the movements of invertebrates and vertebrates are more than sufficient to detach microscopic spores from their conidiophores or sporangiophores, or to dislodge masses of spores from pycnidial and perithecial cirrhi. The intriguing part of the interactions comes from the specificity of some of these animal–fungus interactions: the production of chemical attractants by fungi serving to lure vectors, the behavioral complexity of the animal vectors, the adhesion of spores to the vector surface, fungal survival during digestive passage through mycophagous vectors, and the evolution of these diverse relationships.

VII. Turgor Pressure

A. Ascospore Discharge

Explosive discharge of asci drives some of the fastest movements in nature. The ecological and agricultural importance of the mechanism is apparent from the observation that the Ascomycota is the largest fungal phylum and the most numerous group of plant pathogens. The considerable diversity in ascus structure, ascospore development, and spore morphology among the Ascomycota has significant effects on the dynamics of the discharge process (Webster and Weber 2007). The simplest mechanism operates in pathogens like *Taphrina* that belong to the basal lineage of the Ascomycota called the Taphrinomycotina. For example, in *Taphrina deformans*, that causes peach leaf curl, the ascus is exposed on the infected leaf surface, splits open at its tip, and expels a cloud of infectious spores (Yarwood 1941). In the Saccharomycotina, most of the yeast species engage in passive distribution of ascospores that relies upon digestion of the ascus wall. Exceptions to this include the discharge of needle-shaped spores from *Eremothecium* and *Metschnikowia*, and the slow extrusion of ascospores in *Dipodascus* (Van Heerden et al. 2007). In the Pezizomycotina asci are formed within multicellular fruiting bodies or ascomata. The chasmothecia of the Erysiphales (powdery mildews, e.g., *Phyllactinia*) crack open to expose their explosive asci; the cleistothecia of the Eurotiales (e.g., *Eurotium*) contain non-explosive asci that spill from the ascomata when the wall fragments. Apothecial ascomycetes with operculate asci include *Ascobolus immersus*, a non-pathogenic coprophilous fungus that serves as a model for research on ascospore discharge. The "enormous" asci of this species (up to 1 mm long) project from the ascoma, open via an operculum, and expel clusters of eight spores embedded in mucilage (Fischer et al. 2004). Other species of *Ascobolus* and ascomycetes that form larger apothecia exhibit "puffing," which is the release of a cloud of ascospores (Fig. 6.5). Puffing is caused by a wave of ascus discharge that creates an updraft of air that seems to increase discharge distance (Ingold 1968). Puffing is also seen in the stalked apothecia of *Sclerotinia*, which is an inoperculate apothecial fungus (Harthill and Underhill 1976). In perithecial ascomycetes the ascus apex is thickened in the form of an apical apparatus that operates as a sphincter, maintaining ascus pressure and separating spores as they are discharged. The discharge process in these fungi has been studied in greatest detail in *Podospora* and *Sordaria* (Waker and Harvey 1966; Ingold 1971); pathogens with this type of ascus include species of *Nectria* and *Gibberella* (Trail et al. 2002, 2005). Explosive ascospore discharge is not a universal feature of the perithecial ascomycetes. Many important pathogens exude masses of spores and sticky sap to form cirrhi that protrude from their perithecia. Spores of these fungi are dispersed, secondarily, by insects (Sect. VI). *Ophiostoma* species work in this fashion, their ascus walls dissolving within the perithecium to liberate an ooze of spores. The distribution of the slow-exudation process among groups of perithecial ascomycetes that exhibit explosive spore discharge is consistent with the loss of this "violent" mechanism in multiple lineages (Berbee and Taylor 2001). Finally, many ascomycetes with bitunicate asci engage in a two-stage discharge mechanism in which the outer wall (ectotunica) of the ascus ruptures, allowing the inner wall (endotunica) to elongate, and discharge the spores when its apex ruptures. The bitunicate ascus, also referred to as fissitunicate, is formed in fruiting bodies called pseudothecia that resemble perithecia or other types of ascomata. Important pathogens that form

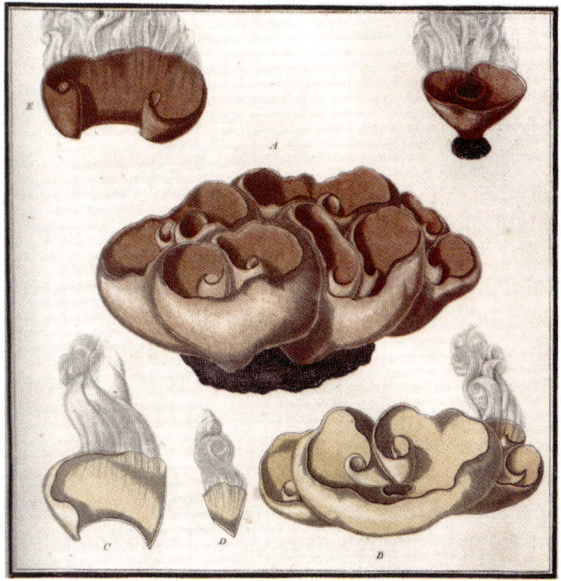

Fig. 6.5. Apothecial puffing in *Peziza* illustrated by Pierre Bulliard (1791). Image courtesy of the Lloyd Library (Cincinnati, Ohio)

bitunicate asci include species of *Cochliobolus*, *Mycosphaerella*, *Pleospora*, and *Venturia*.

The explosive action of the individual ascus has been studied for more than a century and in most species we know that the ascus functions as a pressurized spore gun whose dehiscence spurts spores into the air along with a stream of sap. To understand how different species launch their spores we need to answer a series of questions that include:

1. How much pressure is generated in the ascus?
2. How is this pressure generated?
3. How fast are the spores expelled?
4. How far are they launched?

Until recently, there had been few attempts to measure ascus turgor pressure. Ingold (1939, 1966) estimated osmotic pressures of 1.0–3.0 MPa for the ascus sap of *Ascobolus furfuraceus* and *Sordaria fimicola*. Precise measurements of osmotic pressure reflect the magnitude of ascus turgor under conditions of optimal hydration, but Ingold's estimates appear to be an order of magnitude too high. Fischer et al. (2004) measured ascus turgor of 0.3 MPa in *Ascobolus immersus* by dimpling the ascus wall with a microprobe attached to a miniature strain gauge (Fig. 6.6). These measurements were in agreement with osmotic pressure data for this species. Spectroscopic analysis of ascus sap samples identified polyols as a major osmolyte, which is consistent with histological studies by Ingold (1939) indicating that low molecular weight osmolytes might be generated from glycogen reserves within young asci. The polyol concentration estimated from the spectroscopic analysis indicates that two-thirds of the ascus pressure is probably generated by inorganic ions. Similar work on *Gibberella zeae*, that causes head blight of cereals, identified mannitol as the major sugar in ascus sap (Trail et al. 2002). The use of some osmolytes is more energetically efficient than others. For example, for the same molarity, glycerol (3-C sugar alcohol) and mannitol (6-C sugar alcohol) generate roughly the same osmotic pressures, but glycerol does so for half the carbon investment (Davis et al. 2000). The composition of ascus sap seems to vary within single ascomycete species and may be a feature of colony age (Davis and Money, unpublished data). This interpretation is consistent with the observation that the range of the ascospore discharge mechanism is greatest in young perithecia of *G. zeae* and falls with increasing age (Schmale et al. 2005).

Discharge distance varies from a few millimeters to tens of centimeters among the ascomycetes. Species of *Podospora* are among the record holders, with *P. dicipiens* firing its spores farther than 0.5 m (Walker and Harvey 1966). These spores are shot over a distance of 16 000 times the length of the spore. A comparably impressive cannon would propel a human over a distance of 30 km. In reality, gravity brings a human fired from a circus cannon to the ground just a few meters from the muzzle. Gravity certainly affects the range of the ascospore, but drag from the air acts as a far greater brake (Sect. II). Spores decelerate during flight at a rate that depends on this drag force and the mass (or inertia) of the spore according to Newton's Second Law, $F = ma$. As the size of the spore increases, both the drag and mass increase. The mass however, growing as the cube of the spore size, becomes increasingly significant so that larger spores experience less deceleration than smaller ones and traverse greater distances. For the same ascus turgor pressure and launch speed, larger projectiles travel farther than smaller ones. This principle probably accounts for the evolution of mucilage coats and appendages that connect the spores and increase projectile mass in coprophilous ascomycetes like *Ascobolus* and *Podospora* whose asci exhibit the greatest range. Ingold and Hadland (1959) studied the ballistics of *Sordaria fimicola* by capturing spores on a transparent disc

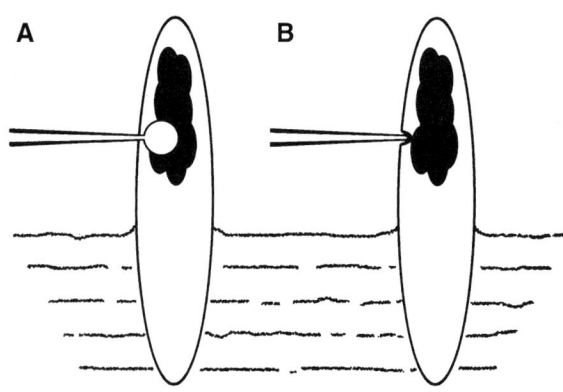

Fig. 6.6. Experimental approaches to measuring ascus turgor pressure in *Ascobolus immersus*. **A** Oil-filled pressure probe micropipet penetrating cell wall and underlying plasma membrane of a mature ascus. In most experiments, the asci deflate upon penetration, but can be re-inflated by oil injection. **B** Indentation of ascus surface with a microprobe attached to a miniature strain gauge. From Fischer et al. (2004), with permission

rotating above a sporulating culture. Spores were shot upward onto the disc, providing a record of discharge from individual asci in the form of short files of spores laid out along the disc. The raw data presented in this paper yield an estimated launch speeds of up to 10.4 m s⁻¹ (Ingold and Hadland made an error in their calculations) which is satisfyingly close to the rough estimate of 10.8 m s⁻¹ from an equation in Buller (1909). The estimated launch speed of the spore clusters in *A. immersus* is 9.3 m s⁻¹ and the maximum range is 30 cm (Buller 1909; Fischer et al. 2004; Fig. 6.7). Trail et al. (2005) estimate launch speeds in excess of 34 m s⁻¹ for *G. zeae*. The difference between *A. immersus* and *G. zeae* is related to the mass of their projectiles; the spore clusters of *A. immersus* have an estimated mass of 1 μg compared with the individual spores of *G. zeae* that weigh no more than 0.2 ng (a 5000-fold difference). The fast launch speed of *G. zeae* spores is necessitated by the overwhelming effect of air viscosity on such small particles, but even at 34 m s⁻¹ they travel only a few millimeters. Future progress in understanding ascospore discharge is dependent upon the use of ultra-high-speed videomicroscopy to analyze the launch of the spores and their flight paths (Sect. VIII).

Turgor-driven mechanisms of conidial discharge are a neglected area of research. This is unfortunate, because these processes are inherently interesting from a biomechanical perspective and occur in fungi that cause significant crop damage. Meredith (1962) was first to suggest that the conidia of *Magnaporthe grisea* might be discharged by "an active discharge mechanism under humid conditions." This process was studied by Ingold (1964) who did not observe discharge, but noted that liberated spores carried a portion of their ruptured stalk cells that served to attach them to their conidiophores. This open-ended pedicel is conspicuous in electron micrographs of *M. grisea* conidia (Fig. 6.8). Ingold suggested that pressure within this cell might induce its rupture and thereby propel the conidium a short distance into the air. Evidence for an active discharge mechanism is obtained from the scatter pattern of conidia released from a sporulating colony of *M. grisea* that is mounted in a vertical orientation (Money and Donofrio, personal observation; Fig. 6.9). The range of this mechanism is very limited, apparently doing little more than separating the spore from its

B. Active Conidial Discharge

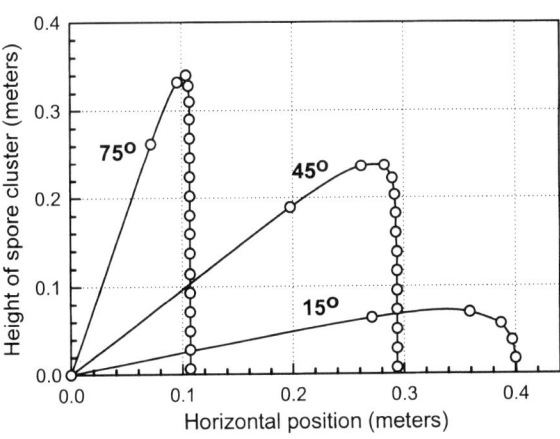

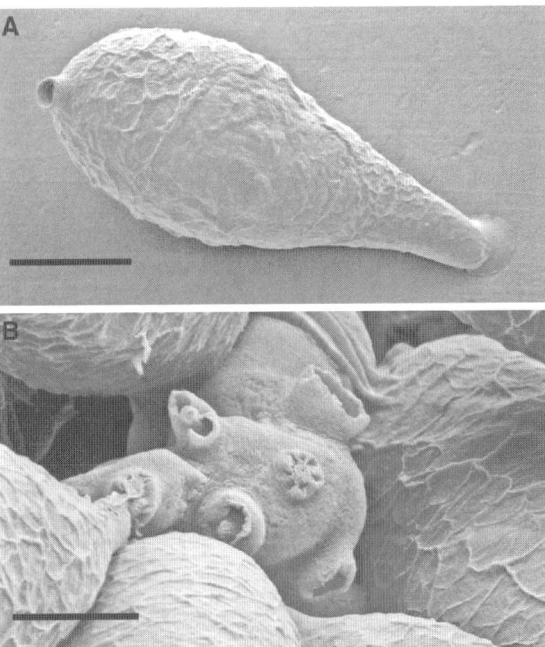

Fig. 6.7. Trajectories for spore clusters of *Ascobolus immersus* predicted for various launch angles. The data points (*circles*) show the position of the spores at 50 ms time intervals; for a launch angle of 15 degrees, the model predicts that the spore cluster will travel a horizontal distance of 40 cm in 0.25 s. From Fischer et al. (2004), with permission

Fig. 6.8. Cryo-SEM images of *Magnaporthe grisea*. **A** Conidium attached to plastic coverslip via spore tip mucilage. **B** Broken pedicels on conidiophore apex. Courtesy of Richard J Howard (DuPont Crop Genetics, Wilmington). *Bars* 5.0 μm

conidiophore, but for a microorganism growing on a surface this disengagement is crucial. Analysis of the discharge process using high-speed video is necessary.

A clearer case of turgor-driven conidial discharge is seen in species of *Nigrospora* (Webster 1952). This genus includes *N. oryzae* that causes ear rot of corn. The globose conidium of *Nigrospora* is subtended by a supporting cell and connecting ampulliform cell. Fluid squirting from the ampulliform cell through a nozzle in the supporting cell is believed to propel the 20 μm diameter conidium over a distance of 1–2 cm. Again, study of this genus using high-speed video microscopy is necessary to clarify this mechanism.

Conidial discharge may also be driven by supporting cells whose walls flex outward under pressure but do not burst. This process has been described in *Epicoccum nigrum*, *Arthrinium cupidatum*, and *Xylosphaera furcata* (Dixon 1965; Webster 1966). The large conidia of entomopathogenic species of *Conidiobolus*, *Erynia*, *Entomophthora* (secondary conidia), and *Furia* are discharged by a similarly

Fig. 6.9. Overnight deposit of *Magnaporthe grisea* conidia discharged from vertically-oriented agar (shadow at *left*) onto surface of coverglass. *Bar* 0.25 mm

rapid pressure-driven eversion of the two-ply septum between the spore and conidiophore apex (Webster and Weber 2007). The range of this mechanism can reach several centimeters. A related process is also involved in discharging rust aeciospores. Rather than septal eversion, pressure boosted by high humidity results in the sudden rounding-off and mutual repulsion of aeciospores, but the mechanical details are unclear. The efficacy of the mechanism is obvious, however, from the shower of spores shot to distances of up to 1.0 cm from mature aecia (Buller 1924; Dodge 1924).

C. Zoospore Discharge

Phytopathogenic oomycetes, including the downy mildew genera *Albugo*, *Pythium*, and *Phytophthora*, and root-rotting opportunists, like *Achlya* and *Aphanomyces*, produce biflagellate zoospores (Dick 2001). The paired flagella of these spores differ in structure and function: one projects ahead of the spore and tripartite filaments (mastigonemes) that ornament its surface serve to pull the spore forward as the flagellum undulates (Cahill et al. 1996); the other is smooth and trails behind, flexing to change the direction of the swimming spore. Variations in the life cycles of these zoosporic pathogens complicate the description of the spore release process. For example, the sporangia of *P. infestans* are dispersed by wind from the tips of branched sporangiophores that develop from the stomata on the infected leaves of its host (Sect. IV.A). When these sporangia land on wet leaves, under cool conditions, they discharge multiple zoospores that can establish new infections.

Zoospore discharge is also characteristic of the phytopathogenic plasmodiophorid slime molds, including *Plasmodiophora brassicae*. These microorganisms produce biflagellate zoospores with a pair of smooth flagella. Mycologists and plant pathologists have also studied zoosporic labyrinthulid slime molds (Labyrinthulomyota) and hyphochytrids (Hyphochytriomycota). Labyrinthulids include *Labyrinthula* species that cause eelgrass wasting disease and rapid blight of turfgrass. Their zoospores are biflagellate, reflecting the group's relationship to the oomycetes. The hyphochytrids infect other water molds and algae, but are not known as agents of plant disease. Hyphochytrid zoospores have a single anteriorly directed flagellum equipped with mastigonemes. In Kingdom Fungi, the uniflagellate zoospores of Chytridiomycota are also released from sporangia.

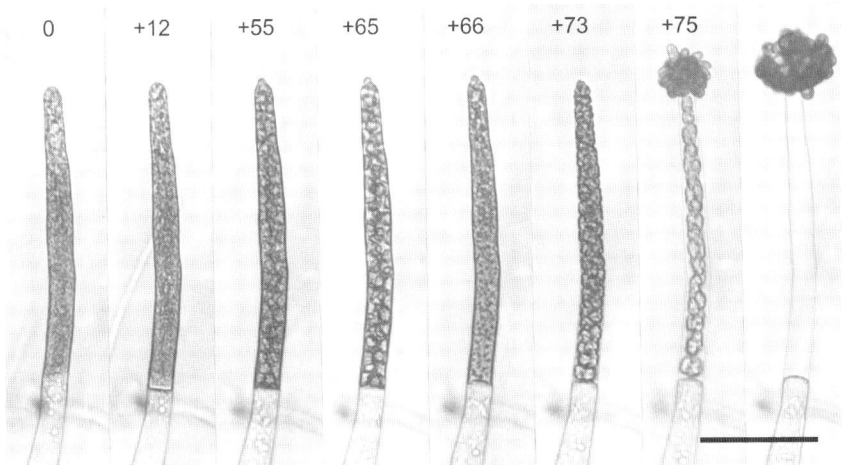

Fig. 6.10. Sporangial development and emptying in *Achlya intricata*. Time(min) is indicated at top of images. Sequence begins with hypha with rounded apex that has stopped extending (0 min). Septation follows (+12 min), then cleavage of the sporangial cytoplasm (+55 min, +65 min), and completion of cleavage (+66 min). Separation of spores from sporangial wall (+73 min) precedes emptying (+75 min). In *Achlya* spp., the spores encyst at the tip of the sporangium; in the related genus *Saprolegnia*, the zoospores swim away upon release. *Bar* = 100 μm

Most of the research on sporangial emptying has concerned Oomycetes, and the biomechanics of this process have been studied in greatest detail in the Saprolegniales, particularly *Achlya* and *Saprolegnia* (Money and Webster 1988, 1989; Money et al. 1988). In these water molds, the sporangium develops from a vegetative hypha whose tip becomes packed with dense cytoplasm before isolation from the subtending mycelium with the formation of a septum (Fig. 6.10). Subsequently, the cytoplasm within the sporangium is differentiated into numerous uninucleate spores. Initially, the *Achlya* sporangium is pressurized to 0.5 MPa, but turgor is lost when the plasma membrane of the sporangium is partitioned by the cleavage mechanism (Fig. 6.11). This results in a dramatic shrinkage of the sporangium as fluid squeezes through the cell wall (measured as current of potassium ions; Money and Brownlee 1987; Thiel et al. 1988) and its pressure falls to less than 1% of its pre-cleavage maximum. At the completion of cleavage, flagellar activity begins with the lashing of the paired flagella (in *Saprolegnia*), the apical papilla is dissolved, and the spores are expelled in less than 1 min. The maximum initial velocity of the spores as they exit the sporangium is 40 μm s⁻¹ (2.4 mm min⁻¹), but the discharge velocity shows a hyperbolic decrease as pressure falls with the expulsion of more and more spores. In *Saprolegnia*, a few spores often remain in the sporangium after the pressure within the sporangium is dissipated; these stragglers usually exit by flagellar action (1 and Webster 1989).

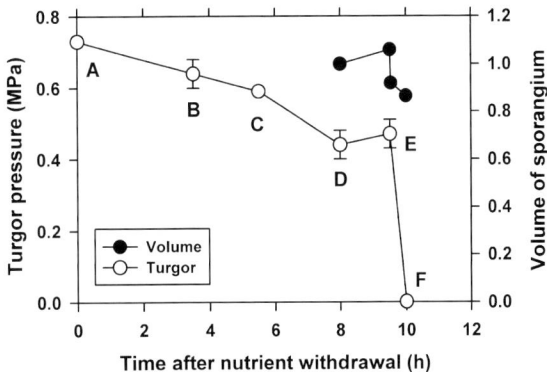

Fig. 6.11. Changes in turgor pressure (*white circles*) and volume (*black circles*) of the sporangial initials and sporangia of *Achlya intricata* during sporangial development, maturation, and spore discharge. *A* Maximum turgor pressure in hyphae growing in nutrient medium. *B* Pressure falls steadily after hyphae are incubated in nutrient-free solution. *C* Hyphae (now referred to as sporangial initials) become charged with cytoplasm; pressure continues to fall. *D* Pressure increases slightly with the formation of a discharge papilla and septum; this is associated with an increase in sporangial volume. *E* At the completion of cleavage, most of the sporangial turgor pressure is lost and volume decreases. *F* A small residual pressure drives the discharge of the spores. Turgor pressure measured with pressure probe according to methods in Money and Harold (1992)

One feature of the sporangial emptying mechanism that is of particular interest is the source of the pressure within the mature sporangium. The plasma membrane of the developing sporangium maintains the differential in osmotic pressure that generates its

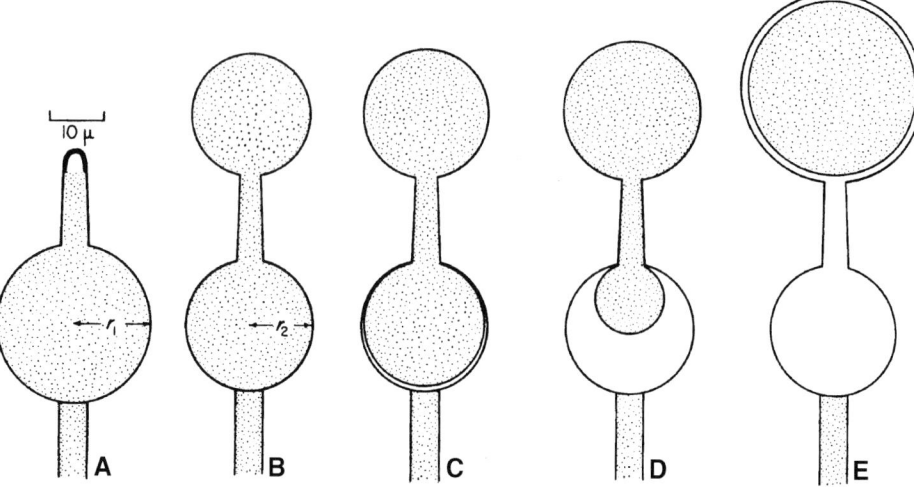

Fig. 6.12. Model for sporangial emptying in *Pythium middletonii*. **A** Sporangium shortly before discharge indicating thickened wall at tip of papilla that will expand into the vesicle. **B** Discharge in progress, sporangium has shrunk with the movement of cytoplasm into the vesicle. **C** Critical point at which the volume of cytoplasm in the vesicle increases due to water uptake causing a small pressure differential between the cytoplasm in the sporangium (high) and vesicle (low). **D** Cytoplasm continues to empty from the sporangium. **E** Vesicle is fully charged. From Webster and Dennis (1967), with permission

turgor pressure. Turgor is lost with the fragmentation of this membrane, but because the osmotic pressure of the individual spores remains high, each spore exports water via the pumping activity of its water expulsion vacuole (contractile vacuole) to avoid pressurization and bursting. The fluid surrounding the spores is derived from the central vacuole of the developing sporangium and the cleavage vacuoles that developed around each of the spore initials. After cleavage, any ions and low molecular weight organic molecules in these compartments are free to diffuse through the cell wall. Only molecules whose molecular size exceeds the wall's porosity are retained (Money and Webster 1988). A very dilute solution of these compounds (no more than a few millimoles) generates the necessary pressure, but the identity of these osmolytes has not been established. The same processes occur in *Aphanomyces euteiches* that causes root rot of legumes (Hoch and Mitchell 1973).

Sporangial emptying in the pathogenic Pythiales and Peronosporales (Oomycota) occurs according to slightly different mechanisms than those described for the Saprolegniales. In *Pythium*, the sporangial cytoplasm is expelled through an exit tube into a vesicle that forms a bag outside the spherical chamber of the sporangium. This vesicle is continuous with the cell wall of the sporangium. Cleavage of the zoospores occurs within this vesicle and the mature spores swim free

from the vesicle when it is lysed. The process of cytoplasmic transfer from the sporangium to the vesicle was studied in detail in *P. middletonii* by Webster and Dennis (1967). The movement of the cytoplasm occurs in two stages (Fig. 6.12). In the first stage, the expansion of the vesicle at the tip of the exit tube provides an escape for the pressurized cytoplasm. At the end of this stage, half of the cytoplasmic volume remains in the spherical sporangium and half in the vesicle. During the second stage, which can occur very slowly, the cytoplasm in the sporangium is shifted into the vesicle. This motion is driven by a differential in surface tension and pressure between the cytoplasm in the sporangium and the vesicle. The second stage requires a "kick start," because there is an instant when the surface area and pressure in the sporangium and vesicle are equal. This is supplied by water influx into the vesicle. Unlike the wall of the sporangium, the vesicle is highly plastic and enlargement caused by water influx lowers its surface tension and hydrostatic pressure, producing the pressure differential that completes the emptying process. The gradient becomes steeper as the sporangial cytoplasm shrinks. This is a beautifully simple mechanism. The same mechanical principles apply to excystment, when oomycete zoospores are expelled from cysts.

In *Phytophthora* and *Albugo*, the zoospores are cleaved within the sporangium and discharge into a vesicle that usually breaks with the continued passage of spores (Fig. 6.13). The vesicle in these genera does not seem to be derived from cell wall materials in the papilla, but is an extension of the plasma membrane within the sporangium. (The same kind

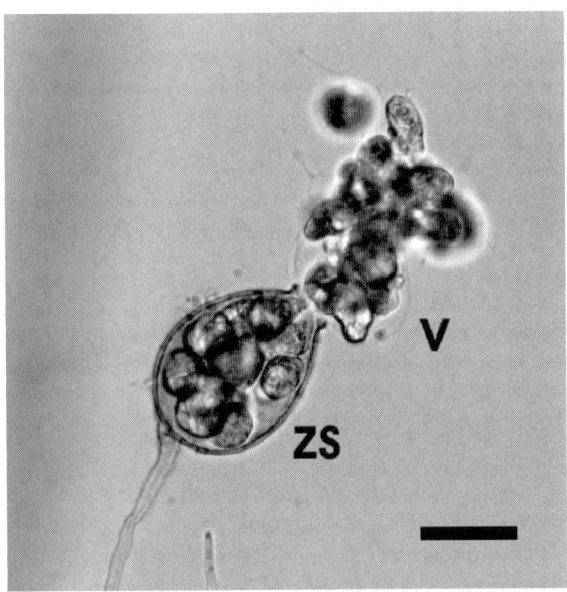

Fig. 6.13. Sporangial emptying in *Phytophthora cinnamomi*. Zoospores are expelled through the open papilla of the zoosporangium (zs). The vesicle (v) is broken and the biflagellate zoospores swim away from the sporangium. *Bar* = 25m. Image kindly supplied by Jenise Bauman (Miami University, Oxford, Ohio)

of cytoplasmic herniation occurs when the wall of a turgid hypha is digested to form protoplasts.) In contrast to the cleavage process in saprolegniaceous water molds, the plasma membrane of the sporangium of *Phytophthora* remains intact and spores are delimited by de novo membrane synthesis (Hyde et al. 1991). This means that the sporangial pressure probably remains high until discharge. In mechanical terms, spore discharge in *Phytophthora* is similar to the discharge of ascospores. Critical experiments on the mechanism were conducted by Gisi (1983), and recent molecular studies are reviewed by Hardham (2005, 2007) and Walker and van West (2007).

There has been very little experimental work on spore discharge mechanisms among the other groups of zoosporic fungi and water molds. The hyphochytrid *Rhizidiomyces apophysatus* (a parasite of straminipiles) discharges an uncleaved cytoplasm into a vesicle generated from wall material that accumulates in a condensed form in the papilla. This process is reminiscent of the sporangial emptying in *Pythium* and probably involves an identical mechanism. Among the Chytridiomycota there is considerable variation in the emptying process. In the simplest cases there is no obvious pressure-driven discharge: the papilla opens, sometimes at the tip of an elongated discharge tube, and the spores swim around in the chamber of the sporangium until they find the exit. In some species the papilla is capped by an operculum that is shed to initiate discharge and in others the spores pour into a vesicle (like *Pythium* and *Rhizidiomyces*).

D. *Pilobolus*

The *Pilobolus* (Zygomycota) "squirt gun" is probably the best known device for spore discharge. Turgor pressure is generated osmotically within a fluid-filled translucent sporangiophore that develops from the herbivore dung on which *Pilobolus* thrives. In the commonest species, *P. kleinii*, a black-pigmented sporangium filled with 30 000–90 000 spores forms at the tip of the sporangiophore. The sporangiophore in *P. kleinii* is usually 2–5 mm in height and its sporangium has a diameter of approximately 0.5 mm. The region of the sporangiophore beneath the sporangium swells to form a vesicle that functions as a lens to direct the phototropic bending of the sporangiophore. Dehiscence occurs along a circular locus of cell wall at the tip of the vesicle, propelling the sporangium and sporangiophore fluid for a horizontal distance of up to 2.5 m (Buller 1934; Ingold 1971). The pressure within the sporangiophore was estimated by Buller (1934) to be 0.55 MPa, but the osmolytes responsible for pressure generation have not been identified using modern methods. Pringsheim and Czurda (1927) estimated the launch speed of the sporangium using an apparatus consisting of a pair of spinning paper discs. To hit the second disc, the sporangia had to pass through a hole in the first disc; the displacement of sporangia on the circumference of the second disc, relative to the hole in the first, was used to estimate a velocity of $14 \, \mathrm{m \, s^{-1}}$. Using an equally ingenious photocell apparatus to record the interruption of light as sporangia passed through two beams of light separated by 3 mm, Page and Kennedy (1964) estimated an average initial velocity of *P. kleinii* sporangia of $10.8 \, \mathrm{m \, s^{-1}}$. In a second paper, Page (1964) captured the discharge process using a high-speed electronic flash.

There are a few similar mechanisms among the fungi, including the processes of conidial discharge in *Basidiobolus* and *Entomophthora* (Webster and Weber 2007). In *Basidiobolus*, the apical part of the conidiophore is discharged along with the spore, so that the projectile resembles a microscopic rocket (Ingold 1971); the spore and this subtending conidiophore tip may separate after launch by the septal eversion process described in Sect. VII.B. In *Entomophthora*, the primary conidium is discharged along with a stream of sap from its conidiophore; the septal eversion mechanism is responsible for discharging the secondary conidium of *Entomophthora* species, illustrating how different mechanisms of spore liberation can operate in a single fungal species.

E. *Sphaerobolus*

Sphaerobolus (Basidiomycota) propels a 1 mm diameter spherical spore-filled projectile called the gleba over distances of up to 6 m (Buller 1933; Ingold 1971). The anatomy of the basidiome of *Sphaerobolus* is unusually complex for a fungal organ. It is spherical, no more than 2 mm in diameter, and is composed of six layers of interwoven hyphae surrounding the central gleba. At maturity, the surface of the organ fractures and opens outward to form a star-shape, with the exposed gleba sitting in its center. In this form, the mature basidiome is best described as a cup within a cup; a slightly flaccid tennis ball, pushed inward on one side to form an unstable dimple, serves as an excellent model for the propulsive mechanism. The gleba is bathed in fluid derived from the innermost tissue layer and is supported by the underlying pair of tissue layers that form an elastic "membrane" called the peridium. The two layers of the peridium consist of a palisade of elongated, radially oriented cells whose long axes point inward, and a backing layer of tangentially oriented thin hyphae. The palisade cells solubilize sugars from glycogen reserves and become pressurized by osmosis (Engel and Schneider 1963). While the peridium is in its concave untriggered form, the exposed ends of the palisade cells are more compressed than their bases. The resulting strain within the walls of these cells is relieved when the peridium flips outward, propelling the gleba into the air at a maximum estimated velocity of 10 m s^{-1}. This is associated with an audible "pop."

Buller (1933) calculated that the power required for glebal discharge is about 1×10^{-4} horsepower, or 0.1 W. *Sphaerobolus* species are saprobes that are not known to cause any plant diseases.

VIII. Surface Tension

The majority of the Basidiomycota utilize the process of ballistospore discharge (ballistospory) that involves the rapid motion of a fluid droplet over the spore surface (Ingold 1971; Webster and Chen 1990; Money 1998; Pringle et al. 2005). The mechanism is responsible for launching basidiospores from the gills, spines, and tube surfaces of mushroom-forming fungi, and is also featured in the life cycles of basidiomycete yeasts and the phytopathogenic rusts and smuts. The wide distribution of ballistospory and the similarity of the mechanism throughout the phylum suggests that it had a single origin in an ancestral group (Thiers 1984). Ballistospores have an asymmetric shape, with a prominent bulge at their base, called the hilar appendix, adjacent to the region of contact with the pointed stalk or sterigma (Fig. 6.14A). A few seconds before discharge, fluid begins to condense on the spore surface in two locations: (a) as a prominent droplet on the hilar appendix (called Buller's drop), and (b) on the adjacent spore surface (called the adaxial drop). Once initiated, Buller's drop increases in diameter for a few seconds, and then, the spore and drop are discharged from the sterigma (Fig. 6.14B). Discharge

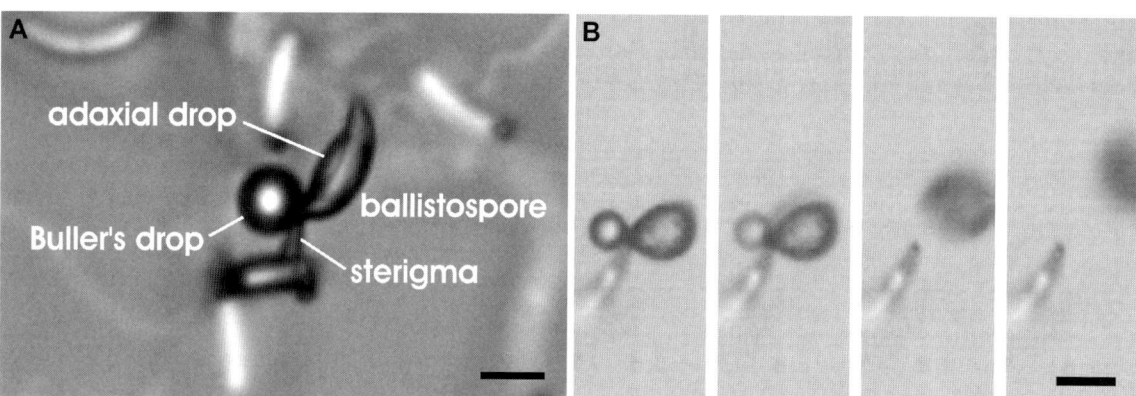

Fig. 6.14. A Ballistospore (also known as secondary sporidium) of the smut *Tilletia caries* photographed immediately before discharge. Hilar appendix is hidden by drop in this image. *Bar* 10 µm. Image kindly supplied by Jessica Stolze-Rybczynski (Miami University, Oxford, Ohio). **B** Ballistospore discharge in the basidiomycete yeast *Itersonilia perplexans* recorded at a frame rate of 100 000 frames s^{-1}. Motion of Buller's drop onto the spore surface is evident in the second frame. *Bar* 10 µm. From Pringle et al. (2005) with permission

occurs when Buller's drop reaches a critical diameter and contacts and coalesces with the adaxial drop on the adjacent spore surface. When Buller's drop moves, mass is redistributed from the hilar appendix in the direction of the free end of the spore; this imparts momentum to the spore, and the spore springs from its sterigma. Fluid movement is powered by the reduction in free energy (surface tension) when the two drops fuse, so we refer to the mechanism as a surface tension catapult (Turner and Webster 1991). The energetics of the mechanism are addressed below.

The fluid accumulation on the hilar appendix (Buller's drop), and on the adjacent spore surface (adaxial drop), occurs by condensation of water from the surrounding air. Condensation is driven by the release of osmolytes onto the spore surface that cause a localized reduction in the chemical activity of water (or water potential) and vapor pressure. Evidence for this mechanism comes from: (a) experiments on the effect of humidity on drop expansion (Webster et al. 1989), (b) mathematical models of the rate of drop expansion (Turner and Webster 1991), and (c) spectroscopic analysis of fluid washed from spore deposits showing high concentrations of mannitol and other osmotically-active compounds (Turner and Webster 1995; Webster et al. 1995). It is not clear how the osmolytes are delivered to the spore surface in a sufficient concentration to act as nuclei for the condensation of water, but electron microscopic studies suggest that these compounds may be prepackaged in the form of a discrete organelle in the cytoplasm of the maturing spore (McLaughlin 1982; McLaughlin et al. 1985).

The basidiomycete yeast *Itersonilia perplexans* causes petal blight of *Chrysanthemum* and also causes a destructive disease called black streak on edible burdock (Fox 2003; Horita et al. 2005). It forms particularly large ballistospores in pure culture and has served as a model for biomechanical research on the discharge mechanism for many years (Webster et al. 1984). The launch of the spore is too fast for analysis using conventional video, but has been captured using ultra-high-speed video cameras (Pringle et al. 2005). The estimated initial velocity of the spore is 0.7 m s^{-1}, which is slow relative to the explosive discharge of ascospores. In terms of acceleration, however, this is an impressive launch: motion of Buller's drop and separation of the spore from the sterigma is completed in less than 10 μs, implying an acceleration of 7000 g. A variety of other basidiomycetes have been studied in the author's laboratory using high-speed video and maximum launch speeds of 1.5 m s^{-1} have been estimated. This is interesting in light of C.T. Ingold's maxim, that "the basidium is a spore-gun of precise range"

(Ingold 1992). The ballistospore discharge mechanism is capable of propelling spores over distances of no more than 1–2 mm; the maximum range for *I. perplexans* is estimated at 1.2 mm, but many spores are propelled over distances of only a few tenths of one millimeter. This must be sufficient to clear the boundary layer on the surface of host. Recall from Sect. III that the boundary layer surrounding leaves varies from 0.1 mm to 9.0 mm in thickness, depending upon leaf size and wind speed (Nobel 1991). If the fungus is positioned so that its spores will fall free from the host surface after discharge, the boundary layer does not present a problem. But spore wastage may be significant for the colony on the upper surface of a petal, for example, that will fall back on the plant after a brief flight. Similar considerations apply to other ballistosporic yeasts, jelly fungi, rusts, and smuts. But for mushroom-forming species, the limited range of the discharge mechanism must be tightly controlled to ensure release into the surrounding air; e.g., spores formed on gills must be propelled over a limited distance so that they do not hit the opposite gill. The control of discharge distance is a complex issue that involves modifications in spore morphology that affect the size of Buller's drop and shall be treated in a separate publication.

Models of the effect of viscous drag upon spores of different sizes show good agreement between the launch speeds measured from video data and the range of the mechanism. For example, *I. perplexans* spores travel 0.4 mm with an initial velocity of 0.7 m s^{-1} (Fig. 6.15). The rapid acceleration of

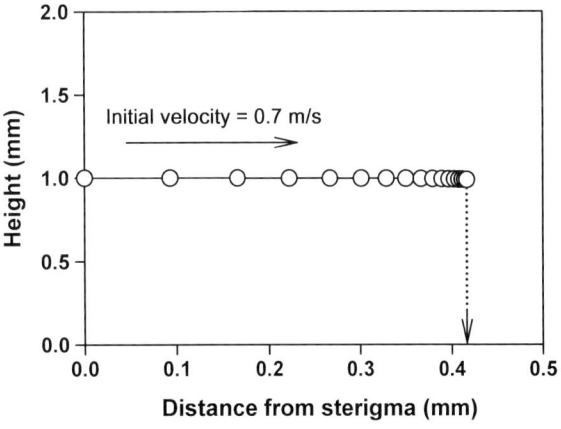

Fig. 6.15. Modeled flight path of ballistospore of *Itersonilia perplexans* predicted from the mass of the spore and associated fluid (1.4 × 10^{-12} kg) and its initial velocity determined by high-speed video (0.7 m s^{-1}). Model for viscous drag based on Fischer et al. (2004). Position of spore is plotted at intervals of 150 μs following horizontal discharge from sterigma at a height of 1 mm from a surface. Note rapid deceleration of spore and complete braking by air viscosity within 3 ms (20 points on the plot), by which time the spore has traveled 0.42 mm. The vertical dotted line shows the subsequent descent of the spore. The terminal velocity of the *I. perplexans* spore and attached fluid is about 6 mm s^{-1}

a stationary spore into air is an impressive feat of microengineering, but it is important to recognize that spore motion is also an example of dramatic deceleration: as soon as the spore is discharged, the viscosity of the air begins to drag it to a halt. At any lower initial velocity, the microscopic particle would make little headway at all through air. The direction of the launch has very little effect upon the distance that the spore is propelled. In other words, the same initial velocity shoots a spore upwards or downwards over the same distance. The influence of gravity is imperceptible until the spore is braked by air viscosity. For spores that are shot horizontally, the typical flight path may be aptly described as a "Wile E. Coyote trajectory," recalling the tragic canine featured in Warner Brothers cartoons. This trajectory is critical to the effectiveness of the mechanism in mushrooms whose spores are propelled from the surface of gills, spines, or tubes.

As explained previously, the current model for ballistospore discharge posits that the energy for spore discharge is derived from the rapid movement of Buller's drop onto the spore surface. The fundamental idea can be tested by calculating the theoretical match between the surface tension in Buller's drop and the energy needed to discharge the ballistospore. Measurements of spore and drop size from *I. perplexans* can be used as an illustration. The spores of this fungus are quite large relative to other basidiospores ($8 \times 15 \mu m$) and spore discharge is associated with the formation of a $5 \mu m$ radius Buller's drop. The surface tension at an air–water interface at $20 °C$ is 7.28×10^{-2} N m^{-1} = 73 mJ m^{-2}. The surface area of the $5 \mu m$ radius drop is 3.1×10^{-10} m^2, and so its surface tension is 2.3×10^{-11} J = 2.3×10^{-4} ergs (1 erg = 10^{-7} J). The mass of the spore and attached fluid is 1.4×10^{-12} kg (using a published density estimate for other basidiospores of 1.2×10^3 kg m^{-3}; Gregory 1973), including the mass of a hemispherical adaxial drop with radius $5 \mu m$. For an initial velocity (v) of 0.7 m s^{-1} (enabling the spore to travel 0.4 mm), kinetic energy ($0.5 mv^2$) = 3.4×10^{-13} J (or kg m^2 s^{-2}) = 3.4×10^{-6} ergs, which is only 1.5% of the energy in Buller's drop.

But this calculation greatly overestimates the energy released by drop movement because the fluid in Buller's drop is carried with the spore when it merges with the adaxial fluid. It is necessary, therefore, to estimate the net change in surface free energy associated with drop merger, because this is all that is available to drive the launch of the spore. To a first approximation, we can model the process as the merger of a sphere and hemisphere to form a larger hemisphere. The sum of the surface free energy in the spherical Buller's drop and the hemispherical adaxial drop is 3.5×10^{-4} ergs. When these bodies merge they will form a hemisphere with radius $7 \mu m$ and the surface tension of the exposed surface is 2.3×10^{-4} ergs (approximately the same as a $5 \mu m$ radius sphere). Therefore, the estimated change in surface free energy associated with the merger of the spherical drop on the hilar appendix with the adaxial

drop is 1.2×10^{-5} ergs, which is four-fold greater than the energy consumed in the launch. It is this energetic saving that powers the movement of Buller's drop and provides the momentum for the launch.

There are two important features of these calculations. First, they demonstrate that the hypothetical link between drop movement and spore discharge makes sense energetically: minuscule droplets of fluid do possess sufficient energy (in the form of surface free energy or surface tension) to discharge spores the appropriate distance. Second, it is clear from the calculations that any change in spore and drop size has a dramatic effect upon the launch speed and discharge distance. For a constant spore size, an increase in the radius of Buller's drop tends to catapult the spore over a greater distance. Conversely, any decrease in drop size reduces the launch speed and range. Calculations show that the *relative* size of the spore and drop are the crucial variables. The morphology of the spore, and/or the hydrophobic nature of the spore surface surrounding the hilar appendix, causes the water condensing in this spot to remain as a discrete drop until the moment of discharge. Because Buller's drop forms on the end of the hilar appendix, longer appendices tend to support larger drops because the drop must expand farther before it contacts and runs over the adjacent spore surface. This is one example of the kind of micromorphological detail that may control the dynamics of the launch process. The remarkable variation in spore morphology among the Basidiomycota may reflect the control of discharge speed and distance necessitated by the evolution of diverse fruiting body forms. The agaricomycete basidome may also enhance spore release by interrupting airflow around their caps (Deering et al. 2001) and by evaporative cooling of the spore-producing tissues (Husher et al. 1999).

The reliance of the ballistospore discharge mechanism upon condensation limits spore discharge to wet environments. Discharge of the ballistosporic stages of pathogenic yeasts, rusts, and smuts can occur only under conditions of high ambient humidity. Mushroom-forming fungi have some degree of control over the humidity of the air between their gills, but even these organisms are restricted to wet habitats and fruit after rainfall. The surface tension catapult mechanism is not found in other fungi, but analogous fluid movement discharges the spores of the protostelid slime mold *Schizoplasmodium cavostelioides* and related species (Olive and Stoianovitch 1966).

IX. Conclusions and Future Research

This chapter examines the diverse mechanisms of spore discharge that operate in phytopathogenic fungi, including passive separation of spores from their parent colonies by wind, vibration, and raindrops, discharge by the explosive formation of cavitation bubbles, and active mechanisms powered by osmotically generated turgor pressure and fluid motion driven by surface tension. A few decades ago, interest in these processes stimulated a great deal of innovative research among mycologists and plant pathologists. Few researchers are investigating these processes today, and yet as this article has identified, our picture of many discharge mechanisms is incomplete. Research on the micromechanical processes that have evolved among the fungi also has some surprising implications for other scientific disciplines. The study of biomimicry is a relatively new area of investigation for engineers whose designs are inspired by natural structures and mechanisms. Investigations on spore discharge may have major implications for future development of machines that operate on the micrometer and nanometer scales. The recent fabrication of a nanoscale machine that is driven by the coalescence of liquid droplets shows that processes fueled by surface tension may have important applications in engineering (Regan et al. 2005). Ballistospore discharge offers a natural model for this kind of device. Atmospheric science is another area of inquiry that is impacted by work on spore discharge. Elbert et al. (2006) found that ballistospore and ascospore discharge is a source of massive quantities of airborne carbohydrates and ions. Using the chemical signature of osmolytes carried in their Buller's drops, they estimate an annual emission of 17 Tg of basidiospores into the atmosphere. Returning to plant pathology, the potential for control methods based upon a better understanding of the mechanisms that create this airborne megatonnage argues for fresh research initiatives using modern tools.

References

Agrios GN (2006) Plant pathology, 5th edn. Academic, New York

Ahimera N, Gisler S, Morgan DP, Michailides TJ (2004) Effects of single-drop impactions and natural and simulated rains on the dispersal of *Botryosphaeria dothidea* conidia. Phytopathology 94:1189–1197

Ainsworth GC (1976) Introduction to the history of mycology. Cambridge University Press, Cambridge

Aylor DE (1975) Force required to detach conidia of *Helminthosporium maydis*. Plant Physiol 55:99–101

Aylor DE (1990) The role of intermittent wind in the dispersal of fungal pathogens. Annu Rev Phytopathol 28:73–92

Aylor DE, Paw U KT (1980) The role of electrostatics in spore liberation by *Drechslera turcica*. Mycologia 72:1213–1219

Aylor DE, Wang Y, Miller DR (1993) Intermittent wind close to the round within a grass canopy. Boundary-Layer Meteorol 66:427–448

Barksdale TH, Asai GN (1961) Diurnal spore release of *Piricularia oryzae* from rice leaves. Phytopathology 51:313–317

Beakes GW, Glocking SL (1998) Injection tube differentiation in gun cells of a *Haptoglossa* species which infects nematodes. Fungal Genet Biol 24:45–68

Berbee ML, Taylor JW (1992) Convergence in ascospore discharge mechanism among Pyrenomycete fungi based on 18S ribosomal RNA gene sequence. Mol Phylogenet Evol 1:59–71

Borkowski M (1972) Zur Anwendbarkeit der Quellungshypothese auf die Sporangienentleerungsweise bei *Achlya*. Z Allg Mikrobiol 12:371–383

Brodie HJ (1975) The bird's nest fungi. University of Toronto Press, Toronto

Buller AHR (1909-1934) Researches on fungi, vols 1–6. Longmans, London

Bulliard P (1791) Histoire des champignons de la France, ou, traité élémentaire renfermant dans un ordre méthodique les descriptions et les figures des champignons qui croissent naturellement en France. Bazan, Paris

Cahill DM, Cope M, Hardham AR (1996) Thrust reversal by tubular mastigonemes: immunological evidence for a role of mastigonemes in forward motion of zoospores of *Phytophthora cinnamomi*. Protoplasma 194:18–28

Chabane S, Sarfati J, Ibrahim-Granet O, Du C, Schmidt C, Mouyna I, Prevost M-C, Calderone R, Latgé JP (2006) Glycosylphosphatidylinositol-anchored Ecm33p influences conidial cell wall biosynthesis in *Aspergillus fumigatus*. Appl Environ Microbiol 72:3259–3267

Davis DJ, Burlak C, Money NP (2000) Osmotic pressure of fungal compatible osmolytes. Mycol Res 104:800–804

Deering R, Dong F, Rambo D, Money NP (2001) Airflow patterns around mushrooms and their relationship to spore dispersal. Mycologia 93:732–736

Dick MW (2001) Straminipilous fungi. Systematics of the Peronosporomycetes including accounts of the marine straminipilous protists, the plasmodiophorids and similar organisms. Kluwer, Dordrecht

Dixon PA (1965) The development and liberation of the conidia of *Xylosphaera furcata*. Trans Br Mycol Soc 48:211–217

Dodge BO (1924) Aecidiospore discharge as related to the character of the spore wall. J Agric Res 27:749–756

Elbert W, Taylor PE, Andreae MO, Pöschl U (2006) Contribution of fungi to primary biogenic aerosols in the atmosphere: active discharge of spores, carbohydrates, and ionorganic ions by Asco- and Basidiomycota. Atmos Chem Phys Discuss 6:11317–11355

Engel H, Schneider JC (1963) Die Umwandlung von Glykogen in Zucker in den Fruchtkörpern von *Sphaerobolus stellatus* (Thode) Pers., vor ihrem Abschluss. Ber Dtsch Bot Ges 75:397–400

Fischer M, Cox J, Davis DJ, Wagner A, Taylor R, Huerta AJ, Money NP (2004) New information on the mechanism of forcible ascospore discharge from *Ascobolus immersus*. Fungal Genet Biol 41:698–707

Fitt BDL, McCartney HA (1986) Spore dispersal in splash droplets. In: Ayres PG, Boddy L (eds) Water, fungi and plants. Cambridge University Press, Cambridge, pp 87–104

Fox RTV (2003) *Chrysanthemum Itersonilia* petal blight. Fungal foes in your garden, 55. Mycologist 17:44

Gisi U (1983) Biophysical aspects of the development of *Phytophthora*. In: Erwin DC, Bartnicki-Garcia S, Tsao PH (eds) *Phytophthora*, its biology, taxonomy, ecology and pathology. American Phytopathological Society, St Paul, pp 109–119

Górny RL, Reponen T, Grinshpun SA, Willeke K (2001) Source strength of fungal spore aerosolization from moldy building material. Atmos Environ 35:4853–4862

Gregory PH (1949) The operation of the puff-ball mechanism of *Lycoperdon perlatum* by raindrops as shown by ultra-high-speed Schlieren cinematography. Trans Br Mycol Soc 32:11–15

Gregory PH (1957) Electrostatic charges on spores of fungi in air. Nature 180:330

Gregory PH (1973) The microbiology of the atmosphere, 2nd edn. Leonard Hill, Plymouth

Gunn R, Kinzer GD (1949) The terminal velocity of fall for water drops in stagnant air. J Meteorol 6:243–248

Hardham AR (2005) *Phytophthora cinnamomi*. Mol Plant Pathol 6:589–604

Hardham AR (2007) Cell biology of plant–oomycete interactions. Cell Microbiol 9:31–39

Harthill WFT, Underhill AP (1976) "Puffing" in *Sclerotinia sclerotiorum* and *S. minor*. NZ J Bot 14:355–358

Heath IB, Steinberg G (1999) Mechanisms of hyphal tip growth: tube dwelling amebae revisited. Fungal Genet Biol 28:79–93

Herwitz SR (2006) Raindrop impact and water flow on the vegetative surfaces of trees and the effects on stem-flow and throughfall generation. Earth Surf Process Land 12:425–432

Hoch HC, Mitchell JE (1973) The effect of osmotic water potentials on *Aphanomyces euteiches* during zoosporogenesis. Can J Bot 51:413–420

Horberg HM (2002) Patterns of splash dispersed conidia of *Fusarium poae* and *Fusarium culmorum*. Eur J Plant Pathol 108:73–80

Horita H, McGovern RJ, Komatsu T, Yasuoka S (2005) Effects of inoculum density, leaf age, termperature, and wetness duration on black streak of edible burdock. J Gen Plant Pathol 71:247–252

Husher J, Cesarov S, Davis C, Fletcher T, Mbuthia K, Richey L, Sparks R, Turpin LA, Money NP (1999) Evaporative cooling of mushrooms. Mycologia 91:351–352

Hyde GJ, Lancelle S, Hepler PK, Hardham AR (1991) Freeze substitution reveals a new model for sporangial cleavage in *Phytophthora*, a result with implications for cytokinesis in other eukaryotes. J Cell Sci 100:735–746

Ingold CT (1939) Spore discharge in land plants. Oxford University Press, Oxford

Ingold CT (1956) A gas phase in viable fungal spores. Nature 177:1242–1243

Ingold CT (1964) Possible spore discharge mechanism in *Pyricularia*. Trans Br Mycol Soc 47:573–575

Ingold CT (1966) Aspects of spore liberation: violent discharge. In: Madelin MF (ed) The fungus spore. Butterworths, London, pp 113–132

Ingold CT (1968) Increased distance of discharge due to puffing in *Ascobolus*. Trans Br Mycol Soc 51:592–594

Ingold CT (1971) Fungal spores: their liberation and dispersal. Oxford University Press, Oxford

Ingold CT (1992) The basidium: a spore gun of precise range. Mycologist 6:111–113

Ingold CT, Hadland SA (1959) The ballistics of *Sordaria*. New Phytol 58:46–57

Jarvis WR (1962) Splash dispersal of spores of *Botrytis cinerea* Pers. Nature 193:599

Kim CK, Min HS, Yoshino R (1990) Epidemiological studies of rice blast disease caused by *Pyricularia oryzae* Cavara (III); diurnal pattern of conidial release and dispersal under the natural conditions. Ann Phytopathol Soc Japan 56:315–321

King AL (1944) The spore discharge mechanism of common ferns. Proc Natl Acad Sci USA 30:155–161

Leach CM (1976) An electrostatic theory to explain violent spore liberation by *Drechslera turcica* and other fungi. Mycologia 68:63–86

Lovell DJ, Parker SR, Van Peteghem P, Webb DA, Welham SJ (2002) Quantification of raindrop kinetic energy for improved prediction of splash-dispersed pathogens. Phytopathology 92:497–503

McCartney HA, Bainbridge A, Legg BJ (1982) Electric charge and the deposition of spores of barley mildew *Erysiphe graminis*. Atmos Environ 16:1133–1143

McLaughlin DJ (1982) Ultrastructure and cytochemistry of basidial and basidiospore development. In: Wells K, Wells EK (eds) Basidium and basidiocarp: evolution, cytology, function, and development. Springer, New York, pp 37–74

McLaughlin DM, Beckett A, Yoon KS (1985) Ultrastructure and evolution of ballistosporic basidiospores. Bot J Linn Soc 91:253–271

Meredith DS (1961) Spore discharge in *Deightoniella torulosa* (Syd.) Ellis. Ann Bot 25:271–278

Meredith DS (1962) Spore discharge in *Cordana musae* (Zimm.) Höhnel and *Zygosporium oscheoides* Mont. Ann Bot 26:233–241

Meredith DS (1963) Violent spore release in some Fungi Imperfecti. Ann Bot 27:39–47

Meredith DS (1965) Violent spore release in *Helminthosporium turcicum*. Phytopathology 55:1099–1102

Milburn JA (1970) Cavitation and osmotic potentials of *Sordaria* ascospores. New Phytol 69:133–141

Money NP (1998) More g's than the space shuttle: ballistospore discharge. Mycologia 90:547–558

Money NP, Brownlee C (1987) Structural and physiological changes during sporangial development in *Achlya intricata* Beneke. Protoplasma 136:199–204

Money NP, Harold FM (1992) Extension growth in the water mold *Achlya*: Interplay of turgor and wall strength. Proc Natl Acad Sci USA 89:4245–4249

Money NP, Webster J (1988) Cell wall permeability and its relationship to spore release in Achlya intricata. Exp Mycol 12:169–179

Money NP, Webster J (1989) The mechanism of sporangial emptying in *Saprolegnia*. Mycol Res 92:45–49

Money NP, Webster J, Ennos R (1988) Dynamics of sporangial emptying in *Achlya intricata*. Exp Mycol 12:13–27

Money NP, Caesar-TonThat T-C, Frederick B, Henson JM (1998) Melanin synthesis is associated with changes in hyphopodial turgor, permeability, and wall rigidity in *Gaeumannomyces graminis* var. *graminis*. Fungal Genet Biol 24:240–251

Moriura N, Matsuda Y, Oichi W, Nakashima S, Hirai T, Sameshima T, Nonomura T, Kakutani K, Kusakari S, Higashi K, Toyoda H (2006) Consecutive monitoring of lifelong production of conidia by individual conidiophores of *Blumeria graminis* f. sp. *hordei* on barley leaves by digital microscopic techniques with electrostatic micromanipulation. Mycol Res 110:18–27

Müller HG (1958) The constricting ring mechanism of two predacious hyphomycetes. Trans Br Mycol Soc 41:341–364

Nobel PS (1991) Physicochemical and environmental plant physiology. Academic, San Diego

Oichi W, Matsuda Y, Nonomura T, Toyoda H (2006) Formation of conidial pseudochains by tomato powdery mildew *Oidium neolycopersici*. Plant Dis 90:915–919

Olive LS, Stoianovitch C (1966) A simple new mycetozoan with ballistospores. Am J Bot 53:344–349

Pady SM, Gregory PH, Kramer CL, Clary R (1969) Periodicity of spore release in *Cladosporium*. Mycologia 61:87–98

Page RM (1964) Sporangium discharge in *Pilobolus*: a photographic study. Science 146:925–927

Page RM, Kennedy D (1964) Studies on the velocity of discharged sporangia of *Pilobolus kleinii*. Mycologia 56:363–368

Paul PA, El-Allaf SM, Lipps PE, Madden LV (2004) Rain splash dispersal of *Gibberella zeae* within wheat canopies in Ohio. Phytopathology 94:1342–1349

Peres NA, Timmer LW, Adaskaveg JE, Correll JC (2005) Lifestyles of *Colletotrichum acutatum*. Plant Dis 89:784–796

Pinckard JA (1942) The mechanism of spore dispersal in *Peronospora tabacina* and certain other downy mildew fungi. Phytopathology 32:505–511

Pringle A, Patek SN, Fischer M, Stolze J, Money NP (2005) The captured launch of a ballistospore. Mycologia 97:866–871

Pringsheim EG, Czurda V (1927) Phototropische und ballistische Probleme bei *Pilobolus*. Jahrb Wiss Bot 66:863–901

Regan BC, Aloni S, Jensen K, Zettl A (2005) Surface-tension-driven nanoelectromechanical relaxation oscillator. Appl Phys Lett 86(123119):1–3

Sache I (2000) Short-distance dispersal of wheat rust spores by wind and rain. Agronomie 20:757–767

Saint-Jean S, Testa A, Madden LV, Huber L (2006) Relationship between pathogen splash dispersal gradient and Weber number of impacting drops. Agric For Meteorol 141:257–262

Saville DBO (1965) Spore discharge in Basidiomycetes: a unified theory. Science 147:165–166

Schmale DG, Arntsen QA, Bergstrom GC (2005) The forcible discharge distance of ascospores of *Gibberella zeae*. Can J Plant Pathol 27:376–382

Skotheim JM, Mahadevan L (2005) Physical limits and design principles for plant and fungal movements. Science 308:1308–1310

Smith RS (1966) The liberation of cereal stem rust uredospores under various environmental conditions in a wind tunnel. Trans Br Mycol Soc 49:33–41

Thiel R, Schreurs WJ, Harold FM (1988) Transcellular ion currents during sporangium development in the water mould *Achlya bisexualis*. J Gen Microbiol 134:1089–1097

Thiers HD (1984) The secotioid syndrome. Mycologia 76:1–8

Trail F, Xu H, Loranger R, Gadoury D (2002) Physiological and environmental aspects of ascospore discharge in *Gibberella zeae* (anamorph *Fusarium graminearum*). Mycologia 94:181–189

Trail F, Gaffoor I, Vogel S (2005) Ejection mechanics and trajectory of the ascospores of *Gibberella zeae* (anamorph *Fusarium graminearum*). Fungal Genet Biol 42:528–533

Tucker K, Stolze JL, Kennedy AH, Money NP (2007) Biomechanics of conidial dispersal in the toxic mold *Stachybotrys chartarum*. Fungal Genet Biol 44:641–647

Turner JCR, Webster J (1991) Mass and momentum transfer on the small scale: How do mushrooms shed their spores? Chem Eng Sci 46:1145–1149

Turner JCR, Webster J (1995) Mushroom spores – the analysis of Buller's drop. Chem Eng Sci 50:2359–2360

Van Heerden A, Van Wyk PWJ, Botes PJ, Pohl CH, Strauss CJ, Nigam S, Kock JLF (2007) The release of elongated, sheathed ascospores from bottle-shaped asci in *Dipodascus geniculatus*. FEMS Yeast Res 7:173–179

Vogel S (1998) Life's devices. The physical world of animals and plants. Princeton University Press, Princeton

Vogel S (2005a) Living in a physical world. II. The bio-ballistics of small projectiles. J Biosci 30:167–175

Vogel S (2005b) Living in a physical world. III. Getting up to speed. J Biosci 30:303–312

Walker CA, van West P (2007) Zoospore development in the oomycetes. Mycol Res 21:10–18

Walker DGA, Harvey R (1966) Studies of the ballistics of ascospores. New Phytol 65:59–74

Webster J (1952) Spore projection in the hyphomycete *Nigrospora sphaerica*. New Phytol 52:229–235

Webster J (1966) Spore projection in *Epicoccum* and *Arthrinium*. Trans Br Mycol Soc 49:339–343

Webster J, Chen C-Y (1990) Ballistospore discharge. Trans Mycol Soc Jpn 31:301–315

Webster J, Dennis C (1967) The mechanism of sporangial discharge in *Pythium middletonii*. New Phytol 66:307–313

Webster J, Weber RWS (2007) Introduction to fungi. Cambridge University Press, Cambridge

Webster J, Davey RA, Ingold CT (1984) Origin of the liquid in Buller's drop. Trans Br Mycol Soc 83:524–527

Webster J, Proctor MCF, Davey RA, Duller GA (1988) Measurement of the electrical charge on some basidiospores and an assessment of two possible mechanisms of ballistospore propulsion. Trans Br Mycol Soc 91:193–203

Webster J, Davey RA, Turner JCR (1989) Vapour as the source of water in Buller's drop. Mycol Res 93:297–302

Webster J, Davey RA, Smirnoff N, Fricke W, Hinde P, Tomos D, Turner JCR (1995) Mannitol and hexoses are components of Buller's drop. Mycol Res 99:833–838

Weston W, Taylor RE (1948) The plant in health and disease. Crosby Lockwood and Son, London

Yarwood CE (1941) Diurnal cycle of ascus maturation of *Taphrina deformans*. Am J Bot 28:355–357

Zoberi MH (1961) Take-off of mould spores in relation to wind speed and humidity. Ann Bot 25:53–64

7 Gene for Gene Models and Beyond: the *Cladosporium fulvum*–Tomato Pathosystem

PIERRE J.G.M. DE WIT[1], MATTHIEU H.A.J. JOOSTEN[1], BART H.P.J. THOMMA[1], IOANNIS STERGIOPOULOS[2]

CONTENTS

I.   Introduction 135
II. The *Cladosporium fulvum*–Tomato
     Pathosystem 136
     A. Infection of Tomato by *C. fulvum*......... 136
     1. The Compatible Interaction.............. 136
     2. The Incompatible Interaction 137
         a) Basal Defense Responses 137
         b) Effectors and Effector-Triggered
             Plant Defense Responses 137
         c) Structure and Function of Effectors
             of *C. fulvum*...................... 138
         d) Evolution of Effector Genes
             of *C. fulvum*...................... 140
         e) Secreted Cysteine-Rich Effectors
             of Other Fungal Pathogens.......... 140
         f) Translocation of Secreted
             Fungal Effectors.................. 140
III. Tomato *Cf* Resistance Genes 143
     A. Overall Structure of *Cf* Resistance Proteins .. 143
     B. Cf-Like Resistance Proteins Effective Against
         Other Extracellular Fungal Plant Pathogens... 144
     C. Evolution of *Cf* Resistance Genes 145
     D. Recognitional Specificity of Cf Proteins ... 146
     E. Interaction Between R Proteins
         and Effectors 146
         1. Indirect Interaction Between
             Cf Proteins and Effectors 146
         2. Direct Interaction Between R
             Proteins and Other Fungal Effectors.... 147
IV. Effector-Triggered Cf-Mediated Defense
     Responses............................. 147
     A. Plant Components Interacting with
         Cf Proteins.......................... 147
     B. Genetic Approaches to Identify
         Defense Signaling Components
         Downstream of Cf Proteins............. 148
     C. Transcript Profiling of Cf-Mediated
         Defense Genes....................... 148
         1. cDNA-AFLP Analysis................ 148
         2. Micro-Array Analysis. 149
     D. Post-Translational Modifications of
         Downstream Plant Defense Components... 149
     E. Induction of Other Cf-Mediated
         Downstream Physiological Responses 149
V.   Conclusions and Future Prospects............ 150
     References 150

I. Introduction

The fungal pathogen *Cladosporium fulvum* (syn. *Passalora fulva*; Braun et al. 2003) causes leaf mould of tomato, a disease first described by Cooke (1883). The fungus presumably originates from South America, the center of origin of the cultivated tomato (*Lycopersicon esculentum*) and related wild species. Worldwide, the fungus causes serious economic losses to commercially grown tomatoes that lack Cladosporium fulvum (*Cf*) resistance genes. Breeding for resistance to *C. fulvum* started in the 1930s in the USA and Canada and *Cf-2*, which originates from *Lycopersicon pimpinellifolium*, was the first resistance gene introduced into a commercial tomato cultivar called Vetomold (Langford 1937, 1948). However, soon after the introduction of this gene, the first reports appeared on the occurrence of *C. fulvum* strains that could overcome *Cf-2* resistance (race 2 strains; Bailey 1950). Various wild species of tomato provide a rich source of different *Cf* genes that have been introgressed into tomato cultivars by breeders (Boukema and Garretsen 1975). Although worldwide many races of the fungus still occur that can overcome several *Cf* genes, leaf mould of tomato is successfully kept under control by the wide deployment of *Cf* genes.

The availability of near-isogenic lines of tomato cultivar Moneymaker (MM) carrying different *Cf* genes (Tigchelaar 1984) and the large collection of complex races that can infect these tomato genotypes (Lindhout et al. 1989) make the *C. fulvum*–tomato pathosystem an ideal model to study gene for gene interactions (Flor 1942; Oort 1944). This model has been very useful for unraveling the molecular basis of communication between host plants and their pathogens and has resulted in the molecular cloning of the first fungal avirulence (*Avr*)

[1] Laboratory of Phytopathology, Wageningen University, and Wageningen Centre for Biosystems Genomics, Binnenhaven 5, 6709 PD Wageningen, The Netherlands; e-mail: pierre.dewit@wur.nl
[2] Laboratory of Phytopathology, Wageningen University, Binnenhaven 5, 6709 PD Wageningen, The Netherlands

Plant Relationships, 2nd Edition
The Mycota V
H. Deising (Ed.)
© Springer-Verlag Berlin Heidelberg 2009

gene (*C. fulvum Avr9*), followed by *Avr* genes from other fungi and oomycetes. To date, many *C. fulvum Avr* genes and their cognate tomato *Cf* genes have been cloned and some studied in detail. In the past decades, several reviews have been written on the *C. fulvum*–tomato interaction (De Wit 1992, 1995a, b, 1997; Joosten and De Wit 1999; Rivas and Thomas 2005; Thomma et al. 2005). In this chapter, we focus on recent developments in this pathosystem.

II. The *Cladosporium fulvum*– Tomato Pathosystem

Cladosporium fulvum is considered an asexual fungus and taxonomists have changed its phylogenetic classification several times (Ciferri 1954; Braun et al. 2003). The name *Fulvia fulva* was proposed (Ciferri 1954) as the fungus has little in common with truly plant pathogenic *Cladosporium* species such as *C. cucumerinum* with respect to pathogenic behavior (*C. fulvum* is biotrophic, *C. cucumerinum* is necrotrophic) and the production of cell wall-degrading enzymes, which are nearly absent in *C. fulvum* and abundantly produced by *C. cucumerinum*. Based on ITS sequences, *C. fulvum* was recently proposed to belong to the Mycosphaerellaceae, which contain taxa with anamorphs and teleomorphs. Currently *C. fulvum* is placed in the anamorph genus *Passalora* and it has now been re-named *Passalora fulva* (Braun et al. 2003). This classification is also supported by the sequences of mating type loci of *C. fulvum* that are highly homologous to those of *Mycosphaerella graminicola* and *M. fijienis* (Waalwijk et al. 2002; Conte-Ferraez et al. 2007; Stergiopoulos et al. 2007b). It is presently unknown whether the mating types genes of *C. fulvum* are functional, but based on cDNA-AFLP-analysis and the polymorphisms observed in eight effector genes in a worldwide collection of strains of *C. fulvum*, it is suggested that sexual reproduction may have occurred in the past (Stergiopoulos et al. 2007a). Indeed, *C. fulvum* now serves as a model for functional studies of pathogens belonging to the Mycosphaerellaceae (Thomma et al. 2005). All members belonging to the Mycosphaerellaceae show a similar infection strategy as they all penetrate stomata and show a biotrophic lifestyle during the early stages of infection, when they proliferate in the extracellular space between mesophyll cells without producing haustoria. Eventually, plants become wilted

and necrotic due to a disfunction of stomata, from which conidiophores emerge bearing high amounts of conidia. Comparative genomics among the Mycosphaerellaceae shows that *C. fulvum* is most related to *M. fijiensis*, the causal agent of the black Sigatoka disease of banana (www.genome.jgi-psf. org/Mfijiensis). Several homologues of *C. fulvum* effectors occur in *M. fijiensis* (Bolton et al. 2008; Stergiopoulos et al. Unpublished data). Currently, the genome of *C. fulvum* is being sequenced and soon the opportunity to compare complete gene sets to other fungal genomes will be available.

A. Infection of Tomato by *C. fulvum*

1. The Compatible Interaction

Most fungal species belonging to the Mycosphaerellaceae are leaf pathogens. They enter leaves through stomata and stay confined to the extracellular space for the major part of their infection cycle (Lazarovits and Higgins 1976a, b; De Wit 1977; Thomma et al. 2005). Although *C. fulvum* is able to penetrate the leaves of different plant species through stomata, fungal growth on all plants, except tomato, is arrested soon after penetration (Fernandez and Heath 1991). On tomato, *C. fulvum* causes a serious disease called leaf mold (Butler and Jones 1949). Real epidemics have not been reported in wild populations of various *Lycopersicon* species, presumably due to the presence of many different *Cf* resistance genes. On cultivated tomato plants that are susceptible to *C. fulvum*, the disease only occurs when environmental conditions are favorable for the fungus; in humid and cool conditions in greenhouses and outdoors. When conidia contact tomato leaves they germinate and produce thin runner hyphae that grow in various directions over the leaf surface and penetrate open stomata. Once inside the leaf, the hyphae increase in diameter. The intercellular hyphae retrieve nutrients from the apoplast where they grow in close contact with mesophyll cells. There is no evidence that the fungus produces significant amounts of cell wall-degrading enzymes during infection (Lazarovits and Higgins 1976a, b; De Wit 1977). Within 10—14 days after penetration, conidiophores emerge from stomata, producing numerous one- to four-celled conidia that are released by wind or water and can re-infect plants. Conidia are believed to survive in the absence of the host plant over a long period of time. Little is known about the nutrients *C. fulvum* retrieves from the apoplast and how this is achieved, but it is suggested that the fungus produces invertases or, alternatively, induces plant invertases to convert apoplastic saccharose into glucose and fructose. These sugars are subsequently taken up and stored by the fungus after converting them into mannitol, which cannot be metabolized by tomato itself (Joosten et al. 1990). Targeted deletion of genes involved in mannitol metabolism in several phytopathogenic fungi, including *C. fulvum*, shows that mannitol can serve both as a carbon source and as a scavenger of reactive oxygen species (Solomon et al. 2007). The apoplast is poor in nitrogen sources, but several amino acids and other nitrogen compounds like γ-amino butyric acid (GABA) occur. GABA concentrations increase significantly during infection (Solomon and Oliver 2002) and can be utilized by *C. fulvum* both as a carbon and as a nitrogen source.

2. The Incompatible Interaction

a) Basal Defense Responses

It is generally accepted that plants show basal defense responses to pathogens and potential pathogens. This basal defense in plants is triggered after recognition of pathogen-associated molecular patterns (PAMPs) and includes callose deposition, cell wall enforcement, non-specific necrosis and accumulation of pathogenesis-related (PR) proteins including chitinases, glucanases and proteases that are all thought to negatively affect the pathogen (Higgins 1982; Joosten et al. 1989; Linthorst et al. 1991; van Esse et al. 2008). Several PAMPs of plant pathogens have been identified, including flagellin, lipopolysaccharide and elongation factor Tu from Gram-negative bacteria, and chitin and β-glucans from fungi and oomycetes, respectively (Nürnberger and Brunner 2002; Fliegman et al. 2004; Kaku et al. 2006; Zipfel et al. 2006; Chinchilla et al. 2007; Miya et al. 2007). PAMPs are recognized by pathogen recognition receptors (PRRs) that recognize a particular domain of a large PAMP molecule which often possesses structural functions that are crucial for a microbe or pathogen to survive (Jones and Dangl 2006; De Wit 2007).

C. fulvum releases several potential PAMPs including cell wall-derived glycoproteins (Lazarovits and Higgins 1979; Lazarovits et al. 1979; De Wit and Kodde 1981a) that induce non-specific

basal defense responses in tomato, irrespective of the *Cf* resistance genes that are present. Hydrophobins that are localized in conidial and hyphal walls of fungi and play distinct roles in fungal development may also function as PAMPs. *C. fulvum* produces six hydrophobins (HCfs), of which HCf-6 is secreted by the hyphae during invasion of tomato (Whiteford et al. 2004). Similar basal defense responses as in tomato are also induced in non-host plant species soon after penetration of stomata by *C. fulvum*, and these can be suppressed by factors of virulent strains from other pathogens, as observed by Fernandez and Heath (1991). They found that, upon injection of conidia into the intercellular spaces of bean leaves, growth of *C. fulvum* in rust-infected leaves exceeded that of control non-rust-infected leaves. Only in its host plant tomato, can *C. fulvum* fully overcome basal defense responses and cause disease (Higgins 1982). How these basal defense responses are suppressed or overcome is still largely unknown, but it is assumed that this is achieved by effectors that are delivered into the apoplast during growth in the host (Thomma et al. 2005; De Wit 2007; Van Esse et al. 2008).

b) Effectors and Effector-Triggered Plant Defense Responses

The gene for gene hypothesis as proposed by Flor (1942) and Oort (1944) states that for every dominant avirulence (*Avr*) gene in the pathogen there is a cognate resistance (*R*) gene in the host. Interaction between their gene products is assumed to activate a hypersensitive response (HR) and resistance. It has been a mystery for a long time why a pathogen would produce avirulence (Avr) factors that enable a resistant host cultivar to recognize the pathogen and trigger an HR that arrests further growth of the pathogen. Due to the clear HR induced by Avrs of *C. fulvum* in resistant tomato cultivars carrying cognate *Cf* resistance genes, many of them could easily be identified (De Wit and Spikman 1982; Thomma et al. 2005). So far, four *Avr* genes of *C. fulvum* have been cloned and they all encode small cysteine-rich proteins that are secreted during infection. They are Avr2, Avr4, Avr4E and Avr9, whose recognition in tomato is mediated by Cf-2, Cf4, Cf-4E and Cf-9, respectively (Thomma et al. 2005). Also, *Ecp* (for extracellular protein) genes encoding Ecp1, Ecp2, Ecp4 and Ecp5 have been cloned whose recognition is mediated by cognate *Cf-Ecp* genes that have not been cloned yet (Laugé et al.

2000; De Kock et al. 2005). Recently, two additional Ecps (Ecp6, Ecp7) were characterized for which no cognate *Cf-Ecp* genes have yet been identified (Bolton et al. 2008). The even number of cysteine residues present in all these Avrs and Ecps of *C. fulvum* are crucial as they form disulfide bridges that provide their stability in the apoplast, which contains many proteases (Rooney et al. 2005; Van Esse et al. 2006, 2008). Although all Avr and Ecp effectors induce an HR in resistant tomato plants carrying the cognate *Cf* genes, their primary function is to facilitate colonization of susceptible plants (Thomma et al. 2005; Van Esse et al. 2007, 2008; Bolton et al. 2008).

c) Structure and Function of Effectors
    of *C. fulvum*

The overall structure of several effector proteins of *C. fulvum* has been determined, but for only a few has the function been elucidated (Laugé and De Wit 1998; Luderer et al. 2002a; Thomma et al. 2005; Van Esse et al. 2007, 2008; Bolton et al. 2008; Fig. 7.1A).

The *Avr2* gene encodes a 58-amino-acid mature protein that contains eight cysteine residues and induces an HR in plants carrying the *Cf-2* resistance gene (Dixon et al. 1996; Luderer et al. 2002b). An additional gene, *Rcr3^pimp* (required for **C**ladosporium fulvum **r**esistance) originating from *L. pimpinellifolium* was cloned that is required for *Cf-2*-mediated HR and resistance (Dixon et al. 2000; Krüger et al. 2002; Rooney et al. 2005). *Rcr3* encodes a cysteine protease (Krüger et al. 2002) and it was shown that Avr2 binds to and inhibits Rcr3, suggesting it plays an offensive role in virulence (Rooney et al. 2005). Indeed, Avr2 was shown to be a genuine virulence factor (Van Esse et al. 2008). Plants expressing the *Avr2* gene are more susceptible to *C. fulvum* strains lacking *Avr2* and other extracellular fungal pathogens, such as *Botrytis cinerea* and *Verticillium dahliae*. Knocking down expression of the *Avr2* gene in *C. fulvum* decreases its virulence on tomato. In addition to Rcr3, Avr2 also inhibits cysteine proteases that include Pip1, aleurain and TDI65 (Shabab et al. 2008; Van Esse et al. 2008). Avr2 does not show significant structural homology to well known cysteine protease inhibitors and seems rather unique (Van't Klooster et al., unpublished data). Structural modification of Rcr3 by Avr2, rather than Rcr3 inhibition, seems to be the cause of triggering Cf-2-mediated defense signaling, as

a natural mutation in the Rcr3 protein (Rcr3^esc) occurs in *L. esculentum* cultivars that causes spontaneous HR-like necrosis in the presence of the *Cf-2* gene in an Avr2-independent manner (Krüger et al. 2002). The Rcr3^esc protein present in *L. esculentum* still shows protease activity and is likely to have a modified tertiary structure as compared to Rcr3^pimp, which is recognized by Cf-2 (that also originates from *L. pimpinellifolium*, see below) in the absence of Avr2 (Krüger et al. 2002). Present studies aim at proving that Avr2, Rcr3 and Cf-2 occur in one receptor complex that triggers Cf-2-mediated defense signaling. Circumvention of Avr2-triggered Cf-2-mediated HR occurs by point mutations, deletions or transposon insertions in the *Avr2* gene (Luderer et al. 2002b; Stergiopoulos et al. 2007).

Avr4 encodes a 135-amino-acid pre-pro-protein, which is C- and N-terminally processed after secretion into the apoplast, resulting in an 86-amino-acid mature protein with eight cysteine residues (Joosten et al. 1994, 1997). The disulfide pattern of Avr4 shows homology to the invertebrate chitin-binding domain of tachycitin (inv ChBD; Shen and Jacobs-Lorena 1999). Binding of Avr4 to chitin was experimentally confirmed and Avr4 was found to protect the cell wall of the fungi *Trichoderma viride* and *Fusarium solani* against deleterious effects of basic plant chitinases in vitro (Van den Burg et al. 2003, 2004). During growth in vitro, *C. fulvum* does not produce Avr4 and its cell wall chitin is not exposed. However, during host colonization, chitin is exposed and the production of Avr4 is induced (Van den Burg et al. 2006). This suggests that Avr4 protects *C. fulvum* in planta against plant chitinases, thereby playing a defensive role in virulence. Indeed, additional experiments demonstrated that strains of *C. fulvum*, in which expression of the *Avr4* gene was knocked down, are less virulent on tomato (Van Esse et al. 2007). In addition, tomato plants expressing Avr4 appeared more susceptible to *C. fulvum* and to other chitin-containing pathogens of tomato (Van Esse et al. 2007). In tomato genotypes that contain the cognate Cf-4 resistance protein, Avr4 triggers Cf-4-mediated HR, and natural isoforms of Avr4 occur that no longer trigger Cf-4-mediated HR but are still able to bind chitin. Most of these Avr4 isoforms lack one cysteine residue that causes Avr4 to become more sensitive to proteases occurring in the host apoplast. Virulent strains producing these unstable Avr4 isoforms are still protected against the deleterious effects of plant chitinases, but the Avr4 protein does not accumulate in the apoplast and will no longer trigger Cf-4-mediated HR (Joosten et al. 1997; Van den Burg et al. 2003).

The Avr4E gene encodes a cysteine-rich protein of 101 amino acids that is secreted into the apoplast of tomato and triggers Cf-4E-mediated HR (Westerink et al. 2004). Various strains of *C. fulvum* have been identified that evade Cf-4E-mediated resistance and these all show two point mutations in the *Avr4E* gene, leading to a stable protein with two amino acid changes (Avr4E^LT) that no longer induces Cf-4E-mediated HR (Westerink et al. 2004). Targeted mutational analysis showed that the amino acid substitution Phe62>Leu is fully sufficient for a complete evasion of Avr4E-triggered Cf-4E-mediated HR, and it is not clear why a double mutation is present in Avr4E alleles of natural strains virulent on Cf-4E plants. Some strains evade Cf-4E-mediated HR by loosing the *Avr4E* gene, similar to strains of *C. fulvum* that evade Cf-9-mediated HR by loosing the *Avr9* gene (see below; Van Kan et al. 1991; Van den Ackerveken et al. 1992). So far, the intrinsic function of the Avr4E protein has not been identified (Westerink et al. 2004).

Avr9 encodes a 63-amino-acid protein that is C- and N-terminally processed both by fungal and plant proteases, leading to a mature 28-amino-acid protein containing six cysteine residues (Van Kan et al. 1991; Van den Ackerveken et al. 1993). The three-dimensional structure of the mature peptide consists of three anti-parallel beta-sheets with two solvent-exposed loops, which are stabilized by three disulfide bridges (Van den Hooven et al. 2001). This overall structure is typical for cystine-knotted peptides which, although structurally related, share very little sequence homology and display very diverse biological functions. An alanine scan of the Avr9 peptide revealed that all six cysteine residues in Avr9 are essential for its overall three-dimensional structure and necrosis-inducing activity (Kooman-Gersmann et al. 1997). Based on its cysteine spacing and overall structure, Avr9 is most homologous to a carboxypeptidase inhibitor (Van den Hooven et al. 2001). However, functional assays did not show carboxypeptidase inhibition by Avr9 (Van den Hooven et al. 2001). Disruption of *Avr9* in *C. fulvum* by homologous recombination did not affect growth in vitro or virulence on susceptible tomato plants, suggesting that *Avr9* is dispensable for full virulence (Marmeisse et al. 1993). Furthermore, all natural strains that no longer induce Cf-9-mediated HR lack the *Avr9* gene, suggesting that the gene can

be lost without a serious fitness penalty. The expression of *Avr9* is induced under nitrogen-limiting conditions in vitro (Van den Ackerveken et al. 1994; Pérez-García et al. 2001; Thomma et al. 2006), which suggests that Avr9 might be involved in nitrogen metabolism. Pérez-García et al. (2001) identified a gene in *C. fulvum*, designated *Nrf1* (for **n**itrogen **r**esponsive **f**actor), which has strong similarity to the nitrogen regulatory proteins of *Aspergillus nidulans*. Although *Nrf1*-deficient strains do not express *Avr9* under nitrogen-limiting conditions in vitro, these strains are still avirulent on *Cf-9* tomato plants, suggesting that Nrf1 is a major, yet not the only, positive regulator of *Avr9* expression (Pérez-García et al. 2001). The expression of all other known *C. fulvum* effector genes was studied under nitrogen-limiting conditions, and they were not significantly induced. Although the expression of all identified *C. fulvum* effectors is specifically induced in planta, the regulators or inducing conditions have not yet been identified (Thomma et al. 2006).

In addition to the Avrs, six effector genes encoding extracellular proteins Ecp1, Ecp2, Ecp4, Ecp5, Ecp6 and Ecp7 have been cloned (Van den Ackerveken et al. 1993; Laugé et al. 2000; Bolton et al. 2008). Like the *Avr* genes, the *Ecp* genes encode proteins that are abundantly secreted by all strains of *C. fulvum* during infection of tomato, suggesting that they are virulence factors. Except for Ecp6, which contains three typical LysM domains (Bolton et al. 2008), they share no, or little sequence homology with other sequences present in databases; all Ecps have an even number of cysteine residues of which most are involved in intramolecular disulfide bridges (Luderer et al. 2002a). *Ecp1* and *Ecp2* were found to contribute to virulence, as their disruption significantly decreased virulence (Laugé et al. 1997). Ecp1 shares some structural homology to **t**umor-**n**ecrosis **f**actor **r**eceptor (TNFR), but whether this homology is functional remains to be shown (Laugé et al. 1997). Although Ecp6 may function in protection of the *C. fulvum* cell wall against host chitinases in a similar fashion as Avr4, Ecp6 may also be involved in binding of chitin fragments that are released by the action of plant chitinases on the fungal cell walls, by its three lysM domains (Bolton et al. 2008). Chitin fragments are PAMPs that elicit plant defense and Ecp6 might scavenge these fragments to prevent elicitation of basal defense responses. Silencing of the *Ecp6* gene compromises virulence of *C. fulvum*

(Bolton et al. 2008). Interestingly, in contrast to most of the *C. fulvum* effectors that are rather unique to this fungus, Ecp6 orthologues occur in many different fungi (Bolton et al. 2008). For the Ecp1, Ecp2, Ecp4 and Ecp5 effectors of *C. fulvum*, tomato accessions have been identified that develop an HR upon inoculation with recombinant potato virus X (PVX) expressing one of these *Ecp* genes or after injection with the encoded proteins (Laugé et al. 1998a, 2000). The responding accessions all carry a single dominant *Cf-Ecp* gene and show HR-associated resistance toward strains that contain the cognate *Ecp* gene (Laugé et al. 2000). All *Cf-Ecp* genes remain to be cloned (de Kock et al. 2005).

A role of the effectors in the virulence of *C. fulvum* and that of other pathogens has also been studied by heterologous expression of their encoding genes in both tomato and *Arabidopsis*, which were subsequently challenged with pathogens other than *C. fulvum* and assayed for altered susceptibility to these pathogens. This showed that expression of *Avr2* and *Avr4* of *C. fulvum* in tomato increases the virulence to the extracellular pathogens *Fusarium oxysporum* f. sp. *lycopersici*, *B. cinerea* and *V. dahliae*, whereas expression of these genes in *Arabidopsis* increases the virulence of pathogens such as *B. cinerea*, *Plectosphaerella cucumerina* and *V. dahliae* (Van Esse et al. 2007, 2008). Similarly, *C. fulvum* Ecp6 was heterologously expressed in *F. oxysporum* f. sp. *lycopersici* that subsequently became more aggressive on tomato (Bolton et al. 2008).

d) Evolution of Effector Genes of *C. fulvum*

In contrast to the Avr genes, few polymorphisms have thus far been found in the Ecp genes in a worldwide collection *C. fulvum* strains obtained from areas were tomatoes are commercially grown (Fig 7.1A; Stergiopoulos et al. 2007a). This might be due to a lack of selection pressure imposed on the pathogen to overcome *Cf-Ecp*-mediated resistance, as *Cf-Ecps* have not yet been introduced in commercial tomato cultivars. Introduction of *Cf-Ecp* genes into newly developed cultivars might provide a good source of resistance for the coming decades.

e) Secreted Cysteine-Rich Effectors of Other Fungal Pathogens

Several other fungal plant pathogens secrete cysteine-rich effectors during infection. The barley pathogen *Rhynchosporium secalis* secretes cysteine-rich necrosis-inducing proteins (Nips) that cause avirulence on barley plants carrying the cognate *R* gene(s) (Van't Slot et al. 2007). The vascular tomato pathogen *F. oxysporum* f. sp. *lycopersici* produces several cysteine-rich secreted in xylem (Six) proteins during growth in the xylem vessels of tomato (Houterman et al. 2007), some of which are both virulence and avirulence factors. Avr3 (=Six1) is required for full virulence of *F. oxysporum* f.sp. *lycopersici* (Rep et al. 2005), as well as for avirulence on tomato plants carrying the resistance gene *I-3* (Rep et al. 2004). Avr1 (=Six 4) is an avirulence factor on tomato plants carrying the *I* or *I-1* resistance gene, but also suppresses resistance mediated by the *I-2* and *I-3* resistance genes, respectively (Houterman et al. 2008). Thus, in the presence of Avr1 both Avr2 and Avr3 do not induce I-2- and I-3-mediated resistance, respectively. Avr1 itself, however, is dispensable for virulence in the absence of *I* or *I-1*.

f) Translocation of Secreted Fungal Effectors

Plant and animal pathogenic bacteria contain six secretory systems of which the type three secretion system (TTSS) appears to be the most important for virulence (Alfano and Collmer 2004; Abramovitch et al. 2006; Bingle et al. 2008). By the TTSS, plant pathogenic bacteria inject many effectors simultaneously into cells of the host plant and without a functional TTSS they only induce the basal defense responses and are not pathogenic. It has now been shown that many TTSS-injected effectors of plant pathogenic bacteria suppress basal plant defense responses, and for several of them the mechanism of suppression has been elucidated (Abromovitch et al. 2006).

Also, fungal plant pathogens translocate effector proteins into the host cells to manipulate basal defense, but the mechanism of their delivery into the plant cells is still unknown. All rusts, powdery and downy mildews produce haustoria during infection that penetrate the host cell wall and invaginate the host plasma membrane, and communication between host and pathogen is assumed to take place in the extrahaustorial matrix (Catanzariti et al. 2007).

The effectors of the flax rust fungus *Melampsora lini* (Dodds et al. 2004; Dodds et al. 2006; Ellis et al. 2007) are most likely important for subverting host defense during the obligate biotrophic phase of infection. The molecular basis of this communication remains unknown. In oomycetes, the amino acid motifs RXLR and EER are present in translocated effectors, as was shown for the *Arabidopsis*

A

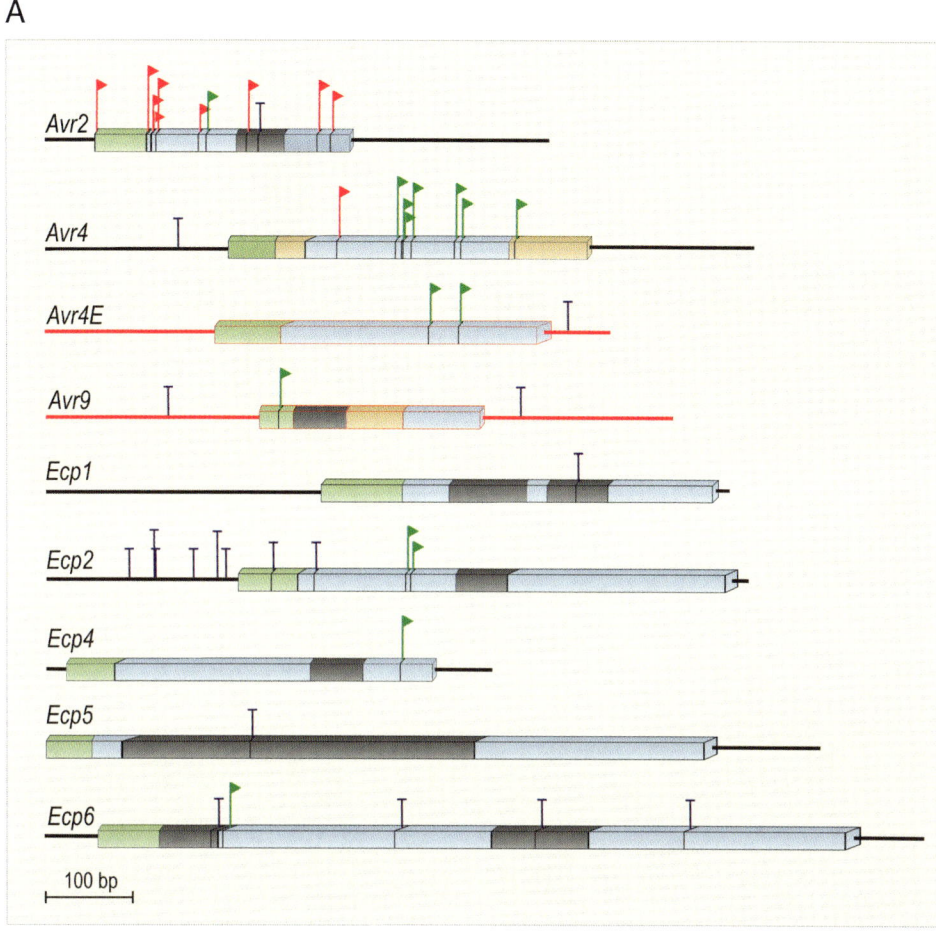

B

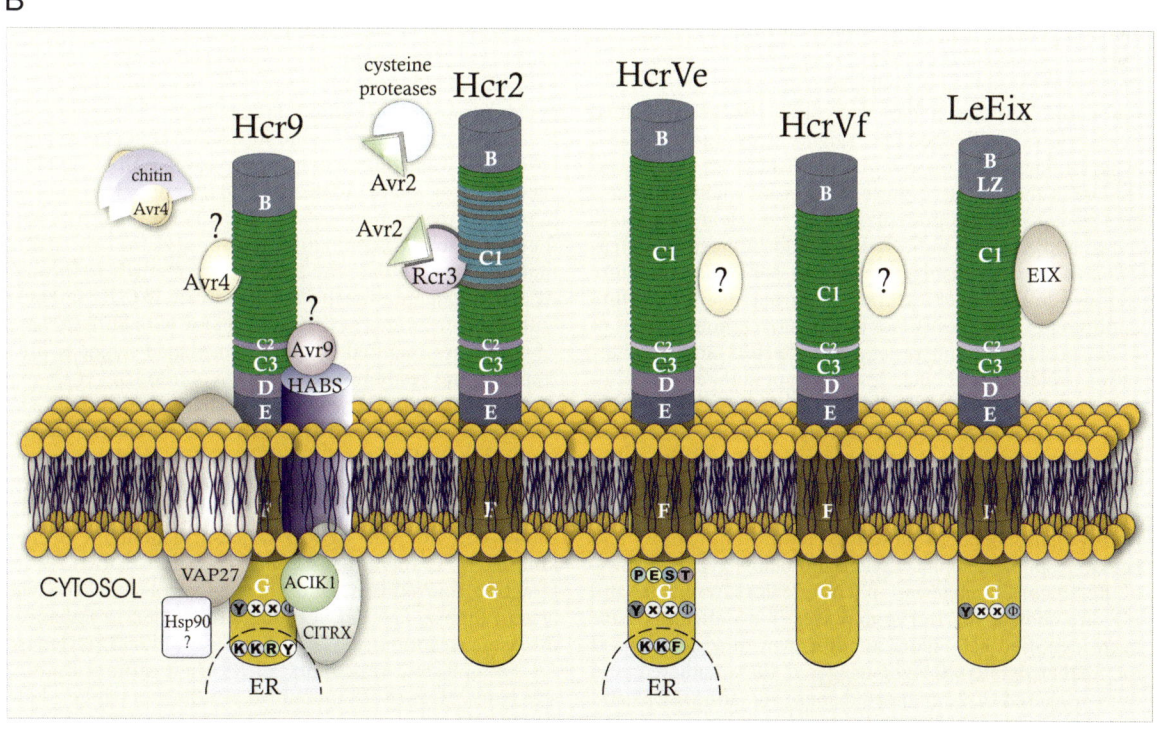

C

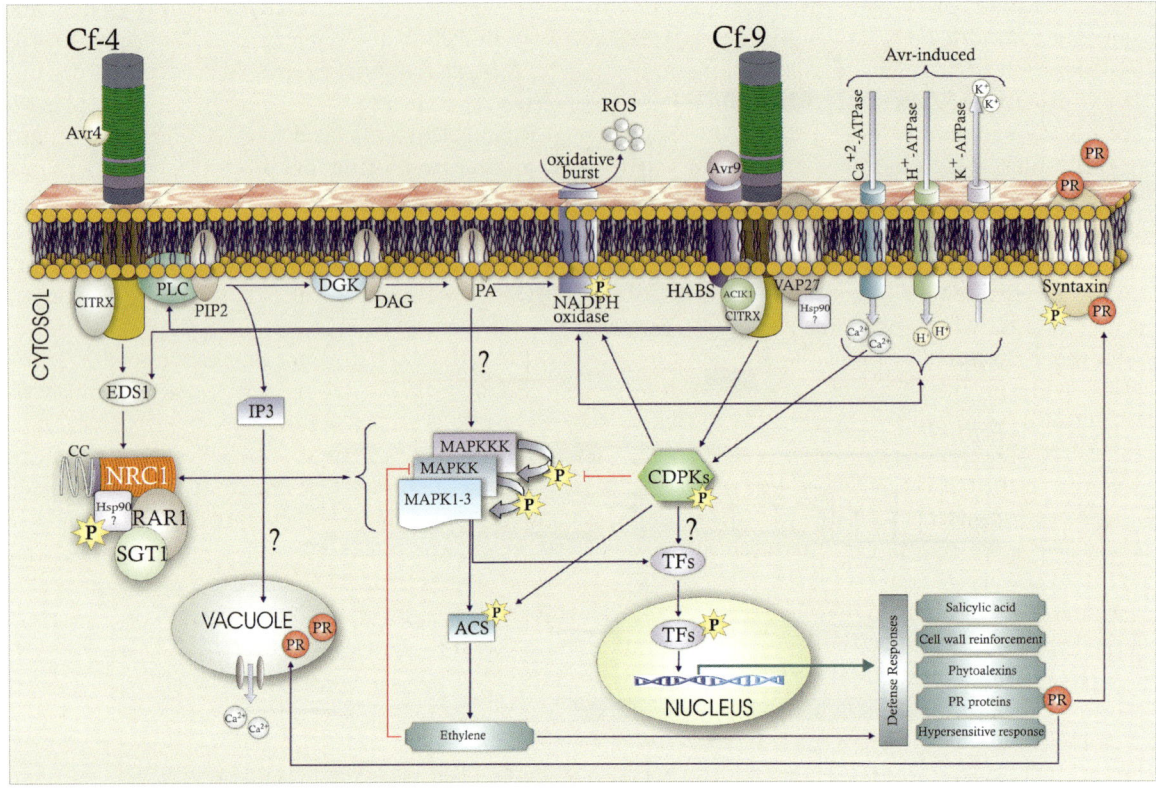

Fig. 7.1. Effectors of *Cladosporium fulvum*, structure of Cf and Cf-like resistance proteins, and Cf-4- and Cf-9-mediated defense signaling upon perception of their cognate effectors. **A** Sequence variation in the *C. fulvum Avr* and *Ecp* effector genes. Open reading frames (*boxed*) are shown with signal sequences (*green*), introns (*gray*), parts that are removed by processing after secretion (*brown*) and the mature proteins (*blue*). Nucleotide changes identified in a worldwide collection of *C. fulvum* strains are indicated as *green flags* when these mutations lead to an amino acid change in the produced protein, *red flags* when mutations lead to a truncated protein, or a *T* for silent mutations. *Avr4E* and *Avr9* are indicated in *red* to indicate that virulent strains occur that lack the entire *Avr* gene. The figure is drawn to scale (see text for further details). **B** Structure of different Cf-like resistance proteins (*Hcr*: **h**omologues of **C**ladosporium fulvum **r**esistance), effectors and Cf-interacting proteins. *Hcr9*: Includes all **h**omologues of the **C**. ful-vum **r**esistance (Hcr) proteins Cf-**9** (Cf-9, Cf-4 presumably also Cf-Ecps). *Hcr2*: Includes all Hcr proteins Cf-**2** (Cf-2, Cf-5). *HcrVe* Includes the Hcr proteins Ve1 and Ve2 active against *Verticillium dahliae*. HcrVf includes the Hcr proteins of the *Vf* region, Vfa1 and Vfa2, active against *Venturia inaequalis*. LeEix includes two Hcr proteins, LeEix1 and LeEix2 responsive to the ethylene-inducible xylanase (EIX) of the fungus *Trichoderma* sp. The effectors Avr4, Avr2 and EIX are shown; for HcrVe and HcrVf no effectors have been identified yet. Figure is not drawn to scale (see text for further details). **C** Avr4-triggered Cf-4-mediated, and Avr9-triggered Cf-9-mediated defense signaling. EDS1 and

NRC1 are required for Cf-4- and Cf-4-mediated resistance. PLC is activated, leading to IP3 and DAG. DAG is converted by DGK into PA, which stimulates NADPH-oxidase causing the accumulation of ROS. IP3 releases Ca²⁺ from the vacuole. SGT1, RAR1 and Hsp90 are in a complex with NRC1 or act just downstream of it. The Hsp90 complex interacts with the MAPK cascade eventually phosphorylating TFs and ACS causing accumulation of ethylene. TFs induce various defense responses including SA and ethylene accumulation, cell wall enforcement, phosphorylation of syntaxin, accumulation of phytoalexins and PR proteins, and the hypersensitive response. Avr9 interacts via the HABS with the Cf-9 protein, stimulating phosphorylation of CDPKs that in turn stimulate the production of ethylene through ACS. Ethylene inhibits the MAPK cascade. ACIK1 interacts with the C-terminus of Cf-9 and CITRX with both Cf-9 and Cf-4 and are required for the Cf-mediated hypersensitive response and resistance. Avr9 and Avr4 also trigger Cf-mediated opening of Ca²⁺, K⁺ and H⁺ channels through Ca²⁺-ATPase, K⁺-ATPase and H⁺-ATPase, respectively (see text for further details). *B, C1, C2, C3, D, E, F, G* Domains present in Hcrs. *ACIK1* Avr9/Cf-9-induced kinase 1. *ACS* 1-Aminocyclopropane-1-carboxylic acid synthase. *Avr* **Av**irulence. *CC* **C**oiled-coil domain. *CDPKs* **C**alcium-**d**ependent **p**rotein **k**inases. *Cf* **C**. fulvum resistance protein. *CITRX* Cf-9-interacting thioredoxin. *DAG* **D**iacylglycerol. *DGK* **D**iacylglycerol kinase. *Ecp* **E**xtracellular **p**rotein. *EDS1* **E**nhanced **d**isease **s**usceptibility-1. *EIX* **E**thylene **i**nducing **x**ylanase. *ER* **E**ndoplasmic **r**eticulum. *HABS* **H**igh **a**ffinity **b**inding site. *Hsp90* **H**eat-**s**hock **p**rotein **90.** *IP3* **I**nositol **trip**hosphate.

downy mildew pathogen *Hyaloperonospora para-sitica* (Rehmany et al. 2005) and the potato patho-gen *Phytophthora infestans* (Whisson et al. 2007). For secretion or targeting to the haustorium of the Avr3a protein of *P. infestans* the RXLR-EER motifs are not required, but they are required for deliv-ery into plant cells where Avr3a is recognized by the R3a resistance protein. RXLR-EER-encoding genes are also transcriptionally upregulated dur-ing infection (Whisson et al. 2007). Bioinformatics analysis identified 425 genes potentially encoding secreted RXLR-EER proteins in the *P. infestans* genome. Functional analyses of this class of pro-teins will be instrumental to determine how Oomycetes manipulate their hosts in order to establish infection (Jiang et al. 2008). Also, the effectors from the rice blast fungus *Magnaporthe grisea* are supposed to act inside the host cell (Jia et al. 2000), as well as the host-selective protein toxin ToxA secreted by the wheat tan spot patho-gen *Pyrenophora tritici-repentis* (Ciuffetti et al. 1997). ToxA is required for full virulence on wheat, as a naturally occurring non-virulent strain of this pathogen becomes virulent when the *ToxA* gene is transferred into this strain. Internalization of ToxA into sensitive wheat mesophyll cells is corre-lated with its activity (Manning and Ciuffetti 2005; Sarma et al. 2005). The solvent-exposed, Arg-Gly-Asp (RGD)-containing loop of ToxA is a candidate for interaction with the plasma membrane, which is a likely prerequisite to toxin internalization (Manning et al. 2008). A similar RGD signature also mediating interaction with the plasma mem-brane is present in the IpiO effector proteins of *P. infestans* (Senchou et al. 2004).

III. Tomato *Cf* Resistance Genes

A. Overall Structure of *Cf* Resistance Proteins

Many *R* genes encode so-called NB-LRRs; pro-teins that are located in the cytoplasm and con-tain either an N-terminal coiled coil (CC) or a Toll interleukin receptor (TIR) domain in addition

to a nucleotide binding (NB) and a C-terminal leucine-rich repeat (LRR) domain (Jones and Dangl 2006; Van Ooijen et al. 2008; Chaps 17, 18). *C. fulvum* resistance genes (*Cf*) encode a distinct class of R proteins containing extracytoplasmic LRRs (eLRRs), a membrane-spanning region and a short cytoplasmic domain without any obvious signaling function (Fritz-Laylin et al. 2005; Rivas and Thomas 2005; Wang et al. 2008). These pro-teins are collectively called receptor-like proteins (RLPs; Kruijt et al. 2005a). Cf proteins contain seven structural domains (Fig. 7.1B). Domain A represents the signal peptide sequence for extra-cellular targeting, domain B contains the mature N-terminus with several cysteine residues, domain C contains a varying number of eLRRs including a "hinge" region with poor homology to the LRR consensus, domain D contains no obvious fea-ture, domain E contains many acidic amino acid residues, domain F represents the membrane-spanning region and domain G consists of a short cytoplasmic tail lacking obvious signaling signatures (Fig. 7.1B). The LRRs of Cf proteins are thought to be involved in direct or indirect recognition of Avrs and Ecps of *C. fulvum* and show homology to the LRRs of other plant RLPs and receptor-like kinases (RLKs; Kruijt et al. 2005; Van der Hoorn et al. 2005; Wang et al. 2008). The *Cf* genes are mem-bers of multigene families in tomato and are des-ignated homologues of Cladosporium resistance genes *Cf-2* and *Cf-9* (*Hcr2* and *Hcr9*, respectively). The *Cf-4*, *Cf-4E*, *Hcr9-9B*, *Cf-9*, *9DC* and *Cf-Ecp* genes are highly homologous and belong to the *Hcr9* gene family (Jones et al. 1994; Parniske et al. 1997; Thomas et al. 1997; Laugé et al. 1998b; Takken et al. 1999; Kruijt et al. 2004; de Kock et al. 2005) and directly or indirectly recognize Avr4, Avr4E, Avr9B (not yet identified), Avr9 (both *Cf-9* and *9DC*) and Ecps, respectively (Kruijt et al. 2005a; Thomma et al. 2005). Similarly, the *Cf-2* and *Cf-5* genes belong to the *Hcr2* gene family (Dixon et al. 1996, 1998) and mediate recognition of Avr2 and Avr5 (not identified yet), respectively. In contrast to the Hcr9 proteins, most of which carry 25 or 27 LRRs, the number of LRRs in the Hcr2 proteins is

KKRY Lys-Lys-Arg-Tyr signature. *KKF* Lys-Lys-Phe sig-nature. *MAPK1–3* Mitogen-activated protein kinase 1–3. *MAPKK* MAPK kinase. *MAPKKK* MAPKK kinase. *NADPH-oxidase* Nicotinamide adenine dinucleotide phosphateH-oxidase. *NRC1* NB-LRR protein required for HR-associated cell death-1. *PA* Phosphatidic acid. *PEST* Pro-Glu-Ser-Thr signature. *PLC* Phospolipase C. *PIP2* Phosphatidyl inositol disphosphate. *PR* Pathogenesis-related. *RAR1* Required for Mla12 resistance-1. *ROS* Reactive oxygen species. *SGT1* Suppressor of G-two allele of Skp-1. *TFs* Transcription fac-tors. *VAP27* Vesicle-associated protein 27. *YXXΦ* Tyr-X-X- Φ signature more inconsistent and varies between 25 and 38 LRRs. The LRRs of Hcr2 proteins can be classified into two groups that alternate to give a second structure level (Dixon et al. 1996, 1998). Mutational analysis showed that at least the number of LRRs determines the specificity of the Cf-2 and Cf-5 proteins (Seear and Dixon 2003). Cf proteins are supposed to be extracellularly anchored to the plasma membrane and all Hcr9 proteins contain a putative dilysine motif at the C-terminus for targeting to the ER, which may suggest that they are ER-localized. However, whereas one study demonstrated that the Cf-9 protein resides in the ER when over-expressed in tobacco and Arabidopsis (Benghezal et al. 2000), another study showed that the functional Cf-9 protein was localized at the plasma membrane upon over-expression in tobacco (Piedras et al. 2000). Finally, mutational analysis of the dilysine motif (KKRY) of Cf-9 showed that the ER retrieval retention motif is not required for Cf-9 function and might be masked by interacting proteins (Van der Hoorn et al. 2001c; Wilson et al. 2005). In addition, Hcr9 proteins appear to contain the mammalian YXXΦ endocytosis signature (Y represents tyrosine, Φ represents a hydrophobic amino acid, X represents any amino acid; Thomas et al. 2005). The YXXΦ signal can stimulate receptor-mediated endocytosis and degradation of mammalian cell-surface receptors (Bonifacino and Traub 2003). The YXXΦ signature in Hcr9 proteins might selectively capture ligands, or regulate responsiveness to these ligands by regulating the concentration of Hcr proteins on the plasma membrane via controlled breakdown. Cf proteins also contain many putative N-glycosylation sites (NxS/T) that may play an important structural role in stabilizing the protein and protection against proteases. The Cf-9 protein contains 22 putative N-glycosylation sites in the eLRR region, of which all but one are glycosylated (Van der Hoorn et al. 2005). Removal of each of the 22 N-linked glycosylation sites revealed that many contribute to Cf-9 activity and that those occurring in the putative α-helices of the LRR modules are essential for its activity.

B. Cf-Like Resistance Proteins Effective Against Other Extracellular Fungal Plant Pathogens

Cf-like proteins (RLPs) have also been shown to confer resistance to other extracellular pathogenic fungi such as *V. dahliae* and *Venturia inaequalis* (Kawchuk et al. 2001; Belfanti et al. 2004; Fig. 7.1B).

In tomato, resistance to race 1 strains of *V. dahliae* and *V. albo-atrum* is conferred by a single dominant *Ve* locus. This locus contains two closely linked genes, *Ve1* and *Ve2*, with inverted orientation (Kawchuk et al. 2001). Expression of the individual *Ve* genes in susceptible potato plants conferred resistance to a race 1 isolate of *V. albo-atrum*. However, transfer of these genes individually into susceptible tomato plants to test their role in resistance to this fungus could not confirm the functionality of both genes. Recent studies suggest that only the *Ve1* gene, but not the *Ve2* gene, is a functional resistance gene (Fradin et al., unpublished data). The deduced primary structures of Ve1 and Ve2 comprise a signal peptide, 38 imperfect LRRs, a membrane-spanning domain and a C-terminal cytoplasmic YXXΦ endocytosis signal (Kawchuk et al. 2001). In addition, a Pro-Glu-Ser-Thr (PEST) sequence, which might be involved in ubiquitination, internalization and degradation of the protein (Rogers et al. 1986), resides at the C-terminus of Ve2 (Fig. 7.1B).

V. inaequalis secretes effectors that induce an HR in resistant apple cultivars (Win et al. 2003). *Vf* resistance has been introgressed from the wild apple species *Malus floribunda*. Two clusters of *RLP* genes with homology to the *Cf* resistance gene family have been identified. The first two of the four *HcrVf* genes (homologues of the **C. fulvum** resistance genes of the *Vf* region, later abbreviated as *Vf*) in both clusters are identical. Transgenic complementation with both genes, under control of the constitutive CaMV 35S promoter, in a susceptible apple cultivar has shown that *HcrVf2* is the functional *R* gene against *V. inaequalis* (Belfanti et al. 2004). Recently, Malnoy et al. (2008) showed that, under control of their own promoter, both *HcrVf1* and *HcrVf2* (now called *Vfa1* and *Vfa2*) confer resistance to *V. inaequalis*.

The PAMP **e**thylene-**i**nducing **x**ylanase (EIX), secreted by the fungus *Trichoderma viride* is recognized in tomato and this recognition is controlled by a single dominant locus comprising three homologous *LeEix* genes, of which two have been cloned (Hania and Avni 1997; Ron and Avni 2004). *LeEix1* and *LeEix2* encode Cf-like RLPs. Common to the Cf and Ve proteins is the presence of the C-terminal YXXΦ endocytosis signal. Both LeEix1 and LeEix2 bind EIX, but only LeEix2 transmits the signal that induces HR. Mutations in the YXXΦ domain of LeEix2 abolish its ability to induce the HR, suggesting that endocytosis plays a crucial role in LeEix-mediated defense signaling (Ron and Avni, 2004). EIX was also shown to interact with a cytoplasmic **s**mall **u**biquitin-related **mo**difier protein (SUMO) in a yeast two-hybrid system, suggesting that EIX is internalized (Hanania et al. 1999). Binding of EIX to LeEix2 may induce receptor-mediated endocytosis, thus allowing LeEix and/or EIX to interact with cytoplasmic proteins such as SUMO (Ron and Avni 2004).

C. Evolution of *Cf* Resistance Genes

Numerous *R* genes have been characterized from different plant species that provide resistance to a variety of different pathogens. *R* genes usually belong to tightly-linked gene families and their evolution is assumed to be driven by selection on allelic variants that originate from mutations and recombination between alleles or different gene family members. R proteins are assumed to be under positive selection in pathogen populations and are expected to be durable when the cognate effector proteins are crucial for the pathogen and the costs to the plant are limited. PRR and R proteins that recognize conserved PAMPs and effectors, respectively, are assumed most durable.

Cf genes have been introgressed into cultivated tomato from different wild relatives, such as *L. pimpinellifolium*, *L. peruvianum* and *L. hirsutum*, and functional screens with eight different effectors of *C. fulvum* show that most wild species respond with an HR and thus contain functional *Cf* genes (Laugé et al. 2000; Kruijt et al. 2005a). *Hcr2* and *Hcr9* genes are most likely derived from a common ancestral gene. Whereas *Hcr2s* map on chromosome 6 (Dickinson et al. 1993), all *Hcr9s* map on chromosome 1 (Jones et al. 1993; Parniske et al. 1997; Haanstra et al. 1999; Takken et al. 1999; Yuan et al. 2002; Kruijt et al. 2004). Subsequent independent evolution probably resulted in the two distinct groups of *Cf* homologues known today (Dixon et al. 1996). The *Cf-2* cluster that originates from *L. pimpinellifolium* comprises three *Hcr2* genes, including two functional *Cf-2* genes (*Cf-2-1*, *Cf-2-2*), which are probably the result of a recent duplication. They encode proteins that differ by only three amino acids and both confer resistance to *C. fulvum* isolates that produce the Avr2 effector (Dixon et al. 1996; Luderer et al. 2002). The *Cf-5* cluster, originating from *L. esculentum* var. *cerasiforme*, contains four *Hcr2* genes of which *Hcr2-5C* is the functional *Cf-5* gene (Dixon et al. 1998). The *Hcr2-5D* gene encodes a protein with two additional LRRs compared to Cf-5, and further differs only by a single amino acid on both flanks of these two LRRs, but does not recognize *C. fulvum* strains that produce Avr5 (Dixon et al. 1998). All *Hcr9* genes are derived from a single *Hcr9* progenitor gene, as shown by Parniske et al. (1997) who isolated the *Cf-0*, *Cf-4* and *Cf-9* gene

clusters. Whereas the *Cf-0* cluster of *L. esculentum* consists of only a single homologue with no known function in *C. fulvum* disease resistance, the *Cf-4* and *Cf-9* clusters both comprise five homologues. Analysis of the individual *Hcr9* genes of the *Cf-0*, *Cf-4* and *Cf-9* clusters showed that positive selection for diversification has acted on nucleotides that encode putative solvent-exposed amino acids in the 17 N-terminal LRRs of Hcr9 proteins (Dixon et al. 1996; Parniske et al. 1997). Sequence exchange between *Hcr9* genes, rather than accumulation of point mutations, is proposed as the mechanism by which novel *Hcr9* genes are generated (Parniske et al. 1997, 1999; Parniske and Jones, 1999). Unequal recombination can alter the number and composition of *Hcr9* genes and might lead to an increase of haplotype variation in the population. In a study on *Cf-9* variation in the *L. pimpinellifolium* population, the *9DC* gene was discovered (Van der Hoorn et al. 2001a; Kruijt et al. 2004). The N-terminal part of the 9DC protein is nearly identical to that of a protein encoded by *Hcr9-9D*, a homologue adjacent to the *Cf-9* gene, whereas the C-terminus is nearly identical to Cf-9, suggesting that intragenic recombination between *Hcr9* genes occurred, leading to a novel *Cf* gene. The complete *9DC* cluster was isolated from a *L. pimpinellifolium* accession and it was shown that the *9DC* cluster is organized like the *Cf-9* cluster (Kruijt et al. 2004).

For each of the *C. fulvum* effectors, except for Avr4, HR-responding plants have been identified in *L. pimpinellifolium* (Laugé et al. 2000). In a genus-wide screen for functional homologues of *Cf-4* and *Cf-9*, comprising eight wild tomato species, many Avr4- and Avr9-responsive tomato plants were identified (Kruijt et al. 2004). From five different species, including *L. hirsutum*, *Hcr9* genes that confer Avr4 responsiveness (*Hcr9-Avr4s*) were isolated and shown to carry features essential for Cf-4 function (Van der Hoorn et al. 2001b; Wulff et al. 2001; Kruijt et al. 2005b). The high sequence conservation of the *Hcr9* genes that mediate recognition of Avr4 an d Avr9 in diverged *Lycopersicon* species suggests that these genes predate *Lycopersicon* speciation and apparently provide a selective advantage in natural *Lycopersicon* populations. This notion is further supported by the identification of two Avr9-responsive *Solanum* species that are closely related to the genus *Lycopersicon* (De Wit, unpublished data). This further suggests that *C. fulvum* is an ancient pathogen of solanaceous plants.

Three resistance genes from *L. pimpinellifolium* that confer recognition of three different *C. fulvum* Ecp effectors (*Cf-Ecp* genes) have been found to co-segregate with *Hcr9* genes. Therefore, the *Cf-Ecp* genes are expected to belong to the *Hcr9* gene family. *Cf-Ecp2* and *Cf-Ecp3* are located at the chromosome 1 *Orion* (*OR*) locus (Haanstra et al. 1999; Yuan et al. 2002), whereas *Cf-Ecp5* maps at the *Aurora* (*AU*) locus of the same chromosome (Haanstra et al. 2000). The *Cf-Ecp2 OR* cluster contains three tandemly repeated *Hcr9* genes (Laugé et al. 2000; De Kock et al. 2005). The *Cf-Ecp3* cluster contains two *Hcr9* genes, whereas the *Cf-0* and *Cf-Ecp5* loci both contain only one *Hcr9* gene (Yuan et al. 2004). Functional analysis of candidate *Cf-Ecp2*, *Cf-Ecp3* and *Cf-Ecp5* genes has yet to be performed. In conclusion, *Cf* genes are generally located in clusters of tandemly repeated homologues, of which a few encode functional *Cf* genes, whereas the others may represent a reservoir of variation that may be exploited in the generation of novel *Cf* genes. Duplications, translocations, intra- and intergenic recombination, gene conversions and point mutations have all been reported, and the major mechanism for generation of novel variation appears to be sequence exchange between the various homologues.

D. Recognitional Specificity of Cf Proteins

Cf-2 and Cf-5 proteins are structurally very similar except for the number of LRRs, of which Cf-2 has 38 and Cf-5 has 32 repeats (Fig 7.1B). The specificity for Avr2 and Avr5 recognition resides in the N-terminus, as determined by swapping these domains between the Cf-2 and Cf-5 proteins (Seear and Dixon 2003). Swaps could delimit the specificity for Avr2 and Avr5 to LRRs 4-27 in the Cf-2 protein and to LRRs 4-21 in the Cf-5 protein. Further proof for the critical role of the N-terminal domain of Cf-2 in Avr2 perception was obtained by studying a Cf-2/Cf-9 chimera containing domains A, B and the 34 N-terminal LRRs of Cf-2 fused to the three C-terminal LRRs and the C-terminal D, E, F and G domains of Cf-9, which conferred Avr2-triggered Rcr3-dependent resistance to *C. fulvum*. This chimeric protein also mediates H_2O_2 production in an Avr2-dependent manner (Krüger et al. 2002; Rivas et al. 2004). Thus, the chimeric Cf-2/Cf-9 protein can also signal through Cf-9 after triggering by Avr2 and in the presence of Rcr3,

proving that the N-terminal domain of Cf-2 is important for Avr2 perception.

Cf-4 and Cf-9 proteins are also much related. Cf-4 has 25 and Cf-9 has 27 LRRs. In wild tomato species several functional homologues of Cf-9 occur that all contain 27 LRRs but differ in many amino acid residues, suggesting that considerable variation is allowed in the LRRs without losing specificity. The number of LRRs is important for specificity as no Cf-9 chimeras with fewer than 27 LRRs provided Avr9-dependent HR and no Cf-4 chimeras with more than 25 LRRs provided Avr4-dependent HR. Thus, LRR copy number is an important determinant in recognitional specificity; variation in LRRs affects spacing between the solvent-exposed residues within LRRs. Analysis of shuffled clones, defined reciprocal swaps and directed mutagenesis of critical amino acid residues varying between the Cf-4 and Cf-9 proteins show that most amino acid residues can be exchanged without affecting specificity. However, three residues in LRR11, 12 and 14 of Cf-4 cannot be exchanged without losing Avr4-dependent activity (Van der Hoorn et al. 2001b), but some of these residues were found to be less crucial in another study (Wulff et al. 2001). However, all crucial amino acids locate in the solvent-exposed β-strand/β-turn domain of the LRRs, indicating that selection pressure is imposed on these residues, as they are supposedly involved in recognition.

E. Interaction Between R Proteins and Effectors

1. Indirect Interaction Between Cf Proteins and Effectors

As soon as the first *Cf* gene (*Cf-9*; Jones et al. 1994) and the cognate *Avr* gene (*Avr9*; Van den Ackerveken et al. 1992) had been cloned, experiments were designed to study whether the encoded proteins would interact directly, as was expected from the biochemical model derived from the gene for gene hypothesis. The mature 28-amino-acid Avr9 protein was labeled with iodine-125 at the N-terminal tyrosine residue and used in binding studies. ^{125}I-Avr9 showed specific, saturable and reversible binding to plasma membranes isolated from Cf-0 or Cf-9 tomato leaves. The dissociation constant was found to be 0.07 nM, and the receptor concentration was 0.8 pmol/mg microsomal protein (Kooman-Gersmann et al. 1996). However, binding kinetics and binding capacity were similar for membranes of the Cf-0 and Cf-9 genotypes. In addition, it was found that all solanaceous plant species tested contained a **high affinity binding site** (HABS) for Avr9, whereas such a HABS could not be identified in non-solanaceous species. The ability of membranes isolated from different solanaceous plant species to bind Avr9 correlated with the presence of members of the *Hcr9* gene family. However, subsequent experiments using different approaches showed no direct interaction between Avr9 and Cf-9. Cf-9 produced in COS cells or insect cells did not bind to ^{125}I-Avr9. Also, Cf-9 protein produced in tobacco did not bind to Avr9 as measured by surface plasmon resonance, BIAcore and surface enhanced laser desorption and ionization, respectively (Luderer et al. 2001). Similarly, for Avr2 no direct interaction with Cf-2 could be shown (Rooney et al. 2005). As mentioned earlier, Avr2 binds and inhibits Rcr3 (which originates from *L. pimpinellifolium*; Rcr3^{pimp}) and it is assumed that Avr2-binding modulates Rcr3^{pimp}, thus enabling the Cf-2 protein to activate an HR. Many additional studies could not show a direct interaction between R proteins and their cognate Avrs, suggesting that direct interaction is the exception rather than the rule (Jones and Dangl 2006). Indirect interaction between an R protein and its cognate effector follows the guard hypothesis (Van der Biezen and Jones 1998), which states that an R protein acts as a guard of a virulence target of a pathogen effector and effector-triggered immunity occurs after modulation of the target by the effector (Jones and Dangl 2006). A few *Cf* genes have been identified that cause autonecrosis when expressed in tobacco in the absence of the cognate Avr factor (Panter et al. 2002; Wulff et al. 2004; Barker et al. 2006a, b). The mechanism behind this phenomenon is not yet understood. Most likely, this activation is not caused by auto-activation but by recognition of a soluble non-self tobacco protein.

2. Direct Interaction Between R Proteins and Other Fungal Effectors

For different R proteins of flax and Avrs of the flax rust fungus *M. lini* direct interactions were observed in the yeast two-hybrid system (Dodds et al. 2006; Ellis et al. 2007). Rusts are obligate biotrophs and form haustoria that penetrate the host cells but remain separated from the cytoplasm by the extrahaustorial matrix. All effector genes from *M. lini* encode small proteins with N-terminal secretion signals for targeting to the ER. They are expected to interact directly with R proteins located in the cytoplasm. In addition, the Avr-Pita protein of *Magnaporthe grisea* directly interacts with its cognate R protein Pi-ta (Jia et al. 2000). *M. grisea* produces invasive hyphae to parasitize its host (Kankanala et al. 2007). Thus, both the *M. lini* and *M. grisea* Avrs interact directly with cognate cytoplasmic R proteins inside host cells to initiate R-mediated defense responses.

As discussed above, the *C. fulvum* effector Avr4 binds chitin, thus shielding the *C. fulvum* cell wall against plant chitinases (Van den Burg et al. 2006; Van Esse et al. 2007). Subsequent studies, including micro-array analysis of Avr4-expressing tomato, showed that Avr4 triggers only a few plant responses in the absence of the Cf-4 resistance protein (Van Esse et al. 2008). It is therefore unlikely that, in addition to the chitin in the *C. fulvum* cell wall, Avr4 has in planta targets. Therefore, it is proposed that the interaction between Cf-4 and Avr4 is direct (Van Esse et al. 2007). The fact that nearly all virulent alleles of *Avr4* represent point mutations rather than frame shifts or deletions and still produce nearly the same protein with only one single amino acid change also suggests that this interaction is direct (Joosten et al. 1994, 1997; Stergiopoulos et al. 2007; Van Esse et al. 2007; Fig. 7.1B).

IV. Effector-Triggered Cf-Mediated Defense Responses

A. Plant Components Interacting with Cf Proteins

Although most Cf and Avr proteins do not interact directly, all effectors trigger Cf-mediated downstream defense signaling eventually leading to HR and resistance (Fig. 7.1C). The Cf-mediated defense signaling pathway is only poorly understood. Both genetic and biochemical approaches have been used to identify downstream defense signaling components (Rivas and Thomas 2005; Joosten and De Wit 1999). Thus far, two components have been identified that interact with a Cf protein in yeast two-hybrid analyses using cDNA libraries of both tomato and tobacco as prey. **Vesicle-associated protein 27** (VAP27) interacts with the EFG domains of Cf-9 (Laurent et al. 2000) and is most likely localized in the plasma membrane. The extracellular C-terminus of VAP27 is highly similar to the mammalian VAP33 protein, which specifically interacts with a vesicle membrane protein and may participate in membrane trafficking. The cytoplasmic N-terminus of VAP27 bears homology to a CC motif present in the major sperm protein (MSP). MSP may be involved in protein–protein interactions and can trigger the movement of membrane vesicles. This suggests that VAP27 may play a role in endocytosis of components of the Cf-9 receptor complex. In addition, the Cf-9 interacting thio-redoxin (CITRX) was identified in a yeast two-hybrid screen for interactors in which the C-terminal 33 amino acids of Cf-9 were used as bait, comprising the cytoplasmic tail and part of the transmembrane domain. CITRX acts as a negative regulator of Cf-9-mediated defense responses but does not interfere with Cf-2-mediated defense responses, as determined by virus-induced gene silencing (VIGS) of *CITRX* (Rivas et al. 2004). CITRX also interacts with the Avr9/Cf-9-induced kinase 1 (ACIK1) protein kinase upon initiation of the Avr9-triggered Cf-9-mediated defense response (Nekrasov et al. 2006).

B. Genetic Approaches to Identify Defense Signaling Components Downstream of Cf Proteins

Tomato plants carrying different *Cf* resistance genes have been mutagenized by EMS, and M2 populations analyzed for loss of resistance either by challenging with a strain of *C. fulvum* carrying the cognate Avr gene, or by direct treatment with the cognate Avr protein and screening for loss of HR. In this way the *Rcr3* gene, discussed earlier, was identified because Cf-2 plants carrying mutations in *Rcr3* could no longer provide Avr2-triggered HR (Krüger et al. 2002; Rooney et al. 2005). Two additional mutations affecting Cf-9-mediated defense, designated *Rcr-1* and *Rcr-3*, were identified but these mutants were not characterized further (Hammond-Kosack et al. 1994). In fact, surprisingly few mutants have been identified by mutational approaches. This could imply that either only few downstream components are required, that many of the downstream components are redundant, or that mutations in many downstream defense signaling components are lethal.

C. Transcript Profiling of Cf-Mediated Defense Genes

1. cDNA-AFLP Analysis

Trancriptional analysis of *Cf-4* tomato leaves or *Cf-9* transgenic tobacco cell cultures allowed identifying downstream defense signaling genes that are either up- or down-regulated after treatment with cognate Avr factors (Durrant et al. 2000; Gabriëls et al. 2006, 2007). Subsequently, up-regulated genes were silenced in order to determine their roles in defense. In this way, several genes were identified and their functions analyzed using VIGS. Using tobacco cell cultures transgenic for *Cf-9* and treated with Avr9, Durrant et al. (2000) found several hundreds of cDNAs that showed differential accumulation, and clones were obtained for 13 Avr9/Cf-9 rapidly elicited (*ACRE*) genes. Some *ACRE* genes were functionally analyzed using VIGS and found to affect plant defense and resistance to *C. fulvum*, including the putative E3 ubiquitin ligase gene *NtCMPG1*. *Nicotiana benthamiana* plants silenced for *NtCMPG1* showed reduced HR after Cf-9/Avr9 elicitation, while over-expression of *NtCMPG1* induced a stronger HR in *Cf-9* tobacco plants after Avr9 infiltration. In tomato, silencing of *CPMG1* decreased resistance to *C. fulvum* (González-Lamothe et al. 2006). Another *ACRE* gene encoded a Ser/Thr protein kinase, ACIK. ACIK1 was found to be required for Cf-9/Avr9- and Cf-4/Avr4-induced HR but not for the HR or resistance induced by other *R/Avr* gene pairs (Rowland et al. 2005). Moreover, VIGS of *LeACIK1* in tomato decreased Cf-9-mediated resistance to *C. fulvum*.

Gabriëls et al. (2006) used *Cf-4/Avr4*-expressing seedlings carrying both the *Cf-4* gene and the cognate *Avr4* gene. Such seedlings express defense genes within 2–4 days after germination and finally mount a systemic HR when grown at room temperature, although at elevated temperatures (33 °C) these seedlings are able to survive (De Jong et al. 2002). *Cf-4/Avr4* transgenic seedlings were grown at 33 °C for some time and subsequently the temperature was shifted to room temperature to establish a synchronized HR. The transcriptomes of the seedlings were compared at 0, 30, 60 and 90 min after temperature shift by cDNA-AFLP analysis with seedlings that express only *Cf-4* or *Avr4*. In this way, several hundreds of differentially expressed genes were identified, designated Avr responsive tomato (*ART*) genes.

After silencing by VIGS, four different *ART* genes clearly affected *Cf-4/Avr4*-induced HR (Gabriëls et al. 2006). Among these was a cDNA of a resistance analogue encoding a CC-NB-LRR designated NB-LRR required for HR-associated cell death (NRC1) that affected *Cf-4*-mediated HR and resistance (Gabriëls et al. 2006).

2. Micro-Array Analysis

Currently, sequencing of the tomato genome is underway and in recent years commercial micro-arrays have become available, allowing the measuring of Cf-mediated transcriptome changes after inoculation with avirulent strains of *C fulvum* or treatment with effectors of plants containing the cognate Cf proteins. Similarly, this type of analysis can be performed in plants after inoculation with virulent strains or in transgenic plants lacking cognate Cf proteins but expressing single effectors to study transcriptional changes induced by virulent strains or single effectors. In this way, expression and co-regulation of genes crucial in Cf-mediated defense and effector-mediated susceptibility have been identified (Van Esse et al. 2007, 2008, unpublished data).

D. Post-Translational Modifications of Downstream Plant Defense Components

A number of early plant defense responses are extremely fast and do not require de novo RNA and protein synthesis, but rather post-translational modifications. These include proteins such as ion channels, mitogen-activated protein kinases (MAPKs) that become (de)phosphorylated and, in turn, may phosphorylate other proteins crucial for defense signaling (Romeis et al. 1999; Stulemeijer et al. 2007; Van Ooijen et al. 2007; Stulemeijer and Joosten 2008). Three MAPKs from tomato, designated LeMPKs, were described to be activated after the temperature shift of *Cf-4/Avr4* seedlings. They were subsequently shown to play a role in HR development and resistance to *C. fulvum* in *Cf-4* plants (Stulemeijer et al. 2007). In addition, phospholipase C (PLC) was shown to be activated, releasing diacylglycerol kinase (DGK)-dependent accumulation of the signaling molecule phosphatidic acid (PA) in Avr4-treated *Cf-4* tobacco cell cultures (De Jong et al. 2004). Similarly, calcium-dependent

protein kinases (CDPKs) were activated in Avr9-treated *Cf-9* tobacco cell cultures (Romeis et al. 2000; Ludwig et al. 2005). NtCDPK2 is required for HR, is activated by phosphorylation, stimulates ethylene synthesis and negatively regulates MAPK signaling. Syntaxin, a protein involved in vesicle transport, was also rapidly phosphorylated in *Cf-9* cells by Avr9 (Heese et al. 2005; Fig. 7.1C).

E. Induction of Other Cf-Mediated Downstream Physiological Responses

Cf-mediated accumulation of compounds such as reactive oxygen species, glutathione, callose, PR proteins and antimicrobial compounds, including phytoalexins, is proposed to play an important role in limiting fungal growth, eventually leading to *C. fulvum* resistance in tomato (De Wit 1992; Hammond-Kosack and Jones 1996; May et al. 1996; Higgins et al. 1998). Accumulation of reactive oxygen species (ROS) such as H_2O_2 occurs in planta and in tomato cell cultures upon treatment with cognate Avr factors (Lu and Higgins 1999; Piedras et al. 1999; Borden and Higgins 2002). Inhibitor studies on the Avr-induced oxidative burst in tomato cell cultures suggested that the majority of the H_2O_2 accumulation originates from NADPH-oxidase (Xing et al. 1997). In addition, Avr-induced activation of Ca^{2+}- and H^+-ATPase in tomato cell cultures and activation of K^+ channels in tomato guard cells were observed (Vera-Estrella et al. 1994; Xing et al. 1997b; Blatt et al. 1999). Treatment of *Cf-4* transgenic tobacco cell cultures with Avr4 activates PLC, resulting in the formation of diacylglycerol (DAG) and inositol 1,4,5 trisphosphate (IP_3), whereas simultaneous activation of DGK gives rise to accumulation of PA (De Jong et al. 2004). PA might stimulate various enzymes such as NADPH-oxidase producing ROS. IP_3 might release Ca^{2+} from the vacuole, that, in addition to Ca^{2+} entering from outside the cell stimulates Ca^{2+}-dependent protein kinases (CDPKs). CDPKs also activate NADPH-oxidase and can negatively affect phosphorylation of MAPkinases. A tomato CDPK was also found to phosphorylate a tomato 1-amino-cyclopropane-1-carboxylic acid synthase (ACS2; Tatsuki and Mori 2001). In tomato plants Avr-triggered Cf-mediated release of salicylic acid (SA) is also induced, but Cf-mediated resistance of tomato to *C. fulvum* seems not to depend on SA, as NahG-transgenic plants accumulating little or no

SA were as resistant to the fungus as control plants (Brading et al. 2000). Also antimicrobial poly-acetylenes and sequiterpenes accumulate faster in incompatible interactions than in compatible ones (De Wit and Flach 1997; De Wit and Kodde 1981b); and the same was observed for the accumulation of glutathione (May et al. 1996) and PR proteins, including chitinases and glucanases (De Wit and Van der Meer 1986; Joosten and De Wit. 1989; Van Kan et al. 1992; Danhash et al. 1993). Figure 7.1C shows a schematic overview of most of the various effector-triggered Cf-mediated physiological responses observed in tomato (and sometimes other host plants, including *Arabidopsis*). However, at present it is difficult to conclude whether these changes are the cause or consequence of resistance, and it is now generally accepted that limitation of fungal growth is not caused by only one component, but most likely by a timely array of antimicrobial enzymes and metabolites (Jones and Dangl 2006; De Wit 2007).

V. Conclusions and Future Prospects

Over 30 years of research on the *C. fulvum*–tomato pathosystem has provided useful general information on the molecular basis of interactions between plants and extracellular fungal pathogens. The intrinsic biological function of a few *C. fulvum* effectors has been elucidated, and they seem to interact indirectly rather than directly with Cf proteins in resistant plant genotypes, with the possible exception of the interaction between Avr4 and Cf-4. Effector-perception by the Cf proteins is still enigmatic. Moreover, effector-triggered Cf-mediated downstream signaling in the *C. fulvum*–tomato pathosystem is still poorly understood (Fig. 7.1C), but the availability of both the *C. fulvum* and the tomato genome sequences in the near future will speed up the discovery of the full secretome of *C. fulvum* and also enable the dissection of the plant defense signaling pathways by combined transcriptome, proteome and functional analyses utilizing RNA-silencing or over-expression of interesting fungal and plant genes. In addition to *C. fulvum*, several other extracellular pathogens secrete cysteine-rich peptides. At the genomic level, *C. fulvum* is most related to *M. fijiensis* and future experiments will reveal whether effectors of *M. fijiensis* have similar functions in virulence as those of *C. fulvum*. Therefore, *C. fulvum* will

remain a versatile model fungus for future studies on extracellular fungal plant pathogens.

References

Abramovitch RB, Anderson JC, Martin GB (2006) Bacterial elicitation and evasion of plant innate immunity. Nat Rev Mol Cell Biol 7:601–611

Alfano JR, Collmer A (2004) Type III secretion system effector proteins: Double agents in bacterial disease and plant defense. Annu Rev Phytopathol 42:385–414

Bailey DL (1950) Studies in racial trends and constancy in *Cladosporium fulvum* Cooke. Can J Res Sect C Bot Sci 28:535–536

Barker CL, Baillie BK, Hammond-Kosack KE, Jones JDG, Jones DA (2006a) Dominant-negative interference with defense signaling by truncation mutations of the tomato *Cf-9* disease resistance gene. Plant J 46:385–399

Barker CL, Talbot SJ, Jones JDG, Jones DA (2006b) A tomato mutant that shows stunting, wilting, progressive necrosis and constitutive expression of defence genes contains a recombinant *Hcr9* gene encoding an autoactive protein. Plant J 46:369–384

Belfanti E, Silfverberg-Dilworth E, Tartarini S, Patocchi A, Barbieri M, Zhu J, Vinatzer BA, Gianfranceschi L, Gessler C, Sansavini S (2004) The *HcrVf2* gene from a wild apple confers scab resistance to a transgenic cultivated variety. Proc Natl Acad Sci USA 101:886–890

Benghezal M, Wasteneys GO, Jones DA (2000) The C-terminal dilysine motif confers endoplasmic reticulum localization to type I membrane proteins in plants. Plant Cell 12:1179–1201

Bingle LE, Bailey CM, Pallen MJ (2008) Type VI secretion: a beginner's guide. Curr Opin Microbiol 11:3–8

Blatt MR, Grabov A, Brearley J, Hammond-Kosack K, Jones JDG (1999) K+ channels of Cf-9 transgenic tobacco guard cells as targets for *Cladosporium fulvum* Avr9 elicitor-dependent signal transduction. Plant J 19:453–462

Bolton MD, Esse HP, Vossen JH, de Jonge R, Stergiopoulos I, Stulemeijer IJE, Van den Berg G, Borrás-Hidalgo O, Dekker HL, de Koster CG, De Wit PJGM, Joosten MHAJ, Thomma BPHJ (2008) The novel *Cladosporium fulvum* Lysine motif effector Ecp6 is a virulence factor with orthologs in other fungal species. Mol Microbiol. doi:10.1111/j.1365-2958.2008.06270.x

Bonifacino JS, Traub LM (2003) Signals for sorting of transmembrane proteins to endosomes and lysosomes. Annu Rev Biochem 72:395–447

Borden S, Higgins VJ (2002) Hydrogen peroxide plays a critical role in the defence response of tomato to *Cladosporium fulvum*. Physiol Mol Plant Pathol 61:227–236

Boukema IW, Garretsen F (1975) Uniform resistance to *Cladosporium fulvum* Cooke in tomato (*Lycopersicon esculentum* Mill). Investigations on F2s and F3s from diallel crosses. Euphytica 24:105–116

Braun U, Crous PW, Groenewald JZ, de Hoog GS (2003) Phylogeny and taxonomy of *Cladosporium*-like hyphomycetes, including *Davidiella* gen. nov. the teleomorph of *Cladosporium* s. str. Mycol Prog 2:3–18

Butler EJ, Jones SG (1949) Plant pathology. Tomato leaf mould, *Cladosporium fulvum* Cooke. Macmillan, London, pp 672–678

Catanzariti AM, Dodds PN, Ellis JG (2007) Avirulence proteins from haustoria-forming pathogens. FEMS Microbiol Lett 269:181–188

Chinchilla D, Zipfel C, Robatzek S, Kemmerling B, Nurnberger T, Jones JDG, Felix G, Boller T (2007) A flagellin-induced complex of the receptor FLS2 and BAK1 initiates plant defence. Nature 448:497–500

Ciferri R (1954) A few critical Italian fungi. Atti 10:237–247

Ciuffetti LM, Tuori RP, Gaventa JM (1997) A single gene encodes a selective toxin causal to the development of tan spot of wheat. Plant Cell 9:135–144

Conte-Ferraez L, Waalwijk C, Canto-Canche BB, Kema GHJ, Crous PW, James AC, Abeln ECA (2007) Isolation and characterisation of the mating type locus of *Mycosphaerella fijiensis*, the causal agent of black leaf streak disease of banana. Mol Plant Pathol 8:111–120

Cooke MC (1883) New American fungi. Grivillea 12:32

Danhash N, Wagemakers CAM, Van Kan JAL, De Wit PJGM (1993) Molecular characterization of 4 chitinase cDNAs obtained from *Cladosporium fulvum*-infected tomato. Plant Mol Biol 22:1017–1029

De Jong CF, Takken FLW, Cai XH, De Wit PJGM, Joosten MHAJ (2002) Attenuation of Cf-mediated defense responses at elevated temperatures correlates with a decrease in elicitor-binding sites. Mol Plant–Microbe Interact 15:1040–1049

De Jong CF, Laxalt AM, Bargmann BOR, De Wit PJGM, Joosten MHAJ, Munnik T (2004) Phosphatidic acid accumulation is an early response in the Cf-4/Avr4 interaction. Plant J 39:1–12

De Kock MJD, Brandwagt BF, Bonnema G, De Wit PJGM, Lindhout P (2005) The tomato Orion locus comprises a unique class of *Hcr9* genes. Mol Breed 15:409–422

De Wit PJGM (1977) A light and scanning electron microscopy study of infection of tomato plants by virulent and avirulent races of *Cladosporium fulvum*. Neth J Plant Pathol 83:109–122

De Wit PJGM (1992) Molecular characterization of gene for gene systems in plant-fungus interactions and the application of avirulence genes in control of plant pathogens. Annu Rev Phytopathol 30:391–418

De Wit PJGM (1995a) *Cf9* and *Avr9*, two major players in the gene for gene game. Trends Microbiol 3:251–252

De Wit PJGM (1995b) Fungal avirulence genes and plant resistance genes: Unraveling the molecular basis of gene for gene interactions, Adv Bot Res 21:147–185

De Wit PJGM (1997) Pathogen avirulence and plant resistance: a key role for recognition. Trends Plant Sci 2:452–458

De Wit PJGM (2007) How plants recognise pathogens and defend themselves. Cell Mol Life Sci 64:2726–2732

De Wit PJGM, Flach W (1979) Differential accumulation of phytoalexins in tomato leaves but not in fruits after inoculation with virulent and avirulent races of *Cladosporium fulvum*. Physiol Plant Pathol 15:257–267

De Wit PJGM, Joosten MHAJ (1999) Avirulence and resistance genes in the *Cladosporium fulvum*–tomato interaction. Curr Opin Microbiol 2:368–373

De Wit PJGM, Kodde E (1981a) Further characterization and cultivar-specificity of glycoprotein elicitors from culture filtrates and cell walls of *Cladosporium fulvum* (Syn *Fulvia fulva*). Physiol Plant Pathol 18:297–314

De Wit PJGM, Kodde E (1981b) Induction of polyacetylenic phytoalexins in *Lycopersicon esculentum* after inoculation with *Cladosporium fulvum* (Syn *Fulvia fulva*). Physiol Plant Pathol 18:143–148

De Wit PJGM, Spikman G (1982) Evidence for the occurrence of race and cultivar-specific elicitors of necrosis in intercellular fluids of compatible interactions of *Cladosporium fulvum* and tomato. Physiol Plant Pathol 21:1–11

De Wit PJGM, Van der Meer FE (1986) Accumulation of the pathogenesis-related tomato leaf protein P14 as an early indicator of incompatibility in the interaction between *Cladosporium fulvum* (Syn *Fulvia fulva*) and tomato. Physiol Mol Plant Pathol 28:203–214

Dikinson MJ, Jones DA, Jones JDG (1993) Close linkage between the Cf-2/Cf-5 and Mi resistance loci in tomato. Mol Plant–Microbe Interact 6:341–347

Dixon MS, Jones DA, Keddie JS, Thomas CM, Harrison K, Jones JDG (1996) The tomato Cf-2 disease resistance locus comprises two functional genes encoding leucine-rich repeat proteins. Cell 84:451–459

Dixon MS, Hatzixanthis K, Jones DA, Harrison K, Jones JDG (1998) The tomato *Cf-5* disease resistance gene and six homologs show pronounced allelic variation in leucine-rich repeat copy number. Plant Cell 10:1915–1925

Dixon MS, Golstein C, Thomas CM, Van der Biezen EA, Jones JDG (2000) Genetic complexity of pathogen perception by plants: The example of *Rcr3*, a tomato gene required specifically by Cf-2. Proc Natl Acad Sci USA 97:8807–8814

Dodds PN, Lawrence GJ, Catanzariti AM, Ayliffe MA, Ellis JG (2004) The *Melampsora lini AvrL567* avirulence genes are expressed in haustoria and their products are recognized inside plant cells. Plant Cell 16:755–768

Dodds PN, Lawrence GJ, Catanzariti AM, Teh T, Wang CIA, Ayliffe MA, Kobe B, Ellis JG (2006) Direct protein interaction underlies gene for gene specificity and coevolution of the flax resistance genes and flax rust avirulence genes. Proc Natl Acad Sci USA 103:8888–8893

Durrant WE, Rowland O, Piedras P, Hammond-Kosack KE, Jones JDG (2000) cDNA-AFLP reveals a striking overlap in race-specific resistance and wound response gene expression profiles. Plant Cell 12:963–977

Ellis JG, Dodds PN, Lawrence GJ (2007) Flax rust resistance gene specificity is based on direct resistance–avirulence protein interactions. Annu Rev Phytopathol 45:289–306

Fernandez MR, Heath MC (1991) Interactions of the non-host French bean plant (*Phaseolus vulgaris*) with parasitic and saprophytic fungi 4. Effect of preinoculation with the bean rust fungus on growth of parasitic fungi nonpathogenic on beans. Can J Bot 69:1642–1646

Fliegmann J, Mithofer A, Wanner G, Ebel J (2004) An ancient enzyme domain hidden in the putative beta-glucan elicitor receptor of soybean may play an active part in the perception of pathogen-associated molecular patterns during broad host resistance. J Biol Chem 279:1132–1140

Flor HH (1942) Inheritance of pathogenicity in *Melampsora lini*. Phytopathology 32:653–669

Fritz-Laylin LK, Krishnamurthy N, Tor M, Sjolander KV, Jones JDG (2005) Phylogenomic analysis of the receptor-like proteins of rice and arabidopsis. Plant Physiol 138:611–623

Gabriels S, Takken FLW, Vossen JH, De Jong CF, Liu Q, Turk SCHJ, Wachowski LK, Peters J, Witsenboer HMA, De Wit PJGM, Joosten MHAJ (2006) cDNA-AFLP combined with functional analysis reveals novel genes involved in the hypersensitive response. Mol Plant–Microbe Interact 19:567–576

Gabriels S, Vossen JH, Ekengren SK, VanOoijen G, Abd-El-Haliem AM, Van den Berg GCM, Rainey DY, Martin GB, Takken FLW, De Wit PJGM, Joosten MHAJ (2007) An NB-LRR protein required for HR signalling mediated by both extra- and intracellular resistance proteins. Plant J 50:14–28

Gonzalez-Lamothe R, Tsitsigiannis DI, Ludwig AA, Panicot M, Shirasu K, Jones JDG (2006) The U-Box protein CMPG1 is required for efficient activation of defense mechanisms triggered by multiple resistance genes in tobacco and tomato. Plant Cell 18:1067–1083

Haanstra JPW, Laugé R, Meijer-Dekens F, Bonnema G, De Wit PJGM, Lindhout P (1999) The Cf-Ecp2 gene is linked to but not part of the Cf4/Cf-9 cluster on the short arm of chromosome 1 in tomato. Mol Gen Genet 262:839–845

Haanstra JPW, Meijer-Dekens F, Laugé R, seethana DC, Joosten MHAJ, De Wit PJGM, Lindhout P (2000) Mapping strategy for resistance genes against *Cladosporium fulvum* on the short arm of Chromosome 1 of tomato: Cf-Ecp5 near the Hcr9 Milky Way cluster. Theor Appl Genet 101:661–668

Hammondkosack KE, Jones DA, Jones JDG (1994) Identification of two genes required in tomato for full Cf-9-dependent resistance to *Cladosporium fulvum*. Plant Cell 6:361–374

Hanania U, Avni A (1997) High-affinity binding site for ethylene-inducing xylanase elicitor on *Nicotiana tabacum* membranes. Plant J 12:113–120

Hanania U, Furman-Matarasso N, Ron M, Avni A (1999) Isolation of a novel SUMO protein from tomato that suppresses EIX-induced cell death. Plant J 19:533–541

Heese A, Ludwig AA, Jones JDG (2005) Rapid phosphorylation of a syntaxin during the Avr9/Cf-9-race-specific signaling pathway. Plant Physiol 138:2406–2416

Higgins VJ (1982) Response of tomato to leaf injection with conidia of virulent and avirulent races of *Cladosporium fulvum*. Physiol Plant Pathol 20:145–146

Higgins VJ, Lu HG, Xing T, Gelli A, Blumwald E (1998) The gene for gene concept and beyond: interactions and signals. Can J Plant Pathol 20:150–157

Houterman PM, Speijer D, Dekker HL, de Koster CG, Cornelissen BJC, Rep M (2007) The mixed xylem sap proteome of *Fusarium oxysporum*-infected tomato plants. Mol Plant Pathol 8:215–221

Houterman PM, Cornelissen BJC, Rep M (2008) Suppression of plant resistance gene-based immunity by fungal effector. Plant Pathol. doi:10.1371/journal.ppat.1000061.g003

Jia Y, McAdams SA, Bryan GT, Hershey HP, Valent B (2000) Direct interaction of resistance gene and avirulence gene products confers rice blast resistance. EMBO J 19:4004–4014

Jiang RHY, Tripathy S, Govers F, Tyler BM (2008) RXLR effector reservoir in two Phytophthora species is dominated by a single rapidly evolving superfamily with more than 700 members. Proc Natl Acad Sci USA 105:4874–4879

Jones DA, Dickinson MJ, Balint-Kurti PJ, Dixon MS, Jones JDG (1993) Two complex resistance loci revealed in tomato by classical and RFLP mapping of the Cf-2, Cf-4, Cf-5 and Cf-9 genes for resistance to *Cladosporium fulvum*. Mol Plant–Microbe Interact 6:348–357

Jones DA, Thomas CM, Hammondkosack KE, Balint-Kurti PJ, Jones JDG (1994) Isolation of the Tomato Cf-9 gene for resistance to *Cladosporium fulvum* by transposon tagging. Science 266:789–793

Jones JDG, Dangl JL (2006) The plant immune system. Nature 444:323–329

Joosten MHAJ, De Wit PJGM (1989) Identification of several pathogenesis-related proteins in tomato leaves inoculated with *Cladosporium fulvum* (Syn *Fulvia fulva*) as 1,3-beta-glucanases and chitinases. Plant Physiol 89:945–951

Joosten MHAJ, De Wit PJGM (1999) The tomato–*Cladosporium fulvum* interaction: a versatile experimental system to study plant-pathogen interactions. Annu Rev Phytopathol 37:335–367

Joosten MHAJ, Hendrickx LJM, De Wit PJGM (1990) Carbohydrate composition of apoplastic fluids isolated from tomato leaves inoculated with virulent or avirulent races of *Cladosporium fulvum* (Syn *Fulvia fulva*). Neth J Plant Pathol 96:103–112

Joosten MHAJ, Cozijnsen TJ, De Wit PJGM (1994) Host-resistance to a fungal tomato pathogen lost by a single base-pair change in an avirulence gene. Nature 367:384–386

Joosten MHAJ, Vogelsang R, Cozijnsen TJ, Verberne MC, De Wit PJGM (1997) The biotrophic fungus *Cladosporium fulvum* circumvents Cf-4-mediated resistance by producing unstable Avr4 elicitors. Plant Cell 9:367–379

Kankanala P, Czymmek K, Valent B (2007) Roles in rice membrane dynamics and plasmadesmata during biotrophic invasion by the blast fungus. Plant Cell 19:706–724

Kaku H, Nishizawa Y, Ishii-Minami N, Akimoto-Tomiyama C, Dohmae N, Takio K, Minami E, Shibuya N (2006) Plant cells recognize chitin fragments for defense signaling through a plasma membrane receptor. Proc Natl Acad Sci USA 103:11086–11091

Kawchuk LM, Hachey J, Lynch DR, Kulcsar F, VanRooijen G, Waterer DR, Robertson A, Kokko E, Byers R, Howard RJ, Fischer R, Prufer D (2001) Tomato Ve disease resistance genes encode cell surface-like receptors. Proc Natl Acad Sci USA 98:6511–6515

Kooman-Gersmann M, Honée G, Bonnema G, De Wit PJGM (1996) A high-affinity binding site for the Avr9 peptide elicitor of *Cladosporium fulvum* is present on plasma membranes of tomato and other solanaceous plants. Plant Cell 8:929–938

Kooman-Gersmann M, Vogelsang R, Hoogendijk ECM, De Wit PJGM (1997) Assignment of amino acid residues of the Avr9 peptide of *Cladosporium fulvum* that determine elicitor activity. Mol Plant–Microbe Interact 10:821–829

Krüger J, Thomas CM, Golstein C, Dixon MS, Smoker M, Tang SK, Mulder L, Jones JDG (2002) A tomato cysteine protease required for Cf-2-dependent disease resistance and suppression of autonecrosis. Science 296:744–747

Kruijt M, Brandwagt BF, De Wit PJGM (2004) Rearrangements in the *Cf-9* disease resistance gene cluster of wild tomato have resulted in three genes that mediate Avr9 responsiveness. Genetics 168:1655–1663

Kruijt M, De Kock MJD, De Wit PJGM (2005a) Receptor-like proteins involved in plant disease resistance – review. Mol Plant Pathol 6:85–97

Kruijt M, Kip DJ, Joosten MHAJ, Brandwagt BF, De Wit PJGM (2005b) The *Cf-4* and *Cf-9* resistance genes against *Cladosporium fulvum* are conserved in wild tomato species. Mol Plant–Microbe Interact 18:1011–1021

Langford AN (1937) The parasitism of *Cladosporium fulvum* Cooke and the genetics of resistance to it. Can J Res Part C 15:108

Langford AN (1948) Autogenous necrosis in interspecific tomato hybrids and its relation to the breeding of tomatoes for resistance to *Cladosporium fulvum*. Phytopathology 38:16–18

Laugé R, De Wit PJGM (1998) Fungal avirulence genes: structure and possible functions. Fungal Genet Biol 24:285–297

Laugé R, Joosten MHAJ, Van den Ackerveken GFJM, Van den Broek HWJ, De Wit PJGM (1997) The in planta-produced extracellular proteins Ecp1 and Ecp2 of *Cladosporium fulvum* are virulence factors. Mol Plant–Microbe Interact 10:725–734

Laugé R, Joosten MHAJ, Haanstra JPW, Goodwin PH, Lindhout P, De Wit PJGM (1998a) Successful search for a resistance gene in tomato targeted against a virulence factor of a fungal pathogen. Proc Natl Acad Sci USA 95:9014–9018

Laugé R, Dmitriev AP, Joosten MHAJ, De Wit PJGM (1998b) Additional resistance gene(s) against *Cladosporium fulvum* present on the Cf-9 introgression segment are associated with strong PR protein accumulation. Mol Plant–Microbe Interact 11:301–308

Laugé R, Goodwin PH, De Wit PJGM, Joosten MHAJ (2000) Specific HR-associated recognition of secreted proteins from *Cladosporium fulvum* occurs in both host and non-host plants. Plant J 23:735–745

Laurent F, Labesse G, De Wit P (2000) Molecular cloning and partial characterization of a plant VAP33 homologue with a major sperm protein domain. Biochem Biophys Res Commun 270:286–292

Lazarovits G, Higgins VJ (1976a) Histological comparison of *Cladosporium fulvum* Race 1 on immune, resistant and susceptible tomato varieties. Can J Bot 54:224–234

Lazarovits G, Higgins VJ (1976b) Ultrastructure of susceptible, resistant, and immune-reactions of tomato to races of *Cladosporium fulvum*. Can J Bot 54:235–249

Lazarovits G, Bhullar BS, Sugiyama HJ, Higgins VJ (1979) Purification and partial characterization of a glycoprotein toxin produced by *Cladosporium fulvum*. Phytopathology 69:1062–1068

Lindhout P, Korta W, Cislik M, Vos I, Gerlagh T (1989) Further identification of races of *Cladosporium fulvum* (*Fulvia fulva*) on tomato originating from the Netherlands, France and Poland. Neth J Plant Pathol 95:143–148

Linthorst HJM, Danhash N, Brederode FT, Van Kan JAL, De Wit PJGM, Bol JF (1991) Tobacco and tomato PR proteins homologous to Win and Pro-Hevein lack the Hevein domain. Mol Plant–Microbe Interact 4:586–592

Lu HG, Higgins VJ (1999) The effect of hydrogen peroxide on the viability of tomato cells and of the fungal pathogen *Cladosporium fulvum*. Physiol Mol Plant Pathol 54:131–143

Luderer R, Rivas S, Nurnberger T, Mattei B, Van den Hooven HW, Van der Hoorn RAL, Romeis T, Wehrfritz JM, Blume B, Nennstiel D, Zuidema D, Vervoort J, De Lorenzo G, Jones JDG, De Wit PJGM, Joosten MHAJ (2001) No evidence for binding between resistance gene product Cf-9 of tomato and avirulence gene product Avr9 of *Cladosporium fulvum*. Mol Plant–Microbe Interact 14:867–876

Luderer R, De Kock MJD, Dees RHL, De Wit PJGM, Joosten MHAJ (2002a) Functional analysis of cysteine residues of Ecp elicitor proteins of the fungal tomato pathogen *Cladosporium fulvum*. Mol Plant Pathol 3:91–95

Luderer R, Takken FLW, De Wit PJGM, Joosten MHAJ (2002b) *Cladosporium fulvum* overcomes Cf-2-mediated resistance by producing truncated Avr2 elicitor proteins. Mol Microbiol 45:875–884

Ludwig AA, Saitoh H, Felix G, Freymark G, Miersch O, Wasternack C, Boller T, Jones JDG, Romeis T (2005) Ethylene-mediated cross-talk between calcium-dependent protein kinase and MAPK signaling controls stress responses in plants. Proc Natl Acad Sci USA 102:10736–10741

Malnoy M, Xu M, Borejsza-Wysocka E, Korban SS, Aldwinckle HS (2008) Two receptor-like genes, *Vfa1* and *Vfa2*, confer resistance to the fungal pathogen *Venturia inaequalis* inciting apple scab disease. Mol Plant–Microbe Interact 21:448–458

Manning VA, Ciuffetti LM (2005) Localization of Ptr ToxA produced by *Pyrenophora tritici*–repentis reveals protein import into wheat mesophyll cells. Plant Cell 17:3203–3212

Manning VA, Hamilton SM, Karplus PA, Ciuffetti LM (2008) The Arg-Gly-Asp-containing, solvent-exposed loop of Ptr ToxA is required for internalization. Mol Plant–Microbe Interact 21:315–325

Marmeisse R, Van den Ackerveken GFJM, Goosen T, De Wit PJGM, Van den Broek HWJ (1993) Disruption of the avirulence gene *Avr9* in two races of the tomato pathogen *Cladosporium fulvum* causes virulence on tomato genotypes with the complementary resistance gene *Cf-9*. Mol Plant–Microbe Interact 6:412–417

May MJ, HammondKosack KE, Jones JDG (1996) Involvement of reactive oxygen species, glutathione metabolism, and lipid peroxidation in the Cf-gene-dependent defense response of tomato cotyledons induced by race-specific elicitors of *Cladosporium fulvum*. Plant Physiol 110:1367–1379

Miya A, Albert P, Shinya T, Desaki Y, Ichimura K, Shirasu K, Narusaka Y, Kawakami N, Kaku H, Shibuya N (2007) CERK1, a LysM receptor kinase, is essential for chitin elicitor signaling in *Arabidopsis*. Proc Natl Acad Sci USA 104:19613–19618

Nekrasov V, Ludwig AA, Jones JDG (2006) CITRX thioredoxin is a putative adaptor protein connecting Cf-9 and the ACIK1 protein kinase during the Cf-9/Avr9-induced defence response. FEBS Lett 580:4236–4241

Nürnberger T, Brunner F (2002) Innate immunity in plants and animals: emerging parallels between recognition of general elicitors and pathogen-associated molecular patterns. Curr Opin Plant Biol 5:318–324

Oort AJP (1944) Onderzoekingen over stuifbrand. II. Overgevoeligheid voor stuifbrand (*Ustilago tritici*). Tijdschr Plantenziekten 50:73–106.

Panter SN, Hammond-Kosack KE, Harrison K, Jones JDG, Jones DA (2002) Developmental control of promoter activity is not responsible for mature onset of Cf-9B-mediated resistance to leaf mold in tomato. Mol Plant–Microbe Interact 15:1099–1107

Parniske M, Jones JDG (1999) Recombination between diverged clusters of the tomato *Cf-9* plant disease resistance gene family. Proc Natl Acad Sci USA 96:5850–5855

Parniske M, HammondKosack KE, Golstein C, Thomas CM, Jones DA, Harrison K, Wulff BBH, Jones JDG (1997) Novel disease resistance specificities result from sequence exchange between tandemly repeated genes at the Cf-4/9 locus of tomato. Cell 91:821–832

Parniske M, Wulff BBH, Bonnema G, Thomas CM, Jones DA, Jones JDG (1999) Homologues of the *Cf-9* disease resistance gene (*Hcr9s*) are present at multiple loci on the short arm of tomato chromosome 1. Mol Plant–Microbe Interact 12:93–102

Perez-Garcia A, Snoeijers SS, Joosten MHAJ, Goosen T, De Wit PJGM (2001) Expression of the avirulence gene *Avr9* of the fungal tomato pathogen *Cladosporium fulvum* is regulated by the global nitrogen response factor Nrf1. Mol Plant–Microbe Interact 14:316–325

Piedras P, Hammond-Kosack KE, Harrison K, Jones JDG (1998) Rapid, Cf-9- and Avr9-dependent production of active oxygen species in tobacco suspension cultures. Mol Plant–Microbe Interact 11:1155–1166

Piedras P, Rivas S, Dröge S, Hillmer S, Jones JDG (2000) Functional c-myc tagged Cf-9 resistance gene products are plasma-membrane localized and glycosylated. Plant J 21:1–8

Rehmany AP, Gordon A, Rose LE, Allen RL, Armstrong MR, Whisson SC, Kamoun S, Tyler BM, Birch PRJ, Beynon JL (2005) Differential recognition of highly divergent downy mildew avirulence gene alleles by *RPP1* resistance genes from two *Arabidopsis* lines. Plant Cell 17:1839–1850

Rep M, Van der Does HC, Meijer M, VanWijk R, Houterman PM, Dekker HL, de Koster CG, Cornelissen BJC (2004) A small, cysteine-rich protein secreted by *Fusarium oxysporum* during colonization of xylem vessels is required for I-3-mediated resistance in tomato. Mol Microbiol 53:1373–1383

Rep M, Meijer M, Houterman PM, Van der Does HC, Cornelissen BJC (2005) *Fusarium oxysporum* evades I-3-mediated resistance without altering the matching avirulence gene. Mol Plant–Microbe Interact 18:15–23

Rivas S, Thomas CM (2005) Molecular interactions between tomato and the leaf mold pathogen *Cladosporium fulvum*. Annu Rev Phytopathol 43:395–436

Rivas S, Rougon-Cardoso A, Smoker M, Schauser L, Yoshioka H, Jones JDG (2004) CITRX thioredoxin interacts with the tomato Cf-9 resistance protein and negatively regulates defence. EMBO J 23:2156–2165

Rogers S, Wells R, Rechsteiner M (1986) Amino-acid-sequences common to rapidly degraded proteins-the PEST hypothesis. Science 234:364–368

Romeis T, Piedras P, Zhang SQ, Klessig DF, Hirt H, Jones JDG (1999) Rapid Avr9- and Cf-9-dependent activation of MAP kinases in tobacco cell cultures and leaves: convergence of resistance gene, elicitor, wound, and salicylate responses. Plant Cell 11:273–287

Romeis T, Piedras P, Jones JDG (2000) Resistance gene-dependent activation of a calcium-dependent protein kinase in the plant defense response. Plant Cell 12:803–815

Ron M, Avni A (2004) The receptor for the fungal elicitor ethylene-inducing xylanase is a member of a resistance-like gene family in tomato. Plant Cell 16:1604–1615

Rooney HCE, Van 't Klooster JW, Van der Hoorn RAL, Joosten MHAJ, Jones JDG, De Wit PJGM (2005) Cladosporium Avr2 inhibits tomato Rcr3 protease required for Cf-2-dependent disease resistance. Science 308:1783–1786

Rowland O, Ludwig AA, Merrick CJ, Baillieul F, Tracy FE, Durrant WE, Fritz-Laylin L, Nekrasov V, Sjolander K, Yoshioka H, Jones JDG (2005) Functional analysis of Avr9/Cf-9 rapidly elicited genes identifies a protein kinase, ACIK1, that is essential for full Cf-9-dependent disease resistance in tomato. Plant Cell 17:295–310

Sarma GN, Manning VA, Ciuffetti LM, Karplus PA (2005) Structure of Ptr ToxA: An RGD-containing host-selective toxin from *Pyrenophora tritici-repentis*. Plant Cell 17:3190–3202

Seear PJ, Dixon MS (2003) Variable leucine-rich repeats of tomato disease resistance genes *Cf-2* and *Cf-5* determine specificity. Mol Plant Pathol 4:199–202

Seeler JS, Dejean A (2003) Nuclear and unclear functions of SUMO. Nature Reviews Molecular Cell Biol 4:690–699

Senchou V, Weide R, Carrasco A, Bouyssou H, Pont-Lezica R, Govers F, Canut H (2004) High affinity recognition of a *Phytophthora* protein by Arabidopsis via an RGD motif. Cell Mol Life Sci 61:502–509

Shabab M, Shido T, Gu C, Kaschani F, Pansuriya T, Chintha R, Harzen A, Colby T, Kamoun S, Van der Hoorn RAL (2008) Fungal effector protein Avr2 targets diversifying defence-related Cys proteases of tomato. Plant Cell 20:1169–1183

Shen ZC, Jacobs-Lorena M (1999) Evolution of chitin-binding proteins in invertebrates. J Mol Evol 48:341–347

Solomon PS, Oliver RP (2002) Evidence that gamma-aminobutyric acid is a major nitrogen source during *Cladosporium fulvum* infection of tomato. Planta 214:414–420

Solomon PS, Waters ODC, Oliver RP (2007) Decoding the mannitol enigma in filamentous fungi. Trends Microbiol 15:257–262

Stergiopoulos I, De Kock MJD, Lindhout P, De Wit PJGM (2007a) Allelic variation in the effector genes of the tomato pathogen *Cladosporium fulvum* reveals different modes of adaptive evolution. Mol Plant–Microbe Interact 20:1271–1283

Stergiopoulos I, Groenewald M, Staats M, Lindhout P, Crous PW, De Wit PJGM (2007b) Mating-type genes and the genetic structure of a world-wide collection of the tomato pathogen *Cladosporium fulvum*. Fungal Genet Biol 44:415–429

Stulemeijer IJE, Joosten MHAJ (2008) Post-translational modification of host proteins in pathogen-triggered defence signalling in plants. Mol Plant Pathol 8:773–784

Stulemeijer IJE, Stratmann JW, Joosten MHAJ (2007) Tomato mitogen-activated protein kinases LeMPK1, LeMPK2, and LeMPK3 are activated during the Cf-4/Avr4-induced hypersensitive response and have distinct phosphorylation specificities. Plant Physiol 144:1481–1494

Takken FLW, Thomas CM, Joosten MHAJ, Golstein C, Westerink N, Hille J, Nijkamp HJJ, De Wit PJGM, Jones JDG (1999) A second gene at the tomato Cf-4 locus confers resistance to *Cladosporium fulvum* through recognition of a novel avirulence determinant. Plant J 20:279–288

Tatsuki M, Mori H (2001) Phosphorylation of tomato 1-amino-cyclopropane-1-carboxylic acid synthase, LEACS2, at the C-terminal region. J Biol Chem 276:28051–28057

Thomas CM, Jones DA, Parniske M, Harrison K, Balint-Kurti PJ, Hatzixanthis K, Jones JDG (1997) Characterization of the tomato *Cf-4* gene for resistance to *Cladosporium fulvum* identifies sequences that determine recognitional specificity in Cf-4 and Cf-9. Plant Cell 9:2209–2224

Thomma BPHJ, Van Esse HP, Crous PW, De Wit PJGM (2005) *Cladosporium fulvum* (syn. *Passalora fulva*), a highly specialized plant pathogen as a model for functional studies on plant pathogenic *Mycosphaerellaceae*. Mol Plant Pathol 6:379–393

Thomma BPHJ, Bolton MD, Clergeot PH, De Wit PJGM (2006) Nitrogen controls in planta expression of *Cladosporium fulvum Avr9* but no other effector genes. Mol Plant Pathol 7:125–130

Tigchelaar EC (1984) Collections of isogenic tomato stocks. Rep Tomato Genet Coop 34:55–57

Van den Ackerveken GFJM, Van Kan JAL, De Wit PJGM (1992) Molecular analysis of the avirulence gene *Avr9* of the fungal tomato pathogen *Cladosporium fulvum* fully supports the gene for gene hypothesis. Plant J 2:359–366

Van den Ackerveken GFJM, Vossen P, De Wit PJGM (1993) The Avr9 race-specific elicitor of *Cladosporium fulvum* is processed by endogenous and plant proteases. Plant Physiol 103:91–96

Van den Ackerveken GFJM, Dunn RM, Cozijnsen AJ, Vossen JPMJ, Van den Broek HWJ, De Wit PJGM (1994) Nitrogen limitation induces expression of the avirulence gene *Avr9* in the tomato pathogen *Cladosporium fulvum*. Mol Gen Genet 243:277–285

Van den Burg HA, Westerink N, Francoijs KJ, Roth R, Woestenenk E, Boeren S, De Wit PJGM, Joosten MHAJ, Vervoort J (2003) Natural disulfide bond-disrupted mutants of Avr4 of the tomato pathogen *Cladosporium fulvum* are sensitive to proteolysis, circumvent Cf-4-mediated resistance, but retain their chitin-binding ability. J Biol Chem 278:27340–27346

Van den Burg HA, Spronk C, Boeren S, Kennedy MAHJ, Vissers JPC, Vuister GW, De Wit PJGM, Vervoort J (2004) Binding of the Avr4 elicitor of *Cladosporium fulvum* to chitotriose units is facilitated by positive allosteric protein–protein interactions – the chitin-binding site of Avr4 represents a novel binding site on the folding scaffold shared between the invertebrate and the plant chitin-binding domain. J Biol Chem 279:16786–16796

Van den Burg HA, Harrison SJ, Joosten M, Vervoort J, De Wit PJGM (2006) *Cladosporium fulvum* Avr4 protects fungal cell walls against hydrolysis by plant chitinases accumulating during infection. Mol Plant–Microbe Interact 19:1420–1430

Van den Hooven HW, Van den Burg HA, Vossen P, Boeren S, De Wit PJGM, Vervoort J (2001) Disulfide bond structure of the Avr9 elicitor of the fungal tomato pathogen *Cladosporium fulvum*: evidence for a cystine knot. Biochemistry 40:3458–3466

Van der Biezen, Jones JDG (1998) Plant disease resistance proteins and the gene for gene concept. Trends Biochem Sci 23:454–456

Van der Hoorn RAL, Kruijt M, Roth R, Brandwagt BF, Joosten MHAJ, De Wit PJGM (2001a) Intragenic recombination generated two distinct *Cf* genes that mediate Avr9 recognition in the natural population of *Lycopersicon pimpinellifolium*. Proc Natl Acad Sci USA 98:10493–10498

Van der Hoorn RAL, Roth R, De Wit PJGM (2001b) Identification of distinct specificity determinants in resistance protein Cf-4 allows construction of a Cf-9 mutant that confers recognition of avirulence protein Avr4. Plant Cell 13:273–285

Van der Hoorn RAL, Van der Ploeg A, De Wit PJGM, Joosten MHAJ (2001c) The C-terminal dilysine motif for targeting to the endoplasmic reticulum is not required for Cf-9 function. Mol Plant–Microbe Interact 14:412–415

Van der Hoorn RAL, Wulff BBH, Rivas S, Durrant MC, Van der Ploeg A, De Wit PJGM, Jones JDG (2005) Structure-function analysis of Cf-9, a receptor-like protein with extracytoplasmic leucine-rich repeats. Plant Cell 17:1000–1015

Van Esse HP, Thomma BPHJ, Van 't Klooster JW, De Wit PJGM (2006) Affinity-tags are removed from *Cladosporium fulvum* effector proteins expressed in the tomato leaf apoplast. J Exp Bot 57:599–608

Van Esse HP, Bolton MD, Stergiopoulos I, De Wit PJGM, Thomma BPHJ (2007) The chitin-binding *Cladosporium fulvum* effector protein Avr4 is a virulence factor. Mol Plant–Microbe Interact 20:1092–1101

Van Esse HP, Van 't Klooster JW, Bolton MD, Yadeta K, VanBaarlen P, Boeren S, Vervoort J, De Wit PJGM and Thomma BPHJ (2008) The *Cladosporium fulvum* virulence protein Avr2 inhibits host proteases required for basal defense. Plant Cell (in press)

Van Kan JAL, Van den Ackerveken GFJM, De Wit PJGM (1991) Cloning and characterization of cDNA of avirulence gene *Avr9* of the fungal pathogen *Cladospo-*

rium fulvum, causal agent of tomato leaf mold. Mol Plant–Microbe Interact 4:52–59

Van Kan JAL, Joosten MHAJ, Wagemakers CAM, Van den Berg-Velthuis GCM, De Wit PJGM (1992) Differential accumulation of messenger-RNAs encoding extracellular and intracellular PR proteins in tomato induced by virulent and avirulant races of *Cladosporium fulvum*. Plant Mol Biol 20:513–527

Van Ooijen G, Van den Burg HA, Cornelissen BJC, Takken FLW (2007) Structure and function of resistance proteins in solanaceous plants. Annu Rev Phytopathol 45:43–72

Van't Slot KAE, Gierlich A, Knogge W (2007) A single binding site mediates resistance- and disease-associated activities of the effector protein NIP1 from the barley pathogen *Rhynchosporium secalis*. Plant Physiol 144:1654–1666

Vera-Estrella R, Barkla BJ, Higgins VJ, Blumwald E (1994) Plant defense response to fungal pathogens – activation of host-plasma membrane H^+-ATPase by elicitor-induced enzyme dephosphorylation. Plant Physiol 104:209–215

Waalwijk C, Mendes O, Verstappen ECP, de Waard MA, Kema GHJ (2002) Isolation and characterization of the mating-type idiomorphs from the wheat septoria leaf blotch fungus *Mycosphaerella graminicola*. Fungal Genet Biol 35:277–286

Wang G, Ellendorff U, Kemp B, Mansfield JW, Forsyth A, Mitchell K, Bastas K, Liu C-M, Woods-Tör E, Zipfel C, De Wit PJGM, Jones JDG, Tör M, Thomma BPHJ (2008) A genome-wide functional investigation into the roles of receptor-like proteins in *Arabidopsis*. Plant Physiol. doi: 10.1104/pp.108.119487

Westerink N, Brandwagt BF, De Wit PJGM, Joosten MHAJ (2004) *Cladosporium fulvum* circumvents the second functional resistance gene homologue at the Cf-4 locus (*Hcr9-4E*) by secretion of a stable avr4E isoform. Mol Microbiol 54:533–545

Whisson SC, Boevink PC, Moleleki L, Avrova AO, Morales JG, Gilroy EM, Armstrong MR, Grouffaud S, van West P,

Chapman S, Hein I, Toth IK, Pritchard L, Birch PRJ (2007) A translocation signal for delivery of oomycete effector proteins into host plant cells. Nature 450:115–116

Whiteford JR, Lacroix H, Talbot NJ, Spanu PD (2004) Stage-specific cellular localisation of two hydrophobins during plant infection by the pathogenic fungus *Cladosporium fulvum*. Fungal Genet Biol 41:624–634

Wilson RJ, Baillie BK, Jones DA (2005) ER retrieval of Avr9 compromises its elicitor activity consistent with perception of Avr9 at the plasma membrane. Mol Plant Pathol 6:193–197

Win J, Greenwood DR, Plummer KM (2003) Characterisation of a protein from *Venturia inaequalis* that induces necrosis in Malus carrying the *V-m* resistance gene. Physiol Mol Plant Pathol 62:193–202

Wulff BBH, Thomas CM, Smoker M, Grant M, Jones JDG (2001) Domain swapping and gene shuffling identify sequences required for induction of an Avr-dependent hypersensitive response by the tomato Cf-4 and Cf-9 proteins. Plant Cell 13:255–272

Wulff BBH, Kruijt M, Collins PL, Thomas CM, Ludwig AA, De Wit PJGM, Jones JDG (2004) Gene shuffling-generated and natural variants of the tomato resistance gene *Cf-9* exhibit different auto-necrosis-inducing activities in Nicotiana species. Plant J 40:942–956

Xing T, Higgins VJ, Blumwald E (1997) Race-specific elicitors of *Cladosporium fulvum* promote translocation of cytosolic components of NADPH oxidase to the plasma membrane of tomato cells. Plant Cell 9:249–259

Yuan YN, Haanstra J, Lindhout P, Bonnema G (2002) The *Cladosporium fulvum* resistance gene *Cf-Ecp3* is part of the Orion cluster on the short arm of tomato chromosome 1. Mol Breed 10:45–50

Zipfel C, Kunze G, Chinchilla D, Caniard A, Jones JDG, Boller T, Felix G (2006) Perception of the bacterial PAMP EF-Tu by the receptor EFR restricts *Agrobacterium*-mediated transformation. Cell 125:749–760

8 The cAMP Signaling and MAP Kinase Pathways in Plant Pathogenic Fungi

Rahim Mehrabi[1], Xinhua Zhao[1], Yangseon Kim[1], Jin-Rong Xu[1]

CONTENTS

I. Introduction.............................. 157
II. *Magnaporthe grisea* 157
   A. Surface Recognition and cAMP Signaling ... 158
   B. *PMK1* is Essential for Appressorium
      Formation and Infectious Growth.......... 159
   C. Mps1 for Penetration, Conidiation,
      and Cell Wall Integrity 160
   D. The Osm1 Pathway Regulates Osmoregulation
      but not Appressorial Penetration........... 161
III. *Ustilago maydis* 161
   A. The cAMP-PKA Pathway Regulates
      Morphogenesis and Infection 161
   B. The Kpp2/Kpp6 MAPK Cascade
      for Mating and Pathogenesis 163
IV. Other Plant Pathogenic Fungi................ 164
   A. Importance of the cAMP-PKA
      Pathway in Pathogenesis.................. 164
   B. *PMK1* Homologs are Well Conserved
      for Plant Infection 165
   C. Homologs of Mps1 MAPK Are Important
      for Plant Infection 166
   D. *HOG1* Homologs Regulate Stress Responses.... 167
V. Conclusions 167
   References 169

Abbreviations: AC, adenylate cyclase; Camp, cyclic AMP; C-PKA, catalytic subunit of PKA; MAPK, mitogen-activated protein kinase; MEK, MAPK kinase; MEKK, MEK kinase; PKA, protein kinase A; R-PKA, regulatory subunit of PKA

I. Introduction

All eukaryotic organisms must be able to sense and respond to extracellular signals for regulating various developmental and differentiation processes. The cyclic AMP (cAMP) signaling and mitogen-activated protein (MAP) kinase pathways are among the best studied signal transduction pathways in eukaryotes. The key components of the cAMP-PKA pathway include the adenylate cyclase (AC) and regulatory and catalytic subunits

[1] Department of Botany and Plant Pathology, Purdue University, West Lafayette, IN 47907, USA; e-mail: jinrong@purdue.edu

of protein kinase A (PKA). In *Saccharomyces cerevisiae*, both small GTPase Ras and trimeric G-protein Gpa2 Gα function upstream from the cAMP-PKA pathway. Adenylate cyclase is activated by Gα subunits in *Schizosaccharomyces pombe* and the model filamentous fungus *Neurospora crassa* (Kays et al. 2000). Although there are no experimental data to determine whether AC is activated by Ras or heterotrimeric G-proteins in most ascomycetous fungal pathogens, phenotype analyses of related mutants and pharmacological studies often suggest the involvement of Gα in the cAMP-PKA pathway (Gronover et al. 2001; Liu and Dean 1997).

A MAP kinase cascade normally consists of a MAP kinase (MAPK), a MAPK kinase (MEK), and a MEK kinase (MEKK). The sequential activation of the MEKK-MEK-MAPK cascade results in the activation of transcription factors and the expression of specific sets of genes in response to extracellular signals. While *Sac. cerevisiae* has five MAPK pathways that regulate mating (pheromone response), invasive growth, cell wall integrity, osmoregulation, and ascospore formation (Bardwell 2004), there are only three well conserved MAPK genes in *N. crassa* and most of plant pathogenic fungi that have been sequenced. Both the cAMP-PKA and MAPK pathways have been characterized in a number of plant pathogenic fungi. Because of advanced studies in the rice blast fungus *Magnaporthe grisea* and the corn smut fungus *Ustilago maydis,* this chapter focuses on these two model fungal pathogens.

II. *Magnaporthe grisea*

Rice blast caused by *M. grisea* is one of the most severe fungal diseases of rice. The fungus forms a specialized infection structure called appressorium. Enormous turgor pressure is generated in appressoria to physically penetrate the plant cuticle and cell wall. After penetration, the penetration peg differentiates into primary and secondary infectious

Plant Relationships, 2nd Edition
The Mycota V
H. Deising (Ed.)
© Springer-Verlag Berlin Heidelberg 2009

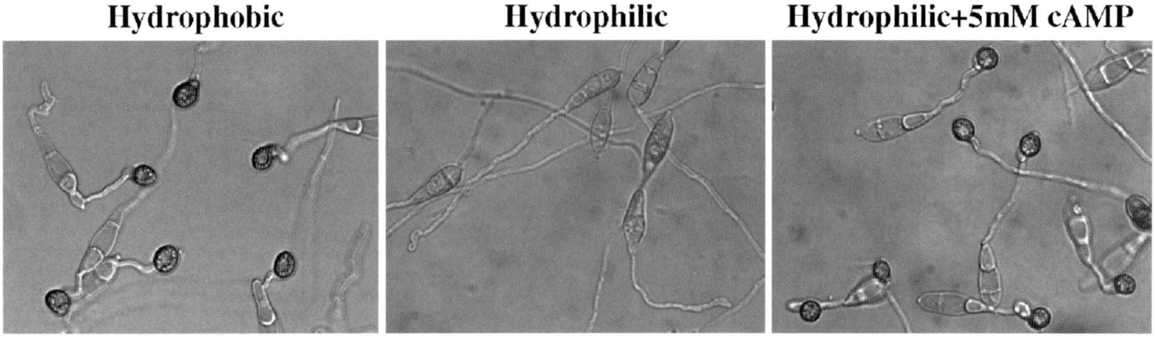

Hydrophobic   **Hydrophilic**   **Hydrophilic+5mM cAMP**

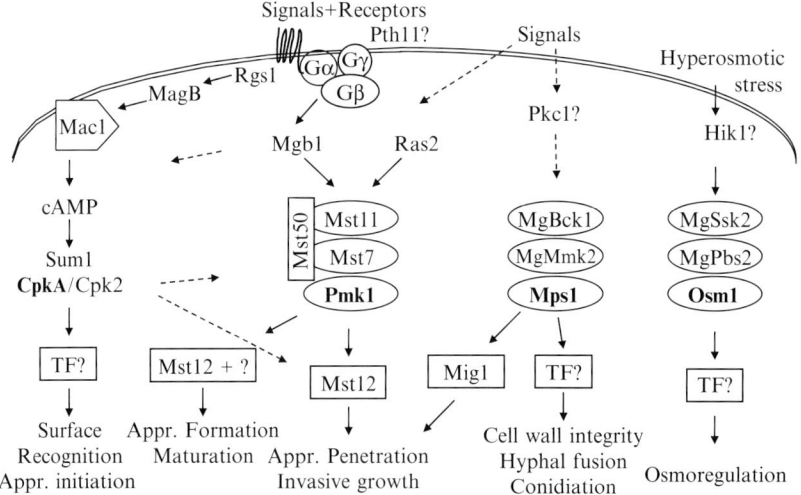

Fig. 8.1. Appressorium formation in *Magnaporthe grisea*. Conidia attached to hydrophobic surfaces germinate and form melanized appressoria. On hydrophilic surfaces, germ tubes fail to differentiate into appressoria. In the presence of 5 mM cAMP, appressorium formation can be observed on nonconducive surfaces

hyphae, which initially grow biotrophically in plant tissues (Kankanala et al. 2007). Eventually, lesions develop on infected rice leaves and the fungus produces more conidia to reinitiate the infection cycle.

A. Surface Recognition and cAMP Signaling

One major factor affecting appressorium formation in *M. grisea* is surface hydrophobicity. On rice leaves or hydrophobic artificial surfaces, the germ tube tip arrests its growth, swells, and eventually forms an appressorium. On hydrophilic surfaces, *M. grisea* forms long, slender germ tubes without tip differentiation. Although the exact surface sensing mechanism is not clear, the secondary intracellular signal is known to be cAMP. In the presence of exogenous cAMP or 3-isobutyl-1-methylxanthine (IBMX), appressorium formation can be induced on hydrophilic surfaces (Fig. 8.1). Deletion of the *MAC1* AC gene (encoding an adenylate cyclase essential for cAMP synthesis) blocks appressorium formation (Choi and Dean 1997). The Δ*mac1* mutant also is defective in the growth of aerial hyphae and conidiation, which can be suppressed by exogenous cAMP or spontaneous mutations in the regulatory subunit of PKA (R-PKA) gene *SUM1* (Fig. 8.2). The *M. grisea* genome has two genes,

Fig. 8.2. The cAMP and MAPK pathways in *M. grisea*. Surface recognition and initiation of appressorium formation is regulated by the cAMP-PKA pathway, which also plays a role in invasive growth in rice plants. The Pmk1 pathway controls appressorium formation, maturation, penetration, and invasive growth. Both Ras2 and Mgb1 may function upstream from the Pmk1 and cAMP signaling pathways but it is not clear how the surface recognition signal is transduced from cAMP signaling to the Pmk1 cascade. The Mps1 MAPK pathway is involved in cell wall integrity, invasive growth, and conidiation. Osm1 is the only MAPK in *M. grisea* that is dispensable for plant infection

CPKA and *CPK2*, that encode catalytic subunits of PKA (C-PKA). *CPKA* is dispensable for appressorium formation but required for appressorial penetration. Lipid and glycogen degradation are delayed in the Δ*cpkA* mutant during appressorium formation (Thines et al. 2000). The Δ*cpkA* mutant is nonpathogenic on healthy plants but can infect through wounds and still responds to exogenous cAMP for appressorium formation on hydrophilic surfaces. Deletion of *CPK2* has no obvious defects in growth, differentiation, and plant infection (Kim and Xu, unpublished data), suggesting that it has no important functions in *M. grisea*. In *U. maydis* and *Aspergillus nidulans*, *CPK2* homologs play only a minor role in the cAMP signaling pathway (Durrenberger et al. 1998; Ni et al. 2005).

Like many other fungal pathogens, *M. grisea* has three Gα (MagA, MagB, and MagC), one Gβ(Mgb1), and one Gγ (Mgg1) subunits. Deletion of *MAGB*, but not the other two Gα genes, significantly reduces appressorium formation and virulence but does not completely block appressorium formation and plant infection. Expression of the dominant active *MAGB*[G42R] allele stimulates appressorium formation on hydrophilic surfaces and mis-scheduled melanization at hyphal tips (Fang and Dean 2000). Unlike *MAGB*, both *MGB1* and *MGG1* are essential for appressorium formation and plant infection (Liang et al. 2006; Nishimura et al. 2003). Deletion of *MGB1* results in a reduced intracellular cAMP level and pleiotropic defects, including fluffy colonies, reduced conidiation, and failure to form appressoria and cause lesions. Exogenous cAMP induces appressorium formation in the Δ*mgb1* mutant but these appressoria are morphologically abnormal and nonfunctional for plant penetration. The *mgg1* disruptant has phenotypes similar to the Δ*mgb1* mutant (Liang et al. 2006). The *M. grisea* genome contains eight putative regulator of G-protein signaling (*RGS*) genes. Recently, one of them, *RGS1*, has been shown to physically interact with all three Gα subunits and functions as a negative regulator of G-proteins. The Δ*rgs1* deletion mutant forms appressoria on hydrophilic surfaces and has elevated levels of intracellular cAMP (Liu et al. 2007).

The G-protein-coupled receptors (GPCRs) regulate different cellular processes by association with heterotrimeric G proteins. A large number of GPCR-like genes has been identified in the *M. grisea* genome (Kulkarni et al. 2005). Twelve of these putative GPCR genes form a subfamily that contains an N-terminal extracellular membrane-spanning (CFEM) domain. One CFEM-GPCR gene, *PTH11*, is implicated in surface recognition and pathogenesis in *M. grisea* (DeZwaan et al. 1999). The Δ*pth11* mutant is nonpathogenic on healthy leaves and significantly reduced in appressorium formation. Exogenous cAMP restores appressorium formation and pathogenicity in the Δ*pth11* mutant (DeZwaan et al. 1999), suggesting that *PTH11* may be involved in surface recognition and regulating Mac1 activities. Recently, a cutinase gene, *CUT2*, was shown to function upstream from the cAMP-PKA pathway (Skamnioti and Gurr 2007). The Δ*cut2* mutant is reduced in conidiation and virulence and forms multiple elongated germ tubes and aberrant appressoria on hydrophobic surfaces. These defects are restored in the *cut2* mutant by addition of synthetic cutin monomers, cAMP, or IBMX. Hydrolytic products generated by Cut2 may serve as one of the ligands recognized by *M. grisea* receptors.

B. *PMK1* is Essential for Appressorium Formation and Infectious Growth

Although surface recognition and early stages of germ tip deformation are regulated by cAMP signaling, late stages of appressorium formation and penetration are regulated by the *PMK1* MAPK gene (Xu and Hamer 1996). The Δ*pmk1* deletion mutant fails to form appressoria but still recognizes hydrophobic surfaces and responds to cAMP for germ tube tip deformation (Xu and Hamer 1996). *PMK1* also is essential for infectious hyphal growth and the Δ*pmk1* mutant is defective in colonizing the host through wounds. In transformants expressing the GFP-*PMK1* fusion construct, a weak GFP signal is detectable in vegetative hyphae, conidia, and germ tubes. Increased expression level and nuclear localization of the GFP-Pmk1 fusion are observed in appressoria and developing conidia (Bruno et al. 2004).

Several upstream components of the *PMK1* pathway, including the MEKK Mst11 and MEK Mst7, and an Ste50 homolog Mst50 (Fig. 8.2), have been characterized in *M. grisea* (Park et al. 2006; Zhao et al. 2005). Like the Δ*pmk1* mutant, the Δ*mst7*, Δ*mst11*, and Δ*mst50* mutants fail to form appressoria and are nonpathogenic. Both Mst11 and Mst50 have one sterile alpha motif (SAM) and one Ras-association (RA) domain. Deletion analyses indicate that SAM but not RA is essential for the function of Mst11 and

Mst50 (Park et al. 2006; Zhao et al. 2005). Mst50 directly interacts with both Mst7 and Mst11 and may function as an adaptor protein for the Mst11-Mst7-Pmk1 cascade. The interaction between Mst50 and Mst11 is mediated by the SAM domain (Zhao et al. 2005). However, the interaction site of Mst50 with Mst7 has not been determined. Mst50 also interacts with Mgb1 and MgCdc42 in yeast two-hybrid assays and may serve as the converging point for multiple upstream signals (Park et al. 2006).

In *M. grisea*, deletion of either *MST20* or *CHM1*, the only two PAK kinase genes in the genome, does not abolish appressorium formation (Li et al. 2004), suggesting that, unlike the yeast mating or filamentation pathway, the Pmk1 MAPK cascade is not activated by PAK kinases. Mst50 and Mst11 both interact with Ras1 and Ras2, two Ras proteins in *M. grisea*. Co-immunoprecipitation (co-IP) assays have shown that Mst11 interacts with Ras2 in vivo, and a dominant active mutation of Ras2 enhances its interaction with Mst11 (Kim and Xu, unpublished data). The $\Delta ras1$ mutant has no defect in plant infection and appressorium formation, but *RAS2* appears to be an essential gene. Expression of a dominant active *RAS2*^{G18V} allele in the wild type but not in the $\Delta pmk1$ or $\Delta mst7$ mutant stimulated abnormal appressorium formation on hydrophilic surfaces (Park et al. 2006). Elevated intracellular cAMP levels and melanized appressorium-like structures on aerial hyphae and conidiophores were observed in transformants expressing the *RAS2*^{G18V} allele (Zhao and Xu, unpublished data). Therefore, *RAS2* may function upstream of both the cAMP-PKA and *PMK1* pathways (Fig. 8.2). Because the $\Delta mgb1$ mutant is defective in appressorium formation and invasive growth (Nishimura et al. 2003) and Mgb1 directly interacts with Mst50, *MGB1* may also be involved in the *PMK1* pathway to regulate appressorial penetration and invasive growth (Nishimura et al. 2003).

MST12 is homologous to yeast *STE12* and encodes a putative transcription factor regulated by Pmk1. The $\Delta mst12$ deletion mutant is nonpathogenic but still forms melanized appressoria. It fails to develop penetration pegs due to cytoskeleton defects in mature appressoria (Park et al. 2004a). *MST12* may function downstream of *PMK1* to regulate genes involved in appressorial penetration and invasive growth, but other transcription factors must exist in *M. grisea* to control appressorium formation (Fig. 8.2). One REMI mutant defective in appressorium formation is disrupted in a putative transcription factor gene *PTH12* (Sweigard et al. 1998). However, the $\Delta pth12$ mutant still occasionally forms appressoria and responds to exogenous cAMP for appressorium formation (Dr. Y. Peng, personal communication). Multiple transcription factors may form heterodimers or a transcription complex to regulate appressorium formation in *M. grisea*.

To identify genes regulated by *PMK1*, subtraction libraries enriched with genes regulated by *PMK1* were constructed (Xue et al. 2002). Two genes identified in this library, *GAS1* and *GAS2*, encode small proteins that are required for full virulence and appressorium function. Comparative analysis of ESTs sequenced from cDNA libraries of wild-type appressoria and $\Delta pmk1$ germlings also have been used to identify genes that are differentially expressed in the $\Delta pmk1$ mutant (Ebbole et al. 2004). Recently, the whole-genome microarray of *M. grisea* was used to identify genes differentially expressed between 24 h appressoria formed by the wild type and germlings of the $\Delta pmk1$ mutant on hydrophobic surfaces (Ding and Xu, unpublished data). In comparison with the wild type, 1803 and 1048 genes were up- and down-regulated, respectively, at least two-fold in the $\Delta pmk1$ mutant. A number of putative transcription factors and *M. grisea*-unique genes were among these putative *PMK1*-regulated genes identified by microarray analysis. The down-regulated genes include *PTH11* and seven genes containing the putative MAPK-docking site (Zhao and Xu 2007).

C. Mps1 for Penetration, Conidiation, and Cell Wall Integrity

The *MPS1* MAPK gene is a homolog of *SLT2* which regulates cell wall integrity in yeast. The $\Delta mps1$ mutant is significantly reduced in aerial hyphal growth and conidiation, but it has no obvious changes in the growth rate (Xu et al. 1998). Unlike Pmk1, Mps1 is dispensable for appressorium formation but essential for appressorial penetration and plant infection. Appressoria formed by the $\Delta mps1$ mutant fail to penetrate and develop infectious hyphae but still elicit plant defense responses, such as papilla formation and autofluorescence. Vegetative hyphae of the $\Delta mps1$ mutant have a weakened cell wall and are hypersensitive to cell wall lytic enzymes (Xu et al. 1998). The $\Delta mps1$ mutant has increased sensitivities to oxidative stress and the plant defensin MsDef1 (Mehrabi and Xu, unpublished data), suggesting that it is defective in overcoming plant defense responses.

Many components of the yeast Pkc1-Slt2 pathway, including *BCK1*, *MMK2*, *RLM1*, *SWI4*, and *SWI6*, have distinct homologs in the *M. grisea* genome (Dean et al. 2005). Mutants deleted of *BCK1* or *MMK2* also were defective in cell wall integrity, conidiation, appressorial penetration, and plant infection (Kim and Xu, unpublished data). However, the Mig1 transcription factor, a homolog of yeast Rlm1, is dispensable for aerial hyphal growth and cell wall integrity in *M. grisea* (Mehrabi et al. 2008). Nevertheless, like the Δ*mps1* mutant, the Δ*rlm1* deletion mutant is nonpathogenic and defective in overcoming plant defense response. To date, no target genes regulated by the Mps1 pathway have been characterized in *M. grisea*. Recently, the *CATB* catalase gene was shown to play a role in strengthening the cell wall during plant penetration (Skamnioti et al. 2007), but it is not clear whether *CATB* is regulated by *MPS1*.

D. The Osm1 Pathway Regulates Osmoregulation but not Appressorial Penetration

Osm1 is the only MAPK that is dispensable for plant infection in *M. grisea* (Dixon et al. 1999). Although the Δ*osm1* mutant is hypersensitive to desiccation and hyperosmotic stress, it has no defect in conidiation, appressorium formation, turgor generation, and penetration. In the presence of 0.4 M NaCl, the *osm1* mutant forms multiple appressoria (Dixon et al. 1999), suggesting that *OSM1* suppresses inappropriate activation of the *PMK1* pathway under hyperosmotic conditions. Deletion of the *PBS2* or *SSK2* homolog or a histidine kinase gene *HIK1* in *M. grisea* also has no effect on appressorium formation and virulence (Motoyama et al. 2005; Zhao et al. 2005). The Δ*mgpbs2* and Δ*mgssk2* mutants, like the Δ*osm1* mutant, have increased sensitivities to hyperosmotic stress. Interestingly, the *hik1* mutant is hypersensitive to high concentrations of sugars but not salts, suggesting that *M. grisea* can recognize different hyperosmotic stresses.

In comparison with the wild type, the Δ*mst11* and Δ*mst50* deletion mutants are also more sensitive to hyperosmotic stresses (Park et al. 2006; Zhao et al. 2005). Transformants of the Δ*mst11* or Δ*mst50* mutant expressing the dominant active *MST7* allele are not fully restored in its osmoregulation defects. Therefore, *MST11* and *MST50* may cross-talk with and play a role in the *OSM1* pathway. Recently, a novel protein encoded by the *MGA1* gene was shown to play a role in glycerol accumulation and lipid metabolism (Gupta and Chattoo 2007). The Δ*mga1* mutant has a reduced glycerol level under hyperosmotic stress conditions and increased sensitivity to hyperosmotic stress. The relationship between the *OSM1* pathway and *MGA1* is not clear but the *mga1* mutant is nonpathogenic.

In summary, the cAMP-PKA pathway regulates surface recognition and the initiation of appressorium formation in *M. grisea*. It may also be involved in turgor generation and appressorial penetration (Fig. 8.2). The *PMK1* pathway is required for appressorium formation, penetration, and invasive growth. For the other two MAPK genes, *MPS1* but not *OSM1* is essential for plant infection. Defects in cell wall integrity and overcoming plant defense responses may be responsible for the loss of pathogenicity in the *mps1* mutant (Fig. 8.2).

III. *Ustilago maydis*

The corn smut fungus *U. maydis* is a facultative biotrophic pathogen that has been extensively studied for signal transduction pathways regulating mating responses and pathogenesis. The unicellular yeast form is nonpathogenic and can grow saprophytically by budding. Fusion of compatible haploid yeast cells leads to the development of dikaryotic hyphae that are obligately biotrophic and pathogenic on corn plants.

A. The cAMP-PKA Pathway Regulates Morphogenesis and Infection

The role of cAMP signaling (Fig. 8.3) in the yeast–hypha growth transition and pathogenic development is well characterized in *U. maydis* (Kahmann and Kämper 2004; Lee et al. 2003). In general, mutants with high PKA activities have a budding phenotype and those with low PKA activities display filamentous growth. However, filamentous development in vitro does not reflect the ability to grow biotrophically in planta or cause tumors. The Δ*uac1* (AC) mutant is nonpathogenic and displays a constitutively filamentous growth phenotype, which can be reversed to budding by exogenous cAMP or suppress mutations in the *ubc1* (R-PKA) gene (Gold et al. 1994). The Δ*adr1* (C-PKA) mutant has similar defects in growth and plant infection with the Δ*uac1* mutant. However, deletion of the *uka1* gene that encodes the other C-PKA in *U. maydis* has no obvious defects in morphology and virulence, indicating (Durrenberger et al. 1998). The Δ*ubc1* mutant is defective in the separation of mother and daughter cells and has altered bud site selection. It has elevated transcription levels for genes involved in phosphate acquisition and storage, suggesting a connection between cAMP signaling and phosphate metabolism (Larraya et al. 2005).

homologs are well conserved in other fungi but none of them is described as a MAPK.

In summary, both cAMP-PKA and pheromone-responsive MAPK pathways regulate mating, filamentous growth, and pathogenesis in *U. maydis* (Fig. 8.3). Nutritional, pheromone, and plant signals must be recognized and integrated to mediate responses to specific stimuli by these pathways. During plant infection, lipids may represent one of the signals that promote and maintain filamentous growth via both the cAMP-PKA and MAPK pathways (Klose et al. 2004).

IV. Other Plant Pathogenic Fungi

A. Importance of the cAMP-PKA Pathway in Pathogenesis

Components of the cAMP-PKA pathway have been characterized in various plant pathogenic fungi (D'Souza and Heitman 2001; Lee et al. 2003). Although deletion of genes encoding the AC, C-PKA, R-PKA, Gα, or Gβ subunits often results in pleiotropic defects, reduced virulence, or loss of pathogenicity is commonly observed in these mutants. In appressorium-forming fungi, the role of the cAMP-PKA pathway in appressorium formation and invasive growth appears to be specific for each fungal pathogen. In *Cochliobolus heterostrophus,* deletion of the *CGA1* Gα gene significantly reduces appressorium formation, but the Δ*cgb11* mutant is blocked in conidiation and appressorium formation (Ganem et al. 2004). Because appressorium formation is not essential for plant infection in *C. heterostrophus*, only the *cgb1* mutant that has cell death-related defects is significantly reduced in virulence. In *Colletotrichum lagenarium*, deletion of the *CPK1* C-PKA, *RPK1* R-PKA, or *CAC1* AC gene has no effect on appressorium formation on plant surfaces (Takano et al. 2001; Yamauchi et al. 2004). However, these mutants are nonpathogenic on intact cucumber plants and defective in appressorial penetration and lipid metabolism during appressorium formation. Unlike the Δ*cpk1* and Δ*cac1* mutants that fail to infect through wounds, the *rpk1* mutant can form lesions on wounded cucumber plants (Takano et al. 2001; Yamauchi et al. 2004). The Δ*rpk1* mutant is more severely reduced in vegetative growth and conidiation than the Δ*cpk1* or Δ*cac1* mutant. In *C. trifolii*, the Δ*Ct-pakc* mutant fails to infect intact alfalfa plants although it has only a slight delay in

conidium germination and appressorium formation (Yang and Dickman 1999). Unlike the *C. lagenarium* Δ*cpk1* mutant, the Δ*Ct-pakc* mutant still colonizes host tissues through wounds, indicating that the role of C-PKA in invasive growth after penetration also is fungus-specific.

In necrotrophic pathogens, including *Cryphonectria parasitica*, *Fusarium oxysporum*, *Botrytis cinerea*, *Alternaria alternata*, and *Sclerotinia sclerotiorum*, the cAMP-PKA pathway is also involved in various differentiation and plant infection processes (see Chap. 2). In *C. parasitica*, deletion of the *CPG-1* Gα gene results in phenotypes that are similar to but more severe than those associated with hypovirus infection, including reduced vegetative growth and loss of virulence and sporulation (Nuss 2005). Deletion or expression a constitutive active allele of *CPG-1* abolishes conidiation. The Δ*cpgb-1* Gβmutant is also reduced in pigmentation and virulence but increased in vegetative growth. One RGS gene, *CPRGS-1*, is implicated in *CPG-1*-mediated in *C. parasitica*. Deletion of *CPRG-1*, results in reduced protein levels of *CPG-1* and *CPGB-1* and loss of pigmentation, sporulation, and virulence (Segers et al. 2004).

In *F. oxysporum*, the Δ*fga2* Gα mutant fails to cause wilt symptoms but the Δ*fgb1* Gβmutant retains a reduced virulence on tomato plants. The Δ*fgb1* mutant also displays an abnormal hyphal growth phenotype that can be partially suppressed by exogenous cAMP (Delgado-Jarana et al. 2005). In *B. cinerea*, mutants with a deleted *BCG1* Gα gene have pleiotropic defects in colony morphology, secretion of extracellular proteases, and plant infection (Gronover et al. 2001). Although conidia germination and penetration of plant tissues is normal, the infection process stops after the formation of primary lesions in the Δ*bcg1* mutant. The Δ*bac* AC mutant has morphological defects similar to the Δ*bcg1* mutant and is defective in sporulation on infected plants (Klimpel et al. 2002). Exogenous cAMP can partially suppress the phenotype of both the Δ*bac* and Δ*bcg1* mutants. However, unlike the Δ*bcg1* mutant, the Δ*bac* mutant still causes spreading soft rot lesions (Klimpel et al. 2002). In *A. alternata*, deletion of the *AGA1* Gα gene does not affect the lesion size but significantly reduces the number of lesions formed on apple leaves. The production of a host-selective toxin, AM toxin, is unaltered in the Δ*aga1* mutant (Yamagishi et al. 2006). In *S. sclerotiorum*, Sac1 adenylate cyclase is important for regulating sclerotium formation and infection cushion formation. The *sac1* mutant is reduced in growth rate and nonpathogenic on intact leaves but it still infects mechanically wounded tissues (Jurick and Rollins 2007).

In the hemibiotroph *Mycospharella graminicola*, both the *MgTPK2* (C-PKA) and *MgBCY1* (R-PKA) genes are dispensable for penetration through stomata and plant tissue colonization (Mehrabi and Kema 2006). However, the Δ*mgtpk2* and Δ*mgbcy1* deletion mutants are defective in pycnidium formation and fail to produce pycnidiospores. In *Stagonospora nodorum*, a close relative of *M. graminicola*, deletion of the *GNA1* Gα gene also blocks pycnidium formation in planta (Solomon et al. 2004). The Δ*gna1* mutant is still pathogenic but has a defect in direct penetration and reduced virulence. In the obligate biotroph *Erysiphe graminis*, treatment with cholera toxin and forskolin (activators of AC) significantly increases the frequency of primary and appressorium germ tubes on cellulose membranes. However, exogenous cAMP affects appressorium formation only and not the emergence of primary germ tubes (Kinane and Oliver 2003).

In some fungi, cAMP signaling also plays a critical role in conidium germination (Osherov and May 2001), which is an important early step in plant infection for fungal pathogens. Conidium germination is delayed in the Δ*cga1* mutant of *C. heterostrophus* but completely blocked in the Δ*cpk1* and Δ*cac1* mutants on artificial surfaces in *C. lagenarium* (Ganem et al. 2004; Yamauchi et al. 2004). The Δ*mgtpk2* mutant of *M. graminicola* is normal in conidium germination on water agar but exhibits an intensive micro-conidiation pattern on rich medium (Mehrabi and Kema 2006). Interestingly, conidium germination is more rapid in the Δ*fgb1* mutant of *F. oxysporum*, which has a reduced level of intracellular cAMP (Jain et al. 2003). The cAMP-PKA pathway is also important for virulence in human pathogens such as *Candida albicans*, *Cryptococcus neoformans*, and *Aspergillus fumigatus*. In *C. neoformans*, capsule formation, melanin production, growth at elevated temperatures, and yeast–hypha transition are well known pathogenicity factors. While the Δ*pka1* C-PKA mutant of serotype A is defective in capsule formation and avirulent, the Δ*pkr1* R-PKA mutant produces enlarged capsules and is hyper-virulent (D'Souza and Heitman 2001). Both Gpa1 Gα and Cac1 AC function upstream of Pkr1-Pka1 and are

involved in melanin production and capsule formation. *GPR1* encodes a GPCR that interacts with Gpa1. The *gpr4* deletion mutant is attenuated in capsule production but retains normal virulence (Xue et al. 2006).

B. *PMK1* Homologs are Well Conserved for Plant Infection

Homologs of *PMK1* have been characterized in over a dozen phytopathogenic fungi (Zhao et al. 2007). In all the appressorium-forming pathogens studied, including *C. heterostrophus*, *C. lagenarium*, *C. gloeosporioides*, and *Pyrenophora teres*, this MAPK pathway is essential for appressorium formation. In *C. lagenarium*, the Δ*cmk1* mutant is reduced in conidiation and defective in conidium germination (Takano et al. 2000). It also fails to colonize wounded tissues. The similarity in phenotypes of the Δ*cmk1* and Δ*cpk1* or Δ*cac1* mutants (Yamauchi et al. 2004) suggest that cAMP signaling and Cmk1 MAPK pathways cooperatively control developmental and infection processes in *C. lagenarium*. In *C. heterostrophus*, *CHK1* is required for hyphal pigmentation, appressorium formation, female fertility, and full virulence (Lev et al. 1999). The *CGB1* Gβ gene is also essential for appressorium formation and may function upstream from *CHK1* (Ganem et al. 2004). Even in the biotroph *C. purpurea* or the hemibiotroph *M. graminicola* that do not form appressoria, the *PMK1* homolog is required for the penetration and colonization of plant tissues (Cousin et al. 2006; Mey et al. 2002b). In *M. graminicola*, *MgFUS3* is also important for hyphal melanization and pycnidium formation. The Δ*mgfus3* mutant fails to colonize the mesophyll tissue through stomata (Cousin et al. 2006).

PMK1 homologs are also important for pathogenesis in several necrotrophic ascomycetes, including *Alternaria brassicicola*, *Bipolaris oryzae*, *B. cinerea*, *C. parasitica*, *Fusarium graminearum*, *F. oxysporum*, and *S. nodorum* (Cho et al. 2007; Zhao et al. 2007). In *F. oxysporum*, the Δ*fmk1* mutant is defective in root attachment and the differentiation of penetration hyphae. It is nonpathogenic on tomato plants but normal in virulence in infection assays with an immunodepressed mouse model (Ortoneda et al. 2004), suggesting different mechanisms are involved in plant and animal infection. Unlike observations in *C. heterostrophus* and *M. grisea*, the *FGB1* Gβ gene appears to function upstream of the cAMP-PKA pathway, but not the Fmk1 MAPK pathway. The Δ*fgb1* mutant has an unaltered Fmk1 phosphorylation level (Delgado-Jarana et al. 2005). In *F. graminearum*, the Δ*gpmk1* mutant is impaired in its colonization of flowering wheat heads and spreads from inoculated florets to neighboring spikelets. Gpmk1 regulates the early induction of extracellular endoglucanase, xylanolytic, and proteolytic activities and is responsible for the overall induction of secreted lipolytic activities (Jenczmionka and Schafer 2005). One of the genes regulated by Gpmk1 is *FGL1*, which encodes a secreted lipase and is an important virulence factor in *F. graminearum*. In *A. brassicicola*, *AMK1* is required for plant penetration and increased expression of several hydrolytic enzymes during plant infection (Cho et al. 2007). The Δ*amk1* disruption mutant is nonpathogenic on intact host plants but can colonize through wounds in the presence of nutrient supplements. It is also defective in conidiogenesis and fails to produce mature conidia.

In *S. nodorum* and *C. parasitica*, deletion of the *PMK1* homolog significantly reduces virulence but does not completely block plant infection. The Δ*mak2* mutant of *S. nodorum* fails to form penetration structures but can enter through natural openings and cause limited necrosis (Solomon et al. 2005). Cankers caused by the Δ*cpmk2* mutant in *C. parasitica* are smaller than those caused by the wild type or a hypovirulent strain (Choi et al. 2005). Unlike the cAMP-PKA pathway, activation of *CpMK2* is not affected by hypovirus infection. However, *CpSTE12* is down-regulated by hypovirus infection. Many genes affected by hypovirus infection are also down-regulated in the Δ*cpste12* mutant (Deng et al. 2007), suggesting that *CpSTE12* functions downstream of the cAMP-PKA pathway.

Unlike its role in plant pathogens, this well conserved MAPK pathway plays no or only a minor role pathogenesis in human pathogens (Alonso-Monge et al. 2006). In *C. neoformans*, the Cpk1 MAPK is important for mating but is dispensable for pathogenesis or switching between the yeast form and hyphal growth. In *C. albicans*, mutants blocked in the Cek1 MAPK pathway are only reduced in virulence and defective in yeast–hypha switching on certain media (Alonso-Monge et al. 2006). In the mycoparasitic fungus *Trichoderma viren*, the Δ*tmkA* mutant is less effective in colonizing the sclerotia of *Rhizoctonia solani* and fails to parasitize *Sclerotium rolfsii* (Mukherjee et al. 2003). However, deletion of the same MAPK gene in another isolate increases the effectiveness of disease control against *R. solani* and *Pythium*.

C. Homologs of Mps1 MAPK Are Important for Plant Infection

Functional characterization of the *MPS1* homologs in a few plant pathogenic fungi, including *C. lagenarium*, *C. purpurea*, *F. graminearum*, *M. graminicola*, and *B. cinerea*, indicates this MAPK pathway is well conserved for regulating plant infection processes (Zhao et al. 2007; Chap. 2). However, the exact function of this MAPK pathway varies among different plant pathogens. While deletion of *MAF1* blocks appressorium formation in *C. lagenarium* (Kojima et al. 2002), like *MPS1* in *M. grisea*, *ChMPS1* is dispensable for appressorium formation but essential for penetration in *C. heterostrophus* (Eliahu et al. 2007). In *C. purpurea*, *CMPK2* is also necessary for plant penetration (Mey et al. 2002a). In contrast, the Δ*mgslt2* mutant of *M. graminicola* is normal in penetrating stomata but fails to colonize plant tissues after penetration (Mehrabi et al. 2006a). In *F. graminearum*, the accumulation of deoxynivalenol, a virulence factor, is greatly reduced in wheat kernels infected by the Δ*mgv1* mutant. In vitro assays showed the Δ*mgv1* mutant is hypersensitive to plant defensin MsDef1 (Ramamoorthy et al. 2007).

In addition to its conserved role in pathogenesis, the *MPS1* homolog has species-specific functions in cell wall integrity, conidiation, and stress responses. Like the *mps1* mutant in *M. grisea*, the *C. purpurea cpmk2*, *F. graminearum mgv1*, and *C. heterostrophus cpmps1* mutants have a weakened cell wall and increased susceptibility to cell wall lytic enzymes or stressors (Eliahu et al. 2007; Hou et al. 2002; Mey et al. 2002a). In contrast, deletion of the *SLT2* homolog has no obvious effect on cell wall integrity in *M. graminicola* and *B. cinerea* (Mehrabi et al. 2006a; Rui and Hahn 2007). However, the Δ*mgslt2* mutant of *M. graminicola* is hypersensitive to several azole fungicides. In *B. cinerea*, the Δ*bmp3* mutant has increased sensitivity to paraquat and phenylpyrrole fungicide fludioxonil but not to the azoles. The *SLT2* homolog is important for conidiation in *C. lagenarium* and *C. purpurea* but not in *F. graminearum* and *M. graminicola* (Hou et al. 2002; Mehrabi et al. 2006a). The Δ*bmp3* mutant produces fewer macroconidia but more microconidia than the wild type (Rui and Hahn 2007). It is also defective in sclerotium formation and has a reduced growth rate on solid media. In *F. graminearum*, Mgv1 is important for normal hyphal growth and essential for hyphal fusion and heterokaryon formation (Hou et al. 2002). These observations indicate that *MPS1* homologs are involved in various important developmental and infection processes in fungal pathogens. Unfortunately, genes regulated by this MAPK pathway have not been characterized in any plant pathogenic fungus. In the human pathogens *C. albicans* and *C. neoformans,* this MAPK pathway is also important for virulence, probably mainly by regulating cell wall integrity, stress responses, and growth at elevated temperatures (Alonso-Monge et al. 2006).

D. *HOG1* Homologs Regulate Stress Responses

Although less studied than the other two MAPKs, the *HOG1* homolog has also been characterized in several fungal pathogens. In general, it plays a critical role in responses to hyperosmotic stress. However, its function in plant infection varies among fungal pathogens. Like the Δ*osm1* mutant, the Δ*osc1* mutant of *C. lagenarium* and the Δ*srm1* mutant of *B. oryzae* are fully pathogenic (Kojima et al. 2004; Moriwaki et al. 2006). In *C. parasitica*, Δ*cpmk1* mutants are reduced in virulence and form smaller cankers than the wild type (Park et al. 2004b). In contrast, Δ*mghog1* mutants of *M. graminicola* and Δ*bcsak1* mutants of *B. cinerea* are nonpathogenic (Mehrabi et al. 2006b; Segmüller et al. 2007). The Δ*mghog1* mutant fails to switch to filamentous growth on water agar and is defective in melanization and the formation of infectious germ tubes. In *B. cinerea*, Δ*bcsak1* mutants are blocked in appressorium formation and plant penetration. In several plant pathogenic fungi, this MAPK pathway is also involved in responses to oxidative stress and UV irradiation (Moriwaki et al. 2006; Zhao et al. 2007). It appears that the HOG pathway is conserved in plant pathogens for regulating responses to a variety of stresses. Differences in infection mechanisms or host defensive responses may contribute to the fungus-specific role of this MAPK in pathogenesis.

In several filamentous ascomycetes, including *N. crassa, C. lagenarium, M. grisea,* and *M. graminicola,* mutants blocked in the HOG pathway are resistant to phenylpyrrole, dicarboximide, and aromatic hydrocarbon fungicides. Treatments with these fungicides stimulate the activation of Hog1 homologs, glycerol accumulation, and cell burst, indicating that fungicidal effects may be related to the over-stimulation of this MAPK pathway (Kojima et al. 2004; Zhang et al. 2002). Even in *C. neoformans*, a basidiomycete, mutants blocked in the HOG pathway are resistant to fludioxonil. The HOG pathway also regulates responses to hyperosmotic, oxidative, and other stresses, and is required for full virulence in both *C. neoformans* and *C. albicans* (Alonso-Monge et al. 2006). In another human pathogen *A. fumigatus*, two closely related MAPKs, SakA and MpkC, are similar to Hog1 and have the TGY phosphorylation motif. While SakA is required for responses to heat, hyperosmotic, and oxidative stresses, MpkC is dispensable for stress responses (Reyes et al. 2006). The Δ*sakA* mutant, but not Δ*mpkC*, is resistant to fludioxonil.

V. Conclusions

Plant pathogenic fungi have a typical cAMP-PKA pathway. Unlike *Sac. cerevisiae*, most fungal pathogens have only two genes encoding catalytic subunits of PKA. In general, one of them plays a major role in cAMP signaling and is important for plant infection, such as *CPKA* in *M. grisea* and *adr1* in *U. maydis* (Durrenberger et al. 1998). The other catalytic subunit, such as *CPK2* in *M. grisea*, is dispensable for pathogenesis. However, the Δ*cpkA* Δ*cpk2* double mutant appears to be nonviable (Kim and Xu, unpublished data) and the *cpkA* mutant still responds to exogenous cAMP, indicating that *CPK2* has overlapping functions with *CPKA* in *M. grisea*. Thus, care must be exercised to interpret the phenotype of mutants deleted of a single gene. This is also true for phenotypic analysis of mutants disrupted in genes encoding Ras or trimeric G-proteins. Most plant pathogenic fungi, except *U. maydis* and *S. nodorum,* have three Gα genes. In *N. crassa* and *C. parasitica*, the deletion of one subunit of trimeric G-proteins affects the expression or protein level of other subunits. To date, the regulatory subunit of PKA is the only known cAMP-binding protein in fungal pathogens. However, it remains possible that filamentous fungi have putative cAMP receptors that are similar to cAR1 or cAR2 of Dictyostelium (Borkovich et al. 2004) or additional intracellular targets. Downstream targets of the cAMP-PKA pathway are also not well studied in phytopathogenic fungi. To better understand the role of cAMP signaling in fungal pathogenesis, it will be important to identify and characterize the signal inputs of and transcriptional networks controlled by this conserved pathway.

Table 8.1. Components of the cAMP-PKA pathway in several phytopathogenic fungi. *AC* Adenylate cyclase, *C-PKA* catalytic subunit of PKA, *Gα* G-protein alpha subunit, *Gβ*G-protein beta subunit, *Gγ* G-protein gama subunit, *R-PKA* regulatory subunit of A, *RGS* regulator of G-protein signaling, *TF* transcription factor

Fungus	Gene	Function	Major biological processes involved
Colletotrichum			
lagenarium	CAC1	AC	Germination, appressorium penetration, pathogenicity
	CPK1	C-PKA	Germination, appressorium penetration, pathogenicity
	RPK1	R-PKA	Growth, conidiation, penetration, pathogenicity
Magnaporthe			
grisea	MAGA	Gα	None
	MAGB	Gα	Appressorium formation, pathogenicity
	MAGC	Gα	Minor role in conidiation
	MGB1	Gβ	Appressorium formation, invasive growth, conidiation
	MGG1	Gγ	Appressorium formation, invasive growth, conidiation
	MAC1	AC	Appressorium formation, pathogenicity, hyphal growth
	SUM1	R-PKA	Uncharacterized
	CPKA	C-PKA	Appressorium penetration, virulence
	RGS1	RGS	Asexual growth, pathogenicity, surface recognition
Cochliobolus			
heterostrophus	CGA1	Gα	Female fertility, appressorium formation, virulence
	CGB1	Gβ	Female fertility, appressorium formation, pathogenicity
Fusarium			
oxysporum	FGB1	Gβ	Pathogenicity, germination, hyphal growth, conidiation
	FGA1	Gα	Pathogenicity, conidiation
	FGA2	Gα	Pathogenicity, conidiation, colony morphology
Mycospharella			
graminicola	MgTPK2	C-PKA	Pathogenicity, pycnidium formation, filamentation
	MgBCY1	R-PKA	Pathogenicity, pycnidium formation, filamentation
	MgGPA1	Gα	Pathogenicity, microcycle conidiation
	MgGPA2	Gα	No role found
	MgGPA3	Gα	Positive regulator of filamentation, pathogenicity
	MgGPB1	Gβ	Anastomosis, pathogenicity
Cryphonectria			
parasitica	CPG-1	Gα	Sporulation, growth rate, pathogenicity, female fertility
	CPG-2	Gα	Growth rate, sporulation
	CPG-3	Gα	Not studied
	CPGB-1	Gβ	Sporulation, pathogenicity, pigmentation
Ustilago maydis	GPA1	Gα	None
	GPA2	Gα	None
	GPA3	Gα	Mating, pathogenicity, cell morphology
	GPA4	Gα	None
	BPP1	Gβ	Dimorphic switch, pheromone response
	ADR1	C-PKA	Dimorphic switch, pathogenicity
	UKA1	C-PKA	None
	UBC1	R-PKA	Morphogenesis, pathogenicity
	UAC1	AC	Dimorphic switch, pathogenicity
	PRF1	TF	Mating and pathogenicity
	HGL1	TF	Dimorphic switch, teliospore formation
	SQL1	TF	Suppression of *gpa3Q206L* phenotype

Most filamentous fungi have three well conserved MEKK-MEK-MAPK cascades. However, the upstream components and downstream transcription factors of these MAPK cascades are not well characterized. Fungal pathogens may have specific receptors for recognizing various host and environmental signals to regulate different plant penetration and colonization processes. Another area that is not well studied in fungal pathogens is the specificity of these three parallel MAPK pathways, which have multiple functions in pathogenesis and fungal differentiation. Therefore, each pathway must have specific scaffold or adaptor proteins, such as Ste5 in *Sac. cerevisiae*, to confer specificities. To date, a number of fungal pathogens have been sequenced, including *M. grisea*, *U. maydis*, *F. graminearum*, and *S. nodorum*. Comparative and functional genomic studies will be useful to identify and characterize downstream targets or transcriptional regulatory networks of these MAPK pathways and better understand their functions in plant infection.

Last, but not least, the interaction between these MAPK pathways and their relationship with the cAMP-PKA or other signaling pathways also needs to be further characterized in plant pathogenic fungi. Several studies have indicated possible cross-talk between MAPK pathways, such as the formation of multiple appressoria by the *M. grisea osm1* mutant in the presence of 0.4 M NaCl (Dixon et al. 1999) and the co-regulation of transcription factor Cmr1 and melanin synthesis by Chk1 and Mps1 in *C. heterostrophus* (Eliahu et al. 2007). The cAMP-PKA and MAPK pathways regulate similar developmental and infection processes or infection-related morphogenesis in a number of fungal pathogens (Kinane and Oliver 2003; Lee et al. 2003). However, molecular mechanisms governing the interactions among different signal transduction pathways are not clear and remain to be illustrated in most plant pathogens. For example, it is not clear how the surface recognition signal mediated by the cAMP-PKA pathway is transduced to the Pmk1 MAPK cascade for regulating appressorium formation in *M. grisea*. In *U. maydis*, both the upstream Ras2 and downstream transcription factor Prf1 are functionally related to the cAMP-PKA and MAPK pathways (Lee and Kronstad 2002). Further characterization of these signal transduction pathways and their interactions in representative phytopathogenic fungi will improve our understanding of regulatory mechanisms important for fungal development and pathogenesis.

Acknowledgements: We thank Drs. Larry Dunkle and Stephen Goodwin for critical reading of the manuscript. We also thank Dr. Youling Peng for communicating with us on unpublished data. Research in J.X.'s laboratory on M. grisea is supported by grants 2005-35319-16073 and 2007-35319-102681 from the National Research Initiative of the USDA Cooperative State Research, Education and Extension Service.

References

Alonso-Monge R, Roman E, Nombela C, Pla J (2006) The MAP kinase signal transduction network in *Candida albicans*. Microbiology 152:905–912

Bardwell L (2004) A walk-through of the yeast mating pheromone response pathway. Peptides 25:1465–1476

Borkovich KA, Alex LA, Yarden O, et al (2004) Lessons from the genome sequence of *Neurospora crassa*: Tracing the path from genomic blueprint to multicellular organism. Microbiol Mol Biol Rev 68:1–108

Brachmann A, Schirawski J, Müller P, Kahmann R (2003) An unusual MAP kinase is required for efficient penetration of the plant surface by *Ustilago maydis*. EMBO J 22:2199–2210

Bruno KS, Tenjo F, Li L, Hamer JE, Xu JR (2004) Cellular localization and role of kinase activity of *PMK1* in *Magnaporthe grisea*. Eukaryot Cell 3:1525–1532

Cho Y, Cramer RA, Kim K, Davis J, Mitchell TK, Figuli P, Pryor BA, Lawrence CB (2007) The Fus3/Kss1 MAP kinase homolog Amk1 regulates the expression of genes encoding hydrolytic enzymes in *Alternaria brassicicola*. Fungal Genet Biol 44:543–553

Choi ES, Chung HJ, Kim MJ, Park SM, Cha BJ, Yang MS, Kim DH (2005) Characterization of the *ERK* homologue CpMK2 from the chestnut blight fungus *Cryphonectria parasitica*. Microbiology 151:1349–1358

Choi WB, Dean RA (1997) The adenylate cyclase gene *MAC1* of *Magnaporthe grisea* controls appressorium formation and other aspects of growth and development. Plant Cell 9:1973–1983

Cousin A, Mehrabi R, Guilleroux M, Dufresne M, Van der Lee T, Waalwijk C, Langin T, Kema GHJ (2006) The MAP kinase-encoding gene *MgFus3* of the non-appressorium phytopathogen *Mycosphaerella graminicola* is required for penetration and in vitro pycnidia formation. Mol Plant Pathol 7:269–278

D'Souza CA, Heitman J (2001) Conserved cAMP signaling cascades regulate fungal development and virulence. FEMS Microbiol Rev 25:349–364

Dean RA, Talbot NJ, Ebbole DJ, Farman M, et al (2005) The genome sequence of the rice blast fungus *Magnaporthe grisea*. Nature 434:980–986

Delgado-Jarana JS, Martinez-Rocha AL, Roldan-Rodriguez R, Roncero MIG, Di Pietro A (2005) *Fusarium oxysporum* G-protein beta subunit Fgb1 regulates hyphal growth, development, and virulence through multiple signaling pathways. Fungal Genet Biol 42:61–72

Deng FY, Allen TD, Nuss DL (2007) Ste12 transcription factor homologue *CpST12* is down-regulated by hypovirus infection and required for virulence and female fertility of the chestnut blight fungus *Cryphonectria parasitica*. Eukaryot Cell 6:235–244

DeZwaan TM, Carroll AM, Valent B, Sweigard JA (1999) *Magnaporthe grisea* Pth11p is a novel plasma membrane protein that mediates appressorium differentiation in response to inductive substrate cues. Plant Cell 11:2013–2030

Dixon KP, Xu JR, Smirnoff N, Talbot NJ (1999) Independent signaling pathways regulate cellular turgor during hyperosmotic stress and appressorium-mediated plant infection by *Magnaporthe grisea*. Plant Cell 11:2045–2058

Durrenberger, F, Wong, K, Kronstad, J.W (1998) Identification of a cAMP-dependent protein kinase catalytic subunit required for virulence and morphogenesis in *Ustilago maydis*. Proc Natl Acad Sci USA 95:5684–5689

Durrenberger F, Laidlaw RD, Kronstad JW (2001) The *hgl1* gene is required for dimorphism and teliospore formation in the fungal pathogen *Ustilago maydis*. Mol Microbiol 41:337–348

Ebbole DJ, Jin Y, Thon M, Pan HQ, Bhattarai E, Thomas T, Dean R (2004) Gene discovery and gene expression in the rice blast fungus, *Magnaporthe grisea*: Analysis of expressed sequence tags. Mol Plant–Microbe Interact 17:1337–1347

Eliahu N, Igbaria A, Rose MS, Horwitz BA, Lev S (2007) Melanin biosynthesis in the maize pathogen *Cochliobolus heterostrophus* depends on two mitogen-activated protein kinases, Chk1 and Mps1, and the transcription factor Cmr1. Eukaryot Cell 6:421–429

Fang EGC, Dean RA (2000) Site-directed mutagenesis of the *MAGB* gene affects growth and development in *Magnaporthe grisea*. Mol Plant–Microbe Interact 13:1214–1227

Ganem S, Lu SW, Lee B, Chou D, Hadar R, Turgeon BG, Horwitz BA (2004) G-protein beta subunit of *Cochliobolus heterostrophus* involved in virulence, asexual and sexual reproductive ability, and morphogenesis. Eukaryot Cell 3:1653–1663

Garrido E, Voss U, Müller P, Castillo-Lluva S, Kahmann R, Perez-Martin J (2004) The induction of sexual development and virulence in the smut fungus *Ustilago maydis* depends on Crk1, a novel MAPK protein. Genes Dev 18:3117–3130

Gold S, Duncan G, Barrett K, Kronstad J (1994) cAMP regulates morphogenesis in the fungal pathogen *Ustilago maydis*. Genes Dev 8:2805–2816

Gronover CS, Kasulke D, Tudzynski P, Tudzynski B (2001) The role of G protein alpha subunits in the infection process of the gray mold fungus *Botrytis cinerea*. Mol Plant–Microbe Interact 14:1293–1302

Gupta A, Chattoo BB (2007) A novel gene *MGA1* is required for appressorium formation in *Magnaporthe grisea*. Fungal Genet Biol 44:1157–1169

Hartmann HA, Kahmann R, Bolker M (1996) The pheromone response factor coordinates filamentous growth and pathogenicity in *Ustilago maydis*. EMBO J 15:1632–1641

Hou Z, Xue C, Peng Y, Katan T, Kistler HC, Xu JR (2002) A mitogen-activated protein kinase gene (*MGV1*) in *Fusarium graminearum* is required for female fertility, heterokaryon formation, and plant infection. Mol Plant–Microbe Interact 15:1119–1127

Jain S, Akiyama K, Kan T, Ohguchi T, Takata R (2003) The G protein beta subunit *FGB1* regulates development and pathogenicity in *Fusarium oxysporum*. Curr Genet 43:79–86

Jenczmionka NJ, Schafer W (2005) The Gpmk1 MAP kinase of *Fusarium graminearum* regulates the induction of specific secreted enzymes. Curr Genet 47:29–36

Jurick WM, Rollins JA (2007) Deletion of the adenylate cyclase (*sac1*) gene affects multiple developmental pathways and pathogenicity in *Sclerotinia sclerotiorum*. Fungal Genet Biol 44:521–530

Kaffarnik F, Müller P, Leibundgut M, Kahmann R, Feldbrugge M (2003) PKA and MAPK phosphorylation of Prf1 allows promoter discrimination in *Ustilago maydis*. EMBO J 22:5817–5826

Kahmann R, Kämper J (2004) *Ustilago maydis*: how its biology relates to pathogenic development. New Phytol 164:31–42

Kankanala P, Czymmek K, Valent B (2007) Roles for rice membrane dynamics and plasmodesmata during biotrophic invasion by the blast fungus. Plant Cell 19:706724

Kays AM, Rowley PS, Baasiri RA, Borkovich KA (2000) Regulation of conidiation and adenylyl cyclase levels by the Galpha protein *GNA-3* in *Neurospora crassa*. Mol Cell Biol 20:7693–7705

Kinane J, Oliver RP (2003) Evidence that the appressorial development in barley powdery mildew is controlled by MAP kinase activity in conjunction with the cAMP pathway. Fungal Genet Biol 39:94–102

Klimpel A, Gronover CS, Williamson B, Stewart JA, Tudzynski B (2002) The adenylate cyclase (BAC) in *Botrytis cinerea* is required for full pathogenicity. Mol Plant Pathol 3:439–450

Klose J, de Sa MM, Kronstad JW (2004) Lipid-induced filamentous growth in *Ustilago maydis*. Mol Microbiol 52:823–835

Kojima K, Kikuchi T, Takano Y, Oshiro E, Okuno T (2002) The mitogen-activated protein kinase gene *MAF1* is essential for the early differentiation phase of appressorium formation in *Colletotrichum lagenarium*. Mol Plant–Microbe Interact 15:1268–1276

Kojima K, Takano Y, Yoshimi A, Tanaka C, Kikuchi T, Okuno T (2004) Fungicide activity through activation of a fungal signaling pathway. Mol Microbiol 53:1785–1796

Kulkarni RD, Thon MR, Pan HQ, Dean RA (2005) Novel G-protein-coupled receptor-like proteins in the plant pathogenic fungus *Magnaporthe grisea*. Genome Biol 2005:6

Larraya LM, Boyce KJ, So A, Steen BR, Jones S, Marra M, Kronstad JW (2005) Serial analysis of gene expression reveals conserved links between protein kinase A, ribosome biogenesis, and phosphate metabolism in *Ustilago maydis*. Eukaryot Cell 4:2029–2043

Lee N, Kronstad JW (2002) *ras2* controls morphogenesis, pheromone response, and pathogenicity in the fungal pathogen *Ustilago maydis*. Eukaryot Cell 1:954–966

Lee N, D'Souza CA, Kronstad JW (2003) Of smuts, blasts, mildews, and blights: cAMP signaling in phytopathogenic fungi. Annu Rev Phytopathol 41:399–427

Lev S, Sharon A, Hadar R, Ma H, Horwitz Benjamin A (1999) A mitogen-activated protein kinase of the corn leaf pathogen *Cochliobolus heterostrophus* is involved in conidiation, appressorium formation, and pathogenicity: Diverse roles for mitogen-activated protein kinase homologs in foliar pathogens. Proc Natl Acad Sci 96:13542–13547

Li L, Xue CY, Bruno K, Nishimura M, Xu JR (2004) Two PAK kinase genes, *CHM1* and *MST20*, have distinct functions in *Magnaporthe grisea*. Mol Plant–Microbe Interact 17:547–556

Liang S, Wang ZY, Liu PJ, Li DB (2006) A G gamma subunit promoter T-DNA insertion mutant – A1-412 of *Magnaporthe grisea* is defective in appressorium formation, penetration and pathogenicity. Chin Sci Bull 51:2214–2218

Liu H, Suresh A, Willard FS, Siderovski DP, Lu S, Naqvi NI (2007) Rgs1 regulates multiple G alpha subunits in *Magnaporthe* pathogenesis, asexual growth and thigmotropism. EMBO J 26:690–700

Liu S, Dean RA (1997) G Protein alpha subunit genes control growth, development, and pathogenicity of *Magnaporthe grisea*. Mol Plant–Microbe Interact 10:1075–1086

Loubradou G, Brachmann A, Feldbrugge M, Kahmann R (2001) A homologue of the transcriptional repressor Ssn6p antagonizes cAMP signalling in *Ustilago maydis*. Mol Microbiol 40:719–730

Martinez-Espinoza AD, Ruiz-Herrera J, Leon-Ramirez CG, Gold SE (2004) MAP kinase and CAMP signaling pathways modulate the pH-induced yeast-to-mycelium dimorphic transition in the corn smut fungus *Ustilago maydis*. Curr Microbiol 49:274–281

Mayorga ME, Gold SE (1998) Characterization and molecular genetic complementation of mutants affecting dimorphism in the Fungus *Ustilago maydis*. Fungal Genet Biol 24:364–376

Mayorga ME, Gold SE (2001) The ubc2 gene of Ustilago maydis encodes a putative novel adaptor protein required for filamentous growth, pheromone response and virulence. Mol Microbiol 41:1365–1379

Mehrabi R, Kema GHJ (2006) Protein kinase A subunits of the ascomycete pathogen *Mycosphaerella graminicola* regulate asexual fructification, filamentation, melanization and osmosensing. Mol Plant Pathol 7:656–677

Mehrabi R, van der Lee T, Waalwijk C, Kema GHJ (2006a) *MgSlt2*, a cellular integrity MAP kinase gene of the fungal wheat pathogen *Mycosphaerella graminicola*, is dispensable for penetration but essential for invasive growth. Mol Plant–Microbe Interact 19:389–398

Mehrabi R, Zwiers L-H, de Waard MA, Kema GHJ (2006b) *MgHog1* regulates dimorphism and pathogenicity in the fungal wheat pathogen *Mycosphaerella graminicola*. Mol Plant–Microbe Interact 19:1262–1269

Mehrabi R, Ding S, Xu JR (2008) The MADS-box transcription factor Mig1 is required for infectious growth in *Magnaporthe grisea*. Eukaryot Cell 7: 791–799.

Mey G, Held K, Scheffer J, Tenberge KB, Tudzynski P (2002a) *CPMK2*, an *SLT2*-homologous mitogen-activated protein (MAP) kinase, is essential for pathogenesis of *Claviceps purpurea* on rye: evidence for a second conserved pathogenesis-related MAP kinase cascade in phytopathogenic fungi. Mol Microbiol 46:305–318

Mey G, Oeser B, Lebrun MH, Tudzynski P (2002b) The biotrophic, non-appressorium-forming grass pathogen *Claviceps purpurea* needs a Fus3/Pmk1 homologous mitogen-activated protein kinase for colonization of rye ovarian tissue. Mol Plant–Microbe Interact 15:303–312

Moriwaki A, Kubo E, Arase S, Kihara J (2006) Disruption of *SRM1*, a mitogen-activated protein kinase gene, affects sensitivity to osmotic and ultraviolet stressors in the phytopathogenic fungus *Bipolaris oryzae*. FEMS Microbiol Lett 257:253–261

Motoyama T, Kadokura K, Ohira T, Ichiishi A, Fujimura M, Yamaguchi I, Kudo T (2005) A two-component histidine kinase of the rice blast fungus is involved in osmotic stress response and fungicide action. Fungal Genet Biol 42:200–212

Mukherjee PK, Latha J, Hadar R, Horwitz B (2003) TmkA, a mitogen-activated protein kinase of *Trichoderma* virens, is involved in biocontrol properties and repression of conidiation in the dark. Eukaryot Cell 2:446–455

Müller P, Katzenberger JD, Loubradou G, Kahmann R (2003) Guanyl nucleotide exchange factor Sql2 and Ras2 regulate filamentous growth in *Ustilago maydis*. Eukaryot Cell 2:609–617

Müller P, Leibbrandt A, Teunissen H, Cubasch S, Aichinger C, Kahmann R (2004) The G beta-subunit-encoding gene *bpp1* controls cyclic-AMP signaling in *Ustilago maydis*. Eukaryot Cell 3:806–814

Ni M, Rierson S, Seo JA, Yu JH (2005) The *pkaB* gene encoding the secondary protein kinase a catalytic subunit has a synthetic lethal interaction with *pkaA* and plays overlapping and opposite roles in *Aspergillus nidulans*. Eukaryot Cell 4:1465–1476

Nishimura M, Park G, Xu JR (2003) The G-beta subunit *MGB1* is involved in regulating multiple steps of infection-related morphogenesis in *Magnaporthe grisea*. Mol Microbiol 50:231–243

Nuss DL (2005) Hypovirulence: mycoviruses at the fungal-plant interface. Nat Rev Microbiol 3:632–642

Osherov N, May GS (2001) The molecular mechanisms of conidial germination. FEMS Microbiol Lett 199:153–160

Park G, Bruno KS, Staiger CJ, Talbot NJ, Xu JR (2004a) Independent genetic mechanisms mediate turgor generation and penetration peg formation during plant infection in the rice blast fungus. Mol Microbiol 53:1695–1707

Park G, Xue C, Zhao X, Kim Y, Orbach M, Xu JR (2006) Multiple upstream signals converge on an adaptor protein Mst50 to activate the *PMK1* pathway in *Magnaporthe grisea*. Plant Cell 18:2822–-2835

Park SH, Choi ES, Kim MJ, Cha BJ, Yang MS, Kim DH (2004b) Characterization of *HOG1* homologue, *CpMK1*, from *Cryphonectria parasitica* and evidence for hypovirus-mediated perturbation of its phosphorylation in response to hypertonic stress. Mol Microbiol 51:1267–1277

Ramamoorthy V, Zhao X, Snyder A, Xu JR, Shah DM (2007) Two mitogen-activated protein kinase signalling cascades mediate basal resistance to antifungal plant defensins in *Fusarium graminearum*. Cell Microbiol 9:1491–1506

Regenfelder E, Spellig T, Hartmann A, Lauenstein S, Bolker M, Kahmann R (1997) G proteins in *Ustilago maydis*: Transmission of multiple signals. EMBO J 16:1934–1942

Reyes G, Romans A, Nguyen CK, May GS (2006) Novel mitogen-activated protein kinase MpkC of *Aspergillus fumigatus* is required for utilization of polyalcohol sugars. Eukaryot Cell 5:1934–1940

Rui O, Hahn M (2007) The Slt2-type MAP kinase Bmp3 of *Botrytis cinerea* is required for normal saprotrophic growth, conidiation, plant surface sensing and host tissue colonization. Mol Plant Pathol 8:173–184

Segers GC, Regier JC, Nuss DL (2004) Evidence for a role of the regulator of g-protein signaling protein *CPRGS-1* in got subunit *CPG-1*-mediated regulation of fungal virulence, conidiation, and hydrophobin synthesis in the chestnut blight fungus *Cryphonectria parasitica*. Eukaryot Cell 3:1454–1463

Segmüller N, Ellendorf U, Tudzynski B, Tudzynski P (2007) BcSak1, a stress-activated mitogen-activated protein kinase, is involved in vegetative differentiation and pathogenicity in *Botrytis cinerea*. Eukaryot Cell 6:211–221

Skamnioti P, Gurr SJ (2007) *Magnaporthe grisea* cutinase2 mediates appressorium differentiation and host penetration and is required for full virulence. Plant Cell 19:2674–2689

Skamnioti P, Henderson C, Zhang ZG, Robinson Z, Gurr SJ (2007) A novel role for catalase B in the maintenance of fungal cell-wall integrity during host invasion in the rice blast fungus *Magnaporthe grisea*. Mol Plant–Microbe Interact 20:568–580

Smith DG, Garcia-Pedrajas MD, Hong W, Yu ZY, Gold SE, Perlin MH (2004) An Ste20 homologue in *Ustilago maydis* plays a role in mating and pathogenicity. Eukaryot Cell 3:180–189

Solomon PS, Tan KC, Sanchez P, Cooper RM, Oliver RP (2004) The disruption of a Galpha subunit sheds new light on the pathogenicity of *Stagonospora nodorum* on wheat. Mol Plant–Microbe Interact 17:456–466

Solomon PS, Waters ODC, Simmonds J, Cooper RM, Oliver RP (2005) The Mak2 MAP kinase signal transduction pathway is required for pathogenicity in *Stagonospora nodorum*. Curr Genet 48:60–68

Sweigard JA, Carroll AM, Farrall L, Chumley FG, Valent B (1998) *Magnaporthe grisea* pathogenicity genes obtained through insertional mutagenesis. Mol Plant–Microbe Interact 11:404–412

Takano Y, Kikuchi T, Kubo Y, Hamer JE, Mise K, Furusawa I (2000) The *Colletotrichum lagenarium* MAP kinase gene *CMK1* regulates diverse aspects of fungal pathogenesis. Mol Plant–Microbe Interact 13:374–383

Takano Y, Komeda K, Kojima K, Okuno T (2001) Proper regulation of cyclic AMP-dependent protein kinase is required for growth, conidiation, and appressorium function in the anthracnose fungus *Colletotrichum lagenarium*. Mol Plant–Microbe Interact 14:1149–1157

Thines E, Weber RWS, Talbot NJ (2000) MAP kinase and protein kinase A-dependent mobilization of triacylglycerol and glycogen during appressorium turgor generation by *Magnaporthe grisea*. Plant Cell 12:1703–1718

Xu JR, Hamer JE (1996) MAP kinase and cAMP signaling regulate infection structure formation and pathogenic growth in the rice blast fungus *Magnaporthe grisea*. Genes Dev 10:2696–2706

Xu JR, Staiger CJ, Hamer JE (1998) Inactivation of the mitogen-activated protein kinase *MPS1* from the rice blast fungus prevents penetration of host cells but allows activation of plant defense responses. Proc Natl Acad Sci USA 95:12713–12718

Xue C, Park G, Choi W, Zheng L, Dean RA, Xu JR (2002) Two novel fungal virulence genes specifically expressed in appressoria of the rice blast fungus. Plant Cell 14:2107–2119

Xue CY, Bahn YS, Cox GM, Heitman J (2006) G protein-coupled receptor Gpr4 senses amino acids and activates the cAMP-PKA pathway in *Cryptococcus neoformans*. Mol Biol Cell 17:667–679

Yamagishi D, Otani H, Kodama M (2006) G protein signaling mediates developmental processes and pathogenesis of *Alternaria alternata*. Mol Plant–Microbe Interact 19:1280–1288

Yamauchi J, Takayanagi N, Komeda K, Takano Y, Okuno T (2004) cAMP-PKA signaling regulates multiple steps of fungal infection cooperatively with Cmk1 MAP kinase in *Colletotrichum lagenarium*. Mol Plant–Microbe Interact 17:1355–1365

Yang ZH, Dickman MB (1999) *Colletotrichum trifolii* mutants disrupted in the catalytic subunit of cAMP-dependent protein kinase are nonpathogenic. Mol Plant–Microbe Interact 12:430–439

Zhang Y, Lamm R, Pillonel C, Lam S, Xu JR (2002) Osmoregulation and fungicide resistance: the *Neurospora crassa OS-2* gene encodes a *HOG1* mitogen-activated protein kinase homologue. Appl Environ Microbiol 68:532–538

Zhao X, Kim Y, Park G, Xu J-R (2005) A mitogen-activated protein kinase cascade regulating infection-related morphogenesis in *Magnaporthe grisea*. Plant Cell 17:1317–1329

Zhao X, Xu JR (2007) A highly conserved MAPK-docking site in Mst7 is essential for Pmk1 activation in *Magnaporthe grisea*. Mol Microbiol 63:881–894

Zhao X, Mehrabi R, Xu J-R (2007) Mitogen-activated protein kinase pathways and fungal pathogenesis. Eukaryot Cell 6:1701–1714

9 The Secretome of Plant-Associated Fungi and Oomycetes

Sophien Kamoun[1]

CONTENTS

I.   Introduction........................... 173
II. Definition of Effectors 173
III. Identification of Secreted Proteins
     and Effectors 174
IV. Classes of Effectors 175
V.   Apoplastic Effectors 175
     A. *Cladosporium fulvum* and *Phytophthora
        infestans* Protease Inhibitors 175
     B. *Phytophthora* Glucanase Inhibitors....... 175
     C. *Cladosporium fulvum* Avr4.............. 176
     D. *Fusarium oxysporum* Six1.............. 176
VI. Cytoplasmic Effectors..................... 176
     A. Flax Rust Effectors AvrL567, AvrM,
        AvrP123, and AvrP4.................... 176
     B. Uromyces fabae Uf-RTP1 176
     C. Leptosphaeria maculans AvrLm1 177
     D. Barley Powdery Mildew Fungus
        AVRa10 and AVRk1..................... 177
     E. Oomycete RXLR Effectors 177
     F. Oomycete Crinklers................... 178
VII. Conclusions 178
     References............................. 179

I. Introduction

The secretome of plant-associated fungi and oomycetes has been the subject of much research since the publication about 10 years ago of the first edition of "*The Mycota: Plant Relationships*". The concept that filamentous microbes require secreted proteins to alter their environment and the organisms they colonize is not particularly novel, but technology has matured to the point where it is nowadays possible to generate catalogs of the complete set of secreted proteins (the secretome) for a given organism. This has been driven by the coming of age of genome sequencing coupled with robust computational predictions of secretion signals. This chapter surveys some of the key concepts and findings that recently emerged from the study of the fungal and oomycete secretome, with an emphasis on effector proteins.

[1] The Sainsbury Laboratory, Colney Lane, Norwich, NR1 3LY, United Kingdom; e-mail: sophien.kamoun@tsl.ac.uk

Filamentous microorganisms, such as fungi and oomycetes, include highly developed plant pathogens that are intimately associated with their host plants and cause a variety of disease pathologies in natural and agricultural plant communities. Until recently, our knowledge of fungal and oomycete pathogenicity was mainly limited to the development of specialized infection structures, secretion of hydrolytic enzymes, and production of toxins (see Chaps. 10, 11). New findings, however, broadened our view of pathogenicity and suggested that filamentous pathogens are sophisticated manipulators of plant cells. It is now well accepted that similar to bacterial pathogens, eukaryotic pathogens secrete an arsenal of effector proteins that modulate plant innate immunity and enable parasitic infection (Birch et al. 2006; Chisholm et al. 2006; Kamoun 2006; O'Connell and Panstruga 2006). Deciphering the biochemical activities of effectors to understand how pathogens successfully colonize and reproduce on their host plants became a driving paradigm in the field of fungal and oomycete pathology.

II. Definition of Effectors

As in previous publications (Kamoun 2003; Torto et al. 2003; Huitema et al. 2004), I continue to use a flexible definition of the term "effectors". I define effectors as molecules that alter host cell structure and function, thereby facilitating infection (virulence factors or toxins) and/or triggering defense responses (avirulence factors or elicitors). The concept of "extended phenotype" (i.e. genes "whose effects reach beyond the cells where they reside") put forward by Richard Dawkins in a classic 1982 book (Dawkins 1999) sums up perfectly this view of effectors. Effectors can be viewed as "parasite genes having phenotypic expression in host bodies and behavior" (Dawkins 1999). Indeed, effectors are the products of genes residing in pathogen

Plant Relationships, 2nd Edition
The Mycota V
H. Deising (Ed.)
© Springer-Verlag Berlin Heidelberg 2009

genomes but actually functioning at the interface with the host plant or even inside plant cells, providing a vivid example of Dawkins "extended phenotype" (Kamoun 2006, 2007).

III. Identification of Secreted Proteins and Effectors

Biochemical, genetic, and bioinformatic strategies, often in combination, have been applied to the identification of secreted proteins from filamentous

pathogens. Traditionally, secreted proteins were identified by biochemical purification followed by genetic analysis. With the advent of genomics, novel strategies emerged. Identification of candidate secreted proteins was facilitated by the fact that in fungi and oomycetes, as in other eukaryotes, most secreted proteins are exported through the general secretory pathway via short, N-terminal, amino acid sequences known as signal peptides (Torto et al. 2003). Signal peptides are highly degenerate and cannot be identified using DNA hybridization or PCR-based techniques. Nonetheless, computational tools, particularly the SignalP program

Table 9.1. Filamentous pathogen effectors discussed in this chapter. *NA* Not applicable

Pathogen species	Effector	Localization in plant tissue[a]	Signal peptide length[b]	Virulence activities
Blumeria graminis f. sp. *hordei*	AVRa10	Cytoplasmic	NA	Enhances infection in susceptible barley plants[c]
	AVRk1	Cytoplasmic	NA	Enhances infection in susceptible barley plants[c]
Cladosporium fulvum	Avr2	Apoplastic	20	Cysteine protease inhibitor, inhibits tomato Rcr3[c]
	Avr4	Apoplastic	18	Contains CBM14 chitin binding domain, protects fungal cell walls from hydrolysis by plant chitinases[c]
	Avr9	Apoplastic	23	Structural similarity to cystine knot carboxypeptidase inhibitor[d]
Fusarium oxysporum f. sp. *lycopersici*	SIX1	Apoplastic (xylem)	21	
Hyaloperonospora parasitica	ATR1	Cytoplasmic	15	
	ATR13	Cytoplasmic	18	
Leptosphaeria maculans	AvrLm1	Probably cytoplasmic	22	
Melampsora lini	AvrL567	Cytoplasmic	23	
	AvrM	Cytoplasmic	28	
	AvrP123	Cytoplasmic	23	Kazal-like protease inhibitor[d]
	AvrP4	Cytoplasmic	28	
Phytophthora infestans	Avr3a	Cytoplasmic	21	Cell death suppressor[c]
	CRN1	Cytoplasmic	17	Elicits cell death in host plants[c]
	CRN2	Cytoplasmic	22	Elicits cell death in host plants[c]
	CRN8	Cytoplasmic	17	Similarity to RD kinase[d], elicits cell death in host plants[d]
	EPI1	Apoplastic	16	Kazal-like serine protease inhibitor, inhibits tomato P69B[c]
	EPI10	Apoplastic	21	Kazal-like serine protease inhibitor, inhibits tomato P69B[c]
	EPIC1	Apoplastic	21	Cystatin-like cysteine protease inhibitor[d]
	EPIC2B	Apoplastic	21	Cystatin-like cysteine protease inhibitor, inhibits tomato PIP1[c]
Phytophthora sojae	Avr1b-1	Cytoplasmic	21	
Uromyces fabae	*Uf*-RTP1	Cytoplasmic	19	Localizes to host nucleus[c]

[a] Cytoplasmic versus apoplastic effectors based on the classification described in the text.
[b] Length in amino acids, based on SignalP ver. 2.0-NN (www.cbs.dtu.dk/services/SignalP-2.0).
[c] Evidence is based on wet laboratory experimental data.
[d] Evidence is based on computational analyses

that was developed using machine learning methods (Nielsen et al. 1999), can assign signal peptide prediction scores and cleavage sites to unknown amino acid sequences with a high degree of accuracy (Menne et al. 2000; Schneider and Fechner 2004). Therefore, with the accumulation of cDNA and genome sequences, lists of candidate secreted proteins can be readily generated using bioinformatics tools (Table 9.1). For instance, Torto et al. (2003) developed PexFinder (with Pex standing for *Phytophthora* extracellular protein), an algorithm based on SignalP ver. 2.0 (Nielsen et al. 1999) to identify proteins containing putative signal peptides from expressed sequence tags (ESTs). PexFinder was then applied to ESTs in *P. infestans* to identify candidate secreted proteins, ultimately leading to the discovery of novel effectors of the RXLR and Crinkler families (for more information about these effectors, see Sect. VI.E, F). Variations on the Torto et al. (2003) approach have been successfully implemented, resulting in the identification of a number of important effectors from fungal pathogens (Dodds et al. 2004; Kemen et al. 2005; Catanzariti et al. 2006).

The overwhelming majority of filamentous pathogen effectors identified to date carry typical signal peptides that can be predicted using SignalP (Kamoun 2007). SignalP ver. 2.0 predictions were also convincingly validated for filamentous pathogens using proteomics (Torto et al. 2003) and a yeast secretion assay (Lee et al. 2006). Similar high degrees of accuracy of SignalP were also reported for other eukaryotes (Menne et al. 2000; Schneider and Fechner 2004). However, it is evident that many secreted proteins do not carry signal peptides and, therefore, cannot be identified using SignalP. One future challenge is to identify alternative secretory pathways, to complete cataloguing the secretomes of filamentous pathogens.

IV. Classes of Effectors

Effectors can be classified in two classes based on their target sites in the host plant (Kamoun 2006, 2007). Apoplastic effectors are secreted into the plant extracellular space, where they interact with extracellular targets and surface receptors. Cytoplasmic effectors are translocated inside the plant cell presumably through specialized structures like infection vesicles and haustoria that invaginate inside living host cells. In the following sections, we review a selection of examples of apoplastic and cytoplasmic effectors from plant pathogenic fungi and oomycetes.

V. Apoplastic Effectors

A. *Cladosporium fulvum* and *Phytophthora infestans* Protease Inhibitors

Effector proteins with inhibitory activities for protection against host proteases have been reported in fungi and oomycetes (Rooney et al. 2005; Kamoun 2006; van den Burg et al. 2006). Among these, *P. infestans* EPI1 and EPI10 are multidomain secreted serine protease inhibitors of the Kazal family that bind and inhibit the pathogenesis-related (PR) protein P69B, a subtilisin-like serine protease of tomato that functions in defense (Tian et al. 2004, 2005). *P. infestans* also secretes the cystatin-like cysteine protease inhibitors EPIC1 and EPIC2B that target PIP1 and other apoplastic cysteine proteases of tomato (Tian et al. 2007). PIP1 is closely related to tomato Rcr3, another apoplastic cysteine protease that is required for Cf-2 mediated resistance to the fungus *Cladosporium fulvm* and is inhibited by the fungal Avr2 protein (Kruger et al. 2002; Rooney et al. 2005). Unlike *P. infestans* cystatin-like EPICs, Avr2 does not have any obvious similarity to other cysteine protease inhibitors.

Other effectors that might function as protease inhibitors include *C. fulvum* Avr9 (van Kan et al. 1991). This protein shows structural similarity to cystine knot carboxypeptidase inhibitors but so far no biochemical data has been reported to support the observed similarity (van den Hooven et al. 2001). The flax rust avirulence protein AvrP123 is a 117-amino-acid protein with similarity to Kazal serine protease inhibitors and might, therefore, also target host proteases (Catanzariti et al. 2006).

B. *Phytophthora* Glucanase Inhibitors

Other secreted proteins with inhibitory activities against host hydrolytic enzymes are *Phytophthora* glucanase inhibitors. The glucanase inhibitors GIP1 and GIP2 are secreted proteins of *P. sojae* that inhibit the soybean endo-β-1,3 glucanase EGaseA (Rose et al. 2002). These inhibitor proteins share significant structural similarity with the trypsin class of serine proteases, but bear mutated catalytic residues and are proteolytically nonfunctional. GIPs are thought to function as counterdefensive molecules that inhibit the degradation of β-1,3/1,6 glucans in the pathogen cell wall and/or the release of defense-eliciting oligosaccharides by host β-1,3 endoglucanases. There is some degree of specificity in inhibition because

GIP1 does not inhibit another soybean endoglucanase, EGaseB. Positive selection has acted on β-1,3 endoglucanases in the plant legume genus *Glycine* and may have been driven by coevolution with glucanase inhibitors in *P. sojae* (Bishop et al. 2004). Four genes with similarity to GIPs have been identified in *P. infestans*, and their ability to inhibit tomato endoglucanases is under investigation (C. Damasceno and J. Rose, personal communication).

C. *Cladosporium fulvum* Avr4

The *Cladosporum fulvum* secreted protein Avr4 is a cysteine-rich protein that contributes to fungal virulence on tomato plants that lack the resistance protein Cf-4 (Joosten et al. 1997). The Avr4 protein has similarity to the chitin-binding domain CBM14 and was shown to bind chitin (van den Burg et al. 2004). Natural variants of Avr4 that are disrupted in their disulfide bridges retain the ability to bind chitin but evade recognition by Cf-4 (van den Burg et al. 2004). Avr4 appears to contribute to virulence by protecting fungal cell walls from hydrolysis by plant chitinases (van den Burg et al. 2006).

D. *Fusarium oxysporum* Six1

Six1 is a 32-kDa cysteine-rich protein secreted by the fungus *Fusarium oxysporum* f. sp. *lycopersici*, a vascular pathogen of tomato (Rep et al. 2004). Within infected tomato plants, Six1 accumulates in the xylem where it is processed by either fungal or plant proteases into a 12-kDa protein (Rep et al. 2004). Six1 mediates avirulence to tomato plants carrying the resistance gene *I-3*, as shown by complementation and knockout experiments (Rep et al. 2004). However, a direct cell death elicitor activity has not been reported for Six1. Presumably, Six1 interacts with a receptor at the surface of the xylem parenchyma cells to trigger hypersensitive cell death, but cloning and functional characterization of the *I-3* gene is needed to help clarify this issue (Rep et al. 2004).

VI. Cytoplasmic Effectors

A. Flax Rust Effectors AvrL567, AvrM, AvrP123, and AvrP4

The haustoria of the flax rust fungus *Melampsora lini* mediate the delivery of effector proteins inside host cells (Dodds et al. 2004; Catanzariti et al. 2006). Four flax rust effectors, AvrL567, AvrM, AvrP123, and AvrP4, were identified among 21 haustorially expressed secreted proteins (HESPs) as having a hypersensitive elicitor activity towards their cognate flax resistance genes (Catanzariti et al. 2006). This work indicated that haustoria are highly enriched in secreted effector proteins and most likely play a role in mediating effector translocation into host cells.

How the flax rust effectors enhance virulence is unknown. As mentioned above, AvrP123 has similarity to Kazal serine protease inhibitors and might target host proteases (Catanzariti et al. 2006). AvrL567 is a highly polymorphic effector that binds flax L5, L6, and L7 resistance proteins in the plant cytoplasm to activate hypersensitivity and defense (Dodds et al. 2004). Diversifying selection has acted on the *AvrL567* gene, and the positively selected sites were shown to alter binding to plant resistance protein receptors, providing evidence that natural selection has acted on modifying binding affinity between pathogen and plant proteins (Dodds et al. 2006).

B. *Uromyces fabae* Uf-RTP1

Uf-RTP1 is a 24-kDa haustorial protein secreted by the rust fungus *Uromyces fabae* (Kemen et al. 2005). This protein is exceptional in possessing, in addition to a classic signal peptide, a bipartite nuclear localization signal (NLS) RQHHKR[X9]HRRHK. Expression of a green fluorescent protein (GFP) and *Uf*-RTP1 fusion protein in tobacco protoplasts demonstrated that the NLS mediates protein accumulation in plant cell nuclei (Kemen et al. 2005). Most interestingly, *Uf*-RTP1 was detected inside infected plant cells, including host nuclei, by immunofluorescence and electron microscopy, providing direct evidence that this protein translocates into host cells during colonization of broad bean plants by the rust fungus (Kemen et al. 2005). This suggests that, similar to some bacterial effectors like *Xanthomonas* AvrBs3 (Lahaye and Bonas 2001), filamentous pathogen effectors also target the host nucleus where they possibly alter host gene expression.

Besides *Uf*-RTP1, *P. infestans* candidate effectors Nuk6, Nuk7, Nuk10 and Nuk12 were recently shown to carry a combination of signal peptide and NLS and to accumulate in plant nuclei in transient expression assays (Kamoun 2006; Kanneganti et al. 2007). Nuclear localization of Nuk6, Nuk7, and Nuk10 was dependent on the plant protein importin-α, suggesting that these effectors exploit the host machinery to localize in the nucleus (Kanneganti et al. 2007). These findings are particularly interesting in view of the emerging evidence that nucleo-cytoplasmic trafficking is important for plant disease resistance response and resistance protein activity (Palma et al. 2005; Shen et al. 2006).

C. *Leptosphaeria maculans* AvrLm1

AvrLm1, a 205-amino-acid protein secreted by the fungal pathogen *Leptosphaeria maculans*, mediates avirulence on *Brassica napus* plants carrying the resistance gene *Rlm1* (Gout et al. 2006). A laborious 10-year positional cloning project was recently capped by the identification of *AvrLm1* in a gene-poor heterochromatin-like region of the *L. maculans* genome (Gout et al. 2006). This 260-kb region is composed essentially of nested long tandem repeat (LTR) retrotransposons and is surrounded by isochores, high GC gene-rich islands. *AvrLm1* was the only predicted gene in the 260-kb region (Gout et al. 2006). The occurrence of *AvrLm1* in such a distinct genome environment is reminiscent of the location of *Magnaporthe oryzae* avirulence gene *Avr-Pita* in a highly unstable telomeric region (Orbach et al. 2000). Also, in the oomycete *P. infestans*, the *Avr3b-Avr10-Avr11* locus displays copy number variation resulting in amplification of multiple truncated copies of a transcription factor-like gene in avirulent strains (Jiang et al. 2006). Localization of effector genes in regions with high genome plasticity is likely to increase genetic and epigenetic variation, perhaps resulting in accelerated evolution (Orbach et al. 2000). In *Plasmodium*, genes for host-translocated effectors are often found near chromosomal ends, possibly because this position favors rapid adaptation to the host (Freitas-Junior et al. 2000; Marti et al. 2004).

The origin of *AvrLm1* remains mysterious. This gene has a lower GC content compared to linked genes, and is absent in virulent races of *L. maculans* as well as isolates of related species (Gout et al. 2006). Although the function of *AvrLm1* was confirmed by transformation into a virulent strain, Rlm1-specific elicitor activity has not been demonstrated yet (Gout et al. 2006). Furthermore, it remains unclear whether AvrLm1 functions in the apoplast or inside plant cells. However, this protein carries only a single cysteine residue and thus differs from apoplastic effectors, which are typically cysteine-rich.

D. Barley Powdery Mildew Fungus AVRa$_{10}$ and AVRk$_1$

The AVR_{a10} and AVR_{k1} genes of the barley powdery mildew fungus *Blumeria graminis* f. sp. *hordei* were first identified as candidate genes by positional cloning, and then shown to trigger hypersensitive cell death when expressed in the cytoplasm of barley cells carrying the *Mla10* and *Mlk1* resistance genes, respectively (Ridout et al. 2006; Chaps. 3, 18). The predicted AVR$_{a10}$ and AVR$_{k1}$ proteins are remarkable among filamentous pathogen effectors in being the only proteins to lack a typical secretion signal peptide (Ridout et al. 2006). How these proteins are secreted by the fungus is unclear but most likely involves alternative secretory pathways. Transient expression experiments of AVR$_{a10}$ and AVR$_{k1}$ in susceptible barley cells indicated that these effectors increase the number of successful infections sites, suggesting a virulence function of an unknown nature (Ridout et al. 2006).

E. Oomycete RXLR Effectors

The oomycete effectors ATR1, ATR13, AVR3a, and AVR1b are defined by an N-terminal motif (arginine, any amino acid, leucine, arginine; RXLR) and are thought to be delivered inside plant cells where they alter host defenses (Rehmany et al. 2005; Birch et al. 2006; Kamoun 2006; Whisson et al. 2007). Genome-wide catalogs of RXLR effectors, generated using computational approaches, unraveled a remarkably complex and divergent set of hundreds of candidate genes. Tyler et al. (2006) reported 350 RXLR effectors each in the genomes of *Phytophthora ramorum* and *P. sojae* using iterated similarity searches. Win et al. (2007) used combinations of motif and hidden Markov model searches to uncover at least 50 candidates in the downy mildew *Hyaloperonaspora parasitica* and more than 200 each in *P. capsici*, *P. infestans*, *P. ramorum*, and *P. sojae*. These large numbers of effectors suggest that oomycetes extensively modulate host processes during infection. They also raise technical challenges for studying RXLR effectors and call for the implementation of high-throughput functional analyses (Torto et al. 2003; Huitema et al. 2004).

RXLR effectors are modular proteins with two main functional domains (Bos et al. 2006; Kamoun 2006). While the N-terminal domain encompassing the signal peptide and conserved RXLR region functions in secretion and targeting, the remaining C-terminal domain carries the effector activity and operates inside plant cells. The RXLR motif defines a domain that functions in delivery of the effector proteins into the host cell (Whisson et al. 2007). Interestingly, the RXLR motif is similar in sequence and position to the plasmodial host translocation (HT)/Pexel motif that functions in delivery of parasite proteins into the red blood cells of mammalian hosts (Hiller et al. 2004; Marti et al. 2004). The RXLR domains of *P. infestans* AVR3a and another RXLR protein PH001D5 are able to mediate the export of the GFP from the *Plasmodium falciparum* parasite to red blood cells (Bhattacharjee et al. 2006), indicating that plant and animal eukaryotic pathogens share similar secretory signals for effector delivery into host cells (Haldar et al. 2006).

The *Phytophthora infestans* RXLR effector Avr3a is able to suppress hypersensitive cell death induced by another *P. infestans* protein, INF1 elicitin, pointing to a possible virulence function (Bos et al. 2006). The occurrence of cell death suppressing effectors has been hypothesized for biotrophic fungal and oomycete pathogens (Panstruga 2003), based on cytological observations of susceptible interactions and the prevalence of cell death suppressors among bacterial type III secretion system (TTSS) effectors (Jamir et al. 2004; Janjusevic et al. 2006). However, to date Avr3a is the only cell death suppressor described in filamentous pathogens. The extent to which other effectors can suppress cell death is a subject of intense investigation.

F. Oomycete Crinklers

The "Crinkler" proteins (CRNs) form a distinct class of secreted proteins that alter host responses and are thought to play important roles in disease progression (Torto et al. 2003; Kamoun 2007; Win et al. 2007). The CRNs were identified following an in planta functional expression screen of candidate secreted proteins of *P. infestans* based on a vector derived from Potato virus X (Torto et al. 2003). Ectopic expression of both genes in *Nicotiana* spp. and in the host plant tomato resultED in a leaf-

crinkling and cell-death phenotype accompanied by an induction of defense-related genes. Torto *et al.* (2003) proposed that the CRNs function as effectors that perturb host cellular processes based on analogy to bacterial effectors, which typically cause macroscopic phenotypes such as cell death, chlorosis, and tissue browning when expressed in host cells. In planta expression of a collection of deletion mutants of *crn2* indicated that this protein activates defense responses in the plant cytoplasm, suggesting they are cytopalsmic effectors (T. Torto and S. Kamoun, unpublished data).

Computational analyses revealed that the CRNs form a complex family of relatively large proteins (about 400–850 amino acids) in *Phytophthora*. Interestingly, the CRNs are defined by a distinct conserved N-terminal motif characterized by the consensus LXLFLAK (Kamoun 2007; Win et al. 2007). In *H. parasitica*, LXLFLAK overlaps with the RXLR motif, resulting in RXLRLFLAK (Win et al. 2007). This finding, along with the observation that the N-terminal region of the CRNs is dispensable for cell death induction in planta, suggests that the LXLFLAK motif contributes to host targeting analogous to RXLR. Thus, similar to RXLR effectors, the CRNs are modular proteins consisting of distinct N-terminal and C-terminal domains. Importantly, the evolutionary history of the CRN family is fundamentally different from that of the RXLR effectors. The CRNs show high rates of gene conversion and a prominent recombination site is present after the conserved N-terminal motifs (Z. Liu and S. Kamoun, personal communication). Therefore, a number of the *P. infestans* CRNs are chimeras that show unique associations between a conserved N-terminal domain and a variety of divergent C-terminal regions. These findings point to an alternative mode of host adaptation by *P. infestans* and suggest that the CRN proteins may fulfill different functions from the RXLR proteins.

VII. Conclusions

With the recent emergence of novel and cheaper DNA sequencing technologies, the flow of cDNA and genome sequences of plant-associated filamentous pathogens has rapidly accelerated. Genome sequences are becoming available for species that represent the diverse lifestyles and phylogenetic spectrum of plant-associated fungi and oomycetes. The available sequences will likely further reinforce the importance of secreted proteins in the associations between filamentous microbes and their host plants. For instance, the genome sequence of the mycorrhizal fungus *Laccaria bicolor* revealed an unexpectedly diverse and complex secretome (F. Martin, personal communication). The extent to which secreted proteins in *L. bicolor* and a diverse range of other filamentous microbes function as effectors that impact host plants is therefore poised to continue to be an exciting topic of research.

Acknowledgements: Research in my laboratory is supported by the Gatsby Charitable Foundation. Parts of this review were adapted from previous publications (Kamoun 2006, 2007).

References

Bhattacharjee S, Hiller NL, Liolios K, Win J, Kanneganti TD, Young C, Kamoun S, Haldar, K (2006) The malarial host-targeting signal is conserved in the Irish potato famine pathogen. PLoS Pathog 2:e50

Birch PR, Rehmany AP, Pritchard L, Kamoun S, Beynon JL (2006) Trafficking arms: oomycete effectors enter host plant cells. Trends Microbiol 14:8–11

Bishop JG, Ripoll DR, Bashir S, Damasceno CM, Seeds JD, Rose JK (2004) Selection on *Glycine* beta-1,3-endoglucanase genes differentially inhibited by a *Phytophthora* glucanase inhibitor protein. Genetics 169:1009–1019

Bos JI, Kanneganti TD, Young C, Cakir C, Huitema E, Win, J, Armstrong MR, Birch PR, Kamoun S (2006) The C-terminal half of *Phytophthora infestans* RXLR effector AVR3a is sufficient to trigger R3a-mediated hypersensitivity and suppress INF1-induced cell death in *Nicotiana benthamiana*. Plant J 48:165–176

Catanzariti AM, Dodds PN, Lawrence GJ, Ayliffe MA, Ellis JG (2006) Haustorially expressed secreted proteins from flax rust are highly enriched for avirulence elicitors. Plant Cell 18:243–256

Chisholm ST, Coaker G, Day B, Staskawicz BJ (2006) Host–microbe interactions: shaping the evolution of the plant immune response. Cell 124:803–814

Dawkins R (1999) The extended phenotype: the long reach of the gene. Oxford University Press, Oxford

Dodds PN, Lawrence GJ, Catanzariti AM, Ayliffe MA, Ellis JG (2004) The *Melampsora lini* AvrL567 avirulence genes are expressed in haustoria and their products are recognized inside plant cells. Plant Cell 16:755–768

Dodds PN, Lawrence GJ, Catanzariti AM, Teh T, Wang CI, Ayliffe MA, Kobe B, Ellis JG (2006) Direct protein interaction underlies gene-for-gene specificity and coevolution of the flax resistance genes and flax rust avirulence genes. Proc Natl Acad Sci USA 103:8888–8893

Freitas-Junior LH, Bottius E, Pirrit LA, Deitsch KW, Scheidig C, Guinet F, Nehrbass U, Wellems TE, Scherf A (2000) Frequent ectopic recombination of virulence factor genes in telomeric chromosome clusters of *P. falciparum*. Nature 407:1018–1022

Gout L, Fudal I, Kuhn ML, Blaise F, Eckert M, Cattolico L, Balesdent MH, Rouxel, T (2006) Lost in the middle of nowhere: the *AvrLm1* avirulence gene of the Dothideomycete *Leptosphaeria maculans*. Mol Microbiol 60:67–80

Haldar K, Kamoun S, Hiller NL, Bhattacharje S, van Ooij C (2006) Common infection strategies of pathogenic eukaryotes. Nat Rev Microbiol 4:922–931

Hiller NL, Bhattacharjee S, van Ooij C, Liolios K, Harrison T, Lopez-Estrano C, Haldar K (2004) A host-targeting signal in virulence proteins reveals a secretome in malarial infection. Science 306:1934–1937

Huitema E, Bos JIB, Tian M, Win J, Waugh ME, Kamoun S (2004) Linking sequence to phenotype in Phytophthora-plant interactions. Trends Microbiol 12:193–200

Jamir Y, Guo M, Oh HS, Petnicki-Ocwieja T, Chen S, Tang X, Dickman MB, Collmer A, Alfano JR (2004) Identification of Pseudomonas syringae type III effectors that can suppress programmed cell death in plants and yeast. Plant J 37:554–565

Janjusevic R, Abramovitch RB, Martin GB, Stebbins CE (2006) A bacterial inhibitor of host programmed cell death defenses is an E3 ubiquitin ligase. Science 311:222–226

Jiang RH, Weide R, van de Vondervoort PJ, Govers F (2006) Amplification generates modular diversity at an avirulence locus in the pathogen *Phytophthora*. Genome Res 16:827–840

Joosten MHAJ, Vogelsang R, Cozijnsen TJ, Verberne MC, de Wit PJGM (1997) The biotrophic fungus *Cladosporium fulvum* circumvents Cf-4 mediated resistance by producing unstable AVR4 elicitors. Plant Cell 9:367–379

Kamoun S (2003) Molecular genetics of pathogenic oomycetes. Eukaryot Cell 2:191–199

Kamoun S (2006) A catalogue of the effector secretome of plant pathogenic oomycetes. Annu Rev Phytopathol 44:41–60

Kamoun S (2007) Groovy times: filamentous pathogen effectors revealed. Curr Opin Plant Biol 10:358–365

Kanneganti T-D, Bai X, Tsai C-W, Win J, Meulia T, Goodin M, Kamoun S, Hogenhout SA (2007) A functional genetic assay for nuclear trafficking in plants. Plant J 50:149–158

Kemen E, Kemen AC, Rafiqi M, Hempel U, Mendgen K, Hahn M, Voegele RT (2005) Identification of a protein from rust fungi transferred from haustoria into infected plant cells. Mol Plant–Microbe Interact 18:1130–1139

Kruger J, Thomas CM, Golstein C, Dixon MS, Smoker M, Tang S, Mulder L, Jones JD (2002) A tomato cysteine protease required for Cf-2-dependent disease resistance and suppression of autonecrosis. Science 296:744–747

Lahaye T, Bonas U (2001) Molecular secrets of bacterial type III effector proteins. Trends Plant Sci 6:479–485

Lee SJ, Kelley BS, Damasceno CM, St John B, Kim BS, Kim BD, Rose JK (2006) A functional screen to characterize the secretomes of eukaryotic pathogens and their hosts in planta. Mol Plant–Microbe Interact 19:1368–1377

Marti M, Good RT, Rug M, Knuepfer E, Cowman AF (2004) Targeting malaria virulence and remodeling proteins to the host erythrocyte. Science 306:1930–1933

Menne KM, Hermjakob H, Apweiler R (2000) A comparison of signal sequence prediction methods using a test set of signal peptides. Bioinformatics 16:741–742

Nielsen H, Brunak S, von Heijne G (1999) Machine learning approaches for the prediction of signal peptides and other protein sorting signals. Protein Eng 12:3–9

O'Connell RJ, Panstruga R (2006) Tete a tete inside a plant cell: establishing compatibility between plants and biotrophic fungi and oomycetes. New Phytol 171:699–718

Orbach MJ, Farrall L, Sweigard JA, Chumley FG, Valent B (2000) A telomeric avirulence gene determines efficacy for the rice blast resistance gene Pi-ta. Plant Cell 12:2019–2032

Palma K, Zhang Y, Li X (2005) An importin alpha homolog, MOS6, plays an important role in plant innate immunity. Curr Biol 15:1129–1135

Panstruga R (2003) Establishing compatibility between plants and obligate biotrophic pathogens. Curr Opin Plant Biol 6:320–326

Rehmany AP, Gordon A, Rose LE, Allen RL, Armstrong MR, Whisson SC, Kamoun S, Tyler BM, Birch PR, Beynon JL (2005) Differential recognition of highly divergent downy mildew avirulence gene alleles by *RPP1* resistance genes from two *Arabidopsis* lines. Plant Cell 17:1839–1850

Rep M, van der Does HC, Meijer M, van Wijk R, Houterman PM, Dekker HL, de Koster CG, Cornelissen BJ (2004) A small, cysteine-rich protein secreted by *Fusarium oxysporum* during colonization of xylem vessels is required for I-3-mediated resistance in tomato. Mol Microbiol 53:1373–1383

Ridout CJ, Skamnioti P, Porritt O, Sacristan S, Jones JD, Brown JK (2006) Multiple avirulence paralogues in cereal powdery mildew fungi may contribute to parasite fitness and defeat of plant resistance. Plant Cell 18:2402–2414

Rooney HC, Van't Klooster JW, van der Hoorn RA, Joosten MH, Jones JD, de Wit PJ (2005) Cladosporium Avr2 inhibits tomato Rcr3 protease required for Cf-2-dependent disease resistance. Science 308:1783–1786

Rose JK, Ham KS, Darvill AG, Albersheim P (2002) Molecular cloning and characterization of glucanase inhibitor proteins: coevolution of a counterdefense mechanism by plant pathogens. Plant Cell 14:1329–1345

Schneider G, Fechner U (2004) Advances in the prediction of protein targeting signals. Proteomics 4:1571–1580

Shen QH, Saijo Y, Mauch S, Biskup C, Bieri S, Keller B, Seki H, Ulker B, Somssich IE, Schulze-Lefert P (2006) Nuclear activity of MLA immune receptors links isolate-specific and basal disease-resistance responses. Science 315:1098–1103

Tian M, Huitema E, da Cunha L, Torto-Alalibo T, Kamoun S (2004) A Kazal-like extracellular serine protease inhibitor from *Phytophthora infestans* targets the tomato pathogenesis-related protease P69B. J Biol Chem 279:26370–26377

Tian M, Benedetti B, Kamoun S (2005) A Second Kazal-like protease inhibitor from *Phytophthora infestans* inhibits and interacts with the apoplastic

pathogenesis-related protease P69B of tomato. Plant Physiol 138:1785–1793

Tian M, Win J, Song J, van der Hoorn R, van der Knaap E, Kamoun S (2007) A *Phytophthora infestans* cystatin-like protein targets a novel tomato papain-like apoplastic protease. Plant Physiol 143:364–377

Torto T, Li S, Styer A, Huitema E, Testa A, Gow NAR, van West P, Kamoun S (2003) EST mining and functional expression assays identify extracellular effector proteins from *Phytophthora*. Genome Res 13:1675–1685

Tyler BM, Tripathy S, Zhang X, Dehal P, Jiang RH, Aerts A, Arredondo FD, Baxter L, Bensasson D, Beynon JL, Chapman J, Damasceno CM, Dorrance AE, Dou D, Dickerman AW, Dubchak IL, Garbelotto M, Gijzen M, Gordon SG, Govers F, Grunwald NJ, Huang W, Ivors KL, Jones RW, Kamoun S, Krampis K, Lamour KH, Lee MK, McDonald WH, Medina M, Meijer HJ, Nordberg EK, Maclean DJ, Ospina-Giraldo MD, Morris PF, Phuntumart V, Putnam NH, Rash S, Rose JK, Sakihama Y, Salamov AA, Savidor A, Scheuring CF, Smith BM, Sobral BW, Terry A, Torto-Alalibo TA, Win J, Xu Z, Zhang H, Grigoriev IV, Rokhsar DS, Boore JL (2006) Phytophthora genome sequences uncover evolutionary origins and mechanisms of pathogenesis. Science 313:1261–1266

van den Burg HA, Spronk CA, Boeren S, Kennedy MA, Vissers JP, Vuister GW, de Wit PJ, Vervoort J (2004) Binding of the AVR4 elicitor of Cladosporium fulvum to chitotriose units is facilitated by positive allosteric protein-protein interactions: the chitin-binding site of AVR4 represents a novel binding site on the folding scaffold shared between the invertebrate and the plant chitin-binding domain. J Biol Chem 279:16786–16796

van den Burg HA, Harrison SJ, Joosten MH, Vervoort J, de Wit PJ (2006) *Cladosporium fulvum* Avr4 protects fungal cell walls against hydrolysis by plant chitinases accumulating during infection. Mol Plant–Microbe Interact 19:1420–1430

van den Hooven HW, van den Burg HA, Vossen P, Boeren S, de Wit PJ, Vervoort J (2001) Disulfide bond structure of the AVR9 elicitor of the fungal tomato pathogen *Cladosporium fulvum*: evidence for a cystine knot. Biochemistry 40:3458–3466

van Kan JAL, van den Ackerveken GFJM, de Wit PJGM (1991) Cloning and characterization of cDNA of avirulence gene *avr9* of the fungal pathogen *Cladosporium fulvum*, causal agent of tomato leaf mold. Mol Plant–Microbe Interact 4:52–59

Whisson SC, Boevink PC, Moleleki L, Avrova AO, Morales JG, Gilroy EM, Armstrong MR, Grouffaud S, van West P, Chapman S, Hein I, Toth IK, Pritchard L, Birch PR (2007) A translocation signal for delivery of oomycete effector proteins into host plant cells. Nature 450:115–118

Win J, Morgan W, Bos J, Krasileva KV, Cano LM, Chaparro-Garcia A, Ammar R, Staskawicz BJ, Kamoun S (2007) Adaptive evolution has targeted the C-terminal domain of the RXLR effectors of plant pathogenic Oomycetes. Plant Cell 19:2349–2369

10 From Tools of Survival to Weapons of Destruction: The Role of Cell Wall-Degrading Enzymes in Plant Infection

Antonio Di Pietro[1], Mª Isabel González Roncero[1], Mª Carmen Ruiz Roldán[1]

CONTENTS

I Introduction.............................. 181
II. Role of CWDEs Produced by Plant
   Pathogenic Fungi......................... 182
   A. Pectinases 182
      1. Polygalacturonases 182
      2. Pectate Lyases.................... 184
      3. Pectin Methylesterases 184
   B. Xylanases............................ 185
   C. Cellulases and β-Glucanases 185
   D. Proteases 186
   E. Cutinases........................... 187
   F. Lipases.............................. 187
III. Regulation of CWDE Expression 188
   A. Mechanism of Substrate Induction....... 188
   B. Signalling Pathways Regulating
      CWDE Expression.................... 190
   C. Effect of Ambient pH on
      CWDE Expression.................... 191
   D. Other Mechanisms Regulating
      CWDE Expression.................... 191
IV. CWDE Inhibitor Proteins from Plants....... 192
   A. Polygalacturonase-Inhibiting
      Proteins............................ 192
   B. Endoxylanase-Inhibiting Proteins........ 192
V. Conclusions and Outlook:
   Genomics and Proteomics
   Applied to the Study of CWDEs
   in Plant Pathogens...................... 193
   References............................. 194

I. Introduction

Cells of higher plants are surrounded by the wall, a resilient and heterogeneous network made up of different classes of polymers, mainly cellulose, xyloglucan, pectin and structural proteins. Current cell wall models propose a structure in which cellulose microfibrils are coated with xyloglucan and embedded in multiple layers of pectic polysaccharides (Carpita and Gibeaut 1993; Ha et al. 1997; Reiter 2002; O'Neill and York

2003). The hemicellulose fraction of some plant families includes non-cellulosic β-glucans such as callose (β-1,3-linked glucan) or mixed-linked β-1,3- β1,4-glucan. These different polymers and proteins form a strong fibrillar network which provides mechanical support but also functions as a dynamic, metabolically active organelle able to accommodate the developmental changes necessary for plant cell growth (Cosgrove 2001).

In addition to providing structural support, the cell wall constitutes an efficient line of defence against microbial invaders. Any micro-organism attempting to colonize a plant must, at some point, contend with the wall. In order to accomplish successful penetration of the cell wall, fungal pathogens employ two different, yet complementary strategies: buildup of mechanical pressure at the hyphal tip (Howard et al. 1991; Bastmeyer et al. 2002) and secretion of **cell wall-degrading enzymes (CWDEs)**.

The ability of plant-infecting fungi to produce a remarkable variety of CWDEs has been traditionally associated with disease (Walton 1994). Besides facilitating penetration, CWDEs are also thought to play a key role in releasing nutrients from insoluble wall polymers, which can then be readily metabolized by the pathogen during growth in planta. However, a major challenge to this traditional view of CWDEs as specific pathogenicity determinants comes from the observation that fungal depolymerases are ubiquitous both among pathogenic and non-pathogenic species (Walton 1994; de Vries and Visser 2001). Not only do saprophytic fungi thrive on dead plant cell wall tissue as well as, or even better than pathogens, but certain CWDEs or the products of their enzymatic activity can even act as avirulence factors by eliciting a potent plant immune response (De Lorenzo et al. 2001; Belien et al. 2006). Thus, the role of CWDEs in fungal plant pathogens remains controversial: are they evolutionary relics from a former saprophytic existence, or highly specialized virulence determinants conferring a selective advantage

[1] Departamento de Genética, Universidad de Córdoba, Campus de Rabanales, Edificio Gregor Mendel, 14071 Córdoba, Spain; e-mail: ge2dipia@uco.es

Plant Relationships, 2nd Edition
The Mycota V
H. Deising (Ed.)
© Springer-Verlag Berlin Heidelberg 2009

during plant infection? Here we provide an update of the current knowledge on CWDEs from pathogenic fungi and present a short outlook on novel approaches that should advance our understanding of their role in plant infection.

II. Role of CWDEs Produced by Plant Pathogenic Fungi

Filamentous fungi secrete a battery of plant cell wall depolymerases which serve in nutrient acquisition and substrate colonization (Walton 1994; de Vries and Visser 2001). Major classes of extracellular CWDEs are pectinases, xylanases, cellulases, glucanases, proteases, cutinases and lipases. Each of these groups can be further classified into subfamilies according to substrate specificity, mode of action and catalytic properties. This review focuses mainly on CWDEs from plant pathogens which have been examined at the molecular level in at least one fungus–plant interaction. In many of these studies, the role of specific CWDEs was tested by targeted knockout of the encoding gene(s), and in some cases a contribution to virulence was demonstrated (summarized in Table 10.1).

A. Pectinases

Pectin-degrading enzymes have been considered as putative pathogenicity determinants for a very long time (Reid 1950) and thus have been studied in more detail than any other class of CWDEs. They are usually among the first extracellular enzymes secreted by pathogens during growth on purified plant cell walls. Moreover, cell-free pectinolytic enzyme preparations from microbial pathogens, when applied to plant tissues, cause symptoms characteristic of plant disease such as tissue maceration or cell death (Collmer and Keen 1986). Pectinases and the products of their enzymatic activity, the oligogalacturonides, can also trigger profound physiological effects in plants, as discussed in Sect. IV (De Lorenzo et al. 2001; Poinssot et al. 2003; Federici et al. 2006).

In spite of this long history and the considerable amount of research devoted to fungal pectinases, our understanding of their role during plant infection is still far from complete. A major limiting factor is the complexity of the fungal pectinolytic systems which makes in-depth analysis both lengthy and tedious. According to their catalytic mechanism, pectin-degrading enzymes can be subdivided into endo- and exo-polygalacturonases, endo- and exo-pectate lyases and pectin methylesterases. On top of this variety, both pathogenic and saprophytic fungi tend to have multiple genes for each class of enzyme, and some of these appear to be functionally redundant (Scott-Craig et al. 1998; Rogers et al. 2000). In spite of these challenges, extensive molecular studies on the role of pectinases have been conducted in a number of fungus–host systems. Some of these studies have been successful in providing evidence for their involvement in plant disease.

1. Polygalacturonases

Polygalacturonases (PGs) are glycoside hydrolases that degrade the pectate matrix by adding water molecules to break the glycosidic bonds. The first gene disruption study to test the role of fungal PG in plant pathogenicity was carried out in the maize pathogen *Cochliobolus carbonum* (Scott-Craig et al. 1990). This fungus produces endoPG, exoPG and PME when grown in culture on pectin. Disruption of *PGN1* encoding the major secreted endoPG caused a 35% decrease of total PG activity in culture, but had no effect on virulence. The residual enzymatic activity detected in the *pgn1* mutant was due to an exoPG. The responsible protein was purified and the encoding gene, *PGX1*, was isolated and deleted both in the *C. carbonum* wild type and in the *pgn1* mutant background (Scott-Craig et al. 1998). Growth of *pgx1* mutants on pectin was reduced by ca. 20% and they were still pathogenic on maize. Even the double *pgn1/pgx1* mutant was still fully pathogenic, despite having less than 1% of total wild-type PG activity. The strain also grew well on pectin and retained PME and mycelium-associated PG activity. Based on these studies, a role of PG in virulence of *C. carbonum* cannot be ruled out yet (Scott-Craig et al. 1998).

Likewise, deletion of *enpg-1* encoding the major extracellular endoPG purified from culture filtrates of the chestnut blight fungus *Cryphonectria parasitica* did not result in a reduction in stem canker formation (Gao et al. 1996). Analysis of PG activity extracted from diseased bark tissue revealed that ENPG-1 was only a minor activity component (less than 5%), while the predominant activity consisted of two acidic PG forms that were absent in culture filtrates. The role of these acidic PGs in virulence of *C. parasitica* remains to be tested.

The vascular root-infecting pathogen *Fusarium oxysporum* secreted at least two endoPGs and two exoPGs during growth on pectin (Di Pietro and Roncero 1996). A strain unable to produce the major extracellular endoPG, PG1, due to a transposon insertion in the encoding gene, was still able to infect muskmelon plants (Di Pietro et al. 1998). Overexpression of the wild-type *pg1* gene in the *pg1*-deficient strain resulted in strongly increased PG activity but had no effect on the level of virulence, suggesting that PG1 is not a major virulence determinant in *F. oxysporum* (Di Pietro and Roncero 1998). Likewise, deletion of a second endoPG gene, *pg5*, or an exoPG gene, *pgx4*, both of which were expressed during early stages of infection, did not affect the ability of *F. oxysporum* to cause wilt disease on tomato plants (Garcia-Maceira et al. 2000, 2001). Since *F. oxysporum* has multiple PG genes, elimination of most or all of these activities will probably be required to conclusively determine the role of PGs in pathogenicity.

The bean pathogen *Colletotrichum lindemuthianum* expresses at least two endoPG genes during interaction with its host plant, *CLPG1* and *CLPG2* (Centis et al. 1996; Dumas et al. 1999). Whereas CLPG1 is secreted to the extracellular medium, CLPG2 accumulates in the fungal cell wall during growth on pectin medium and during appressorium formation (Herbert et al. 2004). Using a monoclonal antibody specific for methyl-esterified galacturonan, extensive pectin dissolution was detected during the development of secondary hyphae (Herbert et al. 2004). The role of *CLPG1* and *CLPG2* in virulence has still not been elucidated.

There are a few examples of single or double deletions of PG genes affecting fungal virulence. Early evidence for the involvement of a specific endoPG in infection was provided for the cotton pathogen *Aspergillus flavus* (Shieh et al. 1997). Deletion of a single endoPG gene, *pecA*, caused a significant reduction in aggressiveness of the mutant compared to a nondisrupted control transformant. Adding the *pecA* gene to a strain naturally lacking the gene significantly increased its virulence, thus providing direct evidence that *pecA* contributes to the invasiveness of *A. flavus* on cotton bolls (Shieh et al. 1997).

The role of PGs in pathogenicity has been extensively studied in *Botrytis cinerea*, a broad host range pathogen that causes diseases in over 200 plant species (van Kan 2006). Like the other species already discussed, *B. cinerea* has multiple endoPG genes whose expression is differentially regulated during growth in vitro and in planta (Wubben et al. 2000; ten Have et al. 2001). Deletion of *Bcpg1*, a gene which was expressed at similar levels on polygalacturonic acid, sucrose, or during infection of tomato leaves, resulted in significantly decreased secondary infectious growth from the lesion beyond the inoculation spot (ten Have et al. 1998). Inactivation of another endoPG gene, *Bcpg2*, resulted in 50–85% reduction in lesion expansion rate (Kars et al. 2005). These results demonstrate that: (a) Bcpg1 and Bcpg2 are not functionally redundant and (b) both PGs are required for full virulence of *B. cinerea*. In a recent study, five *B. cinerea* endoPGs were expressed in the yeast *Pichia pastoris*. BcPG1 and BcPG2 exhibited the strongest leaf-necrotizing activity, leading to plant tissue collapse within a few minutes. A single amino acid substitution in the active site of BcPG2 abolished both its catalytic and necrotizing activity (Kars et al. 2005). Interestingly, a survey of endoPG genes in 34 isolates of *B. cinerea* revealed that *Bcpg1* and *Bcpg2*, but not *Bcpg3* were highly polymorphic when compared with non-pathogenicity loci. These data provide evidence for specialization among individual endoPGs and potential diversification of endoPGs interacting directly with host defences (Rowe and Kliebenstein 2007).

The pectinolytic system of the related necrotroph *Sclerotinia sclerotiorum* has some similarity with that of *B. cinerea* and consists of multiple endoPG genes belonging to distinct phylogenetic groups, which are sequentially expressed during infection (Kasza et al. 2004). Four endo-PG and two exo-PG genes differentially expressed under saprophytic and pathogenic conditions were identified in a *S. sclerotiorum* isolate from *Brassica napus*, but their role in virulence has not yet been addressed (Li et al. 2004).

The above studies confirm the general view that endoPGs are important in virulence of broad host-range necrotrophic pathogens such as *B. cinerea*. In 1984, Cooper suggested that degradation of primary walls was often rapid and extensive during infection by facultative parasites which cause necrosis, and that most evidence implicated endoPGs as major factors in pathogenicity. By contrast infection by ecologically obligate parasites should involve minimal changes to host walls

(Cooper 1984). However, recent evidence indicates that endoPGs also play important roles in highly specialized fungus–plant systems. Simultaneous deletion of two endoPG genes in the biotrophic, organ-specific rye pathogen *Claviceps purpurea* resulted in dramatically reduced virulence (Oeser et al. 2002). Another biotroph, the basidiomycete corn pathogen *Ustilago maydis*, secreted both PGs and PLs as well as other classes of CWDEs during in planta growth, but their exact role in virulence has not been determined yet (Cano-Canchola et al. 2000).

An interesting example of a differential role for a specific endoPG was reported in two morphologically indistinguishable pathogens of citrus: *Alternaria citri*, the causal agent of black rot, and *A. alternata*, which produces brown spot (Isshiki et al. 2001). Whereas the former causes rot by macerating tissues, the latter causes necrotic spots by producing a host-selective toxin. Strikingly, targeted knockout of an endoPG gene displaying 99% nucleotide identity between the two species produced completely different virulence phenotypes. An endoPG mutant of *A. citri* was severely affected in its ability to macerate potato tissue and to cause black rot symptoms on citrus. By contrast, a mutant of *A. alternata* lacking the orthologous gene was not altered in virulence (Isshiki et al. 2001). This outcome indicates that a given CWDE can play differential roles even in taxonomically closely related pathogens, depending on their mode of infection.

2. Pectate Lyases

In contrast to PGs which are hydrolases, pectate lyases (PLs) cleave the glycosidic bonds via a β-elimination mechanism. Extracellular PL activity is often detected during fungal growth on the host plant. A PL from the tomato vascular wilt pathogen *F. oxysporum* f. sp. *lycopersici* was secreted during infection, and expression of the encoding gene was detected during fungal invasion of tomato roots and stems, suggesting a possible role in virulence (Di Pietro and Roncero 1996; Huertas-Gonzalez et al. 1999). However, knockout of *pl1* had no effect on virulence, and zymogram analysis of colony supernatants of *pl1* mutants revealed the presence of a second PL activity band encoded by another, possibly functionally redundant gene (Roncero et al. 2003).

Likewise, targeted knockout of a pectin lyase gene in *Colletotrichum gloeosporioides* (anamorph *Glomerella cingulata*) had no detectable effect on virulence on Capsicum and apple (Bowen et al. 1995). By contrast, deletion of the *pelB* gene in *C. gloeosporioides* resulted in 25% lower PL activity and a 36–45% reduction in decay diameter on avocado fruits (Yakoby et al. 2001). Heterologous expression of the *pelB* gene in *C. magna* increased its virulence on avocado and watermelon seedlings (Yakoby et al. 2000). Taken together, these results indicate that PL is a virulence factor required for host penetration and colonization of *Colletotrichum* species.

The role of PL in virulence of the pea pathogen *Fusarium solani* f. sp. *pisi* (*Nectria haematococca*) has been studied in detail. Antibodies raised against a purified extracellular PL inhibited both the catalytic activity of the enzyme and the ability of the fungus to infect pea stems (Crawford and Kolattukudy 1987). Four genes encoding extracellular PLs were identified in *F. solani* f. sp. *pisi*, two of which (*pelA*, *pelD*) were substrate-inducible, whereas the other two (*pelB*, *pelC*) were constitutively expressed at low levels (Gonzalez-Candelas and Kolattukudy 1992; Guo et al. 1996). Disruption of either *pelA* or *pelD* caused no detectable effect on the virulence of *F. solani*, but inactivation of both genes drastically reduced the capacity of the fungus to infect pea (Rogers et al. 2000). Complementation of the double mutant with the *pelD* gene or supplementation of the infection droplets with purified PLA or PLD enzyme led to complete virulence recovery. This study shows that determination of the role of a specific class of CWDE in pathogenicity may require inactivation of multiple functionally redundant genes present in the species.

3. Pectin Methylesterases

Pectin methylesterases (PMEs) hydrolyse pectin into methanol and polygalacturonic acid. The role of PMEs in fungal virulence is poorly understood. In a recent study, *B. cinerea*, knockout mutants lacking the *Bcpme1* gene which encodes an extracellular PME, showed 75% decreased PME activity, reduced growth on pectin and a strong decrease in virulence on apple, grapevine and *Arabidopsis thaliana* (Valette-Collet et al. 2003). Thus, although *B. cinerea* appears to have multiple PME-encoding genes, *Bcpme1* acts as a key virulence determinant in this fungus. Whether this important role of PME is conserved in other fungal pathogens remains to be determined.

B. Xylanases

Xylan, a heteropolysaccharide with groups of acetyl, 4-O-methyl-D-glucuronosyl and α-arabinofuranosyl residues linked to a backbone of β-1,4-linked xylopyranose units, is the predominant hemicellulose component in plant cell walls and the second most abundant polysaccharide found in nature (Subramaniyan and Prema 2002). Due to the heterogeneous structure of plant xylan, its complete degradation requires the activity of a complex of hydrolytic enzymes with diverse modes of action: β-1,4-endoxylanases, β-xylosidases, α-L-arabinofuranosidases, α-glucuronidases, acetylxylan esterases, and phenolic acid (ferulic and *p*-coumaric acid) esterases (Beg et al. 2001). The most important xylanolytic enzyme is endo-β-1,4-xylanase, which cleaves the main xylan backbone chain (Biely and Tenkanen 1998). Endoxylanases belong mainly to either glycoside hydrolase family 10 (GH10) or glycoside hydrolase family 11 (GH11) (Collins et al. 2005). GH10 endoxylanases typically have a molecular mass higher than 30 kDa and a low pI (Subramaniyan and Prema 2002), whereas GH11 endoxylanases have a lower molecular mass of around 22 kDa and a high pI (Torronen and Rouvinen 1997).

Endoxylanases can act synergistically with other types of xylanolytic enzymes. Acetylxylan esterase hydrolyses the ester linkages of the acetyl groups at position 2 and/or position 3 of the xylose moieties of natural acetylated xylan from hardwood, increasing the accessibility of the xylan backbone for endoxylanase attack. β-Xylosidase enhances the hydrolysis of xylan by endoxylanase by relieving the end-product inhibition of endoxylanases. Finally, the addition of α-arabinofuranosidase to endoxylanase enhances the saccharification of arabinoxylan (Beg et al. 2001).

Because xylan represents a significant proportion of the hemicellulosic fraction in plant cell walls, xylan-degrading enzymes are suggested to act as components of the offensive arsenal of plant pathogens (Belien et al. 2006). Many plant pathogenic fungi secrete endoxylanases when grown in the presence of host cell walls (Cooper et al. 1988; Lehtinen 1993; Ruiz et al. 1997; Wu et al. 1997;

Giesbert et al. 1998; Carlile et al. 2000; Hatsch et al. 2006). *C. carbonum* secreted one major β-xylosidase (Xyp1) and a α-arabinofuranosidase (Arf) during growth on maize cell walls (Ransom and Walton 1997; Wegener et al. 1999). Transcription of a gene encoding a putative α-L-arabinofuranosidase from *F. oxysporum* f. sp. *dianthi* was detected in infected carnation plants (Chacon-Martinez et al. 2004). Differential expression of several GH10 and GH11 endoxylanase genes was reported in *M. grisea, F. oxysporum, C. carbonum, Helminthosporium turcicum, C. purpurea* and *B. cinerea* during infection of their host plants (Apel-Birkhold and Walton 1996; Giesbert et al. 1998; Ruiz-Roldan et al. 1999; Gomez-Gomez et al. 2001; Gomez-Gomez et al. 2002; Brutus et al. 2005; Brito et al. 2006; Wu et al. 2006). While these results suggested the possible involvement of these enzymes in disease, the targeted disruption of structural endoxylanase genes in *C. carbonum*, (Apel et al. 1993; Apel-Birkhold and Walton 1996), *F. oxysporum* (Gomez-Gomez et al. 2001; Gomez-Gomez et al. 2002) and *M. grisea* (Wu et al. 1995), or of a β-xylosidase gene in *C. carbonum* (Wegener et al. 1999) failed to produce any detectable effect on virulence. The occurrence of multiple endoxylanase and β-xylosidase genes in fungi is proposed as the main reason why mutants in individual xylanase genes remain pathogenic (Apel-Birkhold and Walton 1996; Wegener et al. 1999; Gomez-Gomez et al. 2002).

The only experimental evidence so far demonstrating a role for a single endoxylanase gene in virulence was recently obtained in the necrotroph *B. cinerea*. Deletion of *xyn11* encoding a GH11 endoxylanase expressed *in planta* strongly affected the ability of the fungus to infect tomato leaves and grape berries, delaying the appearance of secondary lesions and reducing the lesion size by as much as 70% (Brito et al. 2006).

C. Cellulases and β-Glucanases

The cellulase group includes endo-β-1,4-glucanase, P-glucosidase and cellobiohydrolase, which cooperate in the complete hydrolysis of cellulose to glucose (Lynd et al. 2002). Because of the crystalline nature of native cellulose polymer, its enzymatic degradation is slow. Plant pathologists generally feel that cellulases are not particularly important in pathogenesis, since extensive cellulose degradation typically occurs only late in infection, if at all (Cooper 1984).

C. carbonum grown either on cellulose or maize cell walls secreted several cellobiohydrolase, endoglucanase and β-glucosidase activities that were separable by different fractionation protocols. The *CEL1* gene encoding a putative cellobiohydrolase was highly expressed during growth on cellulose or maize cell walls but not in the presence of sucrose. Unlike most endoglucanases and cellobiohydrolases, Cel1 does not have a putative cellulose binding domain or associated hinge region. A strain of *C. carbonum* lacking *CEL1* was as virulent as the wild type (Sposato et al. 1995). Mutation of a second cellulase gene, *CEL2*, also had no detectable effect on virulence (Ahn et al. 2001).

In the biotrophic rye pathogen *C. purpurea*, transcription of the *cel1* gene encoding a putative cellobiohydrolase which also lacks a substrate binding domain was induced during the first days of infection, suggesting that it may be involved in the penetration and degradation of host cell walls (Müller et al. 1997). In *A. alternata*, the causal agent of black spot disease in persimmon fruits (*Diospyros kaki* L.), production of an endo-1,4-β-glucanase by the fungus was related to host susceptibility (Eshel et al. 2000).

Non-cellulosic β-glucans such as callose (β-1,3-linked glucan) or mixed-linked β-1,3- β1,4-glucan are important components of the plant cell wall, particularly in cereals where they are associated with defense-related structures known as papillae. The role of extracellular β-glucanases was studied in detail in *C. carbonum*. Genes encoding two exo-β-1,3-glucanases and two mixed-linked β-1,3- β1,4-glucanases were cloned and knockout mutants lacking one, two or all four genes were produced (Schaeffer et al. 1994; Gorlach et al. 1998; Nikolskaya et al. 1998; Kim et al. 2001). Total mixed-linked glucanase and β-1,3-glucanase activities in culture filtrates of the quadruple mutants were reduced by 73% and 96%, respectively, but only a modest decrease in growth on β-glucans and no effect on virulence was detected (Kim et al. 2001). Thus, either degradation of β-glucan is not important for penetration or growth of *C. carbonum* in maize leaves, or small amounts of residual enzyme activity are sufficient for the mutants to show virulence. It is possible that unknown β-glucanases which are only expressed in planta can compensate for the loss of the deleted genes.

CWDEs may play unexpected or indirect roles during infection. The maize pathogen *Fusarium verticillioides* produces large amounts of the fumonisin mycotoxin FB1 when grown on starchy kernels, amylopectin, or its hydrolysis product dextrin. A mutant strain carrying a disrupted α-amylase gene was impaired in its ability to produce FB1 on starchy kernels, indicating that hydrolysis of amylopectin is required for FB1 production in planta (Bluhm and Woloshuk 2005).

D. Proteases

Plant cell walls contain different classes of proteins (Carpita and Gibeaut 1993). Degradation of these cell wall proteins by fungal proteases may be critical for successful infection. Proteases catalyse the cleavage of proteins into peptides or free amino acids and are classified according to their mode of action and catalytic site (http://www.merops. ac.uk/merops/merops/htm). The implication of secreted proteases in fungal virulence has been extensively studied in animal pathogens (Monod et al. 2002), but much less is known about their role in plant pathogens.

Production of acidic aspartyl proteases during infection has been reported in two necrotrophic pathogens, *B. cinerea* and *S. sclerotiorum*. Five genes encoding either intra- or extracellular aspartic proteases from *B. cinerea* and two genes encoding putative acidic aspartyl-like proteases from *S. sclerotiorum* were expressed both in liquid culture and in infected plant tissues (Poussereau et al. 2001a, b; ten Have et al. 2004). The role of these different acidic proteases in virulence has not been determined.

Serine proteases, especially those belonging to the subtilisin class, are also associated with fungal pathogenicity. In *Magnaporthe poae*, a fungal pathogen of Kentucky bluegrass, immunoblot analysis detected expression of a subtilisin-like serine proteinase in infected roots, concomitant with increasing severity of disease symptoms (Sreedhar et al. 1999). In *C. carbonum*, three extracellular serine proteases, two of the trypsin type (Alp1a, Alp1b) and one of the subtilisin family (Alp2) were purified and characterized. Mutants lacking both Alp1a and Alp1b showed significantly reduced extracellular protease activities, but their disease phenotype was indistinguishable from that of the wild type (Murphy and Walton 1996). Likewise *prt1*, encoding an extracellular subtilisin-type serine protease from *F. oxysporum*, was expressed during the entire infection cycle, but was not required for virulence on tomato plants (Di Pietro et al. 2001). In the wheat pathogen *S. nodorum*, deletion of *snp1* encoding an extracellular trypsin-like protease expressed during early stages of infection resulted in mutants lacking trypsin activity in vitro and on inoculated wheat leaves, but showing wild-type virulence. These mutants still had 50% of the wild-type alkaline protease activity, which could be attributed to a subtilisin protease and might have compensated for the loss of trypsin activity in the SNP1-deletion mutants (Bindschedler et al. 2003). The presence of a large number of subtilase genes in the genome sequences of the two cereal pathogens *M. grisea* and *Fusarium graminearum* (Dean et al. 2005; Cuomo et al. 2007) provides further support to the idea that this class of proteases may play an important role in pathogenicity.

Interestingly, secreted fungal proteases can also function as determinants of avirulence. *AVR-Pita* encodes a protein of the rice blast fungus *M.* Complementary experiments on conidial separation from short chains of spores in the powdery mildew *Oidium neolycopersici* are described by Oichi et al. (2006). Leach (1976) proposed that electrostatic charges were involved in the release of conidia of *Drechslera turcica*, and electrostatic charges have also been implicated in the discharge of zoospores from oomycete sporangia (Borkowski 1972) and from ballistospores of Basidiomycota (Gregory 1957; Saville 1965). Although spores, like other microscopic particles, respond to electrostatic charge, there is no evidence that charges play a significant role in spore release mechanisms. Aylor and Paw U (1980) argued that naturally-occurring electrostatic forces are too small to detach conidia from their conidiophores. Other studies also show that the charges carried on the surface of fungal spores are insufficient to effect discharge or to affect spore deposition on surfaces (McCartney et al. 1982; Webster et al. 1988).

grisea with features typical of metalloproteases, which interacts in a gene-for-gene fashion with the rice disease resistance gene *Pi-ta* (Orbach et al. 2000). When introduced into virulent fungal strains, *AVR-Pita* specifically conferred avirulence toward rice cultivars containing *Pi-ta* and this avirulence function was absent in mutants altered in residues of the catalytic site, suggesting that metalloprotease activity of AVR-Pita is required for avirulence function (Orbach et al. 2000). It is currently unknown whether AVR-Pita also plays a role in virulence of *M. grisea*.

E. Cutinases

The cuticle forms a hydrophobic coating that covers nearly all aboveground parts of terrestrial plants and constitutes the interface between plant and environment. The main structural component of the plant cuticle is cutin, a waxy polyester consisting of esterified hydroxy and epoxy fatty acids which are n-C_{16} and n-C_{18} types (Kolattukudy

2001). To penetrate the cutin barrier, fungal pathogens use a combination of physical pressure (Howard et al. 1991) and enzymatic degradation by extracellular cutinase, a class of methyl esterase that catalyses the hydrolysis of ester bonds from the cutin fatty acid polymer (Kolattukudy 2001).

The role of cutinase in fungal pathogenicity is subject to debate (Schäfer 1993). A number of gene knockout studies in different pathogens failed to detect an essential role for individual cutinase genes and provided evidence for the presence of multiple functionally redundant cutinase isozymes (Stahl and Schäfer 1992; Sweigard et al. 1992; Yao and Köller 1995; Crowhurst et al. 1997; van Kan et al. 1997). By contrast, disruption of the *cut1* cutinase gene in a *F. solani* f. sp. *pisi* isolate naturally producing low amounts of cutinase resulted in a mutant lacking cutinase activity, in which lesion formation was much less frequent and milder than in the wild type (Rogers et al. 1994). Insertion of *cut1* into *Mycosphaerella*, a pathogen that normally requires wounds to infect, allowed the transgenic strains to penetrate an intact host surface (Dickman et al. 1989). Biochemical studies also support a role for cutinase in virulence. Immunolocalization with antibodies against Cut1 indicated that the cutinase was localized at the growing tip of germinating spores of *F. solani* f. sp. *pisi* (Podila et al. 1995). Application of cutinase antibodies to germinating spores inhibited infection of pea by the fungus (Maiti and Kolattukudy 1979). Cutinase and serine-esterase activity was detected on the surface of uredospores of the obligate parasitic rust fungus *Uromyces viciae-fabae* (Deising et al. 1992). Treatment of living spores with the serine-esterase inhibitor diisopropyl fluorophosphate prevented their adhesion to the leaf surface, whereas addition of cutinase and nonspecific esterases to autoclaved spores restored their ability to adhere to the host cuticle. These findings suggest that, besides penetration, esterase and cutinase may also have a functional role in the process of adhesion (Deising et al. 1992).

F. Lipases

Lipases (triacylglycerol acylhydrolases) are lipolytic serine esterases acting on the carboxyl ester bonds present in acylglycerols to generate organic acids and glycerol (Jaeger and Eggert 2002). Fungal lipases and esterases may be involved in providing carbon sources during plant cell wall degradation, as well as during adhesion to and penetration of the plant surface. Few reports exist on the role of secreted fungal lipases in plant pathogenicity. A purified lipase from *Alternaria brassicicola* was found to cross-react with antibodies against a *B. cinerea* lipase, and addition of the anti-lipase antibodies to a conidial suspension of *A. brassicicola* resulted in a 90% reduction of blackspot lesions on intact cauliflower leaves, but not on leaves from which the surface wax had been removed (Berto et al. 1999). A gene encoding a secreted lipase from *F. solani* f. sp. *pisi* was highly induced by lipidic substrates, repressed by glucose and expressed in planta at different times after inoculation (Nasser Eddine et al. 2001). A secreted lipase of *B. cinerea* was induced during early stages of infection (Commenil et al. 1999). When specific anti-lipase antibodies were added to a conidial suspension of *B. cinerea*, lesion formation was completely suppressed (Commenil et al. 1998). However, targeted disruption of the corresponding gene *lip1* did not affect virulence of *B. cinerea* (Reis et al. 2005). Even double mutants which lacked both *lip1* and the cutinase gene *cutA* and were largely devoid of extracellular esterase activity, still retained full virulence in various host plant systems (Reis et al. 2005).

Conclusive evidence for the role of a single fungal lipase in virulence was recently obtained in *F. graminearum* (Voigt et al. 2005). Disease severity on wheat spikes was strongly reduced when wild-type conidia of the fungus were supplemented with the lipase inhibitor ebelactone B. Moreover, disruption of the lipase gene *FGL1* led to reduced extracellular lipolytic activity in culture and a decrease in virulence on both wheat and maize. The exact role of FGL1 during plant infection is currently unknown (Voigt et al. 2005).

III. Regulation of CWDE Expression

From the wealth of studies reviewed in the previous section, it becomes evident that the main limitation of the structural gene knockout approach is functional redundancy of CWDE genes. To date, there are no reported mutants of phytopathogenic fungi in which all the genes of a given CWDE class have been deleted. One way to circumvent this problem is to target CWDE genes at the transcriptional level. Transcriptional control of CWDE genes is accomplished through specific regulators that act according to a precise programme, in response to environmental and cellular factors such as nutrient availability (Aro et al. 2005). Deletion of these master regulators is expected to abolish the expression of an entire class of CWDE genes, giving rise to strains that are almost completely devoid of one or several CWDE activities (Tonukari et al. 2000).

A. Mechanism of Substrate Induction

In spite of recent advances, our understanding of the regulatory circuits controlling CWDE gene expression in fungal pathogens is still limited. Most fungal wall depolymerases are subject to substrate induction and (carbon) catabolite repression (Cooper and Wood 1973). Substrate induction occurs generally at the transcriptional level and is mostly mediated by monomers which are readily taken up by the fungal cell (in contrast to the insoluble polymer). This raises the question of how the inducing monomers are generated before expression of CWDE genes is activated. The current model suggests that fungi continuously secrete very low amounts of different CWDEs which act as "sensors" of the surrounding environment (Fig. 10.1). Some fungal pathogens have been shown to produce basal depolymerase activities during early infection stages. The extracellular matrix produced by germlings of *B. cinerea* which serves in their attachment to the plant surface, exhibited low amounts of different CWDE activities such as PG, PL, PME, cutinase or cellulase (Doss 1999). Likewise, the surface of uredospores of *U. viciae-fabae* contained low amounts of cutinase and non-specific serine-esterases which were released rapidly upon contact with an aqueous environment (Deising et al. 1992). In the presence of a given substrate such as cutin, monomers are rapidly generated and taken up by the fungal cell, triggering expression of the corresponding depolymerase genes (Fig. 10.1A; Woloshuk and Kolattukudy 1986). In agreement with the "enzyme sensor model", many inducible CWDE genes show low constitutive expression even in the absence of an inducer, thus ensuring the continuous presence of basal amounts of enzyme. In some cases, different genes encoding the same class of depolymerase are regulated differentially: those constitutively expressed at basal level provide the low "sensing"

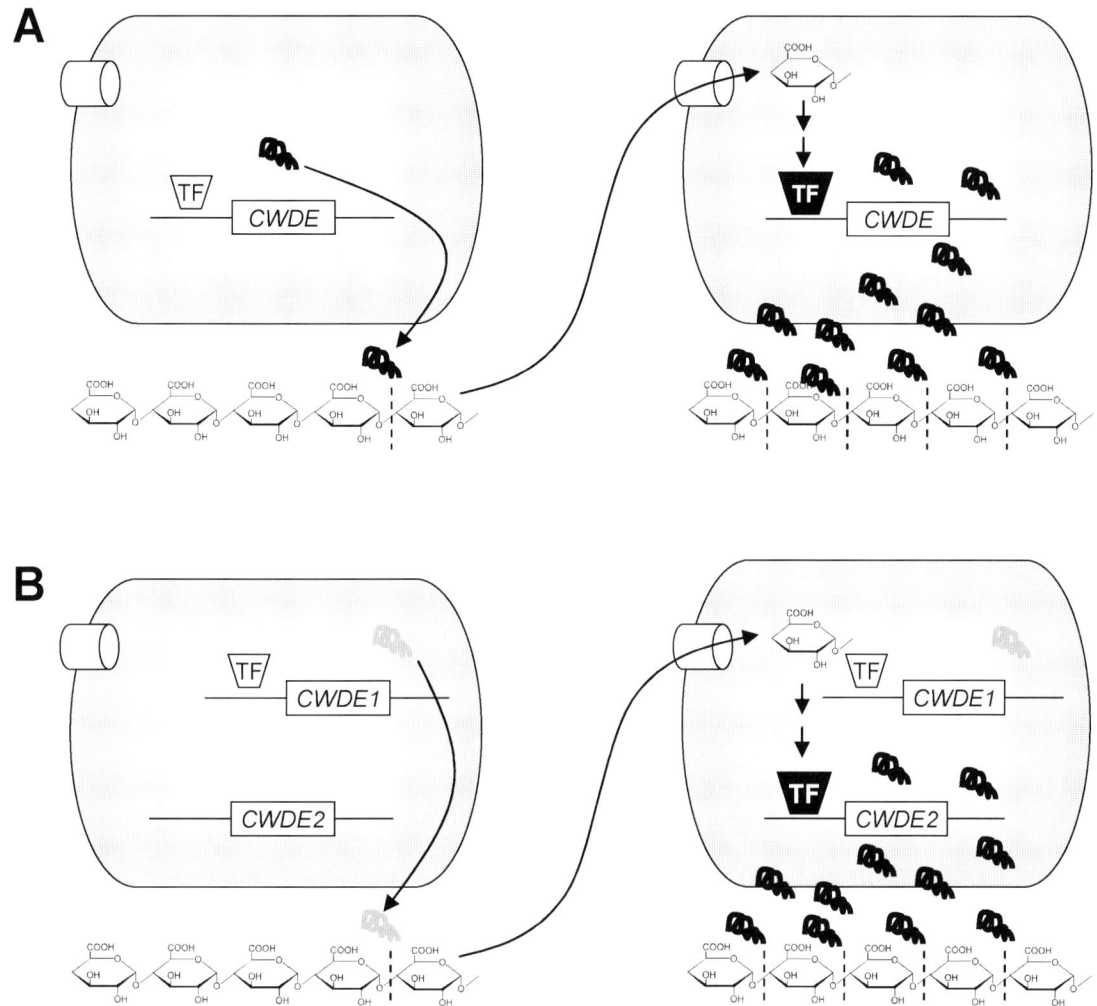

Fig. 10.1. Mechanisms for substrate induction of fungal cell wall-degrading enzymes (CWDEs). **A** Constitutive expression of a CWDE gene at basal levels leads to the presence of low amounts of extracellular enzyme activity. Upon contact with the substrate (e.g. during plant infection), monomers are released and actively taken up by the fungal cell, leading to activation of transcription factor(s) (*TF*), an increase in CWDE gene expression and secretion of large amounts of CWDE. **B** Model similar to **A**, except that constitutive and inducible expression are carried out by two different CWDE genes. In both cases, constitutive and inducible gene expression can be mediated either by the same or by distinct transcription factors. **B** is based on a model formulated by Li et al. (2002)

enzyme activity which releases the monomers from the substrate, thus triggering expression of a second set of highly inducible genes (Fig. 10.1B).

A well studied example that illustrates such regulatory specialization of CWDE genes is provided by three cutinase genes from *F. solani* f. sp. *pisi*: *cut1*, *cut2* and *cut3*. While their amino acid sequences share a high degree of identity, their transcriptional regulation is completely divergent. *cut2* and *cut3* are constitutively expressed at basal levels (Li et al. 2002), whereas expression of *cut1* is strongly induced by cutin monomers (Woloshuk and Kolattukudy 1986). A Cys(6)Zn(2) motif-containing transcription factor, CTF1β, was found to bind a sequence element in the *cut2* promoter, leading to basal expression of *cut2* and continuous production of low levels of cutinase (Li et al. 2002). This activity releases cutin monomers that activate a second zinc finger transcription factor, CTF1α, allowing it to bind a palindromic sequence in the *cut1* promoter and induce transcription of the *cut1* gene (Li and Kolattukudy 1997; Li et al. 2002).

Expression of fungal xylanase genes is usually induced by substrates such as xylan, xylose or arabinose (Ruiz-Roldan et al. 1999; de Vries et al. 2002) and repressed by glucose (Mach et al. 1996; Orejas et al. 1999). Transcriptional activation of xylanolytic genes is mediated by the zinc finger protein XlnR which was originally identified in *Aspergillus niger* (van Peij et al. 1998). Orthologues of XlnR were shown to perform similar functions in other saprophytic species such as *A. oryzae* and *Trichoderma reesei* (Marui et al. 2002; Rauscher et al. 2006). XlnR controls transcription not only of xylanolytic genes, but also of endoglucanase and cellobiohydrolase genes, suggesting a general role in cellulose and hemicellulose degradation and metabolism (van Peij et al. 1998; de Vries and Visser 1999; Rauscher et al. 2006). Recently, the *xlnR* orthologue from the vascular wilt pathogen *F. oxysporum* was cloned (Calero-Nieto et al. 2007). *Fusarium* XlnR had DNA-binding properties highly similar to those of XlnR from *Aspergilli*. Targeted knockout of *xlnR* in *F. oxysporum* resulted in a lack of transcriptional activation of xylanase genes both in culture and during infection of tomato plants, as well as in dramatically reduced extracellular xylanase activity. However, the Δ*xlnR* mutants were still fully pathogenic on tomato plants, demonstrating that XlnR is not an essential virulence determinant in *F. oxysporum* (Calero-Nieto et al. 2007). The maize pathogen *C. carbonum* has two copies of *xlnR*. Mutants lacking both *xlnR* genes showed reduced growth when cultured on xylan but exhibited wild-type virulence, indicating that XlnR is not essential for infection of corn plants by *C. carbonum* (J.D. Walton, personal communication).

B. Signalling Pathways Regulating CWDE Expression

Upstream of the transcription factors, expression of CWDE-encoding genes is controlled by a number of conserved signalling pathways. A mitogen-activated protein kinase (MAPK) cascade, which is required for pathogenicity on plants (Zhao et al. 2007), controls expression of CWDEs in different pathogen species. Mutants of *F. oxysporum* lacking the pathogenicity MAPK showed reduced expression of endoPG and endoPL genes (Di Pietro et al. 2001; Delgado-Jarana et al. 2005). The orthologous MAPK was shown to regulate expression of two cellulase genes in *Cochliobolus heterostrophus* (Lev and Horwitz 2003) as well as different CWDE activities in *F. graminearum* (Jenczmionka and Schäfer 2005) and *A. brassicicola* (Cho et al. 2007). Interestingly, deletion of a *Colletotrichum lindemuthianum* transcription factor similar to Ste12p, which acts downstream of the orthologous MAPK cascade in yeast, resulted in significantly reduced pectinase activity indicating that it is involved in regulation of CWDE expression (Hoi et al. 2007). In agreement with this finding, a cis-acting sequence similar to the Ste12/Tec1 binding element in yeast was required for the transcriptional activation of the *CLPG2* gene in *C. lindemuthianum* by pectin and during appressorium development (Herbert et al. 2002).

Another important group of signalling components, heterotrimeric G protein subunits, often function upstream of a conserved cAMP-regulated pathway. Gα subunits from the two plant pathogens *B. cinerea* and *S. nodorum* were involved in fungal virulence and in activating the expression of extracellular proteases (Schultz Gronover et al. 2001; Solomon et al. 2004). A suppression subtractive hybridization (SSH) approach was used to identify fungal genes whose expression on the host plant was specifically reduced in the *B. cinerea* Gα mutant, leading to the identification of multiple protease and endoxylanase genes (Schulze Gronover et al. 2004; see Chap. 2).

A common feature of many fungal CWDE genes is their repression by carbon catabolites such as glucose (Cooper and Wood 1973; Walton 1994). In *Saccharomyces cerevisiae*, the Snf1 kinase is required for expression of catabolite-repressed genes in conditions of glucose limitation (Hardie et al. 1998). An orthologue of *SNF1* was cloned and deleted in the plant pathogen *C. carbonum* (Tonukari et al. 2000). Growth of *ccsnf1* mutants on complex carbon sources such as xylan, pectin or purified maize cell walls was reduced and the activities of secreted β-1,3-glucanase, pectinase and xylanase were strongly decreased. Accordingly, expression of genes encoding endo-β-1,4-xylanases, an β-arabinosidase, a β-xylosidase, exo- and endoPG, mixed-linked [β-1,3–β-1,4] glucanase and exo-β-1,3-glucanases was downregulated in the mutant. The Δ*ccsnf1* strain was significantly less virulent on maize, indicating that ccSNF1 is required for biochemical processes important in pathogenesis of *C. carbonum* (Tonukari et al. 2000). Knockout of the *snf1* gene in the vascular wilt fungus *F. oxysporum snf1* also resulted in a reduction of CWDE expression and virulence (Ospina-Giraldo et al. 2003).

C. Effect of Ambient pH on CWDE Expression

Ambient pH is emerging as a key factor in controlling both the catalytic activity of CWDEs and transcription of the encoding genes. Fungal CWDE genes often show maximal expression at those pH values that support optimum catalytic activity of the corresponding enzyme. For example, in *C. gloeosporioides* both secretion of PL and expression of the *pelB* gene are regulated by the pH of the host tissue. PL secretion only occurred at values higher than pH 5.8, such as in the pericarp of susceptible fruits, but not at lower values encountered in resistant varieties (Yakoby et al. 2000). *PelB* transcription was activated upon fruit tissue alkalinization by the fungus, which was achieved via secretion of ammonia (Drori et al. 2003; Kramer-Haimovich et al. 2006). Interestingly, addition of 0.1 M ammonium hydroxide at pH 10 to ripening avocado fruits enhanced the activation of quiescent infection and disease symptoms, suggesting that pH alkalinization regulates both *pelB* expression and transition of *C. gloeosporioides* from the quiescent biotrophic stage to the necrotrophic stage (Kramer-Haimovich et al. 2006).

In contrast to PLs, PGs exhibit maximum catalytic activity at acidic ambient pH. Accordingly, endo PG genes from *S. sclerotiorum* and *F. oxysporum* were transcribed maximally under acidic culture conditions (Rollins and Dickman 2001; Caracuel et al. 2003). In *Sclerotinia*, transcripts of genes encoding PGs and other lytic enzymes were restricted to the most acidic zones of the infected tissue, suggesting that progressive acidification of the substrate by the fungus, possibly via secretion of oxalic acid, may be a major strategy for regulating the sequential expression of CWDEs (Cotton et al. 2003).

In filamentous fungi, gene expression by ambient pH is regulated by a conserved signalling cascade whose terminal component, the zinc finger transcription factor PacC, activates alkaline-expressed genes and represses acid-expressed genes (Penalva and Arst 2004). Inactivation of *pacC* in the root pathogen *F. oxysporum* resulted in an acidity-mimicking phenotype: impaired growth at alkaline pH, higher transcript levels of acid-expressed endoPG genes and increased virulence on tomato plants (Caracuel et al. 2003). Conversely, *Fusarium* strains carrying a dominant activating *pacC^c* allele exhibited an the opposite phenotype (alkalinity-mimicking), a decrease in endoPG gene expression and reduced virulence on tomato (Caracuel et al. 2003). Since *F. oxysporum* is also a opportunistic pathogen of mammals, virulence of the *pacC* mutants was assayed on immunodepressed mice. Strikingly, the *pacC* loss-of-function mutant, which was hypervirulent on tomato, was essentially avirulent on mice (Ortoneda et al. 2004). The divergent roles of PacC in virulence on plants and on mammals may be due to differences in ambient pH between the two types of host organisms. In the necrotrophic fungus *S. sclerotiorum*, mutants lacking the orthologous *pac1* gene showed a significant shift in PG gene expression to higher ambient pH and, in sharp contrast to *F. oxysporum*, a dramatically reduced virulence on tomato (Rollins 2003). These contrasting results suggest that requirements for pH-related expression and function of CWDEs during infection may vary considerably in different pathogen–host systems.

D. Other Mechanisms Regulating CWDE Expression

Additional regulatory mechanisms affecting expression of CWDEs have been reported. Deletion of *HDC1* encoding a histone deacetylase of *C. carbonum* gave rise to mutants with normal growth on glucose, but significantly reduced growth on complex carbohydrates (Baidyaroy et al. 2001). Extracellular CWDE activities and expression of the corresponding genes were downregulated in *hdc1* mutant strains, which also had strongly reduced penetration efficiency and virulence. Except for altered conidial morphology, the phenotypes of *hdc1* mutants were similar to those of *C. carbonum* strains mutated in *ccSNF1* (Tonukari et al. 2000). The exact mechanism of regulation of CWDE expression by the histone deacetylase is currently unknown.

IV. CWDE Inhibitor Proteins from Plants

More than 30 years ago, Albersheim and colleagues reported that plant cell walls contained proteins able to reduce the activity of fungal endoPGs, which they termed PG-inhibiting proteins or PGIPs (Albersheim and Anderson 1971). Proteinaceous inhibitors of other classes of fungal CWDEs have since been identified in plants, including inhibitors of pectin lyases (Bugbee 1993), endoxylanases (Debyser et al. 1999) or proteases (Ryan 1990; McLauchlan et al. 1999).

A. Polygalacturonase-Inhibiting Proteins

The inhibitory mechanism of PGIPs has been studied in detail (De Lorenzo et al. 2001; Federici et al. 2006). The interaction between PGIP and endoPG is characterized by high affinity, reversibility and a 1:1 stoichiometry, resulting in up to 99.7% reduction in the catalytic rate of endoPG. Catalytic function is not a prerequisite for PGIP binding, since site-directed mutagenesis of a histidine residue critical for enzymatic and macerating activity of a *F. moniliforme* endoPG did not affect its affinity for PGIP (Caprari et al. 1996).

Fungal PGs with different substrate degradation patterns vary substantially in their susceptibilities to PGIPs. In a comparative study, four PGs which were inhibited by all plant PGIPs tested exhibited an endo/exo mode of substrate cleavage, whereas three PGs showing a classic endo pattern of cleavage were resistant to inhibition by one or more of the PGIPs (Cook et al. 1999). Ambient pH also has a major effect on the PG-PGIP interaction (Favaron et al. 2004; Kemp et al. 2004). Indeed, a recent study suggests that the interaction with PGIP may result in either inhibition or activation of fungal endoPG, depending on the ambient pH (Kemp et al. 2004).

PGIPs are thought to play an important role in plant defence (De Lorenzo et al. 2001). Besides direct inhibition of fungal endoPG, PGIPs favour the accumulation of elicitor-active oligogalacturonides, thus converting a potential virulence factor into a trigger of plant defence mechanisms (Cervone et al. 1989). Moreover, a recent study suggested that endoPG1 from *B. cinerea* can activate grapevine defence reactions in a way that is unrelated to its enzymatic activity (Poinssot et al. 2003).

Supporting a role of PGIPs as part of the plant defence machinery, constitutive expression of pear PGIP in transgenic tomato plants resulted in reduced growth of *B. cinerea* on ripe fruit and a 25% decrease of lesions of macerated tissue (Powell et al. 2000). In an opposite approach, antisense inhibition of *AtPGIP1* gene expression resulted in enhanced susceptibility of transgenic *Arabidopsis* plants to *B. cinerea*, further supporting the idea that PGIP contributes to basal resistance against this fungal pathogen (Ferrari et al. 2006).

B. Endoxylanase-Inhibiting Proteins

An excellent review on the role of endoxylanase inhibitors in plant defence was recently published (Belien et al. 2006). Two distinct classes of endoxylanase inhibitors are reported in cereals: *Triticum aestivum* xylanase inhibitor (TAXI; Debyser et al. 1999) and xylanase inhibitor protein (XIP; McLauchlan et al. 1999). TAXI exhibits inhibitory activity towards fungal and bacterial GH11 family endoxylanases (Gebruers et al. 2004; Belien et al. 2005), whereas XIP typically inhibits GH10 and GH11 endoxylanases from fungi (Juge et al. 2004). This distinct specificity towards endoxylanases of microbial origin, together with the ineffectiveness of TAXI and XIP against endogenous wheat endoxylanases, led to the hypothesis that both types of endoxylanase inhibitors evolved as plant defence mechanisms, in analogy with inhibitors of PGs (De Lorenzo et al. 2001).

Endoxylanases from two phytopathogenic fungi were tested for inhibition by TAXI and XIP. TAXI-I and XIP-I, but not TAXI-II, were able to inhibit the GH11 endoxylanase XynBc1 from *B. cinerea* (Brutus et al. 2005), whereas two GH11 endoxylanases from *F. graminearum* (XylA and XylB) were inhibited by TAXI-I but not by XIP-I (Belien et al. 2005). Based on these findings, a co-evolutionary process was proposed, in which plants evolved different classes of inhibitors to counteract the arsenal of endoxylanases secreted by microbial pathogens, while the latter evolved to produce multiple endoxylanases with distinct sensitivities towards the endoxylanase inhibitors (Belien et al. 2006).

Intriguingly, certain fungal endoxylanases act as potent elicitors of the plant defence response (Dean et al. 1989). When applied to specific cultivars of tobacco and tomato plants, an endoxylanase from

Table 10.1. Fungal cell wall-degrading enzymes with a role in virulence demonstrated by gene knockout

Gene	Enzyme activity	Pathogen/plant host	GenBank no.	Reference
pecA	Endopolygalacturonase	*Aspergillus flavus*/cotton	U05015	Shieh et al. (1997)
Bcpg1	Endopolygalacturonase	*Botrytis cinerea*/tomato, apple	U68715	ten Have et al. (1998)
Bcpg2	Endopolygalacturonase	*B. cinerea*/tomato, bean	U68716	Kars et al. (2005)
cppg1 cppg2	Endopolygalacturonase	*Claviceps purpurea*/rye	Y10165	Oeser et al. (2002)
acpg-1	Endopolygalacturonase	*Alternaria citri*/potato, citrus	AB047543	Isshiki et al. (2001)
pelB	Endopectate lyase	*C. gloeosporioides*/avocado	AF052632	Yakoby et al. (2001)
pelA, pelD	Endopectate lyase	*Fusarium solani* f. sp. *pisi*/pea	M94691 U13050	Rogers et al. (2000)
Bcpme1	Pectin methylesterase	*B. cinerea*/apple, grape	AJ309701	Valette-Collet et al. (2003)
xyn11A	Endoxylanase (GH11)	*B. cinerea*/tomato, grape	DQ057980	Brito et al. (2006)
cut1	Cutinase	*F. solani* f. sp. *pisi*/pea	M29759	Rogers et al. (1994)
FGL1	Lipase	*F. graminearum*/wheat, maize	AY292529	Voigt et al. (2005)

T. viride referred to as the ethylene-inducing xylanase (EIX) induced ethylene production and rapid cell death, two hallmarks of the hypersensitive response (Bailey et al. 1990, 1992). The plant receptor for EIX was recently shown to be a member of a resistance-like gene family in tomato (Ron and Avni 2004).

V. Conclusions and Outlook: Genomics and Proteomics Applied to the Study of CWDEs in Plant Pathogens

The past decade saw important advances in the knowledge of fungal CWDEs and their contribution to pathogenicity. A number of gene knockout studies successfully demonstrated a role of fungal depolymerases in virulence (Table 10.1). Substantial progress was also made in unravelling the regulatory circuits that control CWDE expression in fungi (Aro et al. 2005). In spite of these positive notes, for most pathogen-plant systems the role of depolymerases during infection remains largely unresolved. Besides functional redundancy, a second important setback is our incomplete knowledge on the variety of wall depolymerases made by pathogens, which makes it difficult to identify those enzymes that are crucial for infection. The promising results from recent genomics and proteomics studies on CWDEs in fungal pathogens suggest that these approaches have a strong potential to help fillling this knowledge gap.

Lack of information on the complete set of CWDE genes in a fungal organism represents a severe drawback for mutational studies. Such knowledge not only provides an idea on the complexity of different isozymes expected for each group of depolymerases but may even lead to the identification of novel types of wall depolymerases specific for uncommon linkages in the native wall, whose disruption is critical for successful infection (Walton 1994). Knowing all the CWDEs present in a given pathogen species should also lead to a more rational approach in choosing the most interesting candidates for gene knockout studies. A number of fungal genome sequences were recently published, including those of the cereal pathogens *Magnaporthe grisea* and *Fusarium graminearum* and those of the saprophytes *Aspergillus nidulans* and *Neurospora crassa*. The availability of these sequences provided, for the first time, an insight into the complete arsenal of fungal CWDEs. Comparative analysis between these species revealed a significant expansion of the number of genes encoding cutinases, pectate lyases and subtilisin-like serine proteases in the pathogens, suggesting an implication of these types of CWDEs in pathogenicity (Dean et al. 2005; Cuomo et al. 2007). A more detailed examination of the CWDE-encoding sequences in the genomes of these and other fungal pathogens will most certainly provide novel insights into their regulation and role in pathogenicity.

Proteomics approaches allow the high-throughput analysis of complex protein mixtures with little or no prior fractionation and are therefore particularly well suited to the study of intermixed genomes such as a plant infected by a fungus. A recent analysis of the secretome of *F. graminearum* identified 289 proteins secreted during

growth on different media or during infection
of wheat, a significant proportion of which were
CWDEs (Paper et al. 2007). Among a total of 120
secreted fungal proteins identified with high reli-
ablility in wheat heads, 27 were depolymerases of
cell wall polysaccharides, including pectate lyase, β-
xylosidase, exo-arabinase, cellulase, cellobiohydrolase, endoglucanase, β-1,3-glucosidase,
mixed-linked glucanase, glucuronyl hydrolase, α-
galactosidase and α-amylase, and 13 were proteases.
For several classes of CWDEs (α-galactosidase,
xylanase, cellulase, cellobiohydrolase, subtilase),
multiple paralogues were expressed in planta. This
high degree of genetic redundancy points to the
importance of these enzymes for the fungus, but it
also highlights the difficulty to test their function
by conventional gene disruption (Paper et al. 2007).

A previous proteomic analysis of *F. graminearum* grown in vitro on cell walls of hop (*Humulus
lupulus*, L.) identified 84 extracellular proteins,
including different groups of CWDEs (Phalip et al.
2005). Similar to the study by Paper et al. (2007),
high functional redundancy was detected for some
enzyme classes such as endo-1,4-β-cellulase, endo-
1,4-β-xylanase, 1,4-xylosidase or pectate lyase, with
up to eight different secreted proteins within a single class (Phalip et al. 2005).

A more targeted proteomic approach focused
on the analysis of secreted proteins in the xylem
sap of tomato plants infected with *F. oxysporum*, a vascular wilt pathogen that colonizes the
xylem vessels of the host plant (Houterman et al.
2007). Seven fungal proteins were identified by
mass spectrometry, including a number of small
cysteine-rich proteins and two CWDEs, an exo-
arabinase and a subtilisin-like serine protease.

The results from these and future genomics
and proteomics studies should prove highly useful in directing research efforts towards the most
promising classes of CWDEs/genes for targeted
mutational analysis. Complementary approaches
combining comparative genomics with large-scale
DNA-binding studies such as those performed in
Saccharomyces cerevisiae (Harbison et al. 2004)
should provide new insights into the cellular
mechanisms regulating CWDE expression during
pathogenesis. This knowledge will be essential to
overcome the problem of functional redundancy,
which is likely to remain the major limitation in
furthering our understanding of the role of fungal
CWDEs in plant infection.

References

Ahn JH, Sposato P, Kim SI, Walton JD (2001) Molecular
cloning and characterization of *cel2* from the fungus
Cochlibolus carbonum. Biosci Biotechnol Biochem
65:1406–1411

Albersheim P, Anderson AJ (1971) Proteins from plant cell
walls inhibit polygalacturonases secreted by plant
pathogens. Proc Natl Acad Sci USA 68:1815–1819

Apel PC, Panaccione DG, Holden FR, Walton JD (1993)
Cloning and targeted gene disruption of *XYL1*, a
beta 1,4-xylanase gene from the maize pathogen
Cochliobolus carbonum. Mol Plant–Microbe Interact
6:467–473

Apel-Birkhold PC, Walton JD (1996) Cloning, disruption,
and expression of two endo-beta 1,4-xylanase genes,
XYL2 and *XYL3*, from *Cochliobolus carbonum*. Appl
Environ Microbiol 62:4129–4135

Aro N, Pakula T, Penttila M (2005) Transcriptional regulation of plant cell wall degradation by filamentous
fungi. FEMS Microbiol Rev 29:719–739

Baidyaroy D, Brosch G, Ahn JH, Graessle S, Wegener S,
Tonukari NJ, Caballero O, Loidl P, Walton JD (2001)
A gene related to yeast HOS2 histone deacetylase
affects extracellular depolymerase expression and
virulence in a plant pathogenic fungus. Plant Cell
13:1609–1624

Bailey BA, Dean JF, Anderson JD (1990) An ethylene
biosynthesis-inducing endoxylanase elicits electrolyte
leakage and necrosis in *Nicotiana tabacum* cv Xanthi
leaves. Plant Physiol 94:1849–1854

Bailey BA, Korcak RF, Anderson JD (1992) Alterations in
Nicotiana tabacum L. cv Xanthi cell membrane function
following treatment with an ethylene biosynthesis-
inducing endoxylanase. Plant Physiol 100:749–755

Bastmeyer M, Deising HB, Bechinger C (2002) Force exer-
tion in fungal infection. Annu Rev Biophys Biomol
Struct 31:321–341

Beg Q, Kapoor M, Mahajan L, Hoondal G (2001) Microbial
xylanases and their industrial applications: a review.
Appl Microbiol Biotechnol 56:326–338

Belien T, Van Campenhout S, Van Acker M, Volckaert G
(2005) Cloning and characterization of two endoxylanases from the cereal phytopathogen *Fusarium
graminearum* and their inhibition profile against
endoxylanase inhibitors from wheat. Biochem Biophys Res Commun 327:407–414

Belien T, Van Campenhout S, Robben J, Volckaert G (2006)
Microbial endoxylanases: effective weapons to breach
the plant cell-wall barrier or, rather, triggers of
plant defense systems? Mol Plant–Microbe Interact
19:1072–1081

Berto P, Commenil P, Belingheri L, Dehorter B (1999)
Occurrence of a lipase in spores of *Alternaria brassicicola* with a crucial role in the infection of cauliflower
leaves. FEMS Microbiol Lett 180:183–189

Biely P, Tenkanen M (1998) Enzymology of hemicellulose
degradation. In: Kubicek CP, Harman GE, Ondik KL (ed)
Trichoderma and *Gliocladium*: basic biology, taxonomy
and genetics. Taylor and Francis, London, pp 25–47

Bindschedler LV, Sanchez P, Dunn S, Mikan J, Thangavelu M, Clarkson JM, Cooper RM (2003) Deletion of the SNP1 trypsin protease from *Stagonospora nodorum* reveals another major protease expressed during infection. Fungal Genet Biol 38:43–53

Bluhm BH, Woloshuk CP (2005) Amylopectin induces fumonisin B1 production by *Fusarium verticillioides* during colonization of maize kernels. Mol Plant Microbe Interact 18:1333–1339

Bowen JK, Templeton MD, Sharrock KR, Crowhurst RN, Rikkerink EH (1995) Gene inactivation in the plant pathogen *Glomerella cingulata*: three strategies for the disruption of the pectin lyase gene *pnlA*. Mol Gen Genet 246:196–205

Brito N, Espino JJ, Gonzalez C (2006) The endo-beta-1,4-xylanase xyn11A is required for virulence in *Botrytis cinerea*. Mol Plant Microbe Interact 19:25–32

Brutus A, Reca IB, Herga S, Mattei B, Puigserver A, Chaix JC, Juge N, Bellincampi D, Giardina T (2005) A family 11 xylanase from the pathogen *Botrytis cinerea* is inhibited by plant endoxylanase inhibitors XIP-I and TAXI-I. Biochem Biophys Res Commun 337:160–166

Calero-Nieto F, Di Pietro A, Roncero MI, Hera C (2007a) Role of the transcriptional activator XlnR of *Fusarium oxysporum* in regulation of xylanase genes and virulence. Mol Plant–Microbe Interact 20:977–985

Calero-Nieto F, Hera C, Di Pietro A, Orejas M, Roncero MI (2007b) Regulatory elements mediating expression of xylanase genes in *Fusarium oxysporum*. Fungal Genet Biol (in press)

Cano-Canchola C, Acevedo L, Ponce-Noyola P, Flores-Martinez A, Flores-Carreon A, Leal-Morales CA (2000) Induction of lytic enzymes by the interaction of *Ustilago maydis* with *Zea mays* tissues. Fungal Genet Biol 29:145–151

Caprari C, Mattei B, Basile ML, Salvi G, Crescenzi V, De Lorenzo G, Cervone F (1996) Mutagenesis of endopolygalacturonase from *Fusarium moniliforme*: histidine residue 234 is critical for enzymatic and macerating activities and not for binding to polygalacturonase-inhibiting protein (PGIP). Mol Plant–Microbe Interact 9:617–624

Caracuel Z, Roncero MI, Espeso EA, Gonzalez-Verdejo CI, Garcia-Maceira FI, Di Pietro A (2003) The pH signalling transcription factor PacC controls virulence in the plant pathogen *Fusarium oxysporum*. Mol Microbiol 48:765–779

Carlile AJ, Bindschedler LV, Bailey AM, Bowyer P, Clarkson JM, Cooper RM (2000) Characterization of SNP1, a cell wall-degrading trypsin, produced during infection by *Stagonospora nodorum*. Mol Plant-Microbe Interact 13:538–550

Carpita NC, Gibeaut DM (1993) Structural models of primary cell walls in flowering plants: consistency of molecular structure with the physical properties of the walls during growth. Plant J 3:1–30

Centis S, Dumas B, Fournier J, Marolda M, Esquerre-Tugaye MT (1996) Isolation and sequence analysis of *Clpg1*, a gene coding for an endopolygalacturonase of the phytopathogenic fungus *Colletotrichum lindemuthianum*. Gene 170:125–129

Cervone F, Hahn MG, De Lorenzo G, Darvill A, Albersheim P (1989) Host-pathogen interactions: XXXIII. A plant protein converts a fungal pathogenesis factor into an elicitor of plant defense responses. Plant Physiol 90:542–548

Chacon-Martinez C, Anzola J, Rojas A, Hernandez F, Junca H, Ocampo W, Del Portillo P (2004) Identification and characterization of the alpha-L-arabinofuranosidase B of *Fusarium oxysporum* f. sp *dianthi*. Physiol Mol Plant Pathol 64:201–208

Cho Y, Cramer RA, Jr., Kim KH, Davis J, Mitchell TK, Figuli P, Pryor BM, Lemasters E, Lawrence CB (2007) The Fus3/Kss1 MAP kinase homolog Amk1 regulates the expression of genes encoding hydrolytic enzymes in *Alternaria brassicicola*. Fungal Genet Biol 44:543–553

Collins T, Gerday C, Feller G (2005) Xylanases, xylanase families and extremophilic xylanases. FEMS Microbiol Rev 29:3–23

Collmer A, Keen NT (1986) The role of pectic enzymes in plant pathogenesis. Annu Rev Phytopathol 24:383–409

Commenil P, Belingheri L, Dehorter B (1998) Antilipase antibodies prevent infection of tomato leaves by *Botrytis cinerea*. Physiol Mol Plant Pathol 52:1–14

Commenil P, Belingheri L, Bauw G, Dehorter B (1999) Molecular characterization of a lipase induced in *Botrytis cinerea* by components of grape berry cuticle. Physiol Mol Plant Pathol 55:37–43

Cook BJ, Clay RP, Bergmann CW, Albersheim P, Darvill AG (1999) Fungal polygalacturonases exhibit different substrate degradation patterns and differ in their susceptibilities to polygalacturonase-inhibiting proteins. Mol Plant–Microbe Interact 12:703–711

Cooper R, Wood RKS (1973) Induction of synthesis of extracellular cell-wall degrading enzymes in vascular wilt fungi. Nature 246:309–311

Cooper R (1984) The role of cell wall-degrading enzymes in infection and damage. In: Wood RKS, Jellis GJ (ed) Plant diseases: infection, damage and loss. Blackwell, Oxford, pp 13–27

Cooper R, Longman D, Campbell A, Henry M, Lees P (1988) Enzymic adaptation of cereal pathogens to the monocotyledonous primary wall. Physiol Mol Plant Pathol 32:33–47

Cosgrove DJ (2001) Wall structure and wall loosening. A look backwards and forwards. Plant Physiol 125:131–134

Cotton P, Kasza Z, Bruel C, Rascle C, Fevre M (2003) Ambient pH controls the expression of endopolygalacturonase genes in the necrotrophic fungus *Sclerotinia sclerotiorum*. FEMS Microbiol Lett 227:163–169

Crawford MS, Kolattukudy PE (1987) Pectate lyase from *Fusarium solani* f. sp. *pisi*: purification, characterization, in vitro translation of the mRNA, and involvement in pathogenicity. Arch Biochem Biophys 258:196–205

Crowhurst RN, Binnie SJ, Bowen JK, Hawthorne BT, Plummer KM, Rees-George J, Rikkerink EH, Templeton MD (1997) Effect of disruption of a cutinase gene (*cutA*) on virulence and tissue specificity of *Fusarium solani* f. sp. *cucurbitae* race 2 toward *Cucurbita maxima* and *C. moschata*. Mol Plant Microbe Interact 10:355–368

Cuomo CA, Guldener U, Xu JR, Trail F, Turgeon BG, Di Pietro A, Walton JD, Ma LJ, Baker SE, Rep M, Adam G, Antoniw J, Baldwin T, Calvo S, Chang YL, Decaprio

D, Gale LR, Gnerre S, Goswami RS, Hammond-Kosack K, Harris LJ, Hilburn K, Kennell JC, Kroken S, Magnuson JK, Mannhaupt G, Mauceli E, Mewes HW, Mitterbauer R, Muehlbauer G, Munsterkotter M, Nelson D, O'Donnell K, Ouellet T, Qi W, Quesneville H, Roncero MI, Seong KY, Tetko IV, Urban M, Waalwijk C, Ward TJ, Yao J, Birren BW, Kistler HC (2007) The *Fusarium graminearum* genome reveals a link between localized polymorphism and pathogen specialization. Science 317:1400–1402

De Lorenzo G, D'Ovidio R, Cervone F (2001) The role of polygalacturonase-inhibiting proteins (PGIPs) in defense against pathogenic fungi. Annu Rev Phytopathol 39:313–335

de Vries RP, Visser J (1999) Regulation of the feruloyl esterase (*faeA*) gene from *Aspergillus niger*. Appl Environ Microbiol 65:5500–5503

de Vries RP, Visser J (2001) Aspergillus enzymes involved in degradation of plant cell wall polysaccharides. Microbiol Mol Biol Rev 65:497–522

de Vries RP, van de Vondervoort PJ, Hendriks L, van de Belt M, Visser J (2002) Regulation of the alpha-glucuronidase-encoding gene (*aguA*) from *Aspergillus niger*. Mol Genet Genomics 268:96–102

Dean J, Gamble H, Anderson J (1989) The ethylene biosynthesis-inducing xylanase – its induction in *Trichoderma viride* and certain plant pathogens. Phytopathology 79:1071–1078

Dean RA, Talbot NJ, Ebbole DJ, Farman ML, Mitchell TK, Orbach MJ, Thon M, Kulkarni R, Xu JR, Pan H, Read ND, Lee YH, Carbone I, Brown D, Oh YY, Donofrio N, Jeong JS, Soanes DM, Djonovic S, Kolomiets E, Rehmeyer C, Li W, Harding M, Kim S, Lebrun MH, Bohnert H, Coughlan S, Butler J, Calvo S, Ma LJ, Nicol R, Purcell S, Nusbaum C, Galagan JE, Birren BW (2005) The genome sequence of the rice blast fungus *Magnaporthe grisea*. Nature 434:980–986

Debyser W, Peumans WJ, Van Damme EJM, Delcour JA (1999) *Triticum aestivum* xylanase inhibitor (TAXI), a new class of enzyme inhibitor affecting breadmaking performance. J Cereal Sci 30:39–43

Deising H, Nicholson RL, Haug M, Howard RJ, Mendgen K (1992) Adhesion pad formation and the involvement of cutinase and esterases in the attachment of uredospores to the host cuticle. Plant Cell 4:1101–1111

Delgado-Jarana J, Martinez-Rocha AL, Roldan-Rodriguez R, Roncero MI, Di Pietro A (2005) *Fusarium oxysporum* G-protein beta subunit Fgb1 regulates hyphal growth, development, and virulence through multiple signalling pathways. Fungal Genet Biol 42:61–72

Di Pietro A, Roncero MIG (1996a) Purification and characterization of a pectate lyase from *Fusarium oxysporum* f.sp. *lycopersici* produced on tomato vascular tissue. Physiol Mol Plant Pathol 49:177–185

Di Pietro A, Roncero MIG (1996b) Endopolygalacturonase from *Fusarium oxysporum* f.sp. *lycopersici*: Purification, characterization, and production during infection of tomato plants. Phytopathology 86:1324–1330

Di Pietro A, Roncero MI (1998) Cloning, expression, and role in pathogenicity of *pg1* encoding the major extracellular endopolygalacturonase of the vascular wilt pathogen *Fusarium oxysporum*. Mol Plant–Microbe Interact 11:91–98

Di Pietro A, Garcia-Maceira FI, Huertas-Gonzalez MD, Ruiz-Roldan MC, Caracuel Z, Barbieri AS, Roncero MIG (1998) Endopolygalacturonase PG1 in different formae speciales of *Fusarium oxysporum*. Appl Environ Microbiol 64:1967–1971

Di Pietro A, Garcia-Maceira FI, Meglecz E, Roncero MI (2001) A MAP kinase of the vascular wilt fungus *Fusarium oxysporum* is essential for root penetration and pathogenesis. Mol Microbiol 39:1140–1152

Dickman MB, G.K. P, Kolattukudy PE (1989) Insertion of cutinase gene into a root pathogen enables it to infect intact host. Nature 342:446–448

Doss RP (1999) Composition and enzymatic activity of the extracellular matrix secreted by germlings of *Botrytis cinerea*. Appl Environ Microbiol 65:404–408

Drori N, Kramer-Haimovich H, Rollins J, Dinoor A, Okon Y, Pines O, Prusky D (2003) External pH and nitrogen source affect secretion of pectate lyase by *Colletotrichum gloeosporioides*. Appl Environ Microbiol 69:3258–3262

Dumas B, Centis S, Sarrazin N, Esquerre-Tugaye MT (1999) Use of green fluorescent protein to detect expression of an endopolygalacturonase gene of *Colletotrichum lindemuthianum* during bean infection. Appl Environ Microbiol 65:1769–1771

Eshel D, Ben-Aire R, Dinoor A, Prusky D (2000) Resistance of giberellin-treated persimmon fruit to *Alternaria alternata* arises from reduced ability of the fungus to produce endo-1,4-β-glucanase. Phytopathology 90:1256–1262

Favaron F, Sella L, D'Ovidio R (2004) Relationships among endo-polygalacturonase, oxalate, pH, and plant polygalacturonaseinhibiting protein (PGIP) in the interaction between *Sclerotinia sclerotiorum* and soybean. Mol Plant–Microbe Interact 17:1402–1409

Federici L, Di Matteo A, Fernandez-Recio J, Tsernoglou D, Cervone F (2006) Polygalacturonase inhibiting proteins: players in plant innate immunity? Trends Plant Sci 11:65–70

Ferrari S, Galletti R, Vairo D, Cervone F, De Lorenzo G (2006) Antisense expression of the *Arabidopsis thaliana AtPGIP1* gene reduces polygalacturonase-inhibiting protein accumulation and enhances susceptibility to *Botrytis cinerea*. Mol Plant–Microbe Interact 19:931–936

Gao S, Choi GH, Shain L, Nuss DL (1996) Cloning and targeted disruption of *enpg-1*, encoding the major in vitro extracellular endopolygalacturonase of the chestnut blight fungus, *Cryphonectria parasitica*. Appl Environ Microbiol 62:1984–1990

Garcia-Maceira FI, Di Pietro A, Roncero MI (2000) Cloning and disruption of *pgx4* encoding an in planta expressed exopolygalacturonase from *Fusarium oxysporum*. Mol Plant Microbe Interact 13:359–365

Garcia-Maceira FI, Di Pietro A, Huertas-Gonzalez MD, Ruiz-Roldan MC, Roncero MI (2001) Molecular characterization of an endopolygalacturonase from *Fusarium oxysporum* expressed during early stages of infection. Appl Environ Microbiol 67:2191–2196

Gebruers K, Brijs K, Courtin CM, Fierens K, Goesaert H, Rabijns A, Raedschelders G, Robben J, Sansen S,

Sorensen JF, Van Campenhout S, Delcour JA (2004) Properties of TAXI-type endoxylanase inhibitors. Biochim Biophys Acta 1696:213–221

Giesbert S, Lepping H, Tenberge K, Tudzynski P (1998) The xylanolytic system of *Claviceps purpurea*: Cytological evidence for secretion of xylanases in infected rye tissue and molecular characterization of two xylanase genes. Phytopathology 88:1020–1030

Gomez-Gomez E, Roncero MIG, Di Pietro A, Hera C (2001) Molecular characterization of a novel endo-beta-1,4-xylanase gene from the vascular wilt fungus *Fusarium oxysporum*. Curr Genet 40:268–275

Gomez-Gomez E, Ruiz-Roldan MC, Di Pietro A, Roncero MI, Hera C (2002) Role in pathogenesis of two endo-beta-1,4-xylanase genes from the vascular wilt fungus *Fusarium oxysporum*. Fungal Genet Biol 35:213–222

Gonzalez-Candelas L, Kolattukudy PE (1992) Isolation and analysis of a novel inducible pectate lyase gene from the phytopathogenic fungus *Fusarium solani* f. sp. *pisi* (*Nectria haematococca*, mating population VI). J Bacteriol 174:6343–6349

Gorlach JM, Van Der Knaap E, Walton JD (1998) Cloning and targeted disruption of *MLG1*, a gene encoding two of three extracellular mixed-linked glucanases of *Cochliobolus carbonum*. Appl Environ Microbiol 64:385–391

Guo W, Gonzalez-Candelas L, Kolattukudy PE (1996) Identification of a novel *pelD* gene expressed uniquely in planta by *Fusarium solani* f. sp. *pisi* (*Nectria haematococca*, mating type VI) and characterization of its protein product as an endo-pectate lyase. Arch Biochem Biophys 332:305–312

Ha MA, Apperley DC, Jarvis MC (1997) Molecular rigidity in dry and hydrated onion cell walls. Plant Physiol 115:593–598

Harbison CT, Gordon DB, Lee TI, Rinaldi NJ, Macisaac KD, Danford TW, Hannett NM, Tagne JB, Reynolds DB, Yoo J, Jennings EG, Zeitlinger J, Pokholok DK, Kellis M, Rolfe PA, Takusagawa KT, Lander ES, Gifford DK, Fraenkel E, Young RA (2004) Transcriptional regulatory code of a eukaryotic genome. Nature 431:99–104

Hardie DG, Carling D, Carlson M (1998) The AMP-activated/SNF1 protein kinase subfamily: metabolic sensors of the eukaryotic cell? Annu Rev Biochem 67:821–855

Hatsch D, Phalip V, Petkovski E, Jeltsch JM (2006) *Fusarium graminearum* on plant cell wall: no fewer than 30 xylanase genes transcribed. Biochem Biophys Res Commun 345:959–966

Herbert C, Jacquet C, Borel C, Esquerre-Tugaye MT, Dumas B (2002) A cis-acting sequence homologous to the yeast filamentation and invasion response element regulates expression of a pectinase gene from the bean pathogen *Colletotrichum lindemuthianum*. J Biol Chem 277:29125–29131

Herbert C, O'Connell R, Gaulin E, Salesses V, Esquerre-Tugaye MT, Dumas B (2004) Production of a cell wall-associated endopolygalacturonase by *Colletotrichum lindemuthianum* and pectin degradation during bean infection. Fungal Genet Biol 41:140–147

Hoi JW, Herbert C, Bacha N, O'Connell R, Lafitte C, Borderies G, Rossignol M, Rouge P, Dumas B (2007) Regulation and role of a STE12-like transcription factor from the plant pathogen *Colletotrichum lindemuthianum*. Mol Microbiol 64:68–82

Houterman PM, Speijer D, Dekker HJ, De Koster CG, Cornelissen BJC, Rep M (2007) The mixed xylem sap proteome of *Fusarium oxysporum*-infected tomato plants. Mol Plant Pathol 8:215–221

Howard RJ, Ferrari MA, Roach DH, Money NP (1991) Penetration of hard substrates by a fungus employing enormous turgor pressures. Proc Natl Acad Sci USA 88:11281–11284

Huertas-Gonzalez MD, Ruiz-Roldan MC, Garcia Maceira FI, Roncero MI, Di Pietro A (1999) Cloning and characterization of *pl1* encoding an in planta-secreted pectate lyase of *Fusarium oxysporum*. Curr Genet 35:36–40

Isshiki A, Akimitsu K, Yamamoto M, Yamamoto H (2001) Endopolygalacturonase is essential for citrus black rot caused by *Alternaria citri* but not brown spot caused by *Alternaria alternata*. Mol Plant–Microbe Interact 14:749–757

Jaeger KE, Eggert T (2002) Lipases for biotechnology. Curr Opin Biotechnol 13:390–397

Jenczmionka NJ, Schäfer W (2005) The Gpmk1 MAP kinase of *Fusarium graminearum* regulates the induction of specific secreted enzymes. Curr Genet 47:29–36

Juge N, Payan F, Williamson G (2004) XIP-I, a xylanase inhibitor protein from wheat: a novel protein function. Biochim Biophys Acta 1696:203–211

Kars I, Krooshof GH, Wagemakers L, Joosten R, Benen JA, van Kan JA (2005) Necrotizing activity of five *Botrytis cinerea* endopolygalacturonases produced in *Pichia pastoris*. Plant J 43:213–225

Kasza Z, Vagvolgyi C, Fevre M, Cotton P (2004) Molecular characterization and in planta detection of *Sclerotinia sclerotiorum* endopolygalacturonase genes. Curr Microbiol 48:208–213

Kemp G, Stanton L, Bergmann CW, Clay RP, Albersheim P, Darvill A (2004) Polygalacturonase-inhibiting proteins can function as activators of polygalacturonase. Mol Plant–Microbe Interact 17:888–894

Kim H, Ahn JH, Gorlach JM, Caprari C, Scott-Craig JS, Walton JD (2001) Mutational analysis of beta-glucanase genes from the plant-pathogenic fungus *Cochliobolus carbonum*. Mol Plant Microbe Interact 14:1436–1443

Kolattukudy PE (2001) Polyesters in higher plants. Adv Biochem Eng Biotechnol 71:1–49

Kramer-Haimovich H, Servi E, Katan T, Rollins J, Okon Y, Prusky D (2006) Effect of ammonia production by *Colletotrichum gloeosporioides* on *pelB* activation, pectate lyase secretion, and fruit pathogenicity. Appl Environ Microbiol 72:1034–1039

Lehtinen U (1993) Plant-cell wall degrading enzymes of *Septoria nodorum*. Physiol Mol Plant Pathol 43:121–134

Lev S, Horwitz BA (2003) A mitogen-activated protein kinase pathway modulates the expression of two cellulase genes in *Cochliobolus heterostrophus* during plant infection. Plant Cell 15:835–844

Li D, Kolattukudy PE (1997) Cloning of cutinase transcription factor 1, a transactivating protein containing Cys6Zn2 binuclear cluster DNA-binding motif. J Biol Chem 272:12462–12467

Li D, Sirakova T, Rogers L, Ettinger WF, Kolattukudy PE (2002) Regulation of constitutively expressed and induced cutinase genes by different zinc finger transcription factors in *Fusarium solani* f. sp. *pisi* (*Nectria haematococca*). J Biol Chem 277:7905–7912

Li R, Rimmer R, Buchwaldt L, Sharpe AG, Seguin-Swartz G, Hegedus DD (2004) Interaction of *Sclerotinia sclerotiorum* with *Brassica napus*: cloning and characterization of endo- and exo-polygalacturonases expressed during saprophytic and parasitic modes. Fungal Genet Biol 41:754–765

Lynd LR, Weimer PJ, van Zyl WH, Pretorius IS (2002) Microbial cellulose utilization: fundamentals and biotechnology. Microbiol Mol Biol Rev 66:506–577

Mach RL, Strauss J, Zeilinger S, Schindler M, Kubicek CP (1996) Carbon catabolite repression of xylanase I (*xyn1*) gene expression in *Trichoderma reesei*. Mol Microbiol 21:1273–1281

Marui J, Tanaka A, Mimura S, de Graaff LH, Visser J, Kitamoto N, Kato M, Kobayashi T, Tsukagoshi N (2002) A transcriptional activator, AoXlnR, controls the expression of genes encoding xylanolytic enzymes in *Aspergillus oryzae*. Fungal Genet Biol 35:157–169

McLauchlan WR, Garcia-Conesa MT, Williamson G, Roza M, Ravestein P, Maat J (1999) A novel class of protein from wheat which inhibits xylanases. Biochem J 338:441–446

Monod M, Capoccia S, Lechenne B, Zaugg C, Holdom M, Jousson O (2002) Secreted proteases from pathogenic fungi. Int J Med Microbiol 292:405–419

Müller U, Tenberge KB, Oeser B, Tudzynski P (1997) *Cel1*, probably encoding a cellobiohydrolase lacking the substrate binding domain, is expressed in the initial infection phase of *Claviceps purpurea* on *Secale cereale*. Mol Plant–Microbe Interact 10:268–279

Murphy JM, Walton JD (1996) Three extracellular proteases from *Cochliobolus carbonum*: cloning and targeted disruption of *ALP1*. Mol Plant–Microbe Interact 9:290–297

Nasser Eddine A, Hannemann F, Schäfer W (2001) Cloning and expression analysis of *NhL1*, a gene encoding an extracellular lipase from the fungal pea pathogen *Nectria haematococca* MP VI (*Fusarium solani* f. sp. *pisi*) that is expressed in planta. Mol Genet Genomics 265:215–224

Nikolskaya AN, Pitkin JW, Schaeffer HJ, Ahn JH, Walton JD (1998) EXG1p, a novel exo-beta1,3-glucanase from the fungus *Cochliobolus carbonum*, contains a repeated motif present in other proteins that interact with polysaccharides. Biochim Biophys Acta 1425:632–636

O'Neill MA, York WS (2003) The composition and structure of plant primary walls. In: Rose J (ed) The plant cell wall. Blackwell, Oxford, pp 1–54

Oeser B, Heidrich PM, Müller U, Tudzynski P, Tenberge KB (2002) Polygalacturonase is a pathogenicity factor in the *Claviceps purpurea*/rye interaction. Fungal Genet Biol 36:176–186

Orbach MJ, Farrall L, Sweigard JA, Chumley FG, Valent B (2000) A telomeric avirulence gene determines efficacy for the rice blast resistance gene *Pi-ta*. Plant Cell 12:2019–2032

Orejas M, MacCabe AP, Perez Gonzalez JA, Kumar S, Ramon D (1999) Carbon catabolite repression of the *Aspergillus nidulans xlnA* gene. Mol Microbiol 31:177–184

Ortoneda M, Guarro J, Madrid MP, Caracuel Z, Roncero MI, Mayayo E, Di Pietro A (2004) *Fusarium oxysporum* as a multihost model for the genetic dissection of fungal virulence in plants and mammals. Infect Immun 72:1760–1766

Ospina-Giraldo MD, Mullins E, Kang S (2003) Loss of function of the *Fusarium oxysporum SNF1* gene reduces virulence on cabbage and Arabidopsis. Curr Genet 44:49–57

Paper JM, Scott-Craig JS, Adhikari ND, Cuomo CA, Walton JD (2007) Comparative proteomics of extracellular proteins in vitro and in planta from the pathogenic fungus *Fusarium graminearum*. Proteomics 7:3171–3183

Penalva MA, Arst HN Jr (2004) Recent advances in the characterization of ambient pH regulation of gene expression in filamentous fungi and yeasts. Annu Rev Microbiol 58:425–451

Phalip V, Delalande F, Carapito C, Goubet F, Hatsch D, Leize-Wagner E, Dupree P, Dorsselaer AV, Jeltsch JM (2005) Diversity of the exoproteome of *Fusarium graminearum* grown on plant cell wall. Curr Genet 48:366–379

Poinssot B, Vandelle E, Bentejac M, Adrian M, Levis C, Brygoo Y, Garin J, Sicilia F, Coutos-Thevenot P, Pugin A (2003) The endopolygalacturonase 1 from *Botrytis cinerea* activates grapevine defense reactions unrelated to its enzymatic activity. Mol Plant–Microbe Interact 16:553–564

Poussereau N, Creton S, Billon-Grand G, Rascle C, Fevre M (2001a) Regulation of *acp1*, encoding a non-aspartyl acid protease expressed during pathogenesis of *Sclerotinia sclerotiorum*. Microbiology 147:717–726

Poussereau N, Gente S, Rascle C, Billon-Grand G, Fevre M (2001b) *aspS* encoding an unusual aspartyl protease from *Sclerotinia sclerotiorum* is expressed during phytopathogenesis. FEMS Microbiol Lett 194:27–32

Powell AL, van Kan J, ten Have A, Visser J, Greve LC, Bennett AB, Labavitch JM (2000) Transgenic expression of pear PGIP in tomato limits fungal colonization. Mol Plant Microbe Interact 13:942–950

Ransom RF, Walton JD (1997) Purification and characterization of extracellular beta-xylosidase and alpha-arabinosidase from the plant pathogenic fungus *Cochliobolus carbonum*. Carbohydrate Res 297:357–364

Rauscher R, Wurleitner E, Wacenovsky C, Aro N, Stricker AR, Zeilinger S, Kubicek CP, Penttila M, Mach RL (2006) Transcriptional regulation of *xyn1*, encoding xylanase I, in *Hypocrea jecorina*. Eukaryot Cell 5:447–456

Reid WW (1950) Estimation and separation of the pectinesterase and polygalacturonase of micro-fungi. Nature 166:569

Reis H, Pfiffi S, Hahn M (2005) Molecular and functional characterization of a secreted lipase from *Botrytis cinerea*. Mol. Plant Pathol. 6:257–267

Reiter WD (2002) Biosynthesis and properties of the plant cell wall. Curr Opin Plant Biol 5:536–542

Rogers LM, Flaishman MA, Kolattukudy PE (1994) Cutinase gene disruption in *Fusarium solani* f sp *pisi* decreases its virulence on pea. Plant Cell 6:935–945

Rogers LM, Kim YK, Guo W, Gonzalez-Candelas L, Li D, Kolattukudy PE (2000) Requirement for either a host- or pectin-induced pectate lyase for infection of *Pisum sativum* by *Nectria hematococca*. Proc Natl Acad Sci USA 97:9813–9818

Rollins JA (2003) The *Sclerotinia sclerotiorum pac1* gene is required for sclerotial development and virulence. Mol Plant–Microbe Interact 16:785–795

Rollins JA, Dickman MB (2001) pH signaling in *Sclerotinia sclerotiorum*: identification of a *pacC/RIM1* homolog. Appl Environ Microbiol 67:75–81

Ron M, Avni A (2004) The receptor for the fungal elicitor ethylene-inducing xylanase is a member of a resistance-like gene family in tomato. Plant Cell 16:1604–1615

Roncero MIG, Hera C, Ruiz-Rubio M, García-Maceira FI, Madrid MP, Caracuel Z, Calero F, Delgado-Jarana J, Roldán-Rodríguez R, Martínez-Rocha AL, Velasco C, Roa J, Martín-Urdiroz M, Córdoba D, Di Pietro A (2003) Fusarium as a model for studying virulence in soilborne plant pathogens. Physiol Mol Plant Pathol 62:87–98

Rowe HC, Kliebenstein DJ (2007) Elevated genetic variation within virulence-associated *Botrytis cinerea* polygalacturonase loci. Mol Plant–Microbe Interact 20:1126–1137

Ruiz MC, DiPietro A, Roncero MIG (1997) Purification and characterization of an acidic endo-beta-1,4-xylanase from the tomato vascular pathogen *Fusarium oxysporum* f sp. *lycopersici*. FEMS Microbiol Lett 148:75–82

Ruiz-Roldan MC, Di Pietro A, Huertas-Gonzalez MD, Roncero MI (1999) Two xylanase genes of the vascular wilt pathogen *Fusarium oxysporum* are differentially expressed during infection of tomato plants. Mol Gen Genet 261:530–536

Ryan CA (1990) Protease inhibitors in plants: genes for improving defenses against insects and pathogens. Annu Rev Phytopathol 28:425–449

Scott-Craig JS, Panaccione DG, Cervone F, Walton JD (1990) Endopolygalacturonase is not required for pathogenicity of *Cochliobolus carbonum* on maize. Plant Cell 2:1191–1200

Scott-Craig JS, Cheng YQ, Cervone F, De Lorenzo G, Pitkin JW, Walton JD (1998) Targeted mutants of *Cochliobolus carbonum* lacking the two major extracellular polygalacturonases. Appl Environ Microbiol 64:1497–1503

Schaeffer HJ, Leykam J, Walton JD (1994) Cloning and targeted gene disruption of *EXG1*, encoding exo-beta 1,3-glucanase, in the phytopathogenic fungus *Cochliobolus carbonum*. Appl Environ Microbiol 60:594–598

Schäfer W (1993) The role of cutinase in fungal pathogenicity. Trends Microbiol 1:69–71

Schultz Gronover C, Kasulke D, Tudzynski P, Tudzynski B (2001) The role of G protein alpha subunits in the infection process of the gray mold fungus *Botrytis cinerea*. Mol Plant–Microbe Interact 14:1293–1302

Schulze Gronover C, Schorn C, Tudzynski B (2004) Identification of *Botrytis cinerea genes* up-regulated during infection and controlled by the Galpha subunit BCG1 using suppression subtractive hybridization (SSH). Mol Plant–Microbe Interact 17:537–546

Shieh MT, Brown RL, Whitehead MP, Cary JW, Cotty PJ, Cleveland TE, Dean RA (1997) Molecular genetic evidence for the involvement of a specific polygalacturonase, P2c, in the invasion and spread of *Aspergillus flavus* in cotton bolls. Appl Environ Microbiol 63:3548–3552

Solomon PS, Tan KC, Sanchez P, Cooper RM, Oliver RP (2004) The disruption of a Galpha subunit sheds new light on the pathogenicity of *Stagonospora nodorum* on wheat. Mol Plant–Microbe Interact 17:456–466

Sposato P, Ahn JH, Walton JD (1995) Characterization and disruption of a gene in the maize pathogen *Cochliobolus carbonum* encoding a cellulase lacking a cellulose binding domain and hinge region. Mol Plant–Microbe Interact 8:602–609

Sreedhar L, Kobayashi DY, Bunting TE, Hillman BI, Belanger FC (1999) Fungal proteinase expression in the interaction of the plant pathogen *Magnaporthe poae* with its host. Gene 235:121–129

Stahl DJ, Schäfer W (1992) Cutinase is not required for fungal pathogenicity on pea. Plant Cell 4:621–629

Subramaniyan S, Prema P (2002) Biotechnology of microbial xylanases: enzymology, molecular biology, and application. Crit Rev Biotechnol 22:33–64

Sweigard JA, Chumley FG, Valent B (1992) Disruption of a *Magnaporthe grisea* cutinase gene. Mol Gen Genet 232:183–190

ten Have A, Mulder W, Visser J, van Kan JA (1998) The endopolygalacturonase gene *Bcpg1* is required for full virulence of *Botrytis cinerea*. Mol Plant–Microbe Interact 11:1009–1016

ten Have A, Breuil WO, Wubben JP, Visser J, van Kan JA (2001) *Botrytis cinerea* endopolygalacturonase genes are differentially expressed in various plant tissues. Fungal Genet Biol 33:97–105

ten Have A, Dekkers E, Kay J, Phylip LH, van Kan JA (2004) An aspartic proteinase gene family in the filamentous fungus *Botrytis cinerea* contains members with novel features. Microbiology 150:2475–2489

Tonukari NJ, Scott-Craig JS, Walton JD (2000) The *Cochliobolus carbonum SNF1* gene is required for cell wall-degrading enzyme expression and virulence on maize. Plant Cell 12:237–248

Torronen A, Rouvinen J (1997) Structural and functional properties of low molecular weight endo-1,4-beta-xylanases. J Biotechnol 57:137–149

Valette-Collet O, Cimerman A, Reignault P, Levis C, Boccara M (2003) Disruption of *Botrytis cinerea* pectin methylesterase gene *Bcpme1* reduces virulence on several host plants. Mol Plant–Microbe Interact 16:360–367

van Kan JA (2006) Licensed to kill: the lifestyle of a necrotrophic plant pathogen. Trends Plant Sci 11:247–253

van Kan JA, van't Klooster JW, Wagemakers CA, Dees DC, van der Vlugt-Bergmans CJ (1997) Cutinase A of *Botrytis cinerea* is expressed, but not essential, during penetration of gerbera and tomato. Mol Plant–Microbe Interact 10:30–38

van Peij NN, Gielkens MM, de Vries RP, Visser J, de Graaff LH (1998a) The transcriptional activator XlnR regulates both xylanolytic and endoglucanase gene expression in *Aspergillus niger*. Appl Environ Microbiol 64:3615–3619

van Peij NN, Visser J, de Graaff LH (1998b) Isolation and analysis of *xlnR*, encoding a transcriptional activator co-ordinating xylanolytic expression in *Aspergillus niger*. Mol Microbiol 27:131–142

Voigt CA, Schäfer W, Salomon S (2005) A secreted lipase of *Fusarium graminearum* is a virulence factor required for infection of cereals. Plant J 42:364–375

Walton JD (1994) Deconstructing the cell wall. Plant Physiol 104:1113–1118

Wegener S, Ransom RF, Walton JD (1999) A unique eukaryotic beta-xylosidase gene from the phytopathogenic fungus *Cochliobolus carbonum*. Microbiology 145:1089–1095

Woloshuk CP, Kolattukudy PE (1986) Mechanism by which contact with plant cuticle triggers cutinase gene expression in the spores of *Fusarium solani* f. sp. *pisi*. Proc Natl Acad Sci USA 83:1704–1708

Wu SC, Kauffmann S, Darvill AG, Albersheim P (1995) Purification, cloning and characterization of two xylanases from *Magnaporthe grisea*, the rice blast fungus. Mol Plant–Microbe Interact 8:506–514

Wu SC, Ham KS, Darvill AG, Albersheim P (1997) Deletion of two endo-beta-1,4-xylanase genes reveals additional isozymes secreted by the rice blast fungus. Mol Plant–Microbe Interact 10:700–708

Wu SC, Halley JE, Luttig C, Fernekes LM, Gutierrez-Sanchez G, Darvill AG, Albersheim P (2006) Identification of an endo-beta-1,4-D-xylanase from *Magnaporthe grisea* by gene knockout analysis, purification, and heterologous expression. Appl Environ Microbiol 72:986–993

Wubben JP, ten Have A, van Kan JA, Visser J (2000) Regulation of endopolygalacturonase gene expression in *Botrytis cinerea* by galacturonic acid, ambient pH and carbon catabolite repression. Curr Genet 37:152–157

Yakoby N, Freeman S, Dinoor A, Keen NT, Prusky D (2000a) Expression of pectate lyase from *Colletotrichum gloesosporioides* in *C. magna* promotes pathogenicity. Mol Plant–Microbe Interact 13:887–891

Yakoby N, Kobiler I, Dinoor A, Prusky D (2000b) pH regulation of pectate lyase secretion modulates the attack of *Colletotrichum gloeosporioides* on avocado fruits. Appl Environ Microbiol 66:1026–1030

Yakoby N, Beno-Moualem D, Keen NT, Dinoor A, Pines O, Prusky D (2001) *Colletotrichum gloeosporioides pelB* is an important virulence factor in avocado fruit-fungus interaction. Mol Plant–Microbe Interact 14:988–995

Yao C, Köller W (1995) Diversity of cutinases from plant pathogenic fungi: different cutinases are expressed during saprophytic and pathogenic stages of *Alternaria brassicicola*. Mol Plant–Microbe Interact 8:122–130

Zhao X, Mehrabi R, Xu JR (2007) MAP Kinase pathways and fungal pathogenesis. Eukaryot Cell (in press)

11 Photoactivated Perylenequinone Toxins in Plant Pathogenesis

Margaret E. Daub[1], Kuang-Ren Chung[2]

CONTENTS

I. Introduction............................ 201
II. Perylenequinone Toxins................. 201
   A. Fungal Perylenequinone Toxins 201
   B. Mode of Action of
      Perylenequinone Toxins 203
III. Cercosporin as a Model for
   Understanding Fungal
   Perylenequinone Toxins................. 204
   A. *Cercospora* Fungi 204
   B. Mode of Action of Cercosporin......... 204
   C. Role of Cercosporin in Plant Disease 205
IV. Cercosporin Production.................. 205
   A. Feeding Studies 205
   B. Environmental Regulation
      of Production 205
V. Isolation and Molecular Characterization
   of Biosynthetic Genes.................... 206
   A. Light-Regulated Genes 206
   B. Cercosporin Toxin Biosynthesis
      (CTB) Cluster Genes 206
   C. Demarcation of the CTB
      Gene Cluster 207
   D. Proposed Model for the Cercosporin
      Biosynthetic Pathway 208
   E. Lack of a Role for the CTB Cluster
      Genes in Self-Defense Against
      Cercosporin Toxicity.................. 210
VI. Regulation of Cercosporin Biosynthesis 210
VIII. Fungal Resistance to Cercosporin........... 212
   A. General Photosensitizer Resistance 212
   B. Reductive Detoxification 213
   C. Toxin Export....................... 214
IX. Isolation and Characterization of
   Cercosporin Autoresistance Genes.......... 214
   A. Pyridoxine Biosynthetic Genes 214
   B. Regulatory Genes.................... 214
X. Cercosporin Degradation Genes
   from Bacteria.......................... 215
XI. Conclusions 215
   References............................. 215

I. Introduction

Fungi that parasitize plants have at their disposal a large and diverse set of tools required for successful colonization of their hosts. Among the most effective of these strategies is the production of phytotoxic compounds that play multiple roles in plant disease. These range from toxins that facilitate infection by suppressing normal plant defense pathways to ones that alter normal metabolic processes and symptom expression to toxins that directly kill the cells of the host, allowing for colonization of dead tissue. Among the most intriguing of the well studied plant pathogen toxins are the photoactivated perylenequinones. These substances, the product of the polyketide pathway in ascomycete fungi, are colored compounds that are converted to their toxic state through photoactivation. Although studied most for their involvement in plant pathogenesis, photoactivated perylenequinones have also been recovered from saprophytic species, suggesting that they may have broad functions in fungi. This chapter summarizes the current state of knowledge of the mode of action, biosynthesis, regulation, and understanding of cellular resistance to this group of compounds, with an emphasis on studies done on the perylenequinone cercosporin, produced by members of the genus *Cercospora*, a large and successful group of foliar plant pathogens.

II. Perylenequinone Toxins

A. Fungal Perylenequinone Toxins

All of the perylenequinone toxins that have been characterized are synthesized by fungi in the phylum Ascomycota. Many of the producing fungi are found within the Dothideomycetes, but other classes are also represented (Table 11.1). Many of the identified perylenequinone producers are

[1] Department of Plant Biology, North Carolina State University, Raleigh, NC 27695-7612, USA; e-mail: margaret_daub@ncsu.edu
[2] Citrus Research and Education Center, University of Florida, Lake Alfred, FL 33850, USA; e-mail: krchung@ufl.edu

Plant Relationships, 2nd Edition
The Mycota V
H. Deising (Ed.)
© Springer-Verlag Berlin Heidelberg 2009

Table 11.1. Photoactivated perylenequinones produced by fungi

Perylenequinone	Fungal species	References
Alteichin Alterlosins Altertoxins	*Alternaria alternata* *Alternaria eichorniae* *Stemphylium botryosum*	Davis and Stack (1991), Stack et al. (1986), Stierle and Cardellina (1989)
Cercosporin Isocercosporin Acetylisocercosporin	*Cercospora* species *Scolecotrichum graminis* *Stagnospora convolvuli*	Ahonsi et al. (2005), Daub and Ehrenshaft (2000), Tabuchi et al. (1994)
Calphostin C Cladochrome Ent-isophleichrome Phleichrome	*Cladosporium cucumerinum* *Cladosporium cladosporioides* *Cladosporium herbarum* *Cladosporium phlei*	Arnone et al. (1988), Robeson and Jalal (1992), Weiss et al. (1987), Yoshihara et al. (1975)
Elsinochromes	*Elsinoe* species *Stagnospora convolvuli*	Ahonsi et al. (2005), Weiss et al. (1987)
Hypocrellins Isohypocrellin	*Hypocrella bambusae* *Shiraia bambusicola* *Graphis hematites*	Diwu (1995), Fang et al. (2006), Mathey and Lukins (2001), Weiss et al. (1987)
Hypomycin A	*Hypomyces* spp.	Liu et al. (2001)
Shiraiachromes	*Shiraia bambusicola*	Wu et al. (1989)
Stemphyltoxins	*Stemphylium botryosum*	Davis and Stack (1991)
Other perylenequinones	*Bulgaria inquinans*	Li et al. (2006)

Fig. 11.1. Structures of fungal perylenequinone toxins. All share the same same 3,10-dihydroxy-4,9-perylenequinone chromophore and differ in side-chains

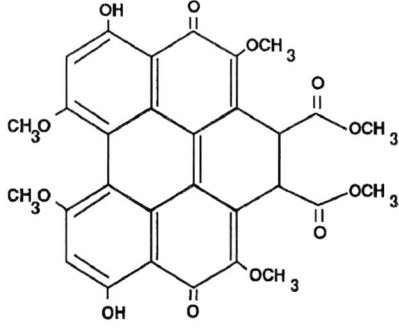

Cercosporin
Cercospora spp.

Elsinochrome A
Elsinoe spp.

Phleichrome
Cladosporium phlei

Hypocrellin A
Hypocrella bambusae

plant pathogens, but these compounds have also been isolated from saprophytic species and a lichen. Many compounds and analogues have been identified, but all share the same 3,10-dihydroxy-4,9-perylenequinone chromophore (Fig. 11.1). Interestingly, the fungal compounds show similarities to compounds isolated from plants and protozoans. These include the photoactive compound hypericin (Giese 1980) that is produced by St. John's wort, a popular medicinal herb used for treatment of depression. In nature, animals that feed on *Hypericum* species develop a disease known as hypericism that is characterized by skin lesions, changes in behavior, and sometimes convulsions and death. Moderate to severe photosensitivity has been reported in humans taking St. John's wort for treatment of depression (Rodriguez-Landa and Contreras 2003). In addition, some species of protozoa produce photoactivated extended quinones that are involved in photomovement responses (Giese 1981).

B. Mode of Action of Perylenequinone Toxins

The fungal perylenequinone toxins fit the classification of a group of compounds known as photosensitizers. Photosensitizer is a name that dates back to the beginning of the twentieth century when a German medical student showed that some common dyes, including xanthenes and acridines, kill paramecia only in the presence of oxygen and visible light (Raab 1900). He coined the term photosensitizer to denote the ability of these compounds to "sensitize" living cells to visible wavelengths of light. Beyond the perylenequinones and synthetic dyes, there are many classes of compounds that have photosensitizing activity. These include numerous compounds isolated from plants such as chlorophylls, coumarins, thiophenes, and acetylenes, compounds important in human health such as riboflavin and porphyrins, and other commonly used dyes such as methylene blue, rose bengal, and acridine orange (Spikes 1989). Due to their toxicity, photosensitizers are being investigated for possible utility as herbicides and insecticides, as antiviral agents, and in photodynamic tumor therapy (Diwu 1995; Heitz and Downum 1995; Hudson and Towers 1991).

Regardless of their structure, photosensitizers share a common mode of action (Fig. 11.2).

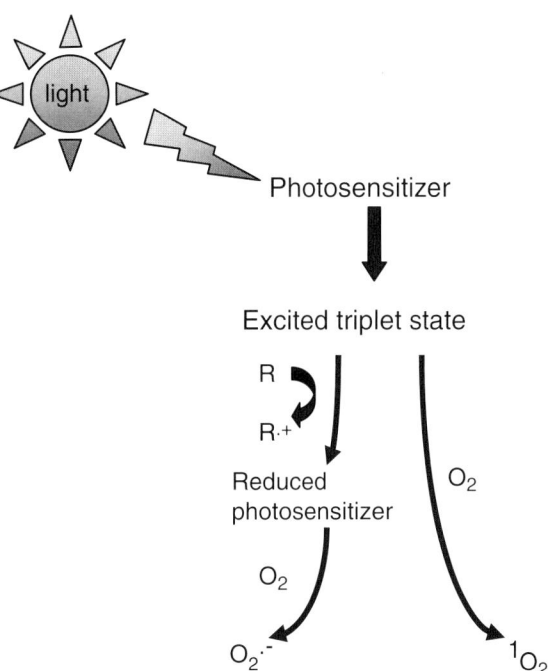

Fig. 11.2. Mode of action of photosensitizers. Upon activation by light, photosensizers are converted to an electronically activated triplet state. The triplet photosensitizer may react in two ways. The triplet sensitizer may react with a reducing substrate (R) to yield a reduced sensitizer molecule; and reaction with oxygen then leads to the generation of superoxide ($O_2^{\cdot-}$). Alternatively, the triplet sensitizer may react directly with oxygen through an energy transfer reaction which results in the production of singlet oxygen ($^1O_2$)

Photosensitizers are colored and absorb visible wavelengths of light. The absorbed light energy causes the photosensitizer to be converted to the energetically activated and long-lived "triplet state". Triplet state photosensitizers may react through radical or energy transfer reactions. In the so-called "type I reaction" the triplet sensitizer reacts through a reducing substrate, leading to the formation of a reduced photosensitizer. The reduced photosensitizer in turn may react with cellular macromolecules such as lipids, proteins, or nucleic acids, causing oxidative damage and leading to the production of lipid peroxides and other free radical compounds. The reduced photosensitizer may also react with oxygen, leading to the production of reduced and free radical forms of oxygen including superoxide ($O_2^{\cdot-}$), hydrogen peroxide (H_2O_2), or the hydroxyl radical ($OH^{\cdot}$). Photosensitizers may also react directly with oxygen via energy transfer, a reaction known as a "type II reaction". These reactions result in the production of the highly toxic singlet oxygen ($^1O_2$). In cells the type of damage resulting from photosensitization is due to where the photosensitizer molecule localizes (Ito 1981); lipid-soluble photosensitizers, for example, generally cause membrane damage whereas those that localize in nuclei damage DNA.

III. Cercosporin as a Model for Understanding Fungal Perylenequinone Toxins

A. *Cercospora* Fungi

A number of important genera of plant pathogenic fungi produce perylenequinone toxins (Table 11.1), but the best studied are members of the genus *Cercospora*, which produce the toxin cercosporin (Fig. 11.1). *Cercospora* species are a group of imperfect fungi that cause damaging leaf spot and blight diseases on an extremely broad range of plant hosts, including economically important crops such as sugar beet, corn, coffee, banana, peanut, soybean, tobacco, and many vegetable and ornamental species. *Cercospora* species have world-wide distribution and are difficult to control due to a lack of high levels of resistance in cultivated varieties. Most *Cercospora* species that have been tested have been shown to produce cercosporin. The toxin was first isolated in the 1950s from *C. kikuchii*, a soybean pathogen (Kuyama and Tamura 1957). It was subsequently isolated from many species as well as from *Cercospora*-infected plants (Daub and Ehrenshaft 2000). Its structure and chemical properties were characterized in the early 1970s (Lousberg et al. 1971; Yamazaki and Ogawa 1972). Cercosporin is red in color and is lipid-soluble, with only limited water-solubility. Secretion of cercosporin from the fungal hyphae in culture results in the accumulation of crystals of cercosporin clustered around the hyphae and in the culture medium, resulting in red pigmentation on the underside of the fungal colony.

B. Mode of Action of Cercosporin

The photosensitizing activity of cercosporin was first documented in 1975. Yamazaki and co-workers (1975) demonstrated that cercosporin was toxic to mice and bacteria and that light and oxygen were required for toxicity. Subsequently Macri and Vianello (1979) documented cercosporin-induced ion leakage from corn, potato, and beet tissues and protection by antioxidants. Studies in our laboratory with tobacco cell cultures showed that the action spectrum for the killing of cells matched the absorption spectrum of cercosporin, with peak activity at 470 nm and 570 nm, the two cercosporin absorption maxima (Daub 1982a). Further, the production of both $^1O_2$ and $O_2^{-\cdot}$ by cercosporin has been confirmed in several studies. Generation of $^1O_2$ in the light has been measured directly from cercosporin and from cercosporin-producing cultures (Bilski et al. 2000, 2002; Dobrowolski and Foote 1983; Leisman and Daub 1992). Production of $O_2^{-\cdot}$ has also been documented when cercosporin is incubated in light in the presence of a reducing substrate such as methionine or urate (Daub and Hangarter 1983; Hartman et al. 1988). Cercosporin is a highly effective producer of $^1O_2$, with a quantum yield (yield of $^1O_2$ per quantum of light absorbed) of 0.81–0.97 (Daub et al. 2000; Dobrowolski and Foote 1983). In support of a primary role of $^1O_2$ in cercosporin toxicity, the killing of cultured tobacco cells by cercosporin could be inhibited by compounds that quench $^1O_2$, including 1.4-diazabicyclo octane (DABCO) and the carotenoid carboxylic acid bixin (Daub 1982a). Free radical forms of active oxygen ($O_2^{-\cdot}$, H_2O_2, OH˙) may also be important in cercosporin toxicity. Sugarbeet transformed for expression of superoxide dismutase showed increased resistance to cercosporin and to *C. beticola* as compared to non-transformed controls, suggesting an important role of $O_2^{-\cdot}$ in cercosporin toxicity (Tertivanidis et al. 2004).

Whether caused by $^1O_2$ or by free radical oxygen species, all evidence to date indicates that cercosporin damages plants by causing a peroxidation of the membrane lipids, leading to membrane breakdown and cell death. Ultrastructural studies of sugar beet leaves infected with *C. beticola* documented membrane damage as a primary ultrastructural symptom (Steinkamp et al. 1979). Cercosporin results in rapid ion leakage from cultured tobacco cells and bursting of tobacco protoplasts, confirming damage to the plasma membrane (Daub 1982b). Further studies documented the production of lipid peroxide products, an increase in the ratio of saturated to unsaturated fatty acids, and a decrease in the fluidity of protoplast membranes, all indicators of lipid peroxidation (Daub and Briggs 1983).

C. Role of Cercosporin in Plant Disease

Both observational and experimental results document an important role for cercosporin in disease development. For example, *Cercospora* infection of coffee, sugar beet, and banana has long been known to require light for symptom development, as symptoms were significantly reduced or failed to develop on shaded leaves (Calpouzos 1966; Calpouzos and Stalknecht 1967; Echandi 1959). On coffee, reduced light exposure was correlated with reduced penetration of stomata, whereas on sugar beet and banana, no effect was noted on penetration, but high light intensities were required for symptom development once the fungus had penetrated the leaf. On banana, the necessity for high light intensities for disease development led to the recommendation in the 1940s that banana be grown under the shade of coconut palms to protect against the yellow Sigatoka disease caused by *Mycosphaerella musicola* (*C. musae*; Thorold 1940).

The importance of cercosporin in plant disease has also been directly tested with cercosporin-deficient mutants. UV-generated mutants of *C. kikuchii* deficient in cercosporin production produced only scattered, pin-point lesions when inoculated onto soybean as compared to the large necrotic and chlorotic lesions produced by wild-type infection (Upchurch et al. 1991). Targeted gene disruption of the polyketide synthase responsible for cercosporin production in *C. nicotianae* resulted in fewer and smaller lesions and a lack of coalescing of lesions that leads to typical blighting symptoms on tobacco (Choquer et al. 2005). Disruption mutants for the cercosporin *CFP* transporter gene in *C. kikuchii* had significantly decreased cercosporin production and symptom expression on soybean (Callahan et al. 1999). Finally, disruption of a MAP kinase gene in *C. zeae-maydis* resulted in decreased cercosporin synthesis and conidiation; these mutants also produced significantly attenuated symptoms on corn.

All of the above studies document a strong correlation between cercosporin production and disease development, indicating that cercosporin is a critical virulence factor in *Cercospora* disease development. Whether cercosporin plays this role via direct toxicity, however, is not clear. Extensive studies described earlier indicate that cercosporin can directly kill host cells by membrane oxidation. However, Dickman et al. (2001) showed that expression of genes from animals that encode anti-apoptotic activity protected tobacco from *C. nicotianae* infection. They hypothesized that colonization by these necrotrophic pathogens resulted from the induction of the host's own apoptotic pathway, allowing for cell death and fungal colonization. Thus it remains to be clarified if cercosporin promotes disease by direct toxicity or by somehow inducing programmed cell death responses in the host.

IV. Cercosporin Production

A. Feeding Studies

The progress toward understanding of the cercosporin biosynthetic pathway and its regulation is still in its infancy compared to significant achievements made on the mycotoxins such as aflatoxin, fumonisin, and trichothecenes, and host-selective toxins such as T-toxin and HC-toxin. The first understanding of cercosporin biosynthesis was derived from the study conducted by Okubo and co-workers (1975) who used nuclear magnetic resonance and mass-spectrometry to analyze culture filtrates after feeding *C. kikuchii* with ^{14}C-labeled acetate. The authors proposed that cercosporin is synthesized from condensation of multiple acetate and malonate molecules via a fungal polyketide pathway. However, no pathway intermediates or enzymes directly pertaining to cercosporin production were identified.

B. Environmental Regulation of Production

In contrast, environmental factors affecting cercosporin production have been well studied. Light is the primary factor, not only for toxicity, but also for cercosporin production (Ehrenshaft and Upchurch 1991). Cercosporin is produced exclusively in the presence of visible light at wavelengths of 400–600 nm, and its production is blocked under complete darkness. Production of cercosporin also is affected by temperature, nutrient conditions, source of carbon or nitrogen, the carbon:nitrogen ratio, and many other physiological factors (Daub and Ehrenshaft 2000). For example, production of cercosporin is markedly inhibited at high temperatures (30 °C; Jenns et al. 1989).

Cercosporin is preferentially synthesized in vegetative cultures, and is repressed when grown on V8 juice agar that induces conidiation. With *C. nicotianae*, we found that a thin layer of Difco potato dextrose agar (PDA) medium (less than 15 ml of medium in a Petri dish) is the best medium to support cercosporin production under light (Chung 2003). Other brands of PDA, PDA made with fresh potatoes, or other synthetic media drastically reduce cercosporin production under light. Recent studies also revealed that micronutrients, including NaCl, KCl, LiCl, ammonium salts, dimethyl sulfoxide (DMSO), citrate buffer, phosphate buffer, casein hydrolysate, arginine, asparagine, glutamine, or yeast extract, markedly suppress cercosporin production (You et al. 2008). It appears that pH has little effect on cercosporin production as addition of citrate and/or phosphate buffers resulted in significant reduction in cercosporin, regardless of the pH values. Cercosporin production is repressed when mannitol and $NaNO_3$ are used as the sole carbon and nitrogen source in a defined medium. In contrast, Zn^{2+}, Fe^{3+}, Co^{2+}, Ca^{2+}, and Mn^{2+} ions had stimulatory effects on cercosporin production. Further, production of cercosporin in axenic culture is highly variable among species and even among isolates of the same species. For example, isolates of *C. nicotianae*, *C. petunia*, and *C. kikuchii* produced maximum quantites of cercosporin when grown on non-buffered PDA (pH 5.6), but accumulated far less cercosporin on phosphate or citrate-buffered PDA (pH 5.6). By contrast, an isolate of *C. beticola* exhibited the opposite pattern, in which *C. beticola* accumulated the highest amounts of cercosporin on phosphate-buffered PDA (pH 5.6). Jenns and co-workers (1989) also documented variation between species and among isolates of the same species in the nutrient requirements that support cercosporin production. Thus, the data accumulated to date indicate that cercosporin production is affected by numerous factors and regulated via complex but interplaying networks.

V. Isolation and Molecular Characterization of Biosynthetic Genes

A. Light-Regulated Genes

Since production of cercosporin is highly induced by light, a differential expression approach was used to identify the first gene, named *CFP* (cercosporin *f*acilitator *p*rotein) required for cercosporin accumulation (Callahan et al. 1999). The CFP protein resembles various membrane transporters belonging to the major facilitator superfamily (MFS transporter) and has been proposed to be responsible for export of cercosporin out of fungal cells. Deletion of the *CFP* gene in *C. kikuchii* created a mutant with dual deficiencies in both cercosporin production and resistance. The actual mechanism by which CFP contributes to cercosporin resistance remains uncertain.

B. Cercosporin Toxin Biosynthesis (CTB) Cluster Genes

Cercosporin is a red pigment and toxin-deficient mutants can be visually selected for lack of red pigmentation of fungal colonies grown on agar plates. To identify cercosporin biosynthetic genes, a strategy using restriction enzyme-mediated insertion (REMI) mutagenesis was used to generate plasmid-tagged mutants that showed various degrees of alterations in cercosporin production in *C. nicotianae* (Chung et al. 2003b). Subsequently, sequences rescued from one of the REMI mutants that is completely defective in cercosporin production led to the identification of *CTB1* (*c*ercosporin *t*oxin *b*iosynthesis 1), a gene encoding a fungal type I polyketide synthase (PKS). As with many fungal polyketide synthases, CTB1 is a large polypeptide containing 2196 amino acids with several conserved functional motifs, including a keto synthase (KS), an acyltransferase (AT), a thioesterase (TE), and two acyl carrier protein (ACP) domains, yet completely lacks ketoreductase (KR), dehydratase (DH), and enoyl reductase (ER) domains (Choquer et al. 2005).

The genes involved in secondary metabolite pathways in filamentous fungi often reside in a cluster (Keller and Hohn 1997; Keller et al. 2005). Further analysis using a chromosome walking strategy recently allowed us to identify a core *CTB* gene cluster containing eight major genes required for cercosporin biosynthesis (Fig. 11.3). In addition to *CTB1*, *CTB2* encodes a putative *O*-methyltransferase (Chen et al. 2007a); *CTB3* encodes a polypeptide containing dual *O*-methyltransferase and flavin adenine dinucleotide (FAD)-dependent monooxygenase domains (Dekkers et al. 2007); *CTB4* encodes a putative MFS membrane transporter with 12 transmenbrane segments (Choquer et al. 2007); *CTB5* encodes a putative oxygen and FAD/FMN-dependent oxidoreductase; *CTB6* encodes a putative NADPH-dependent reductase; *CTB7* encodes a putative FAD/FMN-dependent oxidoreductase; *CTB8* encodes a Zn(II)Cys$_6$ transcriptional activator (Chen et al. 2007a, b). Expression of eight of the *CTB* genes and production of cercosporin were coordinately regulated by light (Fig. 11.3) and by the CTB8 transcriptional activator. Expression of all of the core cercosporin cluster genes was eliminated, or significantly reduced, when the *CTB8* gene was disrupted (Chen et al. 2007a),

Fig. 11.3. A Schematic illustration of the cercosporin toxin biosynthetic (*CTB1-8*) gene cluster with predicted functions in *Cercospora nicotianae*. Arrows indicate the orientation of transcription. Sizes of the coding sequence are indicated. **B** Northern blot analysis indicates differential expression of the *CTB* cluster genes and production of cercosporin (*CR*) by wild-type *C. nicotianae* grown on potato dextrose agar under continuous light (*LT*) or darkness (*DK*). This figure shows data reprinted from Fig. 1 of Chen et al. (2007a), reprinted with permission from Wiley–Blackwell Publishing Ltd

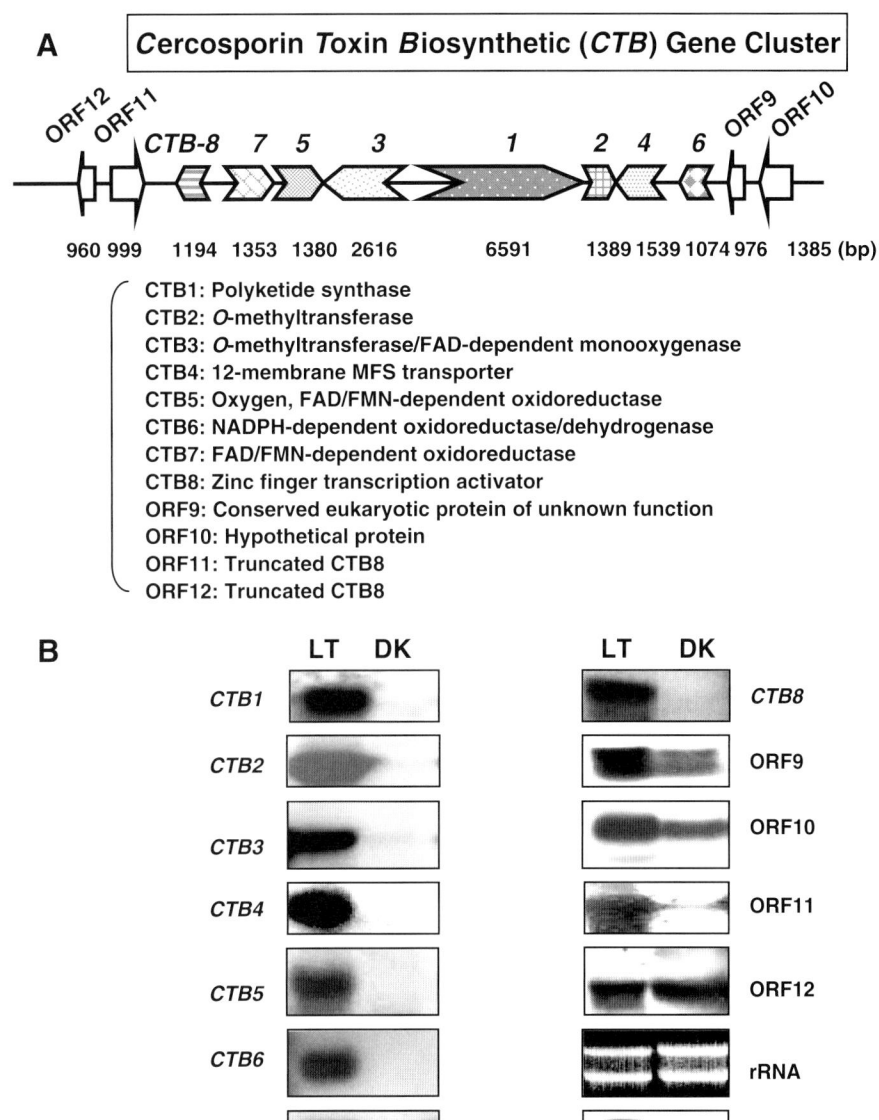

A **Cercosporin *T*oxin *B*iosynthetic (*CTB*) Gene Cluster**

ORF12 ORF11 *CTB-8 7 5 3 1 2 4 6* ORF9 ORF10

960 999 1194 1353 1380 2616 6591 1389 1539 1074 976 1385 (bp)

CTB1: Polyketide synthase
CTB2: *O*-methyltransferase
CTB3: *O*-methyltransferase/FAD-dependent monooxygenase
CTB4: 12-membrane MFS transporter
CTB5: Oxygen, FAD/FMN-dependent oxidoreductase
CTB6: NADPH-dependent oxidoreductase/dehydrogenase
CTB7: FAD/FMN-dependent oxidoreductase
CTB8: Zinc finger transcription activator
ORF9: Conserved eukaryotic protein of unknown function
ORF10: Hypothetical protein
ORF11: Truncated CTB8
ORF12: Truncated CTB8

B

LT DK | LT DK

CTB1 | CTB8
CTB2 | ORF9
CTB3 | ORF10
CTB4 | ORF11
CTB5 | ORF12
CTB6 | rRNA
CTB7 | CR

suggesting that CTB8 plays a crucial role in regulation of the pathway. Expression of the eight genes was also controlled by another transcriptional activator, CRG1, previously shown to regulate cercosporin production and resistance (Chung et al. 2003a). Replacement of each of the *CTB1, 2, 3, 5, 6, 7,* and *8* genes via double cross-over recombination generated mutants that were completely abolished in cercosporin production as a result of specific interruption at the respective genes. In contrast, disruption of the *CTB4* gene created mutants that displayed only partial reduction in cercosporin biosynthesis and secretion (Choquer et al. 2007). Production of cercosporin in each of the *CTB(1–8)* disruptants was restored when a

functional gene cassette was introduced into the respective mutants.

C. Demarcation of the CTB Gene Cluster

In addition to the eight major genes encoding proteins involved in cercosporin production, two open reading frames, ORF9 and ORF10, located near the left end of the cluster were identified that did not encode proteins involved in toxin synthesis (Chen et al. 2007a). The predicted ORF9 protein resembles many conserved eukaryotic proteins of unknown function. The ORF10 protein has similarity to many hypothetical proteins of fungi and proteins containing a GTP-binding motif that are presumably involved in vegetative compatibility in *Podospora anserina*. Two other ORFs were characterized at the right end of the cluster. ORF11 and ORF12 have similarity to CTB8 and many GAL4-like Zn(II)Cys$_6$ transcriptional regulators. However, ORF11 lacks two conserved cysteine residues in the first zinc cluster. ORF12 completely lacks the zinc-binding domain. Northern blot analysis revealed that expression of ORFs located near the two distal ends of the cluster was slightly induced by light, but did not correlate with cercosporin biosynthesis and did not show regulation by CTB8 (Fig. 11.4). Several attempts to disrupt ORF11 failed to identify any cercosporin deficiency mutants, suggesting that ORF11 has no role in cercosporin biosynthesis. Thus, we concluded that the biosynthetic cluster is limited to *CTB1–8* (Chen et al. 2007a).

D. Proposed Model for the Cercosporin Biosynthetic Pathway

A working model that is intended to provide a framework for further investigations of cercosporin biosynthesis is proposed (Fig. 11.5). Similar to the biosynthesis of fatty acids and many fungal polyketides, the functional keto synthase (KS), acyltransferase (AT), thioesterase (TE), and acyl carrier protein (ACP) domains of CTB1 are involved in condensation of acetyl-CoA and malonyl-CoA units (step 1), chain elongation to form a polyketide molecule (step 2), and ring closure in the early stages of cercosporin biosynthesis. The malonyl-CoA subunit is likely attached to the ACP domains of CTB1 by formation of phospho-pantotheine (PPT) as described by Rawlings et al. (1989) and Watanabe and Ebizuka (2004). The AT domain is involved in transferring the acetate unit from acetyl-CoA to the PPT of the ACP domain. The KS domain of CTB1 functions to fuse the malonyl- and acetyl-CoAs by decarboxylation. After each cycle of condensation, the malonyl keto group is reduced. The putative polyketide synthase encoded by CTB1 iteratively catalyzes the synthesis by introducing two carbons in each cycle to form a linear polyketide that is finally released by the function of the TE domain.

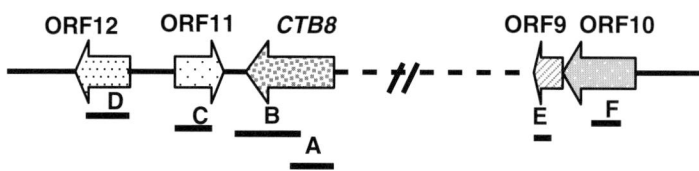

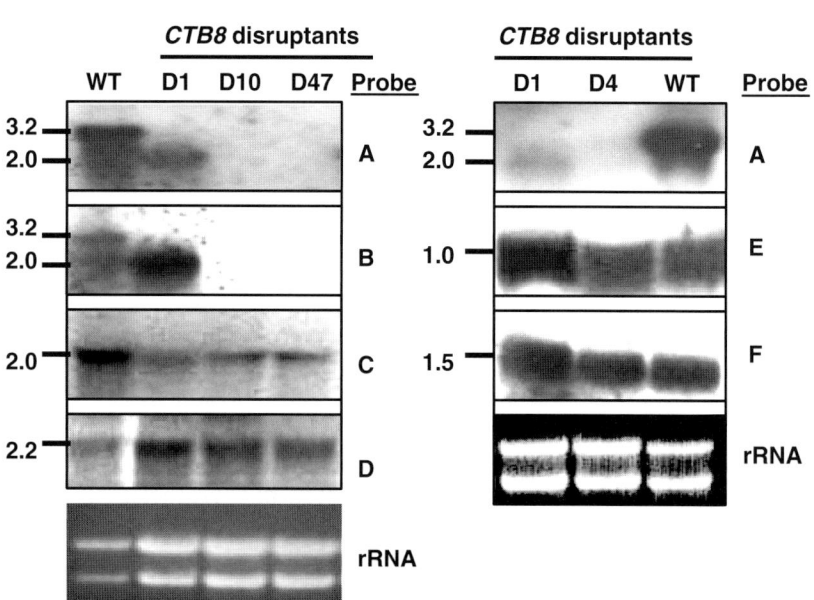

Fig. 11.4. Northern blot analysis of the *CTB8* gene encoding a Zn(II)Cys$_6$ transcriptional activator and four putative genes located near the two distal ends of the *CTB* gene cluster in wild-type (*WT*) and *CTB8* disrupted mutants to demarcate the core *CTB* gene cluster in *C. nicotianae*. The DNA probes corresponding to each gene are shown and sizes of hybridizing bands are indicated in kilobase pairs

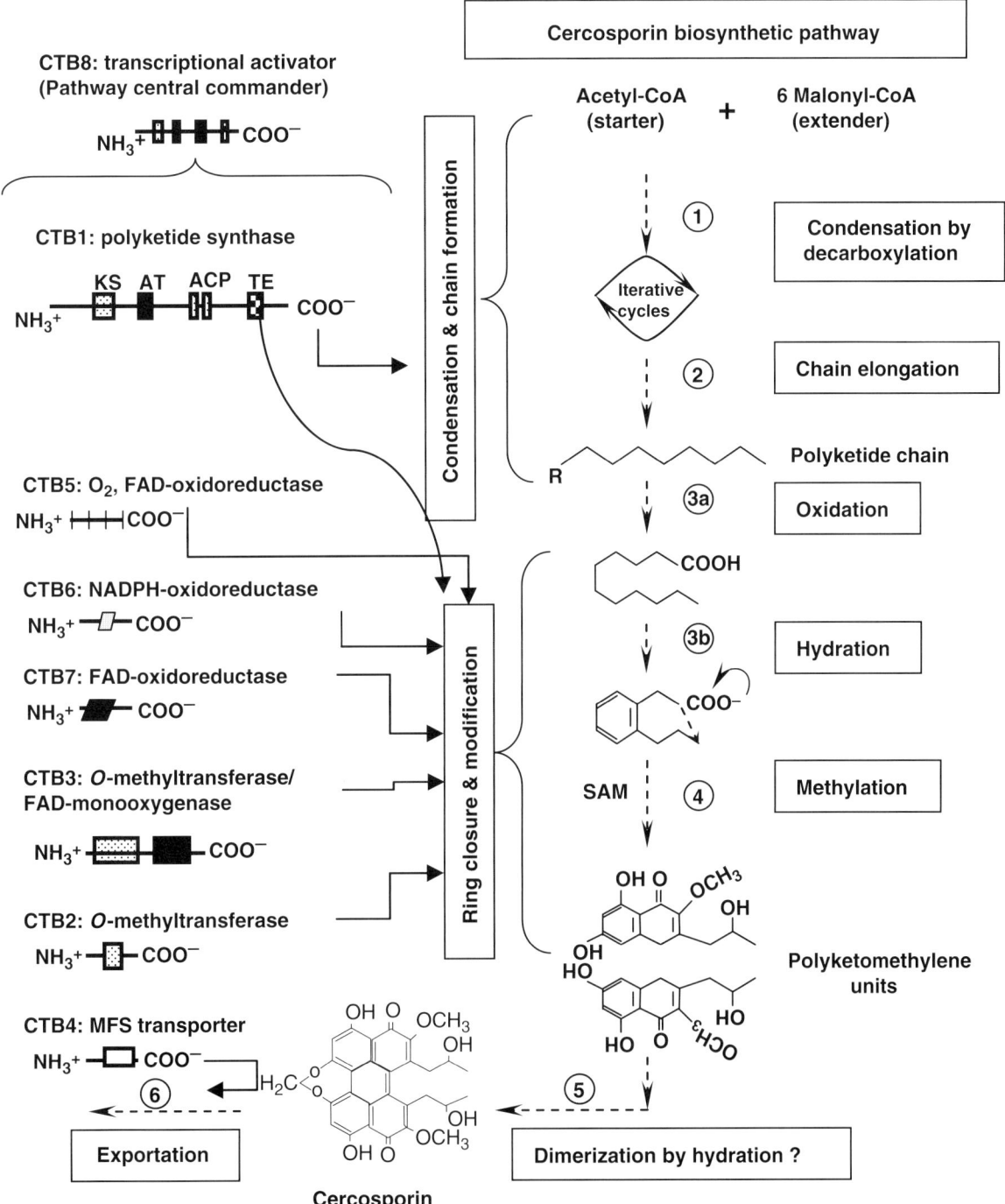

Fig. 11.5. Proposed biosynthetic pathway and involvement of the *CTB* gene products leading to the formation of cercosporin in *Cercospora nicotianae* (see text for details)

Reactions such as oxidation and hydration are required to form aromatic molecules (steps 3a, 3b) during ring closure. The function of the TE domain in the CTB1 may be needed for cyclization of the aromatic ring as proposed for other polyketide compounds (Birch 1967). The oxida-

tion reaction followed by hydration step is likely completed by a monooxygenase domain encoded by *CTB3* (Dekkers et al. 2007) or the translational products of *CTB5*, *CTB6*, and *CTB7*. Once the ring is completed, methyltransferases encoded by *CTB2* and *CTB3* are likely responsible for adding methyl groups to C2 and C11 positions (step 4) to produce polyketomethylene units. Functions of CTB2 and the CTB3 N terminus as an *O*-methyltransferase were not redundant since the *ctb2* and *ctb3* disruptants failed to form any cercosporin via substrate feeding. Finally, two identical polyketomethylene dimers are joined enzymatically or nonenzymatically to form cercosporin with a bilateral symmetrical structure (step 5). The possibility of the *CTB3, 5, 6,* or *7* translational products involved in dimerization cannot be ruled out. Alternatively, dimerization may occur spontaneously as reported for trichothecene biosynthesis in *Fusarium* (McCormick et al. 1990). Once cercosporin is produced, it must be transferred out of the fungal cytoplasm. Exportation of cercosporin out of fungal cells (step 6) is presumably carried out by the function of the MFS transporter encoded by *CTB4* (Choquer et al. 2007). Other membrane transporters, such as *CFP* described earlier, that are not found in the cluster may play a role in transport.

E. Lack of a Role for the CTB Cluster Genes in Self-Defense Against Cercosporin Toxicity

In *Cercospora*, the *CFP* and *CRG1* genes (described in Sects. VI, IX) have been shown to have a dual role in cercosporin resistance and biosynthesis. Disruption of the *CFP* gene encoding a putative cercosporin transporter in *C. kikuchii* yielded a mutant deficient in cercosporin production as well as in resistance (Callahan et al. 1999). Further, disruption mutants for *CRG1*, encoding a putative Zn(II) Cys$_6$ transcription activator, displayed a parallel reduction in both cercosporin production and the ability to tolerate cercosporin toxicity (Chung et al. 2003a, 1999). To date, however, there is no indication that the *CTB* biosynthetic cluster contains genes involved in resistance. The *CTB8* disruptants fail to express the other *CTB* cluster genes, but retain normal resistance to exogenous cercosporin and to other $^1O_2$-generating photosensitizers (Chen et al. 2007a). Further, null mutants of *CTB5, 6,* and *7* also retained normal wild-type levels of resistance against toxicity caused by cercosporin and by other $^1O_2$-generating compounds (Chen et al. 2007b).

VI. Regulation of Cercosporin Biosynthesis

As stated above, expression of the *CTB* cluster genes correlates with environmental conditions conducive for cercosporin production and this regulation is controlled through the CTB8 transcription factor. Clustering of the biosynthetic genes may provide an advantage in coordinated gene regulation at the transcriptional level. As indicated previously, cercosporin biosynthesis is also regulated by another transcription factor, CRG1. The *CRG1* gene was originally identified as a gene involved in cercosporin resistance, but disruption mutants also had a reduction in production (Chung et al. 2003a, 1999). In addition, production of cercosporin appears to be mediated by multiple signal transduction pathways, including the Ca^{2+}/calmodulin signaling (Chung 2003) and MAPK signaling (Shim and Dunkle 2003). Although the role of the G-protein/cAMP/protein kinase A (PKA) signaling in cercosporin biosynthesis remains unknown, these regulatory signaling pathways have been shown to be required for the production of many secondary metabolites by filamentous fungi (Brodhagen and Keller 2006; Yu and Keller 2005). It will be interesting to determine if production of cercosporin is also governed by the G-protein/cAMP/PKA signaling pathway.

Gathering all information available to date about the biosynthetic cluster and the environmental cues and genes involved in cercosporin production, a model illustrating a possible regulatory network for cercosporin biosynthesis is proposed (Fig. 11.6). As with many signal transduction pathways in biological systems, there are likely numerous membrane receptors that receive and respond to a wide range of environmental and physiological signals (such as light, metal ions, nitrogen sources, ammonium, many others). These in turn are proposed to directly or indirectly regulate the transcriptional regulators, CRG1 and CTB8, via signaling transduction pathways. A prior study revealed that *CRG1* is expressed constitutively in cultures grown in the light and in the dark (Chung et al. 1999). However, expression of *CTB8* is differentially regulated by light. It appears that CTB8 is the pathway-specific regulator which regulates the entire cluster of genes (*CTB1–7*) involved in cercosporin biosynthesis and accumulation, whereas CRG1 has a broader

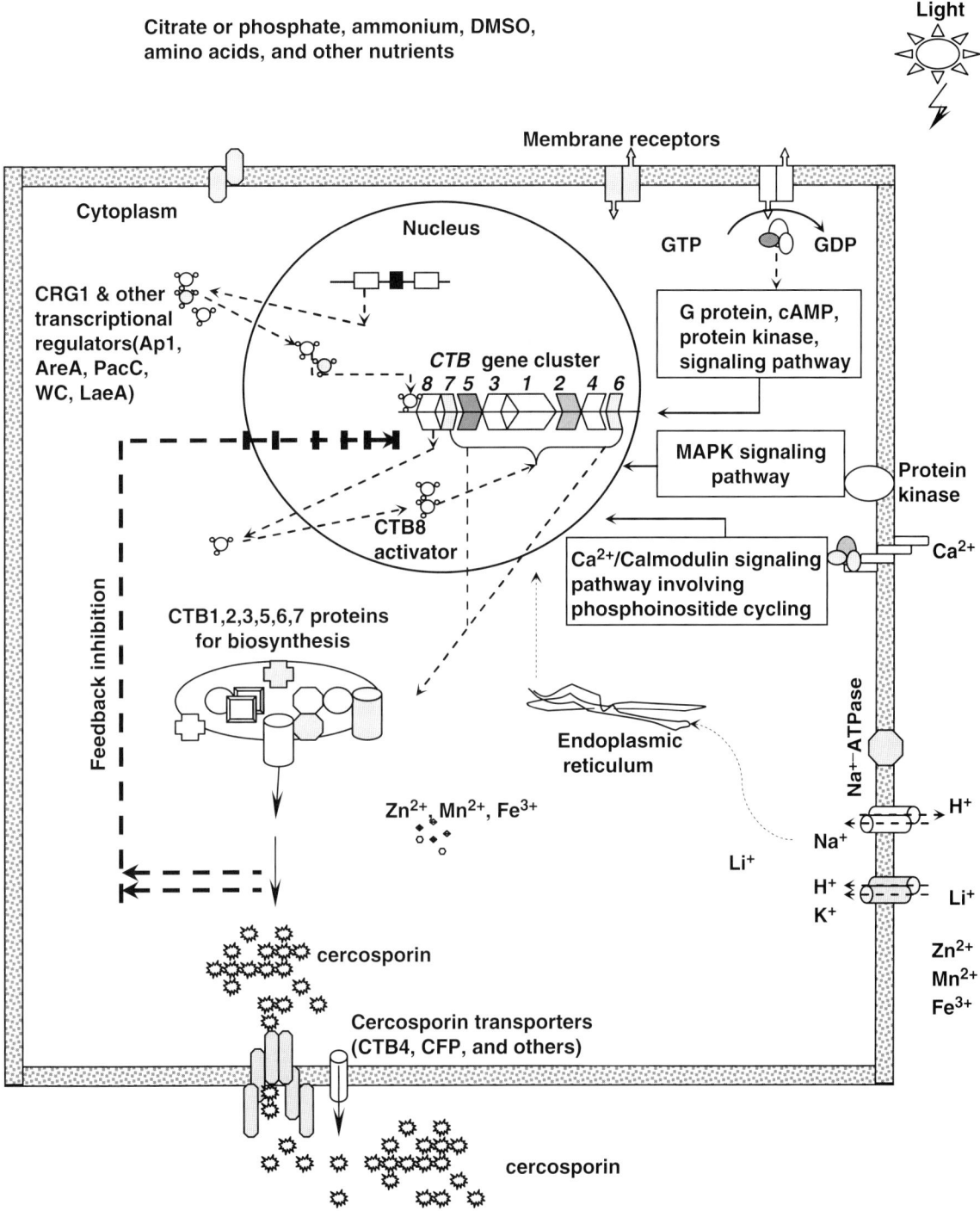

Fig. 11.6. Proposed regulatory pathways in response to diverse environmental signals that lead to trigger or repress production of cercosporin by *C. nicotianae* (see text for details)

regulatory function, including partial control of CTB8 expression (Chen et al. 2007a).

Northern blot analysis of expression of each of the *CTB 1–8* genes in disruptants revealed a feedback inhibition mechanism, in which disruption of one of the biosynthetic genes in the pathway blocked cercosporin production and inhibited expression of the other *CTB* genes in the cluster (Chen et al. 2007a). Such inhibition may not be stringently regulated. It appears that, when one of the biosynthetic genes in the pathway is disrupted, feedback inhibition causes different consequences in expression profiles of other clustering genes or in different mutants.

For example, inactivation of the *CTB1* gene only partially reduced accumulation of the *CTB3* and *CTB8* gene transcripts in four *ctb1* disruptants tested. Expression of the *CTB2* and *CTB4* genes was completely shut down in two *ctb1* disruptants but was unchanged in two other *ctb1* disruptants, likely attributed to "leaky" expression of the genes (Chen et al. 2007a). However, disruption of the *CTB2* gene completely inhibited expression of the *CTB1*, *CTB3*, *CTB4*, and *CTB8* genes (Chen et al. 2007a), whereas the *CTB4*-disrupted mutants retained normal expression of the *CTB1*, *CTB2*, *CTB3*, and *CTB8* genes (Choquer et al. 2007). Expression of *CTB1*, *CTB5*, *CTB6*, *CTB7*, and *CTB8* was significantly reduced or nearly blocked when *CTB5*, *CTB6*, or *CTB7* were inactivated (Chen et al. 2007b). It appears that expression of the *CTB* cluster genes might be regulated collectively as a cluster as well as individually by other transcription factors.

Beyond CRG1, other positive- or negative-acting global regulatory factors, including AreA (nitrogen regulatory protein), AP-1 (oxidative stress-responsive transcription activator), the novel global regulator LaeA, PacC (pH regulatory protein), and WC (light regulatory proteins; Arst and Peñalva 2003; Bok and Keller 2004; Linden et al. 1997; Marzluf 1997; Toone and Jones 1999) may be involved in cercosporin biosynthesis as well.

Analysis of the promoter regions revealed that all *CTB1*–*CTB8* genes have conserved GATA sequences which can serve as the binding sites for the WC1/WC2 light regulatory complex and AreA-type nitrogen regulatory proteins (Ballario et al. 1996; Linden and Macino 1997). None of the *CTB* genes has the CreA (carbon regulatory protein; Dowzer and Kelly 1991) binding site in the promoter regions, suggesting that expression of the *CTB* genes may not be directly responsive to carbon sources. Northern blot studies revealed that the *CTB1*, *CTB2*, *CTB3*, *CTB5*, *CTB6*, *CTB7*, and *CTB8* genes were preferentially expressed in the presence of mannitol and calcium nitrate as the carbon and nitrogen sources, respectively (Chen et al. 2007a). Nitrogen starvation or using ammonium as the sole nitrogen source abolished their expression. Although mannitol increased accumulation of the *CTB1*–*CTB8* gene transcripts, it did not enhance cercosporin production. The promoter regions of *CTB3*, *CTB4*, *CTB6*, *CTB7*, and *CTB8*, but not of *CTB1*, *CTB2*, and *CTB5* contain putative PacC-binding sequences, but there is no direct correlation between cercosporin production and accumulation of the *CTB1–8* gene transcripts under different pH. Unlike other environmental factors, light has a distinct role in expression of the *CTB1–8* genes and accumulation of cercosporin. In addition, all *CTB1*–*CTB8* gene promoters contain CAAT consensus sequences that can be recognized by many other transcriptional regulators, further implying a regulatory complexity in the biosynthetic pathway to form cercosporin.

As discussed above, production of cercosporin was affected by a wide range of environmental factors. Some factors might directly or indirectly regulate *CRG1* and the entire *CTB* cluster genes but others play no role in transcriptional activation. Recent studies examining the expression profiles of the *CRG1*, *CTB1*, and *CTB8* genes indicated that regulation of cercosporin production by KH_2PO_4, DMSO, NH_4NO_3, and LiCl but not Ca^{2+}, Fe^{3+}, Mn^{2+}, and yeast extract appears to occur at the transcriptional level (You et al. 2008). The role of Mn^{2+} and Fe^{3+} in elevating cercosporin production might be due to their serving as cofactors or maintaining the stability of enzymes required for cercosporin regulation and biosynthesis. The negative effect of Na^+, K^+, Li^+ ions and others for cercosporin biosynthesis and accumulation may be attributed, at least in part, to their effect on the function of membrane pumps (Na^+ or Li^+/H^+ antiporters, K^+/H^+ symporters) and/or Na^+-ATPases. This type of regulation might also involve endocytic trafficking systems through the plasma membrane to the endosome in fungi (Herranz et al. 2005). Continued characterization of intermediates accumulated in various *CTB* disrupted mutants will be needed to completely define the cercosporin biosynthetic pathway and to define its complex regulation.

VIII. Fungal Resistance to Cercosporin

A. General Photosensitizer Resistance

Cercosporin shows almost universal toxicity, causing cell death or oxidative damage to plants, bacteria, many fungi, oomycetes, mice, and viruses (Balis and Payne 1971; Daub 1987; Fajola 1978; Hudson et al. 1997; Yamazaki et al. 1975). In spite of the common occurrence of photosensitizers in nature, information on resistance to these agents is limited. Resistance to photosensitizers in animals including insects and protozoa is often due to behavioral, light-avoidance responses (Berenbaum 1987; Giese 1981). Metabolic detoxification of the photosensitizer molecule has also been documented in insects that feed on photosensitizer-containing plants (Li et al. 2003). Compounds are known that effectively quench $^1O_2$ (Bellus 1979; Wilkinson et al. 1995). Although many of these have been identified through chemical studies, some very effective quenchers are found in biological systems. Of these, the most effective are the carotenoid pigments that quench both $^1O_2$ and the triplet state of photosensitizers (Foote et al. 1970; Truscott 1990).

Resistance of a wild rice species ("Louisiana red rice") to cercosporin has been attributed in part to carotenoid content (Batchvarova et al. 1992). Other quenchers present in cells include thiols and the amino acids methionine, histidine, and tryptophan. In addition, antioxidant enzymes such as superoxide dismutase, catalase, and peroxidases would be expected to protect against free radical oxygen species generated by photosensitizers. Data to support their protective role is mixed, however. Sugarbeet plants transformed for expression of superoxide dismutase had increased tolerance to cercosporin (Tertivanidis et al. 2004), but paraquat-resistant tobacco cells that had elevated levels of catalase and superoxide dismutase activity did not show enhanced cercosporin resistance (Hughes et al. 1984). Further, maize plants induce transcripts for superoxide dismutase and catalase in response to cercosporin treatment, but remain sensitive to the toxin (Williamson and Scandalios 1992).

Because of the lack of resistance to cercosporin in most organisms including all host plants studied, we chose to investigate autoresistance to cercosporin in *Cercospora* species and other fungi that produce, and are resistant to, photosensitizers. *Cercospora* fungi produce millimolar concentrations of cercosporin in the light with no measurable decrease in growth and are also resistant to other $^1O_2$-generating photosensitizers such as porphyrins and xanthene dyes (Daub et al. 1992; Jenns et al. 1995). Thus *Cercospora* fungi provide a unique model system for understanding photosensitizer resistance. Early studies eliminated a role for the common defense mechanisms mentioned above. When comparing *Cercospora* mutant and wild-type strains and other fungal species with differing responses to cercosporin, no correlations were found between cercosporin resistance and susceptibility and levels of carotenoids, thiols, reducing substrates, antioxidant activity, or the activity of antioxidant enzymes (Daub 1987; Ehrenshaft et al. 1995; Jenns and Daub 1995; Jenns et al. 1995; Sollod et al.

1992). In addition, as mentioned previously, recent characterization of the *CTB* biosynthetic cluster genes indicated no role for those genes in resistance. By contrast, two potential resistance mechanisms were identified: reductive detoxification of the cercosporin molecule and toxin export out of the cell.

B. Reductive Detoxification

Several lines of evidence support the hypothesis that *Cercospora* species protect themselves against cercosporin in part by a transient reduction of the cercosporin molecule. Reduced cercosporin (Fig. 11.7A), formed by reaction with strong reducing agents such as zinc or dithionite, and its derivatives were shown to be poor generators of $^1O_2$, producing a little as 2% of normal levels, depending on conditions (Daub et al. 2000; Leisman and Daub 1992). The presence of reduced cercosporin in hyphae was confirmed by fluorescence and confocal microscopy, demonstrating that hyphae of resistant fungi (*Cercospora*, *Alternaria*) producing or treated with cercosporin, emitted fluorescence characteristic of reduced cercosporin, whereas sensitive fungi emitted fluorescence characteristic of cercosporin (Daub et al. 1992, 2000). Studies were also conducted to measure the cell surface reducing ability of sensitive and resistant fungi by assaying the ability to reduce tetrazolium dyes spanning a wide range of redox potentials (Sollod et al. 1992). These studies documented that cercosporin-resistant fungi reduce more dyes and dyes with more negative redox potentials than do sensitive fungi. Precise measurement of the redox potential of cercosporin using cyclic voltammetry confirmed a value of −0.14 V, a value within the reducing capability of cells (Clark et al. 1995; Daub et al. 1992). Recent studies with yeast showed that overexpression of a FAD-dependent pyridine nucleotide reductase provided increased resistance to cercosporin and other photosensitizers

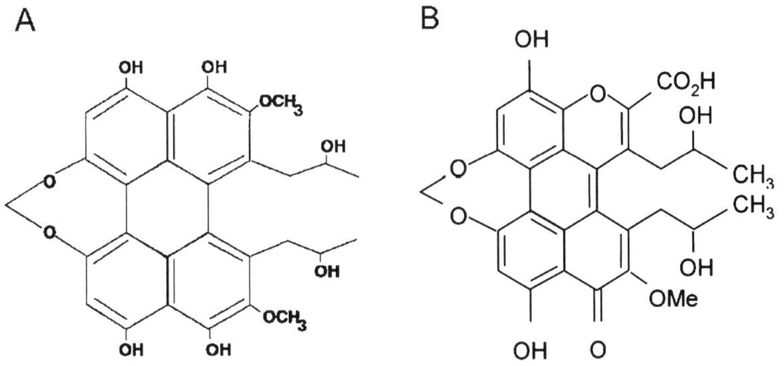

Fig. 11.7. Structures of non-toxic analogues of cercosporin. **A** Reduced cercosporin produced by treating cercosporin with a strong reducing agent. **B** Xanosporic acid produced by *Xanthomonas campestis* pv. *zinniae*

(Ververidis et al. 2001), supporting the hypothesis that reducing ability is linked to cercosporin resistance.

C. Toxin Export

The export of cercosporin out of the hyphae via membrane transport proteins also appears to be an important contributor to cercosporin resistance. Disruption of *CFP* (the gene that encodes the MFS transporter involved in cercosporin resistance, as described earlier; Callahan et al. 1999) resulted in mutants of *C. kikuchii* that were partially sensitive to cercosporin. Further, expression of CFP in *Cochliobolus heterostrophus* increased cercosporin resistance in this normally sensitive fungus (Upchurch et al. 2002). Membrane transport proteins in other systems were also shown to provide cercosporin resistance. For example, the yeast gene *Snq2p* that encodes an ATP-binding cassette (ABC) transporter protein provided resistance to cercosporin when overexpressed (Ververidis et al. 2001). Also, disruption of an MFS transporter (Bcmfs1) in *Botrytis cinerea* resulted in increased sensitivity to cercosporin (Hayashi et al. 2002). These results suggest that membrane transport proteins may enhance cercosporin resistance by transporting cercosporin out of the hyphae.

IX. Isolation and Characterization of Cercosporin Autoresistance Genes

To isolate genes involved in cercosporin resistance, cercosporin-sensitive mutants of *C. nicotianae* were identified and the genes isolated via mutant complementation using a library from the wild type (Jenns and Daub 1995; Jenns et al. 1995). These studies identified three genes required for normal expression of cercosporin resistance. Two of these genes were subsequently shown to encode enzymes in the pyridoxine (vitamin B_6) pathway; the third gene encoded a transcription factor.

A. Pyridoxine Biosynthetic Genes

Five distinct mutants identified by their inability to grow on medium supplemented with cercosporin or with other $^1O_2$-generating compounds were complemented with two different genes from wild-type *C. nicotianae*. Subsequent analysis demonstrated that these genes (*PDX1*, *PDX2*) encoded enzymes in a previously unknown pathway for pyridoxine (vitamin B_6) biosynthesis (Ehrenshaft and Daub 2001; Ehrenshaft et al. 1999). Unlike vitamins C and E, vitamin B_6 vitamers had not previously been shown to play a role in antioxidant defense; rather, the catalytically active form, pyridoxal 5′ phosphate, is a well known co-factor for transamination reactions. Assay for antioxidant properties, however, demonstrated that pyridoxine and its vitamers quench $^1O_2$, $O_2^{-\cdot}$, and have general antioxidant activity (Bilski et al. 2000; Denslow et al. 2005). Further, *Cercospora* fungi have 2- to 3-fold higher mycelial levels of vitamin B_6 than do cercosporin-sensitive fungi, suggesting that elevated B_6 levels may be an important defense mechanism (Herrero and Daub 2007). Studies in other systems support a role for B_6 vitamers in antioxidant defense. For example, this vitamin has been shown to protect against $O_2^{-\cdot}$-mediated eye damage in diabetics and to prevent oxidation of lipids and proteins in blood caused by H_2O_2 and high glucose (Jain and Lim 2001; Jain et al. 2002; Stocker et al. 2003).

To determine if elevated B_6 vitamer levels would protect against cercosporin toxicity in plants, we transformed the *C. nicotianae PDX1* and *PDX2* genes into tobacco and recovered transformed lines with increased expression of both genes (Herrero and Daub 2007). Unfortunately, none of these transgenic lines showed elevated resistance to cercosporin. Molecular analysis of the transgenic plants indicated that most lines did not show any increases in cellular B_6 levels, correlating with a consistent down-regulation of the endogenous tobacco *PDX1* and *PDX2* genes. We concluded that the pathway is tightly regulated at the transcriptional level. Thus, although elevated B_6 levels may protect against cercosporin toxicity, further research is needed to identify strategies to elevate levels of this vitamin in plants.

B. Regulatory Genes

The third gene recovered from our mutant complementation study was *CRG1* (cercosporin *r*esistance *g*ene 1) shown to encode a binuclear zinc cluster transcription factor (Chung et al. 2003a). Mutants for *crg1* show significantly reduced growth on cercosporin-containing medium, but are not inhibited by other $^1O_2$-generating photosensitizers. These mutants also show medium-regulated reductions in cercosporin synthesis. Zinc binuclear transcription factors are an important class of transcriptional regulators in fungi and regulate diverse processes including mycotoxin synthesis and multiple drug resistance (Mamnun et al. 2002; Payne and Brown 1998). We hypothesize that CRG1 regulates genes involved in cercosporin resistance and biosynthesis.

We used suppressive subtractive hybridization to recover libraries of genes that are differentially expressed between the wild-type *C. nicotianae* and a null mutant for *crg1* (Herrero et al. 2007). The libraries recovered are predicted to include genes that are regulated by CRG1, including genes involved in cercosporin resistance. We have identified genes in the libraries that are consistent with the resistance mechanisms previously identified, including several ABC and MFS transporter genes as well as genes involved in reducing activity such as genes predicted to encode oxidoreductases and a quinone reductase. These genes are currently being investigated for a role in cercosporin resistance by gene disruption and over-expression approaches. Any genes identified as playing a role in cercosporin resistance in *Cercospora* will then be tested for the ability to impart cercosporin and *Cercospora* resistance in plants by transformation and expression in host plants such as was done with *PDX1* and *PDX2*.

X. Cercosporin Degradation Genes from Bacteria

Cercospora autoresistance genes are a promising mechanism for engineering resistant plants; however, an alternate approach would be to identify genes encoding proteins that modify or degrade cercosporin to non-toxic breakdown products. This strategy proved effective, for example, in sugarcane where a gene encoding toxin detoxification of the toxin albicidin resulted in resistance against *Xanthomonas albilineans*, the causal agent of leaf scald on susceptible sugarcane (Zhang et al. 1999). Previous studies by Robeson and co-workers (1993) identified bacteria that were capable of degrading cercosporin. Using the methods of that study, we screened bacterial isolates for cercosporin-degrading activity, easily visible as a cleared halo surrounding bacterial colonies growing on cercosporin-containing medium (Mitchell et al. 2002). Among the most active were strains of the zinnia pathogen *Xanthomonas campestris* pv. zinniae (XCZ). Characterization of the reaction led to the identification of the non-toxic breakdown product xanosporic acid (Fig. 11.7B; Mitchell et al. 2003). Mutants of XCZ were isolated that lost the

ability to degrade cercosporin, and complementation of these mutants led to the recovery of a gene encoding a putative oxidoreductase required for cercosporin degradation (Taylor et al. 2006). Southern analysis demonstrated that the presence of this gene correlated with cercosporin degrading activity of different bacterial strains, and quantitative PCR analysis demonstrated that the oxidoreductase gene was significantly upregulated in XCZ in the presence of cercosporin. However, transformation of the oxidoreductase gene into non-degrading *Xanthomonas* strains and into tobacco cells did not impart cercosporin-degradation activity. Thus the oxidoreductase is required but not sufficient for cercosporin degradation and further studies are needed to define additional genes necessary for this process.

XI. Conclusions

Photoactivated perylenequinone toxins play an important role in diseases caused by *Cercospora* species and likely also by a diversity of other fungal pathogens that synthesize these interesting compounds. Significant advances have been made in the understanding of the mode of action of these toxins, their biosynthesis and regulation, and possible mechanisms involved in resistance to their toxic effects. Although efforts thus far to engineer resistance in crop plants to cercosporin and *Cercospora* pathogens have not yet been successful, the strategy of targeting cercosporin in genetic engineering efforts remains a promising strategy for developing crop species resistant to these damaging pathogens.

References

Ahonsi MO, Maurhofer M, Boss D, Defago G (2005) Relationship between aggressiveness of *Stagnospora* sp. isolates on field and hedge bindweeds, and in vitro production of fungal metabolites cercosporin, elsinochrome A and leptosphaerodione. Eur J Plant Pathol 111:203–215

Arnone A, Assante G, Di Modugno V, Merlini L, Nasini G (1988) Perylenequinones from cucumber seedlings infected with *Cladosporium cucumerinum*. Phytochemistry 6:1675–1678

Arst HN, Jr., Peñalva MA (2003) pH regulation in Aspergillus and parallels with higher eukaryotic regulatory systems. Trends Genet 19:224–231

Balis C, Payne MG (1971) Triglycerides and cercosporin from *Cercospora beticola*: fungal growth and cercosporin production. Phytopathology 61:1477–1484

Ballario P, Vittorioso P, Magrelli A, Talora C, Cabibbo A, Macino G (1996) White collar-1, a central regulator of blue light responses in Neurospora, is a zinc finger protein. EMBO J 15:1650–1657

Batchvarova RB, Reddy VS, Bennett J (1992) Cellular resistance in rice to cercosporin, a toxin of *Cercospora*. Phytopathology 82:642–646

Bellus D (1979) Physical quenchers of singlet molecular oxygen. Adv Photochem 11:105–205

Berenbaum MR (1987) Charge of the light brigade: phototoxicity as a defense against insects. In: Heitz JR, Downum KR (eds) Light-activated pesticides. American Chemical Society, Washington, D.C., pp 206–216

Bilski P, Li MY, Ehrenshaft M, Daub ME, Chignell CF (2000) Vitamin B6 (pyridoxine) and its derivatives are efficient singlet oxygen quenchers and potential fungal antioxidants. Photochem Photobiol 71:129–134

Bilski P, Daub ME, Chignell CF (2002) Direct detection of singlet oxygen via its phosphorescence from cellular and fungal cultures. Methods Enzymol 352:41–52

Birch AJ (1967) Biosynthesis of polyketides and related compounds. Science 156:202–206

Bok JW, Keller NP (2004) Global regulation of secondary metabolic gene clusters. Eukaryot Cell 3:527–535

Brodhagen M, Keller NP (2006) Signalling pathways connecting mycotoxin production and sporulation. Mol Plant Pathol 7:285–301

Callahan T, Rose M, Meade M, Ehrenshaft M, Upchurch R (1999) *CFP*, the putative cercosporin transporter of *Cercospora kikuchii*, is required for wild type cercosporin production, resistance, and virulence on soybean. Mol Plant–Microbe Interact 12:901–910

Calpouzos L (1966) Action of oil in the control of plant diseases. Annu Rev Phytopathol 4:369–390

Calpouzos L, Stalknecht GF (1967) Symptoms of Cercospora leaf spot of sugar beets influenced by light intensity. Phytopathology 57:799–800

Chen H, Lee M-H, Daub ME, Chung K-R (2007a) Molecular analysis of the cercosporin biosynthetic gene cluster in *Cercospora nicotianae*. Mol Microbiol 64:755–770

Chen HQ, Lee MH, Chung KR (2007b) Functional characterization of three genes encoding putative oxidoreductases required for cercosporin toxin biosynthesis in the fungus *Cercospora nicotianae*. Microbiology 153:2781–2790

Choquer M, Lahey KA, Chen H-L, Cao L, Ueng PP, Daub ME, Chung KR (2005) The *CTB1* gene encoding a fungal polyketide synthase is required for cercosporin toxin biosynthesis and fungal virulence in *Cercospora nicotianae*. Mol Plant–Microbe Interact 18:468–476

Choquer M, Lee M-H, Bau H-J, Chung K-R (2007) Deletion of a MFS transporter-like gene in *Cercospora nicotianae* reduces cercosporin toxin accumulation and fungal virulence. FEBS Lett 581:489–494

Chung KR (2003) Involvement of calcium/calmodulin signaling in cercosporin toxin biosynthesis by *Cercospora nicotianae*. Appl Environ Microbiol 69:1187–1196

Chung KR, Jenns AE, Ehrenshaft M, Daub ME (1999) A novel gene required for cercosporin toxin resistance in the fungus *Cercospora nicotianae*. Mol Gen Genet 262:382–389

Chung KR, Daub ME, Kuchler K, Schuller C (2003a) The *CRG1* gene required for resistance to the singlet oxygen-generating cercosporin toxin in *Cercospora nicotianae* encodes a putative fungal transcription factor. Biochem Biophys Res Commun 302:302–310

Chung KR, Ehrenshaft M, Wetzel DK, Daub ME (2003b) Cercosporin-deficient mutants by plasmid tagging in the asexual fungus *Cercospora nicotianae*. Mol Gen Genet 270:103–113

Clark RA, Stephens TR, Bowden EF, Daub ME (1995) Electrochemical reduction of the phytotoxin cercosporin. J Electoanal Chem 389:205–208

Daub ME (1982a) Cercosporin, a photosensitizing toxin from *Cercospora* species. Phytopathology 72:370–374

Daub ME (1982b) Peroxidation of tobacco membrane lipids by the photosensitizing toxin, cercosporin. Plant Physiol 69:1361–1364

Daub ME (1987) Resistance of fungi to the photosensitizing toxin, cercosporin. Phytopathology 77:1515–1520

Daub ME, Briggs SP (1983) Changes in tobacco cell membrane composition and structure caused by the fungal toxin, cercosporin. Plant Physiol 71:763–766

Daub ME, Ehrenshaft M (2000) The photoactivated *Cercospora* toxin cercosporin: Contributions to plant disease and fundamental biology. Annu Rev Phytopathol 38:461–490

Daub ME, Hangarter RP (1983) Production of singlet oxygen and superoxide by the fungal toxin, cercosporin. Plant Physiol 73:855–857

Daub ME, Leisman GB, Clark RA, Bowden EF (1992) Reductive detoxification as a mechanism of fungal resistance to singlet-oxygen-generating photosensitizers. Proc Natl Acad Sci USA 89:9588–9592

Daub ME, Li M, Bilski P, Chignell CF (2000) Dihydrocercosporin singlet oxygen production and subcellular localization: a possible defense against cercosporin phototoxicity in *Cercospora*. Photochem Photobiol 71:135–140

Davis VM, Stack ME (1991) Mutagenicity of stemphyltoxin III, a metabolite of *Alternaria alternata*. Appl Environ Microbiol 57:180–182

Dekkers LA, You B-J, Gowda VS, Liao H-L, Lee M-H, Bau H-J, Ueng PP, Chung K-R (2007) The *Cercospora nicotianae* gene encoding dual O-methyltransferase and FAD-dependent monooxygenase domains mediates cercosporin toxin biosynthesis. Fungal Genet Biol 44:444–454

Denslow SA, Walls AA, Daub ME (2005) Regulation of biosynthetic genes and antioxidant properties of vitamin B$_6$ vitamers during plant defense responses. Physiol Mol Plant Pathol 66:244–255

Dickman MB, Park YK, Oltersdorf T, Li W, Clemente T, French R (2001) Abrogation of disease development in plants expressing animal antiapoptotic genes. Proc Natl Acad Sci USA 98:6957–6962

Diwu Z (1995) Novel theraputic and diagnostic applications of hypocrellins and hypericins. Photochem Photobiol 61:529–539

Dobrowolski DC, Foote CS (1983) Chemistry of singlet oxygen 46. Quantum yield of cercosporin-sensitized singlet oxygen formation. Angew Chem 95:729–730

Dowzer CD, Kelly JM (1991) Analysis of the creA gene, a regulator of carbon catabolite repression in *Aspergillus nidulans*. Mol Cell Biol 11:5701–5709

Echandi E (1959) La chasparria de los cafetos causada por el hongo *Cercospora coffeicola* Berk and Cooke. Turrialba 9:54–67

Ehrenshaft M, Daub ME (2001) Isolation of *PDX2*, a second novel gene in the pyridoxine biosynthesis pathway of eukaryotes, archaebacteria, and a subset of eubacteria. J Bacteriol 183:3383–3390

Ehrenshaft M, Upchurch RG (1991) Isolation of light-enhanced cDNA clones of *Cercospora kikuchii*. Appl Environ Microbiol 57:2671–2676

Ehrenshaft M, Jenns AE, Daub ME (1995) Targeted gene disruption of carotenoid biosynthesis in *Cercospora nicotianae* reveals no role for cartenoids in photosensitizer resistance. Mol Plant–Microbe Interact 8:569–575

Ehrenshaft M, Bilski P, Li M, Chignell CF, Daub ME (1999) A highly conserved sequence is a novel gene involved in *de novo* vitamin B6 biosynthesis. Proc Natl Acad Sci USA 96:9374–9378

Fajola AO (1978) Cercosporin, a phytotoxin from *Cercospora* species. Physiol Plant Pathol 79:157–164

Fang LZ, Qing C, Shao HJ, Yang YD, Dong ZJ, Wang F, Zhao W, Yang WQ, Liu JK (2006) Hypocrellin D, a cytotoxic fungal pigment from fruiting bodies of the ascomycete *Shiraia bambusicola*. J Antibiotics 59:351–354

Foote CS, Denny RW, Weaver L, Chang Y, Peters J (1970) Quenching of singlet oxygen. Ann NY Acad Sci 171:139–148

Giese AC (1980) Hypericism. Photochem Photobiol Rev 5:229–255

Giese AC (1981) The photobiology of *Blepharisma*. Photochem Photobiol Rev 6:139–180

Hartman PE, Dixon WJ, Dahl TA, Daub ME (1988) Multiple modes of photodynamic action by cercosporin. Photochem Photobiol 47:699–703

Hayashi K, Schoonbeek HJ, De Waard MA (2002) *Bcmfs1*, a novel major facilitator superfamily transporter from *Botrytis cinerea*, provides tolerance towards the natural toxic compounds camptothecin and cercosporin and towards fungicides. Appl Environ Microbiol 68:4996–5004

Heitz JR, Downum KR (eds) (1995) Light-activated pest control. American Chemical Society, Washington, D.C.

Herranz S, Rodríguez JM, Bussink H-J, Sánchez-Ferrero JC, Arst HNJ, Peñalva MA, Vincent O (2005) Arrestin-related proteins mediate pH signaling in fungi. Proc Natl Acad Sci USA 102:12141–12146

Herrero S, Daub ME (2007) Genetic manipulation of vitamin B-6 biosynthesis in tobacco and fungi uncovers limitations to up-regulation of the pathway. Plant Sci 172:609–620

Herrero S, Amnuaykanjanasin A, Daub ME (2007) Identification of genes differentially expressed in the phytopathogenic fungus *Cercospora nicotianae* between cercosporin toxin-resistant and -susceptible strains. FEMS Microbiol Lett 275:326–337

Hudson JB, Towers GHN (1991) Therapeutic potential of plant photosensitizers. Pharmacol Ther 49:181–222

Hudson JB, Imperial V, Haugland RP, Diwu Z (1997) Antiviral properties of photoactive perylenequinones. Photochem Photobiol 65:352–354

Hughes K, Negrotto D, Daub ME, Meeusen R (1984) Free radical stress response in paraquat-sensitive and resistant tobacco plants. Environ Exp Bot 24:151–157

Ito T (1981) Dye binding and photodynamic action. Photochem Photobiol 33:947–955

Jain AK, Lim G, Langford M, Jain SK (2002) Effect of high-glucose levels on protein oxidation in cultured lens cells, and in crystalline and albumin solution and its inhibition by vitamin B6 and N-acetylcysteine: Its possible relevance to cataract formation in diabetes. Free Radic Biol Med 33:1615–1621

Jain SK, Lim G (2001) Pyridoxine and pyridoxamine inhibit superoxide radicals and prevent lipid peroxidation, protein glycosylation, and (Na$^+$ + K$^+$)-ATPase activity reduction in high glucose-treated human erythrocytes. Free Radic Biol Med 30:232–237

Jenns AE, Daub ME (1995) Characterization of mutants of *Cercospora nicotianae* sensitive to the toxin cercosporin. Phytopathology 85:906–912

Jenns AE, Daub ME, Upchurch RG (1989) Regulation of cercosporin accumulation in culture by medium and temperature manipulation. Phytopathology 79:213–219

Jenns AE, Scott DL, Bowden EF, Daub ME (1995) Isolation of mutants of the fungus *Cercospora nicotianae* altered in their response to singlet-oxygen-generating photosensitizers. Photochem Photobiol 61:488–493

Keller N, Hohn T (1997) Metabolic pathway gene clusters in filamentous fungi. Fungal Genet Biol 21:17–29

Keller NP, Turner G, Bennett JW (2005) Fungal secondary metabolism – from biochemistry to genomics. Nat Rev Microbiol 3:937–947

Kuyama S, Tamura T (1957) Cercosporin. A pigment of *Cercospora kikuchii* Matsumoto et Tomoyasu. I. Cultivation of fungus, isolation and purification of pigment. J Am Chem Soc 79:5725–5726

Leisman GB, Daub ME (1992) Singlet oxygen yields, optical properties, and phototoxicity of reduced derivatives of the photosensitizer cercosporin. Photochem Photobiol 55:373–379

Li PZX, Xu NLJ, Meng DL, Sha Y (2006) A new perylenequinone from the fruit bodies of *Bulgaria inquinans*. J Asian Nat Prod Res 8:743–746

Li WM, Schuler MA, Berenbaum MR (2003) Diversification of furanocoumarin-metabolizing cytochrome P450 monooxygenases in two papilionids: specificity and substrate encounter rate. Proc Natl Acad Sci USA 100:14593–14598

Linden H, Macino G (1997) White collar-2, a partner in blue light signal transduction, controlling expression of light-regulated genes in *Neurospora crassa*. EMBO J 16:98–107

Linden H, Ballario P, Macino G (1997) Blue light regulation in *Neurospora crassa*. Fungal Genet Biol 22:141–150

Liu WZ, Shen YX, Liu XF, Chen YT, Xie JL (2001) A new perylenequinone from *Hypomyces* sp. Chin Chem Lett 12:431–432

Lousberg RJJ, Weiss U, Salmink. CA, Arnone A, Merlini L, Nasini G (1971) The structure of cercosporin, a naturally occurring quinone. Chem Commun 71:1463–1464

Macri F, Vianello A (1979) Photodynamic activity of cercosporin on plant tissues. Plant Cell Environ 2:267–271

Mamnun YM, Pandjaitan R, Mahe Y, Delahodde A, Kuchler K (2002) The yeast zinc finger regulators Pdr1p and Pdr3p control pleiotropic drug resistance (PDR) as homo- and heterodimers in vivo. Mol. Microbiol. 46:1429–1440

Marzluf GA (1997) Genetic regulation of nitrogen metabolism in the fungi. Microbiol Mol Biol Rev 61:17–32

Mathey A, Lukins PB (2001) Spatial distribution of perylenequinones in lichens and extended quinones in quincyte using confocal fluorescence microscopy. Micron 32:107–113

McCormick SP, Taylor SL, Plattner R D, Beremand MN (1990) Bioconversion of possible T-2 toxin precursors by a mutant strain of *Fusarium sporotrichioides* NRRL 3299. Appl Environ Microbiol 56:702–706

Mitchell TK, Chilton WS, Daub ME (2002) Biodegradation of the polyketide toxin cercosporin. Appl Environ Microbiol 68:4173–4181

Mitchell TK, Alejos-Gonzalez F, Gracz HS, Danehower DA, Daub ME, Chilton WS (2003) Xanosporic acid, an intermediate in bacterial degradation of the fungal phototoxin cercosporin. Phytochemistry 62:723–732

Okubo A, Yamazaki S, Fuwa K (1975) Biosynthesis of cercosporin. Agr Biol Chem 39:1173–1175

Payne GA, Brown MP (1998) Genetics and physiology of aflatoxin biosynthesis. Annu Rev Phytopathol 36:329–362

Raab O (1900) The action of fluorescent material on infusorien. Z Biol 39:524–546

Rawlings BJ, Reese PB, Ramer SE, Vederas JC (1989) Comparison of fatty acid and polyketide biosynthesis: stereochemistry of cladosporin and oleic acid formation in *Cladosporium cladosporioides*. J Am Chem Soc 111:3382–3390

Robeson DJ, Jalal MAF (1992) Formation of entisophleichrome by *Cladosporium herbarum* isolated from sugar beet. Biosci Biotechnol Biochem 56:949–952

Robeson JR, Jalal MAF, Simpson RB (1993) Methods for identifying cercosporin-degrading microorganisms. US Patent 5,262,306, Nov 1993

Rodriguez-Landa JF, Contreras CA (2003) A review of clinical and experimental observations about antidepressant actions and side effects produced by *Hypericum perforatum* extracts. Phytomedicine 10:688–699

Shim WB, Dunkle LD (2003) *CZK3*, a MAP kinase kinase kinase homolog in *Cercospora zeae-maydis*, regulates cercosporin biosynthesis, fungal development, and pathogenesis. Mol Plant–Microbe Interact 16:760–768

Sollod CC, Jenns AE, Daub ME (1992) Cell surface redox potential as a mechanism of defense against photosensitizers in fungi. Appl Environ Microbiol 58:444–449

Spikes JD (1989) Photosensitization. In: Smith KC (ed) The science of photobiology, 2nd edn. Plenum, New York, pp 79–110

Stack ME, Mazzola EP, Page SW, Pohland AE, Highet RS, Tempesta MS, Corely DG (1986) Mutagenic perylenequinone metabolites of *Alternaria alternata*: altertoxins I, II, and III. J Nat Prod 49:866–871

Steinkamp MP, Martin SS, Hoefert LL, Ruppel EG (1979) Ultrastructure of lesions produced in leaves of *Beta vulgaris*. Physiol Plant Pathol 15:13–16

Stierle AC, Cardellina JH (1989) Phytotoxins from *Alternaria alternata*, a pathogen of spotted knapweed. J Nat Prod 52:42–47

Stocker P, Lesgards JF, Vidal N, Chalier F, Prost M (2003) ESR study of a biological assay on whole blood: antioxidant efficiency of various vitamins. Biochim Biophys Acta 1621:1–8

Tabuchi H, Tajimi A, Ichihara A (1994) Phytotoxic metabolites isolated from *Scolecotrichum graminis* Fuckel. Biosci Biotech Biochem 58:1956–1959

Taylor TV, Mitchell TK, Daub ME (2006) An oxidoreductase is involved in cercosporin degradation by the bacterium *Xanthomonas campestris* pv. *zinniae*. Appl Environ Microbiol 72:6070–6078

Tertivanidis K, Goudoula C, Vasilikiotis C, Hassioutou E, Perl-Treves R, Tsaftaris A (2004) Superoxide dismutase transgenes in sugarbeets confer resistance to oxidative agents and the fungus *C. beticola*. Transgenic Res 13:225–233

Thorold CA (1940) Cultivation of bananas under shade for the control of leaf spot disease. Trop Agric Trinidad 17:213–214

Toone WM, Jones N (1999) AP-1 transcription factors in yeast. Curr Opin Genet Dev 9:55–61

Truscott TG (1990) New trends in photobiology: the photophysics and photochemistry of the carotenoids. J Photochem Photobiol B 6:359–371

Upchurch RG, Walker DC, Rollins JA, Ehrenshaft M, Daub ME (1991) Mutants of *Cercospora kikuchii* altered in cercosporin synthesis and pathogenicity. Appl Environ Microbiol 57:2940–2945

Upchurch RG, Rose MS, Eweida M, Callahan TM (2002) Transgenic assessment of CFP-mediated cercosporin export and resistance in a cercosporin-sensitive fungus. Curr Genet 41:25–30

Ververidis P, Davrazou F, Diallinas G, Georgakopoulos D, Kanellis AK, Panopoulos N (2001) A novel putative reductase (Cpd1p) and the multidrug exporter Snq2p are involved in resistance to cercosporin and other singlet oxygen-generating photosensitizers in *Saccharomyces cerevisiae*. Curr Genet 39:127–136

Watanabe A, Ebizuka Y (2004) Unprecedented metabolism of chain length determination in fungal aromatic polyketide synthases. Chem Biol 11:1101–1106

Weiss U, Merlini L, Nasini G (1987) Naturally occurring perylenequinones. In: Herz W, Grisebach H, Kirby GW, Tamm CH, (eds) Progress in the chemistry of organic natural products, vol 52. Springer, Vienna, pp 1–71

Wilkinson F, Helman WP, Ross AB (1995) Rate constants for the decay and reactions of the lowest electronically excited singlet state of molecular oxygen in solution. An expanded and revised compilation. J Phys Chem Ref Data 24:663–1021

Williamson JD, Scandalios JG (1992) Differential response of maize catalases and superoxide dismutases to the photoactivated fungal toxin cercosporin. Plant J 2:351–358

Wu H, Lao XF, Wang QW, Lu RR (1989) The shiraiachromes: novel fungal perylenequinone pigments from *Shiraia bambusicola*. J Nat Prod 5:948–951

Yamazaki S, Ogawa T (1972) The chemistry and stereochemistry of cercosporin. Agric Biol Chem 36:1707–1718

Yamazaki S, Okube A, Akiyama Y, Fuwa K (1975) Cercosporin, a novel photodynamic pigment isolated from *Cercospora kikuchii*. Agric Biol Chem 39:287–288

Yoshihara T, Shimanuki T, Araki T, Sakamura S (1975) Phleichrome, a new phytotoxic compound produced by *Cladosporium phlei*. Agric Biol Chem 39:1683–1684

You B-J, Lee M-H, Chung K-R (2008) Production of cercosporin toxin by the phytopathogenic *Cercospora* fungi is affected by diverse environmental signals. Can J Microbiol 54:259–269

Yu J-H, Keller N (2005) Regulation of secondary metabolism in filamentous fungi. Annu Rev Phytopathol 43:437–458

Zhang L, Xu J, Birch RG (1999) Engineered detoxification confers resistance against a pathogenic bacterium. Nat Biotechnol 17:1021–1024

12 Programmed Cell Death in Fungus–Plant Interactions

AMIR SHARON[1], ALIN FINKELSHTEIN[1]

CONTENTS

I.   Introduction 221
II. Apoptosis 222
     A. Apoptosis in Metazoan Organisms........ 222
         1. Apoptotic Pathways 222
         2. Major Classes of Apoptosis-Related
             Proteins.......................... 223
             a) Caspases...................... 223
             b) Bcl-2 Proteins 223
             c) Inhibitor of Apoptosis Proteins.... 223
             d) Mitochondria-Secreted Proteins .. 224
         3. ROI and Apoptosis 224
III. Plant PCD.............................. 224
     A. ROI Production 224
     B. HR and PCD 225
     C. Plant PCD Pathways and Disease........ 225
     D. Effect of Anti-Apoptotic Genes
         on Disease Development.............. 226
         1. Baculovirus Protein p35........... 226
         2. Bcl-2 Proteins.................... 226
         3. BI-1 and Mlo 227
IV. Apoptosis in Yeasts and Fungi............. 228
     A. Apoptosis in Yeasts.................. 228
         1. Induction of Yeast Apoptosis....... 228
         2. Regulation of Yeast Apoptosis
             by Human Bcl-2 Proteins.......... 228
         3. Functional Analysis of Yeast
             Apoptotic Genes 228
             a) Aif1p and AIMD 229
             b) Yca1p 229
             c) HtrA2/Nma111p 229
             d) IAP/Bir1p................... 229
     B. Apoptosis in Filamentous Fungi 229
         1. Morphological and Cytological
             Evidences 229
         2. Genomic Data 230
         3. Functional Characterization
             of Fungal PCD-Related Genes....... 230
         4. Involvement of Fungal Apoptosis
             Pathways in Disease 231
             a) Role of Fungal ROI 231
             b) Role of Fungal PCD 232
V.   Conclusions 232
     References............................. 233

[1] Department of Plant Sciences, Tel Aviv University, Tel Aviv 69978,
Israel; e-mail: amirsh@ex.tau.ac.il, alinf@ex.tau.ac.il

I. Introduction

Apoptosis was originally defined in mammals, where it plays a major role in controlling normal development. Apoptosis and several other forms of programmed cell death (PCD) have since been defined in metazoan as well as in plants, fungi and even in bacteria (Bredesen et al. 2006). For simplicity, when referring to cell death processes in plants and fungi we use the terms apoptosis or PCD throughout this chapter.

Hypersensitive cell death (HR) is a plant resistance response in which spreading of incompatible pathogens is restricted to a small number of plant cells. A hallmark of HR is appearance of small necroses that result from the local death of the host cells. Cell death observed during HR is preceded by enhanced production of reactive oxygen intermediates (ROI) and has typical markers of apoptosis. PCD is also observed during other types of plant–pathogen interactions including non-host as well as in compatible interactions. In some systems PCD is necessary for development of plant resistance, whereas in others it is harmful to the plant. Moreover, various pathogens developed ways to overcome, or even manipulate the plant PCD machinery for their advantage.

Apoptosis is also emerging as an important mechanism in fungi. Yeasts and filamentous fungi undergo cell death with classic markers of metazoan apoptosis during various stages of development. PCD is observed in fungi during vegetative incompatibility, in sexual and asexual reproduction, at stationary phase, and in aged cultures. Homologs of mammalian apoptotic genes have been identified in fungi, supporting a conservation of apoptotic machinery between mammals and fungi. Several recent studies also imply that fungal PCD might be involved in mediating fungus–plant interactions. In this chapter we review the literature on PCD in plants and fungi and the role it may play in mediating fungus–plant interactions.

Plant Relationships, 2nd Edition
The Mycota V
H. Deising (Ed.)
© Springer-Verlag Berlin Heidelberg 2009

II. Apoptosis

A. Apoptosis in Metazoan Organisms

Apoptosis is one of the main types of programmed cell death in multicellular organisms; it involves an orchestrated series of biochemical events leading to a characteristic cell morphology and death. Apoptosis is associated with maintenance of cell homeostasis, elimination of damaged cells, aging and differentiation, as well as the adaptive responses of cells to biotic and abiotic stresses (Danial and Korsmeyer 2004; Green 2005). Apoptosis is distinguished from necrosis by several morphological and cytological characteristics of the dying cells. Necrosis occurs by cell perturbation and usually provokes an inflammatory response and eventually cell lysis. Apoptosis is carried out in an orderly process; extracellular or endogenous signals trigger a chain of cellular responses that lead to non-inflammatory cell death. Apoptotic cells develop typical markers, including cell shrinkage, plasma membrane blabbing, chromatin condensation, specific DNA degradation (resulting in DNA laddering), swelling of the outer mitochondrial membrane, externalization of phosphatidylserine and formation of small vesicles from the cell surface also known as apoptotic bodies. At the end of the apoptotic response, apoptotic bodies are rapidly engulfed by phagocytes and adjacent cells and the cell content is recycled (Gozuacik and Kimchi 2007; Wilfried 2004).

1. Apoptotic Pathways

Apoptosis can follow two general routes, known as the extrinsic and intrinsic pathways. The extrinsic pathway is initiated by extracellular ligands, such as Fas and tumor necrosis factor (TNF), toxins, or other external signals that bind and activate death receptors on the cell membrane. The intrinsic pathway can be activated by cell damage or during specific developmental stages. Mitochondria play a central role in activation and regulation of the intrinsic pathway (Green 1998; Kroemer et al. 2007; Wang 2001). At the biochemical level apoptosis initiated by either pathway can be attributed to the activation of caspases, a group of cysteine proteases that are the executors of apoptosis in mammals and other metazoan organisms (Reed et al. 2004). The intrinsic and extrinsic apoptotic networks include unique as well as common components. In the extrinsic pathway, activated death receptors belonging to the superfamily of cysteine-rich tumor necrosis factor receptor (TNFR) recruit adaptor proteins and pro-caspase 8 molecules, forming a death-inducing signaling complex (DISC) at the plasma membrane. DISC formation activates pro-caspase 8, which then acts to cleave and activate downstream caspases, including caspase 3 (Fig. 12.1). Activation of the intrinsic pathway causes release of certain mitochondria proteins, which associate with and activate downstream components of the apoptotic machinery. Cytochrome *c* that is released to the cytosol binds apoptosis-inducing factor (Apaf1), which in

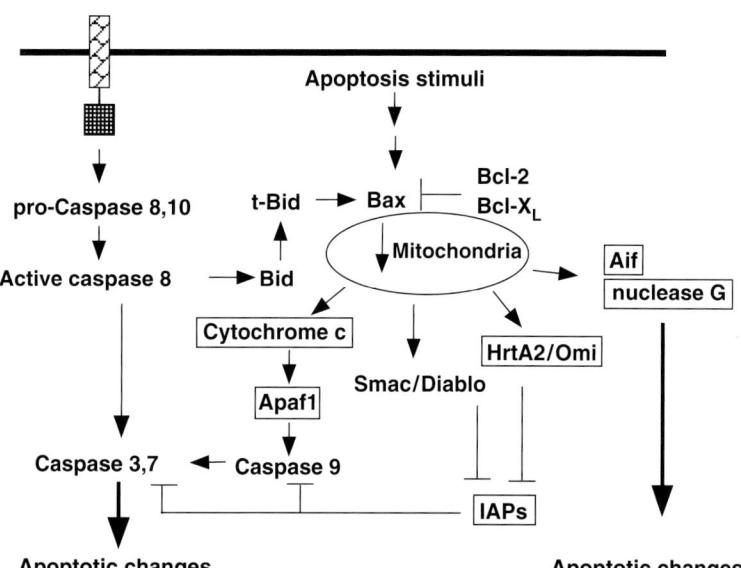

Fig. 12.1. Extrinsic and intrinsic apoptotic pathways. The extrinsic pathway is mediated by membrane death receptors. The intrinsic pathway is mediated by mitochondria. Only major regulators are shown. Homologs of *boxed* proteins have been identified in fungi

turn self–associates and binds procaspase 9, forming the apoptosome complex (Jiang and Wang 2004). Transactivation of procaspase 9 to active caspase 9 follows and the active caspase 9 then cleaves and activates downstream caspases including caspase 3.

Autophagy is another type of cell death (type II cell death) that lacks some of the apoptotic features and does not involve caspases (Gozuacik and Kimchi 2007). The most prominent morphological change observed in autophagy cell death is the appearance of double- or multiple-membrane enclosed vesicles in the cytoplasm that engulf portions of cytoplasm and/or organelles such as mitochondria and endoplasmic reticulum. Nuclear changes, such as chromatin condensation, appear later in autophagic cell death than they do in apoptosis and there is no DNA fragmentation or formation of apoptotic bodies.

2. Major Classes of Apoptosis-Related Proteins

Several classes of protein families and a large number of specific proteins are involved in the regulation of apoptosis. The major groups that are also relevant to the regulation of apoptosis in fungi are briefly described.

a) Caspases

Caspases belong to a family of cysteine- or aspartate-specific proteases (12 members are known in human) which execute the last steps of apoptosis. Caspases exist in cells in the form of proenzymes (zymogens) that are activated by proteolysis at conserved aspartic acid residues, resulting in the formation of a heterodimer containing one large and one small subunit. The active caspase is a tetramer composed of two such heterodimers (Earnshaw et al. 1999; Fuentes-Prior and Salvesen 2004). The apoptotic signaling cascade first activates initiator caspases (e.g., caspases 9 and 8), which then process downstream effector caspases (e.g., caspases 3 and 7). Activation of the effector caspases results in diminished integrity of the actin and intermediate filament networks within the cell, inhibition of protein synthesis, and activation of DNase (Danial and Korsmeyer 2004). **Caspases** have not been identified in plants and fungi. Instead, a class of related proteases called **metacaspases** appears to take the role of caspases in organisms belonging to these kingdoms (Uren et al. 2000).

b) Bcl-2 Proteins

Bcl-2 proteins are major regulators of the intrinsic, mitochondria-dependent apoptotic pathway. Overall identity and homology among protein members of the Bcl-2 family (over 24 in human) is low, except in up to four conserved Bcl-2 homology (BH) domains (BH1–BH4 domains). These BH domains are important for the interaction of Bcl-2 proteins with various Bcl-2 and non-Bcl-2 proteins (Borner 2003; Reed et al. 2004). The Bcl-2 family includes apoptosis-promoting (e.g., Bax, Bak, Bid) and apoptosis-suppressing (e.g., Bcl-2, Bcl-xL) proteins. Bcl-2 proteins are multifunctional and only part of their mechanisms is known. Under normal conditions Bcl-2 proteins are usually linked with the cytoskeleton and during apoptosis they are translocated to mitochondria, where they interact with other proteins and can also form pores in the inner mitochondria membrane (IMM). Anti-apoptotic Bcl-2 proteins also act as antioxidants and block the release of cytochrome c. Neither Bcl-2 nor any of the four BH domains homologs have been identified in fungal or plant genomes.

c) Inhibitor of Apoptosis Proteins

Inhibitor of apoptosis proteins (IAPs) are important regulators of metazoan apoptosis. IAPs block both the mitochondria and death-receptor (Fas) mediated pathways of apoptosis by directly binding to and inhibiting caspases. IAPs are typically characterized by the presence of one to three baculovirus IAP repeat (BIR) domains that mediate the interaction of IAPs with caspases (Dohi et al. 2004; Reed et al. 2004; Verhagen et al. 2001). The XIAP BIR2 domain for example, binds the catalytic groove of effectory caspases 7 and 3, completely filling the active site thus preventing the entry of other substrates (Holcik and Korneluk 2001). Several IAPs also contain a carboxy-terminal RING finger domain which is recognized to have E3 ubiquitin ligase activity (Verhagen et al. 2001). The IAPs are negatively regulated by IAP-binding proteins such as the mitochondria-secreted proteins Omi/HtrA2 and Smac/Diablo that antagonize caspase–IAP interaction. Proteins with BIR domains have been identified in fungi. Usually only a single gene is identified in each fungal genome.

d) Mitochondria-Secreted Proteins

In addition to cytochrome *c*, several other proteins that reside in the mitochondria under normal conditions are translocated to the cytosol or nucleus during apoptosis. Smac/Diablo and HtrA2/Omi are released into the cytosol and facilitate caspase activation by inhibiting IAP proteins. Aif and endonuclease G are translocated to the nucleus and promote caspases-independent DNA degradation (Wang 2001). Fungi and plants have homologs of cytochrome *c*, HtrA2/Omi, Aif, and nuclease G, but not Smac/Diablo.

3. ROI and Apoptosis

Reactive oxygen intermediates (ROI) are produced and accumulate during all forms of PCD (Danial and Korsmeyer 2004; Green 2005). ROI can damage proteins, lipid membranes and DNA, leading to either necrotic or apoptotic cell death, depending on the extent of the damage and duration of the stimulus. In multicellular organisms, ROI could be at the beginning or at the end of death processes. In apoptosis caused by ischemia, ROI accumulate before Bax and caspases. In such cases, ROI scavenging prevents the activation of caspases, indicating that ROI not only cause damage, but also act as a signaling molecules (Maulik et al. 1998; Tan et al. 1998). Treatment with oxidizing compounds, such as menadione, H_2O_2 or depletion of glutathione often results in a dose-dependent apoptotic cell death, suggesting stimulation of apoptosis by ROI. ROI activate opening of the mitochondrial permeability transition pore, which causes swelling, rupture of the outer membrane, and release of mitochondria proteins. These as well as additional experimental evidence show that ROI play important roles not only in induction of apoptosis but also in regulation of responses to various apoptosis-inducing conditions.

III. Plant PCD

Similar to animal cells, plants undergo PCD as part of normal development and in response to external stimuli. Developmentally regulated plant PCD occurs in specific cells or tissues such as during xylogenesis and reproduction, as well as on a global scale during senescence (Greenberg 1996; Pennell and Lamb 1997). PCD is also induced by

various external stimuli including abiotic and biotic stresses. Perhaps the most familiar form of induced plant PCD is associated with pathogen attack. The localized death of a small number of cells triggered by incompatible pathogens is the hallmark of HR resistance. HR is preceded by rapid accumulation of ROI, a response known as oxidative burst (Torres et al. 2006). Treatment of plant tissues with various fungal derived toxins and elicitors also triggers ROI accumulation and PCD (Wang et al. 1996). A cell wall extract of the rice blast fungus *Magnaporthe grisea* elicited rapid H_2O_2 generation and cell death in a rice cell suspension culture (Matsumura et al. 2003). AAL toxin, a sphinganine-analog mycotoxin produced by *Alternaria alternata* induced ROI accumulation and PCD in sensitive tomato tissues (Brandwagt et al. 2000). Nep-1 and Nep-like proteins (NLPs) cause ROI-dependent PCD in a variety of plant species (Bailey et al. 1997; Pemberton and Salmond 2004). Thus, similar to other types of PCD, pathogen-activated plant cell death involves ROI accumulation.

A. ROI Production

Plasma membrane NADPH oxidase and peroxidases are the two most likely sources of plant ROI produced following pathogen recognition. Pharmacological and genetic data now suggest that specific NADPH oxidase genes have pivotal role in the rapid and local accumulation of ROI in response to pathogens (Torres et al. 2005; Chap. 18).

NADPH oxidase, also known as respiratory burst oxidase (Rbo) is a multi-component complex that produces ROI in neutrophils as a defense response against pathogens (Lambeth 2004). The membrane enzyme gp91^{phox}, which is a part of the complex, generates superoxide by transferring an electron to molecular oxygen. Plant Rbo has a similar structure and disruption of the complex affects ROI production and disease resistance (Kawasaki et al. 1999; Torres et al. 2002). Of the ten identified *Arabidopsis* respiratory burst oxidase (*Rbo*) homologous (*AtRboh*) genes, *AtrbohD* and *AtrbohF* are important for ROI production following pathogen attack, as can be determined by the complete blocking of pathogen-induced superoxide formation in an *atrbohd/atrbohf* double mutant (Torrest et al. 2002). Surprisingly, elimination of ROI production has variable effects in different pathosystems: in *atrboh* mutant plants HR is generally reduced and the plants become more susceptible when challenged with avirulent pathogens, whereas compatible pathogens seem to elicit enhanced HR in the *atrboh* mutant and disease is reduced (Torres et al. 2002; Yoshioka et al. 2003).

Above a certain threshold, high ROI levels might have a direct killing effect, as in neutrophils. Indeed, manipulation of ROI scavenging enzymes such as ascorbate peroxidase, glutathione, superoxide dismutase (SOD) or catalase can affect disease in some cases (Jabs et al. 1996; Polidoros et al. 2001). Since ROI are similarly toxic to pathogens and plants, the necrosis associated with HR might represent inevitable plant cell death in areas of high ROI accumulation. In this respect it is important to consider that some pathogens also produce ROI or manipulate host ROI production to their own advantage (see Sect. III.B). Below the toxic levels, ROI may serve as signaling molecules with diverse functions that might not necessarily lead to PCD. Different ROI are formed, which probably have different regulatory effects. The highly reactive oxygen radical produced by Rboh is converted by SOD to a less toxic hydrogen peroxide, which diffuses through membranes and therefore can serve as an efficient second messenger. Outside the cell H_2O_2 can be neutralized by catalases, while inside the cell it can be converted back to a highly reactive hydroxyl radical.

Collectively, these results demonstrated that Rboh proteins are the key producers of ROI during pathogen attack, but the Rboh-generated ROI do not necessarily induce PCD and may have variable effects on fungus–plant interactions.

B. HR and PCD

In a normal HR response, PCD is limited to a small number of cells around pathogen invasion site. Several *Arabidopsis* mutants have been identified in which cell death regulation is impaired (Dietrich et al. 1994; Greenberg and Ausubel 1993). Interestingly, none of the genes defined to date is related to known executors or regulators of metazoan PCD.

Several of the identified genes are transcription factors that were found to be involved in regulation of PCD. The zinc-finger transcription factor Lsd1 negatively regulates the spreading of cell death to cells surrounding infection sites (Dietrich et al. 1997). In *lsd1* mutants, cell death control is impaired, resulting in spreading of cell death to the entire leaf, a phenomenon known as runway cell death (RCD). Additional mutants such as the chloroplast proteins *acd1* and *acd2* also show the RCD phenotype and are highly sensitive to pathogens (Yao and Greenberg 2006). An *lsd1/atrbohd*

double mutant shows enhanced RCD and increased sensitivity to the bacterium Pseudomonas *syringea* compared with *lsd1* plants, while over-expression of *AtrbohD* in *lsd1* mutant plants rescues these phenotypes (Torres et al. 2005). These results demonstrate that ROI are not required for induction of cell death in the *lsd1* mutant plants and that AtrbhoD actually negatively regulates pathogen-induced cell death. Therefore, although AtrbohD is the major generator of superoxide following challenging with a pathogen, the resulting ROI do not activate PCD, but rather limit spreading of PCD to neighboring cells.

Mutation in a related gene, *LOL1*, has opposite effects to the *lsd1* mutation. *LOL1* is a highly conserved *LSD1* paralog that acts as a positive regulator of plant cell death. An *lsd1/lol1* double mutant shows reduced RCD, whereas over-expression of *LOL1* induces HR in the wild type as well as in *lsd1* plants (Epple et al. 2003). These results suggest that Lol1 is a positive regulator of PCD. In wild-type plants it is antagonized by Lsd1, which restricts the activity of Lol1 to a limited number of cells. In *lsd1* plants Lol1 is not blocked, resulting in the RCD phenomenon. Another Lsd1-interacting protein that also enhances PCD was recently characterized. The *Arabidopsis* basic leucine zipper (bZIP) transcription factor AtbZIP10 can be retained outside the nucleus by Lsd1 (Kaminaka et al. 2006). Similar to Lol1, AtbZIP10 is a positive mediator of the *lasd1* RCD phenotype and acts antagonistically to Lsd1 in pathogen-induced HR. Recent data support direct interaction between Lsd1 and the other transcription factors Lol1 and AtbZIP10 (Kaminaka et al. 2006).

These results suggest that Lsd1 might act as an antagonist of PCD by retaining the Lol1 and AtbZIP10 (and possibly other transcription factors) in the cytoplasm, thereby preventing activation of apoptotic process by this group of pro-apoptotic transcription elements. This model resembles the control of cell death in metazoan by the IAP family of zinc-finger proteins, which maintain a threshold for cell death by modulation of caspases. If this analogy is correct, caspase-like activity should be affected in the corresponding *Arabidopsis* mutant plants.

C. Plant PCD Pathways and Disease

Although the identified HR regulators do not show homology to known apoptosis genes, several lines of evidence show that challenging of plants with pathogens activates specific apoptotic pathways and this response directly affects disease development. Pathogen-induced apoptosis is mitochondria-dependent and is at least partly regulated by elements that are similar to apoptosis control elements found in metazoan organisms (del Ponzo and Lam 1998; Lam et al. 2001). Plant mitochondria are involved in generation of ROI, which may activate HR and PCD. The level of mitochondria-derived ROI can be modulated by an alternative oxidase (AOX), an inner mitochondrial membrane (IMM) enzyme that is not found in animal mitochondria (Lam et al. 1999). Interestingly, fungal mitochondria also contain AOX, suggesting possible similarities in regulation of ROI between plants and fungi.

The human mitochondria-associated hexokinase regulates apoptosis by inhibiting cytochrome *c* leakage through interference with opening of permeability transition pores in the mitochondria membrane (Azoulay-Zohar et al. 2004; Pastorino et al. 2002). Silencing of Nicotiana *benthamiana* hexokinase gene *HXK1* caused necrotic lesions on leaves and growth retardation, and cells of the *HXK1*-silenced plants exhibited markers of apoptosis (Kim et al. 2006). Consistent with animal apoptosis, cytochrome *c* was released and caspase-like activity was induced in these plants. Conversely, over-expression of *Hxk1* resulted in increased resistance to H_2O_2-induced PCD and prevented the release of cytochrome *c*.

Changes in phospholipids metabolism can lead to apoptosis in animal cells. The AAL toxin produced by *A. alternate* kills cells of sensitive host plants by inducing apoptotic cell death (Brandwagt et al. 2000). It was shown that administration of AAL toxin to sensitive tissues blocks shingolipid biosynthesis and leads to the accumulation of dihydrosphingosine (DHS), which can induce apoptosis in various systems. AAL-insensitive plants contain the *ASC-1* resistance gene, a homolog of yeast longevity assurance gene (*LAV1*). Asc1 modifies sphingolipid metabolism in AAL-treated cells, thereby preventing accumulation of DHS and induction of apoptosis (Brandwagt et al. 2000; Spassieva et al. 2002). These and other examples show that apoptosis is associated with various types of plant–pathogen interactions and that it is induced and controlled by mechanisms that are similar to those known in animals.

D. Effect of Anti-Apoptotic Genes on Disease Development

1. Baculovirus Protein p35

Further evidence connecting PCD with the plant defense machinery comes from work with plant and non-plant anti-apoptotic genes. p35, an IAP-like protein from baculovirus suppresses apoptosis in mammalian cells by inhibiting caspases (Clem et al. 1996). Expression of the anti-apoptotic baculovirus p35 in plants significantly affected disease caused by various pathogens. Transgenic tomato plants expressing the anti-apoptotic p35 baculovirus protein were protected from AAL toxin-induced cell death and showed enhanced resistance to infection by the tomato fungal pathogens *A. alternate* and *Colletotrichum coccodes* and by the pathogenic bacterium *P. syringea* pv. *tomato* (Lincoln et al. 2002). A p35 binding site mutant clone that is inactive against human caspase 3 did not protect plants from the pathogens or the toxin, indicating the involvement of plant caspase-like proteins in the HR response. Tobacco plants carrying the *N* gene develop a typical HR resistance when challenged with tobacco mosaic virus (TMV). Virus-induced HR and cell death were delayed in transgenic tobacco plants expressing the p35 protein (del Pozo and Lam 2003). The virus was still able to spread and systematically infect the entire plant, suggesting that the p35 protein affected the rate of HR after pathogen challenge. These examples illustrate opposite effects of p35 on disease: it increased plant resistance to fungal and bacterium pathogens, while it broke the HR-based resistance against the virus.

2. Bcl-2 Proteins

Plants and fungi lack any structural homologs of Bcl-2; even proteins with a single BH domain have not been identified in their genomes. Surprisingly however, heterologous expression of Bcl-2 members in plants as well as in yeast can protect (anti-apoptotic members) or induce (pro-apoptotic members) mitochondria-dependent cell death by mechanisms that are similar to those operating in mammalian cells. Expression of anti-apoptotic proteins, the human Bcl-X_L and *Caenorhabditis elegance* Ced-9 (a homolog of the human Bcl-2 protein) enhanced tolerance of tomato plants to infection by a virulent strain of cucumber mosaic virus (Christophel and Santa Cruz 1999). Although cell death and necrosis were delayed or completely prevented, virus reproduction was normal, suggesting that the anti-apoptotic proteins affected plants response after challenging with the virus. The effect was specific to limiting necrotic lesion formation, while other symptoms associated with viral infection were unaffected, again pointing to a specific prevention of virus-induced PCD, but not viral spreading. When expressed in tobacco, these as well as additional anti-apoptotic human and viral proteins provided broad spectrum protection against necrotrophic fungal pathogens (Dickman et al. 2001). Disease was drastically reduced and fungal growth limited in the transgenic plants. DNA laddering that occurred after infection with pathogens in wild-type tobacco plants did not occur in the transgenic plants, suggesting that infection by the necrotrophic pathogens caused PCD, which enhanced plant sensitivity to the pathogens. These results demonstrate that necrotrophic fungi might use plant PCD to their advantage, as distinct from the defensive role of PCD in HR resistance.

3. BI-1 and Mlo

While plants and fungi lack structural homologs of p35 and Bcl-2 proteins, the above examples clearly demonstrate a functional conservation of the apoptotic machinery, which is recognized and activated by these anti-apoptotic metazoan proteins. These results also imply that Bcl-2 structurally unrelated proteins might take over the apoptosis regulatory roles of metazoan Bcl-2 proteins. In a screen for human suppressors of PCD, a new protein called Bax inhibitor-1 (BI-1) was isolated, it blocked Bax-induced PCD in the yeast *Saccharomyces cerevisiae* (Xu and Reed 1998). BI-1 is a membrane protein with six transmembrane helices and it has been suggested that it can form pores in membranes similar to Bcl-2 proteins. Unlike proteins of the Bcl-2 family, BI-1 is highly conserved and homologs are readily identified in all organisms including plants and fungi. Several BI-1 homologs are found in the Arabidopsis genome, AtBI-1 is the most highly related to the mammalian BI-1 proteins (Lam et al. 2001). Over-expression of *AtBI-1* in *Arabidopsis* and tobacco cells blocked Bax, H_2O_2, and salycilic acid-induced cell death (Kawai-Yamada et al. 2003). Rice BI-1 was down-regulated in a cell suspension culture following treatment with a cell wall extract from the rice blast fungus, which

elicits H_2O_2 generation and cell death (Matsumura et al. 2003). Over-expression of the AtBI-1 in the rice cells prevented H_2O_2 generation and the cells showed sustainable survival when challenged with the elicitor. Expression of Barley BI-1 in carrots limited necrosis and restricted spreading of Botrytis *cinerea* (Imani et al. 2006). Thus, expression of BI-1 is suppressed following treatment with elicitors or by challenging with pathogens, and over-expression of BI-1 is associated with enhanced plant resistance both to necrosis-inducing elicitors and to necrotroph pathogens.

Interaction of the biotrophic fungus *Blumeria graminis* f. sp. *hordei* with barley is mediated by the *MLO* gene: *mlo/mol* plants are resistant to the pathogen whereas *MLO* genotypes are sensitive. Resistance of the *mlo* genotypes is characterized by inability of the pathogen to penetrate epidermal cells and by highly restricted HR necroses that are preceded by an oxidative burst (Huckelhoven et al. 2001a). *MLO* genotypes do not develop HR due to suppression of PCD by the defense suppressor Mlo protein and are therefore sensitive to the pathogen (Kim et al. 2002). Expression of BI-1 is up-regulated in barley leaves infected by appropriate *B. graminis* f. sp. *hordei* (Huckelhoven et al. 2001b), and over-expression of BI-1 in *mlo/mol* resistant barley genotypes renders plants sensitive to an otherwise avirulent races of the fungus (Huckelhoven et al. 2003). In addition, over-expression of barley *BI-1* was able to break non-host penetration resistance to the wheat powdery mildew *B. graminis* f. sp. *tritici* (Eichmann et al. 2004). These results demonstrate that BI-1, besides Mlo, is involved in regulation of barley resistance to penetration by the biotrophic powdery mildew fungi. Comparison of BI-1 and Mlo reveals structural and functional similarities. Both genes are similarly expressed in response to pathogen challenge and ROI, as well as during leaf senescence. Over-expression of either gene in barley confers susceptibility to *B. graminis* f. sp. *hordei* and restores susceptibility of *mlo* genotypes. Thus, suppression of the host PCD, either by Mlo or by BI-1, is associated with enhanced sensitivity to biotrophs.

The above examples lead to the conclusion that plant PCD plays a central role in determining the outcome of all types of fungal-plant interactions. A complex system composed of ROI-generating enzymes, a number of transcription-controlling elements, and proteins enhancing and suppressing PCD are involved in the activation or prevention of PCD and in restricting the spread of cell death.

Pathogens cope with plant apoptosis in different ways; depending on the type of interaction, either suppression or activation of plant PCD may be used by pathogens to overcome plant defense and render plants more sensitive.

IV. Apoptosis in Yeasts and Fungi

A. Apoptosis in Yeasts

Because yeasts are unicellular organisms, it was argued that they do not undergo true apoptosis, and until several years ago *S. cerevisiae* was considered an "apoptosis null system" (Frohlich and Madeo 2000). Nevertheless, various reports showed that yeast undergoes PCD accompanied by classic apoptotic markers, including ROI accumulation, nuclear condensation, apoptosis-typical chromatin condensation, DNA cleavage, externalization of phosphatidylserine to the outer leaflet of the plasma membrane, and cytochrome *c* release from mitochondria (Gordeeva et al. 2004; Matsuyama et al. 1999). More recently, yeast orthologues of metazoan apoptotic genes were discovered, including major regulators of apoptosis such as a metacaspase, IAP, HtrA2/Omi and Nuclease G (Frohlich et al. 2007). Analysis of these genes revealed involvement in yeast PCD, providing the final proof that yeast and metazoan apoptosis are two versions of a similar cellular program. Several substantial differences were however observed between yeast and metazoan apoptosis. DNA laddering, the hallmark of metazoan apoptosis could not be determined in yeasts, and many of the apoptosis genes known in metazoan species do not exist in the yeast genome. It is also evident that the gene families for which homologs are found in yeasts (e.g., IAPs, caspase-like genes) are represented by a single yeast homolog (Frohlich et al. 2007).

1. Induction of Yeast Apoptosis

The first observations of yeast PCD were made in a *S. cerevisiae cdc48* mutant strain. Dying cells of this mutant displayed an apoptotic phenotype characterized by mammalian apoptotic markers, such as phosphatidylserine exposure, condensation and margination of chromatin, and DNA fragmentation (Madeo et al. 1997). Soon after this discovery, research showed that exogenous toxic agents such as antifungal compounds, hydrogen peroxide, or acetic acid, as well as various types of stresses can trigger apoptotic-like cell death in *S. cerevisiae* wild-type cells (Madeo et al. 2004). The apoptotic markers were accompanied by ROI generation and translocation of cytochrome *c* into the cytoplasm. Caspase-like enzymatic activity was induced under some conditions, which could be prevented by cycloheximide or antioxidants.

PCD was also noticed under natural conditions, including in aged yeast cultures and during mating (Herker et al. 2004). In aged cultures, the dying old cells accumulate ROI and exhibit markers of apoptosis accompanied by increased caspase activity. Over-expression of *BIR1* (IAP homolog) or deletion of the yeast metacaspase gene *YCA1* blocked PCD and delayed age-induced yeast cell death (Madeo et al. 2002). Pheromone-induced cell death was accompanied by the appearance of apoptotic markers, which were abolished by disruption of the pheromone protein kinase cascade (Pozniakovsky et al. 2005; Severin and Hyman 2002). ROI accumulation was found associated with all types of PCD and is suggested to play a central role in regulating yeast PCD (Frohlich et al. 2007; Madeo et al. 2004).

2. Regulation of Yeast Apoptosis by Human Bcl-2 Proteins

Similar to plants, mammalian Bcl-2 proteins activate the yeast apoptotic machinery. Expression of the pro-apoptotic protein Bax in *S. cerevisiae* caused cell death with typical apoptotic markers and was accompanied by release of cytochrome *c* from mitochondria. Simultaneous expression of the anti-apoptotic Bcl-X_L or Bcl-2 proteins with Bax prevented these effects. Bcl-2 proteins with mutations in the BH domains were inactive, indicating that the intact 3D structure of Bcl-2 proteins is necessary for interaction with the yeast apoptotic machinery. The dimerization-mediating BH3 domain of Bax and targeting of Bax to mitochondrial membranes were both found to be essential for killing both mammalian and yeast cells (Frohlich and Madeo 2000; Gross et al. 2000; Polcic and Forte 2003; Tao et al. 1997).

3. Functional Analysis of Yeast Apoptotic Genes

Homologs of several downstream components of metazoan apoptosis genes were identified in yeasts.

Analysis of these putative yeast apoptosis regulators showed similarity as well as differences in function compared with the mammalian orthologs.

a) Aif1p and AIMD

The human apoptosis-inducing factor (Aif) is secreted from the mitochondria and plays an important role in regulation of apoptosis. The yeast homolog Aif1p is also translocated from mitochondria to the nucleus under apoptosis-inducing conditions and is involved in nuclear degradation. Consistent with the pro-apoptotic role of the human Aif protein, apoptosis is abolished in *aif1Δ* yeast strains (Wissing et al. 2004). Two closely sequence-related genes of "human AIF homologous mitochondrion-associated inducer of cell death" (AMID) are found in yeast: *NDE1* and *NDI1* coding for external NADH dehydrogenase and internal NADH dehydrogenase, respectively. *NDI1* over-expression on glucose media leads to apoptosis and cell cycle arrest in yeast. It is suggested that ROI that are produced as a by-product of the Ndi1p NADH dehydrogenase activity are responsible for cell death. Consistently, deletion of *NDI1* decreased ROI production and prolonged the chronological life span of yeast cells, albeit with loss of survival fitness (Li et al. 2006).

b) Yca1p

Yca1p belongs to the family of metacaspases that are found in fungi and plants (Uren et al. 2000). Disruption of *YCA1* reduced cell death and formation of apoptotic markers in aged cultures. The *yca1Δ* strain accumulated less ROI and had enhanced resistance to oxygen stress, whereas over-expression of *YCA1* had opposite effects and colonies were hypersensitive to apoptotic stimuli (Madeo et al. 2002). These results substantiate the function of Yca1p in apoptosis, although some PCD scenarios do not involve Yca1p.

c) HtrA2/Nma111p

Nma111p is a yeast homolog of the human mitochondria secreted protein HtrA2/Omi, but unlike HtrA2/Omi, Nma111p is a nuclear protein (Fahrenkrog et al. 2004). Under conditions of cellular stress Nma111p aggregates inside the nucleus and is involved in nuclear degradation. Despite these differences in localization, Nma111p is involved in the regulation of apoptosis: *nma111Δ* strains survive better than wild-type cells at elevated temperatures and show no apoptotic hallmarks, whereas over-expression of *NMA111* enhances apoptotic-like cell death (Fahrenkrog et al. 2004).

d) IAP/Bir1p

Yeast has a single *IAP* homolog called *BIR1*. Bir1p is a cytoplasmic and nuclear protein and is a substrate for Nma111p, but unlike human and *Drosophila* IAPs, it does not interact with the metacaspase Yca1p (Walter et al. 2006). Rather than inhibiting caspase-mediated cell death, yeast IAP proteins have roles in cell division and appear to act in a similar way to the IAPs from *C. elegans* and the mammalian IAP survivin (Uren et al. 1999). *bir1Δ* yeast cells are more sensitive to oxidative stress-induced apoptosis while over-expression of *BIR1* reduces apoptotic cell death. The protective effect of Bir1p over-expression can be antagonized in vivo by simultaneous over-expression of Nma111p.

All these data demonstrate the importance of apoptosis in yeasts. The presence of conserved structural homologs of major metazoan regulators of apoptosis imply that apoptosis is regulated in yeasts in a similar way as in metazoans, albeit with some differences. The structural, and even more significantly, the functional conservation of yeast apoptotic proteins suggest that similar conservation can be expected in filamentous fungi. Therefore, the principles of yeast apoptosis are serving as guidelines for the research of apoptosis in filamentous species, including fungal plant pathogens.

B. Apoptosis in Filamentous Fungi

1. Morphological and Cytological Evidences

The best studied form of PCD in filamentous fungi is heterokaryon incompatibility (HI). During interaction between two incompatible hyphae the hyphal fusion cell accumulates apoptotic markers and dies (Esser 2006). Cytological markers of apoptosis were also observed during spore formation or following application of antifungal or apoptosis-inducing compounds in several fungi. In one of the earliest reports, Roze and Linz (1998) showed that *Mucor racemosus* undergoes PCD following treatment with the apoptosis inducing agent lovastatin. In *N. crassa*, apoptotic phenotypes and nuclear degradation were observed during starvation and heterokaryon incompatibility (Esser 2006; Glass and Dementhon 2006). Apoptotic phenotypes were observed in *Aspergillus fumigatus* during entry into stationary phase or on exposure to amphotericin B or H_2O_2 (Mousavi and Robson 2004). In *Aspergillus nidulans*, apoptotic markers and induction of caspase-like activity were observed during sporulation. As in yeasts, DNA laddering could not be detected in filamentous fungi, raising the possibility that fungal PCD might not be true apoptosis. However, most other markers of metazoan apoptosis were detected during fungal PCD, including membrane blabbing, cytoplasm vacuolization, nuclear shrinkage and degradation, and phosphatidylserine translocation from the inner to outer leaflet of the cell membrane. It was proposed that the lack of DNA laddering during PCD in yeast and fungi might be due to differences in heterochromatin structure (Madeo et al. 1997).

ROI seem to be essential mediators of fungal PCD. The oncogenic Ras protein was shown to be important in ROI-mediated PCD in the alfalfa pathogen *Colletotrichum trifolii*. Under nutrient-limiting conditions, the dominant active form of Ras (DARas) caused increased ROI production, abnormal fungal growth and eventual apoptotic-like cell death (Chen and Dickman 2004). These phenotypes were rescued by proline, which reduced ROI levels and suppressed PCD in DARas strains (Chen and Dickman 2005). Proline also protected wild-type *C. trifolii* against various PCD-inducing treatments, such as UV light, salt and oxidative stress. Dihydrosphingosine (DHS) and phosphingosine (PHS), two major sphingoid bases of fungi induced ROI accumulation and cell death with typical markers of apoptosis in *A. nidulans*. Induction of apoptosis by DHS was metacaspase and ROI-independent, but required functional mitochondria (Cheng et al. 2003). Farnesol, a quorum-sensing molecule produced by *Candida albicans*, also induces apoptosis in *A. nidulans* (Semighini et al. 2006). Farnesol-induced apoptosis requires functional mitochondria but is also ROI-dependent, unlike DHS-induced apoptosis which is not blocked by ROI scavengers. The results suggest that redundant PCD pathways might be functional in *A. nidulans*.

2. Genomic Data

In the past few years, over 40 complete fungal genomes have been publicly released, and a similar number of fungi are currently being sequenced (Xu et al. 2006). Homologs of several apoptosis genes are readily identified in the genomes of various fungi. Many of the identified genes are homologs of mitochondria-associated regulators of apoptosis, such as mitochondria-secreted proteins HrtA2, nuclease G, AIF1/AMID, downstream components including metacaspases and IAPs. More than 50 putative human and mouse PCD-associated genes have been identified in Aspergillii (Fedorova et al. 2005). Among these genes are heterokaryon incompatibility genes and species-specific protein families, as well as more conserved core components of the metazoan apoptotic machinery. Interestingly, some of these genes are not present in yeast

genomes. In addition, many of the fungal putative apoptotic genes are structurally more similar to human than to yeast genes (Fedorova et al. 2005), suggesting higher conservation between filamentous fungi and metazoan apoptotic machinery than between yeast and fungi.

3. Functional Characterization of Fungal PCD-Related Genes

Only a handful of fungal PCD-related genes have so far been functionally characterized. Stationary-phase cultures of *A. fumigatus* accumulate apoptotic markers including exposure of phosphatidylserine on the outer leaflet of the cell membrane, elevation of caspase activity. and cell death (Mousavi and Robson 2003, 2004). Two putative metacaspase encoding genes, *CasA* and *CasB*, have been identified and cloned in *A. fumigatus* (Richie et al. 2007). Phosphatidylserine exposure was prevented in a *casa/casb* double mutant but cell viability and metacaspase activity were not altered. The mutant also retained wild-type virulence and showed no difference in sensitivity to various apoptosis-inducing stimuli. Thus, although required for the loss of membrane phospholipids asymmetry at stationary phase, metacaspases seem unrelated to other apoptotic phenotypes or to PCD induced by external conditions, such as oxidative stress and antifungal compounds. These results are in contrast to yeast, in which apoptotic cell death induced by various external stimuli such as acetic acid, H_2O_2, antifungal compounds or starvation depends on the yeast metacaspase Yca1p (Eisler et al. 2004; Silva et al. 2005; Vachova and Palkova 2007).

Genes that regulate age-induced PCD have been characterized in *Podospora anserina*. Cultures of this fungus have a limited life span; after a certain period of growth in a constant rate the culture turns to senescence and dies (Lorin et al. 2006). The final stages of the senescence syndrome require intact mitochondria and are associated with increased ROI production and appearance of apoptotic markers (Gredilla et al. 2006). Several apoptosis regulators were recently isolated and analyzed in *P. anserina*, providing the first molecular proof for the role of mitochondria-dependent PCD in regulation of aging-associated cell death in this fungus. Deletion of a gene encoding mitochondrial fission factor (*PaDNM1*) increased fungal life span and resistance to the apoptosis-inducing compound etoposide (Scheckhuber et al. 2007). The *cyc1-1* mutant carries a single substitution in the *cyc1* gene encoding cytochrome *c*, leading to a splicing defect and loss of function of complex III. The *cyc1-1* mutants produce less ROI, are long lived, and have stabilized mitochondria (Sellem et al. 2007). Caspase-like activity was found to increase in *P. anserina* aging cultures. Two metacaspases, *PaMCA1* and *PaMCA2*, were isolated and knockout mutants were generated and characterized. Knockout of either *PaMCA1* or *PaMCA2* led to 80% or 148% increase of life span, respectively (Hamann et al. 2007). *PaAMID1*, a homolog of mammalian *AMID* was also analyzed as a putative regulator of caspase-independent PCD. Δ*paamid1* knockout mutants had an extended life span, although the effect was somewhat less than in the metacaspase deletion mutants (Hamann et al. 2007). Collectively, these results indicate that increased ROI levels during aging trigger a mitochondria-dependent program that leads to PCD in senescent cultures of *P. anserina*. Similar to situation in *Aspergillus* sp., both caspase-dependent and -independent pathways control PCD in *P. anserina*. Type II autophagy PCD is not part of this machinery since cell death is not prevented in *P. anserina* autophagy mutants and is actually accelerated in aged cultures of autophagy mutants, in contrast to the delayed cell death of mitochondria-dependent PCD mutants (Pinan-Lucarré et al. 2005).

4. Involvement of Fungal Apoptosis Pathways in Disease

a) Role of Fungal ROI

Analysis of PCD genes have not yet been reported in plant pathogenic fungi. Nevertheless, several lines of evidence support a role for fungus-induced plant PCD as well as fungal PCD in regulation of fungus–plant interactions (see Chap. 2). As described above, both plant and fungal PCD are associated with ROI production. In plants, ROI produced by NADPH oxidases regulate HR resistance as well as other types of PCD. Depending on the pathosystem, ROI and resulting PCD may contribute to plant resistance or susceptibility. The necrotroph *B. cinerea* can infect a wide range of plants and there is no known HR resistance against this pathogen. In fact, the degree of disease is correlated with ROI levels: disease was reduced in HR-deficient mutant plants that do not produce ROI, and was enhanced when co-inoculated with an avirulent pathogen, which stimulates ROI production (Govrin and Levine 2000). Furthermore, *Arabidopsis* mutant plants with a delayed or reduced cell death response are generally more resistance to *Botrytis* infection, whereas accelerated cell death mutants are more susceptible (Van Baarlen et al. 2007). Therefore, the HR-like plant PCD that is triggered by *B. cinerea* facilitates plant colonization by this fungus. *B. cinerea* (like other fungi) produces various types of ROI, which may also add to ROI accumulation in and around the infection site. During plant penetration, *B. cinerea* produces significant amounts of H_2O_2 (Tenberge et al. 2002). Functional analysis of hydrogen peroxide generating systems in *B. cinerea* showed that SOD was necessary for H_2O_2 production and for full virulence. *B. cinerea* Δ*bcsod1* mutants produced less H_2O_2 and had significantly reduced pathogenicity (Rolke et al. 2004).

NADPH oxidases (NOX) are major generators of ROI in fungi and are involved in regulating fungal growth and development (Takemoto et al. 2007). The rice blast fungus *M. grisea* undergoes an oxidative burst of its own during plant infection, which is necessary for proper spore germination and for the development of appressoria (Egan et al. 2007). Scavenging of ROI delayed appressoria formation and reduced plant penetration by the fungus. Knockout of one or both *M. grisea NOX* genes resulted in avirulent strains due to inability of appressoria to penetrate the plant cuticle. Mutants were also unable to infect wounded plants indicating a role for *NOX* genes during in planta proliferation. Surprisingly, Δ*nox1* or Δ*nox2* single mutants were unaffected in superoxide production. Moreover, hyphal tips of a Δ*nox1/nox2* double mutant accumulated significantly higher levels of superoxide compared with the wild-type strain, suggesting that ROI can be generated by NOX-unrelated routes during pathogenic development. It also suggests that NOX-generated ROI might suppress other ROI generating pathways. Collectively, these results show that *M. grisea* NOX proteins are necessary for pathogenic development, but additional analysis will be necessary to answer how the fungus-generated ROI affect the plant or fungus–plant communication.

In the endophyte *Epichloe festucae*, mutation in *NOXA*, one of two NADPH oxidase genes present in this fungus, changed interaction with *Lolium perenne* from mutualistic to antagonistic. Plants infected by the *noxa* mutant lost apical dominance, showed precocious senescence, and eventually died (Tanaka et al. 2006). The wild-type endophyte accumulated ROI in the extracellular matrix, which could not be detected in the *noxA* mutant-infected plants. These results show that, in this mutualistic interaction, fungal ROI are suppressors of pathogenic development.

b) Role of Fungal PCD

In *M. grisea*, the spore and germ tube collapse following appressorium formation and it is suggested that the cytoplasm moves from these organs into the incipient appressorium (Howard et al. 1991; Thines et al. 2000). It was recently found that mitosis was necessary for appressoria formation and that the spores died before appressoria-mediated fungal penetration into the plant (Veneault-Fourrey et al. 2006). Blocking of mitosis or deletion of the autophagy gene *MgATG8* prevented appressoria development and spore cell death, respectively. The *mgatg8* autophagy cell death mutants produced appressoria but these appressoria could not penetrate the cuticle and the mutant strains were completely non-pathogenic. This is the first example showing not only that PCD occurs in fungi during plant infection but also that it is necessary for pathogenesis. While demonstrating the importance of fungal cell death in pathogenesis, this result should be considered in light of the more general role of autophagy in fungal growth and development. Deletion of an autophagy gene in *P. anserina* resulted in accelerated cell death, associating autophagy with prevention rather than activation of PCD in this fungus (Pinan-Lucarré et al. 2005). As described, spore PCD was found essential for completion of the early stages of pathogenic development in *M. grisea* (Veneault-Fourrey et al. 2006). It could be expected that PCD would also occur in spores of *C. gloeosporioides*, which produces large melanized appressoria and has a life style similar to *M. grisea*. Surprisingly however, in *C. gloeosporioides* spore cell death does not occur during plant infection and is not required for pathogenicity: blocking of mitosis did not prevent spore germination and appressoria formation, and

spores remained viable long after plant penetration (Nesher et al. 2008).

Barhoom and Sharon (2007) reported on expression of the human Bcl-2 protein in *C. gloeosporioides*. Similar to yeasts, transgenic strains of *C. gloeosporioides* expressing the Bcl-2 protein had prolonged longevity, were protected from Bax-induced cell death, and exhibited enhanced stress resistance. Importantly, the Bcl-2 transgenic strains were hypervirulent to host plants. These results suggest a link between fungal PCD and disease, in which blocking PCD enhances fungal virulence, much in the same way that blocking plant PCD can enhance plant resistance against certain pathogens. Ito et al (2008) recently reported that α-tomatin, the major saponin in tomato, induces apoptotic cell death in the tomato pathogen *Fusarium oxysporum*. This result strengthens the hypothesis that plants might induce fungal PCD, as a counter action to fungal-induced plant PCD.

Thus, although fungal cells undergo PCD during plant colonization, much further research is necessary to determine under which conditions PCD might occur and what might be the role of fungal PCD in disease development. The differences observed between *M. grisea* and *C. gloeosporioides* suggest that substantial variation can be expected between different pathosystems.

V. Conclusions

Apoptotic pathways are composed of hundreds of proteins and are regulated at several levels. Disruption of apoptotic genes or interference with regulation of apoptotic pathways often lead to abnormalities and malignancies in animals. Because of the central role of apoptosis in mammals, many drugs have been developed which affect cell death, either by blocking or promoting apoptosis (Andersen et al. 2005; Fesik 2005). Recent developments in research into plant and fungal apoptosis clearly show that apoptosis is also important for the proper development and adaptation in organisms belonging to these kingdoms. Dying plant and fungal cells have features of mammalian apoptosis, including morphological, cytological and biochemical markers. Overall, the apoptotic machinery present in fungi and plants seems highly similar to mammalian apoptosis machinery and consists of conserved components. Homologs of known apoptosis genes are readily identified in the genomes of plants and fungi, but only homologs of downstream elements of the apoptosis pathways. The missing upstream components are either unidentified due to insufficient sequence conservation, or represent true differences in the regulatory machinery. Functional conservation of "missing" components, such as Bcl-2 proteins in fungi and plants, indicates that plant and fungal PCD machineries are capable of recognizing these mammalian proteins. Therefore functional homologs might exist which lack sequence similarity but have a conserved 3D structure. If such fungal proteins can be identified, they could be attractive targets for future antifungal drugs.

The study of pathogen-induced PCD in plants reveals a complex system with several control levels. PCD can serve as a resistance mechanism in some types of interactions, while in other systems it is harmful to the plant. Manipulation of the plant response, either enhancement (mainly necrotrophs) or suppression (mainly biotrophs) of PCD seems to be a common strategy used by pathogens to weaken the host. Intervention in the host PCD is suggested as an attractive strategy to construct pathogen-resistant plants. This strategy is appealing, but should be treated with caution for a number of reasons. PCD is necessary for normal development of plants and intervention in the delicate regulation of this essential machinery may have unwanted effects on plant growth and development. In addition, any change in the regulation of PCD might have variable and sometimes even opposite effects on plant resistance or susceptibility to different pathogens.

Recent reports highlighted the importance of PCD for normal development and stress adaptation in fungi. While direct proof is still missing, the accumulating evidence indicates that fungal PCD is probably also involved in pathogenic development. The exact way by which fungal PCD might affect fungus–plant interactions probably varies between different pathosystems, similar to the effect of plant PCD. Genome sequences are now available for a growing number of fungal plant pathogens and apoptotic genes can be readily identified in these fungi. Functional analysis of apoptotic genes in several plant pathogens is underway and is expected to be the first step towards elucidating the possible role of fungal apoptosis in fungal-plant interactions. Based on the multiple roles of PCD in other organisms and the importance of PCD in mediating plant responses to pathogens, it can be expected

that central regulators of fungal PCD would also significantly affect fungal pathogenicity.

References

Andersen M, Becker L, Straten P (2005) Regulators of apoptosis: suitable targets for immune therapy of cancer. Nature 4:399–409

Azoulay-Zohar H, Israelson A, Abu-Hamad S, Barmatz V (2004) In self-defence: Hexokinase promotes voltage-dependent anion channel closure and prevents mitochondria-mediated apoptotic cell death. Biochem J 377:347–355

Bailey B, Jennings JC, Anderson JD (1997) The 24-kDa protein from *Fusarium oxysporum* f. sp. *erythroxyli*: Occurrence in related fungi and the effect of growth medium on its production. Can J Microbiol 43:45–55

Barhoom S, Sharon A (2007) Bcl-2 proteins link programmed cell death with growth and morphogenetic adaptations in the fungal plant pathogen *Colletotrichum gloeosporioides*. Fungal Genet Biol 44:32–43

Borner C (2003) The Bcl-2 protein family: sensors and checkpoints for life-or-death decisions. Mol Immunol 39:615–647

Brandwagt B, Mesbah LA, Takken FL, Laurent PL, Kneppers TJ, Hille J, Nijkamp HJ (2000) A longevity assurance gene homolog of tomato mediates resistance to *Alternaria alternata* f. sp. *lycopersici* toxins and fumonisin B1. Proc Natl Acad Sci USA 97:4961–4966

Bredesen D, Rao RV, Mehlen P (2006) Cell death in the nervous system. Nature 433:796–801

Chen C, Dickman, M (2004) Dominant active Rac and dominant negative Rac revert the dominant active Ras phenotype in *Colletotrichum trifolii* by distinct signaling pathways. Mol Microbiol 51:1493–1507

Chen C, Dickman MB (2005) Proline suppresses apoptosis in the fungal pathogen *Colletotrichum trifolii*. Proc Natl Acad Sci USA 102:3459–3176

Cheng J, Park TS, Chio LC, Fischl AS, Ye XS (2003) Induction of apoptosis by sphingoid long-chain bases in *Aspergillus nidulans*. Mol Cell Biol 23:163–177

Christophel L, Santa Cruz S (1999) Bax-induced cell death in tobacco is similar to the hypersensitive response. Proc Natl Acd Sci USA 96:7956–7961

Clem R, Hardwick JM, Miller LK (1996) Anti-apoptotic genes of baculoviruses. Cell Death Differ 3:9–16

Danial N, Korsmeyer S (2004) Cell death: critical control points. Cell 116:205–219

del Pozo O, Lam E (1998) Caspases and programmed cell death in the hypersensitive response of plants to pathogens. Curr Biol 8:1129–1132

del Pozo O, Lam E (2003) Expression of the baculovirus p35 protein in tobacco affects cell death progression and compromises N gene-mediated disease resistance response to tobacco mosaic virus. Mol Plant–Microbe Interact 16:485–494

Dickman MB, Park YK, Oltersdorf T, Li W, Clemente T, French R (2001) Abrogation of disease development in plants expressing animal antiapoptotic genes. Proc Natl Acad Sci USA 98:6957–6962

Dietrich R, Delaney TP, Uknes SJ, Ward ER, Ryals JA, Dangl JL (1994) Arabidopsis mutants simulating disease resistance response. Cell 77:565–577

Dietrich R, Richberg MH, Schmidt R, Dean C, Dangl JL (1997) A novel zinc finger protein is encoded by the *Arabidopsis LSD1* gene and functions as a negative regulator of plant cell death. Cell 88:685–694

Dohi T, Okada K, Xia F, Wilford CE, Samuel T, Welsh K, Marusawa H, Zou H, Armstrong R, Matsuzawa S, Salvesen GS, Reed JC, Altieri DC (2004) An IAP-IAP complex inhibits apoptosis. J Biol Chem 279:34087–34090

Earnshaw W, Martins LM, Kaufmann SH (1999) Mammalian caspases: structure, activation, substrates, and functions during apoptosis. Annu Rev Biochem 68:383–424

Egan M, Wang ZY, Jones MA, Smirnoff N, Talbot NJ (2007) Generation of reactive oxygen species by fungal NADPH oxidases is required for rice blast disease. Proc Natl Acad Sci USA 104:11772–11777

Eichmann R, Schultheiss H, Kogel KH, Huckelhoven R (2004) The barley apoptosis suppressor homologue BAX inhibitor-1 compromises nonhost penetration resistance of barley to the inappropriate pathogen *Blumeria graminis* f.sp. *tritici*. Mol Plant–Microbe Interact 17:484–490

Eisler H, Fröhlich KU, Heidenreich E (2004) Starvation for an essential amino acid induces apoptosis and oxidative stress in yeast. Exp Cell Res 300:345–353

Epple P, Mack AA, Morris VR, Dangl JL (2003) Antagonistic control of oxidative stress-induced cell death in *Arabidopsis* by two related, plant-specific zinc finger proteins. Proc Natl Acad Sci USA 100:6831–3836

Esser K (2006) Heterogenic incompatibility in fungi. In: Esser K (ed) Growth, differentiation and sexuality, 2nd edn. The Mycota. Springer, Heidelberg, pp 141–166

Fahrenkrog B, Sauder U, Aebi U (2004) The *S. cerevisiae* HtrA-like protein Nma111p is a nuclear serine protease that mediates yeast apoptosis. J Cell Sci 117:115–126

Fedorova N, Badger JH, Robson GD, Wortman JR, Nierman WC (2005) Comparative analysis of programmed cell death pathways in filamentous fungi. BMC Genomics 177:1–14

Fesik S (2005) Promoting apoptosis as a strategy for cancer drug discovery. Nature 5:876–885

Frohlich KU, Madeo F (2000) Apoptosis in yeast – a monocellular organism exhibits altruistic behavior. FEBS Lett 473:6–9

Frohlich KU, Fussi H, Ruckenstuhl C (2007) Yeast apoptosis: from genes to pathways. Sem Cancer Biol 17:112–121

Fuentes-Prior P, Salvesen G (2004) The protein structures that shape caspase activity, specificity, activation and inhibition. Biochem J 2004:201–232

Glass LN, Dementhon K (2006) Non-self recognition and programmed cell death in filamentous fungi. Curr Opin Microbiol 9:553–558

Gordeeva A, Labas YA, Zvyagilskaya RA (2004) Apoptosis in unicellular organisms: mechanisms and evolution. Biochemistry 69:1055–1066

Govrin E, Levine A (2000) The hypersensitive response facilitates plant infection by the necrotrophic pathogen *Botrytis cinerea*. Curr Biol 10:751–757

Gozuacik D, Kimchi A (2007) Autophagy and cell death. Curr Top Dev Biol 78:217–245

Gredilla R, Grief J, Osiewacz HD (2006) Mitochondrial free radical generation and lifespan control in the fungal aging model system *Podospora anserina*. Exp Gerontol 41:439–447

Green DR (1998) Apoptotic pathways: the roads to ruin. Cell 94:695–698

Green DR (2005) Apoptotic pathways: ten minutes to dead. Cell 121:671–674

Greenberg J (1996) Programmed cell death: a way of life for plants. Proc Natl Acad Sci USA 93:12094–12097

Greenberg J, Ausubel F (1993) Arabidopsis mutants compromised for the control of cellular damage during pathogenesis and aging. Plant J 4:327–341

Gross A, Pilcher K, Blachly-Dyson E, Basso E, Jockel J, Bassik MC, Korsmeyer SJ, Forte M (2000) Biochemical and genetic analysis of the mitochondrial response of yeast to BAX and Bcl-XL. Mol Cell Biol 20:3125–3136

Hamann A, Brust D, Osiewacz HD (2007) Deletion of putative apoptosis factors leads to lifespan extension in the fungal ageing model *Podospora anserina*. Mol Microbiol 65:948–958

Herker E, Jungwirth H, Lehmann KA, Maldener C, Frohlich KU, Wissing S, Buttner S, Fehr M, Sigrist S, Madeo F (2004) Chronological aging leads to apoptosis in yeast. J Cell Biol 164:501–507

Holcik M, Korneluk R (2001) XIAP, the guardian angel. Nature 2:550–556

Howard RJ, Ferrari MA, Roach DH, Money NP (1991) Penetration of hard substrates by a fungus employing enormous turgor pressures. Proc Natl Acad Sci USA 88:11281–11284

Huckelhoven R, Dechert C, Kogel KH (2001a) Non-host resistance of barley is associated with a hydrogen peroxide burst at sites of attempted penetration by wheat powdery mildew fungus. Mol Plant Pathol 2:199–205

Huckelhoven R, Dechert C, Trujillo M, Kogel KH (2001b) Differential expression of putative cell death regulator genes in near-isogenic, resistant and susceptible barley lines during interaction with the powdery mildew fungus. Plant Mol Biol 47:739–748

Huckelhoven R, Dechert C, Kogel KH (2003) Overexpression of barley BAX inhibitor 1 induces breakdown of mlo-mediated penetration resistance to *Blumeria graminis*. Proc Natl Acad Sci USA 100:5555–5560

Imani J, Baltruschat H, Stein E, Jia G, Vogelsberg J, Kogel KH, Huckelhoven R. (2006) Expression of barley BAX Inhibitor-1 in carrots confers resistance to *Botrytis cinerea*. Mol Plant Pathol 7:279–284

Ito SI, Ihara T, Tamura H, Tanaka S, Ikeda T, Kajihara H, Dissanyake C, Abdel-Mostaal FF, El-Sayed MA (2008) α Tomatin, the major saponin in tomato, induces programmed cell death mediated by reactive oxygen species in the fungal pathogen *Fusarium oxysporum*. FEBS Lett 581:3217–3222

Jabs T, Dietrich RA, Dangl JL (1996) Initiation of runaway cell death in an Arabidopsis mutant by extracellular superoxide. Science 273:1853–1856

Jiang X, Wang X (2004) Cytochrome c-Mediated Apoptosis. Annu Rev Biochem 73:87–106

Kaminaka H, Näke C, Epple P, Dittgen J, Schütze K, Chaban C, Holt BF, Merkle T, Schäfer E, Harter K, Dangl JL (2006) bZIP10-LSD1 antagonism modulates basal defense and cell death in Arabidopsis following infection. EMBO J 25:4400–4411

Kawai-Yamada M, Ohori Y, Uchimiya H (2003) Dissection of Arabidopsis Bax inhibitor-1 suppressing Bax-, hydrogen peroxide-, and salicylic acid-induced cell death. Plant Cell 16:21–32

Kawasaki T, Henmi K, Ono E, Hatakeyama S, Iwano M, Satoh H, Shimamoto K (1999) The small GTP-binding protein Rac is a regulator of cell death in plants. Proc Natl Acad Sci USA 96:10922–10926

Kim M, Panstruga R, Elliott C, Müller J, Devoto A, Yoon HW, Park HC, Cho MJ, Schulze-Lefert P (2002) Calmodulin interacts with Mlo protein to regulate defence against mildew in barley. Nature 416:447–451

Kim M, Lim JH, Ahn CS, Park K, Kim GT, Kim WT, Pai HS (2006) Mitochondria-associated hexokinases play a role in the control of programmed cell death in Nicotiana benthamiana. Plant Cell 18:2341–2355

Kroemer G, Galluzzi L, Brenner C (2007) Mitochondrial membrane permeabilization in cell death. Physiol Rev 87:99–163

Lam E, Pontier D, del Pozo O (1999) Die and let live – programmed cell death in plants. Curr Opin Plant Biol 2:502–507

Lam E, Kato N, Lawton M (2001) Programmed cell death, mitochondria and the plant hypersensitive response. Nature 411:848–853

Lambeth J (2004) NOX enzymes and the biology of reactive oxygen. Nat Rev Immunol 4:181–189

Li W, Sun L, Liang Q, Wang J, Mo W, Zhou B (2006) Yeast AMID homologue Ndi1p Displays Respiration-restricted apoptotic activity and is involved in chronological aging. Mol Biol Cell 17:1802–1811

Lincoln J, Richael C, Overduin B, Smith K, Bostock R, Gilchrist DG (2002) Expression of the antiapoptotic baculovirus p35 gene in tomato blocks programmed cell death and provides broad-spectrum resistance to disease. Proc Natl Acad Sci USA 99:15217–15221

Lorin S, Dufour E, Sainsard-Chanet A (2006) Mitochondrial metabolism and aging in the filamentous fungus Podospora anserina. Biochem Biophys Acta 1757:604–610

Madeo F, Frohlich E, Frohlich KU (1997) A yeast mutant showing diagnostic markers of early and late apoptosis. J Cell Biol 139:729–734

Madeo F, Herker E, Maldener C, Wissing S, Lachelt S, Herlan M, Fehr M, Lauber K, Sigrist SJ, Wesselborg S, Frohlich KU (2002) A caspase-related protease regulates apoptosis in yeast. Mol Cell 9:911–917

Madeo F, Herker E, Wissing S, Jungwirth H, Eisenberg T, Frohlich KU (2004) Apoptosis in yeast. Curr Opin Microbiol 7:655–660

Matsumura H, Nirasawa S, Kiba A, Urasaki N, Saitoh H, Ito M, Kawai-Yamada M, Uchimiya H, Terauchi R (2003) Overexpression of Bax inhibitor suppresses the fungal elicitor-induced cell death in rice (Oryza sativa L) cells. Plant J 33:425–434

Matsuyama S, Nouraini S, Reed JC (1999) Yeast as a tool for apoptosis research. Curr Opin Microbiol 2:618–623

Maulik N, Yoshida T, Das DK (1998) Oxidative stress developed during the reperfusion of ischemic myocardium induces apoptosis. Free Radic Biol Med 24:869–875

Mousavi S, Robson G (2003) Entry into the stationary phase is associated with a rapid loss of viability and an apoptotic-like phenotype in the opportunistic pathogen Aspergillus fumigatus. Fungal Genet Biol 39:221–229

Mousavi SA, Robson GD (2004) Oxidative and amphotericin B-mediated cell death in the opportunistic pathogen Aspergillus fumigatus is associated with an apoptotic-like phenotype. Microbiology 150:1937–1945

Nesher I, Barhoom S, Sharon A (2008) Cell cycle and cell death are not necessary for appressoria formation and plant infection in the fungal plant pathogen Colletotrichum gloeosporioides. Biology BMS. Biol 6:6–9

Pastorino J, Shulga N, Hoek JB (2002) Mitochondrial binding of hexokinase II inhibits Bax-induced cytochrome c release and apoptosis. J Biol Chem 277:7610–7618

Pemberton C, Salmond G (2004) The Nep1-like proteins: a growing family of microbial elicitors of plant necrosis. Mol Plant Pathol 5:353–359

Pennell R, Lamb C (1997) Programmed cell death in plants. Plant Cell 9:1157–1161

Pinan-Lucarré B, Balguerie A, Clavé C (2005) Accelerated cell death in Podospora autophagy mutants. Eukaryot Cell 4:1765–1774

Polcic P, Forte M (2003) Response of yeast to the regulated expression of proteins in the Bcl-2 family. Biochem J 374:393–402

Polidoros A, Mylona PV, Scandalios JG (2001) Transgenic tobacco plants expressing the maize CAT2 gene have altered catalase levels that affect plant-pathogen interactions and resistance to oxidative stress. Transgenic Res 10:555–569

Pozniakovsky Ai, Knorre DA, Markova OV, Hyman AA, Skulachev VP, Severin FF (2005) Role of mitochondria in the pheromone- and amiodarone-induced programmed death of yeast. J Chem Biol 168:257–269

Reed J, Kytbuddin SD, Adam G (2004) The domains of apoptosis: a genomic perspective. Sci STKE 2004(9):1–29

Richie D, Miley MD, Bhabhra R, Robson GD, Rhodes JC, Askew DS (2007) The Aspergillus fumigatus metacaspases CasA and CasB facilitate growth under conditions of endoplasmic reticulum stress. Mol Microbiol 63:591–604

Rolke Y, Liu S, Quidde T, Williamson B, Schouten A, Weltring KM, Siewers V, Tenberge KB, Tudzynski B, Tudzynski P (2004) Functional analysis of H_2O_2-generating systems in Botrytis cinerea: the major Cu-Zn-superoxide dismutase (BCSOD1) contributes to virulence on French bean, whereas a glucose oxidase (BCGOD1) is dispensable. Mol Plant Pathol 5:17–27

Roze LV Linz JE (1998) Lovastatin triggers an apoptosis-like cell death process in the fungus Mucor racemosus. Fungal Genet Biol 25:119–133

Scheckhuber C, Erjavec N, Tinazli A, Hamann A, Nystrom T, Osiewacz HD (2007) Reducing mitochondrial fission results in increased life span and fitness of two fungal ageing models. Nat Cell Biol 9:99–105

Sellem C, Marsy S, Boivin A, Lemaire C, Sainsard-Chanet A (2007) A mutation in the gene encoding cytochrome c1 leads to a decreased ROS content and to a long-lived phenotype in the filamentous fungus Podospora anserina. Fungal Genet Biol 44:648–658

Semighini C, Hornby JM, Dumitru R, Nickerson KW, Harris SD (2006) Farnesol-induced apoptosis in Aspergillus *nidulans* reveals a possible mechanism for antagonistic interactions between fungi. Mol Microbiol 59:753–764

Severin FF, Hyman, AA (2002) Pheromone induces programmed cell death in *S. cerevisiae*. Curr Biol 12:R233–R235

Silva R, Sotoca R, Johansson B, Ludovico P, Sansonetty F, Silva MT, Peinado JM, Corte-Real M (2005) Hyperosmotic stress induces metacaspase- and mitochondria-dependent apoptosis in *Saccharomyces cerevisiae*. Mol Microbiol 58:824–834

Spassieva S, Markham JE, Hille J (2002) The plant disease resistance gene Asc-1 prevents disruption of sphingolipid metabolism during AAL-toxin-induced programmed cell death. Plant J 32:561–572

Takemoto D, Tanaka A, Scott B (2007) NADPH oxidases in fungi: diverse roles of reactive oxygen species in fungal cellular differentiation. Fungal Genet Biol 44:1065–1076

Tan S, Sagara Y, Liu Y, Maher P, Schubert D (1998) The regulation of reactive oxygen species production during programmed cell death. J Cell Biol 141:1423–1432

Tanaka A, Christensen MJ, Takemoto D, Park P, Scott B (2006) Reactive oxygen species play a role in regulating a fungus–perennial ryegrass mutualistic interaction. Plant Cell 18:1052–1066

Tao W, Kurschner C, Morgan JI (1997) Modulation of cell death in yeast by the Bcl-2 family of proteins. J Biol Chem 272:15547–15552

Tenberge K, Beckedorf M, Hoppe B, Schouten A, Solf M, von den Driesch M (2002) In situ localization of AOS in host–pathogen interactions. Microsc Microanal 8:250–251

Thines E, Weber R, Talbot NJ (2000) MAP kinase and protein kinase A-dependent mobilization of triacylglycerol and glycogen during appressorium turgor generation by *Magnaporthe grisea*. Plant Cell 9:1703–1718

Torres M, Dangl JL, Jones JD (2002) Arabidopsis gp91phox homologues AtrbohD and AtrbohF are required for accumulation of reactive oxygen intermediates in the plant defense response. Proc Natl Acad Sci USA 99:517–522

Torres M, Jones JD, Dangl JL (2005) Pathogen-induced, NADPH oxidase-derived reactive oxygen intermediates suppress spread of cell death in *Arabidopsis thaliana*. Nat Genet 37:1130–1134

Torres MA, Jones JD, Dangl JL (2006) Reactive oxygen species signaling in response to pathogens. Plant Physiol 141:373–378

Uren AG, Beilharz T, O'Connell MJ, Bugg SG, van Driel R, Vaux DL, Lithgow, T (1999) Role for yeast inhibitor of apoptosis (IAP)-like proteins in cell division. Proc Natl Acad Sci USA 96:10170–10175

Uren A, O'Rourke K, Aravind LA, Pisabarro MT, Seshagiri S, Koonin EV, Dixit VM (2000) Identification of paracaspases and metacaspases: two ancient families of caspase-like proteins, one of which plays a key role in MALT lymphoma. Mol Cell 6:961–967

Vachova L, Palkova Z (2007) Caspases in yeast apoptosis -like death: facts and artifacts. FEMS Yeast Res 7:12–21

Van Baarlen P, Woltering EJ, Staats M, Van Kan JAL (2007) Histochemical and genetic analysis of host and non-host interactions of Arabidopsis with three *Botrytis* species: an important role for cell death control. Mol Plant Pathol 8:41–54

Veneault-Fourrey C, Barooah M, Egan M, Wakley G, Talbot NJ (2006) Autophagic fungal cell death is necessary for infection by the rice blast fungus. Science 312:580–583

Verhagen A, Coulson EJ, Vaux DL (2001) Inhibitor of apoptosis proteins and their relatives: IAPs and other BIRPs. Genome Biol 2:1–10

Walter D, Wissing S, Madeo F, Fahrenkrog B (2006) The inhibitor-of-apoptosis protein Bir1p protects against apoptosis in *S. cerevisiae* and is a substrate for the yeast homologue of Omi/HtrA2. J Cell Sci 119:1843–1851

Wang H, Li J, Bostock RM, Gilchrist DG (1996) Apoptosis: a functional paradigm for programmed plant cell death induced by a host-selective phytotoxin and invoked during development. Plant Cell 8:375–391

Wang X (2001) The expanding role of mitochondria in apoptosis. Gen Dev 15:2922–2933

Wissing S, Ludovico P, Herker E, Buttner S, Engelhardt SM, Decker T, Link A, Proksch A, Rodrigues F, Corte-Real M, Frohlich KU, Manns J, Cande C, Sigrist SJ, Kroemer G, Madeo F (2004) An AIF orthologue regulates apoptosis in yeast. J Cell Biol 166:969–974

Xu J, Peng YL, Dickman MB, Sharon A (2006) The dawn of fungal pathogen genomics. Annu Rev Phytopathol 44:337–366

Xu Q, Reed J (1998) Bax inhibitor-1, a mammalian apoptosis suppressor identified by functional screening in yeast. Mol Cell 1:337–346

Yao N, Greenberg J (2006) Arabidopsis accelerated cell death2 modulates programmed cell death. Plant Cell 18:397–411

Yoshioka H, Numata N, Nakajima K, Katou S, Kawakita K, Rowland O, Jones JD, Doke N (2003) Nicotiana benthamiana gp91phox homologs NbrbohA and NbrbohB participate in H_2O_2 accumulation and resistance to *Phytophthora infestans*. Plant Cell 15:706–718

13 The Ectomycorrhizal Symbiosis: a Marriage of Convenience

Francis Martin[1], Anders Tunlid[2]

CONTENTS

I. Introduction........................... 237
II. Ectomycorrhiza Morphogenesis............ 239
III. Insights from Sequencing the Genome of the
   Ectomycorrhizal *Laccaria bicolor*........... 240
   A. Sequencing, Assembling and Annotating the
      Genome of *L. bicolor*................... 240
   B. Expanding Multigene Families 241
      1. The Secretome 243
      2. Signal Transduction Pathways........ 244
   C. The Biotrophy–Saprotrophism
      Continuum 246
      1. Carbohydrate-Degrading Enzymes.... 246
      2. Secreted Proteases and Nitrogen
         Acquisition 247
IV. The Ectomycorrhizal Transcriptome........ 248
   A. Alteration in Symbiont Transcriptome 248
   B. Expression of
      Ectomycorrhizae-Specific Genes........ 252
V. Ecological Genomics..................... 252
VI. Future Research......................... 253
   References............................. 253

I. Introduction

Mutualistic ectomycorrhizal symbionts are of major importance as drivers of ecological function and evolutionary processes in forest ecosystems. Within days after their emergence in the upper soil profiles, such as the organic humus and mor layer, most of the short roots of shrubs and trees are colonized by ectomycorrhizal (ECM) fungi (Taylor et al. 2000; Ruess et al. 2003; Adams et al. 2006). In this symbiotic relationship, the running, branching filaments of the soil fungus encounter host root tips, ramify between cells in the root's outer layers, form a sheath around the root and radiate outwards into the surrounding soil and litter, forming extensive intermingling hyphal webs (Horan et al. 1988; Selosse et al. 2006). This transport network then allows the fungus to derive photosynthetically produced sugars from the host and in turn to transfer nitrogen and phosphorus to the plant. Roughly 20% of the carbon assimilated by ectomycorrhizal tree species is estimated to be allocated to their mycorrhizal fungi (Högberg et al. 2001; Leake et al. 2004), whereas ~70% of tree N and P are absorbed via the ectomycorrhizal fungi colonizing the roots (Brandes et al. 1998). This mutualistic association allows host trees to grow efficiently in low-nutrient and marginal environments (Read and Perez-Moreno 2003). Ectomycorrhizal symbiosis is restricted to approximately 8000 plant species, but their ecological importance is amplified by their wide occupancy of terrestrial ecosystems. Trees of the Betulaceae, Cistaceae, Dipterocarpaceae, Fagaceae, Pinaceae, Myrtaceae, Salicaceae and several tribes in the Fabaceae are ectomycorrhizal plants, dominating boreal, temperate, mediterranean and some subtropical forest and woodland ecosystems (Read and Perez-Moreno 2003).

Ectomycorrhizal symbiosis is more recent than the ancient arbuscular mycorrhizal symbiosis involving Glomeromycota (see Chap. 14), which has been around since the earliest stages of land-plant evolution (Brundrett 2002). The oldest known fossil ectomycorrhizae date from 50 million years ago (Mya), but it is now thought that the symbiosis predates this period by some time (~135 Mya), because the Pinaceae and many of the angiosperm families whose current members establish ectomycorrhizal symbiosis were extant well before 50 Mya, along with the major fungal lineages with modern ectomycorrhizal representatives (Brundrett 2002; Moyersoen 2006). The ECM fungi are not a phylogenetically distinct group, but an assemblage of very different fungal species, mainly Basidiomycota, that have independently developed a mutualistic symbiotic lifestyle.

[1] UMR 1136, INRA-Nancy Université,
Interactions Arbres/Microorganismes, INRA-Nancy,
54280 Champenoux, France.
e-mail: fmartin@nancy.inra.fr
[2] Department of Microbial Ecology, Ecology Building,
Lund University, SE 223 62 Lund, Sweden

Plant Relationships, 2nd Edition
The Mycota V
H. Deising (Ed.)
© Springer-Verlag Berlin Heidelberg 2009

The degree to which saprotrophic ability is retained in fungal ECM species is an outstanding issue in the field; and ECM species are placed along a biotrophy–saprotrophy continuum (for a discussion, see Koide et al. 2008). There is currently debate about whether most ectomycorrhizal Basidiomycota are derived from an ectomycorrhizal ancestor, with multiple losses of that lifestyle in many currently saprobic lineages (Hibbett et al. 2000; Lutzoni et al. 2004), or whether the pattern is based strictly on many convergent gains of the mycorrhizal habit with no reversions to saprotrophy (Bruns and Shefferson 2004). The presence of common signalling pathways and developmental gene networks among all mycorrhizal groups would favor the former theory, while the recruitment of novel symbiosis-related gene families in independent lineages would favor the latter.

The underlying mechanisms that shape fungal tissues and build the ectomycorrhizal symbiosis are based on hyphal division, cell-shape changes, hyphal rearrangements and cell death (Kottke and Oberwinkler 1987). The orchestration of these elementary processes depends on constraining genetic programs operating on hyphal behavior (Martin et al. 2007). It is thought that a specific set of yet unknown signalling molecules and more general growth factors (auxins, cytokinins) promote cell divisions and tissue size, whereas regulatory proteins control the orientation of cell divisions, oriented cell rearrangements and hence tissue shape (Martin et al. 2001, 2007). Understanding gene expression and its regulation in response to developmental and environmental cues is one of the greatest challenges for scientists investigating the formation and functioning of mycorrhizal symbioses. Regulatory control of gene expression in the mycobiont and its host plant occurs at pre- and posttranscriptional levels and involves short DNA and RNA signals known as regulatory motifs. An array of signalling molecules (e.g., auxins, ethylene, flavonoids) orchestrates the changes in entangled gene and protein networks (Martin et al. 2001). Fungal auxin regulates the development of the lateral roots, and ethylene and auxin likely interact to control this developmental processes (Rupp and Mudge 1985; Rupp et al. 1989; Jambois et al. 2005). Based on the study of model species, such as *Paxillus involutus* and *Laccaria bicolor*, which develop mycorrhiza using in vitro systems, it is assumed

that such hormone-induced molecular cascades drive the establishment of mycorrhizal symbioses, although more specific "Myc factors" of as yet unknown biochemical structures may also be involved. Unexpectedly, it appears that ontogenic and metabolic programs that lead to the development of symbiosis are mostly driven by the differential expression of pre-existing transcription factors and/or transduction pathways, rather than by the expression of novel symbiosis-specific gene networks (see below; Martin et al. 2008). Elucidating the molecular machinery which leads to partner recognition and accommodation and further establishment of a functional symbiosis is only the first step in a comprehensive understanding of the mutualistic interaction.

The next challenge on the agenda is to insure that the molecular mechanisms identified in vitro or in microcosms (Duplessis et al. 2005; Le Quéré et al. 2005; Morel et al. 2006) are also operating in ECM developed on mature trees in environmental conditions. An important step will be to identify the functions played by the assemblages of mycorrhizal fungi in situ (Read and Perez-Moreno 2003); and the availability of genomic information on the recently sequenced *Populus trichocarpa* (Tuskan et al. 2006) and its mycorrhizal mutualists may significantly support this. The completion of the genome sequences of the ectomycorrhizal *L. bicolor* (Martin et al. 2008) and *Tuber melanosporum* (*Tuber* Genome Consortium, unpublished data) provides an ideal opportunity to identify the key components of interspecific and organism–environment interactions (Whitham et al. 2006). By elucidating, modeling and manipulating patterns of gene expression, we should be able to identify the genetic control points regulating the mycorrhizal response to changing host physiology and environmental cues and to better understand how these symbiotic interactions drive ecosystem function.

Although our understanding of the molecular mechanisms and signalling pathways coupled to symbiosis development and functioning need further refinement, the past few years have brought exciting discoveries in this area. The present paper aims to highlight the most recent work and to illustrate the way functional genomics is altering our thinking about changes in gene expression during the ectomycorrhiza interaction, but also on the ecology of ECM fungi.

II. Ectomycorrhiza Morphogenesis

Signalling processes lead the hyphae of the mycobiont towards the vicinity of susceptible host roots. Only the broad outlines of these signalling processes have been defined and only a limited set of chemical signals produced by either the host or symbiont have been identified so far (Rupp et al. 1989; Ditengou et al. 2000; Lagrange et al. 2001; Martin et al. 2001; Jambois et al. 2005). No specific signal/receptor systems have yet been identified in ECM. Early stages of ectomycorrhiza development have well characterized programmed morphological transitions (Fig. 13.1). Fungal hyphae emerging from soil propagules (spores, sclerotia) or an older mycorrhiza penetrate into the root cap cells. Backwards from the tip, the invasion of root cap cells proceeds inwards until the hyphae reach the epidermal cells (Horan et al. 1988). Morphogenetic changes take place when the hyphæ contact living cortical cells, which is pivotal for initiation of mantle formation and Hartig net construction (Fig. 13.2). The Hartig net is formed by hyphae progressing in the root apoplastic space, where they proliferate and form a finger-like, labyrinthine system (Fig. 13.2; Kottke and Oberwinkler 1987; Massicotte et al. 1987; Bonfante 2001). Abundant membranes of this structure allow ions, metabolites and effector molecules to pass at a high rate between adjacent cells, providing the anatomical basis for intercellular communication and local co-ordination between the symbionts. Numerous mitochondria, lipid bodies, extensive endoplasmic reticulum and dictyosomes with proliferating cisternae are contained within the coenocytic hyphæ of the Hartig net, all of which are illustrations of a highly active anabolic state with highly active biosynthesis of secreted proteins (Kottke and Oberwinkler 1987). Progression from the strongly rhizomorphic outgrowth of the free-living mycelium to the plectenchymatous structure of the ectomycorrhizal sheath and the coenocytic Hartig net hyphæ involves a lack of septation, a loss of apical coherence and intimate juxtaposition of hyphæ (Kottke and Oberwinkler 1987). No root cell penetration is observed, except in senescing ectomycorrhizal tips. After attaching to epidermal cells, hyphae multiply to form a series of layers several hundreds of microns thick which differentiates to form the mature mantle (Fig. 13.2). The hyphae in these structures are encased in an extracellular polysaccharide and proteinacous matrix (Dexheimer and Pargney 1991). Air and water channels that allow the flow of nutrients into the symbiosis innervate these structures, although most of the nutrient transfer probably takes place symplastically (Vesk et al. 2000). In most ECM, an outward network of hyphae prospecting the soil and gathering nutrients irradiates from the outer layers of the mantle. It is

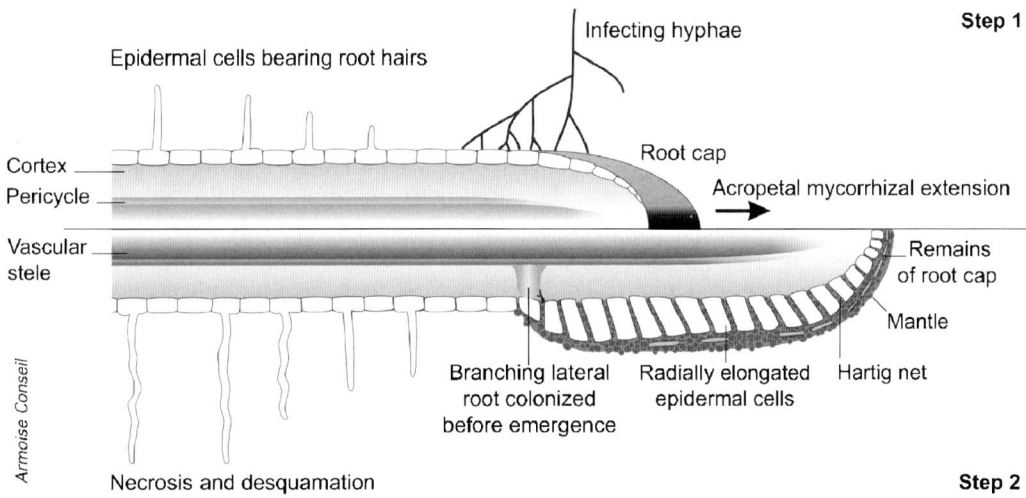

Fig. 13.1. Dynamics of the colonization of the *Eucalyptus* root by the ectomycorrhizal fungus *Pisolithus microcarpus* during symbiosis development. *Step 1* Colonization by the infecting hyphae is initiated in the root cap region and then propagated by an acropetal extension of root and fungal tissues. *Step 2* Epidermal cells, which initiate root hairs in nonmycorrhizal root zones, elongate radially as a specific response to the hyphae. The intercellular hyphal network (*Hartig net*) develops only between these elongated epidermal cells

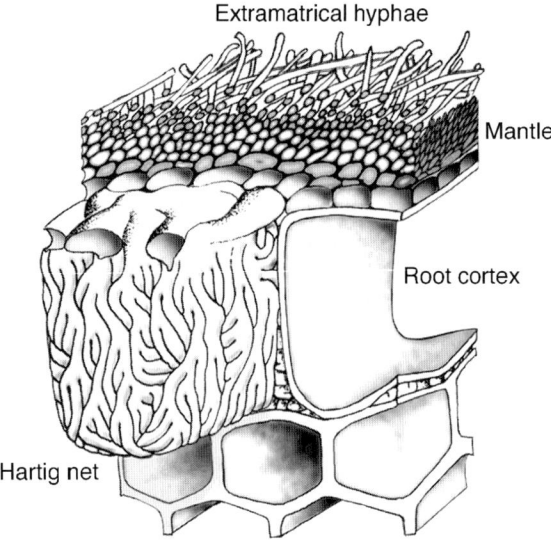

Fig. 13.2. Differentiation of the ectomycorrhizal hyphae during the symbiosis development. Running, branching hyphae ramify between cells of the root's outer layers, forming a finger-like, labyrinthine hyphal system (the Hartig net) and a mantle sheath around the root. The latter hyphal network radiates outwards into the surrounding soil and litter (*Extramatrical hyphae*)

thought that these subterranean hyphal webs connect the trees within the ecosystem (Selosse et al. 2006). Although playing a very similar role in nutrient exchange, this specialized symbiotic structure is anatomically very different from the haustoria of rust fungi, another group of biotrophic fungi (Chap. 4). The haustorium is formed inside the living host cell as a branch of the fungal thallus, but it is not located directly in the cytoplasm. The haustorium is surrounded by an extrahaustorial membrane, representing an extension of the plasmalemma of the plant cell. An extrahaustorial matrix enriched in carbohydrates separates the extrahaustorial membrane from the fungal haustorial wall (Staples 2001; Voegele and Mendgen 2003).

III. Insights from Sequencing the Genome of the Ectomycorrhizal *Laccaria bicolor*

The genomes of a dozen pathogenic fungi have been published (Galagan et al. 2005; Soanes et al. 2007), but no genome of symbiotic fungi is available; therefore, one major goal for the community has been to sequence the genomes of an

ecto- and an endomycorrhizal fungus. In 2004, the Joint Genome Institute (JGI) with the United States Department of Energy started sequencing the genomes of the ectomycorrhizal basidiomycete *L. bicolor* and the endomycorrhizal *Glomus intraradices* (Martin et al. 2004b). In 2006, the French Genoscope Genomics Institute began sequencing the ectomycorrhizal ascomycete *T. melanosporum*. The completion of the *L. bicolor* and *T. melanosporum* genome sequences (Martin et al. 2008; *Tuber* Genome Consortium, unpublished data) provides an unsurpassed opportunity to decipher the key components of ectomycorrhiza development and functioning. In addition, *L. bicolor* is a basidiomycete with a typical basidiocarp form and the availability of its genome will facilitate the characterization of genes involved in the co-ordinated developmental events leading to the formation of basidiocarp tissues. The genome assembly provides a basis for gene discovery, the study of gene organization and evolution, the construction of whole-genome expression oligoarrays and the identification of candidate genes by positional cloning. The following sections summarize what has been learned from *L. bicolor* genome exploration.

A. Sequencing, Assembling and Annotating the Genome of *L. bicolor*

Free-living vegetative mycelium from a single haploid spore was used to prepare high molecular weight DNA for several plasmid and fosmid libraries, with 3-, 8- and 38 kbp inserts, and whole-genome shotgun (WGS) sequencing of paired end sequences (590 million base pairs; Mbp) to 10× coverage (Martin et al. 2008). The shotgun sequence was assembled into 64.9 Mbp of scaffolds. The nuclear genome of *L. bicolor* is bigger than that of previously published fungal genomes, but no evidence for large-scale duplications was observed within the genome, though many examples of short regions of tandem duplication were present within multigene families. The larger size is partly explained by the unusually high number of mobile DNA and repeated sequences, known as transposable elements (TE), that constitute more than 20% of the genome. One unexpected finding is that *L. bicolor* harbors unprecedented transposon diversity – more types than any other fungi studied to date (Martin et al. 2008; Labbé and Quesneville, unpublished data). The most abundant TE are Class 1 elements (e.g., *Copia*, *Gypsy*) which are collectively represented by 962 complete sequences and ~17 000 remnant degraded copies. Class 2 elements, including MITE, *Pogo*, *Ant1/Tc1* and helitrons, account for a total of 5738 sequences, including 355 complete sequences. This abundance of MITE elements and helitrons has not yet been reported in other fungal genomes. Genomic regions of common ancestry between *L. bicolor* and other basidiomycetes appear to have evolved into a mosaic of syntenic blocks that may have diverged between Agaricales species owing to differential insertions of TE and episodes of gene mobilization. In contrast to *Coprinopsis cinerea* (Stajich et al. 2006), TE clusters are not restricted to sub-telomeric or centromeric chromosomal regions, but are distributed randomly in the genome. It appears that TE-rich regions contain a low density of expressed genes.

Using a combination of gene-prediction programs, including FGENESH and EUGENE, 20 614 protein-encoding genes were initially identfed. As inputs, these predictions used coding sequences from UNIPROT (UniProt Consortium 2008) and other basidiomycete genomes, *Cryptococcus neoformans* and *C. cinerea* and cDNA sequences and/or their predicted translations. A fairly deep sequencing of randomly selected cDNA clones (38 900 Sanger ESTs, 180 000 pyrosequencing 454 ESTs) from various tissues (free-living mycelium, ectomycorrhizas, fruiting body) allowed validation of the intron/exon structure of ~40% of genes. In addition, expression of nearly 80% of the predicted genes was detected in free-living mycelium, ectomycorrhizal root tips or fruiting bodies using a custom NimbleGen whole-genome expression oligoarrays (Martin et al. 2008). The remaining genes are likely expressed only at low levels or under conditions not yet analyzed. As of 1 May 2008, the number of gene models is 19 105, with 2565 manually curated predictions. The resulting predictions are distributed through web portals at the JGI *Laccaria* genome browser (http://www.jgi.doe.gov/laccaria) and the INRA *Laccaria* DB (http://www.mycor.nancy.inra.fr/IMGC/LaccariaGenome/). Of these predicted gene models, about 76% showed significant similarity to sequences in protein databases, particularly those from other homobasidiomycetes, the major fungal taxon to which *L. bicolor* belongs. Predicted protein-coding genes were compared with the gene sets of *C. neoformans* (Loftus et al. 2004), *Ustilago maydis* (Kämper et al. 2006), *Phanerochaete chrysosporium* (Martinez

et al. 2004) and *C. cinerea* (Stajich et al. 2006) to derive a common set of gene families attributable to the basidiomycete common ancestor. Compared with other fungal genomes, the *L. bicolor* genome contains both more and larger gene families. Most of them have clear orthologs in other fungi (i.e., genes that evolved from a common ancestor). These include several fungal multigene families coding for membrane, cell wall and secreted proteins. However, up to 1077 families are lineage-specific (see below). Size differences between the genomes of fungal species depend on the relative rates of gene duplication, gene loss, horizontal transfer events and transposable element proliferation. Two evolutionary trends that would result in larger genomes among fungal pathogens are the consistent expansion of certain gene families, as well as the pathogens' apparent affinity for gene acquisition through horizontal transfer (Powell et al. 2008). The expansion of gene families is clearly a major driving process in genome evolution in *L. bicolor*.

The rise of ECM symbiosis has clearly involved multiple phenotype acquisitions and a complex evolutionary history, suggesting that, even within relatively closely related clades, symbiotic, pathogenic and saprotrophic fungi are likely to display many key differences. As stressed above, there is currently debate about whether most ECM are derived from an ectomycorrhizal ancestor, with multiple losses of that habit in current saprotrophic lineages, or whether the mycorrhizal habit results from many convergent gains with no reversions to saprotrophy (Bruns and Shefferson 2004). The presence of common multigene families among all saprobic and mycorrhizal groups would favor the former theory, whereas recruitment of novel lineage-specific, mycorrhiza-regulated gene families in independent lineages would favor the latter. Particular outcomes of duplication events are thus possibly associated with symbiotic life histories in fungi. To date, differential gene gain and loss have not been studied at genomic scales in fungal symbionts, despite the known importance of this phenomenon in virulence in bacterial and fungal pathogens (Powell et al. 2008).

B. Expanding Multigene Families

Duplications of genes or larger chromosome regions in combination with mutations that cause functional divergence of the duplicates are considered to be the

most important mechanism generating evolutionary
novelties including new gene function and expression patterns (Ohno 1970). Gene duplications occur
frequently in eukaryotes. Some of the duplicated
genes are preserved, but the majority were silenced
within a few million years (Lynch and Conery 2000).
The classic model for the origin of functional novelties following gene duplication postulates that gene
duplication creates a redundant locus that is free to
accumulate otherwise deleterious mutations, as long
as the original copy maintains the ancestral function (Ohno 1970). The redundant copy can obtain a
new function by alterations in coding or regulatory
sequences. Alternatively, the original function of the
ancestral gene can be partitioned between the two
daughters (Prince and Pickett 2002).

   Duplications and losses give rise to gene families
containing members with evolutionarily and functionally related sequences (Prince and Pickett 2002).
Recent studies showed that the levels of gene duplication and the sizes of gene families vary greatly
between fungal species (Cornell et al. 2007; Wapinski
et al. 2007; Powell et al. 2008). This is also true when
comparing the genome of the symbiotic fungus
L. bicolor with four other fully sequenced basidiomycetes, including the plant pathogen *U. maydis*, the
saprophytes *C. cinerea* and *P. chrysosporium* and the
human pathogen *C. neoformans*. The species cover
about 550 million years of evolutionary history (Fig.
13.3). Among these species, *L. bicolor* has the largest genome, containing 19 105 predicted protein-
coding genes, as compared with 13 500 in *C. cinerea*,
10 000 in *P. chrysosporium*, 7300 in *C. neoformans*
and 6,500 in *U. maydis* (Martinez et al. 2004; Loftus
et al. 2005; Kämper et al. 2006; Martin et al. 2008).
In total, the five basidiomycete genomes contain 56
400 predicted proteins that cluster into 7350 families
(Martin et al. 2008). The number and mean size
of these families vary among the genomes and are
related to the gene content, being largest in *L. bicolor*,
followed by *C. cinerea*, *P. chrysosporium*, *C. neoformans* and *U. maydis* (Fig. 13.4; Martin et al. 2008). Furthermore, the taxonomic distribution of the families
differs. Some of the multigene families (24%) are
present in all five genomes, 49% are shared between
any two, three or four species and the remaining 27%
are present in only one of the five analyzed species.
The largest number of such lineage-specific families
is found in the *L. bicolor* genome (1077 families).
C. cinerea contains 399, *P. chrysosporium* 249,
C. neoformans 173 and *U. maydis* 71 lineage-specific
families (Rajaskekar et al. unpublished data).

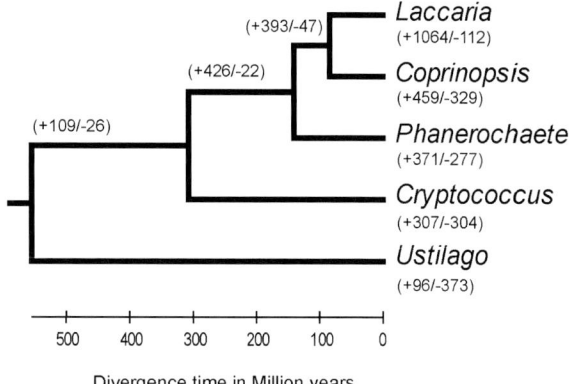

Fig. 13.3. Phylogeny of basidiomycetes with sequenced
genomes. The linearized phylogenetic trees of *Laccaria
bicolor*, *Coprinopsis cinerea*, *Phanerochaete chrysosporium*,
Cryptococcus neoformans and *Ustilago maydis* were constructed using 18S rDNA sequences. *Aspergillus niger* was
used to root the tree and the divergence time between *A.
niger* and *U. maydis* (Berbee and Taylor 2001) was used to
date the nodes in the tree. The *numbers* on the branches
show the numbers of expanded (+) or contracted (−) protein families along the lineages (from Rajashekar et al.
unpublished dara)

   A small number (87) of the basidiomycete
protein families are restricted to the symbiont
L. bicolor and the pathogenic biotroph *U. maydis*.
The majority of these families are small in size, containing less than three members. Only a small fraction of the proteins in families shared by *L. bicolor*
and *U. maydis* display significant sequence similarity to proteins in other fungal genomes, including
ascomycetes. Furthermore, only a limited number
of them contain known protein domains (Rajashekar
et al. unpublished data). Notably, four of the 87 families
contain sequences predicted to encode cysteine-rich,
small secreted proteins (SSPs) that could have a role
as effector molecules in manipulating the host defence
system (see further below).

   It can be expected that a significant part of the
observed differences in protein family sizes is the
result of random processes and they did not evolve
as adaptations to specific environments (Lynch
2007). The variation that could be accounted for by
random processes can be inferred using likelihood
methods and estimations on the divergence time
between the compared taxa (Hahn et al. 2005).
When this method is applied for analyzing the evolution of basidiomycete protein families, the largest number of expanded protein families is found
along the branch leading to *L. bicolor* (Fig. 13.3). In
total, 1064 protein families have expanded in size
along the *L. bicolor* branch, as compared with 96

Fig. 13.4. Expansion of protein families in *Laccaria bicolor*. Relationships between genome size and number of protein families (*A*) and between genome size and protein family sizes (*B*) in *L. bicolor, Coprinopsis cinerea, Phanerochaete chrysosporium, Cryptococcus neoformans* and *Ustilago maydis* predicted proteins were clustered into families using the TRIBE-MCL algorithm (Enright et al. 2002). Reprinted from Martin et al. (2008)

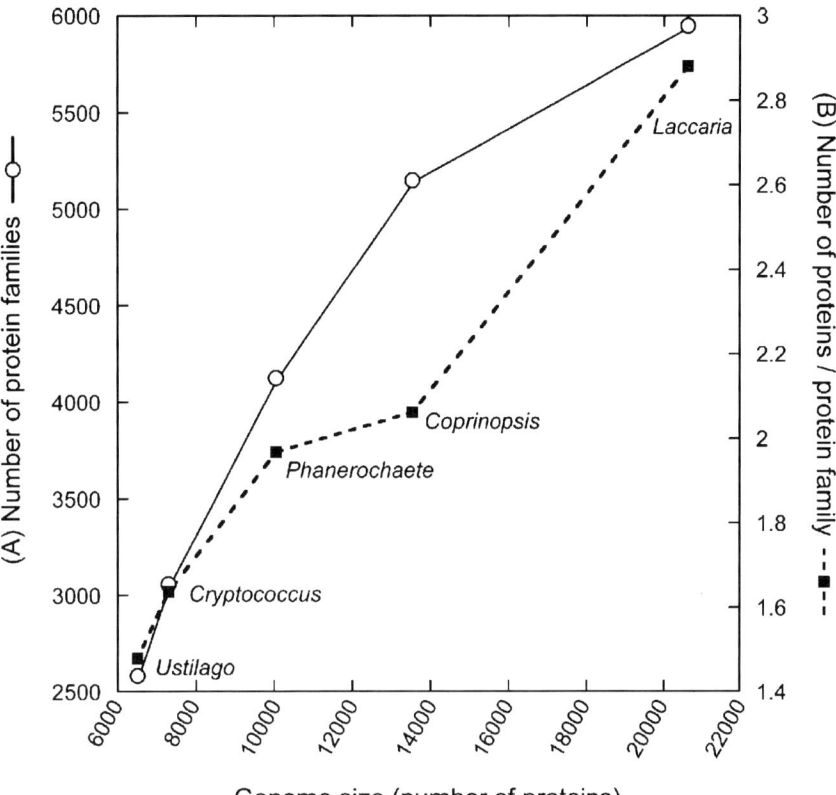

Genome size (number of proteins)

families along the *U. maydis* branch. Conversely, the numbers of contracted and extinct protein families are lower in *L. bicolor* than in the other sequenced basidiomycetes; 112 protein families have contracted and 363 are extinct along the *L. bicolor* branch. In comparison, 373 protein families have contracted and 2871 are extinct along the *U. maydis* branch.

The functions of a majority of expanding families are not known (Martin et al. 2008). However, a few of them contain members with the protein kinase and small GTPase domains, suggesting that they have a role in signal transduction pathways. Below, we discuss some of the expanded and contracted gene families coding for unknown and known proteins.

1. The Secretome

Proteins that are secreted from cells into the extracellular media and onto the cell surface represent a major class of molecules involved in intercellular communication in multicellular organisms and in the molecular cross-talk between organisms.

This class of proteins is referred to as the secretome (Chaffin et al. 1998). Many secreted proteins in fungi are expressed only during specific stages of development. The ECM hyphae colonizing root surfaces secrete various types of extracellular material, much of which is composed of chitosans, β-1,3-glucans and proteins (Dexheimer and Pargney 1991; Tagu and Martin 1996). Although the precise mechanisms that govern cell–cell interactions are not fully defined, a number of specific cell surface molecules have been identified as key elements in the interaction. In ectomycorrhiza, fungal attachment to the epidermal cells involves a polysaccharide mucigel and the secretion of oriented fibrillar materials, containing polysaccharides and glycoproteins, within which the whole of the sheath eventually becomes embedded (Lapeyrie et al. 1989). Interestingly, most of the cell surface CAP proteins involved in the biosynthesis of the polysaccharide capsule of the pathogenic basidiomycete *C. neoformans* (Moyrand et al. 2007) occur in the genome of *L. bicolor* (F. Martin, unpublished data).

A layer of extracellular fibrillar polymers, containing ConA-recognized glycoproteins, is present in the extracellular matrix of free-living mycelia of *L. bicolor* (Lei et al. 1991) and *P. tinctorius* (Lei et al. 1990; Lapeyrie et al. 1989) even before interaction with the root. However, at the contact sites between hyphæ and root surface an increased secretion of these extracellular fibrillar polymers takes place in compatible ectomycorrhizal associations. In contrast, isolates of *P. tinctorius* with delayed symbiosis development do not secrete this fibrillar material (Lei et al. 1990), suggesting that this fibrillar material can improve contact or adhesion and lead to a better colonization. In addition to hydrophobins (Duplessis et al. 2001, 2005), prominent symbiosis-regulated acidic polypeptides (SRAPs) accumulate in cell walls of the ectomycorrhizal gasteromycete *Pisolithus microcarpus* upon symbiosis formation on *Eucalyptus globulus* root tips (Laurent et al. 1999). The 32-kDa SRAPs showed no significant homology with known proteins, but the central part of their predicted protein sequence contains an Arg-Gly-Asp (RGD) cell attachment motif. These proteins are found mainly associated to the flocculent material covering the hyphal surface (Laurent et al. 1999) in the vicinity of hydrophobin proteins (Tagu et al. 2001). Owing to the key role played by extracellular and surface proteins in ectomycorrhiza development, the whole repertoire of predicted secreted proteins in the *L. bicolor* genome was analyzed in depth (Martin et al. 2008).

Of the 2930 proteins predicted to be secreted by *L. bicolor*, most (67%) cannot be functionally annotated and 82% of these predicted proteins are specific to *L. bicolor*. Within this set, an impressive number of sequences was predicted to encode secreted proteins with fewer than 300 amino acids. Of these SSPs, most belong to multigene families scattered all over the genome. Transcript profiling revealed that the expression of several SSP genes is specifically induced upon the symbiotic interaction. Five of the 20 most highly upregulated fungal transcripts in poplar and Douglas fir ectomycorrhizas code for SSP. The mycorrhiza-induced cysteine-rich SSP (MISSP) showing the highest induction in ectomycorrhizas, so-called MISSP7, was detected by immunofluorescence microscopy in the hyphal mantle layers ensheating the root tips, but the protein mainly accumulated in the finger-like, labyrinthine branch hyphal system (Hartig net; Fig. 13.2). Recently derived genome sequences of two *Phytophthora* species encode about 350 secreted proteins in each species carrying the host uptake

RXLR signal, suggesting that they are taken up by and act inside the host cell (Tyler et al. 2006). The genome sequence of the maize pathogenic fungus *U. maydis* also identified a large number of secreted proteins, some of which demonstrated virulence functions (Kämper et al. 2006). Even with the currently small sample of sequenced genomes of plant biotrophs, it is remarkable that a large number of proteins secreted during host infection do not exhibit a clear similarity to previously characterized proteins. The question is whether the diversity represents adaptation of the fungus towards distinct host species. If so, this may reflect differences in virulence/symbiosis targets or convergent evolution of different effectors towards the same host targets. The genome sequences of closely related species within the same genus that colonize the same host species will be of great interest.

As a large set of SSP genes shows significant changes in gene expression during fruiting body formation (Martin et al. 2008), they may not have been recruited only for host-plant colonization. They likely play a role in the formation of complex multicellular tissues, such as the Hartig net and mycorrhizal mantle and also mushroom primordia and specialized tissues. Elucidating the variability in the symbiosis-related secretome between strains of the same species (e.g., rapid nucleotide polymorphism) may be indicative of host–symbiont co-evolution and/or diversifying selection under this selective pressure.

2. Signal Transduction Pathways

Protein phosphorylation performed by protein kinases is of major importance in controlling cellular processes like morphological changes, cell cycle transitions, and stress responses (Manning et al. 2002; Westfall et al. 2004; Chaps. 2, 8). For example, protein kinases are key components in the signalling transduction pathways of *U. maydis* that regulate the morphological changes accompanying a successful colonization and infection of host plants (Klosterman et al. 2007). Likewise, several steps in the infection of plants by *Magnaporthe oryzae*, including attachment of conidia to rice leaves, germination and development of appressoria, are controlled by protein kinase signalling pathways (Xu et al. 2007).

Protein kinases comprise a superfamily of proteins that share the conserved protein kinase domain. Based on sequence similarities, the families can be classified into various subgroups (ePK families), with presumably similar functions (Manning et al. 2002). The five basidiomycete genomes sequenced so far contain 41 families with the protein kinase domain (Rajashekar et al. 2008). The largest of these families, denoted family-2, includes 512 proteins, and this family has been significantly expanded in *L. bicolor*. The *L. bicolor* genome has 140 proteins in family-2 containing the protein kinase motif, the C. *cinerea*

genome 109 proteins, *P. chrysosporium* 107, *C. neoformans* 84 and *U. maydis* 72 proteins.

Of the 140 *L. bicolor* family-2 proteins, 90 can be associated to five ePK families including AGC (cyclic-nucleotide, calcium-phospholipid-dependent kinases, ribosomal S6-phosphorylating kinases, G protein-coupled kinases), CAMK (calmodulin-regulated kinases), CMGC [cyclin-dependent kinases, mitogen-activated protein (MAP) kinases], STE (including many

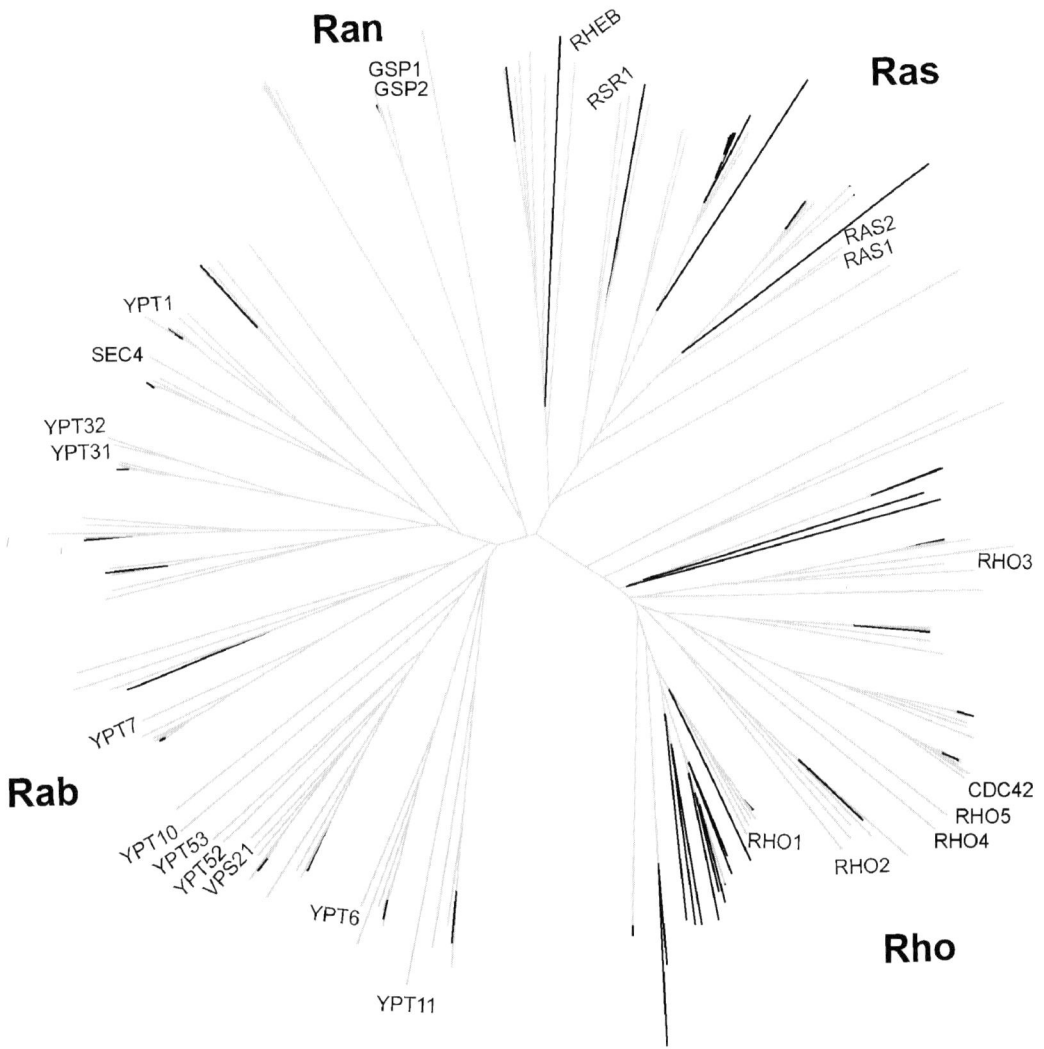

Fig. 13.5. Phylogeny of Ras-like GTPases from five basidiomycetes. Protein sequences from the genomes of *Laccaria bicolor*, *Coprinopsis cinerea*, *Phanerochaete chrysosporium*, *Cryptococcus neoformans* and *Ustilago maydis* were clustered into families using the TRIBE-MCL algorithm (Enright et al. 2002). Shown is a neighbor-joining tree of protein sequences in one of the largest gene families (family-6) containing members of the Ras family of small GTPases. The number of

proteins present in this family was 55 in *L. bicolor* (terminal branches are in *black*), 29 in *C. cinerea*, 27 in *P. chrysosporium*, 24 in *C. neoformans* and 25 in *U. maydis*. Also included are 23 Ras GTPases from *Saccharomyces cerevisiae* (gene names are indicated). The GTPases can be divided into four subgroups: the Ras, Rho, Rab and Ran GTPases, respectively. In *L. bicolor*, there has been a major expansion of Ras GTPases in the Ras and Rho subgroups, respectively (from Rajashekar et al. 2008) kinases functioning in the MAP kinase cascades) and "others" (consisting of kinases not classified into any of the above groups). Sequences of the remaining 50 *L. bicolor* proteins in family-2 do not cluster into the subgroups present in *Saccharomyces cerevisiae* (i.e., they have no yeast orthologs). Expansion of *L. bicolor* protein kinases has occurred both within the ePK family "others", but mainly in clusters lacking yeast orthologs. Of the expanded protein kinases, 24 are paralogs that arose by recent duplication events, i.e., after the separation of *L. bicolor* and *C. cinerea*. The majority of these paralogs are expressed above background levels; and several of them are differentially expressed in mycorrhizal rot tips, fruiting bodies and mycelium (Rajashekar et al. 2008).

Another large and expanded gene family in *L. bicolor* with a presumed role in signal transduction is the Ras family of small GTPases. Based on sequence similarity, the Ras small GTPases can be divided into five subfamilies, the Ras, Rho, Rab, Ran and Arf (Takai et al. 2001; Wennerberg et al. 2005). Phylogenetic analysis shows that expansion of Ras GTPases in *L. bicolor* mainly occurred within the Ras and Rho subfamilies (Fig. 13.5). The *L. bicolor* genome contains 13 genes in the Ras subfamily as compared with two in *C. cinerea*, *P. chrysosporium*, *C. neoformans* and *U. maydis* (Rajashekar et al. unpublished data).

The two *ras* genes (*ras1*, *ras2*) of *U. maydis* and *C. neoformans* are well characterized and have been shown to regulate morphogenesis, growth, mating and pathogenicity (Alspaugh et al. 2000; Lee and Kronstad 2002; Waugh et al. 2002; Müller et al. 2003). Earlier studies identified one *ras* gene (*Lbras*) in *L. bicolor* (Sundaram et al. 2001). Observations suggest that *Lbras* might have a function in vesicular transport (Sundaram et al. 2003). The function of the additional 12 *ras* genes identified in the *L. bicolor* genome is not yet known. Data from microarray expression experiments showed that five of them are expressed above background levels and at different levels in mycorrhizal root tips, fruiting bodies and mycelium (Rajashekar et al. 2008).

The expansion of rho GTPases in *L. bicolor* genome mainly occurred by paralogs of the *RHO1* GTPase of *S. cerevisiae* (Rajashekar et al. 2008). *RHO1* encodes an essential small GTPase in the Rho subfamily of Ras-like GTPases in yeast. Like other Rho-type GTPases, RHO1 is involved in the

establishment of cell polarity, playing a role in reorganizing the actin cytoskeleton (Madden and Snyder 1998). Apart from a *RHO1* ortholog, the *L. bicolor* genome contains 12 paralogs located in a sister clade to the *RHO1* GTPases. Seven of these paralogs were expressed at various levels in fruiting bodies, mycorrhizal root tips and mycelium of *L. bicolor*; and the remaining five could not be analyzed due to cross-hybridization.

Protein kinases and Ras-like GTPases are known to play important roles in linking responses of membrane receptors and signalling molecules to numerous downstream targets so that they can co-ordinately activate molecular processes required for specific cellular responses (Ridley 2001; Manning et al. 2002). The extensive expansions of protein kinases and Ras-like GTPases in the genome of *L. bicolor*, and the divergence of their expression patterns, suggest that these proteins have important roles in controlling the morphological and physiological changes accompanying the establishment of a functional mycorrhizal association.

C. The Biotrophy–Saprotrophism Continuum

The extensive extramatrical mycelium of the ECM fungi is ideally placed for nutrient acquisition in the top soil horizons at the humus/mineral soil interface, where most of the local pools is sequestered in organic form (Lindahl et al. 2007). ECM fungi are known to secrete carbohydrate-degrading enzymes (Lindahl and Taylor 2004), but saprotrophic fungi are more efficient than ECM fungi in colonizing and utilizing fresh litter. In contrast, ECM fungi are efficient in mobilizing organic N in well degraded organic matter and humus (Read and Perez-Moreno 2003). Although the *L. bicolor* genome contains numerous genes coding for key hydrolytic enzymes, such as glucanases, chitinases and proteases, we also observed an extreme reduction in the number of enzymes involved in the degradation of plant cell wall (PCW) oligo- and polysaccharides (Martin et al. 2008) confirming, at the genomic level, the low saprotrophic ability of ECM fungi.

1. Carbohydrate-Degrading Enzymes

Glycoside hydrolases (GH), glycosyltransferases (GT), polysaccharide lyases (PL), carbohydrate esterases and their ancillary carbohydrate-binding modules were identified using the CAZy classification (http://www.cazy.org/). A comparison of the *L. bicolor* candidate CAZymes with other sequenced Basidiomycetes and two ascomycetous phytopathogens confirmed the adaptation of its enzyme repertoire for symbiosis and revealed the strategy used for the interaction with the host. The reduction in the number of GH acting on PCWs affects almost all families, culminating in the complete absence of several key families. For instance, only one candidate cellulase (GH5) is found in the genome and there are no cellulases of the families GH6 and GH7, which are very abundant in saprotrophic fungi, such as the white rot fungus *P. chrysosporium* and the leaf decay fungus *C. cinerea*. Similarly, a reduction or loss of hemicellulose- and pectin-degrading enzymes was also noted. These observations suggest that the arsenal of *L. bicolor* PCW-degrading enzymes underwent massive gene losses as a result of the adaptation to a symbiotic lifestyle. This species is now unable to use many PCW polysaccharides as a carbon source, including those found in planta, in leaf litter and in soil. The remaining small set of secreted CAZymes with potential action on plant polysaccharides (e.g. the unique GH5 cellulase and GH28-polygalacturonases) is probably required for colonizing the root apoplastic space. The up-regulation of the single GH5 cellulase in ectomycorrhiza (Martin et al. 2008) supports this hypothesis. Colonization of the root apoplastic space may also involve mycobiont expansins, showing an increased transcript concentration in symbiotic tissues (Martin et al. 2008).

To survive until establishing a mycorrhizal association with its host, *L. bicolor* appears to have the capacity to degrade non-plant oligo- and polysaccharides (e.g., of animal or bacterial origin), which is suggested by its retention of CAZymes from families GH79, PL8, PL14 and GH88 (Martin et al. 2008). The closest functionally characterized homologs of *L. bicolor* PL8 and PL14 are, respectively, the mollusk alginate lyases of *Haliotis discus* and bacterial chondroitin and xanthan lyases. Alginate and xanthan can be found as component of bacterial exopolysaccharides (EPS). In the absence of know fungal acidic endogenous polysaccharides, these enzymes could be involved in the control of bacterial populations, such as the mycorrhiza-helper *Pseudomonas fluorescens,* living in interaction with *L. bicolor* (Deveau et al. 2007).

Compared with other sequenced basidiomycetes, *L. bicolor* possesses a rich set of CAZymes,

allowing the degradation of the fungal cell wall and in particular β-glucans, chitin/chitosan and α-1,3-glucans. A comparably complex set is found in the leaf decayer *C. cinerea*, although the relative abundances of active CAZyme families are different in these two species (Martin et al. 2008). *C. cinerea* and *L. bicolor* are both able to form a complex mushroom basidiocarp and this rich set of CAZymes active on fungal cell wall may be involved in cell wall remodeling during the development of fruiting bodies. Up to 40% of the corresponding genes are up- or down- regulated during developmental processes requiring cell wall alterations, such as the formation of fruiting bodies or mycorrhiza (Martin et al. 2008).

2. Secreted Proteases and Nitrogen Acquisition

The main pools of soil N are humified matter, plant litter, live and dead microbial materials and inorganic ions. Soil organic matter contains N predominantly as proteins, but amino acids, amino sugars, peptides and chitin could represent a significant proportion of this pool. The current view is that, in ecosystems where the rate of microbial mineralization is slow, such as in boreal forests and heathlands, woody plants rely on ecto- or ericoid mycorrhizal fungal symbioses to break down soil proteins, whereas in ecosystems with high microbial activity and mineralization rates, such as grasslands, temperate and tropical forests, herbaceous and woody plants use mostly inorganic nitrogen (Read 1991; Read and Perez-Moreno 2003). However, there is evidence that (nonmycorrhizal) plants, such as *Arabidopsis thaliana* and *Hakea actites,* can assimilate protein without assistance from soil organisms (Paungfoo-Lonhienne et al. 2008). A major contributing factor in organic N acquisition by ECM fungi is the continuous growth of the mycelial network into patchily distributed soil resources, which it absorbs and transports for storage in the hyphal mantle ensheathing the colonized roots. Absorption of soil protein N requires their enzymatic degradation to peptides and amino acids. ECM fungi secrete proteinases when grown on animal proteins (casein, gelatin, albumin) and protein fractions from beech forest litter as the substrate (El-Badaoui and Botton 1989; Nygren et al. 2007). The ECM hyphal network colonizing the soil might therefore be expected to express a wide diversity of proteolytic enzymes.

The *L. bicolor* genome encodes 116 secreted proteases (Martin et al. 2008), including extracellular fungalysin metalloproteases (Lilly et al. 2008), a relatively large hydrolytic capacity, compared with other sequenced saprotrophic basidiomycetes, such as *C. cinerea* and *C. neoformans*. The ability to secrete aspartyl-, metallo- and serine-proteases may play a role in degradation of decomposing litter and confirms that *L. bicolor* can use nitrogen of animal origin, as suggested previously (Klironomos and Hart 2001). Several of the secreted proteases are likely involved in the developmental remodeling of multicellular structures as the expression of several secreted proteases is up- or down-regulated in fruiting bodies and ectomycorrhizal root tips (Martin et al. 2008).

Many ECM fungi, such as telephoroid species, likely play a significant role in the degradation of organic matter, but a significant degradation of organic matter requires a high capacity to attack plant cell wall polysaccharides. As this capacity is dramatically reduced in *L. bicolor* (Martin et al. 2008), it seems unlikely that this species should be able to degrade litter or wood, or be able to support its carbon metabolism through saprotrophic activities. However, chitinases, secreted proteases and hydrolytic enzymes acting on bacterial, insect and fungal polysaccharides may allow this fungus to fulfill some of its metabolic needs. It should be kept in mind that large sections of the subterranean ECM hyphal network exploring extensive soil regions are located far away from their ectomycorrhizal root tips. These scavenger hyphae likely express some autonomous physiological activities and would benefit from secreting these hydrolyzing enzymes.

IV. The Ectomycorrhizal Transcriptome

A. Alteration in Symbiont Transcriptome

The development of ectomycorrhizal tissues proceeds through the differentiation of a fungal mantle surrounding the roots, inwardly directed hyphae forming the Hartig net around the cortical root cells and externally directed hyphae forming the extramatrical mycelium in the soil (Kottke and Oberwinkler 1987; Massicotte et al. 1987). These stages vary greatly in cell/hypha shape, tissue morphology, biochemical properties and

transcriptional activity. Insights into the genes which are differentially expressed during the development of these tissues and their specialized metabolic activities have been obtained in a number of different systems, including the *Pisolithus–Eucalyptus*, *Paxillus–Betula*, *Laccaria–Populus*, *Laccaria–Pseudotsuga* and *Tuber–Tilia* symbioses (Table 13.1). These experiments were done using several different methods (cDNA arrays, suppression subtractive hybridizations; more recently, whole-genome expression oligoarrays), growth conditions (defined agar media, peat–soil systems) and experimental setups (e.g., time of interactions, selection of fungal strains and host species, amendment of nutrients).

Results from the transcript profiling of ectomycorrhizal fungi are summarized in several recent reviews (Martin 2007; Martin et al. 2004a, 2007). Briefly, studies showed that a moderate part of the mycobiont transcriptome (5–20% of the analyzed transcripts) is differentially expressed in symbiotic tissues, as compared with free-living mycelium. In *L. bicolor*, genome-wide transcript profiling showed that 413 and 665 transcripts were differentially expressed in Douglas fir and Poplar ectomycorrhizal root tips, respectively (Fig. 13.6; Kohler et al. 2008). This means that only 3% of the total gene set is affected by symbiosis development. A large part of these regulated genes, typically more than 50%, displays no significant sequence homologies to genes with known function and several belong to lineage-specific gene families (Martin et al. 2008). The products of the remaining genes are predicted to be involved in a number of different cellular functions including carbon and nitrogen metabolism, respiration, protein synthesis, regulation of morphology, cell wall formation and secretome, transport, defence and stress reactions and cellular signalling (Table 13.1). Overall, the emergent induced gene networks suggest that the fungal symbiotic tissues are characterized by the activation of general cellular growth functions and processes that facilitate accommodation of the mycobiont with the newly colonized root. In two of the symbiotic systems, *Pisolithus–Eucalyptus* and *Paxillus–Betula*, the patterns of gene regulation associated with the development of ectomycorhizal tissues were examined in more detail (Duplessis et al. 2005; Le Quéré et al. 2005). Notably, several of the

Table 13.1. Transcriptional profiling of ectomycorrhizal fungi

Organisms	Methods	Experiments	Genes regulated[a]	Reference
Pisolithus/Eucalyptus	cDNA array 486 probes	Comparisons of gene expressions in ECM rot tips (sheath) and free-living mycelium/plant roots. Organisms grown on defined medium using an agar plate assay.	66 genes (ESTs) Cell wall formation Lipid metabolism Signalling pathways Protein synthesis and degradation Defence and stress reactions	Voiblet et al. (2001)
Pisolithus/Eucalyptus	cDNA array 1345 fungal probes 193 plant probes	Changes in gene expression during the development of the ECM root tissue (contact, sheath formation, Hartig net formation, mature mycorrhizae). Comparisons of ECM tissue with free-living mycelium and plant roots. Organisms grown on defined medium using an agar plate assay.	30–221 genes (ESTs) Carbon metabolism Amino acid biosynthesis Respiration Defence and stress reactions Cytoskeleton organization Cell wall formation Protein synthesis and degradation Signalling pathways	(Duplessis et al. 2005)
Paxillus/Betula	cDNA microarray 1048 fungal probes 1021 plant probes	Comparisons of gene expressions in ECM rot tips (sheath and Hartig net) and free-living mycelium/plant roots. Organisms grown on defined medium using a cellophane-covered agar plate assay.	93 genes (ESTs) Carbon metabolism Lipid metabolism Defence and stress reactions Signalling pathways Cellular transport Protein synthesis and degradation Cell wall formation	Johansson et al. (2004)
Paxillus/Betula	cDNA microarray 1075 fungal probes 1074 plant probes	Changes in gene expression during the development of the ECM root tissue (contact, sheath formation, Hartig net formation, mature mycorrhizae). Comparisons of ECM tissue with free-living mycelium and plant roots. Organisms grown on defined medium using an agar plate assay.	251 genes (ESTs) Carbon metabolism Respiration Defence and stress reactions Signalling pathways Cytoskeleton organization Cell wall formation	Le Quéré et al. (2005)
Paxillus/Betula	cDNA microrray 1075 fungal probes 1074 plant probes	Comparison of gene expression in ECM root tissues, extramatrical mycelium and rhizomorphs. Organisms grown in peat microcosms with ammonium sulphate nutrient patches.	337 genes (ESTs) Nitrogen metabolism Carbon metabolism Signalling pathways Cytoskeleton organization Vesicular transport Protein degradation 66 genes (ESTs)	Wright et al. (2005)

continued

Table 13.1. continued

Organisms	Methods	Experiments	Genes regulated[a]	Reference
Paxillus/Betula	cDNA microrray 1075 fungal probes 1074 plant probes	Comparison of gene expression in three strains of *P. involutus* that differ in host preferences. Strains grown in association with *Betula pendula* on using an agar plate assay.	Signalling pathways Defence and stress reactions Cell wall formation 65 genes (ESTs)	Le Quéré et al. (2004)
Paxillus/Betula	cDNA array 1230 fungal and plant probes	Comparison of gene expression of in ECM root tissues and extramatrical mycelium when grown on peat in a microcosm system.	Nitrogen metabolism Lipid metabolism Cellular transport Protein synthesis and degradation Defence and stress reactions 58 cDNA clones (ESTs)	Morel et al. (2005)
Tuber/Tilia	Suppressive subtractive hybridization (SSH)	Comparison of gene expression in mycelium interacting (without direct contact) with host plant (pre-infection phase) and in free-living mycelium. Northern blot experiments when organisms growing in a peat-moss vermiculite medium.	Defence and stress reactions Protein synthesis and degradation Cell wall formation General metabolism 51 genes (ESTs)	Menotta et al. (2004)
Tuber	cDNA array 2062 fungal probes	Changes in gene expression in mycelium in response to nitrogen starvation. Fungus growing on a synthetic solid medium.	Nitrogen metabolism Carbon metabolism Cell wall modification Stress response Small secreted proteins (SSPs)	Montanini et al. (2006)
Laccaria/Pseudotsuga *Laccaria/Populus*	Whole-genome oligoarray 22 294 gene models	Comparisons of gene expression in ECM root tips (mature mycorrhizae), free-living mycelium and fruiting bodies. EM synthesized in pots with peat and vermiculite.	Transporters Carbohydrate-degrading enzymes Secreted proteases	Martin et al. (2008)

[a] Number of genes, with cellular and molecular functions known to be regulated.

Fig. 13.6. Three Venn diagrams (**A**, **B**,**C**) showing changes in gene expression in *Laccaria bicolor* upon development of ectomycorrhizal roots tips and fruiting body. The numbers of transcript with an increased (**B**) or decreased (**C**) concentration in poplar or Douglas fir ectomycorrhizal tips (*ECM*) or in fruiting body (*FB*) are shown (adapted from Martin et al. 2008)

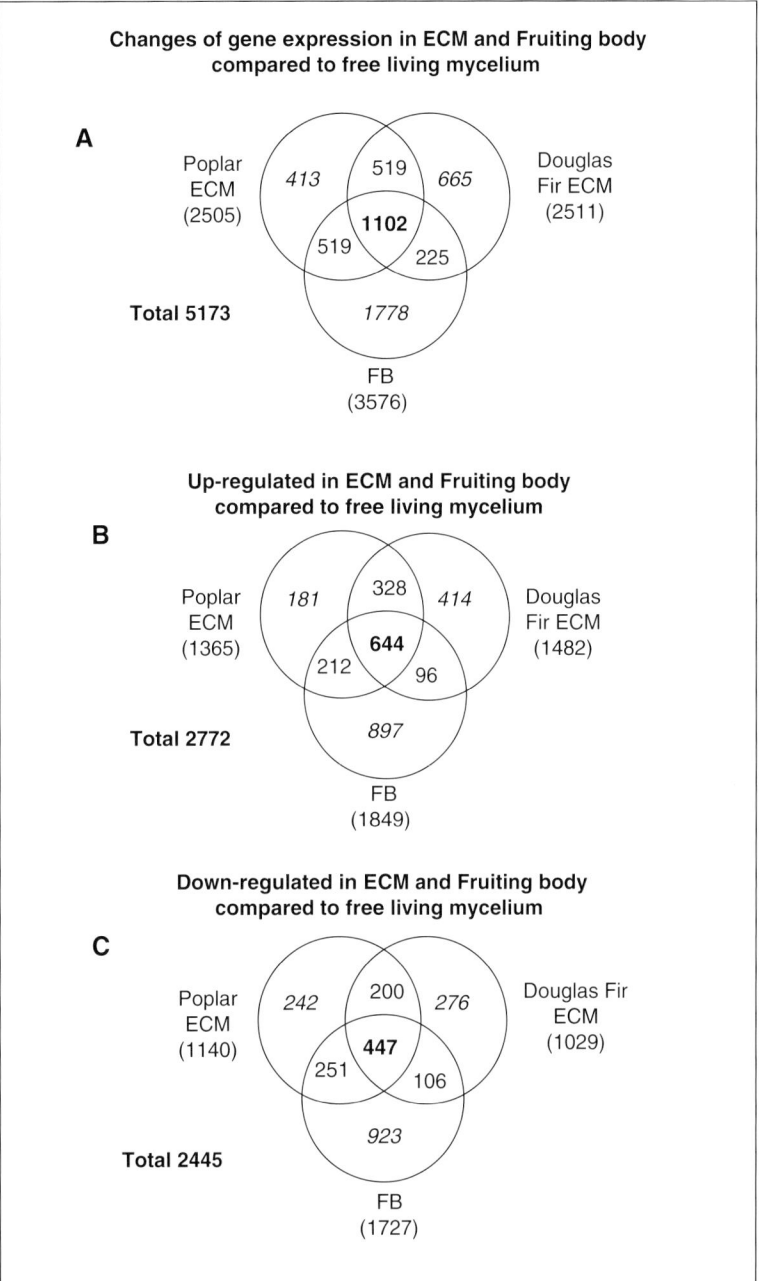

cellular functions were similarly regulated in both systems, suggesting that common genetic networks are activated during the development of fungus–plant symbiotic tissues (Martin et al. 2007).

ECM fungi can differ markedly in their ability to form symbioses with different host plants (Molina and Trappe 1982; Cairney 1999). Le Quéré et al. (2004) showed that such differences could affect the responses of plant-induced gene expression in closely related strains of *P. involutus*. Thus,

3.4% of the analyzed transcripts were found to be differentially expressed in compatible and incompatible strains. The observed differences are most likely due to changes in promoter elements or levels of transcription factors because all the differentially expressed genes were found in the same copy numbers in compatible and incompatible strains.

The extramatrical mycelium of many ECM fungi can proliferate extensively when growing in soils, and it efficiently prospects for nutrient resources, often at considerable distances from the plant roots. The hyphal network can differentiate into several, spatially separated and functionally specialized tissues (Cairney 1996). Thus, when encountering a nutrient-rich patch, the mycelium could branch extensively and form a dense network of aggregated hyphae that is adapted for efficient uptake of nutrients. Behind the patches, the hyphae might aggregate to form rhizomorphs that connect the mycelial front and the mycorrhizal root tips. Studies of the *Paxillus–Betula* association developed in soil microcosms showed that global gene expression profiles differ considerably between hyphal networks aggregating on dense sources of nutrients, rhizomorphs translocating assimilated nutrients and mycorrhizal root tips (Morel et al. 2005; Wright et al. 2005). Distinct expression patterns of genes implicated in metabolic and cellular activities indicate a functional specialization of the various compartments forming the hyphal web.

B. Expression of Ectomycorrhizae-Specific Genes

Transcript profiling based on cDNA microarrays have revealed a moderate change in the expression levels (see above). For example, when comparing expression levels in symbiotic and free-living mycelium, the changes rarely exceed eight-fold up- or down-regulation (Duplessis et al. 2005; Le Quéré et al. 2005; Morel et al. 2005; Wright et al. 2005; Heller et al. 2008). It has been reported that the development of *Pisolithus–Eucalyptus*, *Paxillus–Betula*, *Pisolithus–Populus*, *Laccaria–Populus*, *Laccaria–Pseudotsuga*, *Laccaria–Pinus* and *Tuber–Tilia* ectomycorrhizal symbioses does not induce the expression of ectomycorrhiza-specific genes (coding for ectomycorrhizins; Martin et al. 2007). This apparent lack of ectomycorrhiza-specific genes and moderate induction of symbiosis-regulated genes were recently challenged in experiments using a *L. bicolor* whole-genome expression oligoarray (Martin et al. 2008). This array contained probes for the whole set of predicted gene models. In ectomycorrhizal root tips of Douglas fir or poplar plantlets, the most highly up-regulated *L. bicolor* transcripts displayed values (ectomycorrhizae *vs.* free-living mycelium) exceeding 1000-fold; but only six of them were not detected in the free-living vegetative mycelium or in the fruiting body, confirming that fungal ectomycorrhizins are rare. As stressed in Sect. III.B.1, many of these ectomycorrhiza-specific transcripts are coding for SSPs (e.g., the mycorrhiza-induced small secreted protein of 7 kDa, MiSSP7) that were not detected because their size was below the cut-off value for EST sequencing and were thus lacking from cDNA spotted arrays (Duplessis et al. 2005; Le Quéré et al. 2005; Morel et al. 2005; Wright et al. 2005). The apparent paucity of ectomycorrhiza-specific genes and the moderate induction of symbiosis-regulated genes are striking and suggest that ontogenic and metabolic programs that lead to the development of symbiosis are driven by the differential expression/activity of pre-existing transcription factors and/or transduction pathways, rather than by the expression of symbiosis-specific arrays of genes.

V. Ecological Genomics

A central challenge in the evolutionary and ecological genomics of mycorrhizal symbioses is to identify the genetic basis of adaptive traits that allow the fungus (and its host plant) to survive and reproduce in natural environments. This challenge is made less daunting by the increasing number of genome sequences (*P. trichocarpa*, *L. bicolor*, *T. melanosporum*) and genetic resources (ESTs, genetic maps, cDNA arrays, oligonucleotide arrays) that have become available in the past 5 years. Access to these resources and genetic linkage maps for species of interest, such as *P. trichocarpa* and *L. bicolor*, will facilitate the identification of genes underlying traits of interest by means of (expressed) quantitative trait loci (QTL) mapping and through association studies in natural populations. Such approaches are now widely used by evolutionary biologists to investigate the genetic basis of several traits in higher plants (de Meaux et al. 2006). The mechanistic basis for genotypic variation in fungal responses to the environment and host plant genotype has long been enigmatic, and genotype-specific patterns of gene expression should provide a uniquely integrated view of the anabolic, catabolic and signalling processes that may combine to regulate the performance of mycobionts and their associated host plants.

Given the apparent generality of intraspecific variation in mycorrhizal fungi (Debaud et al. 1995), it is important to understand whether and how these subpopulations are functionally distinct. A given tree may interact with dozens of genotypes of the same fungal ECM species (e.g., *L. bicolor, L. amethystina*; Gherbi et al. 1999; Selosse et al. 1999) and a significant symbiosis phenotypic variability should be expected. At the level of DNA sequence, a substantial fraction of substitutions are silent in terms of amino acid sequence, and others may be nonsynonymous but functionally neutral. However, two individuals that differ by a few percent in their genetic sequence will inevitably have at least minor functional differences, such as optimal temperature or pH for the activity of some hydrolytic enzymes involved in nutrient acquisition or transporters controlling the symbiotic exchanges of nutrients. It should be quite informative to identify the source(s), diversity and range of functionality associated with these nucleotide sequence variabilities by fully resequencing a large number of these genes coding for mycorrhiza-regulated functions and understanding how their individual abundances fluctuate. Even small differences in optimal conditions may have profound effects on the physiological fitness of the symbiosis.

VI. Future Research

The complete genome sequence of *L. bicolor*, a generalist ECM fungus, provides key insights into the symbiotic fungal lifestyle. These should be confirmed by the on-going analysis of the *T. melanosporum* genome. When combined with whole-genome transcript profiling (Table 13.1), genome exploration has revealed potential symbiotic factors along with genes involved in a biotrophic lifestyle, such as a large set of SSPs, signalling proteins and transporters. *L. bicolor* possesses expanded multi-gene families associated with the hydrolysis of soil polymers that were previously only known in the most potent decomposers in the soil. These are mostly secreted degrading proteases, chitinases and glucanases likely involved in nutrient acquisition from decomposed litter. This indicates hitherto unexpected saprotrophic capabilities of *L. bicolor*, thereby giving trees a more direct symbiosis-driven access to nutrients that are otherwise unavailable within complex molecules in the soil. However, the presence of only limited numbers of hydrolytic enzymes able to act on plant cell wall polysaccharides implies that the mycobiont mainly relies on its host for its carbon

requirements. *L. bicolor* also has an unusually large number of genes compared with other fungi examined to date; and those gene families which have undergone expansion may explain how it can simultaneously fulfil its role as a symbiont and as a facultative saprotroph. The *L. bicolor* genome encodes most of the features needed for biotrophy, including most notably a battery of effector-type proteins by which communication with its plant partner is likely established and maintained. But it also has a very restricted arsenal of hydrolytic enzymes that act on plant cell walls, such as cellulases and pectinases, which are characteristic of plant pathogens.

Although a large number of genes that encode structural proteins, enzymes, transporters, transcription factors, components of signalling pathways and potential secreted effectors have been shown to be regulated in the symbiotic tissues of ectomycorrhizal fungi, the external signals that trigger these changes are not yet known. Such signals might include nutrient and carbon levels, the abiotic microenvironment and specific cues from the host plant. Furthermore, little is known about the membrane receptors and signalling pathways that are required for activating the observed cellular responses. Owing to the fact that the transcriptional regulation of the fungal genes is moderate upon symbiosis development, there is a urgent need to investigate the post-translational regulations (e.g., phosphorylation, protein isoforms) at a genomic scale.

Acknowledgements. We would like to thank A. Kohler, D. Ahren T. Thampson and S. Duplessis for their efficient support, and former and current PhD students and postdocs for their collaboration. We would like to acknowledge the members of the *Laccaria* Genome Consortium for their oustanding efforts. Investigations carried out in F.M.'s laboratory were supported by grants from INRA (Project '*Genome Sequencing of Poplar and Associated Micro-Organisms*'), the European Commission Network of Excellence EVOLTREE, the Génoscope Genomics Institute (Project '*ForEST*') and the Région Lorraine. A.T. is supported by the Swedish Research Council. The genome sequencing and analysis of *Laccaria bicolor* was funded by grants from the U.S. Department of Energy, INRA, Région Lorraine, the European Network of Excellence EVOLTREE and the Swedish Research Council.

References

Adams F, Reddell P, Webb MJ, Shipton WA (2006) Arbuscular mycorrhizas and ectomycorrhizas on *Eucalyptus grandis* (Myrtaceae) trees and seedlings in native forests of tropical north-eastern Australia. Aust J Bot 54:271–281

Alspaugh JA, Cavallo LM, Perfect JR, Heitman J (2000) RAS1 regulates filamentation, mating and growth at high temperature of *Cryptococcus neoformans*. Mol Microbiol 36:352–365

Berbee ML, Taylor JW (2001) Fungal molecular evolution: gene trees and geologic time. In: McLaughlin DJ, McLaughlin EG, Lemke PA (eds) The Mycota, vol. VII.B. Springer, Berlin, pp 229–245

Bonfante P (2001) At the interface between mycorrhizal fungi and plants: the structural organization of cell wall, plasma membrane and cytoskeleton. In "The Mycota IX Fungal Associations", Hock ed., Springer-Verlag Berlin, Heidelberg, 45–61

Brandes B, Godbold DL, Kuhn AJ, Jentschke G (1998) Nitrogen and phosphorus acquisition by the mycelium of the ectomycorrhizal fungus *Paxillus involutus* and its effect on host nutrition. New Phytol 140:735–743

Brundrett MC (2002) Coevolution of roots and mycorrhizas of land plants. New Phytol 154:275–304

Bruns TD, Shefferson RP (2004) Evolutionary studies of ectomycorrhizal fungi: recent advances and future directions. Canadian Journal of Botany, 82(8): 1122–1132

Cairney JWG (1996) Physiological heterogeneity within fungal mycelia: an important concept for a functional understanding of the ectomycorrhizal symbiosis. New Phytol 134:685–695

Cairney JWG (1999) Intraspecific physiological variation: implications for understanding functional diversity in ectomycorrhizal fungi. Mycorrhiza 9:125–135

Chaffin WL, Lopez-Ribot JL, Casanova M, Gozalbo D, Martinez JP (1998) Cell wall and secreted proteins of *Candida albicans*: identification, function, and expression. Microbiol Mol Biol Rev 62:130–180

Cornell MJ, Alam I, Soanes DM, et al (2007) Comparative genome analysis across a kingdom of eukaryotic organisms: specialization and diversification in the fungi. Genome Res 17:1809–1822

Debaud JC, Marmeisse R, Gay G (1995) Intraspecific genetic variation in ectomycorrhizal fungi. In: Varma AK, Hock B (eds) Mycorrhiza: structure, molecular biology and function. Springer, Heidelberg, pp 79–113

de Meaux J, Pop A, Mitchell-Olds T (2006) Cis-regulatory evolution of chalcone-synthase expression in the genus *Arabidopsis*. Genetics 174:2181–2202

Deveau A, Palin B, Delaruelle C, Peter M, Kohler K, Pierrat Jc, Sarniguet A, Garbaye J, Martin F, Frey-Klett P (2007) The mycorrhiza helper *Pseudomonas fluorescens* BBc6R8 has a specific priming effect on the growth, morphology and gene expression of the ectomycorrhizal fungus *Laccaria bicolor* S238N. New Phytol 175:743–755

Dexheimer J, Pargney JC (1991) Comparative anatomy of the host-fungus interface in mycorrhizas. Experientia 47:312–320

Ditengou FA, Béguiristain T, Lapeyrie F (2000) Root hair elongation is inhibited by hypaphorine, the indole alkaloid from the ectomycorrhizal fungus *Pisolithus tinctorius*, and restored by indole-3-acetic acid. Planta 211:722–728

Duplessis S, Sorin C, Voiblet C, Palin B, Martin F, Tagu D (2001) Cloning and expression analysis of a new hydrophobin cDNA from the ectomycorrhizal basidiomycete *Pisolithus*. Curr Genet 39:335–339

Duplessis S, Courty PE, Tagu D, Martin F (2005) Transcript patterns associated with ectomycorrhiza development in *Eucalyptus globulus* and *Pisolithus microcarpus*. New Phytol 165:599–611

El-Badaoui K, Botton B (1989) Production and characterization of exocellular proteases in ectomycorrhizal fungi. *Ann Sci For* 46:728s–730s

Enright AJ, Van DS, Ouzounis CA (2002) An efficient algorithm for large-scale detection of protein families. Ncl Acids Res 30:1575–1584

Galagan JE, Henn MR, Li-Jun MA, Cuomo CA, Birren B (2005) Genomics of the fungal kingdom:Insights into eukaryotic biology. Genome Res 15:1620–1631

Gherbi H, Delaruelle C, Selosse MA, Martin F (1999) High genetic diversity in a population of the ectomycorrhizal basidiomycete *Laccaria amethystina* in a 150-year-old beach forest. Mol Ecol 8:2003–2013

Hahn MW, De BT, Stajich JE, Nguyen C, Cristianini N (2005) Estimating the tempo and mode of gene family evolution from comparative genomic data. Genome Res 15:1153–1160

Heller G, Adomas A, Li G, Osborne J, van Zyl L, Sederoff R, Finlay RD, Stenlid J, Asiegbu FO (2008) Transcriptional analysis of *Pinus sylvestris* roots challenged with the ectomycorrhizal fungus *Laccaria bicolor*. BMC Plant Biol 8:19

Hibbett DS, Gilbert LB, Donoghue MJ (2000) Evolutionary instability of ectomycorrhizal symbioses in basidiomycetes. *Nature* 407: 506–508

Högberg P, Nordgren A, Buchmann N, Taylor AFS, Ekblad A, Högberg MN, Nyberg G, Ottosson-Löfvenius M, Read DJ (2001) Large-scale forest girdling shows that current photosynthesis drives soil respiration. Nature 411:789–792

Horan DP, Chilvers GA, Lapeyrie FF (1988) Time sequence of the infection process in eucalypt ectomycorrhizas. New Phytol 109:451–458

Jambois A, Dauphin A, Kawano T, Ditengou FA, Bouteau F, Legué V, Lapeyrie F (2005) Competitive antagonism between IAA and indole alkaloid hypaphorine must contribute to regulate ontogenesis. Physiologia Plant 123:120–129

Johansson T, Le Quéré A, Ahrén D, et al (2004) Transcriptional responses of *Paxillus involutus* and *Betula pendula* during formation of ectomycorrhizal root tissue. Mol Plant–Microb Interact 17:202–215

Kämper J, Kahmann R, Bolker M, et al (2006) Insights from the genome of the biotrophic fungal plant pathogen *Ustilago maydis*. Nature 444:97–101

Klironomos JN, Hart MM (2001) Animal nitrogen swap for plant carbon. Nature 410:651–652

Klosterman SJ, Perlin MH, Garcia-Pedrajas M, Covert SF, Gold SE (2007) Genetics of morphogenesis and pathogenic development of *Ustilago maydis*. Adv Genet 57:1–47

Kohler A, Deveau A, Delaruelle C, Peter M, Brokstein PB, Lindquist E, Aerts A, Grigoriev I, Martin F (2008) The *Laccaria* transcriptome: EST resource, whole-genome expression array and comparative analysis upon development of fruit body and ectomycorrhizas. New Phytol (in press)

Koide RT, Sharda JN, Herr JR, Malcolm GM (2008) Ectomycorrhizal fungi and the biotrophy–saprotrophy continuum. New Phytologist, 178, 230–233

Kottke I, Oberwinkler F (1987) The cellular structure of the Hartig net: coenocytic and transfer cell-like organization. Nord J Bot 7:85–95

Lagrange H, Jay-Allemand C, Lapeyrie F (2001) Rutin, the phenolglycoside from eucalyptus root exudates, stimulates *Pisolithus* hyphal growth at picomolar concentrations. New Phytol 149:349–355

Lapeyrie F, Lei J, Malajczuk M, Dexheimer J (1989) Ultrastructural and biochemical changes at the pre-infection stage of mycorrhizal formation by two isolates of *Pisolithus tinctorius*. Ann Sci For 46s:754s–757s

Laurent P, Voiblet C, Tagu D, De Carvalho D, Nehls U, De Bellis R, Balestrini R, Bauw G, Bonfante P, Martin F (1999) A novel class of ectomycorrhiza-regulated cell wall polypeptides in *Pisolithus tinctorius*. Mol Plant–Microbe Interact 12:62–871

Leake JR, Johnson D, Donnelly DP, Muckle GE, Boddy L, Read DJ (2004) Networks of power and influence: the role of mycorrhizal mycelium in controlling plant communities and agroecosystem functioning. Can J Bot 82:1016–1045

Lei J, Lapeyrie F, Malajczuk N, Dexheimer J (1990) Infectivity of pine and eucalypt isolates of *Pisolithus tinctorius* (Pers.) Coker and Couch on roots of *Eucalyptus urophylla* S.T. Blake in vitro. II. Ultrastructural and biochemical changes at the early stage of mycorrhiza formation. New Phytol 116:115–122

Lei J, Wong KKY, Piché Y (1991) Extracellular concanavalin A-binding sites during early interactions between *Pinus banksiana* and two closely related genotypes of the ectomycorrhizal basidiomycete *Laccaria bicolor*. Mycol Res 95:357–363

Le Quéré A, Schützendübel A, Rajashekar B, et al (2004) Divergence in gene expression related to variation in host specificity of an ectomycorrhizal fungus. Mol Ecol 13:3809–3819

Le Quéré A, Wright DP, Söderström B, Tunlid A, Johansson T (2005) Global patterns of gene regulation associated with the development of ectomycorrhiza between birch (*Betula pendula* Roth.) and *Paxillus involutus* (Batsch) Fr. Mol Plant–Microbe Interact 18:659–673

Lee N, Kronstad JW (2002) *ras2* controls morphogenesis, pheromone response, and pathogenicity in the fungal pathogen *Ustilago maydis*. Eukaryot Cell 1:954–966

Lilly WW, Stajich JE, Pukkila PJ, Wilke SK, Inoguchi N, Gathman AC (2008) An expanded family of fungalysin extracellular metallopeptidases of *Coprinus cinerea*. Mycol Res 112:389–398

Lindahl BD, Taylor AFS (2004) Occurrence of N-acetylhexosaminidase-encoding genes in ectomycorrhizal basidiomycetes. New Phytol 164:193–199

Lindahl BD, Ihrmark K, Boberg J, Trumbore SE, Högberg P, Stenlid J, Finlay RD (2007) Spatial separation of litter decomposition and mycorrhizal nitrogen uptake in a boreal forest. New Phytol 173:611–620

Loftus BJ, Fung E, Roncaglia P, et al (2005) The genome of the basidiomycetous yeast and human pathogen *Cryptococcus neoformans*. Science 307:1321–1324

Lynch M (2007) The fragility of adaptive hypotheses for the origins of organismal complexity. Proc Natl Acad Sci USA 104:8597–8604

Lynch M, Conery JS (2000) The evolutionary fate and consequences of duplicate genes. Science 290:1151–1155

Lutzoni F, Kauff F, Cox CJ, McLaughlin D, Celio G, Denti-nger B, Padamsee M, Hibbett D, James TY, Baloch E, Grube M, Reeb V, Hofstetter V, Schoch C, Arnold AE, Miadlikowska J, Spatafora J, Johnson D, Hambleton S, Crockett M, Shoemaker R, Sung GH, Lücking R (2004) Assembling the fungal tree of life: progress, classification, and evolution of subcellular traits. Am J Bot 91:1446–1480

Madden K, Snyder M (1998) Cell polarity and morphogenesis in budding yeast. Annu Rev Microbiol 52:687–744

Manning G, Plowman GD, Hunter T, Sudarsanam S (2002) Evolution of protein kinase signaling from yeast to man. Trends Biochem Sci 27:514–520

Martin F (2007) Fair trade in the underworld: the ectomycorrhizal symbiosis. In: Howard RJ, Gow NAR (eds) Biology of the fungal cell. The Mycota VIII. Springer, Berlin, pp 291–308

Martin F, Duplessis S, Kohler A, Tagu D (2004a) Exploring the transcriptome of the ectomycorrhizal symbiosis. In: Kumar S, Fladung M (eds) Molecular genetics and breeding of forest trees. Haworth's Food Products, New York, pp 81–109

Martin F, Tuskan GA, DiFazio SP, Lammers P, Newcombe G, Podila GK (2004b) Symbiotic sequencing for the *Populus* mesocosm. New Phytol 161:330–335

Martin F, Kohler A, Duplessis S (2007) Living in harmony in the wood underground: ectomycorrhizal genomics. Curr Opin Plant Biol 10:204–210

Martin F, Aerts A, Ahrén D, et al (2008) Symbiosis insights from the genome of the mycorrhizal basidiomycete *Laccaria bicolor*. Nature 452:88–92

Martin F, Duplessis S, Ditengou F, Lagrange H, Voiblet C, Lapeyrie F (2001) Developmental cross talking in the ectomycorrhizal symbiosis: signals and communication genes. New Phytol 151:145–154

Martinez D, Larrondo LF, Putnam N, et al (2004) Genome sequence of the lignocellulose degrading fungus *Phanerochaete chrysosporium* strain RP78. Nat Biotechnol 22:695–700

Massicotte HB, Peterson RL, Ackerley CA, Ashford AE (1987) Ontogeny of *Eucalyptus pilularis–Pisolithus tinctorius* ectomycorrhizae. II. Transmission electron microscopy. Can J Bot 65:1940–1947

Menotta M, Amicucci A, Sisti D, Gioacchini AM, Stocchi V (2004) Differential gene expression during pre-symbiotic interaction between *Tuber borchii* Vittad. and *Tilia americana* L. Curr Genet 46:158–165

Molina R, Trappe JM (1982) Patterns of ectomycorrhizal host specificity and potential among pacific northwest conifers and fungi. For Sci 28:423–458

Montanini B, Gabella S, Abba S, et al (2006) Gene expression profiling of the nitrogen starvation stress response in the mycorrhizal ascomycete *Tuber borchii*. Fungal Genet Biol 43:630–641

Morel M, Jacob C, Kohler A, et al (2005) Identification of genes differentially expressed in extraradical mycelium and ectomycorrhizal roots during *Paxillus involutus–Betula pendula* ectomycorrhizal symbiosis. Appl Environ Microbiol 71:382–391

Moyersoen B (2006) Pakaraimaea dipterocarpacea is ecto-mycorrhizal, indicating an ancient Gondwanaland origin for the ectomycorrhizal habit in Dipterocar-paceae. New Phytol 172:753–762

Moyrand F, Fontaine T, Janbon G (2007) Systematic capsule gene disruption reveals the central role of galactose metabolism on *Cryptococcus neoformans* virulence. Mol Microbiol 64:771–781

Müller P, Katzenberger JD, Loubradou G, Kahmann R (2003) Guanyl nucleotide exchange factor Sql2 and Ras2 regulate filamentous growth in *Ustilago maydis*. Eukaryot Cell 2:609–617

Nygren CMR, Edqvist J, Elfstrand M, Heller G, Taylor AFS (2007) Detection of extracellular protease activity in different species and genera of ectomycorrhizal fungi. Mycorrhiza 17:241–248

Ohno S (1970) Evolution by gene duplication. Springer, Berlin

Paungfoo-Lonhienne C, Lonhienne TGA, Rentsch D, Robinson N, Christie M, Webb RI, Gamage HK, Carroll BJ, Schenk PM, Schmidt S (2008) Plants can use protein as a nitrogen source without assist-ance from other organisms. Proc Natl Acad Sci USA 18:4524–4529

Powell AJ, Conant GC, Brown DE, Carbone I, Dean RA (2008) Altered patterns of gene duplication and dif-ferential gene gain and loss in fungal pathogens. BMC Genomics 9:147. doi:10.1186/1471-2164-9-147

Prince VE, Pickett FB (2002) Splitting pairs: the diverging fates of duplicated genes. Nat Rev Genet 3:827–837

Rajashekar B, Kohler A, Johansson T, Martin F, Tunlid A, Ahrén D (2008) Expansion of signal pathways in the symbi-otic fungus *Laccaria bicolor* – evolution of nucleotide sequences and expression patterns of protein kinase and RAS GTPase gene families. New Phytol (in press)

Read DJ (1991) Mycorrhizas in ecosystems. Experientia 47:376–391

Read DJ, Perez-Moreno J (2003) Mycorrhizas and nutrient cycling in ecosystems - a journey towards relevance?. New Phytol 157:475–492

Ridley AJ (2001) Rho family proteins: coordinating cell responses. Trends Cell Biol 11:471–477

Ruess RW, Hendrick RL, Burton AJ, Pregitzer KS, Svein-bjornsson B, Allen ME, Maurer GE (2003) Coupling fine root dynamics with ecosystem carbon cycling in black spruce forests of interior Alaska. Ecol Monogr 73:643–662

Rupp LA, Mudge KW (1985) Ethephon and auxin induce mycorrhiza-like changes in the morphology of root organ cultures of Mugo pine. Physiol Plant 64:316–322

Rupp LA, Mudge KW, Negm FB (1989) Involvement of eth-ylene in ectomycorrhiza formation and dichotomous branching of roots of mugo pine seedlings. Can J Bot 67:477–482

Selosse MA, Martin F, Bouchard D, Le Tacon F (1999) Struc-ture and dynamics of experimentally introduced and naturally occurring *Laccaria* sp. discrete genotypes in a Douglas fir plantation. Appl Environ Microbiol 65:2006–2014

Selosse MA, Richard F, He X, Simard SW (2006) Mycor-rhizal networks: des liaisons dangereuses? Trends Ecol Evol 21:621–628

Soanes DM, Richards TA, Talbot NJ (2007) Insights from sequencing fungal and Oomycete genomes: what can

we learn about plant disease and the evolution of pathogenicity? Plant Cell 19:3318–3326

Stajich JE, Birren B, Burns C, Casselton LA, Dietrich F, Fargo DC, Gathman AC, James TY, Kamada T, Lilly WW,Ma L-J, Muraguchi H, Palmerini H, Rehmeyer C, Wilke S, Zolan M, Pukkila PJ (2006) Genomic analysis of *Copri-nus cinereus*. Proc Int Symp Mushroom Sci 2006:59–74

Staples RC (2001) Nutrients for a rust fungus: the role of haustoria. Trends Plant Sci 6:496–498.

Sundaram S, Kim SJ, Suzuki H, et al (2001) Isolation and characterization of a symbiosis-regulated *ras* from the ectomycorrhizal fungus *Laccaria bicolor*. Mol Plant–Microbe Interact 14:618–628

Sundaram S, Brand JH, Hymes MJ, Hiremath ST, Podila GK (2003) Isolation and analysis of a symbiosis-regulated and Ras-interacting vesicular assemby protein gene from the ectomycorrhizal fungus *Laccaria bicolor*. New Phytol 161:529–538

Tagu D, Martin F (1996) Molecular analysis of cell wall pro-teins expressed during the early steps of ectomycor-rhiza development. New Phytol 133:73–85

Tagu D, De Bellis R, Balestrini R, De Vries OMH, Piccoli G, Stocchi V, Bonfante P, Martin F (2001) Immuno-localization of the hydrophobin HYDPt-1 from the ectomycorrhizal basidiomycete *Pisolithus tinctorius* during colonization of *Eucalyptus globulus* roots. New Phytol 149:127–135

Takai Y, Sasaki T, Matozaki T (2001) Small GTP-binding proteins. Physiol Rev 81:153–208

Taylor AFS, Martin F, Read DJ (2000) Fungal diversity in ecto-mycorrhizal communities of Norway spruce [*Picea abies* (L.) Karst.] and beech (*Fagus sylvatica* L.) along north-south transects in Europe. Ecol Stud 142:343–365

Tuskan GA, Difazio S, Jansson S, et al (2006) The genome of black cottonwood, *Populus trichocarpa*. Science 313:1596–1604

Tyler BM, Tripathy S, Zhang X, et al. (2006) *Phytoph-thora* genome sequences uncover evolutionary origins and mechanisms of pathogenesis. Science 313:1261–1266

UniProt Consortium (2008) The universal protein resource (UniProt). Nucleic Acids Res 36:D190–D195

Vesk PA, Ashford AE, Markovina AL, Allaway WG (2000) Apoplasmic barriers and their significance in the exodermis and sheath of *Eucalyptus pilularis-Pisolithus tinctorius* ectomycorrhizas. New Phytol 145:333–346

Voegele RT, Mendgen K (2003) Rust haustoria: nutrient uptake and beyond. New Phytol 159:93–100

Voiblet C, Duplessis S, Encelot N, Martin F (2001) Identi-fication of symbiosis-regulated genes in *Eucalyptus globulus-Pisolithus tinctorius* ectomycorrhiza by differential hybridization of arrayed cDNAs. Plant J 25:181–191

Wapinski I, Pfeffer A, Friedman N, Regev A (2007) Natural history and evolutionary principles of gene duplica-tion in fungi. Nature 449:54–61

Waugh MS, Nichols CB, DeCesare CM, et al (2002) Ras1 and Ras2 contribute shared and unique roles in physiology and virulence of *Cryptococcus neoformans*. Microbiol-ogy 148:191–201

Wennerberg K, Rossman KL, Der CJ (2005) The Ras super-family at a glance. J Cell Sci 118:843–846

Westfall PJ, Ballon DR, Thorner J (2004) When the stress of your environment makes you go HOG wild. Science 306:1511–1512

Whitham TG, Bailey JK, Schweitzer JA, Shuster SM, Bangert RK, LeRoy CJ, Lonsdorf EV, Allan GJ, DiFazio FP, Potts BM, Fischer DG, Gehring CA, Lindroth RL, Marks JC, Hart SC, Wimp GM, Wooley SC (2006) A framework for community and ecosystem genetics: from genes to ecosystems. Nat Rev Genet 7:510–523

Wright DP, Johansson T, Le Quéré A, Söderström B, Tunlid A (2005) Spatial patterns of gene expression in the extramatrical mycelium and mycorrhizal root tips formed by the ectomycorrhizal fungus *Paxillus involutus* in association with birch (*Betula pendula* Roth.) seedlings in soil microcosms. New Phytol 167:579–596

Xu JR, Zhao X, Dean RA (2007) From genes to genomes: a new paradigm for studying fungal pathogenesis in *Magnaporthe oryzae*. Adv Genet 57:175–218

14   Establishment and Functioning of Arbuscular Mycorrhizas

Paola Bonfante[1], Raffaella Balestrini[1], Andrea Genre[1], Luisa Lanfranco[1]

CONTENTS

I.   Introduction............................ 259
II. A Brief Overview of the Colonization Process
     of Plant Roots by Glomeromycota 260
III. Signalling in the Rhizosphere.............. 261
IV. Mycorrhiza-Defective Mutants: a Powerful
     Tool to Decipher the Sym Pathway 263
V.   From the Rhizosphere to the Epidermis 264
     A. Deeper in the Root: Functioning Mycorrhiza .. 266
VI. The Molecular Basis of Nutrient Exchanges .. 267
     A. Plant P Uptake 267
     B. Fungal N Uptake and Delivery.......... 268
     C. The Carbon Cost 270
VII. Conclusions and Perspectives 271
     References............................. 271

I. Introduction

Located at the interface with the soil, plant roots are the preferred niche for many soil fungi that live in the rhizosphere as saprotrophs or are directly associated to the photosynthetic plants as symbionts. Among the latter, arbuscular mycorrhizal (AM) fungi represent a vital component in plant ecosystems: they have a widespread distribution in very diverse environments (Smith and Read 2008) and are present in more than 80% of the land plants, from liverworts to ferns, from gymnosperms to angiosperms. The absence of a strict host specificity is considered a direct consequence of the long coevolutionary history which dates back to the Ordovician (460 million years ago; Remy et al. 1994), when the first land plants, without true roots but with rhizomes, were subjected to mineral nutrient deprivation and forced to associate with heterotrophic soil microbes (Brundrett 2002). The symbiotic nature of the interaction between

plant roots and AM fungi is therefore based on nutritional exchanges, where AM fungi supply the plants with nutrients and in turn receive photosynthetic carbohydrates. Together with the root nodules, arbuscular mycorrhizas (AMs) are considered to be the most important symbioses "that help feed the world" (Marx 2004), since in addition to their role of biological fertilizers, AM fungi increase plant production and confer resistance against biotic and abiotic stresses.

On the basis of their ribosomal gene sequences, AM fungi are grouped in the phylum Glomeromycota, which is a sister clade to Asco- and Basidiomycota (Schüßler et al. 2001). Their uniqueness is mirrored not only by this taxonomical status, but also by their biological traits:

1. They are obligate plant biotrophs, being unculturable in the absence of their host (Declerck et al. 2005).
2. Their concept of species is poorly defined, which reflects a high degree of variability in their functionalities (van der Heijden and Scheublin 2007).
3. Their spores and hyphae possess thousands of syncytial nuclei, making classic genetics approaches unsuitable (Pawlowska 2005).

As a consequence, stable transformation protocols are not yet available, even though Helber and Requena (2007) were successful in introducing fluorescent reporters, like DsRed and GFP, in the AM fungus *Glomus intraradices*. Unlike other fungi, a complete genome analysis of an AM fungus is not yet available, even though the completion of the sequencing project of *G. intraradices* selected for its small genome is not far away (http://darwin.nmsu.edu/ fungi/). The availability of genomics information will pave the way for new scientific approaches aimed at deciphering the particular biological features of AM fungi as well as their lifestyle, which is so closely dependent on their host plant.

---

[1] Dipartimento di Biologia Vegetale, Università di Torino and Istituto Protezione Piante-CNR, Viale Mattioli 25, 10125 Torino, Italy; e-mail: p.bonfante@ipp.cnr.it, r.balestrini@ipp.cnr.it, andrea.genre@unito.it, luisa.lanfranco@unito.it

Plant Relationships, 2nd Edition
The Mycota V
H. Deising (Ed.)
© Springer-Verlag Berlin Heidelberg 2009

In spite of these experimental constraints, AMs are currently part of the mainstream of biology. A combination of genetics, DNA technologies, genomics and cell biology, at the moment mostly applied to the plant side, offers new opportunities to reveal the secrets that allow the establishment and functioning of these ubiquitous beneficial interactions. Many of these aspects are carefully documented in reviews (Harrison 2005; Paszkowski 2006; Bucher 2007; Genre and Bonfante 2007; Reinhardt 2007), in book chapters (Gianinazzi Pearson et al. 2007; Martin et al. 2007) and in special issues (Puhler and Strack 2007).

The main purpose of this chapter is to focus on the events that allow AM fungi to move from the rhizosphere towards the root surface and eventually colonize the root cortex, where the main symbiotic functions are located. This process requires the accomplishment of different events: first, signalling and partner recognition, which lead to a reciprocal compatibility, then, a second set of events which permit the development of the intra-root fungal structures and the functioning of the symbiosis with its multiple benefits. In addition to the improved mineral nutrition, AM fungi in fact play crucial ecological roles, as they have an impact on both the composition, succession and biodiversity of plant species (van der Hejden and Sanders 2002) and on soil structure as a result of improved soil aggregation and increased organic matter (Rillig and Mummey 2006).

As an introduction to the most recent cellular and molecular data, Sect. II summarizes the main features of the colonization process.

II. A Brief Overview of the Colonization Process of Plant Roots by Glomeromycota

The way in which the 150 so far identified species of Glomeromycota colonize more than 200 000 plant species is quite amazing, since – irrespectively of such a huge biodiversity – the colonization mechanism is quite conserved throughout the plant taxa. For this reason, the detailed morphological descriptions of AM roots dating back to the 1970s were quite similar (Bonfante 1984). Less acknowledged information from this data set stated that AM fungi can colonize plant tissues irrespective of their ploidy, since both gametophytic and

sporophytic tissues are involved in the symbiosis (as demonstrated in ferns) and irrespective of the nature of the organ (not only roots, but also rhizomes are colonized, as in *Aglaephyton*; Bonfante and Genre 2008). Interestingly, rhizobial colonization of stem tissues is reported in the literature (Goormachtig et al. 2004), although this is limited to tropical legumes and often left aside in mainstream research on nodulation. This suggests that both symbioses can be considered as current forms of previous interactions where organ specificity was not a rule. However, so-called leaf pathogens (e.g. *Magnaporthe*) can also colonize roots (Sesma and Osbourn 2004) with a developmental programme that largely overlaps their better known behaviour above the ground. Altogether, plant-interacting fungi may be more versatile than generally considered, with regards to their target organs.

As obligate biotrophic microbes, AM hyphae germinating from their large asexual spores can only grow for a few days in the absence of a host plant. Such presymbiotic hyphae, upon recognition of the host, in turn develop infection units that colonize the root epidermis and cortex, in the case where an angiosperm root is considered.

During this presymbiotic phase, AM spores germinate spontaneously in the soil (asymbiotic stage) and develop a germ tube. This grows using the carbon stored in the spore and contacts the epidermal cells, forming appressorium-like hyphopodia (Genre and Bonfante 2007). Following this event, a hyphal peg is produced which enters and crosses the epidermal cell, avoiding direct contact between the fungal wall and host cytoplasm thanks to a surrounding membrane of host origin. Building the interface (Bonfante 2001) therefore starts upon this first penetration event.

Once the fungus overcomes the epidermal layer, it grows inter- and intracellularly all along the root in order to spread fungal structures. During this action, some Glomeromycota (i.e. *Glomus* species) develop structures called vesicles that fill the cellular spaces, probably acting as a storage pool. Only when the fungus reaches the cortical layers, does a peculiar branching process start which leads to highly branched arbuscules (like small trees). These are the key structures of the symbiosis and are considered the site of nutrient exchange (Paszkowski 2006). The development of such massive arbuscules leads to an impressive change in the architecture of the host cell: the nucleus moves from the periphery to the centre of the cell, the vacuole is fragmented, plastids change their organization and a new apoplastic space, based on membrane proliferation, is built around all the arbuscule branches (Fig. 14.1). Arbuscules are ephemeral structures with a life cycle of 4–5 days (Toth and Miller 1984) since, after that time, the arbuscule tips collapse and the plant cell regains its previous organization (Bonfante 1984). However, the whole root colonization process is non-synchronous and early structures, e.g. hyphopodia, coexist with mature symbiotic organs such as arbuscules. It is claimed that, after an arbuscule collapses, the same host cell can undergo a new colonization.

The widespread distribution of AMs, together with their constant structural features (fungal morphogenesis, which is plant-controlled and limited to two major plant tissues, i.e. epidermis, cortex), strongly suggests the existence of a com-mon molecular and genetic determinant which operates across the different plant taxa. The Arum and Paris terminology (Fig. 14.2) introduced by Gallaud (1905) at the start of the past century and proposed by Smith and Smith (1997) and Dickson et al. (2007) to better differentiate among coloni-zation patterns could represent a good starting point to identify the role of plant *versus* fungus genotype.

III. Signalling in the Rhizosphere

The rhizosphere is a dynamic environment in which microbes develop, interact with each other and take advantage of the organic matter released by the root, therefore playing a crucial role in the

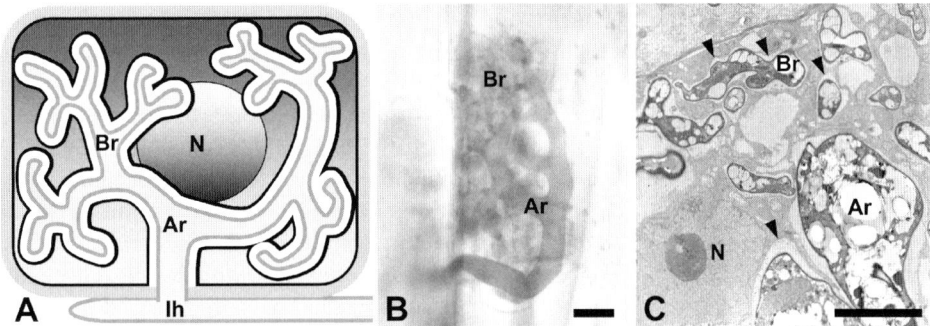

Fig. 14.1. Arbuscule and interface compartment. **A** Model showing the organization of the interface compartment around intracellular fungal structures in an arbusculated cell: a thin sheath of plant cell wall materials (*white*), coated by the perifungal membrane (*black line*), surrounds the whole arbuscule (*grey*). *Ar* Arbuscule, *Br* branches, *Ih* intercellular hypha, *N* nucleus. **B** Optical microscopy image of an arbuscule inside a root cortical cell. *Bar* 10 μm. **C** Transmission electron microscopy image showing the interface space (*arrowheads*) around intracellular fungal structures. *Bar* 5 μm

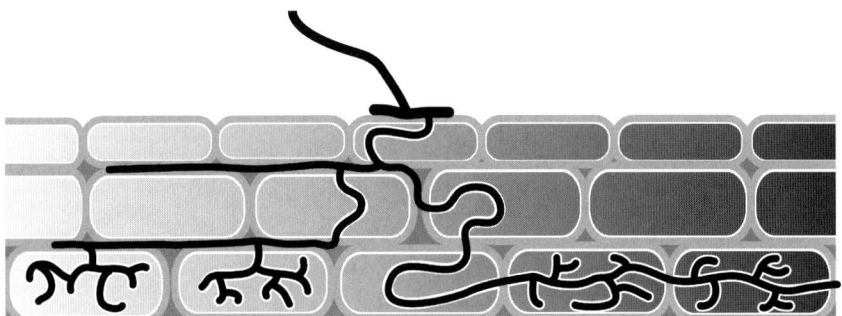

Fig. 14.2. Scheme showing the main characteristic traits of Paris- and Arum-type colonization patterns in arbuscular mycorrhizas. The Paris type (*left*) is charac-terized by intercalary arbuscules, frequent intracellular coils and limited intercellular development. Arum-type mycorrhizas (*right*) usually have terminal arbuscules, limited intracellular coils and abundant intercellular hyphae. Nonetheless, a whole range of intermediate patterns is present in nature, which depend on the plant/fungus combination

field of mycorrhizal research (Martin et al. 2007). Signalling molecules produced early in the interaction between symbiotic fungi and their host plants elicit responses in the partners as the first step in the cascade of events leading to contact at the host surface and eventual symbiosis. Identifying the signalling molecules that are active in the rhizosphere and which regulate the information flow between mycorrhizal fungi and host roots is currently an area of intense research.

As far as AM fungi are concerned, important breakthroughs have been made concerning the mechanisms that govern signalling and recognition with their hosts. Among the root exudates which may act as plant signals in the rhizosphere (for a careful review, see Gianinazzi et al. 2007), a plant molecule identified as a strigolacton, 5-deoxy-strigol (Akyama et al. 2005) causes a specific phenotype in AM fungi, i.e. an extensive branching of the germinating mycelium, similar to that usually shown in the vicinity of host roots, prior to the formation of the hyphopodium (Giovannetti et al. 1993). Interestingly, strigolactones are known to stimulate the seed germination of parasitic plants and to inhibit shoot branching, acting therefore as a plant hormone (Gomez-Roldan et al. 2008). Besserer et al. (2006) found that a strigolactone from a monocotyledonous plant, *Sorghum,* strongly and rapidly stimulated the proliferation of the AM fungus *Gigaspora rosea,* at concentrations as low as 10^{-13} M. Within 1 h of treatment, the density of mitochondria in the fungal cells increased, and their shape and movement changed dramatically. Strigolactones derive from the carotenoid pathway and belong to the group of apocaroteinoids that are identified as the yellow pigments often present in mature mycorrhizas (Walter et al. 2007): these new results suggest that apocarotenoids play important and distinct roles in mycorrhizas.

However, plants perceive diffusible fungal signals, even in the absence of physical contact. Due to the many analogies between Rhizobium/legumes and AM symbioses (Hirsh and Kapulnik 1998), the hypothesis that Myc factors akin to Nod factors may be produced by AM fungi has been developed since 1994 (LaRue and Weeden 1994). The presence of such Myc factor(s), whose chemical nature is still unknown, was demonstrated in experiments where a membrane prevented physical contact between the two partners, while allowing signal exchange. The factor(s) induces the activation of the nodulation-inducible gene

ENOD11 in transformed roots of *Medicago truncatula* (Kosuta et al. 2003), other genes involved in signal transduction (Weidmann et al. 2004) and the development of lateral roots (Olah et al. 2005), suggesting that the plant is timely informed of fungal growth in the rhizosphere.

However, how plant and fungus perceive and transduce these symbiotic signals is not fully understood at the moment. In analogy with the *Rhizobium*–legume symbiosis (Oldroyd and Downie 2006), calcium is often hypothesized to be involved in AM signal transduction. Ca^{2+} is the most common intracellular messenger that couples an array of extracellular stimuli to specific physiological responses. Transient elevations of cytosolic free Ca^{2+} concentrations ($[Ca^{2+}]_{cyt}$) can be recorded in cells undergoing biotic or abiotic stresses, and encode information which is subsequently transduced in a cascade of cellular events (Sanders et al. 2002). By using aequorin-transformed soybean cells treated with the germination medium of *Gigaspora margarita* and *Glomus intraradices* spores, a rapid and significant increase in cytosolic Ca^{2+} level was induced and dissipated within 30 min (Navazio et al. 2007; Fig. 14.3). The medium induced neither the accumulation of reactive oxygen species nor programmed cell death and did not induce any Ca^{2+} transient in aequorin-transformed *Arabidopsis* cells. These data indicate that fungal signals are released in the rhizosphere even before any plant-derived cue can influence the fungus. The transient Ca^{2+} elevation can be considered as an initial event in the response cascade leading to the integration of the symbionts. In addition, the absence of defence responses in

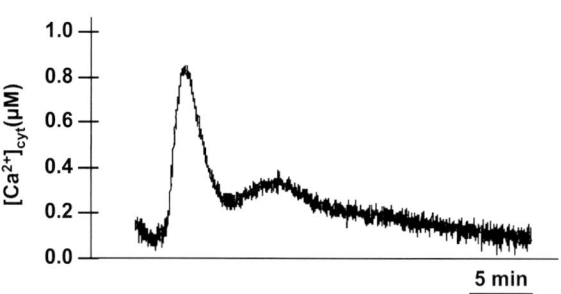

Fig. 14.3. Transient variation of cytosolic Ca^{2+} concentration in soybean cultured cells expressing aequorin, when challenged with the culture medium of germinated *Gigaspora margarita* spores (kindly provided by Dr. Lorella Navazio, University of Padova, Italy)

the host and Ca²⁺ transients in the non-host plant suggests that hosts perceive AM fungi as compatible partners from the very beginning.

The identification of the Myc factor(s) surely represents a crucial challenging step. Its availability as a defined molecule(s) will allow the hypotheses so far advanced to be tested, the fungus-related signalling pathway to be identified and a potential key to understand the absence of host-specificity to be obtained (Fig. 14.4).

IV. Mycorrhiza-Defective Mutants: a Powerful Tool to Decipher the Sym Pathway

The perception of fungal signals at the cell periphery and their subsequent transduction into a cascade of events is expected to lead to the activation of specific genes which permit the formation of a functional mycorrhiza. Such a hypothesis has been tested in AMs, using plant mutants which are unable to form symbioses. These mutant lines in fact provide a powerful tool to identify genetically defined steps in the development of the symbiotic interaction. The genetic dissection of AM development was pioneered through the isolation of pea mutants impaired in AM symbiosis. These mutants, initially identified through their altered root nodule symbiosis with *Rhizobium*, were also affected in AM symbiosis (Duc et al. 1989). This finding demonstrated an overlap in the genetic programmes for the two symbiotic interactions

and opened a new operational scenario, leading to the discovery and identification of the so-called "*sym*" genes that are essential for the establishment of both microbial symbioses (Parniske 2008; Oldroyd and Downie 2006). The analysis of mutants that are defective in nodule and AM formation in legumes allowed the identification of a number of components of the signal perception and transduction pathway which are shared by both symbioses. Among the most important common components is a leucine-rich-repeat receptor-like kinase (LRR-RLK), which is thought to transduce the perception of the Nod and Myc factor, through its intracellular kinase domain, leading to the activation of K⁺ ion channels, coded by *DMI1* in *M. truncatula* and by the *Castor* and *Pollux* couple in *Lotus japonicus*. This symbiotic cascade, which also includes new nucleoporin members (Kanamori et al. 2006), leads to the so-called electrochemical prelude, including membrane depolarization, calcium influx and calcium spiking. These prolonged oscillations of intracellular calcium concentration are well characterized in legume–rhizobium interactions (Oldroyd and Downie 2006) and in mycorrhizal roots (Kosuta et al. 2008). A calcium- and calmodulin-dependent protein kinase coded by *DMI3* acts downstream of the calcium spiking and interprets the calcium signal inducing the transcription of symbiosis-specific genes (Oldroyd and Downie 2006 and references therein). Models illustrating such a sym pathway are often modified as new components are added to this complex puzzle (Fig. 14.5). These genetics data strongly suggest that the sym pathway includes

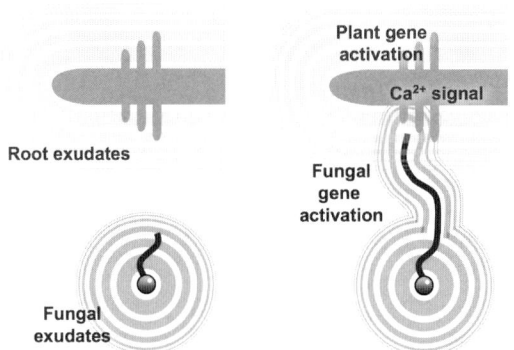
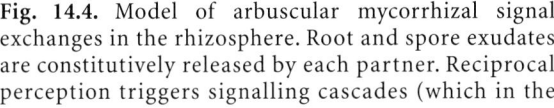
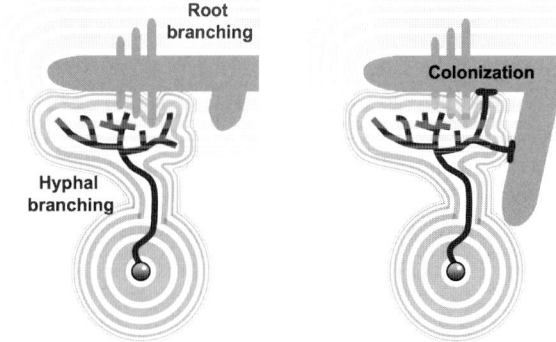

Fig. 14.4. Model of arbuscular mycorrhizal signal exchanges in the rhizosphere. Root and spore exudates are constitutively released by each partner. Reciprocal perception triggers signalling cascades (which in the plant involves calcium) and leads to the regulation of plant and fungal genes. Responses include root and hyphal branching, which in turn increases the chances of direct interaction and successful colonization

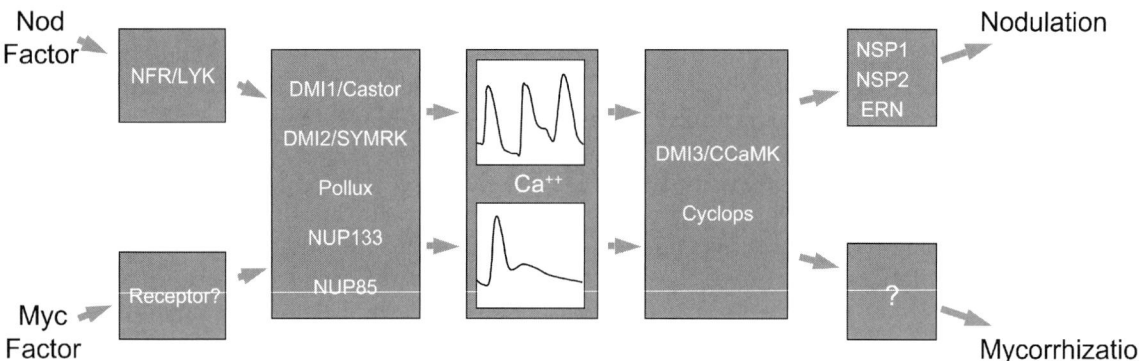

Fig. 14.5. Model of the nodulation and mycorrhization signaling pathways as defined in the model legumes *Medicago truncatula* and *Lotus japonicus*. NFR/LYK genes are candidates for the Nod-factor receptor. DMI1/Castor, DMI2/SYMRK, Pollux, NUP133 and NUP85 are common to both pathways and located upstream the calcium signal. DMI3/CCaMK, also required for both mycorrhization and nodulation, act downstream of this calcium response. Finally, NSP1, NSP2 and ERN represent a nodulation-specific branch downstream calcium spiking. Mycorrhiza-specific genes are very likely to act both upstream and downstream of the common SYM pathway

components which are conserved for both symbioses, while others are specialized. In addition, the phenotypic characterization of mutants (Novero et al. 2002, Demchenco et al. 2004) and some cellular traits, like nuclear movement (Genre et al. 2005), and molecular responses, like *ENOD11* induction (Kosuta et al. 2003) persisting in the mutant *DMI3*, suggest that parallel signalling pathways operate to establish a mycorrhizal root (Parniske 2008).

V. From the Rhizosphere to the Epidermis

Root responses to AM fungi can be described according to the cell type involved, but the notion that epidermal and cortical cells may be programmed differently is a relatively new concept (Genre and Bonfante 2005). The root epidermis is the first barrier to all soil micro-organisms during their colonization process and it is therefore likely to be the site of recognition mechanisms as well as accommodation/defence responses (Parniske 2008). As already mentioned, mycorrhizal mutant plants indicate the importance of epidermal cells in the early steps of AM establishment. Rather than a passive barrier, epidermal cells should be considered as an active checkpoint where signal exchanges and a strong control over root colonization occur (Novero et al. 2002; Demchenko et al. 2004). Direct evidence of this emerged from the recent description of pre-penetration responses, which design the track that the fungus

will follow through the host cell a few hours before cell penetration (Genre et al. 2005). In creating an intracellular niche, the AM interaction resembles all other cases of endo-symbioses, where the guest organisms are confined to specialized membrane-bordered spaces. The impact of AM fungi on root cell contents promoted investigations into the role of the plant cytoskeleton in cell colonization (Takemoto and Hardham 2004). Whilst increases in the complexity of microtubule (Genre and Bonfante 1997; Blancaflor et al. 2001) and microfilament (Genre and Bonfante 1998) organization are striking in arbuscule-colonized cortical cells, evidence for the pivotal role of the cytoskeleton, even during the early stages of fungal accommodation, was first provided from studies of the *L. japonicus Ljsym4-2* mutant. When an AM fungus achieves epidermal cell penetration of mutant roots, the absence of a correct cytoskeletal response leads to plant cell death and colonization arrest (Genre and Bonfante 2002). Taken as a whole, host cell reorganization can be explained by the need on the one hand to preserve cell integrity and on the other to optimize reciprocal compatibility. In particular, the repositioning of the cytoskeleton, endoplasmic reticulum and Golgi bodies, all of which are involved in membrane proliferation and cell wall deposition, is most likely related to the novel construction of the interface compartment.

Live cell imaging of epidermal cells, the most easily accessible through direct microscopic observation, provides an important means of understanding the mechanisms of interface construction in AM interactions. The first description of *in vivo* epidermal cell responses to fungal contact was recently obtained through *in vivo* imaging of GFP-labelled cell components in mycorrhizal root organ cultures using confocal microscopy (Genre et al. 2005).

The main discovery from this research concerns the observation of a novel, ephemeral apparatus, the pre-penetration apparatus (PPA), which is considered to be responsible for assembling the interface compartment. The PPA is organized in the epidermal cell as soon as the hyphopodium develops on its surface, and it remains visible for a few hours, until fungal penetration occurs (Fig. 14.6). Hyphopodium contact triggers the repositioning of the plant cell nucleus at the contact site within a couple of hours. This repositioning is accompanied by a reorganization of the cytoskeleton and the assembly of endoplasmic reticulum (ER) patches. The nucleus then starts to migrate towards the cell wall that faces the cortex. This movement is accomplished in 2–4 h and leads to complete development of the PPA, resulting in a column of cytoplasm that links the nucleus to the contact site and contains a very high density of cytoskeletal fibres and endoplasmic reticulum cisternae. When nuclear migration terminates, a fine membranous thread is present in the middle of the PPA. Only then does fungal penetration

occur and it exactly follows the route traced by the PPA, which then starts to disassemble about 6–8 h after hyphopodium contact.

These observations unambiguously demonstrate an active control by the plant cell over the infection process. Fungal development is arrested at the hyphopodium stage until the PPA is completed. In addition, intracellular fungal growth is confined to the pre-formed compartment. The homology between the PPA and the *Rhizobium*-induced infection thread is a visible sign of the evolutionary relationship between the two symbioses. The existence of the PPA also leads to several new questions about AM interactions (Smith et al. 2006), especially concerning the specificity of such a response to AM fungi or to any root-penetrating micro-organism, or the nature of the local signal(s) triggering cell polarization and PPA orientation. These observations would in fact suggest the existence of a specific fungal signal that orientates plant responses by shifting them towards fungal accommodation and, as a first requirement, interface assembly.

The activation of at least one gene (*MtENOD11*) together with PPA formation (Genre et al. 2005), prompted an additional question concerning whether other transcripts could be connected to this transient structure. Using PPA as a cellular marker to identify the root areas that actively respond to early

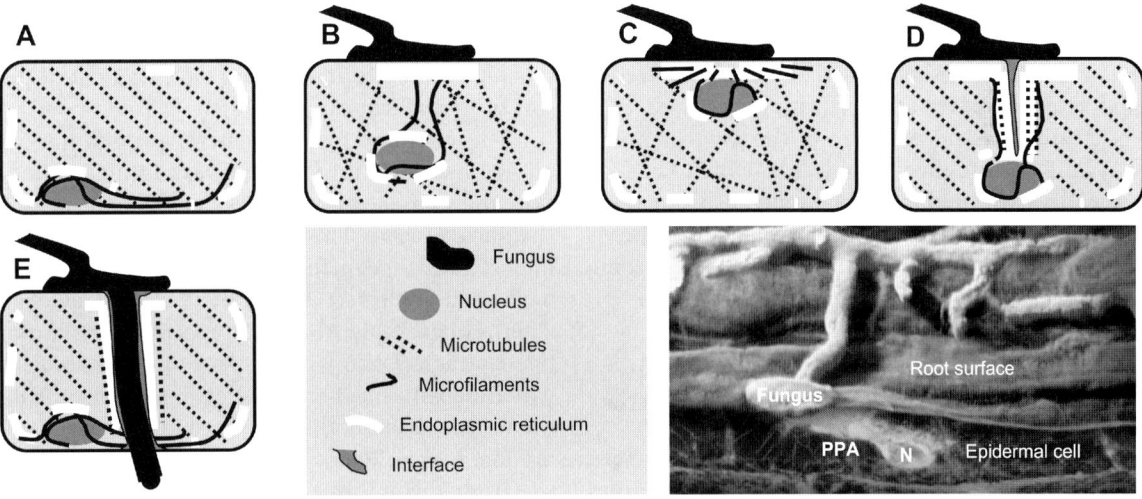

Fig. 14.6. Diagram showing the development of the prepenetration apparatus (*PPA*, steps **A–E**) in root epidermal cells contacted by an arbuscular mycorrhizal fungus. Hyphopodium adhesion (**B, C**) triggers cytoskeleton reorganization, nuclear repositioning at the contact site and local accumulation of endoplasmic reticulum (ER). A second nuclear migration across the cell (**D**) is associated with the formation of a trans-cellular column of cytoplasm, containing cytoskeletal bundles, ER and a central membranous thread. Finally (**E**), fungal penetration occurs, with the penetrating hypha closely following the PPA track across the cell lumen. The *inset* shows a composition of confocal images showing PPA assembly in an epidermal cell underneath the hyphopodium contact site

fungal contact, Siciliano et al. (2007) constructed an SSH library where genes were identified as differentially expressed in *Medicago* roots after the fungal contact *versus* control roots. The differential expression was confirmed for 15 genes by reverse Northern blot analysis. Their expression profile was analysed by real-time PCR and compared both with that of non-inoculated *M. truncatula* and with that of *dmi3-1* mutant plants, which – as already mentioned – lack a Ca^{2+}/calmodulin-dependent protein kinase, do not establish mycorrhizas (Lévy et al. 2004) and never produce PPAs (Genre et al. 2005). The experiments demonstrated that at least two novel genes (encoding for expansin-like and nodulin-like proteins) could be used as good markers of PPA formation. These are significantly up-regulated in the root fragments that contain such structures, compared with the control and the *dmi3-1* mutant. The mRNA of one of them, the expansin-like protein, was preferentially detected in epidermal cells following fungal contact, using *in situ* hybridization. As a second important finding, an Avr9/Cf-9 rapidly elicited protein 264 was found to be up-regulated in the *dmi3-1* mutant plant, suggesting that this gene is under negative control of DMI3 and leading to new perspectives on the mechanisms which control compatibility in AM interactions.

As a counterpart of the plant strigolactones, perceived by AM fungi as signals of the presence of a potential host, the soluble Myc factors and/or the fungal signals involved in the physical contact with the epidermal cells are considered to be the main actors of the signalling dialogue that precedes and accompanies root colonization.

To understand whether, at the contact point, AM fungi release effector molecules which – like the microbe-associated molecular patterns from pathogen microbes – bind to specific plant receptors is one of the key questions involved in deciphering communication in the mycorrhizal symbiosis.

A. Deeper in the Root: Functioning Mycorrhiza

The establishment of a functional mycorrhiza requires co-ordinated developmental programmes from both partners. Once the root epidermis is crossed following colonization patterns that mirror the plant anatomy (Genre and Bonfante 2007), the fungus is ready to spread the infection by forming intercellular hyphae and eventually intracellular, highly branched structures called arbuscules,

which are usually located in the inner cortical cells. These structures, which give their name to the symbiosis, are considered the preferential site of the nutrient exchanges between the partners. In arbuscule-containing cells, the fungus comes into intimate contact with the host, even though they are still physically separated by a surrounding membrane of host origin, the periarbuscular membrane (Bonfante 2001). The creation of such a functional interface is one of the landmarks that clearly define the symbiotic nature of plant/AM association.

Despite the crucial role of arbuscules in AM symbiosis, the cellular and molecular bases of arbuscule morphogenesis (formation, differentiation, senescence, dismantling) are still largely unknown. The cortical cell responses that precede and accompany arbuscule development have been recently outlined (Genre et al. 2008). Such results suggest that the PPA mechanism is conserved and modulated in cortical cells: as a cytoplasmic column, the cortical PPAs anticipate the track followed by the advancing fungus and elicit the host cell reorganization that is observed in the presence of developed arbuscules. Metabolic activities at the arbuscule interface seem to be crucial to sustain the colonization process. In fact, a *Medicago truncatula* inorganic phosphate (Pi) transporter (MtPT4) described as exclusively expressed during AM symbiosis (see next paragraph) and located in the periarbuscular membrane (Harrison et al. 2002) is not only essential for the acquisition of Pi delivered by the AM fungus but it is also required to maintain arbuscule life and sustain development of the AM fungus (Javot et al. 2007a).

The complex morphogenetic programme of an AM fungus, culminating in arbuscule production, therefore seems to be under the control of a specific cell type. This provides a rationale to the important changes in fungal metabolic pathways that have been described (Pfeffer et al. 1999; Bago et al. 2003). For example, AM fungi present quite distinctive characteristics with respect to C metabolism which depend on their location inside or outside the root (Pfeffer et al. 1999; Bago et al. 2002, 2003; Trépanier et al. 2005).

High-throughput genetic, molecular and biochemical analyses provided important insights into many aspects concerning the functioning of AM symbiosis over the past decade. Transcriptomic approaches revealed major changes in gene expression that accompany the formation of AM symbiosis and in particular identified a wide spectrum of plant genes induced or repressed in response to AM fungi (Balestrini and Lanfranco 2006). Most of these studies focused on the late stages of the colonization. Spatial expression patterns revealed the exclusive expression of a number of genes in the cortical cells containing arbuscules, suggesting that a mycorrhiza-specific signal, operating in cell autonomous fashion, activates the expression of these genes. However, other genes are expressed both in arbuscule-containing cells and in cortical cells in the vicinity of arbusculated cells, suggesting the presence of a second mobile signal(s) acting in the colonized region of the roots (Harrison 2005).

Advanced technologies, such as *in vitro* culture systems and methods based on polymerase chain reaction, offered the opportunity of conducting a molecular investigation of the fungal partner too, thus overcoming the problem of the limited availability of biological material. However, knowledge on fungal transcriptional profiling during the root colonization phase is limited. Due to the limited amount of fungal biomass inside the root, investigations performed on RNA from mycorrhizal roots could only identify a few highly constitutively expressed fungal genes (Balestrini and Lanfranco 2006).

VI. The Molecular Basis of Nutrient Exchanges

Since one of the main aims of this chapter is to provide a link between signalling events leading to the establishment of symbiosis and its activities, only those functions related to nutrient exchanges are considered. Important modifications in secondary metabolism, as reviewed by Walter et al. (2007), are not discussed here since a clear function to the symbiosis has not yet been identified.

A. Plant P Uptake

Mycorrhizal plants acquire Pi directly from the soil through plant specific phosphate transporters (PTs; the direct uptake pathway) or through uptake and transport systems of the fungal symbiont (the mycorrhizal uptake pathway). It was demonstrated that, in most cases, there is a preferential uptake via fungal hyphae (Smith et al. 2003) and

that the phosphate is eventually delivered by the arbuscules to the cortical cells. These cells are particularly efficient in Pi uptake, due to the induction of mycorrhiza-specific or mycorrhiza-inducible PTs (Bucher 2007). Current data suggest that Pi, taken up by the extraradical mycelium from soil solutions, is translocated through the AM fungal hyphae as polyphosphate (poly-Pi) and, after hydrolysis in the arbuscule, the Pi is exported from the AM fungus to the periarbuscular space (Javot et al. 2007b). The import of Pi across the periarbuscular membrane into the cell is then mediated by plant transporters. The study of plant PTs and of their functioning thus represents a key point to fully understand this process. In recent years, several PT genes from different plants were described and their expression characterized in mycorrhizal roots with the identification of Pht1 transporters induced in AM symbiosis (for reviews, see Bucher 2007; Javot et al. 2007b). These can be divided into two categories: those expressed strictly in response to AM symbiosis (mycorrhiza-specific) and those strongly induced by AM symbiosis but having a basal expression in non-mycorrhizal roots (mycorrhiza-inducible). All the mycorrhizal plants examined possess at least one mycorrhiza specific PT, with a variation in number according to species (Javot et al. 2007a). On this basis, plant PTs are currently considered markers of a functional symbiosis (for a review, see Bucher 2007). However, other members of the PT family, which are responsible for the direct uptake pathway, can show a down-regulation during AM symbiosis (Versaw et al. 2002; Hohnjec et al. 2005).

Recent data, based on the expression analysis of laser-dissected cells, demonstrate that the five so far identified *LePT* genes in tomato (Nagy et al. 2005) are differentially expressed in three cortical cell populations during interaction with *G. mosseae* (Balestrini et al. 2007). All tomato PT genes are consistently expressed inside arbusculated cells, suggesting that plants guarantee maximum P uptake through activation of the whole gene family. On the fungal side, the presence of *GmosPT* transcripts in the arbusculated cells offers a sound confirmation of the results obtained by Benedetto et al. (2005) on the whole mycorrhizal root. The efflux of phosphate probably occurs in competition with its uptake and the fungus might exert control over the amount of phosphate delivered to the plant. The discovery that five plant and one fungal PT genes are consistently expressed inside arbusculated cells provides a new scenario for plant–fungus nutrient exchanges.

In spite of the number of mycorrhiza-inducible Pi transporters identified to date, protein localization in the interaction between AM fungi and host plants is not so clearly understood. It would be interesting to investigate whether the presence of five mRNAs in arbusculated tomato cells mirrors a differential localization of PT proteins along the plasma membrane of a single arbusculated cell. An asymmetric distribution of PTs can be hypothesized: the myc-specific transporters could be associated to the periarbuscular membrane, as already demonstrated for MtPT4 (Harrison et al. 2002), while the constitutively expressed LePT1 and LePT2 could be confined to the peripherical plasma membrane (Balestrini et al. 2007; Fig. 14.7).

Apart from expression profiles, knock-out mutants of PT genes add to their functional characterization. The first functional study of a mycorrhiza-inducible Pi transporter, LePT4 from tomato, revealed considerable redundancy between the PT proteins at this interface in solanaceous species and showed that LePT4 was dispensable (Nagy et al. 2005). Consequently, multiple gene mutants for *LePT3*, *LePT5* and *LePT4* are needed to obtain a more comprehensive understanding of the molecular regulation of this process. The role of LePT4 in tomato was recently investigated through the molecular and physiological characterizations of a loss-of-function homozygous mutant *lept4* (Xu et al. 2007). The data provide evidence that LePT4 is involved in Pi acquisition in both uptake pathways, i.e. direct or mycorrhizal routes, and suggest that the loss of the LePT4 function is only partially compensated by other members of the family.

Due to absence or a reduced level of gene redundancy, studies on legumes lead to more consistent results. Javot et al. (2007a), using transgenic lines in which the only AM-specific PT gene, *MtPT4*, is down regulated by RNA interference (RNAi) or knock-out mutants obtained from classic mutagenesis, clearly demonstrated that MtPT4 is essential for symbiotic Pi transport. Similar results were obtained using an RNAi approach for *Lotus japonicus* AM-inducible PT (*LjPT3*; Maeda et al. 2006).

Further important information provided by these studies is that the intraradical fungal colonization is reduced, indicating that P transport is critical for the development of AM symbiosis (Maeda et al. 2006; Javot et al. 2007a). It is sug-

gested that the import of Pi by MtPT4 serves as a signal to permit a continuous development of the arbuscule and consequently sustain fungal existence within the root.

The molecular signals responsible for the activation of the PT genes in an arbuscule specific manner are still unknown. Lysophospholipids were recently demonstrated to act as diffusible signals in controlling the activation of a P transporter gene associated with arbuscule development (Drissner et al. 2007). Interestingly, a role of MicroRNA399 in Pi-signalling was recently reported (Aung et al. 2006; Bari et al. 2006) and in silico analysis suggests that Pht1 Pi transporters are potential candidates for regulation by members of the MicroRNA399 family (Chiou et al. 2006). As Pht1 members are induced in mycorrhizal roots, it will be interesting to see whether gene expression regulation through microRNAs occurs in AM symbiosis.

B. Fungal N Uptake and Delivery

Although Pi acquisition receives more attention, important advances in the understanding of nitrogen movement in AMs have been made in recent years. AM fungi directly take up ammonium, nitrate and amino acids (Hawkins et al. 2000). Several studies that mainly focused on inorganic N led to a model that is already supported by experimental evidence.

The first step in N acquisition requires the activity of specific transporters located at the interface between the soil and the extraradical mycelium. Although AM fungi are able to take up both NO_3^- and NH_4^+, a clear preference for NH_4^+ has been demonstrated (Lopez-Pedrosa et al. 2006). The only gene involved in this process and characterized so far is a high-affinity NH_4^+ transporter gene (*GintAMT1*; Lopez-Pedrosa et al. 2006). *GintAMT1* is expressed in the extraradical mycelium in particular when NH_4^+ is present in the surrounding environment at micromolar concentrations.

The glutamine synthetase/glutamate synthase (GS/GOGAT) cycle is possibly responsible for subsequent NH_4^+ assimilation in AM extraradical hyphae (Johansen et al. 1996; Breuninger et al. 2004), although the involvement of glutamate dehydrogenase has not been experimentally excluded.

Inorganic N taken up by the extraradical mycelium is then incorporated into amino acids

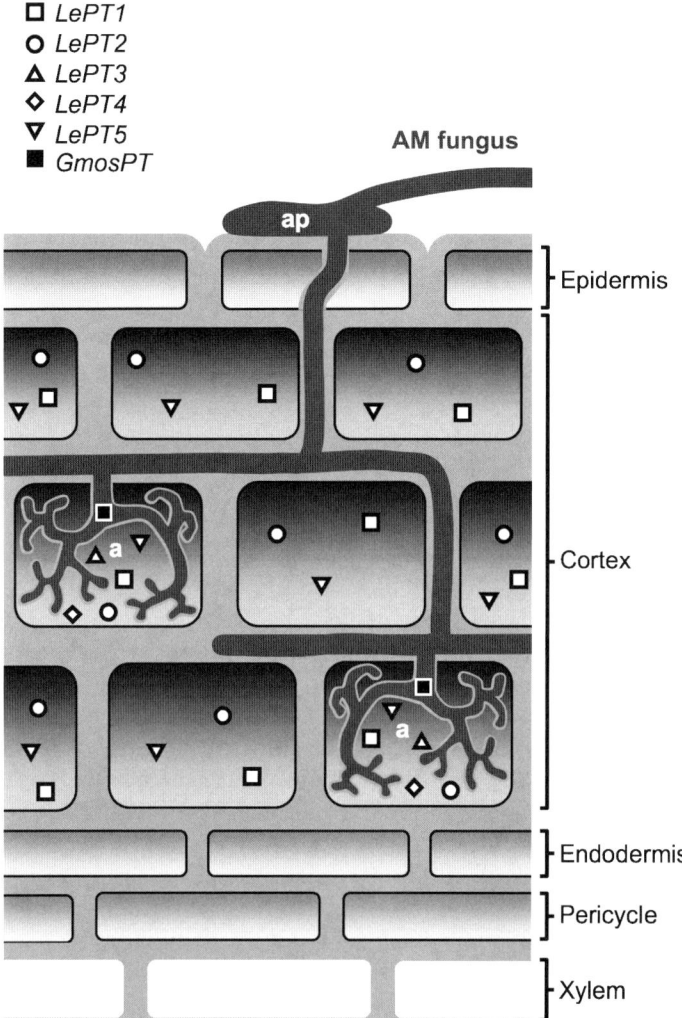

Fig. 14.7. Schematic view of the distribution of plant and fungal phosphate transporter transcripts in an arbuscule-containing cell. *a* Arbuscule, *Hp* hyphopodium

and translocated to the intraradical mycelium, mainly as arginine since this is the predominant free amino acid in the external hyphae (Jin et al. 2005).

Ammonium, produced by the catabolic branch of the urea cycle, is the most likely form of N transferred from fungus to plant. This also ensures that the fungus does not lose any C skeleton. This proposed mechanism requires that enzymes for N assimilation are expressed differently in the extraradical and intraradical mycelia. In fact, quantitative real-time PCR assays show that a gene of primary nitrogen assimilation

(glutamine synthase) is preferentially expressed in the extraradical hyphae, whereas genes involved in arginine breakdown (urease accessory protein, ornithine amino transferase) and NH_4^+ transfer (ammonium transporter) are more highly expressed in the intraradical mycelium (Govindarajulu et al. 2005). Biochemical analyses of enzymatic activities of glutamine synthetase, argininosuccinate synthetase, arginase and urease in extraradical mycelium and AM roots also support the role of arginine as a key component in N translocation in the AM mycelium (Cruz et al. 2007).

Little is known about the organic nitrogen uptake capability of AM fungi. Hawkins et al. (2000) showed that AM hyphae are able to take up glycine and glutamic acid and transport nitrogen from these sources to the plant roots. Furthermore, AM symbiosis was shown to both enhance decomposition of and increase nitrogen capture from organic patches (Hodge et al. 2001). Recent molecular investigations led to the identification of an amino acid permease in *Glomus mosseae* (Cappellazzo et al. 2008) that is exclusively expressed in the external hyphae. In addition, an untargeted approach based on the construction of a subtractive cDNA library showed that *G. intraradices* extraradical structures respond to organic N limitation with a transcriptional activation of genes involved in different functional categories and supported the role of N as a signalling molecule in AM fungi (Cappellazzo et al. 2007).

It is not known how the plant cell takes up NH_4^+ released from the fungus or whether arbuscules are directly involved in this process. Both targeted approaches (Hildebrandt et al. 2002) and transcriptome profiling in different plant species (Frenzel et al. 2005; Guimil et al. 2005; Hohnjec et al. 2005) revealed that the expression of a number of genes coding for ammonium and nitrate transporters is enhanced in AM roots. These findings suggest mechanisms that support not only ammonium uptake but also nitrate acquisition during AM symbiosis. However, some members of the nitrate transporter gene family can also be downregulated during AM symbiosis (Burleigh 2001). A similar situation of up- or down-regulation of genes belonging to the same family is also reported for Pi transporters (see above).

There is a clear need to investigate the spatial and temporal distribution of these transporters in more detail in order to verify to which extent similarities with mycorrhiza-specific Pi transporters can be found. Phosphate and nitrogen are transferred by similar mechanisms in AM roots. Poly-P or Arg are stored in vacuoles and Pi or NH_4^+ are translocated along hyphae from the extra- to the intra-radical mycelium and released to the plant apoplast.

C. The Carbon Cost

During the symbiotic phase, the C metabolism of both partners is significantly modified and this is mirrored at the level of gene expression. Intraradical fungal structures (presumably the arbuscules) are known to take up photosynthetically fixed plant C as hexoses (Pfeffer et al. 1999; Douds et al. 2000). With the exception of the gene recently described in the glomeromycotan *Geosiphon pyriforme* (Schüßler et al. 2006) which forms symbiosis with a cyanobacterium, no transporter responsible for the uptake of hexose released by host cells has so far been characterized in AM fungi. Since this is a crucial point of symbiosis, the *Geosiphon* gene could be helpful for identifying the corresponding genes in AM fungi.

This C flux requires plant sucrose-cleaving enzymes to provide the intraradical mycelium with hexoses. Enhanced gene expression of sucrose synthase was observed in AM colonized roots (Hohnjec et al. 2003; Ravnoskov et al. 2003), leading to the hypothesis that sucrose synthase is involved in generating a sink strength. More recently, it was been shown that both transcripts for an apoplastic invertase and activity of apoplastic invertases are increased near fungal structures and in phloem cells (Schaarschmidt et al. 2006). In addition, reduced apoplastic invertase activity leads to a diminished mycorrhization level (Schaarschmidt et al. 2007a). The invertase activity in the whole plant therefore seems to influence the extent of fungal colonization, probably by providing the fungus with a higher hexose level (Schaarschmidt et al. 2007b).

What is the fate of hexose once it has entered the AM fungus? Triacylglycerols (TAG) and glycogen, the main C storage compounds, are synthesized by the fungus inside the root and then transferred to the extraradical mycelium (Bago et al. 2002, 2003). The results of gene expression studies performed on a putative acyl-coenzyme A dehydrogenase (Bago et al. 2002) and a glycogen synthase (Bago et al. 2003) were consistent with the results obtained with ^{13}C labelling assays. In contrast, genes coding for enzymes of the glyoxylate cycle (isocitrate lyase and malate synthase) were actively expressed in the extraradical mycelium where the fungus, unable to take up hexoses, converts lipids into carbohydrates via this metabolic pathway (Lammers et al. 2001).

Lipid metabolism seems to be a crucial aspect of AM fungi biology. It was in fact recently demonstrated that *de novo* fatty acid synthesis only occurs in the intraradical mycelium and not in germinating spores or in the extraradical mycelium

(Trépanier et al. 2005). This regulation could be one of the reasons of the obligate biotrophism of AM fungi.

This short overview of the metabolic pathways activated in AM roots demonstrates the intimate interdependency of the two partners: on one hand, the fungus has a strong impact on the most important metabolic pathways involved in mineral nutrition of the colonized roots while, on the other hand, the fungus changes its metabolism, depending on the stage of interaction and uses P, N and C not only as nutrients but also as signals for its morphogenesis.

VII. Conclusions and Perspectives

AMs were first described more than 100 years ago (Gallaud 1905) and became objects of experimental research in the 1960s, but only from the 1980s onwards were they acknowledged as a central topic of research that crosses multiple scientific areas. However, in spite of the efforts of the scientific community, our knowledge of the mechanisms that control AM symbiosis and its ecological success is still quite rough. Based on findings with legume model plants, we learn that such mechanisms are multiple and operate according to well defined time-scales: signalling and recognition between partners may take from only a minute to a few hours, successful contacts with cellular responses and gene activation require hours, root colonization and nutrient exchange which – as late events – may take days or weeks. Our current hypothesis is that the plant is the key actor, but we should be aware that this view may be due to our lack of knowledge of fungal biology. For example, at cell level, we lack knowledge about the fungal secretome or the role of fungal molecules in the establishment of compatibility. The metabolic changes described in mycorrhizal roots upon fungal colonization (Güimil et al. 2005; Küster et al. 2007; Liu et al. 2007) and correlated with nutrient exchanges represent a unique aspect of mycorrhizal symbiosis. It is not obvious how to identify models that can be used for comparison purposes. In pathogenic systems, nutrient flow is mostly unidirectional and limited in time while, in *Rhizobium* symbiosis, microbes are located within a nodule, which is a specific niche that is very different from a root. A further level of complexity is given by the variability in the performance of different plant species or cultivars following mycorrhizal colonization. The identification of the mechanisms that regulate such events will be a challenging objective for a better exploitation of AMs in ecology and agricultural programmes.

Acknowledgements. Contributions to this chapter have partly been funded by MIUR projects (FIRB, PRIN 2006-08, CEVIOBEM), by the University of Torino (60% projects), by CNR grants to P.B. and by Regione Piemonte (CIPE B74 project) to L.L.

References

Akiyama K, Matsuzaki K, Hayashi H (2005) Plant sesquiterpenes induce hyphal branching in arbuscular mycorrhizal fungi. Nature 435:824–827

Aung K, Lin SI, Wu CC, Huang YT, Su CL, Chiou TJ (2006) *Pho2*, a phosphate overaccumulator, is caused by a nonsense mutation in a microRNA399 target gene. Plant Physiol 141:1000–1011

Balestrini R, Lanfranco L (2006) Fungal and plant gene expression in arbuscular mycorrhizal symbiosis. Mycorrhiza 16:509–524

Balestrini R, Gómez-Ariza J, Lanfranco L, Bonfante P (2007) Laser microdissection reveals that transcripts for five plant and one fungal phosphate transporter genes are contemporaneously present in arbusculated cells. Mol Microbe–Plant Interact 20:1055–1062

Bago B, Zipfel W, Williams RC, Jun J, Arreola R, Pfeffer PE, Lammers PJ, Shachar-Hill Y (2002) Translocation and utilization of fungal storage lipid in the arbuscular mycorrhizal symbiosis. Plant Physiol 128:108–124

Bago B, Pfeffer PE, Abubaker J, Jun J, Allen JW, Brouillette J, Douds DD, Lammers PJ, Shachar-Hill Y (2003) Carbon export from arbuscular mycorrhizal roots involves the translocation of carbohydrate as well as lipid. Plant Physiol 131:1496–1507

Bari R, Pant BD, Stitt M, Scheible WR (2006) Pho2, microRNA399, and PHR1 define a phosphate-signaling pathway in plants. Plant Physiol 141:988–999

Benedetto A, Magurno F, Bonfante P, Lanfranco L (2005) Expression profiles of a phosphate transporter gene (*GmosPT*) from the endomycorrhizal fungus *Glomus mosseae*. Mycorrhiza 15:620–627

Besserer A, Puech-Pagès V, Kiefer P, Gomez-Roldan V, Jauneau A, Roy S, Portais JC, Roux C, Bécard G, Séjalon-Delmas N (2006) Strigolactones stimulate arbuscular mycorrhizal fungi by activating mitochondria. PLoS Biol 4:1239–1247

Blancaflor EB, Zhao L, Harrison MJ (2001) Microtubule organization in root cells of *Medicago truncatula* during development of an arbuscular mycorrhizal symbiosis with *Glomus versiforme*. Protoplasma 217:154–165

Bonfante P (1984) Anatomy and morphology. In: Powell CL, Bagyaraj DJ (eds) VA mycorrhizas. CRC, Boca Raton, pp 5–33

Bonfante P (2001) At the interface between mycorrhizal fungi and plants: the structural organization of cell wall, plasma membrane and cytoskeleton. In: Esser K, Hock B (eds) The Mycota IX. Springer, Berlin, pp 45–61

Bonfante P, Genre A (2008) Plants and arbuscular mycorrhizal fungi: an evolutionary–developmental perspective. Trends Plant Sci 13:492–498

Breuninger M, Requena N (2004) Recognition events in AM symbiosis: analysis of fungal gene expression at the early appressorium stage. Fungal Genet Biol 41:794–804

Brundrett MC (2002) Coevolution of roots and mycorrhizas of land plants. New Phytol 154:275–304

Bucher M (2007) Functional biology of plant phosphate uptake at root and mycorrhiza interfaces. New Phytol 173:11–26

Burleigh SH (2001) Relative quantitative PCR to study nutrient transport processes in arbuscular mycorrhizas. Plant Sci 160:899–904

Cappellazzo G, Lanfranco L, Bonfante P (2007) A limiting source of organic nitrogen induces specific transcriptional responses in the extraradical structures of the endomycorrhizal fungus *Glomus intraradices*. Curr Genet 51:59–70

Cappellazzo G, Lanfranco L, Fitz M, Wipf D, Bonfante P (2008) Characterization of an amino acid permease from the endomycorrhizal fungus *Glomus mosseae*. Plant Physiol 147:429–437

Chiou TJ, Aung K, Lin SL, Wu CC, Chiang SF, Su CL (2006) Regulation of phosphate homeostasis by microRNA in *Arabidopsis*. Plant Cell 18:412–421

Cruz C, Egsgaard H, Trujillo C, Ambus P, Requena N, Martins-Loução MA, Jakobsen J (2007) Enzymatic evidence for the key role of arginine in nitrogen translocation by arbuscular mycorrhizal fungi. Plant Physiol 144:782–792

Declerck S, Strullu DG, Fortin JA (2005) *In vitro* culture of mycorrhizas. In: Varma A (ed) Soil biology, vol 4. Springer, Berlin, pp 388–400

Demchenko K, Winzer T, Stougaard J, Parniske M, Pawlowski K (2004) Distinct roles of *Lotus japonicus* *SYMRK* and *SYM15* in root colonization and arbuscule formation. New Phytol 163:381–392

Dickson S, Smith FA, Smith SE (2007) Structural differences in arbuscular mycorrhizal symbioses: more than 100 years after Gallaud, where next? Mycorrhiza 17:375–393

Douds DD, Pfeffer PE, Shachar-Hill Y (2000) Application of *in vitro* methods to study carbon uptake and transport by AM fungi. Plant Soil 226:255–261

Drissner D, Kunze G, Callewaert N, Gehrig P, Tamasloukht N, Boller T, Felix G, Amrhein N, Bucher M (2007) The signal induction plant phosphate transporters in the arbuscular mycorrhizal symbiosis is lysophosphatidylcholine. Science 318:265–268

Duc G, Trouvelet A, Gianinazzi-Pearson V, Gianinazzi S (1989) First report of non-mycorrhizal plant mutants (Myc⁻) obtained in pea (*Pisum sativum* L.) and fababean (*Vicia faba* L.). Plant Sci 60:215–222

Frenzel A, Manthey K, Perlick AM, Meyer F, Pühler A, Krajinski F, Küster H (2005) Combined transcriptome profiling reveals a novel family of arbuscular

mycorrhizal-specific *Medicago truncatula* lectin genes. Mol Plant–Microbe Interact 18:771–782

Gallaud I (1905) Etudes sur les mycorrhizes endotrophs. Rev Gen Bot 17:5–50

Genre A, Bonfante P (1997) A mycorrhizal fungus changes microtubule orientation in tobacco root cells. Protoplasma 199:30–38

Genre A, Bonfante P (1998) Actin *versus* tubulin configuration in arbuscule containing cells from mycorrhizal tobacco roots. New Phytol 140:745–752

Genre A, Bonfante P (2002) Epidermal cells of a symbiosis-defective mutant of *Lotus japonicus* show altered cytoskeleton organisation in the presence of a mycorrhizal fungus. Protoplasma 219:43–50

Genre A, Bonfante P (2005) Building a mycorrhizal cell: how to reach compatibility between plants and arbuscular mycorrhizal fungi. J Plant Interact 1:3–13

Genre A, Bonfante P (2007) Check-in procedures for plant cell entry by biotrophic microbes. Mol Plant–Microbe Interact 20:1023–1030

Genre A, Chabaud M, Timmers T, Bonfante P, Barker DG (2005) Arbuscular mycorrhizal fungi elicit a novel intracellular apparatus in *Medicago truncatula* root epidermal cells before infection. Plant Cell 17:3489–3499

Genre A, Chabaud M, Faccio A, Barker DG, Bonfante P (2008) Prepenetration apparatus assembly precedes and predicts the colonization patterns of arbuscular mycorrhizal fungi within the root cortex of both *Medicago truncatula* and *Daucus carota*. Plant Cell 20:1407–1420

Gianinazzi-Pearson V, Séjalon-Delmas N, Genre A, Jeandroz S, Bonfante P (2007) Plants and arbuscular mycorrhizal fungi: cues and communication in the early steps of symbiotic interactions. Adv Bot 46:181–219

Giovannetti M, Sbrana C, Avio L, Citernesi AS, Logi C (1993) Differential hyphal morphogenesis in arbuscular mycorrhizal fungi during pre-infection stages. New Phytol 125:587–593

Gomez-Roldan V, Fermas S, Brewer PB, Puech-Pages V, et al (2008) Strigolactone inhibition of shoot branching. Nature 455:189–194

Goormachtig S, Capoen W, James EK, Holsters M (2004) Switch from intracellular to intercellular invasion during water stress-tolerant legume nodulation. Proc Natl Acad Sci USA 101:6303–6308

Govindarajulu M, Pfeffer PE, Jin HR, Abubaker J, Douds DD, Allen JW, Bucking H, Lammers PJ, Shachar-Hill Y (2005) Nitrogen transfer in the arbuscular mycorrhizal symbiosis. Nature 435:819–823

Güimil S, Chang H-S, Zhu T, Sesma A, Osbourn A, Roux C, Ioannidis V, Oakeley EJ, Docquier M, Descombes P, Briggs SP, Paszkowski U (2005) Comparative transcriptomics of rice reveals an ancient pattern of response to microbial colonization. Proc Natl Acad Sci USA 102:8066–8070

Harrison MJ (2005) Signaling in the arbuscular mycorrhizal symbiosis. Annu Rev Microbiol 59:19–42

Harrison MJ, Dewbre GR, Liu J (2002) A phosphate transporter from *Medicago truncatula* involved in the acquisition of phosphate released by arbuscular mycorrhizal fungi. Plant Cell 14:2413–2429

Hawkins HJ, Johansen A, George E (2000) Uptake and transport of organic and inorganic nitrogen by arbuscular mycorrhizal fungi. Plant Soil 226:275–285

Helber N, Requena N (2007) Expression of the fluorescence markers DsRed and GFP fused to a nuclear localization signal in the arbuscular mycorrhizal fungus *Glomus intraradices*. New Phytol (in press)

Hildebrandt U, Schmelzer E, Bothe H (2002) Expression of nitrate transporter genes in tomato colonized by an arbuscular mycorrhizal fungus. Physiol Plant 115:125–136

Hirsh AM, Kapulnik Y (1998) Signal transduction pathways in mycorrhizal associations: comparisons with the *Rhizobium*-legume symbiosis. Fungal Genet Biol 23:205–212

Hodge A, Campbell CD, Fitter AH (2001) An arbuscular mycorrhizal fungus accelerates decomposition and acquires nitrogen directly from organic material. Nature 413:297–299

Hohnjec N, Perlick AM, Puhler A, Kuster H (2003) The *Medicago truncatula* sucrose synthase gene *MtSucS1* is activated both in the infected region of root nodules and in the cortex of roots colonized by arbuscular mycorrhizal fungi. Mol Plant–Microbe Interact 16:903–915

Hohnjec N, Vieweg MF, Pühler A, Becker A, Küster H (2005) Overlaps in the transcriptional profiles of *Medicago truncatula* roots inoculated with two different *Glomus* fungi provide insights into the genetic program activated during arbuscular mycorrhiza. Plant Physiol 137:1283–1301

Javot H, Penmetsa VR, Terzaghi N, Cook DR, Harrison MJ (2007a) A *Medicago truncatula* phosphate transporter indispensable for the arbuscular mycorrhizal symbiosis. Proc Natl Acad Sci USA 104:1720–1725

Javot H, Pumplin N, Harrison MJ (2007b) Phosphate in the arbuscular mycorrhizal symbiosis: transport properties and regulatory roles. Plant Cell Environ 30:310–322

Jin H, Pfeffer PE, Douds DD, Piotrowski E, Lammers PJ, Shachar-Hill Y (2005) The uptake, metabolism, transport and transfer of nitrogen in an arbuscular mycorrhizal symbiosis. New Phytol 168:687–696

Johansen A, Finlay RD, Olsson PA (1996) Nitrogen metabolism of external hyphae of the arbuscular mycorrhizal fungus *Glomus intraradices*. New Phytol 133:705–712

Kanamori N, Madsen LH, Radutoiu S, Frantescu M, Quistgaard EM, Miwa H, Downie JA, James EK, Felle HH, Haaning LL, Jensen TH, Sato S, Nakamura Y, Tabata S, Sandal N, Stougaard J (2006) A nucleoporin is required for induction of Ca2+ spiking in legume nodule development and essential for rhizobial and fungal symbiosis. Proc Natl Acad Sci USA 103:359–364

Kosuta S, Chabaud M, Lougnon G, Gough C, Dénarie J, Barker DG, Bécard G (2003) A diffusible factor from arbuscular mycorrhizal fungi induces symbiosis-specific MtENOD11 expression in roots of *Medicago truncatula*. Plant Physiol 131:1–11

Kosuta S, Asledine S, Sun J, Miwa H, Morris RJ, Downie JE, Oldroyd GE (2008) Differential and chaotic calcium signatures in the symbiosis signalling pathway of legumes. Proc Natl Acad Sci USA 105:9823–9828

Küster H, Vieweg MF, Manthey K, Baier MC, Hohnjec N, Perlick AM (2007) Identification and expression regulation of symbiotically activated legume genes. Phytochemistry 68:8–18

Lammers PJ, Jun J, Abubaker J, Arreola R, Gopalan A, Bago B, Hernandez-Sebastia C, Allen JW, Douds DD, Pfeffer PE, et al (2001) The glyoxylate cycle in an arbuscular mycorrhizal fungus: gene expression and carbon flow. Plant Physiol 127:1287–1298

LaRue, TA, Weeden NF (1994) The symbiosis genes of the host. Proc Eur Nitrogen Fix Conf 1:147

Lévy J, Bres C, Geurts R, Chalhoub B, Kulikova O, Duc G, Journet EP, Ane JM, Lauber E, Bisseling T, et al (2004) A putative Ca2+ and calmodulin-dependent protein kinase required for bacterial and fungal symbioses. Science 303:1361–1364

Liu J, Maldonado-Mendoza I, Lopez-Meyer M, Cheung F, Town CD, Harrison MJ (2007) Arbuscular mycorrhizal symbiosis is accompanied by local and systemic alterations in gene expression and an increase in disease resistance in the shoots. Plant J 50:529–544

López-Pedrosa A, González-Guerrero M, Valderas A, Azcón-Aguilar C, Ferrol N (2006) *GintAMTi* encodes a functional high-affinity ammonium transporter that is expressed in the extraradical mycelium of *Glomus intraradices*. Fungal Genet Biol 43:102–110

Maeda D, Ashida K, Iguchi K, Chechetka SA, Hijikata A, Okusako Y, Deguchi Y, Izui K, Hata S (2006) Knockdown of an arbuscular mycorrhiza-inducible phosphate transporter gene of *Lotus japonicus* suppresses mutualistic symbiosis. Plant Cell Physiol 47:807–817

Martin F, Perotto S, Bonfante P (2007) Mycorrhizal fungi: a fungal community at the interface between soil and roots. In: Pinton R, Varanini Z, Nannipieri P (eds) The rhizosphere, 2nd edn. CRC, Boca Raton, pp 201–236

Marx J (2004) The roots of plant–microbe collaborations. Science 304:234–239

Nagy R, Karandashov V, Chague V, Kalinkevich K, Tamasloukht M, Xu G, Jakobsen I, Levy AA, Amrhein N, Bucher M (2005) The characterization of novel mycorrhiza-specific phosphate transporters from *Lycopersicon esculentum* and *Solanum tuberosum* uncovers functional redundancy in symbiotic phosphate transport in solanaceous species. Plant J 42:236–250

Navazio L, Moscatiello R, Genre A, Novero M, Baldan B, Bonfante P, Mariani P (2007) A diffusible signal from arbuscular mycorrhizal fungi elicits a transient cytosolic calcium elevation in host plant cells. Plant Physiol 144:673–681

Novero M, Faccio A, Genre A, Stougaard J, Webb KJ, Mulder L, Parniske M, Bonfante P (2002) Dual requirement of the *LjSym4* gene for mycorrhhizal development in epidermal and cortical cells of *Lotus japonicus* roots. New Phytol 154:741–749

Olah B, Brière C, Bécard G, Dénarie' J, Gough C (2005) Nod factors and a diffusible factor from arbuscular mycorrhizal fungi stimulate lateral root formation in *Medicago truncatula via* the DMI1/DMI2 signalling pathway. Plant J 44:195–207

Oldroyd GED, Downie JA (2006) Nuclear calcium changes at the core of symbiosis signalling. Curr Opin Plant Biol 9:351–357

Parniske M (2008) Arbuscular mycorrhiza: the mother of plant root endosymbioses. Nature 6:763–775

Paszkowski U (2006) A journey through signaling in arbuscular mycorrhizal symbioses (Tansley review). New Phytol 172:35–46

Pawlowska TE (2005) Genetic processes in arbuscular mycorrhizal fungi. FEMS Microbiol Lett 251:185–192

Pfeffer PE, Douds DD, Becard G, Shachar-Hill Y (1999) Carbon uptake and the metabolism and transport of lipids in arbuscular mycorrhiza. Plant Physiol 120:587–598

Puhler A, Strack D (2007) Molecular basics of mycorrhizal symbioses. Phytochemistry 68:6–7

Ravnskov S, Wu Y, Graham JH (2003) Arbuscular mycorrhizal fungi differentially affect expression of genes coding for sucrose synthases in maize roots. New Phytol 157:539–545

Reinhardt D (2007) Programming good relations – development of the arbuscular mycorrhiza symbiosis. Curr Opin Plant Biol 10:98–105

Remy W, Taylor TN, Hass H, Kerp H (1994) Four hundred million year old vesicular arbuscular mycorrhizae. Proc Nat Acad Sci USA 91:11841–11843

Rillig MC, Mummey DL (2006) Mycorrhizas and soil structure (Tansley review). New Phytol 171:41–53

Sanders D, Pelloux J, Brownlee C, Harper JF (2002) Calcium at the crossroads of signaling. Plant Cell Suppl 2002:S401–S417

Schaarschmidt S, Roitsch T, Hause B (2006) Arbuscular mycorrhiza induces gene expression of the apoplastic invertase LIN6 in tomato (*Lycopersicon esculentum*) roots. J Exp Bot 57:4015–4023

Schaarschmidt S, Gonzalez MC, Roitsch T, Strack D, Sonnewald U, Hause B (2007a) Regulation of arbuscular mycorrhization by carbon. The symbiotic interaction cannot be improved by increased carbon availability accomplished by root-specifically enhanced invertase activity. Plant Physiol 143:1827–1840

Schaarschmidt S, Kopka J, Ludwig-Müller J, Hause B (2007b) Regulation of arbuscular mycorrhization by apoplastic invertases: enhanced invertase activity in the leaf apoplast affects the symbiotic interaction. Plant J 51:390–405

Schüßler A, Schwarzott D, Walker C (2001) A new fungal phylum, the *Glomeromycota*: phylogeny and evolution. Mycol Res 105:1413–1421

Schüßler A, Martin H, Cohen D, Fitz M, and Wipf D (2006) Characterization of a carbohydrate transporter from symbiotic glomeromycotan fungi. Nature 444:933–936

Sesma A, Osbourn AE (2004) The rice leaf blast pathogen undergoes developmental processes typical of root-infecting fungi. Nature 431:582–586

Siciliano V, Genre A, Balestrini R, Cappellazzo G, deWit PJGM, Bonfante P (2007) Transcriptome analysis of arbuscular mycorrhizal roots during development of the pre-penetration apparatus. Plant Physiol 144:1455–1466

Smith AF, Smith SE (1997) Structural diversity in (vesicular)-arbuscular mycorrhizal symbioses. New Phytol 137:373–388

Smith SE, Read DJ (2008) Mycorrhizal symbiosis, 3rd edn. Academic, London

Smith SE, Smith AF, Jakobsen I (2003) Mycorrhizal fungi can dominate phosphate supply to plants irrespective of growth responses. Plant Physiol 133:16–20

Smith SE, Barker SJ, Zhu Y-G (2006) Fast moves in arbuscular mycorrhizal symbiotic signalling. Trends Plant Sci 11:369–371

Takemoto D, Hardham AR (2004) The cytoskeleton as a regulator and target of biotic interactions in plants. Plant Physiol 136:3864–3876

Toth R, Miller RM (1984) Dynamics of arbuscule development and degeneration in *Zea mays* mycorrhiza. Am J Bot 7:449–460

Trépanier M, Bécard G, Moutoglis P, Willemot C, Gagné S, Avis TJ, Rioux JA (2005) Dependence of arbuscular mycorrhizal fungi on their plant host for palmitic acid synthesis. Appl Environ Microbiol 71:5341–5347

van der Heijden MGA, Sanders IR (2002) Mycorrhizal ecology. Studies in ecology, vol 157. Springer, Heidelberg

van der Heijden MAG, Scheublin TR (2007) Functional traits in mycorrhizal ecology: their use for predicting the impact of arbuscular mycorrhizal fungal communities on plant growth and ecosystem functioning. New Phytol 174:244–250

Versaw WK, Chiou TJ, Harrison MJ (2002) Phosphate transporters of *Medicago truncatula* and arbuscular mycorrhizal fungi. Plant Soil 244:239–245

Walter MH, Floss DS, Hans J, Fester T, Strack D (2007) Apocarotenold biosynthesis in arbuscular mycorrhizal roots: contributions from methylerythritol phosphate pathway isogenes and tools for its manipulation. Phytochemistry 68:130–138

Weidmann S, Sanchez L, Descombin J, Chatagnier O, Gianinazzi S, Gianinazzi-Pearson V (2004) Fungal elicitation of signal transduction-related plant genes precedes mycorrhiza establishment and requires the *dmi3* gene in *Medicago truncatula*. Mol Plant–Microbe Interact 17:1385–1393

Xu GH, Chague V, Melamed-Bessudo C, Kapulnik Y, Jain A, Raghothama KG, Levy AA, Silber A (2007) Functional characterization of *LePT4*: a phosphate transporter in tomato with mycorrhiza-enhanced expression. J Exp Bot 58:2491–2502

15   Epichloë Endophytes: Clavicipitaceous Symbionts of Grasses

Christopher L. Schardl[1], Barry Scott[2], Simona Florea[1], Dongxiu Zhang[1]

CONTENTS

I.   Introduction........................... 275
II. Symbiosis and the Clavicipitaceae 276
  A. Symbiota........................... 276
  B. Life Cycles of *Epichloë* and
    Neotyphodium spp. 277
  C. Evolution of the Epichloae 278
  D. Host Compatibility 279
III. Ecological and Agronomic Roles of
  Endophytes........................... 284
  A. Host Fitness Effects 284
    1. Resistance to Biotic Stresses 284
    2. Drought Tolerance 285
  B. Diversity and Plasticity of Endophyte
    Benefits 286
  C. Ecosystem Effects..................... 286
IV. Endophyte Metabolites.................. 287
  A. Ergot Alkaloids....................... 287
    1. Ergot Alkaloid Activities 287
    2. Ergot Alkaloid Biosynthesis 290
    3. Genetics of Ergot Alkaloid
      Biosynthesis 291
  B. Lolitrems........................... 292
  C. Loline Alkaloids...................... 294
  D. Peramine........................... 297
V.   Genome and Transcriptome of
  Epichloë festucae 298
VI. Concluding Remarks..................... 299
  References............................ 300

I. Introduction

The modern approach to biology emphasizes the workings and system integration of individual organisms, with microbial infections usually considered in a disease context, yet it is benign and mutualistic symbioses that actually dominate the biosphere. Even our own healthy bodies host complex microbial consortia (Gill et al. 2006). The ecological importance of lichens (fungi hosting green algae or cyanobacteria; DePriest 2004) and the reliance of corals on zooxanthellate algae (Baker 2003) are clear. Root nodules, representing symbioses of legumes with rhizobia (*Rhizobium* spp. and related bacteria; Doyle 1994), provide a large portion of fixed nitrogen on which much of the biosphere relies. Even more ubiquitous are the mycorrhizae, which serve a key nutritional role in the vast majority of land plants (Strack et al. 2003; Chaps. 13, 14). These symbioses are readily apparent to the unaided eye. For example, lichens coat rocks and tree trunks in much of earth's wilderness, and algae bestow their bright colors upon vast coral reefs. More sophisticated microbiological techniques must be employed to visualize endophytic and epiphytic microbes, and often the structures observed need much further investigation to determine if they represent benign symbionts or latent plant pathogens.

Perhaps because of the difficulties inherent in studying benign epiphytes and endophytes, the particularly stable and well characterized symbioses of grasses with members of the fungal family Clavicipitaceae have become the favorite model for these lifestyles. Most species in a clade of the Clavicipitaceae (Spatafora et al. 2007) maintain systemic and long-term symbioses with plants that are largely or completely asymptomatic. What is intriguing is that most of these symbioses defy the usual categorization as mutualistic, commensal or antagonistic, because their effects on host fitness can vary with environment, ecological circumstance, and the stage of host development (Clay 1990; Saikkonen et al. 1998). Many of these symbionts, such as *Epichloë*, *Atkinsonella*, *Parepichloë*, and several *Balansia* spp. (White 1997; White and Reddy 1998; White et al. 1995, 1997), fruit on host inflorescences or florets, which consequently fail to develop and set seed. Others, such as *Myriogenospora* spp. and some *Balansia* spp., fruit on leaves or nodes, but still may more or less suppress host flowering and seed production (Clay 1990; Clay et al. 1989; White and Glenn 1994). Ironically,

[1] Department of Plant Pathology, 201F Plant Science Building, 1405 Veterans Drive, Lexington, KY 40546-0312, USA; e-mail: schardl@uky.edu
[2] Institute of Molecular BioSciences, Massey University, Palmerston North 5321, New Zealand

Plant Relationships, 2nd Edition
The Mycota V
H. Deising (Ed.)
© Springer-Verlag Berlin Heidelberg 2009

Neotyphodium spp. that are asexual derivatives of *Epichloë* species actually depend on the host seeds as their only means of dispersal (Freeman 1904; Sampson 1935, 1937; Schardl and Clay 1997).

II. Symbiosis and the Clavicipitaceae

A. Symbiota

Characteristics of epiphytic and endophytic Clavicipitaceae beg us to consider the basic concept of symbiosis. The word *symbiosis* was coined by H. Anton de Bary to represent any extensive physiological interaction (living together) between organisms from different taxa (differently named organisms). Such a definition does not presuppose any relative benefit or detriment to either partner (Lewis 1985). For example, infections causing disease in the host can be regarded as symbioses, though the term is rarely applied to diseases.

It is noteworthy that de Bary was a very early proponent of the germ theory, characterized cereal rust life cycles, and proposed that the microbe *Phytophthora infestans* caused late blight of potato.

More typically, the term is applied to interactions where there is an apparent host benefit. If it can be demonstrated that both partners benefit (increase their fitness), such a symbiosis is **mutualistic**. In contrast, if one benefits to the detriment of the other, the relationship is **antagonistic**. The term **parasitic**, often applied to these, is avoided here because of its ambiguity; any symbiosis where one partner relies on the other for nutrition can be called parasitic. Arguably the most interesting symbioses are those in which relative benefits and detriments vary or are difficult to determine. When a symbiosis varies with conditions, time, or development, it is **pleiotropic** (Michalakis et al. 1992). *Epichloë* spp. live in such pleiotropic symbioses with host grasses (Fig. 15.1). When the plant is in the vegetative state, the fungus causes no disease symptoms, and in some well studied systems, it is apparent that the fungus provides a number of fitness enhancements (Schardl 2001). But after floral initiation and during bolting, the fungus may fruit on some or all of the developing inflorescences. On any tiller exhibiting the fruiting structure (stroma), the inflorescence beneath it stops developing. Typically, *Epichloë* species in pleiotropic symbioses are also vertically transmissible in the seeds on asymptomatic but infected tillers.

Several characteristics of the clavicipitaceous endophytes in general, and *Epichloë* species in particular, appear ideally suited for the evolution of highly beneficial mutualists. One such characteristic is the very long-term systemic symbioses they establish. Most or all aerial portions of host plants are thoroughly colonized throughout the life of the plant. Some of the symbionts, such as the *Myriogenospora* spp., *Parepichloë* spp., and *Balansia hypoxylon* (= *Atkinsonella hypoxylon*), are epiphytes that grow on leaf surfaces and between leaf primordia (Clay 1994; White and Glenn 1994; White and Reddy 1998). Others, such as *Balansia epichloë*, *Balansia henningsiana*, *Balansia obtecta*, and *Epichloë* spp. are endophytes growing between the plant cells (Phelps et al. 1993; White and Reddy 1998; White et al. 1995, 1997). Either way, the plant is the sole source of nutrition for the symbiont, and protection of that food source can benefit the fungus as well as the host. Many Clavicipitaceae produce toxins, and certain neurotoxins such as ergot alkaloids and indolediterpenes are produced by some – but not all – *Claviceps*, *Balansia*, and *Epichloë* spp. (Cole et al. 1977; Glenn and Bacon 1997; Vining 1973).

A third characteristic predisposing evolution of mutualism is the ability to transmit vertically by infecting host seeds without damaging them (Fig. 15.1). Certain *Epichloë* species, and all of their asexual derivatives (*Neotyphodium* spp.), are efficiently transmitted by systemic infections of the seeds of their grass hosts (Freeman 1904; Sampson 1933, 1935; Schardl et al. 2004). Among the other Clavicipitaceae inhabiting monocots, only *B. hypoxylon* is known to have a similar trait; this fungus transmits in cleistogamous seeds, but rarely in open pollinated seeds, of *Danthonia* spp. (Poaceae; Clay 1994). Surprisingly, vertical transmission seems to be of little importance in *B. hypoxylon* populations, and the fungus relies heavily on its sexual cycle and horizontal transmission (Kover et al. 1997). In contrast, a clavicipitaceous epiphyte recently discovered on leaves of the dicot *Ipomoea asarifolia* (Convolvulaceae), and implicated as the source of ergot alkaloids in those plants, appears like many *Neotyphodium* spp. to be strictly seed-transmitted (Steiner et al. 2006).

In the *Epichloë* spp., vertical and horizontal transmission are linked to their asexual and sexual life cycles, respectively (Fig. 15.1). Sacrificing the

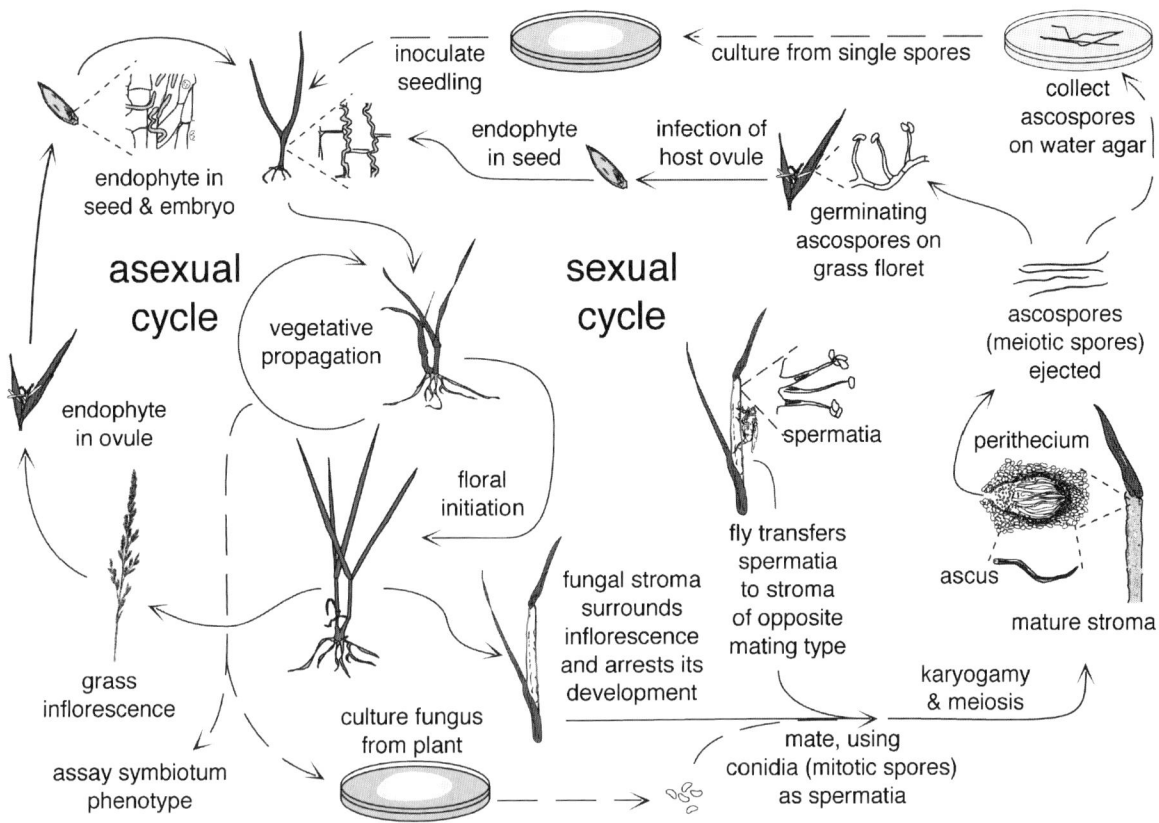

Fig. 15.1. Life cycles and artificial mating strategy for *Epichloë festucae*. In the asexual cycle (*left*) the fungus is vertically transmitted, whereas in the sexual cycle (*right*) it is horizontally transmitted. *Dashed arrows* indicate laboratory manipulations for experimental crosses

sexual life cycle poses no immediate problem to these fungi because they are so efficiently transmitted in the seeds that their survival and dissemination is assured perhaps for thousands of host generations or more. This appears to be why strictly asexual derivatives of *Epichloë* species – classified in form genus *Neotyphodium* – are extremely common (Clay and Leuchtmann 1989; Moon et al. 2004). Together the *Epichloë* and *Neotyphodium* spp. constitute an evolutionarily coherent group of endophytes whose interactions with grasses span a continuum that includes antagonistic, pleiotropic, mutualistic, and probably also commensal symbioses (Clay and Schardl 2002; Müller and Krauss 2005). In this review we use the informal term **epichloë** (plural, **epichloae**) to refer to any *Epichloë* or *Neotyphodium* species.

The list of reported endophyte benefits to grass fitness is extensive. Among the best documented are activities against insects, mammals, and nematodes, enhanced competitiveness, improved

survival and recovery following water deficit, and improved acquisition of mineral phosphate (Bazely et al. 1997; Clay and Schardl 2002; Malinowski and Belesky 2000; Morse et al. 2002; Schardl et al. 2004; Zabalgogeazcoa et al. 2006).

B. Life Cycles of *Epichloë* and *Neotyphodium* spp.

Figure 15.1 depicts the combined life cycle of *E. festucae* and its host grass, and this applies to all pleiotropic grass-*Epichloë* spp. symbioses. The fungal asexual life cycle is similar to that mapped out in great detail by Freeman (Freeman 1904) for *Neotyphodium occultans* in *Lolium temulentum* and to the other life cycles of asexual epichloae. This simply involves infection of almost all aerial tissues of the host plant. Importantly for the stability and heritability of the symbiosis, any new vegetative or reproductive tillers and the embryo are infected at extremely high efficiency (Freeman 1904; Philipson and Christey 1986). All epichloae are restricted to the intercellular spaces of the host (Philipson and Christey 1986) and there are no reports of their hyphae breaching the cell walls or forming specialized feeding structures. Thus, it is necessary for these endophytes to grow on the relatively nutrient-poor apoplastic fluids, plus material liberated from host cell walls and matrices by the action of hydrolytic enzymes (Bryant et al. 2007; Li et al. 2005; Lindstrom et al. 1993; Moy et al. 2002). In natural symbioses, there is no ultrastructural indication of host cell reaction to the adjacent hyphae (Christensen et al. 2002). However, epichloë endophytes appear, albeit inconsistently, to cause changes in plant structure and resource allocation, such as increased tillering, root growth, and stolon number and length (Belesky et al. 1989; de Battista et al. 1990; Pan and Clay 2003, 2004).

Studies of *Lolium perenne* (perennial ryegrass) symbiota with the endophyte *Neotyphodium lolii* indicate coordinated growth of fungal hyphae with the host tissues, such that fungal growth ceases when host tissue stops expanding (Tan et al. 2001). Interestingly, even in fully expanded leaves the hyphae remain metabolically active, as indicated by reporter gene expression (Herd et al. 1997; Tan et al. 2001). Until recently, it was puzzling how the endophyte growth keeps pace with host tissue expansion. Assuming the paradigm of fungal growth by hyphal tip extension, the rate of growth would have to be very fast, which is especially surprising considering that endophyte cultures on rich medium grow much slower than many saprophytic fungi (Herd et al. 1997; Tan et al. 2001). Conceivably, hyphae may be broken by the strain imposed when adjacent plant cells expand, and the broken ends may then grow and reconnect. In a recent study, however, intercalary hyphae were observed to elongate alongside the expanding leaf cells, suggesting a mode of growth quite distinct from the tip-extension paradigm (Christensen et al. 2008).

The *Epichloë* sexual cycle begins with the production of a stroma bearing the female (receptive hyphae) and male structures (sterigma bearing spermatia; Fig. 15.1; White 1997). Ironically, the sexual life cycle is directly antagonistic to the asexual life cycle, because fungal stroma production suppresses host floral development and prevents production of seeds on the affected tiller. Consequently, in the more antagonistic symbioses where the *Epichloë* species tend to produce stromata on all reproductive

tillers, there is no opportunity for them to transmit vertically. In some cases, such as greenhouse-grown *L. perenne* with *Epichoë typhina* or *Bromus erectus* with *Epichloë bromicola*, inflorescences have been observed to escape symptom development and produce seeds, yet the endophyte fails to transmit in those seeds (Chung and Schardl 1997a; Leuchtmann and Schardl 1998). Since the seed-producing tillers are nevertheless infected, it remains a mystery why *E. typhina* fails to transmit vertically, yet can cause horizontal transmission to seeds of the same host species (Chung and Schardl 1997a).

The sexual cycle (Fig. 15.1) depends on a third symbiont. Anthomyiid flies belonging to any of several *Botanophila* spp. (Leuchtmann 2007; Rao et al. 2005) have a mutualistic relationship with *Epichloë* spp. These flies feed upon the stromata, and also mediate cross-fertilization of opposite mating types (Bultman et al. 1995). The adult female fly eats mycelia and conidia and, at some point in her forays, starts laying eggs on stromata. After oviposition, the fly defecates frass that is highly enriched in viable conidia that serve as spermatia. While performing a dance, she uses her ovipositor to streak the frass thoroughly over the stroma, fertilizing the entire fruiting body. This results in development of perithecia and ascospores on which the growing larvae feed. Nevertheless, enough ascospores develop and are ejected to ensure dissemination of the fungus.

Conidia grown in culture can also serve as spermatia in experimental matings (Chung and Schardl 1997a; Fig. 15.1). In approx. 4–6 weeks after fertilization, mature ascospores can be collected by placing the stroma under an overturned water–agar plate and allowing the spores to eject onto the agar surface. The ascospore germinates to produce a palisade of conidiogenous cells, which give rise to conidia, and the conidia may give rise to hyphae or may iteratively germinate to form more conidiogenous cells (Bacon and Hinton 1988). For genetic analysis, it is advisable to isolate single-conidium-derived cultures. These can then be reintroduced into host plants by seedling inoculation (Latch and Christensen 1985) to evaluate phenotypes relevant to the symbiosis. A genetic analysis with this approach revealed a locus for loline alkaloid production (Wilkinson et al. 2000), demonstrating that the sexual stage is amenable to Mendelian analysis.

C. Evolution of the Epichloae

Symbiotic *Epichloë* spp. (*sensu stricto*) and *Neotyphodium* spp. are associated only with subfamily Pooideae, and molecular phylogenetic evidence suggests that the epichloae originated with this large subfamily of grasses (Schardl et al. 2008). The sexual *Epichloë* species are capable of fruiting on these host plants and suppressing seed production (choke disease), much like their relatives in the Clavicipitaceae. What sets them apart, however, is that the *Epichloë* species are seed-transmitted with extraordinary efficiency in most of their hosts (Table 15.1; Leuchtmann and Schardl 1998; Leuchtmann et al. 1994; Schardl and Leuchtmann 1999; Siegel et al. 1984). Since choke disease expression is directly antagonistic to vertical transmission, the latter necessitates some suppression of the former. This is evident in the pleiotropic nature of many of these symbioses (Fig. 15.1). Often the great majority of reproductive tillers develop normally despite the epichloë infection, and these bear the seeds that serve to disseminate not only the grass, but also the endophyte (the **shared diaspore**; Dawkins 1989).

It would appear that loss of the sexual cycle is a simple and logical process in the evolution of the asexual *Neotyphodium* spp. However, extensive phylogenetic analysis has indicated that many have a complex history of evolution as interspecific hybrids (Table 15.2). In a survey of asexual endophytes, including 32 distinct genotypes from various grasses, 20 were polyploid or heteroploid with genome constituents of two or three *Epichloë* spp. ancestors (Moon et al. 2004). The others appeared to be haploid, each derived directly from a sexual *Epichloë* species. In mating tests using cultured endophytes as males (Fig. 15.1), several of the nonhybrids, but none of the hybrids, were capable of mating as males to stromata of their closest sexual relatives (Brem and Leuchtmann 2003; Moon et al. 2004). A survey of endophytes from ten grass species in Argentina indicated that all were interspecific hybrids except for three out of six isolates from *Bromus setifolius* (Gentile et al. 2005; Tables 15.1, 15.2). Most of the South American endophytes were hybrids between *E. typhina* and *E. festucae*, though ancestors of one of the two isolates from *Phleum commutatum* were related to the three extant species, *E. typhina*, *E. baconii*, and *E. amarillans*.

Conceivably, interspecific hybrids could arise either by a sexual or parasexual process. Although sexual hybridization is well known in plants and animals, results of extensive mating tests argue against sexual epichloë hybrids (Leuchtmann and Schardl 1998; Leuchtmann et al. 1994; Schardl and Leuchtmann 1999). Generally, there is a failure to produce ascospores. In the one case – an *E. festucae* × *E. baconii* cross – ascospores were produced in abundance, but failed to eject. In that case, two ascospore progeny were rescued in culture, and were observed to have segregated their genetic markers in Mendelian fashion (Schardl and Leuchtmann 1999). This was quite different from the typical hybrid *Neotyphodium* spp. in nature, which tend to have multiple copies of genes that are single-copy in their ancestral *Epichloë* spp. (Leuchtmann and Clay 1990; Moon et al. 2004). Therefore, it appears likely that hybrid *Neotyphodium* spp. are derived from hyphal (cellular) fusion events. The ability to fuse hyphae is common among fungi, although in many the resulting heterokaryons are killed unless they happen to share alleles at multiple vegetative compatibility loci. No such heterokaryon incompatibility system seems to operate in *Epichloë* spp. (Chung and Schardl 1997b). The opportunity for hyphal fusion can be afforded only if a plant is superinfected with two different endophyte species. A likely means for this occurrence is diagrammed in Fig. 15.2, based on the observation that ascospores can mediate infections of florets and, consequently, seeds (Chung and Schardl 1997a). Following superinfection and hyphal fusion, karyogamy (fusion of cell nuclei) must occur, since the hybrids apparently have polyploid or heteroploid nuclei and are not heterokaryotic (Schardl and Craven 2003; Schardl et al. 1994; Tsai et al. 1994).

D. Host Compatibility

Most of the epichloae have a very limited host range. The narrow host ranges of most asexual species are expected, since their only documented means of dispersal is by vertical transmission. Among the most intensely studied systems are perennial ryegrass with *N. lolii* and *Lolium arundinaceum* (tall fescue) with *Neotyphodium coenophialum*. In both cases, maintenance of the symbiosis appears to be exclusively by vertical transmission. Surprisingly, hybrid endophytes fitting the morphological and molecular phylogenetic species, *Neotyphodium tembladerae*, are found in a broad range of pooid grasses in South America (Table 15.2). It is unclear whether this is due to numerous hybridization events involving similar *Epichloë* genotypes, or an indication of horizontal transmission of *N. tembladerae* (Gentile et al. 2005). The broad host distribution of *N. tembladerae* is

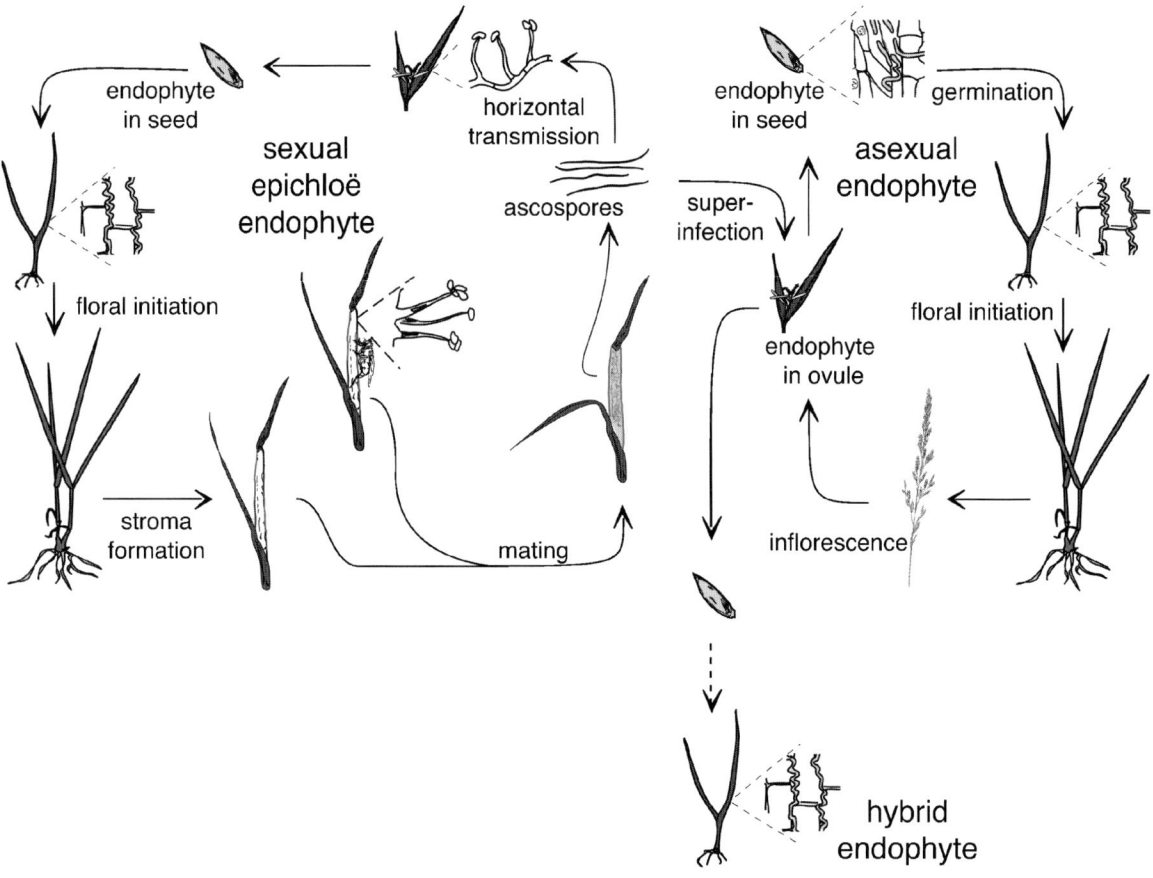

Fig. 15.2. Proposed scenario for the origin of hybrid *Neotyphodium* spp.

exceptional, and other asexual and sexual epichloae tend to exhibit specificity for host species, genera, or groups of related tribes (Tables 15.1, 15.2).

Key features of compatible interactions between epichloë endophytes and grasses are the apparent lack of a host defense response and a highly regulated growth pattern of the endophyte within the grass host (Christensen et al. 2002). The lack of a visible host defense response could be due to an inability of the host to perceive any signals indicative of the presence of the endophyte, for example by a complete lack of elicitors (molecules that could trigger the plant's defense response), or by active protective mechanisms that mask the presence of the fungus (van den Burg et al. 2006). Alternatively, one might speculate that the endophyte actively suppresses host defense responses, perhaps like microbial plant pathogens, by secreting effectors into the apoplastic space or into the host cell itself (De Wit et al. 2002; Chap. 7).

Clearly the host has the potential to respond, as highlighted by the generation of an incompatible response when endophytes are transferred from their natural hosts to alternative hosts (Christensen 1995; Koga et al. 1993). Two types of incompatible responses have been observed. In synthetic associations between *N. coenophialum* and perennial ryegrass the plant intercellular matrix, which maintains contact between the host cell walls and the hyphae, became electron dense, suggesting that the endophyte has elicited a host defense response (Koga et al. 1993). This change may be responsible for the vacuolation, reduced vigor and death of the hyphae in these associations. The second type of response observed is necrosis of the apical meristem and stunted tillers, in synthetic associations of tall fescue with either *N. uncinatum* or *N. lolii* × *E. typhina* (*Neotyphodium* sp. LpTG-2), an asexual interspecific hybrid endophyte of perennial ryegrass (Christensen 1995).

Table 15.1. *Epichloë* and nonhybrid *Neotyphodium* spp., their hosts, and symbiotum phenotypes

Endophyte species	Host species	Host tribe	Stromata	Seed transmission[a]
Epichloë amarillans	*Agrostis hiemalis*	Agrostideae	+	+
E. amarillans	*Ag. hiemalis*	Agrostideae	–	+
E. amarillans	*Agrostis perennans*	Agrostideae	+	+
E. amarillans	*Sphenopholis obtusata*	Agrostideae	+	+
E. bromicola	*Elymus virginicus*	Hordeeae	–	+
E. amarillans	*Sphenopholis nitida*	Agrostideae	+	(–)
E. amarillans	*Sph. obtusata*	Agrostideae	+	(–)
E. amarillans	*Sph. obtusata*	Agrostideae	–	+
E. baconii	*Agrostis stolonifera*	Agrostideae	+	–
E. baconii	*Agrostis tenuis*	Agrostideae	+	–
E. baconii	*Calamagrostis villosa*	Agrostideae	+	–
E. brachyelytri	*Brachyelytrum erectum*	Brachyelytreae	+	+
E. bromicola	*Bromus erectus*	Bromeae	+	–
E. bromicola	*Bromus benekenii*	Bromeae	–	+
E. bromicola	*Bromus ramosus*	Bromeae	–	+
E. bromicola	*Hordelymus europaeus*	Hordeeae	–	+
E. bromicola	*Hordeum brevisubulatum*	Hordeeae	–	+
E. clarkii	*Holcus lanatus*	Poeae	+	–
E. elymi	*Elymus canadensis*	Hordeeae	+	+
E. elymi	*Elymus hystrix*	Hordeeae	+	+
E. elymi	*Elymus villosus*	Hordeeae	+	+
E. elymi	*Elymus virginicus*	Hordeeae	+	+
E. elymi	*Bromus purgans*	Bromeae	–	+
E. festucae	*Festuca rubra* subsp. *commutata*	Poeae	+	+
E. festucae	*F. rubra* subsp. *rubra*	Poeae	+	+
E. festucae	*Festuca pulchella*	Poeae	–	+
E. festucae	*Festuca longifolia*	Poeae	+	+
E. festucae	*Koeleria cristata*	Agrostideae	+	(+)
E. festucae	*Lolium giganteum*	Poeae	+	+
E. festucae	*Lolium perenne*	Poeae	–	+
E. glyceriae	*Glyceria striata*	Meliceae	+	-
Epichloë sp.	*Achnatherum sibiricum*	Stipeae	+	+
Epichloë sp.	*Agropyron ciliare*	Hordeeae	+	+
Epichloë sp.	*Holcus mollis*	Poeae	+	+
E. sylvatica	*Brachypodium sylvaticum*	Brachypodieae	+	–
E. sylvatica	*Bp. sylvaticum*	Brachypodieae	–	+
E. typhina	*Anthoxanthum odoratum*	Poeae	+	(–)
E. typhina	*Arrhenatherum elatius*	Agrostideae	+	(–)
E. typhina	*Brachypodium pinnatum*	Brachypodieae	+	(–)
E. typhina	*Dactylis glomerata*	Poeae	+	–

continued

Table 15.1. continued

Endophyte species	Host species	Host tribe	Stromata	Seed transmis- sion[a]
E. typhina	Lolium perenne	Poeae	+	–
E. typhina	Puccinellia distans	Poeae	+	(–)
E. typhina	Phleum pratense	Agrostideae	+	(–)
E. typhina	Poa nemoralis	Poeae	+	+
E. typhina	Poa pratensis	Poeae	+	(–)
E. typhina	Poa silvicola	Poeae	+	(–)
E. typhina	Poa trivialis	Poeae	+	(–)
E. yangzii	Roegneria kamoji	Hordeeae	+	+
Neotyphodium aotearoae	Echinopogon ovatus	Agrostideae	–	+
N. gansuense	Achnatherum inebrians	Stipeae	–	+
Neotyphodium sp. ex. E. typhina[b]	Bromus setifolius	Bromeae	–	+
N. huerfanum ex.E. typhina[b]	Festuca arizonica	Poeae	–	+
N. lolii ex. E. festucae[b]	L. perenne	Poeae	–	+
Neotyphodium sp. ex. E. festucae[b]	Festuca obtusa	Poeae	–	+
N. typhinum ex. E. typhina[b]	Lolium edwardii	Poeae	–	+
Neotyphodium sp. ex. E. typhina[b]	Poa ampla	Poeae	–	+
Neotyphodium sp. ex. E. typhina[b]	Poa sylvestris	Poeae	–	+

[a] Symbols in parentheses indicate phenotypes that await verification
[b] Probable ancestral *Epichloë* species

Table 15.2. Inferred evolutionary relationships of hybrid *Neotyphodium* spp.

Neotyphodium species	Host grass	Host tribe	Geographic origin[a]	Closest non hybrid groups[b]
N. australiense	Echinopogon ovatus	Agrostideae	AU	Efe, ETC
N. chisosum	Achnatherum eminens	Stipeae	NAM	Eam, Ebo, ETC
N. coenophialum	Lolium arundinaceum	Poeae	EU, NAF	Efe, ETC, LAE
N. funkii	Achnatherum robustum	Stipeae	NAM	Eel, Efe
N. guerinii	Melica ciliata	Meliceae	EU	ETC, Nga
N. melicicola	Melica decumbens	Meliceae	SAF	Efe, Nao
N. occultans	Lolium canariense, L. multiflorum, L. persicum, L. remotum, L. rigidum, L. subulatum, L. temulentum, L. × hybridium	Poeae	EU	Ebo, LAE

continued

Table 15.2. continued

Neotyphodium species	Host grass	Host tribe	Geographic origin[a]	Closest non hybrid groups[b]
N. siegelii	Lolium pratense	Poeae	EU	Ebo, Efe
N. tembladerae	Phleum commutatum	Agrostideae	SAM	Efe, ETC
N. tembladerae	Bromus auleticus, Bs. setifolius	Bromeae	SAM	Efe, ETC
N. tembladerae	Melica stuckertii	Meliceae	SAM	Efe, ETC
N. tembladerae	Festuca argentina, F. hieronymi, F. magellanica, F. superba, Poa huecu, P. rigidifolia	Poeae	SAM	Efe, ETC
N. tembladerae	Festuca arizonica	Poeae	NAM	Efe, ETC
N. uncinatum	Lolium pratense	Poeae	EU	Ebo, ETC
Neotyphodium sp.	Bromus auleticus	Bromeae	SAM	Efe, ETC
Neotyphodium sp. FaTG-2	Lolium sp.	Poeae	EU, NAF	Efe, LAE
Neotyphodium sp. FaTG-3	Lolium sp.	Poeae	EU, NAF	ETC, LAE
Neotyphodium sp.	Festuca altissima	Poeae	EU	Ebo, ETC
Neotyphodium sp.	Festuca paradoxa	Poeae	NAM	Eam, ETC
Neotyphodium sp.	F. superba, P. rigidifolia	Poeae	SAM	Efe, ETC
Neotyphodium sp.	Hordelymus europaeus	Triticeae	EU	Ebo, ETC
Neotyphodium sp.	Hordeum bogdanii	Triticeae	EU	Eam, Eel
Neotyphodium sp.	H. bogdanii	Triticeae	EU	Ebo, ETC
Neotyphodium sp.	Hordeum brevisubulatum	Triticeae	EU	Ebo, ETC
Neotyphodium sp. LpTG-2	Lolium perenne	Poeae	EU	Efe, ETC
Neotyphodium sp.	Phleum commutatum	Agrostideae	SAM	Eam, Eba, ETC
Neotyphodium sp.	Poa autumnalis	Poeae	NAM	Eel, ETC

[a] Abbreviations: *AU* Australasia, *EU* Eurasia, *NAM* North America, *NAF* North Africa, *SAM* South America, *SAF* South Africa

[b] Abbreviations: *Eam Epichloë amarillans, Eba E. baconii, Ebo E. bromicola, Eel E. elymi, Efe E. festucae, ETC E. typhina* complex (including *E. typhina, E. clarkii, E. sylvatica*), *LAE Lolium*-associated endophyte clade, *Nao Neotyphodium aotearoae, Nga N. gansuense*

A highly regulated endophyte growth pattern within the grass host is also crucial for maintaining a compatible interaction. In symbiotic associations between *Epichloë festucae* or *N. lolii* and perennial ryegrass the hyphae grow by tip growth within the leaf and axillary bud primordia, but after colonization of the leaf expansion zone, the hyphae become attached to plant cell walls and further elongate by intercalary extension (Christensen et al. 2002, 2008; Tan et al. 2001). This pattern of growth allows the fungal endophyte to synchronize its growth with that of the grass host and avoid mechanical shear as the leaf cells are displaced away from the leaf expansion zone. However, this highly regulated pattern of hyphal growth breaks down in symbiotic associations between an NADPH oxidase (NoxA) mutant of *E. festucae* and perennial ryegrass (Tanaka et al. 2006). In *noxA* mutant associations the hyphae show a hyperbranched pattern of growth that results in a dramatic increase in fungal biomass in both meristematic and mature leaf tissue. Cells become highly vacuolated and frequently distorted. Plants infected with the *noxA* mutant lose apical dominance, become severely

stunted and show precocious senescence. These results demonstrate that endophyte production of reactive oxygen species (ROS) is critical for maintaining the mutualistic interaction between *E. festucae* and perennial ryegrass (Tanaka et al. 2006).

Synthesis of ROS by specific NADPH oxidases has a key role in both defense and cellular differentiation processes in plants, animals and fungi (Aguirre et al. 2005; Bedard and Krause 2007; Takemoto et al. 2007; Torres and Dangl 2005).

The most well studied member of this group of enzymes is the mammalian gp91^{phox} (also known as Nox2), which is responsible for the phagocytic **oxidative burst**, a hallmark of the mammalian defense response to microbial pathogens (Lambeth 2004). Generation of ROS by the phagocytic NADPH oxidase requires formation of a multi-enzyme complex composed of the catalytic subunit gp91^{phox}, an adaptor protein p22^{phox}, regulatory subunits p40^{phox}, p47^{phox}, p67^{phox}, and the small GTPase Rac2 (Diebold and Bokoch 2001; Lambeth 2004). Among these components, just p67^{phox} and Rac2 are necessary and sufficient for catalytic activation of gp91phox in a cell-free system (Diebold and Bokoch 2001). Genes for homologues of both p67phox and Rac2 – designated NoxR and RacA, respectively – were recently identified in *E. festucae* (Takemoto et al. 2006). Targeted disruption of *noxR* and *racA* in *E. festucae* resulted in plant interaction phenotypes similar to that observed for the *noxA* mutant, demonstrating that these two genes are also essential for the mutualistic symbiosis (Takemoto et al. 2006; Tanaka 2008).

The demonstration that NoxA, NoxR, and RacA are all required to maintain a mutualistic symbiotic interaction between *E. festucae* and perennial ryegrass led to a working model for regulation of endophyte growth within the plant (Takemoto et al. 2006, 2007). This model proposes that hyphal tip growth and branching is controlled by localized bursts of ROS catalyzed by NoxA, following recruitment of NoxR and RacA from the cytosol to the plasma membrane in response to signaling from the grass host.

III. Ecological and Agronomic Roles of Endophytes

A. Host Fitness Effects

1. Resistance to Biotic Stresses

During the extreme North American drought of the 1930s, agronomy professors from the University of Kentucky were introduced to tall fescue by a local farmer in Menifee County, Kentucky, and immediately appreciated its heartiness and potential to hold soil on steep hillsides even in stressful climactic conditions (Stuedemann and Hoveland 1988). They developed the first tall fescue cultivar, Kentucky-31, unaware as they selected and bred the plants that an inconspicuous symbiont passed to every succeeding generation. Tall fescue was widely adopted for pastures, forage, soil conservation, strip-mine reclamation, and eventually for turf. Unfortunately, as its use expanded, conflicting opinions about animal performance on the grass became so acrimonious as to be dubbed **the fescue wars** (Buckner et al. 1979). **Tall fescue toxicosis** eventually became widely recognized and associated with ergot alkaloids produced by the endophyte, *N. coenophialum* (Bacon et al. 1977; Lyons et al. 1986; Watson et al. 2004; see Sect. IV.A). The discovery of the tall fescue endophyte shed light on the cause of another well known malady, **ryegrass staggers**, affecting sheep and other livestock on perennial ryegrass pastures in New Zealand and Australia (Gallagher et al. 1984). This syndrome is attributed to indole-diterpene alkaloids (such as lolitrems) produced by strains of the perennial ryegrass endophyte, *N. lolii* (Bluett et al. 2005; Tor-Agbidye et al. 2001; see Sect. IV.B). Although it is technically feasible to eliminate the endophytes from commercial cultivars, this solution has not gained widespread acceptance because of the multitude of endophyte benefits to the host grasses. But recently, novel genotypes of *N. coenophialum* and *N. lolii* that lack the ergot alkaloids and lolitrems were introduced into cultivars (Bluett et al. 2005; Watson et al. 2004). These endophytes are not as thoroughly studied as those in the original tall fescue and perennial ryegrass cultivars, to which have been ascribed numerous enhancements of host fitness.

The most obvious benefits of the endophytes have been the antiherbivore effects of their alkaloids. Four classes of alkaloids are produced by epichloae: ergot alkaloids, indole-diterpenes, lolines and peramine (Bush et al. 1997; see Sect. IV). All have anti-insect activity, and ergot alkaloids and indole-diterpenes are known for neurotropic activities in mammals. No known epichloë strain produces all four alkaloid classes, although the tall fescue and perennial ryegrass endophyte strains associated with livestock toxicoses each produce three of the alkaloid classes: the original 'Kentucky-31' strain of *N. coenophialum* produces lolines, peramine,

and ergot alkaloids, and the original 'Nui' strain of *N. lolii* produces lolitrems, peramine, and ergot alkaloids (Siegel et al. 1990). Some endophytes of *Achnatherum robustum* (North America), and *Neotyphodium gansuense* in *Achnatherum inebrians* ('drunken horse grass'; Central Asia), produce narcotic forms of ergot alkaloids in such abundance as to effectively prevent grazing by horses and other large herbivores (Faeth et al. 2006; Li et al. 2007). This effect may be responsible for the growing dominance of *Ach. inebrians* and degradation of grasslands on the Tibetan plateau (Li et al. 2007).

The epichloë endophytes are best known as defensive mutualists because they deter vertebrate and invertebrate herbivores and enhance competitiveness of host grasses under insect pressure (Clay et al. 1993; Clay and Schardl 2002). In keeping with this scenario, herbivore pressure has been found to accelerate the increase in endophyte infection frequency in tall fescue over a 54-month field study (Clay et al. 2005). The ergot and indole-diterpene alkaloids may provide a natural protection of the grasses from grazing mammals and birds (Cheplick and Clay 1988; Madej and Clay 1991; Petroski et al. 1992), but their activities against invertebrate herbivores may also be of importance (Potter et al. 2008). However, the more important anti-insect alkaloids are peramine and lolines (Prestidge et al. 1985; Riedell et al. 1991; Schardl et al. 2007; Tanaka et al. 2005; see Sects. IV.C, IV.D). It is highly likely that there are other endophyte metabolites active against insects and other invertebrates (Ball et al. 2006).

Endophytes in tall fescue and perennial ryegrass have also been observed to reduce parasitism by the root knot nematode, *Meloidogyne marylandi*, and to reduce reproduction of the migratory nematodes, *Pratylenchus scribneri* and *Pratylenchus zeae* (Elmi et al. 2000; Kimmons et al. 1990; Panaccione et al. 2006a; Timper et al. 2005). Mechanisms of antagonism to nematodes remain to be elucidated. Given the paucity of endophyte hyphae in host roots, the activities against these root parasites would seem to require long-distance transport of an antibiotic metabolite or induction of a plant defense.

2. Drought Tolerance

Endophyte protection from drought stress has also been documented, particularly for *N. coenophialum* in tall fescue (for a review, see Malinowski and

Belesky 2000). One effect that may contribute is an endophyte-induced alteration of root architecture (Malinowski et al. 1999) and increase in root biomass (de Battista et al. 1990). This, and increased tillering, may be due to indole-3-acetic acid (IAA; auxin; Yue et al. 2000) or other hormonally active metabolites of the endophyte. The increase in average root hair length, as well as an increase in phenolic exudates, may also be important factors in the endophyte-enhanced ability of tall fescue plants to acquire mineral phosphate (Malinowski and Belesky 1999; Malinowski et al. 1999).

Analysis of $^{13}C/^{12}C$ ratios in tall fescue and perennial ryegrass indicate that the effect of endophyte infection is to reduce water use efficiency during water deficit (Eerens et al. 1998; Johnson and Tieszen 1993). The underlying mechanism is unclear, but may involve alterations of stomatal conductance (Malinowski and Belesky 2000).

Osmotic adjustment is a common mechanism of enhanced drought tolerance. In tall fescue, endophyte infection is reported to significantly enhance osmotic adjustment in some experiments (Elmi et al. 2000). This effect may be attributable to greater increases of protective osmolites such as carbohydrates (Richardson et al. 1992; Richardson et al. 1993). In many organisms, L-proline (Pro) levels increase in response to salt or water-deficit stress, but reports to date indicate similar or lower levels of Pro in endophyte-infected versus endophyte-free tall fescue or meadow fescue (*Lolium pratense*) plants (Elbersen and West 1996; Malinowski and Belesky 2000). Interestingly, lolines produced by *N. coenophialum* in tall fescue, or *N. uncinatum* in meadow fescue, are derived from Pro and another amino acid, L-homoserine (Hse; Schardl et al. 2007). The molar concentrations of loline alkaloids are comparable to those of Pro in stressed plants (Hong et al. 2000). Therefore, loline biosynthesis should deplete the Pro pool. If there is some compensation by increased Pro synthesis, the combined loline alkaloid and Pro levels may help enhance osmotic adjustment in these symbiota.

Another reported endophyte effect is to increase the water retention of leaf sheathes (Elbersen and West 1996). This may result from increased cell wall elasticity as well as osmotic adjustment. The tall fescue endophyte is reported to decrease the bulk modulus of tissue elasticity (White et al. 1992). Perhaps hydrolytic enzymes secreted by endophytes (Bryant et al. 2007; Li et al. 2005; Lindstrom et al. 1993; Moy et al. 2002) may help loosen the cell walls.

Taken as a whole, the large literature base suggests a multitude of mechanisms by which endophytes may enhance drought tolerance, particularly in tall fescue. However, as discussed below, extensive studies of other grasses have amply demon-strated that, whereas some host or endophyte gen-otypes positively affect drought tolerance, others have less beneficial or even negative effects.

B. Diversity and Plasticity of Endophyte Benefits

Though the literature tends to emphasize the benefits and anti-herbivore activities of epichloë endophytes, it is possible that some are commensal or antagonistic, at least in some of the environ-ments and hosts in which they are found.

The sexual *Epichloë* species are clearly capable of horizon-tal spread (Chung and Schardl 1997a, Leyronas and Ray-nal 2008; Rao and Baumann 2004), so would be expected to have less selective pressure for host protection. How-ever, even in the case of the asexual endophytes of *Fes-tuca arizonica*, no antiherbivore effect has been observed, and there is little indication of other benefits (Saikkonen et al. 1999). This is particularly surprising given the very high frequency of endophyte infection in populations of *F. arizonica*. An endophyte enhancement of plant growth under water deficit is associated with greater biomass production and leaf net photosynthesis in the endophyte-infected plants compared to endophyte-free plants (Morse et al. 2002). Nevertheless, a test of long-term survival of *F. arizonica* has given no indication of an endophyte ben-efit (Faeth and Hamilton 2006). Maintenance of vertically transmitted symbionts at high levels in host populations need not require mutualism in all environmental circum-stances, but may be a metapopulation effect (Saikkonen et al. 2002). Indeed, several studies of perennial ryegrass-endophyte associations suggest beneficial or detrimental effects depending on host or endophyte genotype and environmental conditions (Cheplick et al. 2000; Hesse et al. 2003, 2004, 2005). Another possible effect of endophytes, to increase phenotypic plasticity of the plant (Cheplick 1997), has received little study, but could be very important for enhancing adaptability.

C. Ecosystem Effects

Endophyte effects on herbivores can have dra-matic implications for ecosystem structure. In a 4-year field study comparing succession in fields planted in tall fescue with and without endophyte, species diversity was much richer in endophyte-free plots (Clay and Holah 1999). Exclusion of small mammals and insecticidal treatments reduced this effect, indicating that small herbiv-ores were largely responsible for the difference in species richness (Rudgers et al. 2007). Differential

herbivory by voles (*Microtus* spp.), for example, had a major effect on the succession from grass-land to woodland, because the voles exhibited 65% greater consumption of tree seedlings in plots with endophyte-infected compared to endophyte-free tall fescue.

If endophytes protect against insects, they should affect the abundance of herbivorous insects such as aphids (Homoptera: Aphididae) as well as their primary and secondary parasitoids (Hymenoptera). The expected effects were indeed observed in a study of these trophic levels asso-ciated with *Lolium multiflorum* (Italian ryegrass) with and without endophyte (Omacini et al. 2001). The lower food quality in a monoculture stand of the endophyte-infected grass was reflected in a lower abundance of the aphid, *Rhopalosiphum padi*, compared to an endophyte-free Italian rye-grass stand. Population densities of another (but far less abundant) aphid, *Metopolophium festucae*, were not significantly affected. The rate of primary parasitism of *R. padi* showed little difference, but the rate of secondary parasitism was far higher, in endophyte-free plots. Endophyte status also sig-nificantly affected the distribution of the species of hymenoptera that parasitize these two aphids. Italian ryegrass typically possesses the endo-phyte, *N. occultans*, which produces loline alka-loids (TePaske et al. 1993). The lolines, in turn, are active against *R. padi* (Wilkinson et al. 2000), thus accounting for the endophyte effect reducing qual-ity of the plants as food for this aphid.

Studies of hyperparasitism of *Spodoptera fru-giperda* (Lepidoptera: Noctuidae; fall armyworm), when fed on tall fescue, indicated a negative effect of *N. coenophialum* on the pupal mass (but not the developmental rate) of *Euplectrus* spp. parasitoids (Hymenoptera: Eulophidae; Bultman et al. 1997). This effect may be attributable to loline alkaloids (Bultman et al. 1997). Similarly, in most tall fescue cultivars the endophyte enhanced nematode (*Het-erorhabditis bacteriophora*) infection and conse-quent mortality of larvae of the Japanese beetle, *Popillia japonica* (Coleoptera: Scarabaeidae; Gre-wal et al. 1995). This effect was mimicked by including ergotamine (a feeding deterrent) in the larval diets, suggesting that starving predisposes the larvae to nematode infection. *Microctonus hyperodae* (Hymenoptera: Braconidae), a hyper-parasite of Argentine stem weevil, *Listronotus bonariensis* (Coleoptera: Curculionidae), was also affected by endophytes (Bultman et al. 2003). The weevils were fed perennial ryegrass with strains of *N. lolii* differing in alkaloid profiles. Some but not all strains of *N. lolii* affected parasitoid development rate and survival, and the effect on survival correlated with the presence of ergovaline.

Significant endophyte effects on other consortia of invertebrates have also been reported. For example, spider diversity was lower in endophyte-infected tall fescue plots (Finkes et al. 2006). Even the Collembola detritivore assemblage was affected, such that more Hypogastruridae were present in litter of tall fescue with endophyte, and more Isotomidae were present in tall fescue litter without endophyte (Lemons et al. 2005).

IV. Endophyte Metabolites

A. Ergot Alkaloids

1. Ergot Alkaloid Activities

Since ancient times the neurotoxic ergot alkaloids have impacted human lives in many ways. Several fungi in the family Clavicipitaceae are known to produce ergot alkaloids. Among these are the plant-pathogenic *Claviceps* spp., infamous for causing ergotism ('St. Anthony's fire') because their sclerotia (ergots) can contain large concentrations of ergot alkaloids and can contaminate certain cereals (particularly rye). But several endophytic Clavicipitaceae, such as *Epichloë*, *Neotyphodium*, and *Balansia* spp., also produce ergot alkaloids and can be important causes of ergot-alkaloid toxicity to livestock (Bacon et al. 1975, 1977). Cattle and horses that graze on tall fescue pastures infested with *N. coenophialum* can suffer from fescue toxicosis, symptoms of which are similar to those caused by *C. purpurea*, and are presented as loss of appetite, poor weight gain, hyperthermia, agalactia, reduced reproductive capability, fat necrosis, dry gangrene, convulsions, and even death (Raisbeck et al. 1991; Thompson and Stuedemann 1993). The potential for toxicosis is higher in heavily grazed fields where the animals consume lower portions of the grass plants (Cross 2003) where the ergot alkaloids tend to be more concentrated (Spiering et al. 2002a). Also particularly dangerous to livestock are developing seed heads because ergot alkaloids increase as seeds develop, perhaps due to increasing endophyte biomass (Rottinghaus et al. 1991). (These tissues also present an increased risk of poisoning by *C. purpurea*, which specifically parasitizes

florets.) Though linked to livestock health problems, ergot alkaloids (including the simpler clavines) are highly varied in structure and biological activities. Activities of some ergot alkaloids against insects and microbes (Clay and Cheplick 1989; Eich and Pertz 1999), for example, suggest that some ergot alkaloids might provide important benefits to host plants quite apart from deterring mammalian grazers.

Ergot alkaloids include compounds with the tetracyclic ergoline ring system, related tricyclic compounds such as the biosynthetic precursor, chanoclavine I, and the more elaborate ergopeptines (Schardl et al. 2006; Figs. 15.3, 15.4, Table 15.3). Those ergot alkaloids that are precursors of paspalic acid and lysergic acid, as well as spur products derived from such precursors, are referred to as **clavines**. The biosynthesis of ergot alkaloids has been studied intensively in *Claviceps* spp., and most steps have been identified (Figs. 15.3, 15.4; Gröger and Floss 1998). Clusters of genes known or suspected to specify these steps have been identified in *C. purpurea* and *C. fusiformis* (Haarmann et al. 2005; Lorenz et al. 2007). The presence of related genes in *Epichloë* and *Neotyphodium* spp. (Fleetwood et al. 2007; Panaccione et al. 2001, 2003; Wang et al. 2004), *Balansia obtecta* (Wang et al. 2004), and the clavicipitaceous symbiont of *I. asarifolia* (Steiner et al. 2006) indicates that a similar pathway should be responsible for ergot alkaloids in those organisms. It should be noted that some Clavicipitaceae, and other fungi such as *Aspergillus fumigatus* (Panaccione and Coyle 2005) produce clavine end products, and that these may also be potent neurotoxins.

The ergopeptines (Table 15.3) are especially toxic to mammals. Results of feeding animals with purified or synthesized ergopeptines suggest that fescue toxicosis is mainly due to the ergopeptine, ergovaline (Cross 2003; Spiers et al. 1995). However, the roles of other lysergic acid derivatives and clavines (precursors) have not been fully evaluated due to the technical difficulties involved in alkaloid purification. Hill et al. (2001) observed faster transport across the ruminant gastric membranes of lysergic acid and lysergol, compared to two ergopeptines, ergotamine and ergocryptine. On this basis, they suggested that the simpler lysergic acid derivatives might be principally responsible for toxicosis in cattle.

The antiherbivore effects of ergot alkaloids have been assessed in perennial ryegrass with *N. lolii* × *E. typhina* Lp1, or Lp1 mutants in the *dmaW* gene or *lpsA* gene. The *dmaW* mutant is interrupted at the determinant step of the clavine- and ergot-alkaloid

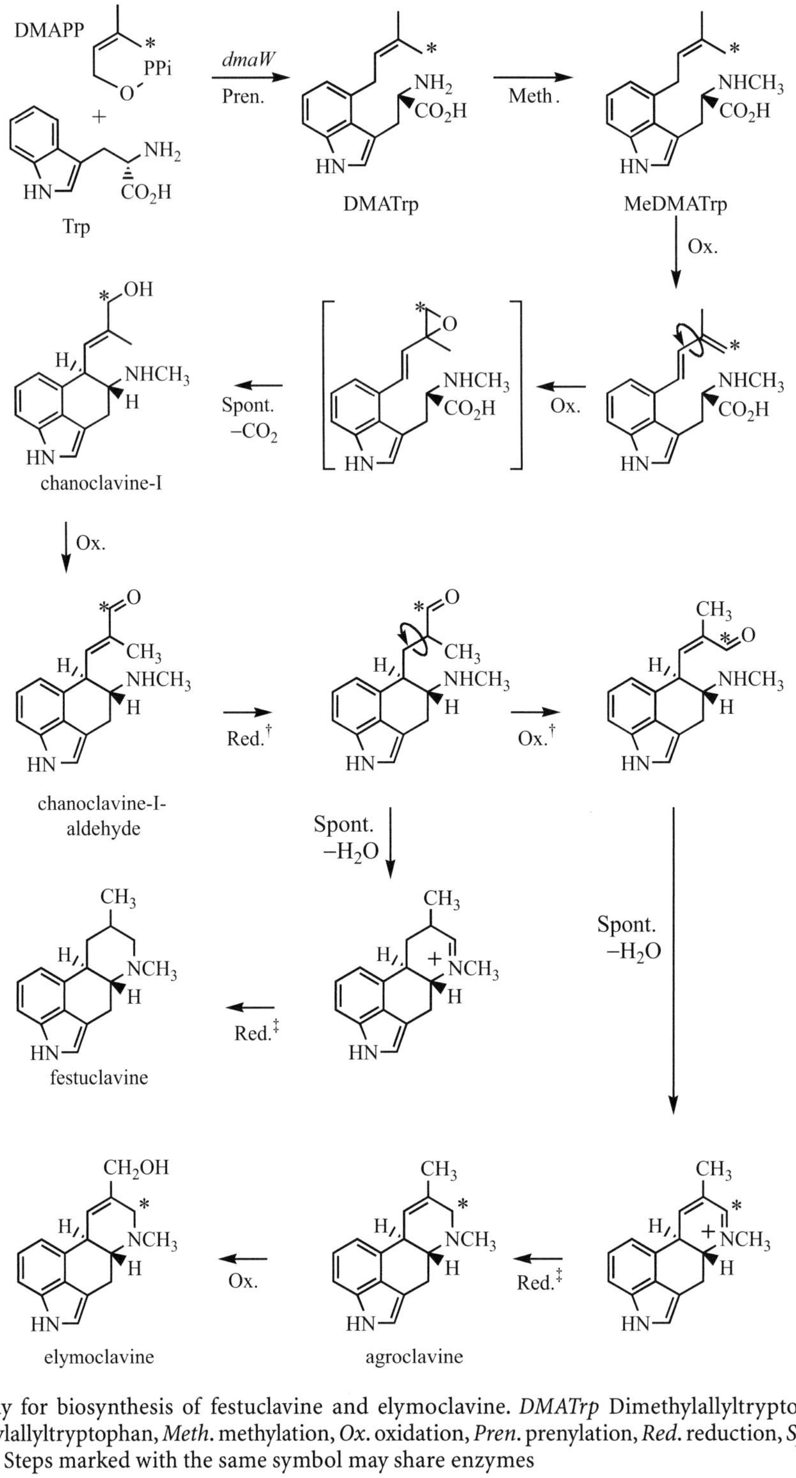

Fig. 15.3. Pathway for biosynthesis of festuclavine and elymoclavine. *DMATrp* Dimethylallyltryptophan, *MeDMATrp* N-methyl-dimethylallyltryptophan, *Meth.* methylation, *Ox.* oxidation, *Pren.* prenylation, *Red.* reduction, *Spont.* spontaneous, *Trp* L-tryptophan. Steps marked with the same symbol may share enzymes

pathway, whereas the *lpsA* mutant synthesizes clavines and lysergic acid, but not the lysergyl amides or ergopeptines. Rabbit preference and satiety tests with Lp1 and the mutants identified roles of clavines and ergopeptines (Panaccione et al. 2006b). Hay derived from plants possessing the *dmaW* mutant was strongly preferred over plants without endophyte, indicating that in the absence of clavine and ergot alkaloids the endophyte made perennial ryegrass more palatable to the rabbits. Preference for the

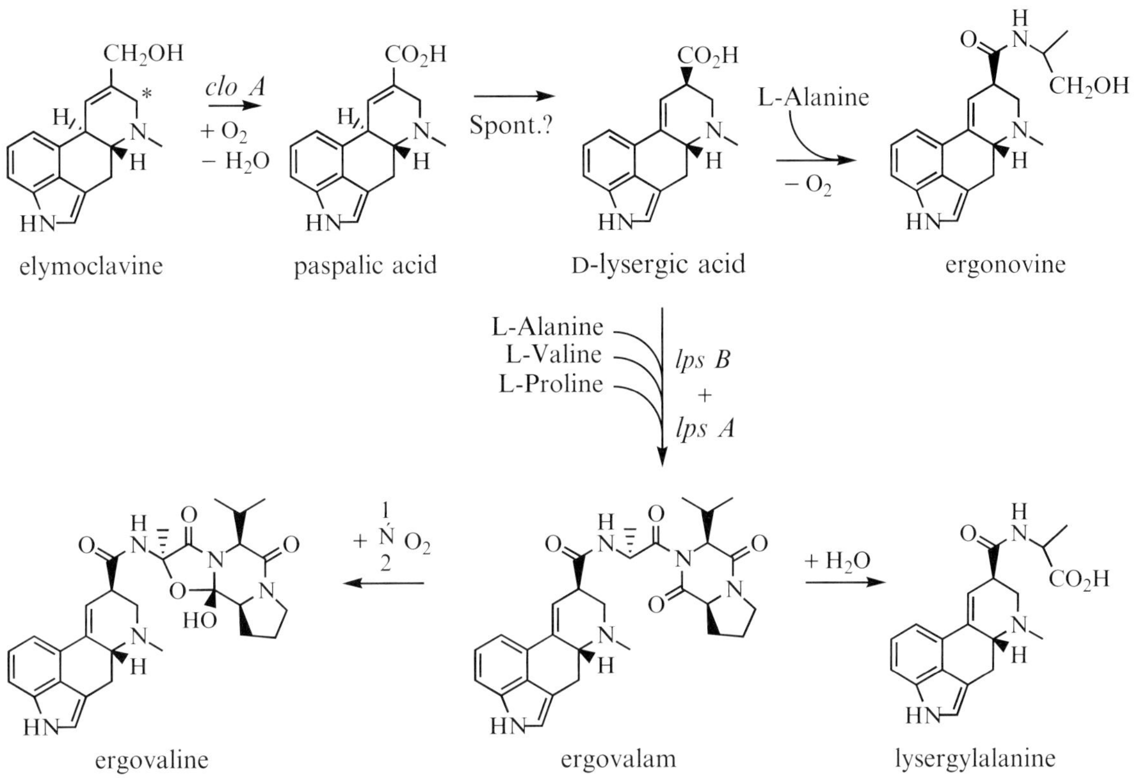

Fig. 15.4. Biosynthetic pathway to ergovaline, and possible pathways to ergonovine and lysergylalanine

Table 15.3. Ergopeptines which differ at the amino acid (*AA*) positions in the tripeptide moiety. For each entry, AA3 is L-proline

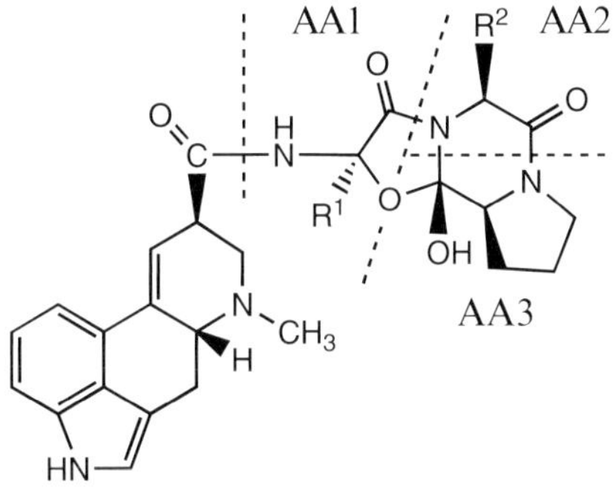

	AA2			
AA1	L-Valine	L-Phenylalanine	L-Leucine	L-Isoleucine
L-Alanine	Ergovaline	Ergotamine	Ergosine[a]	β-Ergosine
L-Valine	Ergocornine	Ergocristine	Ergokryptine	β-Ergokryptine
L-2-Aminobutyric acid	Ergonine	Ergostine	Ergoptine	β-Ergoptine

[a] Ergobalansine, when AA3 is L-alanine

dmaW mutant over the *lpsA* mutant indicated that clavines deterred feeding, but the lack of significant preference of the *dmaW* mutant over the wild type is difficult to explain. Satiety tests suggested that lysergyl amides – most likely the prevalent ergopeptine, ergovaline – reduce appetite in rabbits.

Table 15.4. Ergot alkaloid synthesis genes identified in *Claviceps purpurea* and *Epichloë festucae*

Gene	Gene size (amino acids)	Putative function	Likely cofactors
dmaW	448	DMATrp synthase	Ca^{2+} ?
lpsB	1308	LPS subunit 2	4'-phosphopantetheine, ATP
cloA	507	Elymoclavine oxygenase	Heme-Fe
lpsA1	3585	LPS 1	4'-phosphopantetheine, ATP, Fe
lpsA2	3524	LPS 1	4'-phosphopantetheine, ATP
easA	369	Reductase/dehydrogenase (OYE)	$FMNH_2$
easC	521	Catalase	Heme-M
easD	261	Reductase/dehydrogenase	NAD^+ or NADH
easE	551	Reductase/dehydrogenase	FAD
easF	359	Methyltransferase	AdoMet
easG	257	Reductase/dehydrogenase	NAD^+
easH1	314	Oxygenase/hydroxylase	Fe
easH2	154	Hydroxylase	Fe

The same perennial ryegrass plants were used to study effects on the insect *Agrotis ipsilon* (sod webworm) of Lp1 and the *dmaW* and *lpsA* mutants of Lp1. Results indicated a significant effect of ergot alkaloid – probably ergovaline – on the insect neonates (Potter et al. 2008). Both significant antibiosis and antixenosis were attributable to the lysergyl amides, suggesting that ergovaline has insecticidal and deterrent activity. Nevertheless, the mutant endophytes retained significant activity against this insect, probably due to their production of peramine, discussed in section IV.D.

Elimination of ergot alkaloid production by the *dmaW* mutation in Lp1, as well as the mutation of *lpsA*, had no significant effect on endophyte-conferred resistance to *P. scribneri* (Panaccione et al. 2006a).

2.   Ergot Alkaloid Biosynthesis

Investigation of the ergot alkaloid biosynthesis pathway (Figs. 15.3, 15.4) began when Mothes et al. (1958) injected radiolabeled tryptophan (^{14}C) into rye ears infected with *C. purpurea* and observed incorporation of the label into the lysergic acid moiety of ergonovine. They proposed that formation of the ergoline ring system involves condensation of tryptophan with an isoprene unit. Birch et al. (1960) confirmed this hypothesis when they observed that the label from [^{14}C]mevalonate incorporates into the ergoline ring. The synthesis of dimethylallyltryptophan (DMATrp) from dimethylallyl diphosphate (DMAPP) and L-tryptophan was demonstrated by Heinstein et al. (1971), who proposed that DMATrp synthase catalyzes the first step in the pathway (Fig. 15.3). In additional

studies with *Claviceps* spp. the activity of DMATrp N-methyl transferase was detected in cell free extracts (Otsuka et al. 1980). Two subsequent oxidations allow the C ring to close in a reaction that is coupled to decarboxylation, giving chanoclavine I (Kozikowski et al. 1993). Subsequently, in a series of oxidation and reduction steps collectively called chanoclavine cyclase, chanoclavine I is oxidized to the aldehyde, epimerized, and a C-N bond is formed to close the D ring, resulting in agroclavine or the more reduced festuclavine (Gröger and Floss 1998; Kozikowski et al. 1988; Schardl et al. 2006). Oxidation of agroclavine to elymoclavine is NADPH-dependent and is proposed to be catalyzed by a cytochrome P450 (Maier et al. 1988). Subsequent oxidation of elymoclavine to paspalic acid is catalyzed by the CloA cytochrome P450 (Haarmann et al. 2006), and the isomerization of paspalic acid to lysergic acid may be spontaneous (Haarmann et al. 2006).

The ergopeptines (Table 15.3, Fig. 15.4) are complex alkaloids derived from lysergic acid and three amino acids, assembled by lysergyl peptide synthetase (LPS), which is composed of two subunits, LPS1 and LPS2 (Panaccione et al. 2001; Tudzynski et al. 1999). LpsA belongs to the nonribosomal peptide synthetase (NRPS) family, and the NRPS have modular configurations with each module catalyzing the addition of an amino acid or other carboxylic acid substituent. Within a typical module is an adenylation (A) domain that specifies the substituent, a thiolation domain (T) also known as peptidyl carrier protein domain, and a condensation (C) domain that links the substituent on the T domain to the next substituent in the chain. The LPS2 subunit activates lysergic acid, and LPS1 activates the three amino acids (Riederer et al. 1996). A final oxidation by an unknown enzyme completes the cyclol ring that is peculiar to ergopeptines.

Various ergopeptines have been identified that have similar structure but differ at the amino acid positions in the tripeptide molecule (Table 15.3). Among the clavicipitaceous endophytes, *B. obtecta* produces ergobalansine (Powell et al. 1990), whereas several *Neotyphodium* and *Epichloë* spp. produce mainly ergovaline and small amounts of aci-ergovaline (the 2 epimer), ergosine, ergonine, and didehydroergovaline (not yet fully characterized; Lyons et al. 1986; Shelby et al. 1997). Most of the ergopeptines are formed by linking lysergic acid to two hydrophobic amino acids and L-proline, except that ergobalasine has L-alanine rather than L-proline in the third amino acid position (Powell et al. 1990). Variations in the ergopeptines are due to variations in the LPS1 subunit (Haarmann et al. 2008).

In addition to the complex ergopeptines such as ergovaline, other lysergyl amides found in some endophyte-infected grasses are the simplest, ergine (lysergylamide), the medicinally important, ergonovine, and lysergylalanine (Fig. 15.4; Li et al. 2004; Miles et al. 1996; Panaccione et al. 2003). Interestingly, disruption of the *lpsA* gene in the endophyte isolate, Lp1, eliminated ergine and lysergylalanine from the profile (neither the wild-type Lp1 nor the mutant produced ergonovine; Panaccione et al. 2001, 2003). It is conceivable that lysergylalanine arises by hydrolytic removal of the intermediate from the first module of LPS1; ergine could arise by hydrolysis at any of several stages.

3. Genetics of Ergot Alkaloid Biosynthesis

Purification of DMATrp synthase (Gebler and Poulter 1992) and LPS (Riederer et al. 1996) made possible the cloning and molecular characterization of genes encoding these enzymes (Correia et al. 2003; Panaccione et al. 2001; Tsai et al. 1995; Tudzynski et al. 1999; Wang et al. 2004). These genes are located in clusters with other genes likely to be involved in the pathway in *C. purpurea* (Haarmann et al. 2005), *C. fusiformis* (Lorenz et al. 2007), and *N. lolii* (Fleetwood et al. 2007). There is a confusing array of names for orthologous genes identified in the

different species, so a systematic set of names was recently adopted for the genes in the *EAS* (ergot alkaloid synthesis) clusters. The fully characterized genes are named according to the enzyme activities of the proteins they encode, whereas genes with putative functions or that remain to be characterized are named *easA* to *easH* (Schardl et al. 2006). To date, those genes with established roles are *dmaW* for DMATrp synthase, *lpsA* for LPS1, *lpsB* for LPS2, and *cloA* for the cytochrome P450 that catalyzes oxidation of elymoclavine to paspalic acid (Haarmann et al. 2006; Lorenz et al. 2007).

Based on the *dmaW* sequence from *C. fusiformis,* several homologues of the gene were identified in *C. purpurea, N. coenophialum, N. lolii × E. typhina* Lp1 and *B. obtecta* (Tudzynski et al. 1999; Wang 2000). The role of *dmaW* as the first pathway step was demonstrated by gene disruption in isolate Lp1 (Wang et al. 2004). The disrupted mutant failed to produce ergovaline and chanoclavine I, providing evidence for the role of DMATrp synthase in the pathway. Chromosome walking identified other *EAS* genes closely linked to *dmaW* in *C. purpurea* (Tudzynski et al. 1999). Among these was *lpsA*, which was functionally characterized by gene disruption in Lp1 (Panaccione et al. 2001). Plants symbiotic with the *lpsA* mutant had clavines, lysergic acid, and 6,7-secolysergine, while ergovaline and other lysergyl amides were absent from these symbiota. Two other genes, *lpsB* and *lpsC,* predicted to encode monomodular peptide synthetases, were localized in the cluster (Haarmann et al. 2005). Correia et al. (2003) confirmed by sequence and gene disruption that *lpsB* encoded the protein responsible for the activation of D-lysergic acid, the LPS2 subunit. Analysis of its alkaloid profile revealed that the *lpsB* mutant failed to produce ergopeptines but accumulated D-lysergic acid. Further chromosome walking along the *EAS* cluster in *C. purpurea* identified *cloA,* predicted to encode a cytochrome P450 heme-binding protein (Haarmann et al. 2006). Gene disruption demonstrated that CloA is required to convert elymoclavine to D-lysergic acid. Further analyses of the cluster identified another seven *EAS* genes (*easA, easC, easD, easE, easF, easG, easH*), and bioinformatic analysis suggested putative functions for each (Table 15.4; Haarmann et al. 2005; Schardl et al. 2006; Tudzynski et al. 2001). It was suggested that *easF* is the gene encoding *N*-methyltransferase (dependent on AdoMet) responsible for methylation of DMATrp, and *easA* was suggested to encode an enzyme analogous to old yellow enzymes (with $FMNH_2$ as a cofactor) and to be involved in the epimerization of chanoclavine aldehyde. These and other functions of *EAS* genes remain to be tested experimentally.

Recently, Fleetwood et al., (2007) isolated and characterized the *lpsB* gene from *N. lolii*. As in *C. purpurea*, functional analysis of the gene revealed its involvement in the biosynthesis of ergovaline. For this, a mutated version of the gene was used to disrupt its homolog in *E. festucae* (the sexual ancestor of *N. lolii*). Chromosome walking and sequence comparisons were carry out to identify other genes linked to *lpsB* in *N. lolii*. This approach provided a mechanism for the identification of five more putative *EAS* genes with homologues in *Claviceps* spp., *easA*, *easE*, *easH*, *easF*, and *easG* (Table 15.4). Due to the phylogenetic relationships among members of the family Clavicipitaceae, the presence of an ergot alkaloid cluster in epichloë endophytes is not surprising and suggests that all use a similar biosynthetic pathway. One approach to address ergot alkaloid poisoning such as tall fescue toxicosis is to screen for strains lacking these genes, or to specifically manipulate or knock out genes and eliminate or alter ergot alkaloid profiles (Panaccione et al. 2001, 2003; Wang et al. 2004).

B. Lolitrems

Lolitrems are an important subgroup of indole-diterpene metabolites that are relatively abundant in leaf sheath tissue and seeds of perennial ryegrass and some other grasses containing *E. festucae*, *N. lolii* or other epichloae. The most abundant indole-diterpene found in *N. lolii*–perennial ryegrass symbiota is lolitrem B (Fig. 15.5). The link between lolitrem B and the mammalian mycotoxin disorder, ryegrass staggers, led to a systematic analysis of the indole-diterpene profile in *N. lolii*-infected perennial ryegrass seed, resulting in the isolation and chemical identification of paspaline, paspaline B, 13-desoxypaxilline, terpendole M, lolicine A and B, lolilline, lolitriol, and lolitrems A, E, N and F (for a review, see Saikia et al. 2008). The structural similarity between paxilline – a metabolite produced in abundance by *Penicillium paxilli* – and these diverse indole-diterpenes suggests that the more complex indole-diterpenes found in grass endophytes are derived from either paxilline or proximate precursors of paxilline, such as paspaline.

A cluster of at least six genes was shown to be required for paxilline biosynthesis in *P. paxilli* (Saikia et al. 2006, 2007; Young et al. 2001). Four genes, *paxG* (encoding a geranylgeranyl diphosphate synthase), *paxB* (unknown function), *paxM* (FAD-dependent monooxygenase), and *paxC* (prenyl transferase) were shown to be required for paspaline synthesis, the first stable indole-diterpene intermediate for paxilline biosynthesis (Saikia et al. 2006). Two additional genes, *paxP* (P450 monooxygenase) and *paxQ* (P450 monooxygenase) are required to convert paspaline to paxilline (Saikia et al. 2007). The isolation and functional analysis of these genes from *P. paxilli* has allowed the isolation of homologues from the epichloae (Young et al. 2005, 2006).

Using a combination of PCR, suppression subtractive hybridization and chromosome walking, a complex lolitrem-biosynthesis locus (*LTM*), composed of at least ten genes organized in three mini-clusters, was cloned and characterized from *N. lolii* and *E. festucae* (Young et al. 2005, 2006). The first cluster contains three genes, *ltmG*, *ltmM* and *ltmK*, two of which appear to be functional orthologues of *P. paxilli paxG* and *paxM*. In *P. paxilli*, *paxG* and *paxM* are required for early steps in paxilline biosynthesis. The right-hand side of *LTM* cluster one in *N. lolii* strain Lp19 is flanked by a 17-kb relic of a retrotransposon sequence comprised of one retro-element (Tahi) with a second retro-element (Rua) inserted within (Young et al. 2005). This block of retrotransposon sequence is absent from *E. festucae* strain Fl1. Instead *ltmK* is linked directly to a polyketide synthetase pseudogene that is also present in *N. lolii* strain Lp19. The presence of this pseudogene and additional AT-rich sequence in both strains suggests that *ltmK* defines the right-hand boundary of the *LTM* gene cluster.

The *LTM* cluster two contains five genes, *ltmP*, *ltmQ*, *ltmF*, *ltmC*, and *ltmB*, four of which appear to be homologues of the functionally characterized *paxP*, *paxQ*, *paxC*, and *paxB* genes from *P. paxilli* (McMillan et al. 2003; Saikia et al. 2006; Young et al. 2001). Cluster two is separated from cluster one by a block of AT-rich retrotransposon relic sequence of approximately 35 kb and 32 kb, respectively, in strains Lp19 and Fl1.

The third *LTM* cluster contains just two genes, *ltmE* and *ltmJ*, that appear to be unique to epichloae, hence unique to lolitrem biosynthesis. A 16-kb AT-rich sequence separates cluster two from cluster three. The left-hand side of cluster three is composed of additional AT-rich sequence. Whether *ltmE* defines the left-hand boundary of the *LTM* locus remains to be determined, as no additional linked sequence has been cloned for analysis.

By analogy to the proposed functions of PaxG, PaxM, PaxB, and PaxC in *P. paxilli* (Saikia et al. 2006), the gene products of *ltmG*, *ltmM*, *ltmB*, and *ltmC* are likely to catalyze identical steps in *N. lolii* and *E. festucae* (Young et al. 2005, 2006). The fact that both *ltmM* and *ltmC* have been shown by complementation experiments to be functional orthologues of *paxM* and *paxC* supports this hypothesis. The inability of a symbiotum containing an *E. festucae ltmM* mutant to synthesize lolitrem B or any other indole-diterpenes confirmed that *ltmM* is required for lolitrem B biosynthesis (Young et al. 2005). LtmG is proposed to catalyze the synthesis of geranylgeranyl diphosphate, the first step in lolitrem biosynthesis. GGPP then condenses with indole-3-glycerol phosphate to form 3-geranylgeranylindole, a linear intermediate shown, by application of radiolabeled compound to cultures, to be incorporated into paxilline

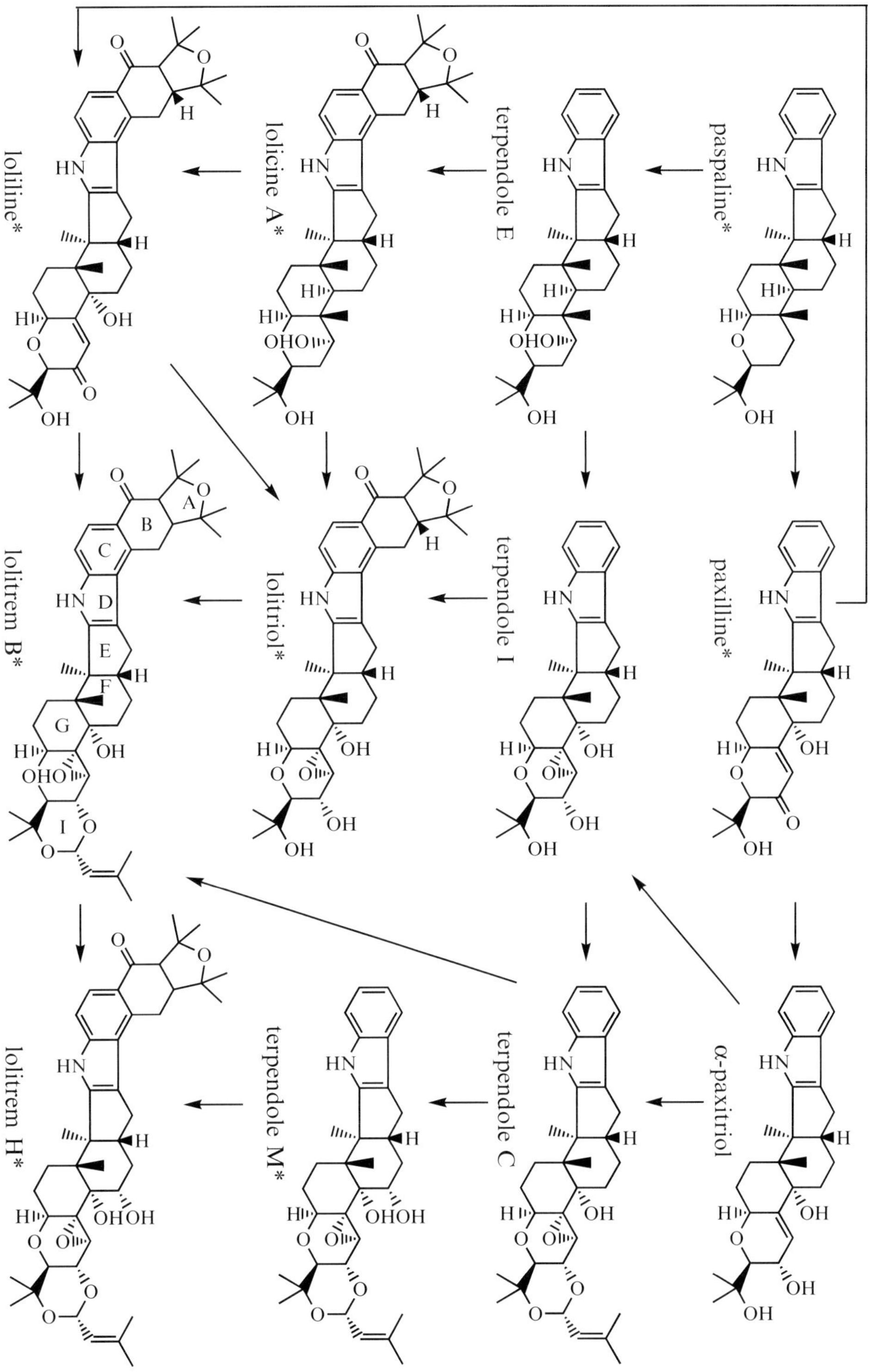

Fig. 15.5. Proposed lolitrem biosynthesis network

(Fueki et al. 2004). LtmM is proposed to catalyze the epoxidation of the two terminal alkenes of the geranylgeranyl moiety, which is then cyclized by LtmC to paspaline.

A comparison of the structure of lolitrem B with paxilline would suggest that LtmP and LtmQ, homologues of PaxP and PaxQ, catalyze analogous biosynthetic steps in *N. lolii* and *E. festucae*.

LtmP is proposed to catalyze the demethylation of C-12 of paspaline and subsequent hydroxylation of C-10, and LtmQ is proposed to hydroxylate the C-13 position of the paspaline ring (McMillan et al. 2003; Young et al. 2006).

Formation of the A- and B-rings of lolitrem B requires prenylation at positions 20 and 21 in the indole ring of paspaline (Fig. 15.5). A candidate enzyme for one or both of these prenylations is LtmE, given the domain structure of this protein, which appears to be a fusion of two prenyl trans-ferases (Young et al. 2006). Two additional catalytic steps complete the oxidation and closure of ring-A of lolitrem B, and the predicted P450 monooxyge-nase, LtmJ, is a candidate enzyme for both of these steps. At least two additional catalytic steps are required to form the epoxide between C-11 and C-12 of paspaline and to oxidize and prenylate C-10 to allow formation of ring-I of lolitrem B (Fig. 15.5). Candidate enzymes that may catalyze the latter reactions are LtmF and LtmK.

The chemical diversity of indole-diterpenes identified in *N. lolii*-infected perennial ryegrass seed indicates that the A and B rings can form independently of the I ring (Gatenby et al. 1999; Munday-Finch et al. 1998). This observation led these authors to propose that lolitrem biosynthesis is modular, proceeding by way of a metabolic grid rather than a linear pathway (Fig. 15.5). Chemical analysis of the metabolites that accumulate in *E. festucae*-perennial ryegrass symbiota containing deletions of each of the Fl1 *LTM* genes will identify the major and minor biosynthetic pathways that comprise this metabolic grid.

C. Loline Alkaloids

Loline alkaloids were first reported from the plant *Lolium temulentum*, a host of a fungal endophyte (Hofmeister 1892) now known as *Neotyphodium occultans* (Moon et al. 2000). Since their discovery, lolines have been found in broad range of genera and tribes of the endophyte-infected cool-season grass subfamily, Pooideae (Siegel et al. 1990; TePaske et al. 1993). They have no apparent toxicity to livestock, and instead, they are insecticidal, anti-invertebrate, or a feeding deterrent to a broad range of insects (Schardl et al. 2007). Activities of lolines have been demonstrated against the large milkweed bug (*Oncopeltus faciatus*; Hemiptera; Yates 1989), Japanese beetle (*Popilla japonica*,

Coleoptera; Patterson et al. 1991), European corn borer (*Ostrinia nubilalis*; Lepidoptera), fall army-worm (*Spodoptera frugiperda*; Lepidoptera), horn fly (*Haematobia irritans*; Diptera), bird cherry oat aphid (*Rhopalosiphum padi*; Homoptera) and greenbug aphid (*Schizapus graminum*; Dougherty et al. 1998; Riedell et al. 1991; Wilkinson et al. 2000). A recent review (Schardl et al. 2007) presents a detailed history of loline alkaloids and their bio-logical activities. Because of their bioprotective activities against insects, loline alkaloids have attracted great interest as natural plant protectants.

Lolines are saturated 1-aminopyrrolizidines with an oxygen bridge that links carbons C-2 and C-7 of the pyrrolizidine A and B rings (Fig. 15.6; Bush et al. 1993; Schardl et al. 2007). The substituents on the exo-1-amine group deter-mine the seven most common loline alkaloids in grass-endophyte symbiota: loline, norloline, *N*-methylloline, *N*-formylloline, *N*-acetylloline, *N*-acetylnorloline, and *N*-formylnorloline. Lolines with other moieties on the 1-amine, reported from *Adenocarpus* spp. (plant family Fabaceae), include *N*-propionylnorloline, *N*-isobutyrylnorloline, and *N*-isovalerylnorloline (Aasen and Culvenor 1969; Powell and Petroski 1992; Veen et al. 1992). The ether linkage, characteristic of the lolines, is very unusual in a natural compound because it cross-links two aliphatic bridgehead carbons.

Loline alkaloids are normally extracted with organic solvents under basic conditions from freeze-dried grass tissues or from *N. uncinatum* cultures (Blankenship et al. 2001, 2005). Differ-ent lolines can be separated by liquid column chromatography, paper chromatography, or thin layer chromatography on silica gel or alumina plates. Capillary gas chromatography (GC), NMR, EI (electron ionization)-MS (mass spectrometry), and GC-MS are used for detection and identification of different lolines (Schardl et al. 2007). These alkaloids have been iden-tified in the following natural symbiota: *Lolium* spp. with *N. occultans*, tall fescue with *N. coenophialum*, meadow fescue with *N. uncinatum*, meadow fescue with *N. siegelii*, *Poa autum-nalis* with *Epichloë elymi* × *E. typhina* PauTG-1, *Achnatherum robustum* with a *Neotyphodium* sp. (possibly *N. funkii*), *L. giganteum* with *E. festucae*, a *Lolium* sp. with *Epichloë festucae* × *E. typhina* FaTG-3, *Echinopogon ovatus* with *N. aotearoae*, and *Agrostis hiemalis* with *Epichloë amarillans* (Schardl et al. 2007; TePaske et al. 1993). It had long been debated whether the grass or the endophyte was the loline producer (Porter 1994), until production was demonstrated in *N. uncinatum* cultures grown in defined minimal media (Blankenship et al. 2001). The pathway for biosynthesis of lolines was largely eluci-dated by application to *N. uncinatum* cultures of specific isotope-labeled compounds hypothesized to be precur-sors and intermediates (Blankenship et al. 2005; Faulkner et al. 2006). When L-[5-^{13}C]ornithine (Orn) was applied to

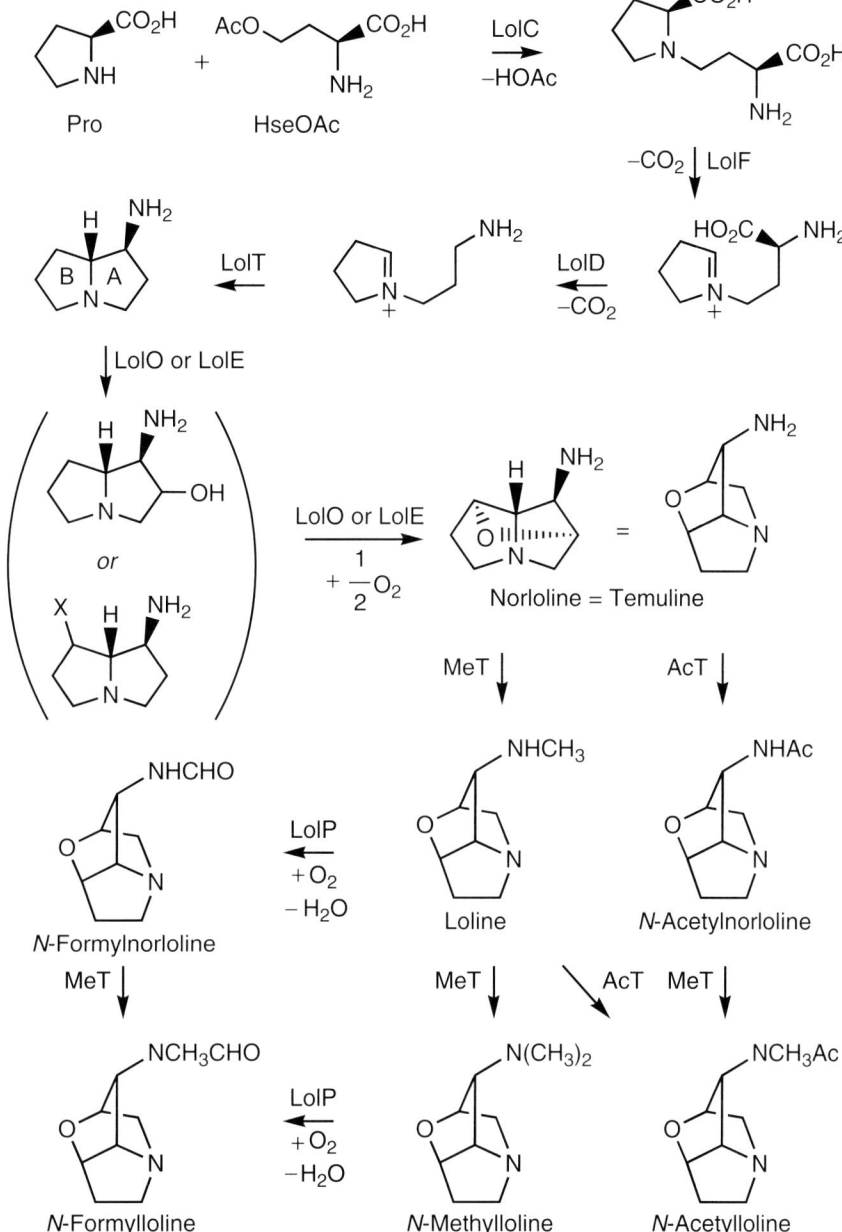

Fig. 15.6. Proposed loline alkaloid biosynthetic pathway. Possible roles of *LOL* gene products are indicated, as are the steps involving an as yet uncharacterized methyltransferase (*MeT*) and acetyltransferase (*AcT*). *HseOAc* O-acetyl-L-homoserine, *Pro* L-proline

cultures, *N*-formylloline was labeled only in the C-5 atom, and not at C-8, as determined by NMR and GC-MS. This result excluded the possibility that putrescine was a loline precursor because the symmetrical structure of putrescine would have dictated that the ¹³C label should distribute equally to both C-5 and C-8 in the lolines (Blankenship et al. 2005). When L-[¹⁵N, U-¹³C]Pro was applied, GC-MS analysis of the *N*-formylloline product showed ¹⁵N in the ring nitrogen, and ¹³C-labeled carbons 5, 6, 7, and 8 (Blankenship et al. 2005). Pro was therefore identified as the contributor

for the B ring. Similarly, L-homoserine (Hse) was identified as the A ring donor, by experiments in which L-[4-¹³C] aspartate, L-[¹⁵N]aspartate, [¹⁵N]Hse or [4,4-²H₂]Hse was applied to the cultures. Results from application of [4,4-²H₂]Hse (Faulkner et al. 2006) or L-[¹⁵N, U-¹³C]methionine (Blankenship et al. 2005) indicated that incorporation of the atoms from Hse did not proceed via methionine, which is a common source (via decarboxylated *S*-adenosylmethionine) of aminopropyl substitutents. Retention of both ²H atoms from [4,4-²H₂]Hse suggested that the first step in the pathway was γ-substitution reaction condensing the aminopropyl moiety of Hse (probably via O-acetylhomoserine) with Pro, and probably catalyzed by a γ-type pyridoxal phosphate (PLP)-containing enzyme.

Application of chemically synthesized ^{2}H-labeled potential intermediates gave evidence that after condensation of Pro and Hse, the carboxyl group from Pro is removed oxidatively (Fig. 15.6; Faulkner et al. 2006; Schardl et al. 2007). Subsequent steps remove the Hse carboxyl group and cyclize the A ring. The ether bridge is apparently formed after A-ring closure, thereby giving the loline core. Among the decorations of the 1-amine are acylation, methylation and formylation. Both the N-methyl and N-formyl carbons are derived from C-6 of methionine (Blankenship et al. 2005), indicating that they are probably donated by S-adenosylmethionine. Therefore, formation of N-formylloline, the predominant loline alkaloid in many symbiota, requires an oxygenation of one of the N-methyl groups of N-methylloline (Spiering et al. 2008).

Among the first loline biosynthesis genes identified was lolC, predicted to encode a γ-type pyridoxal phosphate (PLP) enzyme related to the *Aspergillus nidulans cysD* O-acetylhomoserine (thiol) lyase (Spiering et al. 2002b, 2005). The *lolC* gene was identified among those up-regulated during loline alkaloid synthesis in *N. uncinatum* cultures (Spiering et al. 2002b), and also as an amplified fragment length polymorphism (AFLP) marker in a Mendelian analysis of *E. festucae* isolates that differed in loline-producing capability (Spiering et al. 2002b; Wilkinson et al. 2000). Confirmation that *lolC* is a loline biosynthesis gene was obtained by an RNA interference (RNAi) experiment, whereby reduced *lolC* gene expression was associated with decreased loline production (Spiering et al. 2005).

The expected product of the LolC reaction, N-(3-amino-3-carboxypropyl) proline (NACPP) was synthesized in dideuterated and tetradeuterated form, and applied to loline-producing *N. uncinatum* cultures (Faulkner et al. 2006). Both labeled forms of NACPP showed incorporation into N-formylloline. Several observations suggested that NACPP is toxic to the bacterial or fungal cells. When *A. nidulans alcA* promoter was linked to *lolC* coding sequence and transformed into an *A. nidulans cysD* mutant, the transformants not only failed to complement the *cysD* function, but also lost detectable *lolC* mRNA expression after one subculture (Spiering et al. 2005). Likewise, repeated attempts to clone full-length *lolC* cDNA into *Escherichia coli* failed. Another observation was that *N. uncinatum* fed 1 mM or 2 mM labeled NACPP grew more slowly than unfed cultures, and feeding 4 mM NACPP almost completely inhibited growth and loline production (Faulkner et al. 2006). These observations suggest that loline biosynthesis steps subsequent to NACPP are necessary to detoxify the LolC product.

The *lolC* gene is one of nine genes identified in the 25-kb *LOL* cluster, at least seven of which are predicted to encode enzymes for various biosynthetic steps (Spiering et al. 2005).

Two homologous gene clusters, *LOL1* and *LOL2*, are present in *N. uncinatum*. Of these, the *LOL2* cluster is very similar to the single *LOL* cluster in *N. coenophialum* (Kutil et al. 2007; Spiering et al. 2005). *LOL* clusters are also present in the other epichloë strains known to produce lolines: *N. siegelii* ATCC 74483 from meadow fescue, *E. festucae* CBS 102475 from *L. gigantea*, *E. festucae* × *E. typhina* isolate Tf18 from a *Lolium* sp., and *Neotyphodium* sp. PauTG-1 from *Poa autumnalis* (Kutil et al. 2007; Spiering et al. 2005). In contrast, the *LOL* genes are undetectable in genomes of loline nonproducers such as *E. festucae* CBS 102477, *E. typhina* ATCC 200736, *N. lolii* isolate 138, and *N. lolii* × *E. typhina* isolate Lp1 (Spiering et al. 2002b).

The predicted *LOL* gene products (Spiering et al. 2005) fit well to the proposed pathway based on loline alkaloid biosynthesis studies (Blankenship et al. 2005; Faulkner et al. 2006; Fig. 15.6). After the LolC-catalyzed condensation, decarboxylation of the prolyl moiety would be catalyzed by one of the oxidative enzymes, perhaps the flavine-containing monooxygenase LolF. Then decarboxylation of the Hse moiety would probably be catalyzed by an α-type PLP enzyme, such as LolD, which is closely related to ornithine decarboxylase. The next step is proposed to be the A-ring closure, catalyzed by another α-type PLP enzyme, LolT. Then, oxygenation and oxidation at carbons C-2 and C-7 is required to form the ether bridge. The predicted epoxidase/hydroxylase LolE (with a sequence closely related to a *Penicillium decumbens* epoxidase, and containing the signatures of the enzyme family Fe^{II}/2-oxoglutarate-dependent oxygenase) may be involved in this step. Also, another 2-oxoglutarate-dependent oxygenase encoded by *LolO* may participate in ether bridge formation. Genes encoding enzymes for N-methylation and N-acetylation have not been identified in the cluster, but also there also are no obvious roles for *lolA* and *lolU* gene products. The conversion of N-methylloline to N-formylloline requires *lolP*, predicted to encode a cytochrome P450 monooxygenase (Spiering et al. 2008).

The loline alkaloids can be exceptionally abundant in symbiota, depending on the endophyte and host species. Tall fescue with *N. coenophialum* typically has several milligrams of lolines per gram of plant dry mass. Similar or higher levels are found in meadow fescue with *N. uncinatum* or *N. siegelii*, and total loline levels can increase to 20 mg/g dry mass within 1–2 weeks after these plants are clipped (Craven et al. 2001). This response to clipping is in keeping with an ecological role of lolines as inducible defenses against insect herbivores (Bultman et al. 2004). The possible roles and regulatory mechanisms of *LOL* gene expression, Lol enzyme activities, and metabolic flux of amino acids into lolines will be of interest in future studies of endophyte–plant defensive mutualisms.

D. Peramine

Peramine, the only known natural pyrrolopyrazine (Fig. 15.7), is a potent insect feeding deterrent that is uniquely synthesized by epichloae (Clay and Schardl 2002; Lane et al. 2000; Rowan 1993; Rowan et al. 1990, 1986). Taxonomically, peramine is the most widely distributed of the epichloë bioprotective metabolites identified to date, and has been detected in symbiota containing *E. typhina*, *E. festucae*, *E. amarillans*, *E. bromicola*, *E. elymi,* as well as *N. lolii*, *N. coenophialum*, and many other asexual epichloae (Clay and Schardl 2002; Lane et al. 2000; Siegel et al. 1990).

The lipophilic ring system and the hydrophilic guanidinium group of peramine are novel structural features not reported in any other insect feeding deterrent. The ability of epichloae to synthesize peramine and other bioprotective metabolites is proposed to constitute a major ecological benefit for the symbiota (Lane et al. 2000; Schardl 1996). Peramine is a potent feeding deterrent against both larvae and adults of Argentine stem weevil, *Listronotus bonariensis*, a major pest of perennial ryegrass (Rowan et al. 1990). Protection from insect herbivory may provide strong selective pressure for maintenance of the fungal biosynthetic genes (Schardl 1996).

Based on the structure of peramine, it was suggested to be derived from Pro and arginine via a diketopiperazine intermediate (Rowan 1993; Rowan et al. 1986). This led to the hypothesis that peramine synthetase would be a two-module nonribosomal peptide synthetase (NRPS), containing a methylation domain. This was recently verified by cloning a NRPS gene, designated *perA* (peramine synthetase), from *E. festucae* (Tanaka et al. 2005). The cloning strategy used, as a probe, an RT-PCR product representing a portion of an NRPS gene preferentially expressed in planta. The inferred

PerA sequence indicated a domain structure consistent with the expected functions of an enzyme required for the synthesis of peramine. This structure consists of one module with adenylation (A), thiolation (T), and condensation (C) domains, followed by a second module containing adenylation (A), methylation (M), thiolation (T), and reductase (R) domains. Perennial ryegrass symbiota containing a *perA* deletion mutant lacked detectable levels of peramine but otherwise had a wild-type symbiotic interaction phenotype. Introduction of a wild-type copy of *perA* into the mutant restored the ability to synthesize peramine, confirming that *perA* encodes peramine synthetase. Furthermore, symbiota containing the wild-type peramine-producing strain deterred feeding by *Listronotus bonariensis*, but the *perA* mutant did not, confirming that peramine is the metabolite responsible for *L. bonariensis* feeding deterrence.

No additional candidate genes for peramine biosynthesis were identified adjacent to the *E. festucae perA*. In fact, a comparison of the corresponding genome region in *Fusarium graminearum* demonstrated remarkable conservation of gene structure, order and orientation, with the exception of *perA*, which is absent from the genome of *F. graminearum*. The presence of duplicate 12-bp repeats on either side of *perA* may be a 'footprint' of a former recombination event at this locus.

These results taken together led Tanaka et al. (2005) to propose that PerA alone is responsible for peramine biosynthesis (Fig. 15.7). The condensation domain of PerA is proposed to catalyze formation of the peptide bond linking Δ1-pyrroline-5-carboxylate (the immediate precursor of Pro) with arginine. The methylation domain of PerA is proposed to catalyze the *N*-methylation of the α-amine of the arginine moiety. The reductase domain is proposed to reduce the thioester and cyclize the dipeptide to form an iminium ion that is concomitantly released. Deprotonation of this intermediate and oxidation of the pyrroline ring would give rise to peramine (Fig. 15.7). To test the validity of this proposed biosynthetic scheme will require feeding studies with isotopically labeled substrates as has been used for dissecting the loline biosynthesis pathway, described above. This approach may only be feasible if epichloae can be identified that are naturally derepressed for peramine biosynthesis or culture conditions under which they synthesize peramine can be identified.

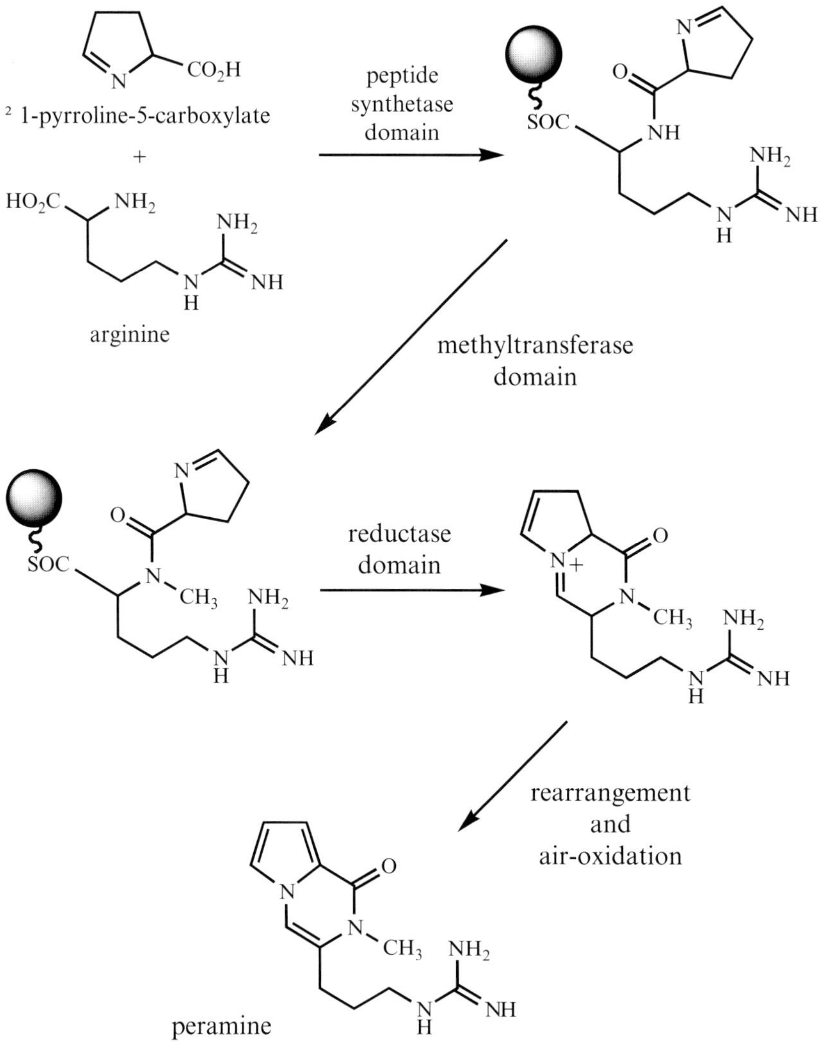

Fig. 15.7. Proposed peramine biosynthesis pathway. Each step is proposed to occur on a domain of peramine synthetase

V. Genome and Transcriptome of *Epichloë festucae*

As the understanding of endophyte biology, ecology, and agronomic importance, as well as the tools for endophyte research, increased at a growing pace, the advantage to sequencing the entire genome sequence for an epichloë endophyte became evident. The main challenge was the choice of a model endophyte. In the United States, the tall fescue endophyte, *N. coenophialum*, is of greatest economic importance. However, due to its hybrid nature, many of its genes are duplicated and its genome is large (approx. 57 Mb) compared to nonhybrid endophytes (approx. 29 Mb; Kuldau et al. 1999). These characteristics would make molecular genetic studies as well as genome sequencing considerably

slower for *N. coenophialum* than for nonhybrids. In New Zealand and Australia, *N. lolii* is particularly important as the common symbiont of perennial ryegrass. This endophyte appears to be a nonhybrid derived from *Epichloë festucae* (Kuldau et al. 1999). Compared with its ancestral species, *N. lolii* is characteristically slow growing (Zhang et al. 2006). For this reason researchers in New Zealand focused on *E. festucae* as a model (Young et al. 2006). In fact, *E. festucae* proved much easier to transform and manipulate by gene knockouts than other endophytes tried to date. Furthermore, as one of the ancestors of *N. coenophialum* (Tsai et al. 1994), the *E. festucae* genome sequence should contribute substantially to the study of *N. coenophialum*. Other advantages to *E. festucae* are its compatibility with many *Lolium* and *Festuca* species (Leuchtmann et al.

1994), and its sexual cycle, which makes it amenable to Mendelian genetics (Wilkinson et al. 2000). Finally, its ability to participate in pleiotropic symbioses (Leuchtmann et al. 1994) makes it a good model both for pathogenic *Epichloë* spp. and for vertically transmitted *Neotyphodium* spp.

An *E. festucae* isolate, designated E2368, was generated in the genetic study of loline alkaloid production (Wilkinson et al. 2000), and was chosen for genome sequencing (U. Hesse, C.L. Schardl, B.A. Roe and M.L. Farman, unpublished data). The nuclear DNA was enriched by isopycnic ultracentrifugation in the presence of bisbenzimide to separate mitochondrial from nuclear DNA. Then the nuclear DNA was sheared to fragments of approximately 4 kb, which were size-selected and inserted into a plasmid vector to create a library of clones. The inserted DNA in approximately 70 000 clones was sequenced from both ends to give a four-fold redundant coverage of the genome sequence. In addition, pyrosequencing of small, PCR-amplified fragments (conducted by B.A. Roe's group at the University of Oklahoma, Norman) resulted in a 20-fold coverage. Computer analysis of overlapping reads assembled the sequences into 2880 contigs totaling 27.4 Mb. The assembled genome is available by ftp and as a blastable database on the University of Oklahoma web site.

In a parallel study, two clone libraries of normalized cDNAs were created; one from stromata formed by *E. festucae* on meadow fescue, and the other from asymptomatic inflorescences on the same plants. In total, 71 015 of these clones were sequenced. The numbers of high-quality sequence reads (expressed sequence tags; ESTs) are given in Table 15.5.

Those EST sequences that matched *E. festucae* genome sequences were assembled into 9866 unigenes, which mapped to 7678 locations on a nonredundant *E. festucae* genome assembly. In addition, the *E. festucae* unigenes were combined with gene predictions (using FGENESH trained to *Fusarium*) to construct a total of 11 035 gene models. These were distributed in 9440 gene locations on the *E. festucae* genome, with the difference attributable to alternatively spliced variants. (The remaining ESTs, assumed to be from the host plant, assembled into 39 517 unigenes.) The assembled ESTs and the genome to which they are mapped are available through a web-based browser (www.endophyte.uky.edu).

VI. Concluding Remarks

Although grass-endophyte symbioses are not as prevalent in nature as the well known and well studied mycorrhizae, nitrogen-fixing nodules, coral–alga symbiota or lichens, a convergence of factors fueled comparably intensive investigations of endophytes and their effects. The agronomic importance of endophytes that contribute to fitness, and in some cases, toxicity of pasture grasses has driven most of the research on the two best-studied symbiota, *N. coenophialum* in tall fescue and *N. lolii* in perennial ryegrass. Studies of *E. festucae* in various *Festuca* and *Lolium* spp. gained prominence recently because of its evolutionary and chemotypic relationship to these other endophytes, and its tractability for Mendelian and molecular genetics; and it is likely to garner even more interest because of its recently completed genome sequence. Nevertheless, focus on these model symbiota is far from exclusive, given the broad diversity of pooid grasses with epichloë endophytes and the global distribution of native symbiota. In fact, an emergent theme is the diversity of endophyte chemotypes and ecological roles, as well as the diversity of symbiotic manifestations including mutualism, antagonism, and perhaps some commensalisms. Improved understanding of the pasture grass endophytes is enhancing programs to develop novel symbiota that take advantage of grass fitness enhancements while minimizing livestock toxicosis. Meanwhile, what is learned about the diverse symbiota provides insight into evolution both of mutualistic and pathogenic systems, genomic evolution of fungi, secondary metabolite roles and biosynthetic pathways, physiology of grass defenses and stress resistance, and broader ecological effects of grass–fungus symbioses.

Table 15.5. Expressed sequence tags (EST) from cDNA libraries of meadow fescue–*E. festucae* symbiota

High-quality reads	Stromata[a]	Inflorescences[a]	Other[b]	Total
5′ end	43 739	24 016	2740	70 495
3′ end	37 154	0	0	37 154

[a] From normalized cDNA libraries
[b] From nonnormalized cDNA libraries

References

Aasen AJ, Culvenor CCJ (1969) Abnormally low vicinal coupling constants for O-CH-CH in a highly strained five-membered-ring ether. Identity of loline and festucine. Aust J Chem 22:2021–2024

Aguirre J, Rios-Momberg M, Hewitt D, Hansberg W (2005) Reactive oxygen species and development in microbial eukaryotes. Trends Microbiol 13:111–118

Bacon CW, Hinton DM (1988) Ascosporic iterative germination in *Epichloë typhina*. Trans Br Mycol Soc 90:563–569

Bacon CW, Porter JK, Robbins JD (1975) Toxicity and occurrence of *Balansia* on grasses from toxic fescue pastures. Appl Microbiol 29:553–556

Bacon CW, Porter JK, Robbins JD, Luttrell ES (1977) *Epichloë typhina* from toxic tall fescue grasses. Appl Environ Microbiol 34:576–581

Baker AC (2003) Flexibility and specificity in coral-algal symbiosis: diversity, ecology, and biogeography of *Symbiodinium*. Annu Rev Ecol System 34:661–689

Ball OJP, Coudron TA, Tapper BA, Davies E, Trently D, Bush LP, Gwinn KD, Popay AJ (2006) Importance of host plant species, *Neotyphodium* endophyte isolate, and alkaloids on feeding by *Spodoptera frugiperda* (Lepidoptera : Noctuidae) larvae. J Econ Entomol 99:1462–1473

Bazely DR, Vicari M, Emmerich S, Filip L, Lin D, Inman A (1997) Interactions between herbivores and endophyte-infected *Festuca rubra* from the Scottish islands of St. Kilda, Benbecula and Rum. J Appl Ecol 34:847–860

Bedard K, Krause KH (2007) The NOX family of ROS-generating NADPH oxidases: physiology and pathophysiology. Physiol Rev 87:245–313

Belesky DP, Stringer WC, Hill NS (1989) Influence of endophyte and water regime upon tall fescue accessions. I. growth characteristics. Ann Bot 63:495–503

Birch AJ, McLoughlin BJ, Ho S (1960) The biosynthesis of the ergot alkaloids. Tetrahedron Lett 7:1–3

Blankenship JD, Spiering MJ, Wilkinson HH, Fannin FF, Bush LP, Schardl CL (2001) Production of loline alkaloids by the grass endophyte, *Neotyphodium uncinatum*, in defined media. Phytochemistry 58:395–401

Blankenship JD, Houseknecht JB, Pal S, Bush LP, Grossman RB, Schardl CL (2005) Biosynthetic precursors of fungal pyrrolizidines, the loline alkaloids. Chembiochem 6:1016–1022

Bluett SJ, Thom ER, Clark DA, Macdonald KA, Minnee EMK (2005) Effects of perennial ryegrass infected with either AR1 or wild endophyte on dairy production in the Waikato. NZ J Agric Res 48:197–212

Brem D, Leuchtmann A (2003) Molecular evidence for host-adapted races of the fungal endophyte *Epichloë bromicola* after presumed host shifts. Evolution 57:37–51

Bryant MK, May KJ, Bryan GT, Scott B (2007) Functional analysis of a β-1,6-glucanase gene from the grass endophytic fungus *Epichloë festucae*. Fungal Genet Biol 44:808–817

Buckner RC, Powell JB, Frakes RV (1979) Historical development. Agronomy 20:1–8

Bultman TL, White JF, Jr., Bowdish TI, Welch AM, Johnston J (1995) Mutualistic transfer of *Epichloë* spermatia by *Phorbia* flies. Mycologia 87:182–189

Bultman TL, Borowicz KL, Schneble RM, Coudron TA, Bush LP (1997) Effect of a fungal endophyte on the growth and survival of two *Euplectrus* parasitoids. Oikos 78:170–176

Bultman TL, McNeill MR, Goldson SL (2003) Isolate-dependent impacts of fungal endophytes in a multitrophic interaction. Oikos 102:491–496

Bultman TL, Bell G, Martin WD (2004) A fungal endophyte mediates reversal of wound-induced resistance and constrains tolerance in a grass. Ecology 85:679–685

Bush LP, Fannin FF, Siegel MR, Dahlman DL, Burton HR (1993) Chemistry, occurrence and biological effects of saturated pyrrolizidine alkaloids associated with endophyte-grass interactions. Agric Ecosyst Environ 44:81–102

Bush LP, Wilkinson HH, Schardl CL (1997) Bioprotective alkaloids of grass-fungal endophyte symbioses. Plant Physiol 114:1–7

Cheplick GP (1997) Effects of endophytic fungi on the phenotypic plasticity of *Lolium perenne* (Poaceae). Am J Bot 84:34–40

Cheplick GP, Clay K (1988) Acquired chemical defences in grasses: the role of fungal endophytes. Oikos 52:309–318

Cheplick GP, Perera A, Koulouris K (2000) Effect of drought on the growth of *Lolium perenne* genotypes with and without fungal endophytes. Funct Ecol 14:657–667

Christensen MJ (1995) Variation in the ability of *Acremonium* endophytes of *Lolium perenne*, *Festuca arundinacea* and *F. pratensis* to form compatible associations in the three grasses. Mycol Res 99:466–470

Christensen MJ, Bennett RJ, Schmid J (2002) Growth of *Epichloë/Neotyphodium* and p-endophytes in leaves of *Lolium* and *Festuca* grasses. Mycol Res 106:93–106

Christensen MJ, Bennett RJ, Ansari HA, Koga H, Johnson RD, Bryan GT, Simpson WR, Koolaard JP, Nickless EM, Voisey CR (2008) Epichloë endophytes grow by intercalary hyphal extension in elongating grass leaves. Fungal Genet Biol (in press)

Chung K-R, Schardl CL (1997a) Sexual cycle and horizontal transmission of the grass symbiont, *Epichloë typhina*. Mycol Res 101:295–301

Chung KR, Schardl CL (1997b) Vegetative compatibility between and within *Epichloë* species. Mycologia 89:558–565, 976

Clay K (1990) Comparative demography of three graminoids infected by systemic, clavicipitaceous fungi. Ecology 7:558–570

Clay K (1994) Hereditary symbiosis in the grass genus *Danthonia*. New Phytol 126:223–231

Clay K, Cheplick GP (1989) Effect of ergot alkaloids from fungal endophyte-infected grasses on fall armyworm (*Spodoptera frugiperda*). J Chem Ecol 15:169–182

Clay K, Holah J (1999) Fungal endophyte symbiosis and plant diversity in successional fields. Science 285:1742–1744

Clay K, Leuchtmann A (1989) Infection of woodland grasses by fungal endophytes. Mycologia 81:805–811

Clay K, Schardl C (2002) Evolutionary origins and ecological consequences of endophyte symbiosis with grasses. Am Nat 160:S99–S127

Clay K, Cheplick GP, Marks S (1989) Impact of the fungus *Balansia henningsiana* on *Panicum agrostoides*: frequency of infection, plant growth and reproduction and resistance to pests. Oecologia 80:374–380

Clay K, Marks S, Cheplick GP (1993) Effects of insect herbivory and fungal endophyte infection on competitive interactions among grasses. Ecology 74:1767–1777

Clay K, Holah J, Rudgers JA (2005) Herbivores cause a rapid increase in hereditary symbiosis and alter plant community composition. Proc Natl Acad Sci USA 102:12465–12470

Cole RJ, Dorner JW, Lansden JA, Cox RH, Pape C, Cunfer B, Nicholson SS, Bedell DM (1977) Paspalum staggers: isolation and identification of tremorgenic metabolites from sclerotia of *Claviceps paspali*. J Agric Food Chem 25:1197–1201

Correia T, Grammel N, Ortel I, Keller U, Tudzynski P (2003) Molecular cloning and analysis of the ergopeptine assembly system in the ergot fungus *Claviceps purpurea*. Chem Biol 10:1281–1292

Craven KD, Blankenship JD, Leuchtmann A, Hignight K, Schardl CL (2001) Hybrid fungal endophytes symbiotic with the grass *Lolium pratense*. Sydowia 53:44–73

Cross DL (2003) Ergot alkaloid toxicity. In: White JF Jr, Bacon CW, Hywel-Jones NL, Spatafora JW (eds) Clavicipitalean fungi: evolutionary biology, chemistry, biocontrol and cultural impacts. Marcel–Dekker, New York, pp 475–494

Dawkins R (1989) The selfish gene, 2nd edn. Oxford University Press, Oxford

de Battista JP, Bouton JH, Bacon CW, Siegel MR (1990) Rhizome and herbage production of endophyte-removed tall fescue clones and populations. Agronomy J 82:651–654

De Wit PJGM, Brandwagt BF, van den Burg HA, Cai X, Van der Hoorn RAL, De Jong CF, Van Klooster J, De Kock MJD, Kruijt M, Lindhout WH, Luderer R, Takken FL, Westerink N, Vervoort J, Joosten MHAJ (2002) The molecular basis of coevolution between *Cladosporium fulvum* and tomato. Antonie Van Leeuwenhoek 81:409–412

DePriest PT (2004) Early molecular investigations of lichen-forming symbionts: 1986–2001. Annu Rev Microbiol 58:273–301

Diebold BA, Bokoch GM (2001) Molecular basis for Rac2 regulation of phagocyte NADPH oxidase. Nat Immunol 2:211–215

Dougherty CT, Knapp FW, Bush LP, Maul JE, Van Willigen J (1998) Mortality of horn fly (Diptera: Muscidae) larvae in bovine dung supplemented with loline alkaloids from tall fescue. J Med Entomol 35:798–803

Doyle JJ (1994) Phylogeny of the legume family: an approach to understanding the origins of nodulation. Annu Rev Ecol System 25:325–349

Eerens JPJ, Lucas RJ, Easton S, White JGH (1998) Influence of the endophyte (*Neotyphodium lolii*) on morphology, physiology, and alkaloid synthesis of perennial ryegrass during high temperature and water stress. NZ J Agric Res 41:219–226

Eich E, Pertz H (1999) Antimicrobial and antitumor effects of ergot alkaloids and their derivatives. In: Kren V, Cvak L (eds) Ergot: the genus *Claviceps*. Harwood, Amsterdam, pp 441–449

Elbersen HW, West CP (1996) Growth and water relations of field-grown tall fescue as influenced by drought and endophyte. Grass Forage Sci 51:333–342

Elmi AA, West CP, Robbins RT, Kirkpatrick TL (2000) Endophyte effects on reproduction of a root-knot nematode (*Meloidogyne marylandi*) and osmotic adjustment in tall fescue. Grass Forage Sci 55:166–172

Faeth S, Hamilton C (2006) Does an asexual endophyte symbiont alter life stage and long-term survival in a perennial host grass? Microb Ecol 52:748–755

Faeth S, Gardner D, Hayes C, Jani A, Wittlinger S, Jones T (2006) Temporal and spatial variation in alkaloid levels in *Achnatherum robustum*, a native grass infected with the endophyte *Neotyphodium*. J Chem Ecol 32:307–324

Faulkner JR, Hussaini SR, Blankenship JD, Pal S, Branan BM, Grossman RB, Schardl CL (2006) On the sequence of bond formation in loline alkaloid biosynthesis. Chembiochem 7:1078–1088

Finkes LK, Cady AB, Mulroy JC, Clay K, Rudgers JA (2006) Plant–fungus mutualism affects spider composition in successional fields. Ecol Lett 9:344–353

Fleetwood DJ, Scott B, Lane GA, Tanaka A, Johnson RD (2007) A complex ergovaline gene cluster in epichloe endophytes of grasses. Appl Environ Microbiol (in press)

Freeman EM (1904) The seed fungus of *Lolium temulentum* L., the darnel. Phil Trans R Soc Lond B 196:1–27

Fueki S, Tokiwano T, Toshima H, Oikawa H (2004) Biosynthesis of indole diterpenes, emindole, and paxilline: involvement of a common intermediate. Org Lett 6:2697–2700

Gallagher RT, Hawkes AD, Steyn PS, Vleggaar R (1984) Tremorgenic neurotoxins from perennial ryegrass causing ryegrass staggers disorder of livestock: structure elucidation of lolitrem B. J Chem Soc Chem Commun 1984:614–616

Gatenby WA, Munday-Finch SC, Wilkins AL, Miles CO (1999) Terpendole M, a novel indole-diterpenoid isolated from *Lolium perenne* infected with the endophytic fungus *Neotyphodium lolii*. J Agric Food Chem 47:1092–1097

Gebler JC, Poulter CD (1992) Purification and characterization of dimethylallyltryptophan synthase from *Claviceps purpurea*. Arch Biochem Biophys 296:308–313

Gentile A, Rossi MS, Cabral D, Craven KD, Schardl CL (2005) Origin, divergence, and phylogeny of epichloë endophytes of native Argentine grasses. Mol Phylogenet Evol 35:196–208

Gill SR, Pop M, DeBoy RT, Eckburg PB, Turnbaugh PJ, Samuel BS, Gordon JI, Relman DA, Fraser-Liggett CM, Nelson KE (2006) Metagenomic analysis of the human distal gut microbiome. Science 312:1355–1359

Glenn AE, Bacon CW (1997) Distribution of ergot alkaloids within the family Clavicipitaceae. In: Hill NS, Bacon CW (eds) *Neotyphodium*/grass interactions. Plenum, New York, pp 53–56

Grewal SK, Grewal PS, Gaugler R (1995) Endophytes of fescue grasses enhance susceptibility of *Popillia japonica* larvae to an entomopathogenic nematode. Entomol Exp Appl 74:219–224

Gröger D, Floss HG (1998) Biochemistry of ergot alkaloids – achievements and challenges. Alkaloids Chem Biol 50:171–218

Haarmann T, Machado C, Lübbe Y, Correia T, Schardl CL, Panaccione DG, Tudzynski P (2005) The ergot alkaloid gene cluster in *Claviceps purpurea*: extension of the cluster sequence and intra species evolution. Phytochemistry 66:1312–1320

Haarmann T, Ortel I, Tudzynski P, Keller U (2006) Identification of the cytochrome P450 monooxygenase that bridges the clavine and ergoline alkaloid pathways. Chembiochem 7:645–652

Haarmann T, Lorenz N, Tudzynski P (2008) Use of a non-homologous end joining deficient strain (?ku70) of the ergot fungus *Claviceps purpurea* for identification of a nonribosomal peptide synthetase gene involved in ergotamine biosynthesis. Fungal Genet Biol (in press)

Heinstein PF, Lee S-L, Floss HG (1971) Isolation of dimethylallylpyrophosphate: tryptophan dimethylallyl transferase from the ergot fungus (*Claviceps* spec.). Biochem Biophys Res Commun 44:1244–1251

Herd S, Christensen MJ, Saunders K, Scott DB, Schmid J (1997) Quantitative assessment of in planta distribution of metabolic activity and gene expression of an endophytic fungus. Microbiology 143:267–275

Hesse U, Schöberlein W, Wittenmayer L, Förster K, Warnstorff K, Diepenbrock W, Merbach W (2003) Effects of Neotyphodium endophytes on growth, reproduction and drought-stress tolerance of three *Lolium perenne* L. genotypes. Grass Forage Sci 58:407–415

Hesse U, Hahn H, Andreeva K, Förster K, Warnstorff K, Schöberlein W, Diepenbrock W (2004) Investigations on the influence of Neotyphodium endophytes on plant growth and seed yield of *Lolium perenne* genotypes. Crop Sci 44:1689–1695

Hesse U, Schoberlein W, Wittenmayer L, Förster K, Warnstorff K, Diepenbrock W, Merbach W (2005) Influence of water supply and endophyte infection (*Neotyphodium* spp.) on vegetative and reproductive growth of two *Lolium perenne* L. genotypes. Eur J Agron 22:45–54

Hill NS, Thompson FN, Stuedemann JA, Rottinghaus GW, Ju HJ, Dawe DL, Hiatt EE (2001) Ergot alkaloid transport across ruminant gastric tissues. J Anim Sci 79:542–549

Hofmeister F (1892) The active constituents of *Lolium temulentum*. Arch Exp Pathol Pharmakol 30:203–230

Hong Z, Lakkineni K, Zhang Z, Verma DPS (2000) Removal of feedback inhibition of ?1-pyrroline-5-carboxylate synthetase results in increased proline accumulation and protection of plants from osmotic stress. Plant Physiol 122:1129–1136

Johnson RC, Tieszen LL (1993) Carbon isotope discrimination, water relations, and gas exchange in temperate grass species and accessions. In: Ehleringer JR, Hall AE, Farquhar GD (eds) Stable isotopes and plant carbon–water relations. Academic, San Diego, pp 281–296

Kimmons CA, Gwinn KD, Bernard EC (1990) Nematode reproduction on endophyte-infected and endophyte-free tall fescue. Plant Dis 74:757–761

Koga H, Christensen MJ, Bennett RJ (1993) Incompatibility of some grass/*Acremonium* endophyte associations. Mycol Res 97:1237–1244

Kover PX, Dolan TE, Clay K (1997) Potential versus actual contribution of vertical transmission to pathogen fitness. Proc R Soc Lond B 264:903–909

Kozikowski AP, Wu J-P, Shibuya M, Floss HG (1988) Probing ergot alkaloid biosynthesis: identification of advanced intermediates along the biosynthetic pathway. J Am Chem Soc 110:1970–1971

Kozikowski AP, Chen C, Wu J-P, Shibuya M, Kim C-G, Floss HG (1993) Probing ergot alkaloid biosynthesis: intermediates in the formation of ring C. J Am Chem Soc 115:2482–2488

Kuldau GA, Tsai H-F, Schardl CL (1999) Genome sizes of *Epichloë* species and anamorphic hybrids. Mycologia 91:776–782

Kutil BL, Greenwald C, Liu G, Spiering MJ, Schardl CL, Wilkinson HH (2007) Comparison of loline alkaloid gene clusters across fungal endophytes: predicting the co-regulatory sequence motifs and the evolutionary history. Fungal Genet Biol 44:1002–1010

Lambeth JD (2004) NOX enzymes and the biology of reactive oxygen. Nat Rev Immunol 4:181–189

Lane GA, Christensen MJ, Miles CO (2000) Coevolution of fungal endophytes with grasses: the significance of secondary metabolites. In: Bacon CW, White JF Jr (eds) Microbial endophytes. Marcel Dekker, New York, pp 341–388

Latch GCM, Christensen MJ (1985) Artificial infections of grasses with endophytes. Ann Appl Biol 107:17–24

Lemons A, Clay K, Rudgers JA (2005) Connecting plant-microbial interactions above and belowground: a fungal endophyte affects decomposition. Oecologia 145:595–604

Leuchtmann A (2007) *Botanophila* flies on *Epichloë* host species in Europe and North America: no evidence for co-evolution. Entomol Exp Appl 123:13–23

Leuchtmann A, Clay K (1990) Isozyme variation in the *Acremonium/Epichloë* fungal endophyte complex. Phytopathology 80:1133–1139

Leuchtmann A, Schardl CL (1998) Mating compatibility and phylogenetic relationships among two new species of *Epichloë* and other congeneric European species. Mycol Res 102:1169–1182

Leuchtmann A, Schardl CL, Siegel MR (1994) Sexual compatibility and taxonomy of a new species of *Epichloë* symbiotic with fine fescue grasses. Mycologia 86:802–812

Lewis DH (1985) Symbiosis and mutualism: crisp concepts and soggy semantics. In: Boucher DH (ed) The biology of mutualism. Oxford University Press, New York, pp 29–39

Leyronas C, Raynal G (2008) Role of fungal ascospores in the infection of orchardgrass (*Dactylis glomerata*) by *Epichloe typhina* agent of choke disease. J Plant Pathol 90:15–21

Li C, Nan ZB, Paul VH, Dapprich PD, Liu Y (2004) A new *Neotyphodium* species symbiotic with drunken horse grass (*Achnatherum inebrians*) in China. Mycotaxon 90:141–147

Li C, Zhang X, Li F, Nan Z, Schardl CL (2007) Disease and pest resistance of endophyte infected and non-infected drunken horse grass. In: Popay A, Thom ER (eds) Proceedings of the sixth international symposium on fungal endophytes of grasses. New Zealand Grassland Association, Dunedin, pp 111–114

Li HM, Crouch JA, Belanger FC (2005) Fungal endophyte *N*-acetylglucosaminidase expression in the infected host grass. Mycol Res 3:363–373

Lindstrom JT, Sun S, Belanger FC (1993) A novel fungal protease expressed in endophytic infection of Poa species. Plant Physiol 102:645–650

Lorenz N, Wilson EV, Machado C, Schardl C, Tudzynski P (2007) Comparison of ergot alkaloid biosynthesis gene clusters in *Claviceps* species indicate loss of late pathway steps in evolution of *C. fusiformis*. Appl Environ Microbiol (in press)

Lyons PC, Plattner RD, Bacon CW (1986) Occurrence of peptide and clavine ergot alkaloids in tall fescue grass. Science 232:487–489

Madej CW, Clay K (1991) Avian seed preference and weight loss experiments: the role of fungal endophyte-infected tall fescue seeds. Oecologia 88:296–302

Maier W, Schumann B, Gröger D (1988) Microsomal oxygenases involved in ergoline alkaloid biosynthesis of various *Claviceps* strains. J Basic Microbiol 28:83–93

Malinowski DP, Belesky DP (1999) *Neotyphodium coenophialum*–endophyte infection affects the ability of tall fescue to use sparingly available phosphorus. J Plant Nutr 22:835–853

Malinowski DP, Belesky DP (2000) Adaptations of endophyte-infected cool-season grasses to environmental stresses: mechanisms of drought and mineral stress tolerance. Crop Sci 40:923–940

Malinowski DP, Brauer DK, Belesky DP (1999) The endophyte *Neotyphodium coenophialum* affects root morphology of tall fescue grown under phosphorus deficiency. J Agron Crop Sci 183:53–60

McMillan LK, Carr RL, Young CA, Astin JW, Lowe RGT, Parker EJ, Jameson GB, Finch SC, Miles CO, McManus OB, Schmalhofer WA, Garcia ML, Kaczorowski GJ, Goetz M, Tkacz JS, Scott B (2003) Molecular analysis of two cytochrome P450 monooxygenase genes required for paxilline biosynthesis in *Penicillium paxilli,* and effects of paxilline intermediates on mammalian maxi-K ion channels. Mol Genet Genomics 270:9–23

Michalakis Y, Olivieri I, Renaud F, Raymond M (1992) Pleiotropic action of parasites: how to be good for the host. Trends Ecol Evol 7:59–62

Miles CO, Lane GA, Di Menna ME, Garthwaite I, Piper EL, Ball OJP, Latch GCM, Allen JM, Hunt MB, Bush LP, Min FK, Fletcher I, Harris PS (1996) High levels of ergonovine and lysergic acid amide in toxic *Achnatherum inebrians* accompany infection by an *Acremonium*-like endophytic fungus. J Agric Food Chem 44:1285–1290

Moon CD, Scott B, Schardl CL, Christensen MJ (2000) The evolutionary origin of Epichloë endophytes from annual ryegrasses. Mycologia 92:1103–1118

Moon CD, Craven KD, Leuchtmann A, Clement SL, Schardl CL (2004) Prevalence of interspecific hybrids amongst asexual fungal endophytes of grasses. Mol Ecol 13:1455–1467

Morse LJ, Day TA, Faeth SH (2002) Effect of Neotyphodium endophyteinfection on growth and leaf gas exchange of Arizona fescue under contrasting water availability regimes. Environ Exper Bot 48:257–268

Mothes KV, Weygand F, Gröger D, Grisebach H (1958) Untersuchungen zur Biosynthese der Mutterkorn-Alkaloide. Z Naturforsch Teil B 13b:41–44

Moy M, Li HJM, Sullivan R, White JF, Belanger FC (2002) Endophytic fungal β-1,6-glucanase expression in the infected host grass. Plant Physiol 130:1298–1308

Müller CB, Krauss J (2005) Symbiosis between grasses and asexual fungal endophytes. Curr Opin Plant Biol 8:450–456

Munday-Finch SC, Wilkins AL, Miles CO (1998) Isolation of lolicine A, lolicine B, lolitriol, and lolitrem N from *Lolium perenne* infected with *Neotyphodium lolii* and evidence for the natural occurrence of 31-epilolitrem N and 31-epilolitrem F. J Agric Food Chem 46:590–598

Omacini M, Chaneton EJ, Ghersa CM, Müller CB (2001) Symbiotic fungal endophytes control insect host-parasite interaction webs. Nature 409:78–81

Otsuka H, Quigley FR, Gröger D, Anderson JA, Floss HG (1980) In vivo and in vitro evidence for *N*-methylation as the second pathway-specific step in ergoline biosynthesis. Planta Med 40:109–119

Pan JJ, Clay K (2003) Infection by the systemic fungus *Epichloë glyceriae* alters clonal growth of its grass host, *Glyceria striata*. Proc R Soc Lond B 270:1585–1591

Pan JJ, Clay K (2004) *Epichloë glyceriae* infection affects carbon translocation in the clonal grass *Glyceria striata*. New Phytol 164:467–475

Panaccione DG, Coyle CM (2005) Abundant respirable ergot alkaloids from the common airborne fungus *Aspergillus fumigatus*. Appl Environ Microbiol 71:3106–3111

Panaccione DG, Johnson RD, Wang JH, Young CA, Damrongkool P, Scott B, Schardl CL (2001) Elimination of ergovaline from a grass-Neotyphodium endophyte symbiosis by genetic modification of the endophyte. Proc Natl Acad Sci USA 98:12820–12825

Panaccione DG, Tapper BA, Lane GA, Davies E, Fraser K (2003) Biochemical outcome of blocking the ergot alkaloid pathway of a grass endophyte. J Agric Food Chem 51:6429–6437

Panaccione DG, Kotcon JB, Schardl CL, Johnson RD, Morton JB (2006a) Ergot alkaloids are not essential for endophytic fungus-associated population suppression of the lesion nematode, *Pratylenchus scribneri*, on perennial ryegrass. Nematology 8:583–590

Panaccione DG, Cipoletti JR, Sedlock AB, Blemings KP, Schardl CL, Machado C, Seidel GE (2006b) Effects of ergot alkaloids on food preference and satiety in rabbits, as assessed with gene-knockout endophytes in perennial ryegrass (*Lolium perenne*). J Agric Food Chem 54:4582–4587

Patterson CG, Potter DA, Fannin FF (1991) Feeding deterrency of alkaloids from endophyte-infected grasses to Japanese beetle grubs. Entomol Exp Appl 285–289

Petroski R, Powell RG, Clay K (1992) Alkaloids of *Stipa robusta* (sleepygrass) infected with an *Acremonium* endophyte. Nat Toxins 1:84–88

Phelps RA, Morgan-Jones G, Owsley MR (1993) Systematic and biological studies in the Balansieae and related anamorphs. VII. Host-pathogen relationship of *Eragrostis capillaris* and *Balansia epichloë*. Mycotaxon 49:117–127

Philipson MN, Christey MC (1986) The relationship of host and endophyte during flowering, seed formation, and germination of *Lolium perenne*. NZ J Bot 24:125–134

Porter JK (1994) Chemical constituents of grass endophytes. In: Bacon CW, White JF Jr (eds) Biotechnology of endophytic fungi of grasses. CRC, Boca Raton, pp 103–123

Potter DA, Stokes JT, Redmond CT, Schardl CL, Panaccione DG (2008) Contribution of ergot alkaloids to suppression of a grass-feeding caterpillar assessed with gene-knockout endophytes in perennial ryegrass. Entomol Exp Appl (in press)

Powell RG, Petroski RJ (1992) The loline group of pyrrolizidine alkaloids. In: Pelletier SW (ed) Alkaloids: chemical and biological perspectives. Springer, New York, pp 320–338

Powell RG, Plattner RD, Yates SG, Clay K, Leuchtmann A (1990) Ergobalansine, a new ergot-type peptide alkaloid isolated from *Cenchrus echinatus* (sandbur grass) infected with *Balansia obtecta*, and produced in liquid cultures of *B. obtecta* and *Balansia cyperi*. J Nat Prod Lloydia 53:1272–1279

Prestidge RA, Lauren DR, van der Zujpp SG, di Menna ME (1985) Isolation of feeding deterrents to Argentine stem weevil in cultures of endophytes of perennial ryegrass and tall fescue. NZ J Agric Res 28:87–92

Raisbeck MF, Rottinghaus GE, Kendall JD (1991) Effects of naturally occurring mycotoxins on ruminants. In: Smith JE, Henderson RS (eds) Mycotoxins and animal foods. CRC, Boca Raton, pp 647–677

Richardson MD, Chapman GW, Hoveland CS, Bacon CW (1992) Sugar alcohols in endophyte infected tall fescue under drought. Crop Sci 32:1060–1064

Richardson MD, Hoveland CS, Bacon CW (1993) Photosynthesis and stomatal conductance of symbiotic and nonsymbiotic tall fescue. Crop Sci 33:145–149

Riedell WE, Kieckhefer RE, Petroski RJ, Powell RG (1991) Naturally occurring and synthetic loline alkaloid derivatives: insect feeding behavior modification and toxicity. J Entomol Sci 26:122–129

Riederer B, Han M, Keller U (1996) D-lysergyl peptide synthetase from the ergot fungus *Claviceps purpurea*. J Biol Chem 271:27524–27530

Rao S, Baumann D (2004) The interaction of a *Botanophila* fly species with an exotic *Epichloë* fungus in a cultivated grass: fungivore or mutualist? Entomol Exp Appl 112:99–105

Rao S, Alderman SC, Takeyasu J, Matson B (2005) The *Botanophila–Epichloe* association in cultivated *Festuca* in Oregon: evidence of simple fungivory. Entomol Exp Appl 115:427–433

Rottinghaus GE, Garner GB, Cornel CN, Ellis JL (1991) HPLC method for quantitating ergovaline in endophyte-infested tall fescue: seasonal variation of ergovaline levels in stems with leaf sheaths, leaf blades, and seed heads. J Agric Food Chem 39:112–115

Rowan DD (1993) Lolitrems, peramine and paxilline: mycotoxins of the ryegrass/endophyte interaction. Agric Ecosyst Environ 44:103–122

Rowan DD, Hunt MB, Gaynor DL (1986) Peramine, a novel insect feeding deterrent from ryegrass infected with the endophyte *Acremonium loliae*. J Chem Soc Chem Commun 12:935–936

Rowan DD, Dymock JJ, Brimble MA (1990) Effect of fungal metabolite peramine and analogs on feeding and development of Argentine stem weevil (*Listronotus bonariensis*). J Chem Ecol 16:1683–1695

Rudgers JA, Holah J, Orr SP, Clay K (2007) Forest succession suppressed by an introduced plant-fungal symbiosis. Ecology 88:18–25

Saikia S, Parker EJ, Koulman A, Scott B (2006) Four gene products are required for the fungal synthesis of the indole diterpene paspaline. FEBS Lett 580:1625–1630

Saikia S, Parker EJ, Koulman A, Scott B (2007) Defining paxilline biosynthesis in *Penicillium paxilli*: functional characterization of two cytochrome P450 monooxygenases. J Biol Chem 282:16829–16837

Saikia S, Nicholson MJ, Young C, Parker EJ, Scott B (2008) The genetic basis for indole-diterpene chemical diversity in filamentous fungi. Mycol Res (in press)

Saikkonen K, Faeth SH, Helander M, Sullivan TJ (1998) Fungal endophytes: a continuum of interactions with host plants. Annu Rev Ecol System 29:319–343

Saikkonen K, Helander M, Faeth SH, Schulthess F, Wilson D (1999) Endophyte–grass–herbivore interactions: the case of Neotyphodium endophytes in Arizona fescue populations. Oecologia 121:411–420

Saikkonen K, Ion D, Gyllenberg M (2002) The persistence of vertically transmitted fungi in grass metapopulations. Proc R Soc Lond B 269:1397–1403

Sampson K (1933) The systemic infection of grasses by *Epichloë typhina* (Pers.) Tul. Trans Br Mycol Soc 18:30–47

Sampson K (1935) The presence and absence of an endophytic fungus in *Lolium temulentum* and *L. perenne*. Trans Br Mycol Soc 19:337–343

Sampson K (1937) Further observations on the systemic infection of *Lolium*. Trans Br Mycol Soc 21:84–97

Schardl CL (1996) *Epichloë* species: fungal symbionts of grasses. Annu Rev Phytopathol 34:109–130

Schardl CL (2001) *Epichloë festucae* and related mutualistic symbionts of grasses. Fungal Genet Biol 33:69–82

Schardl CL, Clay K (1997) Evolution of mutualistic endophytes from plant pathogens. In: Carroll GC, Tudzynski P (eds) Plant relationships. The Mycota, vol V.B. Springer, Berlin, pp 221–238

Schardl CL, Craven KD (2003) Interspecific hybridization in plant-associated fungi and oomycetes: a review. Mol Ecol 12:2861–2873

Schardl CL, Leuchtmann A (1999) Three new species of *Epichloë* symbiotic with North American grasses. Mycologia 91:95–107

Schardl CL, Leuchtmann A, Tsai HF, Collett MA, Watt DM, Scott DB (1994) Origin of a fungal symbiont of perennial ryegrass by interspecific hybridization of a mutualist with the ryegrass choke pathogen, *Epichloë typhina*. Genetics 136:1307–1317

Schardl CL, Leuchtmann A, Spiering MJ (2004) Symbioses of grasses with seedborne fungal endophytes. Annu Rev Plant Biol 55:315–340

Schardl CL, Panaccione DG, Tudzynski P (2006) Ergot alkaloids – biology and molecular biology. Alkaloids Chem Biol 63:45–86

Schardl CL, Grossman RB, Nagabhyru P, Faulkner JR, Mallik UP (2007) Loline alkaloids: currencies of mutualism. Phytochemistry 68:980–996

Schardl CL, Craven KD, Speakman S, Stromberg A, Lindstrom A, Yoshida R (2008) A novel test for host–symbiont codivergence indicates ancient origin of fungal endophytes in grasses. Syst Biol 57:483–498

Shelby RA, Olsovska J, Havlicek V, Flieger M (1997) Analysis of ergot alkaloids in endophyte-infected tall fescue by liquid chromatography electrospray ionization mass spectrometry. J Agric Food Chem 45:4674–4679

Siegel MR, Johnson MC, Varney DR, Nesmith WC, Buckner RC, Bush LP, Burrus PB, II., Jones TA, Boling JA (1984) A fungal endophyte in tall fescue: incidence and dissemination. Phytopathology 74:932–937

Siegel MR, Latch GCM, Bush LP, Fannin FF, Rowan DD, Tapper BA, Bacon CW, Johnson MC (1990) Fungal endophyte-infected grasses: alkaloid accumulation and aphid response. J Chem Ecol 16:3301–3315

Spatafora JW, Sung GH, Sung JM, Hywel-Jones NL, White JF (2007) Phylogenetic evidence for an animal pathogen origin of ergot and the grass endophytes. Mol Ecol 16:1701–1711

Spiering MJ, Davies E, Tapper BA, Schmid J, Lane GA (2002a) Simplified extraction of ergovaline and peramine for analysis of tissue distribution in endophyte-infected grass tillers. J Agric Food Chem 50:5856–5862

Spiering MJ, Wilkinson HH, Blankenship JD, Schardl CL (2002b) Expressed sequence tags and genes associated with loline alkaloid production in the grass endophyte *Neotyphodium uncinatum*. Fungal Genet Biol 36:242–254

Spiering MJ, Moon CD, Wilkinson HH, Schardl CL (2005) Gene clusters for insecticidal loline alkaloids in the grass-endophytic fungus *Neotyphodium uncinatum*. Genetics 169:1403–1414

Spiering MJ, Faulkner JR, Machado C, Zhang D-X, Grossman RB, Schardl CL (2008) Role of the LolP cytochrome P450 in biosynthesis of *N*-formylloline. Fungal Genet Biol (in press)

Spiers DE, Zhang Q, Eichen PA, Rottinghaus GE, Garner GB, Ellersieck MR (1995) Temperature-dependent responses of rats to ergovaline derived from endophyte-infected tall fescue. J Anim Sci 73:1954–1961

Steiner U, Ahimsa-Müller MA, Markert A, Kucht S, Groß J, Kauf N, Kuzma M, Zych M, Lamshöft M, Furmanowa M, Knoop V, Drewke C, Leistner E (2006) Molecular characterization of a seed transmitted clavicipitaceous fungus occurring on dicotyledoneous plants (Convolvulaceae). Planta 224:533–544

Strack D, Fester T, Hause B, Schliemann W, Walter MH (2003) Arbuscular mycorrhiza: biological, chemical, and molecular aspects. J Chem Ecol 29:1955–1979

Stuedemann JA, Hoveland CS (1988) Fescue endophyte: history and impact on animal agriculture. J Prod Agric 1:39–44

Takemoto D, Tanaka A, Scott B (2006) A p67Phox-like regulator is recruited to control hyphal branching in a fungal-grass mutualistic symbiosis. Plant Cell 18:2807–2821

Takemoto D, Tanaka A, Scott B (2007) NADPH oxidases in fungi: diverse roles of reactive oxygen species in fungal cellular differentiation. Fungal Genet Biol 44 (in press)

Tan YY, Spiering MJ, Scott V, Lane GA, Christensen MJ, Schmid J (2001) In planta regulation of extension of an endophytic fungus and maintenance of high metabolic rates in its mycelium in the absence of apical extension. Appl Environ Microbiol 67:5377–5383

Tanaka A, Tapper BA, Popay A, Parker EJ, Scott B (2005) A symbiosis expressed non-ribosomal peptide synthetase from a mutualistic fungal endophyte of perennial ryegrass confers protection to the symbiotum from insect herbivory. Mol Microbiol 57:1036–1050

Tanaka A, Christensen MJ, Takemoto D, Park P, Scott B (2006) Reactive oxygen species play a role in regulating a fungus-perennial ryegrass mutualistic association. Plant Cell 18:1052–1066

Tanaka A, Takemoto D, Hyon G-S, Park P, Scott B (2008) NoxA activation by the small GTPase RacA is required to maintain a mutualistic symbiotic association between *Epichloë festucae* and perennial ryegrass. Mol Microbiol 68:1165–1178

TePaske MR, Powell RG, Clement SL (1993) Analyses of selected endophyte-infected grasses for the presence of loline-type and ergot-type alkaloids. J Agric Food Chem 41:2299–2303

Thompson FN, Stuedemann JA (1993) Pathophysiology of fescue toxicosis. Agric Ecosyst Environ 44:263–281

Timper P, Gates RN, Bouton JH (2005) Response of *Pratylenchus* spp. in tall fescue infected with different strains of the fungal endophyte *Neotyphodium coenophialum*. Nematology 7:105–110

Tor-Agbidye J, Blythe LL, Craig AM (2001) Correlation of endophyte toxins (ergovaline and lolitrem B) with clinical disease: fescue foot and perennial ryegrass staggers. Vet Hum Toxicol 43:140–146

Torres MA, Dangl JL (2005) Functions of the respiratory burst oxidase in biotic interactions, abiotic stress and development. Curr Opin Plant Biol 8:397–403

Tsai H-F, Liu JS, Staben C, Christensen MJ, Latch G, Siegel MR, Schardl CL (1994) Evolutionary diversification of fungal endophytes of tall fescue grass by hybridization with *Epichloë* species. Proc Natl Acad Sci USA 91:2542–2546

Tsai H-F, Wang H, Gebler JC, Poulter CD, Schardl CL (1995) The *Claviceps purpurea* gene encoding dimethylallyltryptophan synthase, the committed step for ergot alkaloid biosynthesis. Biochem Biophys Res Commun 216:119–125

Tudzynski P, Hölter K, Correia T, Arntz C, Grammel N, Keller U (1999) Evidence for an ergot alkaloid gene cluster in *Claviceps purpurea*. Mol Gen Genet 261:133–141

Tudzynski P, Correia T, Keller U (2001) Biotechnology and genetics of ergot alkaloids. Appl Microbiol Biotechnol 57:593–605

van den Burg HA, Harrison SJ, Joosten MH, Vervoort J, de Wit PJ (2006) *Cladosporium fulvum* Avr4 protects fungal cell walls against hydrolysis by plant chitinases accumulating during infection. Mol Plant–Microbe Interact 19:1420–1430

Veen G, Greinwald R, Canto P, Witte L, Czygan FC (1992) Alkaloids of *Adenocarpus hispanicus* (Lam.) DC varieties. Z Naturforsch Tiel C 47:341–345

Vining LC (1973) Physiological aspects of alkaloid production by *Claviceps* species. In: Vanek Z, Hostálek Z, Cudlín J (eds) Genetics of industrial microorganisms II. Actinomycetes and fungi. Elsevier, Amsterdam, pp 405–419

Wang J (2000) *dmaW* encoding tryptophan dimethylallyltransferase in ergot alkaloid biosynthesis from clavicipitaceous fungi. Dissertation, University of Kentucky

Wang J, Machado C, Panaccione DG, Tsai H-F, Schardl CL (2004) The determinant step in ergot alkaloid biosynthesis by an endophyte of perennial ryegrass. Fungal Genet Biol 41:189–198

Watson RH, McCann MA, Parish JA, Hoveland CS, Thompson FN, Bouton JH (2004) Productivity of cow-calf pairs grazing tall fescue pastures infected with either the wild-type endophyte or a nonergot alkaloid-producing endophyte strain, AR542. J Anim Sci 82:3388–3393

White J Jr (1997) Perithecial ontogeny in the fungal genus *Epichloë*: An examination of the clavicipitalean centrum. Am J Bot 84:170–178

White J Jr, Glenn AE (1994) A study of two fungal epibionts of grasses: Structural features, host relationships, and classification in the genus *Myriogenospora* (Clavicipitales). Am J Bot 81:216–223

White JF Jr, Sharp LT, Martin TI, Glenn AE (1995) Endophyte–host associations in grasses. XXI. Studies on the structure and development of *Balansia obtecta*. Mycologia 87:172–181

White JF Jr, Reddy PV (1998) Examination of structure and molecular phylogenetic relationships of some graminicolous symbionts in genera *Epichloë* and *Parepichloë*. Mycologia 90:226–234

White JF Jr, Reddy PV, Glenn AE, Bacon CW (1997) An examination of structural features and relationships in *Balansia* subgenus *Dothichloë*. Mycologia 89:408–419

White RH, Engelke MC, Morton SJ, Johnson-Cicalese JM, Ruemmele BA (1992) *Acremonium* endophyte effects on tall fescue drought tolerance. Crop Sci 32:1392–1396

Wilkinson HH, Siegel MR, Blankenship JD, Mallory AC, Bush LP, Schardl CL (2000) Contribution of fungal loline alkaloids to protection from aphids in a grass–endophyte mutualism. Mol Plant–Microbe Interact 13:1027–1033

Yates SG, Fenster JC, Bartelt RJ (1989) Assay of tall fescue seed extracts, fractions, and alkaloids using the large milkweed bug. J Agric Food Chem 37:354–357

Young C, McMillan L, Telfer E, Scott B (2001) Molecular cloning and genetic analysis of an indole-diterpene gene cluster from *Penicillium paxilli*. Mol Microbiol 39:754–764

Young CA, Bryant MK, Christensen MJ, Tapper BA, Bryan GT, Scott B (2005) Molecular cloning and genetic analysis of a symbiosis-expressed gene cluster for lolitrem biosynthesis from a mutualistic endophyte of perennial ryegrass. Mol Genet Genomics 274:13–29

Young CA, Felitti S, Shields K, Spangenberg G, Johnson RD, Bryan GT, Saikia S, Scott B (2006) A complex gene cluster for indole-diterpene biosynthesis in the grass endophyte *Neotyphodium lolii*. Fungal Genet Biol 43:679–693

Yue Q, Miller CJ, White JF Jr, Richardson MD (2000) Isolation and characterization of fungal inhibitors from *Epichloë festucae*. J Agric Food Chem 48:4687–4692

Zabalgogeazcoa I, Ciudad AG, de Aldana BR, Criado BG (2006) Effects of the infection by the fungal endophyte *Epichloë festucae* in the growth and nutrient content of *Festuca rubra*. Eur J Agron 24:374–384

Zhang NX, Scott V, Al-Samarrai TH, Tan YY, Spiering MJ, McMillan LK, Lane GA, Scott DB, Christensen MJ, Schmid J (2006) Transformation of the ryegrass endophyte *Neotyphodium lolii* can alter its in planta mycelial morphology. Mycol Res 110:601–611

16 Lichen-Forming Fungi and Their Photobionts

ROSMARIE HONEGGER[1]

CONTENTS

I. Introduction........................... 307
II. Lichen-Forming Fungi 309
   A. Species Concepts 309
   B. Intrathalline Diversity................. 310
   C. Gains and Losses of Lichenization 311
   D. Fossil Lichens 312
III. Lichen Photobionts 312
   A. Photobiont Diversity.................. 312
   B. Recognition and Specificity 313
   C. Symbiotic Propagules and Potential
      Vectors............................ 313
   D. Photobiont-Derived Mobile
      Carbohydrates 314
IV. Functional Thallus Anatomy 315
   A. Main Building Blocks 315
   B. Mycobiont–Photobiont Interactions........ 316
   C. Wall Surface Hydrophobicity and Gas
      Exchange.......................... 317
   D. Hydrophilic Wall Components.......... 318
V. Water Relations and Ecology.............. 318
   A. Reversible Cytoplasmatic Cavitation....... 318
   B. Foliicolous Lichens 318
   C. Lichens of Extreme Climates 319
   D. Poikilodyric Water Relations 319
   E. Ice Nucleation Sites 322
VI. Uptake of Elements 322
   A. Ion Uptake and Accumulation.......... 322
   B. Radionuclides....................... 323
   C. Sources of Fixed Nitrogen 323
VII. Secondary Metabolism 324
   A. Mycobiont-Derived Lichen Compounds..... 324
   B. Economic Perspectives 324
VIII. Conclusions and Outlook.................. 325
   References............................. 326

I. Introduction

Lichens are the symbiotic phenotype of nutritionally specialized fungi, ecologically obligate biotrophs which acquire fixed carbon from a population of minute photobiont cells (Honegger 1991). Lichen-forming fungi (also referred to as lichen mycobionts) are, like plant or animal pathogens or mycorrhizal fungi, a polyphyletic, taxonomically diverse group of nutritional specialists, but are otherwise normal representatives of their fungal classes. They differ from non-lichenized taxa by their manifold adaptations to symbiosis with a population of minute photobiont cells. Lichenization is a successful nutritional strategy, almost 20% of all fungal species being lichenized (Kirk et al. 2001). More than 10% of terrestrial ecosystems are lichen-dominated; these are the sites where vascular plants are at their physiological limits: high alpine, arctic, antarctic and desert ecosystems.

In contrast to the parasitic fungi of cyanobacteria or green algae, such as the necrotrophic *Athelia arachnoidea*, which destroys free-living aerophilic algae, lichens and mosses and is the sexual state of the plant pathogenic *Rhizoctonia carotae* (Adams and Kropp 1996), lichen-forming fungi do not kill their photoautotrophic partner. As ecologically obligate biotrophs they are found in nature almost exclusively in the symbiotic state, exceptions being germ tubes derived either from sexual spores or from aposymbiotic vegetative propagules in search of a compatible photobiont. The majority of lichen-forming ascomycetes are physiologically facultatively biotrophic, i.e. can be axenically cultured in the aposymbiotic state (Ahmadjian 1993; Crittenden et al. 1995).

Among the diverse fungal interactions with cyanobacteria or algae only extracellular symbioses are referred to as lichens in which the fungal partner is either quantitatively predominant or at least an equal partner and, in the vast majority of taxa, is the exhabitant. Neither non-parasitic fungal endobionts of seaweeds, such as the ubiquitous *Mycophycias ascophylli* (= *Mycosphaerella ascophylli*) on the intertidal brown algae *Ascophyllum nodosum* and *Pelvetia canaliculata* (Fucales; Kohlmeier and Kohlmeier 1972, 1998; Deckert and Garbary 2005), nor the enigmatic *Geosiphon pyriforme* with its *Nostoc* endocyanosis are considered as lichens. *G. pyriforme* is the only Glomeromycete known so far which does not form mycorrhizae

---

[1] Institute of Plant Biology, University of Zürich, Zollikerstrasse 107, 8008 Zürich, Switzerland; e-mail: rohonegg@botinst.unizh.ch

Plant Relationships, 2nd Edition
The Mycota V
H. Deising (Ed.)
© Springer-Verlag Berlin Heidelberg 2009

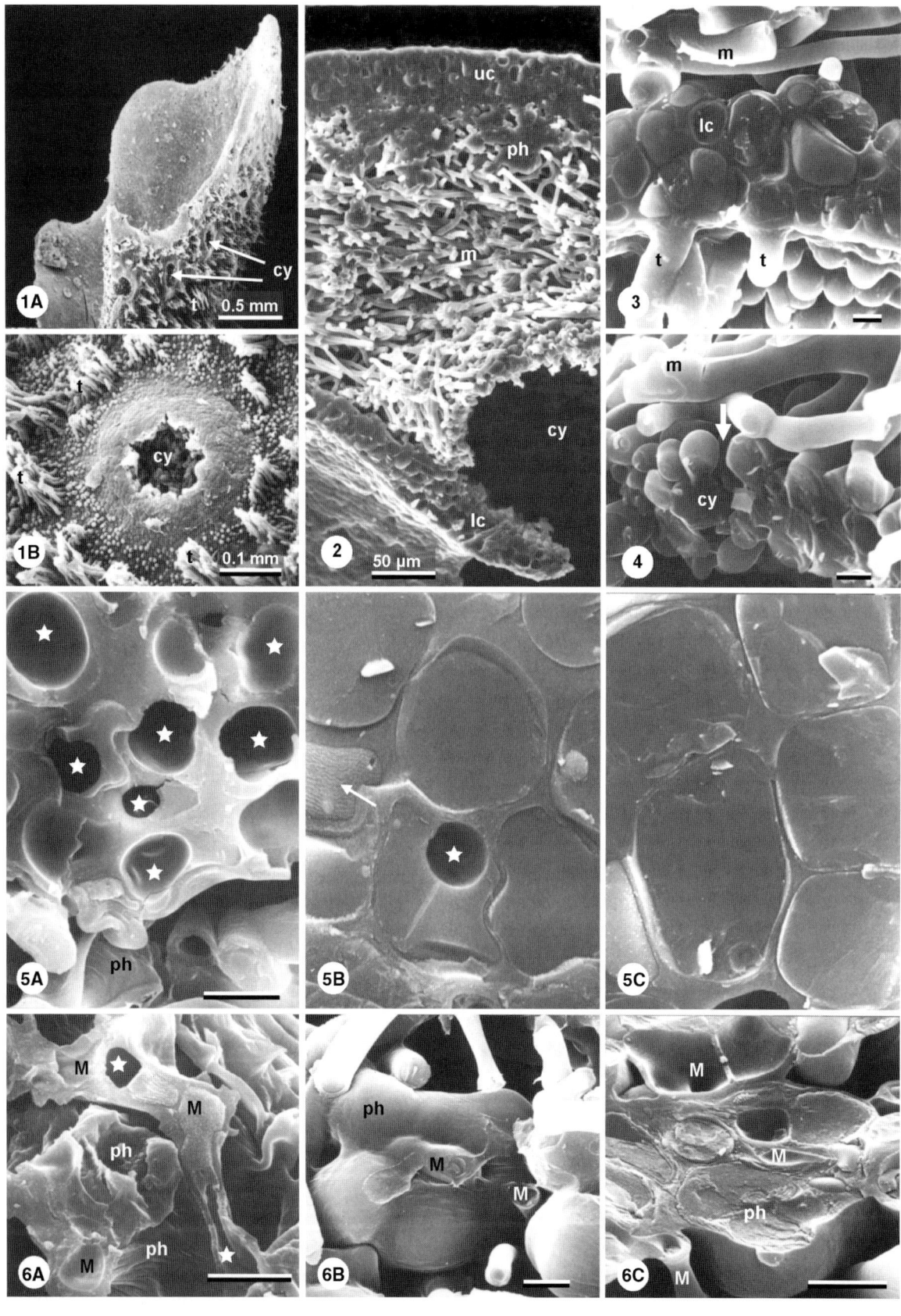

with plant roots, but incorporates filaments of the diazotrophic cyanobacterium *N. punctiforme* in a phagocytosis-like process (Mollenhauer 1992; Kluge et al. 1994; Mollenhauer et al. 1996; Schüssler et al. 1995, 2001; Schüssler and Kluge 2001; Wolf 2003). The *Nostoc* filament is subsequently housed in a large, sausage-shaped, membrane-bound vesicle, functionally a phagosome, within the fungal protoplast. In response to the successful acquisition of a cyanobiont the fungus differentiates an approximately 2 mm long, ovoid, translucent unicellular "bladder" at the soil surface, which receives its greyish-green colour from the phycocyanin pigments of the *Nostoc* cells. These divide and achieve a larger cell size while living in the membrane-bound vesicle within the bladder of *G. pyriforme* than they do in the non-symbiotic state. Only heterocyst-free filaments are phagocytosed, but heterocysts differentiate among the dividing cells of the symbiotic filament, which provides the fungal partner with fixed carbon and fixed nitrogen (Mollenhauer 1992; Kluge et al. 1994; Schüssler and Kluge 2001; Schüssler 2002). In contrast to cyanelles of Glaucophytes, which have lost their genetic autonomy since parts of their genome are incorporated in the nucleus of their heterotrophic exhabitant, *N. punctiforme* cells retain their genetic autonomy in symbiosis with *Geosiphon*.

II. Lichen-Forming Fungi

A. Species Concepts

Species names of lichens refer to the fungal partner; the photoautotrophic symbionts have their own names and phylogenies (Greuter et al. 2000). About 14 000 species of lichen-forming fungi are so far described, approximately 99% of them being ascomycetes classified in the subphylum Pezizomycotina (Kirk et al. 2001; Eriksson 2006a). Traditionally lichen-forming ascomycetes are classified at the suprageneric level on the basis of features related to sexual reproduction such as ascomal ontogeny, ascus structure and function, ascospore structure and pigmentation, and at the subgeneric level mostly on the basis of morphological, anatomical and chemical characters of the symbiotic phenotype (morpho- and chemospecies). As lichen-forming ascomycetes do not routinely and reliably differentiate sexual reproductive structures under laboratory conditions the biological

Figs. 16.1–16.6. Thallus anatomy, hyphal polymorphism and multifunctionality and drought stress-induced structural alterations at the cellular level in the foliose macrolichen *Sticta sylvatica* (Peltigerales; cynobacterial photobiont: *Nostoc* sp.), as observed in conventionally prepared (16.1) and in cryofixed, frozen-hydrated specimens with low temperature scanning electron microscopy (16.2–16.6). Bars 5 μm (unless otherwise stated) **Fig. 16.1. A** Lateral view of a thallus lobe with smooth upper and tomentose (*t*) lower surface with numerous cyphellae (*cy*); these are specialized aeration pores. **B** Laminal view of cyphella (*cy*) and the tomentum (*t*) **Fig. 16.2.** Vertical cross-section (fracture) of the fully hydrated thallus with conglutinate upper cortex (*uc*), photobiont layer (*ph*), gas-filled medullary layer (*m*) and conglutinate lower cortex (*lc*). Part of a cyphella (*cy*) is fractured. Note: free water within the thallus is confined to the symplast and the apoplast. The medullary and photobiont layers are gas-filled, even at water saturation **Fig. 16.3.** Detail of the lower cortex (*lc*), adjacent medullary layer (*m*) and tomentum (*t*), revealing multiple transitions in fungal growth and changes in wall surface properties: filamentous medullary hyphae (= aerial hyphae of the thalline interior) with a hydrophobic wall surface layer switch to apolar, globose growth and secrete hydrophilic β-glucans when participating in the lower cortex (*lc*). Globose cells may switch back to filamentous, polar growth and form the hair-like hyphae of the tomentum (*t*) with a hydrophilic wall surface (→ water uptake) **Fig. 16.4.** Detail of a vertically fractured cyphella (*cy*): medullary hyphae (*m*) with a hydrophobic cell wall surface layer switch from filamentous/polar to globose/apolar growth but retain the wall surface hydrophobicity. The *arrow* points to pores between the globose cells of the cyphella which facilitate gas exchange **Fig. 16.5.** Cross-fractures of the upper cortex at different levels of hydration (A–C). **A** Desiccated state (<20% water, dry weight). *Asterisks* refer to cytoplasmic gas bubbles within the strongly condensed cytoplasm of cavitated fungal cells. Most fungal and also cyanobacterial (*ph*) cells reveal irregular outlines due to severe shrinkage. **B** Rehydration phase, 17 s after the addition of a water droplet to the thallus surface. The top cells are already ovoid in shape, their cytoplasmic gas bubble has disappeared and their plasmamembrane appears smooth. The lower cells still have irregular outlines, their cytoplasmic gas bubble is vanishing within the rehydrating cytoplasm. The *arrow* points to the plasmamembrane of a not yet fully hydrated fungal cell with drought-stress-induced, fine foldings. **C** Fully hydrated cortical cell. Note the dramatic changes in cell volume between **A** and **C** **Fig. 16.6.** The mycobiont-photobiont interface in the desiccated (**A**) and fully hydrated state (**B, C**). The mycobiont (*M*) forms intragelatinous protrusions within the gelatinous sheath of the cyanobacterial colony (*ph*). *Asterisks* refer to coplasmic gas bubbles in desiccated, fractured fungal cells. Same magnification used in **A** and **C**. Modified from Honegger (1997)

Fungal sensing and correction of an upside down position by means of growth processes

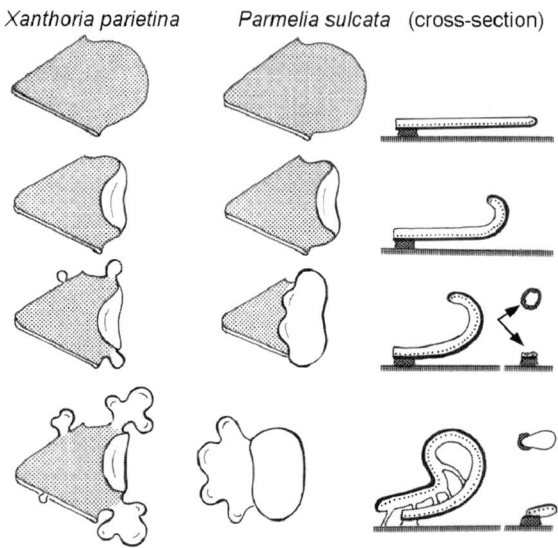

Xanthoria parietina   *Parmelia sulcata*   (cross-section)

Fig. 16.7. Diagram illustrating fungal reactions to spatial disturbance in two foliose species, *Xanthoria parietina* and *Parmelina sulcata*. Dissected thallus lobes were fixed with a small amount of Super Glue Gel in an inverted position to the substratum and incubated outdoors for several months. Many lobes broke off in a hailstorm; the *arrows* point to remains where only a few algal cells and medullary hyphae were left. These regenerated into new lobes with normal upside up position. From Honegger (1995), with permission of the publisher

species concept is not applicable, in which a species is defined as "a set of actually or potentially interbreeding populations" (Mayr 1942, 2000). With the advent of tools for inferring molecular phylogenies the phylogenetic species concept is now applicable in lichenology, phylospecies being defined as "the smallest diagnosable cluster of individual organisms within which there is a parental pattern of ancestry and descent" (Cracraft 1983).

The phylospecies concept allows characterization of cryptic species among lichenized fungi, which cannot or hardly be recognized on the basis of morphological characters; examples are the sulfur-yellow wolf lichens (genus *Letharia*) in western North America. Based on morphological criteria this genus was assumed to consist of one sympatric species pair, comprising the sexually reproducing *L. columbiana* and the vegetatively dispersing *L. vulpina*. With molecular techniques *L. columbiana* was resolved into five species: the vegetatively dispersing *L. lupina* and the sexually reproducing *L. barbata, L. lucida, L. gracilis* and *L. rugosa* (Kroken and Taylor 2001).

In many taxa of lichen-forming fungi the morpho- and phylospecies delimitations are largely in parallel. However, numerous interesting exceptions have been found, some of them referring to convergence in morphologically similar species, which had formerly been classified within the same genus based on morphological criteria, but revealed distinct molecular phylogenies. Examples are representatives of the former genus *Baeomyces,* now being split into *Baeomyces* (Baeomycetaceae) and *Dibaeis* (Icmadophilaceae) within the Lecanoromycetes (Platt and Spatafora 1999; Eriksson 2006a), or representatives of the former genus *Siphula,* a group of sterile (anamorphic) lichens, now being split into the genera *Siphula* (Icmadophilaceae) and *Parasiphula* (Coccotremataceae) within the Lecanoromycetes (Grube and Kantvilas 2006). With molecular tools the taxonomic affiliation of many other sterile taxa has been elucidated; examples are representatives of the anamorphic genera *Lepraria, Leproloma, Leprocaulon* or *Thamnolia,* which are now included in the Stereocaulaceae, Lecanoraceae and Icmadophilaceae (Platt and Spatafora 2000; Ekman and Tønsberg 2002).

Fewer than 50 species of homobasidiomycetes are distinctly lichenized; these belong to the Agaricales (*Lichenomphalia, Semiomphalia*), Atheliales (*Dictyonema*), Cantharellales (*Multiclavula*) and Polyporales (*Lepidostroma*), as summarized by Nelsen et al. (2007).

B.   Intrathalline Diversity

Most investigators assume lichens to be a dual or, in the case of cephalodiate species, a triple symbiosis, formed by a fungal partner in association with a green alga and/or a cyanobacterium. However, beside these main partners are innumerable pro- and eukaryotic epi- and endobionts of lichen thalli, including either symptomless fungal endophytes or parasitic, often devastating lichenicolous fungi, or lichenivorous invertebrates, even gall-producers being among the latter two groups of nutritional specialists (Hawksworth 1988a; Hawksworth and Honegger 1994; Diederich 2003; Lawrey and Diederich 2003). Epibiotic bacterial films are extremely common and widespread in taxonomically diverse lichens from all types of environments, their taxonomic affiliation and potential physiological impact on lichen symbiosis being currently investigated (Cardinale et al. 2006). Future investigations have to show whether there are any lichen-beneficial bacterial partners, analogous to the mycorrhization helper bacteria (Garbaye 1994; Aspray et al. 2006). In morphologically simple, microfilamentous thalli, but also within the tomentum and other hairy or felty appendages of surfaces of morphologically complex species or within hollow-perforate thalli of reindeer lichens (*Cladonia* spp.), a wide array of other organisms are found, including aerophilic cyanobacteria, green algae, diatoms and other algae, tardigrades, pollen grains, etc. (Lakatos et al. 2004; Honegger, unpublished data), their impact on lichen symbiosis being largely unknown. Consequently, contaminating DNA derived from epi- or endobionts often causes problems in molecular studies of lichen mycobionts and photobionts when samples are used as collected in nature. Lichens are not individuals and certainly not plants, as assumed even in the twenty-first century by many biologists, but are consortia with an unknown number of participants.

C. Gains and Losses of Lichenization

Until recently lichenization was assumed to be a nutritional strategy, which was repeatedly and independently gained in diverse groups of asco- and basidiomycetes. Recent multilocus analyses of

The functional anatomy of internally stratified thalli of lichenized ascomycetes

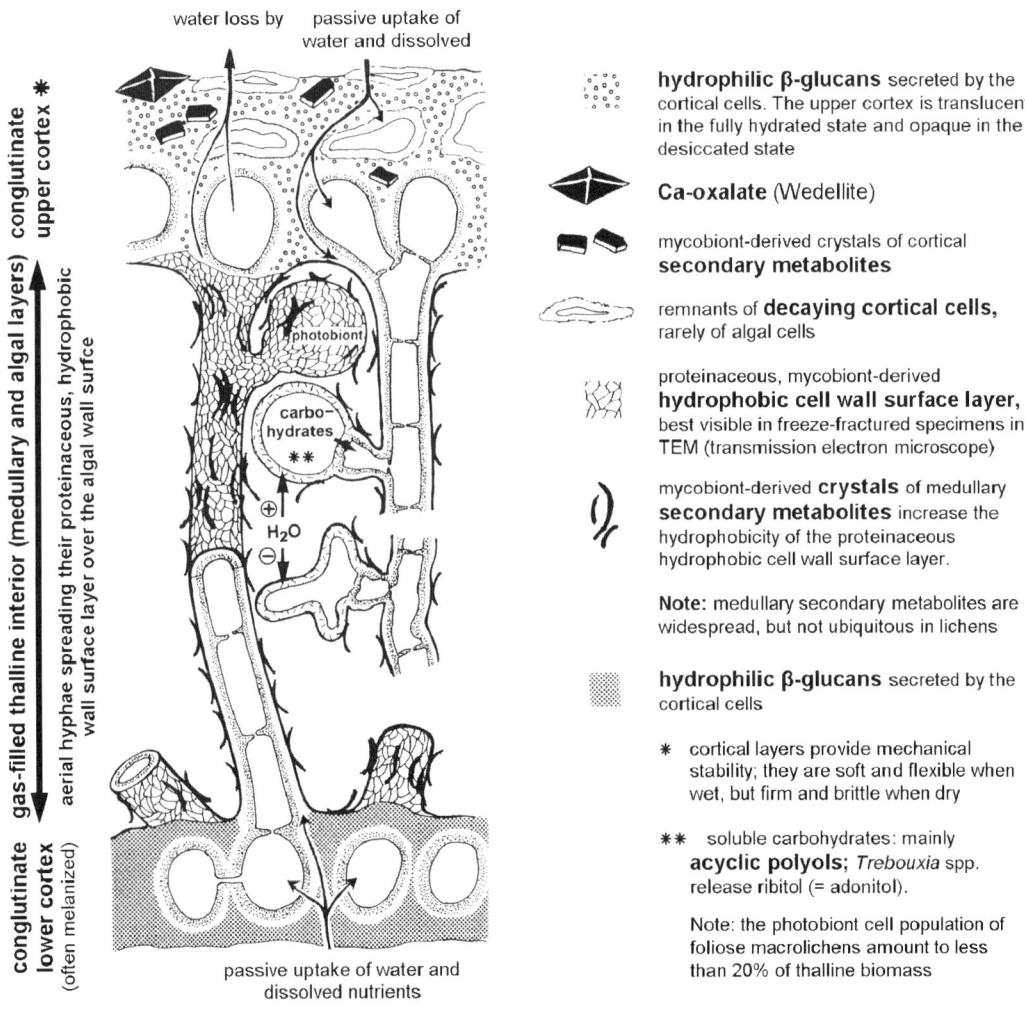

Fig. 16.8. Diagram summarizing the functional morphology and anatomy of foliose macrolichens (from Honegger 1997)

fungal phylogenies refer to multiple independent gains among homobasidiomycetes, but to fewer gains and multiple losses of lichenization in the subphylum Pezizomycotina among the ascomycetes. Thus several groups of non-lichenized ascomycetes derive from lichenized ancestors (Lutzoni et al. 2001, 2004; James et al. 2006). Molecular phylogenies suggest that the switch from the lichenized to the saprotrophic or parasitic lifestyle was relatively recently performed by Eurotiales, indicating that some of the biotechnologically most important taxa of filamentous fungi such as *Aspergillus* spp. derive from lichenized ancestors (Lutzoni et al. 2001, 2004; James et al. 2006). Accordingly, lichenization can be a transient mode of nutrition.

D. Fossil Lichens

Lichenization was speculated to be a very old fungal life style (Galloway 1996; Eriksson 2005, 2006b), but molecular phylogenies of extant lichen-forming fungi apparently do not support this view and fossil records are very scarce. Lichen-like fossils were discovered in marine phosphorites of the Doushantuo Formation in South China (approx. 600×10^6 years before present; MaBP; Yuan et al. 2005) and as epiphytes in the famous Early Devonian Rhynie Chert beds in Scotland (approx. 460 MaBP). The latter fossil, described as *Winfrenatia reticulata*, was assumed to be formed by a zygomycete-like fungus in association with a cyanobacterium, the association being morphologically dissimilar from extant lichens (Taylor et al. 1995). As no lichenized taxa are known among extant zygomycetes some investigators hesitate to interpret *Winfrenatia* as lichen. However, molecular phylogenies of ascomycetes show that lichenization was repeatedly lost (Lutzoni et al. 2001, 2004; James et al. 2006). As cyanobacteria and aerophilic algae were colonizing aquatic and terrestrial habitats long before the advent of vascular plants, lichenization as a nutritional strategy might have repeatedly and independently evolved and been lost. Loss of lichenization in favour of saprotrophism might have occurred in parallel with the advent of vascular plants, which made a wide array of new nutritious substrata available.

Morphologically advanced, foliose or fruticose lichens resembling extant taxa of lichen-forming ascomycetes are known from Tertiary deposits (65–1.5 MaBP; MacGinitie 1937; Peterson 2000).

Amber, fossilized tree resins from the Old World (Baltic amber, 55–35 MaBP) or New World (Dominican amber, 20–15 MaBP) contain extremely well preserved organisms, including lichens (Mägdefrau 1957; Poinar et al. 2000; Rikkinen 2003), but only a few have so far been recognized.

III. Lichen Photobionts

A. Photobiont Diversity

About 85% of lichen-forming fungi associate with green algae (often referred to as chlorolichens; Ahmadjian 1993), about 10% with cyanobacteria (cyanolichens; Ahmadjian 1993) and about 4%, the so-called cephalodiate species, simultaneously with both (Fig. 16.12c–e; Friedl and Büdel 2008). Cyanobacterial partners of such triple symbioses are housed in either external or internal cephalodia, gall-like structures in which the fungal partner creates microaerobic conditions to facilitate cyanobacterial nitrogen fixation (Fig. 16.12c–e; see SEM micrographs in Honegger 2001). Cyanobacterial photobionts in mature cephalodia were shown to comprise a higher proportion of heterocysts and thus a higher rate of dinitrogen fixation than in the free-living state (Millbank 1976; Englund 1977). Only one species each of brown algae (Phaeophyta) and Xanthophyceae were found as lichen photobionts (Tschermak-Woess 1988): *Petroderma maculiforme*, brown algal photobiont of the marine, intertidal *Verrucaria tavaresi* (Sanders et al. 2004, 2005), and *Heterococcus caespitosus*, xanthophycean photobiont of freshwater *Verrucaria* species (Thüs 2004).

Less than 150 species of lichen photobionts are so far described. In less than 5% of lichen-forming fungi was the photoautotrophic partner ever identified at species level with conventional or molecular techniques. In the majority of associations the generic level of the photobiont is known; examples are the Parmeliaceae (approx. 2200 spp. in more than 80 genera), Physciaceae (approx. 490 spp. in more than 20 genera; Kirk et al. 2001), all being symbiotic with unicellular green algae of the genus *Trebouxia* (Friedl 1989; Dahlkild 2001; Helms 2003; Piercey-Normore 2006). However, new photobiont taxa are still being discovered in association with morphologically less advanced, crustose or microfilamentous species of lichen-forming fungi. The highest photobiont diversity is found among the Verrucariaceae, a family comprising marine, freshwater and terrestrial species from temperate to tropical areas (Thüs 2004; Nyati et al. 2007).

Some photobionts of lichen-forming ascomycetes are common and widespread in nature, i.e. outside lichen thalli. This applies mainly for representatives of the Trentepohliaceae, very common photobionts in tropical lichens. Contradictory views are found in the literature concerning the abundance of free-living representatives of the genus *Trebouxia* (Trebouxiophyceae), the most common green algal photobionts of lichen-forming ascomycetes in temperate and especially in extremely cold and/or dry habitats. *Trebouxia* species were postulated to be non-existent outside lichen thalli (Ahmadjian 1988, 1993) and to depend on a carbohydrate supply from the fungal partner for their own nutrition (Ahmadjian 2002). Other investigators reported on the occurrence of free-living *Trebouxia* cells in natural ecosystems (Tschermak-Woess 1978; Mukhtar et al. 1994), and (Sanders 2001, 2005; Sanders and Lücking 2002) used long-term colonization experiments with microscopy slides which were exposed in natural habitats and regularly examined to demonstrate that free-living *Trebouxia* cells are commonly available. Phycologists refer to *T. arboricola*, type species of the genus and photobiont of innumerable species of lichen-forming ascomycetes, as a common and widespread aerophilic alga (Ettl and Gärtner 1995; Graham and Wilcox 2000; Rindy and Guiry 2003). In the near future this problem will be addressed with molecular tools applied to environmental samples. Nevertheless, in most ecosystems the most common aerophilic algae are not compatible lichen photobionts, but no key metabolic features have so far been identified by which lichen photobionts differ from non-symbiotic algae.

B. Recognition and Specificity

As lichen-forming fungi do not routinely express their symbiotic phenotype under sterile culturing conditions, the range of compatible photoautotrophic partners per fungal species cannot be experimentally studied in resynthesis experiments; instead the taxonomic affiliation of the cyanobacterial or algal partner is investigated in samples collected in nature. Thus the mechanisms underlying recognition and specificity are still poorly understood in lichen symbiosis.

Most of the morphologically advanced lichens of temperate climates so far investigated associate with genotypes of one or few related algal or cyanobacterial species (Kroken and Taylor 2000; Dahlkild et al. 2001; Helms et al. 2001; Rikkinen et al. 2002; Helms 2003; Opanowicz and Grube 2004; Piercey-Normore 2004, 2006). A higher diversity of algal partners was detected in foliose-umbilicate *Umbilicaria* species in the Antarctic (Romeike et al. 2002). Morphologically less advanced crustose species of lichen-forming

ascomycetes of temperate regions accept a wider range of photobiont species from the same genus (Helms 2003; Blaha 2006). Fungal and algal phylogenies in the crustose genus *Chaenotheca* suggest multiple green algal switches, leading to a wide array of taxonomically diverse photobionts from different genera, each being associated with one of the closely related fungal partners (Tibell 2001).

C. Symbiotic Propagules and Potential Vectors

A high percentage of lichen-forming fungi disperse efficiently by means of vegetative symbiotic propagules (Fig. 16.9): soredia, isidia, blastidia, but also thallus fragments comprise both symbionts and grow into new lichen thalli upon landing on a suitable substratum in an acceptable environment (Fig. 16.9a). Dispersal via thallus fragments is the predominat mode in reindeer lichens (*Cladonia* spp.), which cover thousands of square kilometres of arctic tundras and are only very rarely seen with sexual reproductive structures; in dry weather their tips get brittle and break off very easily upon trampling, even by small animals. The soredia of many lichens carry large amounts of mycobiont-derived crystalline secondary metabolites at their surfaces and thus are highly hydrophobic. This facilitates anemochory (wind dispersal), but also zoochory since soredia adhere electrostatically to the cuticle of invertebrates (insects, spiders, mites; Stubbs 1995) or to the feathers, fur and extremities of vertebrates. Short- and long-distance transport is also provided by the innumerable invertebrates and vertebrates which use lichen fragments to camouflage either their own body (insects, slugs; Fig. 16.16a–b) or their nests (birds, squirrels, etc.; Gressitt 1965; Richardson 1975; Gerson and Seaward 1977; Scharf 1978; Seyd and Seaward 1984; Brodo et al. 2001; Allgaier 2007).

Lichenivorous invertebrates, such as snails or the ever-present mites (Acari) which feed and live on and between lichens, most likely also disperse both partners of lichen symbiosis via their faecal pellets. With vital stains applied to the contents of faecal pellets of lichenivorous snails, viable photobiont cells were microscopically visualized (Fröberg et al. 2001). Culturing and molecular techniques showed that the ascospores of *Xanthoria parietina* and cells of its *Trebouxia* photobiont survived the gut passage in lichenivorous oribatid mites (Meier et al. 2002). Such faecal pellets are commonly seen as dark dots on the surface of lichens, together with grazing marks. Acarine faecal pellets are dispersed over short distances by their

Examples of vegetative symbiotic propagules in lichens

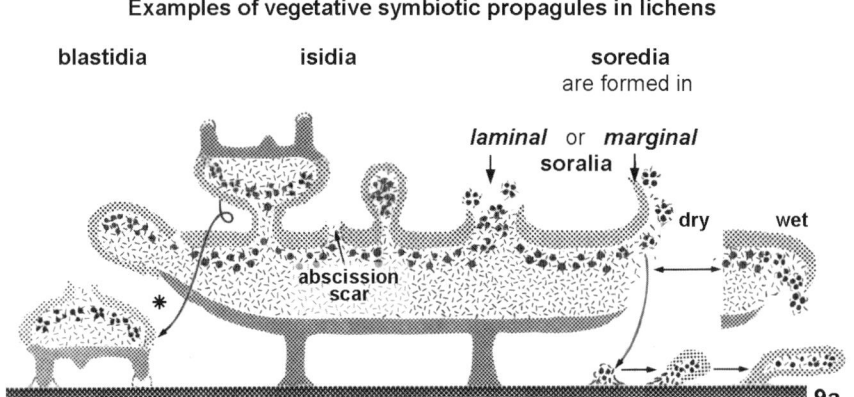

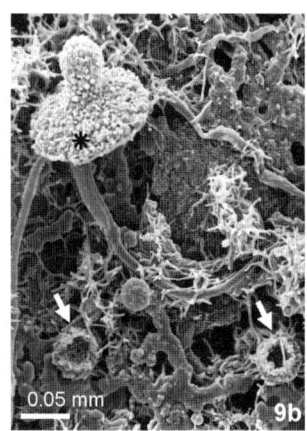

Fig. 16.9. Vegetative symbiotic propagules. **A** Diagram summarizing the most common types. *Asterisk* refers to an uncommon type of isidium (as observed in *Parmelina pastillifera*; Honegger 1987). This is an example of a remarkable level of fungal differentiation in the vegetative thallus: juvenile thalli with strongly melanized lower cortex differentiate in an inverted position on the non-growing part of the mother thallus. With their hydrophilic mucilage at the tips of future rhizinae the propagule is top-heavy and falls on the correct side upon detachment (Honegger 1997).

B Scanning electron micrograph of a dummy-shaped symbiotic propagule of the crustose, foliicolous, tropical *Bacidina scutellifera*. *Asterisk* indicates a detached propagule (in lit. referred to as an isidium) which, upon landing on a suitable substratum, germinates and develops a new thallus. These symbiotic propagules grow like minute mushrooms on the surface of the crustose thallus. *Arrows* point to the now hollow basal structures whence propagules had detached upon completion of their development. Specimen collected and identified by Edit Farkas

producers and by rain. It is highly probable that faecal pellets and mites are dispersed by birds and by wind over short or long distances.

D. Photobiont-Derived Mobile Carbohydrates

Lichens were the first fungal symbiosis with photoautotrophic partners in which the mobile photosynthate fraction was qualitatively identified which moves from the photoautotroph to the heterotroph. The mobile carbohydrates were characterized with either the ^{14}C isotope trapping technique (also termed inhibition technique), as designed by Smith and co-workers (for reviews, see Hill 1976; Smith 1980; Smith and Douglas 1987; Fig. 16.11), or with ^{13}C nuclear magnetic resonance spectroscopy (NMR) techniques (Lines et al. 1989). Green algal lichen photobionts provide their fungal partners with a group-specific acyclic sugar alcohol (polyol); examples are ribitol (as released by the most common and widespread green lichen photobionts of the genera *Trebouxia*, *Coccomyca*, *Dictyochloropsis*, *Myrmecia*, all representatives of Trebouxiophyceae), erythritol (by *Trentepohlia* and *Phycopeltis* spp., Ulvophyceae) and sorbitol (*Heterococcus*, Xanthophyceae; *Trochiscia* spp., Chlorophyceae). Cyanobacterial photobionts release glucose.

Lichen photobionts photosynthesize when the thalli are sufficiently hydrated. For photobionts of lichens in extreme climates only short periods per day are suitable for photosynthesis; at many days of the year either a negative balance or no metabolic activity at all is achieved (Kappen 1988). In inhibition technique experiments the rate of fixed carbon moving from the photoautotroph to the fungal partner was around 20–40% of the total fixed C during 3 h in *Peltigera* species with a cyanobacterial photobiont (*Nostoc* sp.; see example in Fig. 16.12a–h). This is the exception, since the translocation rates in all other interactions with cyanobacteria and especially with green algae were so small (less than 5%, often only 1–2% of the total fixed C in 3 h at full hydration) that the symbiotic system could not function and persist in nature.

In this type of experiment fragments of foliose lichens, horizontally dissected along the algal layer, were placed on an incubation medium containing dissolved, non-radioactive carbohydrates; these were taken up by the fungal partner until presumed saturation of its demands, thus inhibiting further uptake. Radioactively labelled CO_2, added to the incubation chamber, was photosynthesized by the photobiont cell population and radioactive, mobile photosynthates were assumed not to be taken up by the mycobiont, but released into the incubation medium whence they could be recovered for subsequent analysis

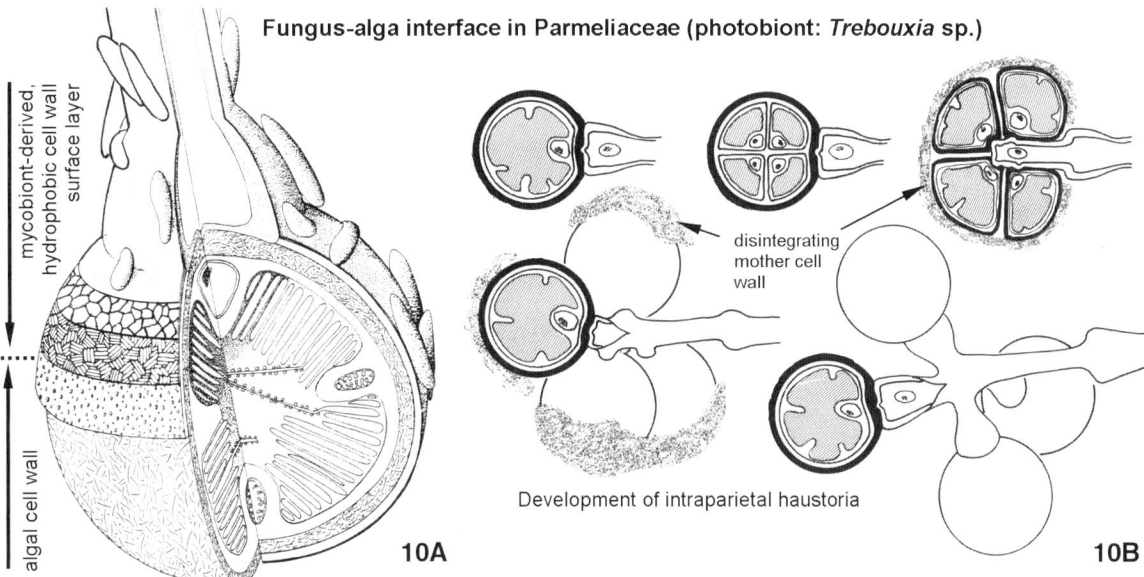

Fig. 16.10. Diagrams of the mycobiont-photobiont interface (**A**) and its ontogeny (**B**) in Parmeliaceae, a large group of morphologically advanced, foliose or fruticose macrolichens with *Trebouxia* species as green algal photobiont. A very similar situation is found in foliose Physciaceae and Teloschistaceae (in the latter with no mycobiont-derived secondary metabolites in the thalline interior). A Line drawing by Sibylle Erni. **B** After Honegger (1985), modified

(Hill 1976; Smith 1980; Smith and Douglas 1987). In ^{13}C NMR studies applied to *Xanthoria calcicola* (photobiont: *Trebouxia arboricola*) a distinctly higher rate of translocation was detected; and a fast transformation of photobiont-derived carbohydrates into fungus-specific compounds such as mannitol was evident (Lines et al. 1989). The low recovery of photobiont-derived fixed carbon in inhibition technique experiments is due to the complexity of the fungal–algal interface, especially to the mycobiont-derived hydrophobic wall surface seal (see below), which prevents radioactively labelled soluble compounds from leaking out into the non-radioactive incubation medium. Consequently only those soluble compounds were recovered from the incubation medium which had been leaking out of dissected or otherwise damaged cells along the cut edges of the lichen fragment under investigation (Honegger 1997; Fig. 16.11).

IV. Functional Thallus Anatomy

A. Main Building Blocks

The majority of lichen-forming fungi form more or less inconspicuous crustose thalli on or within the peripheral layers of the substratum, a wide range of surfaces being colonized, however with distinct preferences (Figs. 16.15, 16.19, 16.20, 16.22). Only about 25% of lichen-forming fungi differentiate leaf-like (foliose) or shrubby (fruticose),

erect or pendulous 3D thalli, which rise above the substratum. In these morphologically advanced lichens the quantitatively predominant fungal partner amounts to 80–90% of thalline biomass (Figs. 16.2, 16.8); it competes for space above ground and controls the photobiont cell population in the thalline interior. The fungal partner reacts upon spatial disturbance, i.e. when thalli are fixed in an upside-down position to the substratum, by means of correcting growth processes (Honegger 1995; Fig. 16.7).

The internally stratified thalli of morphologically advanced lichen-forming fungi are the product of an astonishing hyphal polymorphism, the main building blocks being either filamentous or globose hyphae with either hydrophilic or hydrophobic wall surfaces (Figs. 16.1–16.4, 16.8). The thalli are typically built up by two functionally and anatomically different areas:

1. A hydrophilic, conglutinate zone (usually as a peripheral cortical layer and/or as internal strands) which absorbs water, provides mechanical stability to the whole thallus and transmits light in the fully hydrated state. These conglutinate zones are brittle when dry, but very elastic in the fully hydrated state. The rate of light transmission through the peripheral cortex depends

Investigating the photobiont-derived, mobile carbohydrates with the *Inhibition Technique (= Isotope trapping technique)* according to D.C. Smith and coworkers (reviews: Hill 1976; Smith and Douglas 1987)

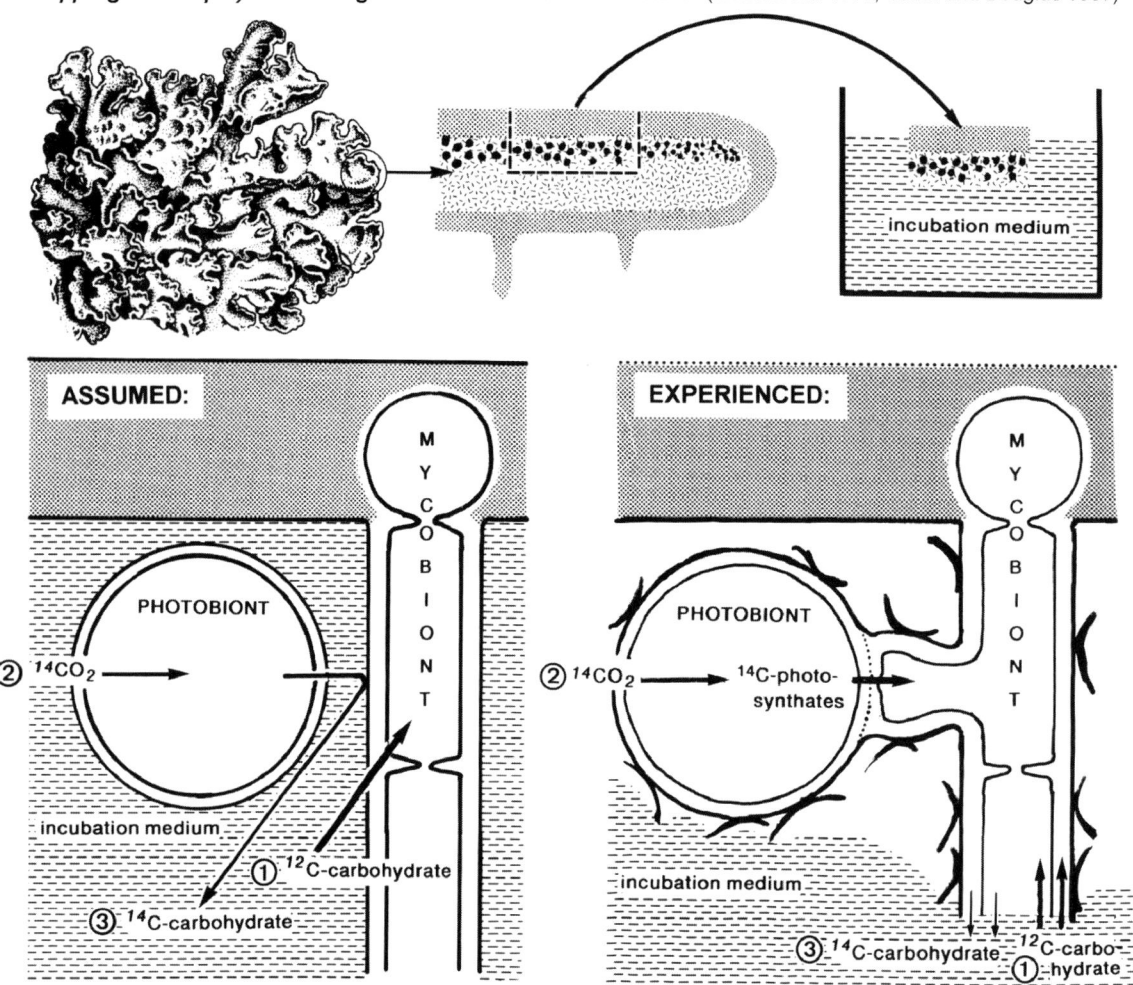

Fig. 16.11. Diagram illustrating the design and outcome of experiments using the inhibition technique. See text for detailed explanations (from Honegger 1997)

on the taxonomic affiliation of the mycobiont and on its habitat: it is high in shady places in species with no cortical secondary metabolites, but low (approx. 44%) in sun-exposed sites, due either to the deposition of crystalline secondary metabolites, oxalates or melanin, or to the formation of a superficial pruina from dead cell material (Dietz et al. 2000; Solhaug et al. 2003; Nybakken et al. 2004; McEvoy 2006, 2007).

2. A system of loosely interwoven aerial hyphae with hydrophobic wall surface layers (usually as a medullary layer in the thalline interior) which remains air-filled at any level of thalline hydration (Figs. 16.2–4, 16.6a–c, 16.8). The photobiont cell population is actively positioned, maintained and

controlled by the fungal partner at the periphery of the air-filled medullary layer (Fig. 16.10b).

B. Mycobiont–Photobiont Interactions

Each photobiont cell in lichen thalli is contacted by a specialized fungal hypha whence it derives water and dissolved mineral nutrients. Intragelatinous fungal protrusions grow into the gelatinous sheaths of cyanobacterial colonies (*Nostoc* sp., Fig. 16.6a–c; Honegger 1991). Fungal hyphae or appressoria adhere to green algal cells with enzymatically non-degradable, sporopollenin-like wall surface layers (*Coccomyxa* or *Elliptochloris* spp.; Honegger and Brunner 1981; Brunner and Honegger, 1985; Honegger 1991). Transparietal ("intracellular") haustoria, as found in morphologically simple crustose lichens, penetrate the algal cell wall (Honegger 1984, 1985, 1986a). In contrast intraparietal haustoria, as typically found in morphologically advanced taxa of lichen-forming ascomycetes, enter but do not pierce the cellulosic cell wall of the unicellular green algal photobiont (predominantly *Trebouxia* spp.; Honegger 1984, 1985, 1986a; Fig. 16.10a–b). Accordingly appressoria, haustoria or intragelatinous protrusions are not only the sites where the fungal partner mobilizes fixed carbon, but also the very place where the green algal or cyanobacterial cells are provided with water and micronutrients via the mycobiont, since the photoautotrophic partner is hidden in the interior of the heterotrophic fungal thallus and has no direct access to water and mineral nutrients.

C. Wall Surface Hydrophobicity and Gas Exchange

Beside appressoria, haustoria and functionally related fungal structures, an inconspicuous structural element plays a key role at the mycobiont–photobiont interface of lichens: a thin, hydrophobic wall surface layer, provided by the fungal partner, spreads over the photobiont wall surface, thus ensheathing the cell–cell interface with a hydrophobic coat (Fig. 16.10a; Honegger 1984, 1986b, 1991; Scherrer et al. 2002; Trembley et al. 2002b). The astonishing wall surface hydrophobicity of the algal and medullary layers of lichen thalli was first described by Goebel (1926), who assumed secondary metabolites of fungal origin crystallized on hyphal surfaces in the medullary layer and created this water repellency. However, by far not all morphologically advanced lichen-forming fungi secrete secondary metabolites in their medullary and algal layers in the thalline interior, yet their hyphal and algal surfaces are hydrophobic. In transmission electron microscopy (TEM) of freeze-fractured samples the fracture plane runs along this hydrophobic discontinuity. At higher magnification a distinct rodlet pattern is resolved in replicas of this thin wall surface layer, which spreads from the fungus over the surface of the algal wall or glucan sheath of cyanobacterial colonies at the contact site (Figs. 16.8, 16.10a). The same type of rodlet layer is found on the surface of aerial hyphae and conidia in a wide range of non-lichenized fungi (Wösten 2001).

Rodlet layers of fungi are built up by hydrophobins, very peculiar, secreted fungal proteins which, due to their special properties, escaped the attention of biochemists until the 1990s. Hydrophobins are approximately 100 amino acids long and reveal very little sequence homology, except eight cysteine residues in conserved pattern (X_{25-85}–C–X_{5-8}–C–C–X_{17-39}–C–X_{6-23}–C–X_{5-6}–C–C–X_{6-18}–C–X_{2-13}) and a distinct hydropathy profile (Wösten et al. 1993, Wessels 1996; Wösten and Wessels 1997; Wösten 2001; Kershaw et al. 2005; Linder et al. 2005). Hydrophobin monomers self-assemble in vivo and in vitro at hydrophilic–hydrophobic interfaces, such as the hyphal surface or the surface of an aqueous solution, to an amphiphilic protein film, which carries the rodlet pattern at its hydrophobic surface (Wösten et al. 1993); such films can not be solubilized using conventional protocols for protein solubilization. Hydrophobins play important roles not only as surfactants, but also in the establishment of mutualistic or antagonistic fungal interactions with plants or invertebrates, their expression often being developmentally regulated (Martin et al. 1995; Talbot et al. 1996; Whiteford and Spanu 2002; Mankel et al. 2002; Kershaw et al. 2005; Linder et al. 2005).

Although rodlet layers were seen in diverse taxa of lichen-forming fungi they could be solubilized and characterized in only a few species. Thorough investigations were conducted in the hydrophobins of vegetative thalli of *Xanthoria* species (lichenized ascomycetes; Scherrer et al. 2000, 2002; Scherrer and Honegger 2003) and in lichenized fruiting bodies of the tropical *Dictyonema glabratum* (syn. *Cora pavonia*, lichenized basidiomycete; Trembley et al. 2002a, b). Only one hydrophobin was found in each of the *Xanthoria* species examined, but three in *Dictyonema glabratum*; these shared 54–66% amino acid homology (Trembley et al. 2002a). In both types of lichen the hydrophobin genes are developmentally expressed (Scherrer et al. 2002; Trembley et al. 2002b). Hydrophobin genes evolve rapidly and thus are excellent phylogenetic markers among closely related species (Scherrer and Honegger 2003).

Mycobiont-derived secondary metabolites crystallize on and partly within the thin, hydrophobic wall surface protein film in the medullary and algal layers, thus enhancing its water repellency (Honegger 1986b). Consequently the rodlet pattern is no longer visible; instead, an irregularly tessellated surface layer, which may carry additional, amorphous overlays, are resolved with TEM techniques (Figs. 16.8, 16.10a; Honegger 1984, 1985, 1986b, 1991). So far it has not been possible to isolate and characterize the rodlet layer of lichens with medullary secondary compounds (Jos Wessels and co-workers, personal communication; Honegger, Scherrer and Haisch, unpublished data); *Dictyonema glabratum* and the different *Xanthoria* species have no secondary metabolites at their hyphal surfaces in the thalline interior. This applies also to cyanobacterial *Peltigera* spp. (dog lichens), where the rodlet layer was isolated, but all attempts to sequence the amino acids failed (Hugelshofer, Scherrer and Honegger, unpublished data).

A pre-requisite for successful gas exchange in thalli of lichenized asco- and basidiomycetes is a gas-filled photobiont layer. A hydrophobic wall surface lining in the algal and medullary layers, as provided by hydrophobin proteins, partly in combination with mycobiont-derived crystalline secondary metabolites, prevents the accumulation of free water on wall surfaces. Moreover it forces the passive water flow from the thallus surface to the algal layer during rehydration and vice versa during desiccation to move within the cell wall underneath the hydrophobic surface layer, i.e. in the apoplastic continuum between the partners (Fig. 16.8). Ideally the role of the hydrophobins in the functioning of the symbiosis would be tested with hydrophobin-deficient mutants, as elegantly performed by Talbot et al. (1996) in the rice blast fungus *Magnaporthe grisea*. This ultimate test is missing since re-lichenization is not routinely achieved under laboratory conditions by combining aposymbiotically cultured fungal and algal isolates, and hydrophobin-deficient mutants are not available.

D.   Hydrophilic Wall Components

Hyphal wall surface properties may change from hydrophilic to hydrophobic within the range of microns, as demonstrated in veins and medullary hyphae of representatives of the *Peltigera aphthosa* group (Honegger and Hugelshofer 2000). Representatives of this genus of foliose, dorsiventrally organized lichens do not differentiate a lower cortex; instead, they have a system of interconnected veins at their lower surface whence either rhizinae or a felty tomentum arise, which anchor in the soil or moss cover (Fig. 16.12a, d, f–h). Rhizinae, tomentum and veins have hydrophilic wall surfaces and absorb water like wicks. Between their mostly parallel-running hyphae water moves passively and very quickly over long distances due to capillary forces, as shown with dyes applied to the tip of single rhizinae (Fig. 16.12d). Medullary hyphae with hydrophobic wall surfaces due to a rodlet layer are connected to the veins. At this point water is no longer creeping along the hydrophilic wall surfaces of hyphae of the vein, but enters the cell wall and moves slowly underneath the hydrophobic

wall surface lining of medullary hyphae towards the photobont cells. The relatively large molecules of dyes are retained in the vein and do not enter the fungal apoplast (Honegger and Hugelshofer, unpublished data).

The area of the fungal cell wall, in which long-distance solute translocation occurs, is differentiated in many taxa of lichen-forming ascomycetes as a thick overlay on the cell wall proper. This zone of the hyphal wall is subjected to massive shrinkage during desiccation. Lichenin, a class of $(1{\rightarrow}3)(1{\rightarrow}4)$-$\beta$-glucans, was located with immunocytochemical techniques in this hydrophilic cell wall area overlying the cell wall proper in hyphae of the medullary and algal layers and in the conglutinate extracellular material of the cortex in *Cetraria islandica* ("Icelandic moss"; Honegger and Haisch 2001). Water, as passively absorbed, is retained in this hydrophilic wall layer before being lost during desiccation. Thus the thick, hydrophilic middle and the thin, hydrophobic surface layers of hyphal walls in the thalline interior fulfil different, yet centrally important functions in thalline water relations (Honegger 2007).

V.   Water Relations and Ecology

A.   Reversible Cytoplasmatic Cavitation

The majority of lichen-forming fungi and their photobionts are poikilohydric organisms, their water relations being subjected to constant changes between full hydration and desiccation (<20% water per dry weight). During drought stress events both partners of the symbiosis shrink dramatically (Brown et al. 1987) and fungal cells cavitate (Figs. 16.5a, 16.6c, 16.12a). However, this cytoplasmatic cavitation is reversible; the cavitation bubble disappears during rehydration (Fig. 16.5b; Honegger and Peter 1994; Scheidegger 1994; Schroeter and Scheidegger 1995). Only a few lichens from continuously moist, temperate or subtropical habitats have so far been found which do not tolerate desiccation (Green et al. 1991; Nyati et al. 2007).

B.   Foliicolous Lichens

More than 10% of terrestrial ecosystems are lichen-dominated; these are the sites where vascular plants are at their physiological limits and thus exert little or no competitive pressure. Accordingly lichens are not a dominant element in tropical rainforests, where they occupy special ecological niches such as the surface of long living leaves.

An astonishing biodiversity was found among the comparatively short-living foliicolous lichens (Lücking and Matzer 2001; Lücking et al. 2003; Herrera-Campos et al. 2004; Lücking 2008), which use leaves as a substratum; they may have a shading effect, but do not penetrate the leaf cuticle.

On single dicotyledonous leaves up to 48 species of foliicolous lichens were found, on one palm leaf even 81 species (Lücking and Matzer 2001). Many of these foliicolous lichen-forming fungi find their photoautotrophic partner within the foliicolous algal community (e.g. *Phycopeltis* spp.; Sanders 2001; Sanders and Lücking 2002), others disperse by means of vegetative symbiotic propagules (Sanders 2002). Some of these crustose foliicolous species produce the most bizarre symbiotic propagules among lichenized fungi (see example in Fig. 16.9b), others differentiate unique conidia-bearing structures termed hyphophore: erect, stalked peltate asexual sporophores (Vežda 1975; for a definition, see Kirk et al. 2001; Lücking 2008). Older leaves of wild and economic plants such as palm trees, coffee, cocoa, rubber, etc. may carry a luxuriant epibiont community, comprising lichens, liverworts and free-living algae. Growers dislike these leaf-dwelling communities and suspect they harm the trees (Hawksworth 1988b). However, foliicolous lichens were shown to grow on plastic materials which had been deposited in rainforests (Sipman 1994; Sanders 2002), since they are in need of space, not nutrition.

C. Lichens of Extreme Climates

Lichens dominate thousands of square kilometres of arctic tundras and cover rock surfaces from sea level to the high alpine areas around the globe, provided that sufficient humidity is available. Wind-dispersed vagrant lichens, which roll up under drought stress and unroll upon rehydration, are important elements of cold and dry alpine ecosystems, e.g. in the Andes (Pérez 1997a, b), and arid steppe and desert ecosystems around the globe (Rogers 1971; Rosentreter 1993). Lichens, often inconspicuous crustose taxa (Fig. 16.22), are integral parts of the ecologically very important, but highly vulnerable soil crust communities of arid lands worldwide. Together with free-living cyanobacteria and desiccation-tolerant bryophytes, they stabilize soils and thus prevent desertification by reducing wind erosion (Eldridge 1996; Belnap and Lange 2003; Lalley and Viles 2005, 2006; Muscha and Hilda 2006; McCune and Rosentreter 2007). In addition, soil crust organisms play an essential role as precursors of the colonization of steppe and desert soils by plants.

D. Poikilodydric Water Relations

Key features of the impressive environmental stress tolerance of lichen-forming fungi and their photobionts are their poikilohydric water relations and the ability of both partners to survive temperature extremes unharmed in a state of dormancy. Lichens were believed to retain their viability for long periods of time after storage at room temperature, as concluded from respiratory activity upon rehydration; however, culturing experiments showed that several species of lichen-forming fungi and their photobionts die after a maximum of 2 years storage at room temperature (Yamamoto et al. 1998), but full viability was retained by the fungal and algal partners for more than 10 years in desiccated specimens which had been kept at −20 °C (Honegger 2003). In nature lichen-forming fungi were often assumed to protect their photoautotrophic partner from desiccation.

However, there is no evidence and no need for this type of protection; on the contrary, only in the desiccated, physiologically inactive state can fungal and algal or cyanobacterial cells withstand temperature extremes. Desiccated thalli of the golden-yellow wall lichen, *Xanthoria parietina* (a common and widespread species of temperate climates) grew normally after cryoimmobilization in subcooled liquid nitrogen (ca. −200 °C), freeze-fracturing, sputter-coating with a gold/palladium alloy and examination under high-vacuum conditions on the cold table of a low-temperature scanning electron microscope at an acceleration voltage of 60 kV; fungal and algal cells of hydrated thalli did not survive this treatment (Honegger 1995, 1998). Measurements of chlorophyll fluorescence and vital staining experiments refer to survival of a 16-day exposure to space conditions in desiccated *Rhizocarpon geographicum* (map lichen) and *Xanthoria elegans* (golden sunburst lichen) and their *Trebouxia* photobionts (Sancho et al. 2007). However, this does not prove that lichens could be metabolically active in space.

Differences during rehydration were detected concerning the type of water required by green algal and cyanobacterial photobionts, as elegantly shown in photosymbiodemes (Green et al. 1993, 2002): these are either green algal or cyanobacterial lichens formed by the same fungal partner. Photosymbiodemes may be morphologically largely identical, except for their colouration (Fig. 16.12c), or distinctly different. Most members of photosymbiodemes are referred to under different species names or, in the case of morphologically distinct phenotypes, under different genus names; this problem remains to be solved by lichen taxonomists (Jörgensen 1998). Green algal lichens reactivate their metabolic activities after drought stress events at very low water contents; elevated nocturnal moisture contents of the air, dew, or fog are sufficient for partial rehydration (Kappen 1988; Lange et al. 1990, 2006). In contrast cyanobacterial lichens require liquid water for rehydration.

Accordingly green algal lichens dominate extreme habitats such as fog deserts (Namib, Atacama, Negev) and are the predominant species in alpine, arctic and antarctic ecosystems and in semi-arid soil crust communities (Lange et al. 1997), whereas cyanobacterial lichens are most abundant in continuously moist forest ecosystems (Green et al. 1993; Sillett and Antoine 2004). Astonishing exceptions are the gelatinous cyanobacterial lichens of the genus *Collema* which, together with free-living cyanobacteria, are important elements of arid soil crust communities. The thick, hydrophilic gelatinous sheaths of the *Nostoc* colonies in gelatinous lichens absorb water in the rainy season, but the fungal and cyanobacterial partners survive the dry periods in a state of dormancy (Lange et al. 1998; see images of sections of desiccated and hydrated *Collema* sp. in Honegger 2001).

Cold temperature extremes and the availability of water are the limiting factors in polar and alpine lichens. Many species use ice as a water source and photosynthetic activity was measured at subzero temperatures, at low photon flux densities and in an only partly hydrated state (Kappen 1988, 2000; Kappen and Schroeter 1997). Covered by snow, ideally with ice windows due to continuous melting and freezing processes above darkly pigmented thalli, arctic and antarctic lichens often live as in a miniature greenhouse, well protected from harsh climates, their metabolic activities resulting in a positive balance as regards to fixed carbon. Thus, long-term observations showed that lichens of the maritime antarctic grew faster than previously assumed (Lewis Smith 1995).

Desiccating lichens shrink without wilting and change their consistency and often their colour; the latter applies for those species which derive their colouration from the photoautotrophic partner, not from secondary fungal metabolites in the peripheral cortical layer. Examples are the vivid green shades of *Peltigera*, *Solorina* and *Lobaria*

Fig. 16.12. *Peltigera* species. **A** *P. membranacea* desiccated thallus. Veins and rhizinae on the lower surface are visible at curved margins. **B** Same thallus in the fully hydrated state; dark grey colour provided by the photobiont (*Nostoc punctiforme*), which is visible through the translucent upper cortex (same magnification in **A**, **B**). **C** Photosymbiodeme: the same fungal partner either with a cyanobacterium (*Nostoc* sp.; dark brown *P. malacea*) or with a green alga as primary photobiont (*Coccomyxa* sp.; green *P. britannica*, growing out of the cyanobacterial morph), described under two different species names. *Arrows* point to external cephalodia on *P. britannica* in which the same *Nostoc* sp. is incorporated as in the cyanobacterial morph. See text for further details. **D, E** Cross-sections of *P. aphthosa* with external cephalodium (*ce*), seen under a dissecting microscope (**D**) or in a stained semithin section in bright field microscopy (**E**). *m* Medullary layer (hydrophobic hyphal surfaces), *ph* photobiont layer with minute, green *Coccomyxa* cells, *t* filthy tomentum (hydrophilic wall surfaces), *uc* upper cortex (translucent when wet). *Asterisk* in **E** refers to the peripheral cortex around the cephalodium. **F–H** *Peltigera membranacea*: solute translocation in rhizinae and veins at the lower thallus surface visualized with dyes (erythrosine or methylene blue, respectively). *Arrows* in **F** point to three rhizinae to each of which a drop of dye had been added, as shown in **G**. The dye was moving fast over considerable distances (**H**) by means of capillary forces along the hydrophilic wall surfaces of the parallel hyphae of the veins. Water, but not the large molecules of the dyes entered the cell walls of medullary hyphae and slowly moved underneath the hydrophobic wall surface layer (rodlet layer) towards the algal layer **Fig. 16.13.** *Xanthoria parietina* in the field, natural size. **A** Shade form with low parietin (anthraquinone) content, growing in the shade of a hedge on an iron fence. **B** Sun-exposed thalli; all except one mutant thallus (*asterisk*) reveal a high parietin content **Fig. 16.14.** Production of secondary metabolites by sterile-cultured, aposymbiotic *Xanthoria parietina* on different carbon sources. The inoculum was cultured on Millipore filters, then dissected and incubated for two months **Fig. 16.15.** Crustose lichens on granitic rock: *Ophioparma ventosum* with different secondary products in the ivory-greenish vegetative thallus (divaricatic acid, a *para*-depside, other compounds) than in the apothecia (*ac*), whose blood-red colour is due to the quinone haemoventosin. For rock weathering activities of this species, see Bjelland and Thorset (2002). *Arrow* points to the juvenile thallus of a map lichen (*Rhizocarpon* sp.) growing on quartz. The non-lichenized fungal prothallus is black due to strongly melanized cell walls; the yellow, photobiont-harbouring areoles synthesize rhizocarpic acid for UV protection **Fig. 16.16.** A bagworm (*Narycia duplicella*, Psychidae, Lepidoptera; det. Peter Hättenschwiler) feeds on the sorediate *Xanthoria ulophyllodes* and fixes soredia as camouflage to its silk bag (**A**), where they develop into small lobules (**B**) **Fig. 16.17.** Sterile cultured, aposymbiotic *X. fallax*. Yellow anthraquinones crystallize on the colonies, but also in and on the agar medium at considerable distances **Fig. 16.18.** Thallus cross-section of *X. parietina* in UV light. Red autofluorescence of the globose green algal cells (*Trebouxia* sp.) and intense yellow anthraquinone autofluorescence show in the upper (*uc*) and lower cortex (*lc*); no secondary fungal metabolites are synthesized in the medullary (*m*) and algal layer (*ph*) **Fig. 16.19.** *Porpidia flavocaerulescens* from Alaska derives its rusty colour from Fe^{3+} ions, which it accumulates at the periphery of the vegetative thallus, but not in the black apothecia **Fig. 16.20.** The crustose *Candelariella vitellina* overgrowing chrysotile fibres in a former asbestos mine; specimen collected by Sergio Favero-Longo **Fig. 16.21.** Antibiotic properties of *Usnea ceratina* (**A**) and usnic acid (**B**), tested on *Bacillus subtilis*. Diameters of the inhibition cores in **B** depend on usnic acid concentration **Fig. 16.22.** Soil crust lichens in the Namib desert. The inconspicuous, crustose *Lecidella crustuliniforme* (asterisk) covers and stabilizes the ground. *Arrow* points to a pebble overgrown by the orange *Caloplaca elegantissima*. About natural size; courtesy of Reinhard Berndt species with green algal photobionts, or the turquoise-grey to blackish shades of *Peltigera*, *Sticta* and *Lobaria* species with cyanobacterial photobionts (Fig. 16.12a, b). Hardly any colour changes during desiccation are seen in innumerable grey to dark brown or black Parmeliaceae, Physciaceae, Umbilicariaceae, or in intensely yellow representatives of Teloschistaceae (*Xanthoria,* many *Caloplaca* and *Teloschistes* species) with golden-yellow anthraquinones in their peripheral cortex, or in the sulfur-yellow *Letharia vulpina* (wolf lichen) and related species with vulpinic acid at their thallus periphery.

E. Ice Nucleation Sites

The fate of fully hydrated antarctic lichens during a steep temperature decline was investigated by low-temperature scanning electron microscopy using defined cooling rates, with actinic fluorescence and gas exchange being measured in parallel experiments (Schroeter and Scheidegger 1995). At temperatures around 0 °C to −6 °C water is confined to the symplast and apoplast of lichen-forming fungi and their photobiont, the hydrophobic wall surface lining preventing the accumulation of free water in intercellular spaces within the thalline interior. At temperatures below the freezing point free water was shown to accumulate on the cell wall surfaces and to crystallize in the intercellular space due to very powerful ice nucleating sites; simultaneously the fungal and algal cells shrivelled and the fungal cells cavitated, as happens under drought stress. This impressive mode of lowering cellular water contents prevents the formation of ice crystals within the cell, which would have a devastating effect on cellular membrane systems and destroy compartmentalization within the cell, leading to cell death. Biological ice nucleation sites were characterized in a range of lichens (Kieft 1988; Kieft and Ahmadjian 1989; Kieft and Ruscetti 1990, 1992; Warren and Wolber 1991).

VI. Uptake of Elements

A. Ion Uptake and Accumulation

During their lifetime lichen thalli may accumulate relatively high amounts of nutrient and non-nutrient elements, radionuclides included, which are passively absorbed over the thallus surface. These elements reach the thalli either via wet or dry deposition, particulate deposition (e.g. from smelters, vehicle exhaust, industrial releases) or by substrate weathering (Richardson 1995; Purvis and Halls 1996; Haas and Purvis, 2006; Purvis and Pawlik-Skowronska 2008). Some lichen-forming fungi enhance weathering processes, much to the concern of rupestrian archaeologists and curators of cultural heritage (Cifferi et al. 2000; Bjelland and Thorset 2002; Bjelland et al. 2002; StClair and Seaward 2004). Many fungal cell wall components such as different glucans, mannans, glycoproteins, etc. reveal a strong binding affinity for cations, which they immobilize via ion exchange or chelatinization processes. Moreover, a wide range of polyphenolic secondary metabolites form insoluble complexes with cations (Purvis and Halls 1996; Haas and Purvis 2006). Potentially harmful elements are thus accumulated, often in astonishing quantities within thalli, without affecting the metabolism of the mycobiont and photobiont. Metallophyte lichens colonize metal-rich substrata such as iron monuments or fences (Haas and Purvis 2006) without suffering from competitive pressure. Some lichen species derive their colouration from metal accumulations (Fig. 16.19); examples are rust-brown thalli of epilithic *Lecidea* spp. due to iron oxide accumulation (Noeske et al. 1970) and the slightly turquoise tinge in otherwise brownish thalli of *Lecanora vinetorum* on wooden posts in vineyards due to the frequent application of copper-containing pesticides (up to 12 times per growing season); these thalli may contain up to 5000 ppm of copper (Poelt and Huneck 1968). Due to their ability to bind a wide range of elements, some investigators suggest the use of lichens as a bioremediation tool (Bennett and Wright 2004; Ates et al. 2007).

With their ability to bioaccumulate metals and other elements lichen thalli are valuable archives, which allow monitoring of changes in pollutant and other element deposition over long periods of time (Purvis et al. 2003, 2007a, b; Walker et al. 2003, 2006a, b). By investigating the contents of lead and several other elements in herbarium samples and freshly collected specimens of the foliose *Flavoparmelia baltimorensis* from Plummers Island (15 km from the city centre of Washington, D.C.) and from two other sites (21 km and 120 km apart), covering the span from 1907 to 1992, the massive increase in lead deposition (mainly from automobile exhaust) was impressively documented, peaking in the 1970s and declining with the advent of catalytic converter technology, irrespective of the continuously growing level of private traffic (Lawrey 1993).

B. Radionuclides

Lichens accumulate nuclear fallout and thus were annalists of early atmospheric testing of nuclear bombs and of the Chernobyl accident (Richardson 1975; Feige et al. 1990; Sawidis and Heinrich 1992; Biazrov 1994). As "primary producers" in the food chains of subarctic and arctic tundra ecosystems, especially in lichen-dominated reindeer and caribou pastures, radioactively contaminated lichens, the main winter food of reindeer and caribou, led to a massive accumulation of radionuclides in domesticated and wild animals and finally in Scandinavian Sàmi and North American Eskimos (Richardson 1975; Strand et al. 1992; Tveten et al. 1998; Mehli et al. 2000; Bostedt 2001).

C. Sources of Fixed Nitrogen

Different demands for fixed nitrogen are evident among different taxa of lichen-forming ascomycetes. Cyanobacterial lichens receive fixed nitrogen (as ammonium ion) from their diazotrophic photobiont, whereas green algal lichens take fixed nitrogen from the environment.

Nitrophilous green algal lichens such as the golden yellow *Xanthoria* species and related Teloschistaceae develop on eutrophicated substrates, indicating the sites where animals congregate, breed or rest. *Xanthoria* species on coastal rocks survive the deposition of seagull excreta, which are otherwise highly corrosive on a wide range of materials, iron included. Consequently their thalline nitrogen contents may reach the values of lichens with cyanobacterial photobionts (up to 28 mg g^{-1} dry weight; Palmquist et al. 2002). The luxuriant populations of endemic, epiphytic *Xanthoria* species near the coast in South Africa and Namibia benefit from ammonia emissions derived from coastal bird colonies or Cape fur seal colonies. This also applies, with high probability, for the lichen fields which cover wide areas of the Namib desert (Theobald et al. 2006); these use fog from the Atlantic as the sole source of water for most of the year (Kappen 1988). By far the highest N contents in green algal lichens (up to 48 mg g^{-1}

dry weight; Palmquist et al. 2002) were recorded in *Mastodia tessellata* (syn. *Turgidosculum turgidum*), which lives in the coastal Antarctic near and in penguin benches above the shoreline. Its green algal partner, the thalloid *Prasiola crispa*, is highly desiccation- and osmotolerant and occurs free-living at the same sites as the lichenized form (Bock et al. 1996).

The majority of green algal lichens have low to very low thalline N contents, the lowest values being found in oligotrophic epiphytic and mat-forming terrestrial species (1–10 mg dry weight; Palmquist et al. 2002). The mat-forming, green algal lichen communities of arctic tundras were shown to be N-limited, fixed nitrogen being recycled within the stand (Ellis et al. 2005). Oligotrophic lichens disappear in response to elevated N input from natural or anthropogenic sources and are replaced by less sensitive or even nitrophilic species. In Scandinavian tundras increased grazing pressure, trampling damage and N input due to the doubling of reindeer herds within the past 25 years resulted in a massive decline in lichen cover and an increased competitive pressure by vascular plants, as documented by remote sensing techniques (Nordberg and Allard 2002; Johansen and Karlsen 2005).

In cyanobacterial lichens the highest N contents (up to 50 mg g^{-1} dry weight) were detected in species with *Nostoc* photobionts (Palmquist et al. 2002). *Nostoc* species as the sole photobiont provide their mycobionts with fixed carbon and nitrogen. In most of the triple symbioses of cephalodiate lichens, i.e. thalli with a green algal primary photobiont and *Nostoc* colonies harboured in cephalodia (e.g. *Peltigera britannica* or *P. aphthosa*; Fig. 16.12c–e) the fungal partner creates microaerobic conditions in the cephalodia, thus enhancing heterocyst frequency and nitrogenase activity as compared with the free-living state of the *Nostoc* colonies (Griffiths et al. 1972; Hitch and Millbank 1975; Millbank 1976; Englund 1977).

Differences in chitin contents were observed in cell walls of cyanobacterial lichens (*Peltigera* spp.) and green algal lichens, as measured in cell wall fractions derived from either entire thalli collected in nature (Boissière 1987) or sterile cultured lichen-forming ascomycetes (Honegger and Bartnicki-Garcia 1991). Cell walls of the green algal lichens *Lasallia pustulata*, *Cladonia caespiticia*, *C. macrophylla* and *Physcia stellaris* comprised between 3.2% and 6.6% chitin; walls of the cyanobacterial *Peltigera canina* had 13.0%. All Peltigerales associate with cyanobacterial photobionts of the genus *Nostoc*, either as the sole partner or as a cephalodial partner (Fig. 16.12a–e) and thus can satisfy their higher demands for fixed nitrogen compared with green algal lichens. Moreover N-enriched leachates were detected in soil around healthy thalli of *Peltigera* spp. (Knowles et al. 2006).

VII. Secondary Metabolism

A. Mycobiont-Derived Lichen Compounds

More than 800 mycobiont-derived secondary metabolites, mostly polyphenolics, were detected in lichen thalli (Huneck and Yoshimura 1996; Huneck 2001). The majority are restricted to lichen-forming fungi, but some belong to classes with a wide distribution among organisms; examples are the anthraquinones, being produced by representatives of bacteria, lichenized and non-lichenized fungi, invertebrates and plants (Culberson 1969; Pankewitz et al. 2007). Many non-lichenized fungi which are currently used in biotechnology as producers of economically interesting secondary metabolites derive from lichenized ancestors, as shown by large-scale phylogenetic analyses (Lutzoni et al. 2001, 2004; James et al. 2006); examples are *Aspergillus* spp. and other representatives of the Eurotiales.

Secondary compounds are released by the cells of lichen-forming fungi in a yet unknown, possibly glycosylated form into the fungal wall, whence they are passively translocated by the main fluxes of solutes within the apoplastic continuum during desiccation and rehydration events. They finally reach their sites of crystallization, either the gelatinous matrix of the conglutinate peripheral cortical layers, or the hydrophobic hyphal wall surfaces in the medullary and algal layers. Secondary compounds of fungal origin reach the photobiont cell population by passive translocation and crystallize on the algal wall surfaces, which are covered by the same, mycobiont-derived hydrophobic wall surface layer as the contacting fungal cells (Figs. 16.8, 16.10a). In their crystallized form the secondary metabolites of lichen-forming fungi are almost insoluble in aqueous systems at low or neutral pH ranges. Thus large amounts of secondary metabolites may accumulate in lichen thalli (usually 1–4%, up to 10% of thallus dry weight). Fairly often the secondary compounds accumulating in the peripheral cortical layers are different from those in the medullary and algal layers (Figs. 16.8, 16.10a). Many species have the same secondary metabolites in both the vegetative thallus and the ascomata (Fig. 16.13a–b); others synthesize different compounds in their ascomata and in the thallus (Fig. 16.15).

In sterile cultures of aposymbiotic lichen-forming fungi secondary metabolites often diffuse out into the agar medium and crystallize at a considerable distance from the mycelium (Fig. 16.17). For a long time most secondary lichen compounds were assumed to be formed exclusively in symbiosis, the exact role of the photobiont being unknown. However, with improved culturing techniques (e.g. application of osmotic stress) the secondary metabolism was triggered in a wide range of aposymbiotic lichen mycobionts (Fig. 16.14; Yamamoto et al. 1985, 1993; Culberson and Armaleo 1992; Stocker-Wörgötter 2001).

In nature secondary compounds fulfil various functions in lichen thalli (Lawrey 1986). They increase the hydrophobicity of hyphal surfaces within the thallus (Figs. 16.8, 16.10) or at the thallus periphery, as observed in some extremely hydrophobic, mainly crustose lichens (Honegger 2004; Shirtcliffe et al. 2006). A positive correlation between thallus surface hydrophobicity and tolerance to SO_2 pollution is postulated (Hauck and Huneck 2007; Hauck et al. 2008). A wide range of cortical secondary lichen metabolites are autofluorescent, i.e. absorb UV light and transmit longer wavelengths to the algal layer in the thalline interior (Fig. 16.18) and thus protect the photobiont cell population from irradiation damage (Rundel 1978; Lawrey 1986; Gauslaa and Uvstedt 2003; Solhaug et al. 2003). Various investigators demonstrated quantitative differences in secondary product contents per lichen species, depending on the intensity of UV exposure (Fig. 16.13a–b; Buffoni Hall et al. 2002; Solhaug et al. 2003; Nybakken et al. 2004; McEvoy et al. 2006, 2007), often with distinct seasonal variations (Bjerke et al. 2002, 2005; Armaleo et al. 2008). However, mutants occur with low secondary metabolite production irrespective of full illumination (Fig. 16.13b; Honegger, unpublished data).

B. Economic Perspectives

Many secondary lichen compounds reveal antibiotic properties (Fig. 16.21a, b); they might protect the long-living lichen thalli from microbial invasion. Although bacterial films are commonly seen on thallus surfaces, the interior of healthy, actively growing lichens is free from bacterial contaminants, irrespective of the presence or absence of medullary lichen compounds. After first screens in the post-World War II period (Barry 1946; Stoll et al. 1947) the interest in lichen-derived antibiotics was largely lost since biotechnologists preferred, for obvious reasons, the fast-growing non-lichenized fungal and bacterial producers of antibiotics. Only in folk medicine are formulations available with mild antibiotic properties due to lichens such as *Usnea* spp. (e.g. in the widely traded Usneasan; trademark registered by Vogel). The advent of bacterial pathogens resistant to conventional antibiotics triggered a renewed interest in lichen-derived antibiotics, some of which inhibit a wide range of bacterial pathogens, resistant strains included (Elo et al. 2007). Analgesic, antiproliferative, or antiinflammatory activities were detected in lichen-derived compounds (Müller 2001; Oksanen 2006). One of the pharmaceutically most promising lichen metabolites is usnic acid (Cochietto et al. 2002; Elo et al. 2007), which occurs not only in beard lichens (*Usnea* spp.) but also in a wide range of other taxa. However, usnic acid attracted the interest of gastroenterologists since it is strongly hepatotoxic when ingested in higher quantities as a presumed weight loss agent ("fat burner"; Durazo et al. 2004; Han 2004).

The widespread *para*-depside atranorin and related secondary lichen metabolites are allergenic, causing contact dermatitis in sensitized persons (Mitchell and Champion 1965; Bernard et al. 2003). This allergic reaction, known as woodcutters disease, is caused by contact with minute lichen fragments, but also by perfumes which contain lichen extracts (mousse de chêne, from *Evernia prunastri*, or mousse d'arbre from *Pseudevernia furfuracea*; Dahlquist and Fregert 1980; Thune et al. 1982; Buckley et al. 2000; Johansen et al. 2002a, b). These lichen-derived compounds provide men's cosmetics, currently a booming and increasing market, with the scent of wilderness and adventure. Since allergies against lichen compounds, as contained in perfumes, have steadily increased from circa 1980 onwards (Buckley et al. 2000), possibly with the increasing use of cosmetics, natural lichen resinoids are currently being replaced by synthetic, non-allergenic compounds in the perfume industry (Roman Kaiser, Givaudan SA, personal communication).

Beside polyphenolic secondary metabolites the water- and alkali-soluble fractions of lichen thalli, i.e. the manifold types of cell wall polysaccharides, attracted the interest of pharmacologists. Some lichen-derived glucans, chemically similar to lentinan or schizophyllan from non-lichenized basidiomycetes, have stong immunomodulating or tumor-inhibiting activities (Olafsdottir and Ingolfsdottir 2001; Olafsdottir et al. 2003; Omarsdottir et al. 2006, 2007). As large-scale harvesting of lichens for pharmaceutical applications should be avoided for conservational reasons, various investigators are currently improving cultural techniques for lichen-forming fungi (Brunauer and Stocker-Wörgötter 2005). However, considering the immensely slow growth of these oligotrophic organisms in pure culture, the characterization of polyketide synthetase genes in lichen-forming fungi and subsequent heterologous expression in fast-growing non-lichenized filamentous fungi seems to be a more promising approach (Miao et al. 2001; Grube and Blaha 2003; Opanowicz et al. 2006). As crystalline secondary metabolites of lichen-forming fungi are difficult to dissolve in aqueous systems new formulations have to be developed (Kristmundsdottir et al. 2005).

VIII. Conclusions and Outlook

In the past decade fascinating insights were gained into the phylogeny of lichen-forming fungi and their photobionts (e.g. tree of life project) and also in the biology of lichen symbiosis, by means of molecular techniques. In the near future population genetics will be intensely explored with molecular tools applied at the subspecific level. In ecological studies a very high diversity was found in habitats which had previously been only partially investigated, such as the foliicolous lichens of tropical rainforests. The importance of lichens and other biota in biological soil crust communities for soil stabilization in arid zones became evident in studies on the prevention of desertification and consequent severe sand storm events. Remote sensing techniques facilitate monitoring of these highly vulnerable communities (Chen et al. 2005; Schultz 2006; Zhang et al. 2007). As in the past the ability of lichens to bioaccumulate elements of various origins, pollutants and radionuclides included, will be of continued interest in monitoring studies. The renewed interest in lichen-derived compounds with pharmaceutical potential and the discovery of promising substances necessitates intense genetic studies, the adaptation of culturing techniques and a search for suitable heterologous expression systems.

Lichens respond to global warming. Psychrophilic species disappear in zones with milder climates and retreat into colder areas, but species favouring milder climates invade new areas (van Herk and Aptroot 2002). As shown in a series of field experiments in the climatically milder part of the arctic, lichens would be continuously outcompeted by angiosperms when temperatures continue to rise and more nutrients become available due to atmospheric input and increased soil mineralization in melting permafrost (Cornelissen et al. 2001; Rustad et al. 2001). The ecological, social and economic consequences are difficult to predict.

Lichen-forming fungi lack all the characteristics of model organisms in main stream biology such as fast growth, in vitro sexual reproduction and short reproductive cycles, etc. and thus are not among the favourites of molecular or developmental biologists. Accordingly, the state of knowledge in molecular genetics of lichen symbiosis is extremely low. Nevertheless, with their taxonomic diversity, their immense ecological significance in extreme climates, their impressive lifestyles and the level of complexity of their symbiotic interactions, lichen-forming fungi will always attract the interest of biologists and also fascinate naturalists by their intrinsic beauty.

Acknowledgements. My sincere thanks are due to the Swiss National Science Foundation for generous financial support (grant 3100A0-116597), to Sibylle Erni for drawing Fig. 16.10a, to Peter Hättenschwiler for identifying the bagworm with lichen camouflage in Fig. 16.16a, b, to Reinhard Berndt for Fig. 16.22 and to my husband, Thomas G. Honegger, for continuous and patient computer support.

References

Adams GC, Kropp BR (1996) *Athelia arachnoidea*, the sexual state of *Rhizoctonia carotae*, a pathogen of carrot in cold storage. Mycologia 88:459–472

Ahmadjian V (1988) The lichen alga *Trebouxia* – does it occur free-living? Plant Syst Evol 158:243–247

Ahmadjian V (1993) The lichen symbiosis. Wiley, New York

Ahmadjian V (2002) *Trebouxia*: reflections on a perplexing and controversial lichen photobiont. In: Seckbach J (ed) Symbiosis: mechanisms and model systems. Kluwer, Dordrecht, pp 373–383

Allgaier C (2007) Active camouflage with lichens in a terrestrial snail, *Napaeus* (N.) *barquini* Alonso and Ibáñez, 2006 (Gastropoda, Pulmonata, Enidae). Zool Sci 24:869–876

Armaleo D, Zhang Y, Cheung S (2008) Light might regulate divergently depside and depsidone accumulation in the lichen *Parmotrema hypotropum* by affecting thallus temperature and water potential. Mycologia 100:55–576

Aspray T, Jones E, Whipps J, Bending G (2006) Importance of mycorrhization helper bacteria: cell density and metabolite localization for the *Pinus sylvestris/Lactarius rufus* symbiosis. FEMS Microbiol Ecol 56:25–33

Ates A, Yildiz A, Yildiz N, Calimli A (2007) Heavy metal removal from aqueous solution by *Pseudevernia furfuracea* (L.) Zopf. Ann Chim 97:385–93

Barry VC (1946) Anti-tubercular compounds. Nature 158:863–865

Belnap J, Lange OL (2003) Biological soil crusts: structure, function and management. Ecological studies, vol 150. Springer, Berlin

Bennett JP, Wright DM (2004) Element content of *Xanthoparmelia scabrosa* growing on asphalt in urban and rural New Zealand. Bryologist 107:421–428

Bernard G, Giménez-Arnau E, Rastogi SC, Heydorn S, Johansen JD, Menné T, Goossens A, Andersen K, Lepoittevin JP (2003) Contact allergy to oak moss: search for sensitizing molecules using combined bioassay-guided chemical fractionation, GC-MS, and structure–activity relationship analysis. Arch Dermatol Res 295:229–35

Biazrov LG (1994) The radionuclides in lichen thalli in Chernobyl and East Urals areas after nuclear accidents. Phyton 34:85–94

Bjelland T, Thorset IH (2002) Comparative studies of the lichen–rock interface of four lichens in Vingen, western Norway. Chem Geol 192:81–98

Bjelland T, Sæbø L, Thorseth IH (2002) The occurrence of biomineralization products in four lichen species growing on sandstone in western Norway. Lichenologist 34:429–440

Bjerke JW, Lerfall K, Elvebakk A (2002) Effects of ultraviolet radiation and PAR on the content of usnic and divaricatic acids in two arctic–alpine lichens. Photochem Photobiol Sci 1:678–685

Bjerke JW, Elvebakk A, Domínguez E, Dahlback A (2005) Seasonal trends in usnic acid concentrations of Arctic, alpine and Patagonian populations of the lichen *Flavocetraria nivalis*. Phytochemistry 66:337–344

Blaha J, Baloch E, Grube M (2006) High photobiont diversity associated with the euryoecious lichen-forming ascomycete *Lecanora rupicola* (Lecanoraceae, Ascomycota). Biol J Linn Soc 88:283–293

Bock C, Jacob A, Kirst GO, Leibfritz D, Mayer A (1996) Metabolic changes of the antarctic green alga *Prasiola crispa* subjected to water stress investigated by in vivo ^{31}P NMR. J Exp Bot 47:241–249

Boissière J-C (1987) Ultrastructural relationship between the composition and the structure of the cell wall of the mycobiont of two lichens. Progress and problems in lichenology in the eighties. Bibl Lichenol 35:117–132

Bostedt G (2001) Reindeer husbandry, the Swedish market for reindeer meat, and the Chernobyl effects. Agric Econ 26:217–226

Brodo IM, Sharnoff SD, Sharnoff S (2001) Lichens of North America. Yale University Press, New Haven

Brown DH, Rapsch S, Beckett A, Ascaso C (1987) The effect of desiccation on cell shape in the lichen *Parmelia sulcata* Taylor. New Phytol 105:295–299

Brunauer G, Stocker-Wörgötter E (2005) Culture of lichen fungi for future production of biologically active compounds. Symbiosis 38:187–201

Brunner U, Honegger R (1985) Chemical and ultrastructural studies on the distribution of sporopollenin-like biopolymers in 6 genera of lichen phycobionts. Can J Bot 63:2221–2230

Buckley DA, Wakelin SH, Seed PT, Holloway D, Rycroft RJ, White IR, McFadden JP (2000) The frequency of fragrance allergy in a patch-test population over a 17-year period. Br J Dermatol 142:279–283

Buffoni Hall RS, Bornman JF, Björn LO (2002) UV-induced changes in pigment content and light penetration in the fruticose lichen *Cladonia arbuscula* ssp. *mitis*. J Photochem Photobiol B 66:13–20

Cardinale M, Puglia AM, Grube M (2006) Molecular analysis of lichen-associated bacterial communities. FEMS Microbiol Ecol 57:484–495

Chen J, Zhang MY, Wang L, Shimazaki H, Tamura M (2005) A new index for mapping lichen-dominated biological soil crusts in desert areas. Remote Sensing Environ 96:165–175

Ciferri O, Tiano P, Mastromei G (2000) Of microbes and art: the role of microbial communities in the degradation and protection of cultural heritage. Kluwer/Plenum, New York

Cocchietto M, Skert N, Nimis PL, Sava G (2002) A review on usnic acid, and interesting natural compound. Naturwissenschaften 89:137–146

Cornelissen JHC, Callaghan TV, Alatalo JM, Michelsen A, Graglia E, Hartley AE, Hik DS, Hobbie SE, Press MC, Robinson CH, Henry GHR, Shaver GR, Phoenix GK, Jones DG, Jonasson S, Chapin FS, Molau U, Neill C, Lee JA, Melillo JM, Sveinbjörnsson B, Aerts R (2001) Global change and arctic ecosystems: is lichen decline a function of increases in vascular plant biomass? J Ecol 89:984–994

Cracraft J (1983) Species diversity, biogeography, and the evolution of biotas. Am Zool 34:33–47

Crittenden PD, Porter N (1991) Lichen-forming fungi: potential sources of novel metabolites. Trends Biotechnol 9:409–414

Crittenden PD, David JC, Hawksworth DL, Campbell FS (1995) Attempted isolation and success in the culturing of a broad spectrum of lichen-forming and lichenicolous fungi. New Phytol 130:267–297

Culberson CF (1969) Chemical and botanical guide to lichen products. University of North Carolina Press, Chapel Hill

Culberson CF, Armaleo D (1992) Induction of a complete secondary-product pathway in a cultured lichen fungus. Exp Mycol 16:52–63

Dahlkild A, Kallersjo M, Lohtander K, Tehler A (2001) Photobiont diversity in the Physciaceae (Lecanorales). Bryologist 104:527–536

Dahlquist I, Fregert S (1980) Contact allergy to atranorin in lichens and perfumes. Contact Dermatitis 6:111–119

Deckert RJ, Garbary DJ (2005) Ascophyllum and its symbionts. VI. Microscopic characterization of the Ascophyllum nodosum (Phaeophyceae), Mycophycias ascophylli (Ascomycetes) symbiotum. Algae 20:225–232

Diederich P (2003) New species and new records of American lichenicolous fungi. Herzogia 16:41–90

Dietz S, Büdel B, Lange OL, Bilger W (2000) Transmittance of light through the cortex of lichens from contrasting habitats. Bibl Lichenol 75:171–182

Durazo FA, Lassman C, Han SHB, Saab S, Lee NP, Kawano M, Saggi B, Gordon S, Farmer DG, Yersiz H, Goldstein RLI, Ghobrial M, Busuttil RW (2004) Fulminant liver failure due to usnic acid for weight loss. Am J Gastroenterol 99:950–952

Ekman S, Tønsberg T (2002) Most species of Lepraria and Leproloma form a monophyletic group closely related to Stereocaulon. Mycol Res 106:1262–1276

Eldridge DJ (1996) Distribution and floristics of terricolous lichens in soil crusts in arid and semi-arid New South Wales, Australia. Aust J Bot 44:581–599

Ellis CJ, Crittenden PD, Scrimgeour CM, Ashcroft CJ (2005) Translocation of ^{15}N indicates nitrogen recycling in the mat-forming lichen Cladonia portentosa. New Phytol 168:423–434

Elo H, Matikainen J, Pelttar E (2007) Potent activity of the lichen antibiotic (+)-usnic acid against clinical isolates of vancomycin-resistant enterococci and methicillin-resistant Staphylococcus aureus. Naturwissenschaften 94:465–468

Englund B (1977) The physiology of the lichen Peltigera aphthosa, with special reference to the blue-green phycobiont (Nostoc sp.). Physiol Plant 41:298–304

Eriksson O (2005) Ascomyceternas ursprung och evolution – Protolichenes-hypotesen. Svensk Mykol Tidskr 26:22–29

Eriksson OE (2006a) Outline of Ascomycota. Myconet 12:1–82

Eriksson OE (2006b) Notes on ascomycete systematics. Myconet 12:83–101

Ettl H, Gärtner G (1995) Syllabus der Boden-, Luft- und Flechtenalgen. Fischer, Stuttgart

Feige GB, Niemann L, Jahnke S (1990) Lichens and mosses – silent chronists of the Chernobyl accident. Bibl Lichenol 38:63–77

Friedl T (1989) Systematik und Biologie von Trebouxia (Microthamniales, Chlorophyta) als Phycobiont der Parmeliaceae (lichenisierte Ascomyceten). Dissertation, University of Bayreuth

Friedl T, Büdel B (2008) Photobionts. In: Nash TH (ed) Lichen Biology, 2nd edn. Cambridge University Press, Cambridge, pp 9–26

Fröberg L, Berg CO, Baur A, Baur B (2001) Viability of lichen photobionts after passing through the digestive tract of a land snail. Lichenologist 33:543–545

Galloway DJ (1996) Lichen biogeography. In: Nash TH (ed) Lichen biology. Cambridge University Press, Cambridge, pp 199–216

Garbaye J (1994) Helper bacteria: a new dimension to the mycorrhizal symbiosis. New Phytol 128:197–210

Gauslaa Y (2005) Lichen palatability depends on investments in herbivore defence. Oecologia 143:94–105

Gauslaa Y, Ustvedt EM (2003) Is parietin a UV-B or a blue-light screening pigment in the lichen Xanthoria parietina? Photochem Photobiol Sci 2:424–432

Gerson U, Seaward MRD (1977) Lichen–invertebrate associations. In: Seaward MRD (ed) Lichen ecology, Academic, London, pp 69–119

Goebel K (1926) Die Wasseraufnahme der Flechten. Ber Dtsch Bot Ges 44:158–161

Graham LE, Wilcox LW (2000) Algae. Prentice Hall, New Jersey

Green TGA, Kilian E, Lange OL (1991) Pseudocyphellaria dissimilis: a desiccation-sensitive, highly shade-adapted lichen from New Zealand. Oecologia 85:498–503

Green TGA, Büdel B, Heber U, Meyer A, Zellner A, Lange OL (1993) Differences in photosynthetic performance between cyanobacterial and green algal components of lichen photosymbiodemes measured in the field. New Phytol 125:723–731

Green TGA, Schlensog M, Sancho LG, Winkler JB, Broom FD, Schroeter B (2002) The photobiont determines the pattern of photosynthetic activity within a single lichen thallus containing cyanobacterial and green algal sectors (photosymbiodeme). Oecologia 130:191–198

Gressitt JL (1965) Flora and fauna on backs of large Papuan moss-forest weevils. Science 150:1833–1835

Greuter W, McNell J, Barrie FR, Burdet H-M, Demoulin V, Filgueras TS, Nicolson DH, Silva PC, Skog JE, Trehane P, Turland NJ, Hawksworth DL (2000) International Code of Botanical Nomenclature (St Louis Code). Regnum Vegetabile 138. Koeltz, Königstein

Griffiths HB, Greenwood AD, Millbank JW (1972) The frequency of heterocysts in the *Nostoc* phycobiont of the lichen *Peltigera canina* Willd. New Phytol 71:11–13

Grube M, Blaha J (2003) On the phylogeny of some polyketide synthase genes in the lichenized genus *Lecanora*. Mycol Res 107:1419–1426

Grube M, Kantvilas G (2006) *Siphula* represents a remarkable case of morphological convergence in sterile lichens. Lichenologist 38:241–249

Haas JR, Purvis OW (2006) Lichen biogeochemistry. In: Gadd GM (ed) Fungi in biogeochemical cycles. Cambridge University Press, Cambridge, pp 343–376

Han D, Matsumaru K, Rettori D, Kaplowitz N (2004) Usnic acid-induced necrosis of cultured mouse hepatocytes: inhibition of mitochondrial function and oxidative stress. Biochem Pharmacol 67:439–451

Hauck M, Huneck S (2007) Lichen substances affect metal adsorption in *Hypogymnia physodes*. J Chem Ecol 33:219–223

Hauck M, Jürgens SR, Brinkmann M, Herminghaus S (2008) Surface hydrophobicity causes SO_2 tolerance in lichens. Ann Bot 101:531–539

Hawksworth DL (1988a) The variety of fungal-algal symbioses, their evolutionary significance, and the nature of lichens. Bot J Linn Soc 96:3–20

Hawksworth DL (1988b) Effects of algae and lichen-forming fungi on tropical crops. In: Agnihotri VP, Sarbhoy KA, Kumar D (eds) Perspectives of mycopathology. Malhotra, New Delhi, pp 76–83

Hawksworth DL, Honegger R (1994) The lichen thallus: symbiotic phenotype and its responses to gall producers. In: Williams MC (ed) Plant galls: organisms, interactions. Clarendon, Oxford, pp 77–98

Helms G (2003) Taxonomy and symbiosis in associations of Physciaceae and *Trebouxia*. Dissertation, University of Göttingen

Helms G, Friedl T, Rambold G, Mayrhofer H (2001) Identification of photobionts from the lichen family Physciaceae using algal-specific ITS rDNA sequencing. Lichenologist 33:73–86

Herrera-Campos MA, Lücking R, Perez R, Campos A, Colin PM, Pena AB (2004) The foliicolous lichen flora of Mexico. V. Biogeographical affinities, altitudinal preferences, and an updated checklist of 293 species. Lichenologist 36:309–327

Hill DJ (1976) The physiology of lichen symbiosis. In: Brown DH, Hawksworth DL, Bailey RH (eds) Lichenology: progress and problems. Academic, London, pp 457–496

Hitch CJB, Millbank JW (1975) Nitrogen metabolism in lichens. VI. The blue-green phycobiont content, heterocyst frequency and nitrogenase activity in *Peltigera* species. New Phytol 74:473–476

Holfeld H (1998) Fungal infections of the phytoplankton: seasonality, minimal host density, and specificity in a mesotrophic lake. New Phytol 138:507–517

Honegger R (1984) Cytological aspects of the mycobiont-phycobiont relationship in lichens. Haustorial types, phycobiont cell wall types, and the ultrastructure of the cell wall surface layers in some cultured and symbiotic myco- and phycobionts. Lichenologist 16:111–127

Honegger R (1985) Fine structure of different types of symbiotic relationships in lichens. In: Brown DH (ed) Lichen physiology and cell biology. Plenum, New York, pp 287–302

Honegger R (1986a) Ultrastructural studies in lichens. I. Haustorial types and their frequencies in a range of lichens with trebouxioid phycobionts. New Phytol 103:785–795

Honegger R (1986b) Ultrastructural studies in lichens. II. Mycobiont and photobiont cell wall surface layers and adhering crystalline lichen products in four Parmeliaceae. New Phytol 103:797–808

Honegger R (1991) Functional aspects of the lichen symbiosis. Annu Rev Plant Physiol Plant Mol Biol 42:553–578

Honegger R (1995) Experimental studies with foliose macrolichens: fungal responses to spatial disturbance at the organismic level and to spacial problems at the cellular level. Can J Bot 73:569–578

Honegger R (1996) Growth and regenerative capacity in the foliose lichen *Xanthoria parietina* (L.) Th. Fr. (Teloschistales, Ascomycotina). New Phytol 133:573–581

Honegger R (1997) Metabolic interactions at the mycobiont-photobiont interface in lichens. In: Carroll GC, Tudzynski P (eds) Plant relationships, vol V.A. Springer, Heidelberg, pp 209–221

Honegger R (1998) The lichen symbiosis – what is so spectacular about it? Lichenologist 30:193–212

Honegger R (2001) The symbiotic phenotype of lichen-forming ascomycetes. In: Hock B (ed) Fungal associations, vol IX. Springer, Berlin, pp 165–188

Honegger R (2003) The impact of different long-term storage conditions on the viability of lichen-forming ascomycetes and their green algal photobiont, *Trebouxia* spp. Plant Biol 5:324–330

Honegger R (2004) Fine structure of the interaction of *Leprocaulon microscopicum* with its integrated green algal photobiont, *Dictyochloropsis symbiontica*. Bibl Lichenol 88:201–210

Honegger R (2007) Water relations in lichens. In: Gadd GM, Watkinson SC, Dyer P (eds) Fungi in the environment. Cambridge University Press, Cambridge, pp 185–200

Honegger R, Bartnicki-Garcia S (1991) Cell wall structure and composition of cultured mycobionts from the lichens *Cladonia macrophylla*, *Cladonia caespiticia*, and *Physcia stellaris* (Lecanorales, Ascomycetes). Mycol Res 95:905–914

Honegger R, Brunner U (1981) Sporopollenin in the cell wall of Coccomyxa and Myrmecia phycobionts of various lichens: an ultrastructural and chemical investigation. Can J Bot 59:2713–2734

Honegger R, Haisch A (2001) Immunocytochemical location of the $(1 \rightarrow 3) (1 \rightarrow 4)$-beta-glucan lichenin in the lichen-forming ascomycete *Cetraria islandica* (Icelandic moss). New Phytol 150:739–746

Honegger R, Hugelshofer G (2000) Water relations in the *Peltigera aphthosa* group visualized with LTSEM techniques. Bibl Lichenol 75:113–126

Huneck S (2001) New results on the chemistry of lichen substances. Springer, Vienna

Huneck S, Yoshimura I (1996) Identification of lichen substances. Springer, Berlin

James TY, Kauff F, Schoch CL, Matheny PB, Hofstetter V, Cox CJ, Celio G, Gueidan C, Fraker E, Miadlikowska Y, Lumbsch HT, Rauhut A, Reeb V, Arnold AE, Amtoft A, Stajich JE, Hosaka K, Sung G-H, Johnson D, O'Rourke B, Crockett M, Binder M, Curtis JM, Slot JC, Wang Z, Wilson AW, Schüssler A, Longcore JE, O'Donnell K, Mozley-Standridge S, Porter D, Letcher PM, Powell MJ, Taylor JW, White MM, Griffith GW, Davies DR, Humber RA, Morton JB, Sugiyama J, Rossman AY, Rogers JD, Pfister DH, Hewitt D, Hansen K, Hambleton S, Shoemaker RA, Kohlmeyer J, Volkmann-Kohlmeyer B, Spotts RA, Serdani M, Crous PW, Hughes KW, Matsuura K, Langer E, Langer G, Untereiner WA, Lücking R, Büdel B, Geiser DM, Aptroot A, Diederich P, Schmitt I, Schultz M, Yahr R, Hibbett DS, Lutzoni F, McLaughlin DJ, Spatafora JW, Vilgalys R (2006) Reconstructing the early evolution of Fungi using a six-gene phylogeny. Nature 443:818–822

Johansen B, Karlsen SR (2005) Monitoring vegetation changes on Finnmarksvidda, northern Norway, using Landsat MSS and Landsat TM/ETM+ satellite images. Phytocoenologia 35:969–984

Johansen JD, Heydorn S, Menné T (2002) Oak moss extracts in the diagnosis of fragrance contact allergy. Contact Dermatitis 46:157–161

Johansen JD, Bernard G, Giménez-Arnau E, Lepoittevin JP, Bruze M, Andersen KE (2006) Comparison of elicitation potential of chloroatranol and atranol-2 allergens in oak moss absolute. Contact Dermatitis 54:192–195

Jörgensen PM (1998) What shall we do with the blue-green counterparts? Lichenologist 30:351–356

Kappen L (1988) Ecophysiological relationships in different climatic regions. In: Galun M (ed) CRC handbook of lichenology, vol 2. CRC, Boca Raton, pp 37–100

Kappen L (2000) Some aspects of the great success of lichens in Antarctica. Antarct Sci 12:314–324

Kappen L, Schroeter B (1997) Activity of lichens under the influence of snow and ice (18th symposium on polar biology). Proc NIPR Symp Polar Biol 10:163–168

Kershaw M, Thornton C, Wakley G, Talbot N (2005) Four conserved intramolecular disulphide linkages are required for secretion and cell wall localization of a hydrophobin during fungal morphogenesis. Mol Microbiol 56:117–125

Kieft T (1988) Ice nucleation activity in lichens. Appl Environ Microbiol 54:1678–1681

Kieft T, Ahmadjian V (1989) Biological ice nucleation activity in lichen mycobionts and photobionts. Lichenologist 21:335–362

Kieft T, Ruscetti T (1990) Characterisation of biological ice nuclei from a lichen. J Bacteriol 172:3519–3623

Kieft T, Ruscetti T (1992) Molecular sizes of lichen ice nucleation sites determined by gamma-radiation inactivation analysis. Cryobiology 29:407–413

Kirk PM, Cannon PF, David JC, Stalpers JA (2001) Ainsworth and Bisby's dictionary of the fungi, 9th edn. CAB International, Oxford

Kluge M, Mollenhauer D, Mollenhauer R (1994) *Geosiphon pyriforme* (Kützing) von Wettstein, a promising system for studying endocyanoses. Prog Bot 55:130–141

Knowles RD, Pastor J, Biesboer DD (2006) Increased soil nitrogen associated with dinitrogen-fixing, terricolous lichens of the genus *Peltigera* in northern Minnesota. Oikos 114:37–48

Kohlmeyer J, Kohlmeyer E (1972) Is *Ascophyllum nodosum* lichenized? Bot Mar 15:109–112

Kohlmeyer J, Volkmann-Kohlmeyer B (1998) *Mycophycias*, a new genus for the mycobionts of *Apophlaea*, *Ascophyllum* and *Pelvetia*. Systema Ascomycetum 16:1–7

Kristmundsdóttir T, Jónsdóttir E, Ogmundsdóttir HM, Ingólfsdóttir K (2005) Solubilization of poorly soluble lichen metabolites for biological testing on cell lines. Eur J Pharm Sci 24:539–543

Kroken S, Taylor JW (2000) Phylogenetic species, reproductive mode, and specificity of the green alga *Trebouxia* forming lichens with the fungal genus *Letharia*. Bryologist 103:645–660

Kroken S, Taylor JW (2001) A gene genealogical approach to recognize phylogenetic species boundaries in the lichenized fungus *Letharia*. Mycologia 93:38–53

Kytöviita M-M, Crittenden PD (2007) Growth and nitrogen relations in the mat-forming lichens *Stereocaulon paschale* and *Cladonia stellaris*. Ann Bot 100:1537–1545

Lakatos M, Lange-Bertalot H, Büdel B (2004) Diatoms living inside the thallus of green algal lichen *Coenogonium linkii* in neotropic lowland rain forests. J Phycol 40:70–73

Lalley JS, Viles HA (2005) Terricolous lichens in the northern Namib Desert of Namibia: distribution and community composition. Lichenologist 37:77–91

Lalley JS, Viles HA (2006) Do vehicle track disturbances affect the productivity of soil-growing lichens in a fog desert? Funct Ecol 20:548–556

Lange O, Meyer A, Zellner H, Ullmann I, Wessels D (1990) Eight days in the life of a desert lichen: water relations and photosynthesis of *Teloschistes capensis* in the coastal fog zone of the Namib desert. Madoqua 17:17–30

Lange OL, Belnap J, Reichenberger H, Meyer A (1997) Photosynthesis of green algal soil crust lichens from arid lands in southern Utah, USA: role of water content on light and temperature responses of CO_2 exchange. Flora 192:1–15

Lange OL, Belnap J, Reichenberger H (1998) Photosynthesis of the cyanobacterial soil-crust lichen *Collema tenax* from arid lands in southern Utah, USA: role of water content on light and temperature responses of CO_2 exchange. Funct Ecol 12:195–202

Lange OL, Green TGA, Melzer B, Meyer A, Zellner H (2006) Water relations and CO_2 exchange of the terrestrial lichen *Teloschistes capensis* in the Namib fog desert: measurements during two seasons in the field and under controlled conditions. Flora 201:268–280

Lawrey JD (1986) Biological role of lichen substances. Bryologist 89:111–122

Lawrey JD (1993) Lichens as monitors of pollutant elements at permanent sites in Maryland and Virginia. Bryologist 96:339–341

Lawrey JD, Diederich P (2003) Lichenicolous fungi: interactions, evolution, and biodiversity. Bryologist 106:80–120

Lewis Smith RI (1995) Colonization by lichens and the development of lichen-dominated communities in the maritime Antarctic. Lichenologist 27:473–483

Linder MB, Szilvay GR, Nakari-Setälä T, Penttilä ME (2005) Hydrophobins: the protein-amphiphiles of filamentous fungi. FEMS Microbiol Rev 29:877–896

Lines CEM, Ratcliffe RG, Rees TAV, Southon TE (1989) A ^{13}C NMR study of photosynthate transport and metabolism in the lichen *Xanthoria calcicola* Oxner. New Phytol 111:447–482

Lücking R (2008) Foliicolous lichenized fungi. Flora neotropica, monograph 103. New York Botanical Garden Press, New York, 867 pp

Lücking R, Matzer M (2001) High foliicolous lichen alpha-diversity on individual leaves in Costa Rica and Amazonian Ecuador. Biodivers Conserv 10:2139–2152

Lücking R, Wirth V, Ferraro LI, Caceres MES (2003) Foliicolous lichens from Valdivian temperate rain forest of Chile and Argentina: evidence of an austral element, with the description of seven new taxa. Global Ecol Biogeogr 12:21–36

Lutzoni F, Pagel M, Reeb V (2001) Major fungal lineages are derived from lichen symbiotic ancestors. Nature 411:937–940

Lutzoni F, Kauff F, Cox C, McLaughlin D, Celio G, Dentinger B, Padamsee M, Hibbett D, James T, Baloch E, Grube M, Reeb V, Hofstetter V, Schoch C, Arnold A, Miadlikowska J, Spatafora J, Johnson D, Hambleton S, Crockett M, Shoemaker R, Sung G-H, Lücking R, Lumbsch T, O'Donnell K, Binder M, Diederich P, Ertz D, Gueidan C, Hansen K, Harris R, Hosaka K, Lim Y-W, Matheny B, Nishida H, Pfister D, Rogers J, Rossman A, Schmitt I, Sipman H, Stone J, Sugiyama J, Yahr R, Vilgalys R (2004) Assembling the fungal tree of life: progress, classification, and evolution of subcellular traits. Am J Bot 91:1446–1480

MacGinitie H (1937) The flora of the Weaverville beds of Trinity County, California, with descriptions of the plant-bearing beds. Eocene flora of western America, vol 465. Carnegie Institution, Washington, pp 83–151

Mägdefrau K (1957) Flechten und Moose im baltischen Bernstein. Ber Dtsch Bot Ges 9:433–435

Mankel A, Krause K, Kothe E (2002) Identification of a hydrophobin gene that is developmentally regulated in the ectomycorrhizal fungus *Tricholoma terreum*. Appl Environ Microbiol 68:1408–1413

Martin F, Laurent P, Decarvalho D, Burgess T, Murphy P, Nehls U, Tagu D (1995) Fungal gene expression during ectomycorrhiza formation. Can J Bot 73:S541–S547

Mayr E (1942) Systematics and the origin of species. Columbia University Press, New York

Mayr E (2000) The biological species concept. In: Wheeler QD, Meier R (eds) Species concepts and phylogenetic theory – a debate. Columbia University Press, New York, pp 17–29

McCune B, Rosentreter R (2007) Biotic soil crust lichens of the Columbia Basin. Monogr N Am Lichenol 1:1–105

McEvoy M, Solhaug KA, Gauslaa Y (2006) Ambient UV irradiation induces a blue pigment in *Xanthoparmelia stenophylla*. Lichenologist 38:285–289

McEvoy M, Gauslaa Y, Solhaug KA (2007) Changes in pools of depsidones and melanins, and their function, during growth and acclimation under contrasting natural light in the lichen *Lobaria pulmonaria*. New Phytol 175:271–282

Mehli H, Skuterud L, Mosdøl A, Tønnessen A (2000) The impact of Chernobyl fallout on the Southern Saami reindeer herders of Norway in 1996. Health Phys 79:682–690

Meier FA, Scherrer S, Honegger R (2002) Faecal pellets of lichenivorous mites contain viable cells of the lichen-forming ascomycete *Xanthoria parietina* and its green algal photobiont, *Trebouxia arboricola*. Biol J Linn Soc 76:259–268

Miao V, Coeffet-LeGal M-F, Brown D, Sinnemann S, Donaldson G, Davies J (2001) Genetic approaches to harvesting lichen products. Trends Biotechnol 19:349–355

Millbank JW (1976) Aspects of nitrogen metabolism in lichens. Lichenology: progress and problems. Academic, London, pp 441–455

Mitchell JC, Champion RH (1965) Human allergy to lichens. Bryologist 68:116–118

Miura S, Yokota A (2006) Isolation and characterization of cyanobacteria from lichen. J Gen Appl Microbiol 52:365–374

Mollenhauer D (1992) *Geosiphon pyriforme*. In: Reisser W (ed) Algae and symbioses. Biopress, Bristol, pp 339–351

Mollenhauer D, Mollenhauer R, Kluge M (1996) Studies on initiation and development of the partner association in *Geosiphon pyriforme* (Kütz.) v. Wettstein, a unique endocytobiotic system of a fungus (Glomales) and the cyanobacterium *Nostoc punctiforme* (Kütz.) Hariot. Protoplasma 193:3–9

Mukhtar A, Garty J, Galun M (1994) Does the lichen alga *Trebouxia* occur free-living in nature: further immunological evidence. Symbiosis 17:247–253

Müller K (2001) Pharmaceutically relevant metabolites from lichens. Appl Microbiol Biotechnol 56:9–16

Muscha JM, Hilda AL (2006) Biological soil crusts in grazed and ungrazed Wyoming sagebrush steppe. J Arid Environ 67:195–207

Nelsen MP, Lücking R, Umana L, Trest MT, Will-Wolf S, Chaves JL, Gargas A (2007) *Multiclavula ichthyiformis* (Fungi: Basidiomycota: Cantharellales: Clavulinaceae), a remarkable new basidiolichen from Costa Rica. Am J Bot 94:1289–1296

Noeske O, Läuchli A, Lange OL, Vieweg GH, Ziegler H (1970) Konzentration und Lokalisierung von Schwermetallen in Flechten der Erzschlackenhalden des Harzes. Dtsch Bot Ges Neue Folge 4:67–79

Nordberg M-L, Allard A (2002) A remote sensing methodology for monitoring lichen cover. Can J Remote Sensing 28:262–274

Nyati S, Beck A, Honegger R (2007) Fine structure and phylogeny of green algal photobionts in the microfilamentous genus *Psoroglaena* (Verrucariaceae, lichenforming ascomycetes). Plant Biol 9:390–399

Nybakken L, Solhaug KA, Bilger W, Gauslaa Y (2004) The lichens *Xanthoria elegans* and *Cetraria islandica* maintain a high protection against UV-B radiation in Arctic habitats. Oecologia 140:211–216

Oksanen I (2006) Ecological and biotechnological aspects of lichens. Appl Microbiol Biotechnol 73:723–734

Olafsdottir ES, Ingólfsdottir K (2001) Polysaccharides from lichens: structural characteristics and biological activity. Planta Med 67:199–208

Olafsdottir ES, Omarsdottir S, Smestad Paulsen B, Wagner H (2003) Immunologically active O6-branched (1→3)-beta-glucan from the lichen *Thamnolia vermicularis* var. *subuliformis*. Phytomedicine 10:318–324

Omarsdottir S, Petersen BO, Paulsen BS, Togola A, Duus JØ, Olafsdottir ES (2006) Structural characterisation of novel lichen heteroglycans by NMR spectroscopy and methylation analysis. Carbohydr Res 341:2449–2455

Omarsdottir S, Freysdottir J, Olafsdottir ES (2007) Immunomodulating polysaccharides from the lichen *Thamnolia vermicularis* var. *subuliformis*. Phytomedicine 14:179–184

Opanowicz M, Grube M (2004) Photobiont genetic variation in *Flavocetraria nivalis* from Poland (Parmeliaceae, lichenized Ascomycota). Lichenologist 36:125–131

Opanowicz M, Blaha J, Grube M (2006) Detection of paralogous polyketide synthase genes in Parmeliaceae by specific primers. Lichenogist 38:47–54

Palmqvist K, Dahlman L, Valladares F, Tehler A, Sancho LG, Mattsson J-E (2002) CO_2 exchange and thallus nitrogen across 75 contrasting lichen associations from different climate zones. Oecologia 133:295–306

Pankewitz F, Zöllmer A, Gräser Y, Hilker M (2007) Anthraquinones as defensive compounds in eggs of Galerucini leaf beetles: biosynthesis by the beetles. Arch Insect Biochem Physiol 662:98–108

Peñas MM, Aranguren J, Ramírez L, Pisabarro AG (2004) Structure of gene coding for the fruit body-specific hydrophobin Fbh1 of the edible basidiomycete *Pleurotus ostreatus*. Mycologia 96:75–82

Pérez FL (1997a) Geoecology of erratic globular lichens of *Catapyrenium lachneum* in a high Andean Paramo. Flora 192:241–259

Pérez FL (1997b) Geoecology of erratic lichens of *Xanthoparmelia vagans* in an equatorial Andean Paramo. Plant Ecol 129:11–28

Peterson E (2000) An overlooked fossil lichen (Lobariaceae). Lichenologist 32:298–300

Piercey-Normore MD (2004) Selection of algal genotypes by three species of lichen fungi in the genus *Cladonia*. Can J Bot 82:947–961

Piercey-Normore M (2006) The lichen-forming ascomycete *Evernia mesomorpha* associates with multiple genotypes of *Trebouxia jamesii*. New Phytol 169:331–344

Platt JL, Spatafora JW (1999) A re-examination of generic concepts of baeomycetoid lichens based on phylogenetic analyses of nuclear SSU and LSU Ribosomal DNA. Lichenologist 31:409–418

Platt JL, Spatafora JW (2000) Evolutionary relationships of nonsexual lichenized fungi: molecular phylogenetic hypotheses for the genera *Siphula* and *Thamnolia* from SSU and LSU rDNA. Mycologia 92:475–487

Poelt J, Huneck S (1968) *Lecanora vinetorum* nova spec., ihre Vergesellschaftung, ihre Ökologie und ihre Chemie. Oesterr Bot Z 115:411–422

Poinar G, Peterson E, Platt J (2000) Fossil *Parmelia* in New World amber. Lichenologist 32:263–269

Purvis OW, Halls C (1996) A review of lichens in metal-enriched environments. Lichenologist 28:571–601

Purvis OW, Pawlik-Skowrońska B (2008) Lichens and metals. In: Avery S, Stratford M, van West P (eds) Stress in yeasts and filamentous fungi. Elsevier, Amsterdam, pp 175–200

Purvis OW, Chimonides J, Din V, Erotokritou L, Jeffries T, Jones GC, Louwhoff S, Read H, Spiro B (2003) Which factors are responsible for the changing lichen floras of London? Sci Tot Environ 310:179–189

Purvis OW, Chimonides PDJ, Jeffries TE, Jones GC, Rusu A-M, Read H (2007) Multi-element composition of historical lichen collections and bark samples, indicators of changing atmospheric conditions. Atmos Environ 41:72–80

Rademaker M (2000) Allergy to lichen acids in a fragrance. Australas J Dermatol 41:50–51

Richardson DHS (1975) The vanishing lichens. David and Charles, Newton

Richardson DHS (1995) Metal uptake in lichens. Symbiosis 18:119–127

Rikkinen J (2003) Calicioid lichens from European Tertiary amber. Mycologia 95:1032–1036

Rikkinen J, Oksanen I, Lohtander K (2002) Lichen guilds share related cyanobacterial symbionts. Science 297:357

Rindi F, Guiry MD (2003) Composition and distribution of subaerial algal assemblages in Galway City, western Ireland. Cryptogam Algol 24:245–267

Rogers RW (1971) Distribution of the lichen *Chondropsis semiviridis* in relation to its heat and drought resistance. New Phytol 70:1069–1077

Romeike J, Friedl T, Helms G, Ott S (2002) Genetic diversity of algal and fungal partners in four species of *Umbilicaria* (lichenized ascomycetes) along a transect of the Antarctic peninsula. Mol Biol Evol 19:1209–1217

Rosentreter R (1993) Vagrant lichens in North America. Bryologist 96:333–338

Rundel PW (1978) The ecological role of secondary lichen substances. Biochem Syst Ecol 6:157–170

Rustad LE, Campbell J, Marion GM, Norby RJ, Mitchell MJ, Hartley AE, Cornelissen JHC, Gurevitch J, and GCTE NEWS (2001) A meta-analysis of the response of soil respiration, net N mineralization and aboveground plant growth to experimental ecosystem warming. Oecologia 126:543–562

Sancho LG, de la Torre R, Horneck G, Ascaso C, de los Rios A, Pintado A, Wierzchos J, Schuster M (2007) Lichens survive in space: results from the 2005 LICHENS Experiment. Astrobiology 7:443–454

Sanders WB (2001) Preliminary light microscope observations of fungal and algal colonization and lichen thallus initiation on glass slides placed near foliicolous lichen communities within a lowland tropical forest. Symbiosis 31:85–94

Sanders WB (2002) In situ development of the foliicolous lichen *Phyllophiale* (Trichotheliaceae) from propagule germination to propagule production. Am J Bot 89:1741–1746

Sanders WB (2005) Observing microscopic phases of lichen life cycles on transparent substrata placed in situ. Lichenologist 37:373–382

Sanders WB, Lücking R (2002) Reproductive strategies, relichenization and thallus development observed in situ in leaf-dwelling lichen communities. New Phytol 155:425–435

Sanders WB, Moe RL, Ascaso C (2004) The intertidal marine lichen formed by the pyrenomycete fungus *Verrucaria tavaresiae* (Ascomycotina) and the brown alga *Petroderma maculiforme* (Phaeophyceae): thallus organization and symbiont interaction. Am J Bot 91:511–522

Sanders WB, Moe RL, Ascaso C (2005) Ultrastructural study of the brown alga *Petroderma maculiforme* (Phaeophyceae) in the free-living state and in lichen symbiosis with the intertidal marine fungus *Verrucaria tavaresiae* (Ascomycotina). Eur J Phycol 40:353–361

Sawidis T, Heinrich G (1992) Cesium-137 monitoring using lichens and mosses from northern Greece. Can J Bot 70:140–144

Scharf CS (1978) Birds and mammals as passive transporters for algae found in lichens. Can Field Nat 92:70–71

Scheidegger C (1994) Low temperature scanning electron microscopy: the location of free and perturbed water and its role in the morphology of the lichen symbionts. Crypt Bot 4:290–299

Scherrer S, Honegger R (2003) Inter- and intraspecific variation of homologous hydrophobin (H1) gene sequences among *Xanthoria* spp. (lichen-forming ascomycetes). New Phytol 158:375–389

Scherrer S, De Vries OMH, Dudler R, Wessels JGH, Honegger R (2000) Interfacial self-assembly of fungal hydrophobins of the lichen-forming ascomycetes *Xanthoria parietina* and *X. ectaneoides*. Fungal Genet Biol 30:81–93

Scherrer S, Haisch A, Honegger R (2002) Characterization and expression of XPH1, the hydrophobin gene of the lichen-forming ascomycete *Xanthoria parietina*. New Phytol 154:175–184

Schroeter B, Scheidegger C (1995) Water relations in lichens at subzero temperatures: structural changes and carbon dioxide exchange in the lichen *Umbilicaria aprina* from continental Antarctica. New Phytol 131:273–285

Schultz C (2006) Remote sensing the distribution and spatiotemporal changes of major lichen communities in the central Namib Desert. Dissertation, University of Kaiserslautern

Schüssler A (2002) Molecular phylogeny, taxonomy, and evolution of *Geosiphon pyriformis* and arbuscular mycorrhizal fungi. Plant Soil 244:75–83

Schüssler A, Kluge M (2001) *Geosiphon pyriforme*, an endocytosymbiosis between fungus and cyanobacteria, and its meaning as a model system for arbuscular mycorrhizal research. In: Hock B (ed) Fungal associations, vol IX. Springer, Heidelberg, pp 151–161

Schüssler A, Schnepf E, Mollenhauer D, Kluge M (1995) The fungal bladders of the endocyanosis *Geosiphon pyriforme*, a *Glomus*-related fungus: cell wall permeability indicates a limiting pore radius of only 0.5 nm. Protoplasma 185:131–139

Schüssler A, Schwarzott D, Walker C (2001) A new fungal phylum, the Glomeromycota: phylogeny and evolution. Mycol Res 105:1413–1421

Seyd EL, Seaward MRD (1984) The association of oribatid mites with lichens. Zool J Linn Soc 80:369–420

Shirtcliffe NJ, Brian Pyatt F, Newton MI, McHale G (2006) A lichen protected by a super-hydrophobic and breathable structure. J Plant Physiol 163:1193–1197

Sillett SC, Antoine ME (2004) Lichens and bryophytes in forest canopies. In: Lowman MD, Rinker BH (eds) Forest canopies. Elsevier, Amsterdam, pp 151–174

Sipman HJM (1994) Foliicolous lichens on plastic tape. Lichenologist 26:311–312

Smith DC (1980) Mechanisms of nutrient movement between lichen symbionts. In: Cook CB, Pappas PW, Rudolph ED (eds) Cellular interactions in symbiosis and parasitism. Ohio State University Press, Columbus, pp 197–227

Smith DC, Douglas AE (1987) The biology of symbiosis. Arnold, London

Solhaug KA, Gauslaa Y, Nybakken L, Bilger W (2003) UV-induction of sun-screening pigments in lichens. New Phytol 158:91–100

StClair L, Seaward M (2004) Biodeterioration of stone surfaces: lichens and biofilms as weathering agents of rocks and cultural heritage. Kluwer, Dordrecht

Stocker-Wörgötter E (2001) Experimental lichenology and microbiology of lichens: culture experiments, secondary chemistry of cultured mycobionts, resynthesis, and thallus morphogenesis. Bryologist 104:576–581

Stoll A, Brack A, Renz J (1947) Die antibakterielle Wirkung der Usninsäure auf Mykobakterien und andere Mikroorganismen. Experientia 3:115–117

Strand P, Selnaes TD, Bøe E, Harbitz O, Andersson-Sørlie P (1992) Chernobyl fallout: internal doses to the Norwegian population and the effect of dietary advice. Health Phys 63:385–392

Stubbs CS (1995) Dispersal of soredia by the oribatid mite, *Humerobates arborea*. Mycologia 87:454–458

Talbot NJ, Kershaw MJ, Wakley GE, Devries OMH, Wessels JGH, Hamer JE (1996) MPG1 encodes a fungal hydrophobin involved in surface interactions during infection-related development of *Magnaporthe grisea*. Plant Cell 8:985–999

Taylor TN, Hass H, Remy W, Kerp H (1995) The oldest fossil lichen. Nature 378:244

Theobald MR, Crittenden PD, Hunt AP, Tang YS, Dragosits U, Sutton MA (2006) Ammonia emissions from a Cape fur seal colony, Cape Cross, Namibia. Geophys Res Lett 33:L03812

Thune P, Solberg Y, McFadden N, Staerfelt F, Sandberg M (1982) Perfume allergy due to oak moss and other lichens. Contact Dermatitis 8:396–400

Thüs H (2002) Taxonomie, Verbreitung und Ökologie silicoler Süßwasserflechten im außeralpinen Mitteleuropa. Bibl Lichenol 85:1–214

Tibell L (2001) Photobiont association and molecular phylogeny of the lichen genus *Chaenotheca*. Bryologist 104:191–198

Trembley ML, Ringli C, Honegger R (2002a) Differential expression of hydrophobins DGH1, DGH2 and DGH3 and immunolocalization of DGH1 in strata of the lichenized basidiocarp of *Dictyonema glabratum*. New Phytol 154:185–195

Trembley ML, Ringli C, Honegger R (2002b) Hydrophobins DGH1, DGH2, and DGH3 in the lichen-forming basidiomycete *Dictyonema glabratum*. Fungal Genet Biol 35:247–259

Tschermak-Woess E (1978) *Myrmecia reticulata* as a phycobiont and free-living – free-living *Trebouxia* – the problem of *Stenocybe septata*. Lichenologist 10:69–79

Tschermak-Woess E (1988) The algal partner. In: Galun M (ed) Handbook of lichenology, vol 1. CRC, Boca Raton, pp 39–92

Tveten U, Brynildsen LI, Amundsen I, Bergan TDS (1998) Economic consequences of the Chernobyl accident in Norway in the decade 1986–1995. J Environ Radioact 41:233–255

van Herk CM, Aptroot A, van Dobben HF (2002) Long-term monitoring in the Netherlands suggests that lichens respond to global warming. Lichenogist 34:141–154

Vežda A (1975) Foliikole Flechten aus Tanzania (Ost-Afrika). Folia Geobot Phytotaxon 10:383–432

Walker TR, Young SD, Crittenden PD, Zhang H (2003) Anthropogenic metal enrichment of snow and soil in north-eastern European Russia. Environ Poll 121:11–21

Walker TR, Crittenden PD, Young SD, Prystina T (2006a) An assessment of pollution impacts due to the oil and gas industries in the Pechora basin, north-eastern European Russia. Ecol Indic 6:369–387

Walker TR, Habeck JO, Karjalainen TP, Virtanen T, Solovieva N, Jones V, Kuhry P, Ponomarev VI, Mikkola K, Nikula A, Patova E, Crittenden PD, Young SD, Ingold T (2006) Perceived and measured levels of environmental pollution: interdisciplinary research in the Subarctic lowlands of northeast European Russia. Ambio 35:220–228

Warren G, Wolber P (1991) Molecular aspects of microbial ice nucleation. Mol Microbiol 5:239–243

Wessels JGH (1996) Fungal hydrophobins: proteins that function at an interface. Trends Plant Sci 1:9–15

Whiteford J, Spanu P (2002) Hydrophobins and the interactions between fungi and plants. Mol Plant Pathol 3:391–400

Wolf E (2003) Partnererkennung und Inkorporation des Photobionten bei der Pilz-Endocyanose *Geosiphon pyriforme*. Dissertation, University of Darmstadt

Wösten HAB (2001) Hydrophobins: multipurpose proteins. Annu Rev Microbiol 55:625–646

Wösten HAB, Wessels JGH (1997) Hydrophobins, from molecular structure to multiple functions in fungal development. Mycoscience 38:363–374

Wösten HAB, DeVries OMH, Wessels JGH (1993) Interfacial self-assembly of a fungal hydrophobin into a hydrophobic rodlet layer. Plant Cell 5:1567–1574

Yamamoto Y, Mizugichi R, Yamada Y (1985) Tissue cultures of *Usnea rubescens* and *Ramalina yasudae* and production of usnic acid in their cultures. Agric Biol Chem 49:3347–3348

Yamamoto Y, Miura Y, Higuchi M, Kinoshita Y, Yoshimura I (1993) Using lichen tissue cultures in modern biology. Bryologist 96:384–393

Yamamoto Y, Kinoshita Y, Takahagi T, Kroken S, Kurokawa T, Yoshimura I (1998) Factors affecting discharge and germination of lichen ascospores. J Hattori Bot Lab 85:267–278

Yuan X, Xiao S, Taylor TN (2005) Lichen-like symbiosis 600 million years ago. Science 308:1017–1020

Zhang YM, Chen J, Wang L, Wang XQ, Gu ZH (2007) The spatial distribution patterns of biological soil crusts in the Gurbantunggut Desert, Northern Xinjiang, China. J Arid Environ 68:599–610

Plant Response

17   Signal Perception and Transduction in Plants

Wolfgang Knogge[1], Justin Lee[1], Sabine Rosahl[1], Dierk Scheel[1]

CONTENTS

I. Introduction 337
II.   MAMP Perception 338
   A. MAMPs of Oomycete or Fungal Origin ... 338
   B. Pattern Recognition Receptors 338
     1. β-Glucan Receptor – Enzymatic Ligand
       Amplification and Optimization 338
     2. Chitin Receptor – Heterodimerization
       of LysM RLP and RLK 339
     3. EIX Receptor – RLP-Mediated
       Endocytosis 340
III.   Signal Transduction 341
   A. Calcium 341
     1. Monitoring Changes in Ca^{2+} Levels
       During Pathogen Attack 342
     2. Ca^{2+} Transients by Fungal/Oomycete-
       Derived Elicitors 342
     3. Calcium Signal Transduction: Sensors
       and Targets...................... 342
     4. Source of Calcium and Identity of
       Elicitor-Activated Channels/Pumps 344
   B. Reactive Oxygen Species................ 344
   C. Nitric Oxide 345
   D. MAPK Cascades....................... 346
     1. Activation of MAPKs During Defense.. 346
     2. Evidence for the Importance of MAPK
       Cascades in Disease Resistance....... 347
   E. Other Components in Signaling Systems 348
     1. Jasmonic Acid 348
     2. Salicylic Acid 351
     3. Cross-Talk 352
IV.   Conclusion 352
   References............................ 353

I.   Introduction

The plant immune system can be activated by two different types of signals, by microbial signatures and by features signifying malfunctioning of plant processes. In other words, plants respond to signals indicating 'non-self' or to signals specifying 'disturbed self'. Perception of these signals is mediated by two different types of receptors: a class of membrane-resident receptors that identify extracellular pathogen-derived molecules and a class of mainly intracellular receptors that recognize the presence or the activity of pathogen-derived effector molecules inside the host cell. Extracellular ligands can be evolutionarily conserved, broadly occurring molecules of functional importance for the microbe although without being specifically intended for the interaction with a host and, hence, cannot easily be modified without loss of functionality. These molecules that are absent from the potential host have been termed microbe-associated molecular patterns (MAMPs; Mackey and McFall 2006) or pathogen-associated molecular patterns (PAMPs; Medzhitov and Janeway 1997; Nürnberger et al. 2004). In the following, the former more general term is used, because plant-recognized PAMPs can also be found in non-pathogenic microbes. MAMPs are recognized on the surface of plant cells by specific pattern recognition receptors (PRRs; Nürnberger and Kemmerling 2006).

The other class of molecules serving as defense triggers is secreted by pathogens with the purpose to specifically manipulate the host physiology (*bona fide* virulence factors). These effectors may act as external ligands of plant resistance protein-associated transmembrane perception systems. More frequently, however, these effectors are transmitted into the host cells, where they either interact directly with resistance proteins or they inflict modifications on host targets, which are detected by resistance proteins, usually of the NB-LRR type (cf. guard hypothesis of resistance protein function; van der Biezen and Jones 1998). For defense-inducing products originating from such indirectly recognized effectors, the term microbe-induced molecular patterns (MIMPs) is proposed (Mackey and McFall 2006).

The current view of the plant immune system and its evolution was outlined in a recent review article as a four-phased model (Jones and Dangl

[1] Leibniz Institute of Plant Biochemistry, Department of Stress and Developmental Biology, Weinberg 3, 06120 Halle (Saale), Germany; e-mail: wknogge@ipb-halle.de, jlee@ipb-halle.de, srosahl@ipb-halle.de, dscheel@ipb-halle.de

Plant Relationships, 2nd Edition
The Mycota V
H. Deising (Ed.)
© Springer-Verlag Berlin Heidelberg 2009

2006). Most plants are resistant to most invading pathogens due to a basic resistance strategy, in which conserved MAMPs are recognized by PRRs and pathogen development is prevented by MAMP-triggered immunity (MTI; in Jones and Dangl (2006) termed PAMP-triggered immunity; PTI). To get access to the plant food market and to allow microbe accommodation, pathogens need to avoid recognition or suppress its consequences. For this purpose, they secrete effectors that interfere with MTI, thus causing effector-triggered susceptibility (ETS), formerly called basic susceptibility (see Chap. 9). Once the plant evolves a receptor (resistance protein) to specifically recognize one of these effectors directly or through its activity, the consequence is effector-triggered immunity (ETI), formerly called cultivar-specific resistance. Meanwhile, hundreds of plant genes encoding putative resistance proteins of the NB-LRR type have been identified in plant genomes (Meyers et al. 2003). In the next phase, the pathogen 'learns' to avoid or to suppress ETI, but selection can also produce new resistance gene specificities, resulting in re-established ETI. This chapter focuses on the plant perception of MAMPs from fungal and Oomycete pathogens and on signaling molecules that are involved in the intracellular signal transduction leading to plant immunity (MTI). Further details on ETI (with an emphasis on bacteria-plant interactions) have been recently reviewed (Abramovitch et al. 2006; Chisholm et al. 2006; DeYoung and Innes 2006; Jones and Dangl 2006).

II.   MAMP Perception

A.   MAMPs of Oomycete or Fungal Origin

Most of our knowledge on MAMP perception originates from studying bacterial MAMPs. For instance, highly conserved parts of the protein building block of bacterial flagellin are recognized by the innate immune system of many plant species and animals (Zipfel and Felix 2005). Plants and animals also have perception systems for lipopolysaccharides, the major structural components of the outer membrane of Gram-negative bacteria (Zipfel and Felix 2005). Some MAMPs are less widely recognized. For instance, the most conserved motif of bacterial cold-shock proteins, the RNA-binding motif, serves as a MAMP in members of the *Solanaceae* (Felix and Boller 2003). In

contrast, the *Brassicaceae* are able to perceive the N-terminus of *elongation factor Tu (EF-TU)*, the most abundant and highly conserved protein in the bacterial cytoplasm (Kunze et al. 2004).

Fungi and Oomycetes are also characterized by the presence of surface-localized or secreted MAMPs. Typical cell wall components, such as Oomycete β-glucans and fungal chitin, have long been recognized as inducers ('general elicitors') of plant defense (Ayers et al. 1976; Hadwiger and Beckman 1980). Two additional cell wall proteins were characterized as Oomycete MAMPs: a *Phytophthora* transglutaminase with its conserved Pep-13 epitope (Brunner et al. 2002) and a cellulose-binding elicitor lectin protein (CBEL; Gaulin et al. 2006). Also secreted proteins, such as Oomycete lipid transfer proteins (elicitins), necrosis and ethylene-inducing protein 1 (Nep1) from *Fusarium oxysporum* (Bailey 1995) and its structural homologues in various Oomycetes, fungi and bacteria (Nep1-like proteins, NLPs; Pemberton and Salmond 2004; Qutob et al. 2006), as well as a fungal endopolygalacturonase (Poinssot et al. 2003) and ethylene-inducing xylanase (EIX; Bailey et al. 1990) were described as ligands in MAMP perception. Finally, the typical fungal sterol, ergosterol (Granado et al. 1995), as well as fungus-specific sphingolipids, cerebroside A and C (Koga et al. 1998), need to be mentioned in this context as well. All these components are not found in higher eukaryotes and, hence, represent molecular signatures that characterize putative microbial plant invaders. Although a variety of different fungal and Oomycete MAMPs was shown to trigger defense reactions in plants, knowledge on the corresponding receptors and the biochemical mechanisms linking receptor activation and intra-cellular signaling has remained sparse with only very few exceptions. Three PRRs involved in the perception of different fungal or Oomycete cell wall components and a secreted fungal protein, respectively, are treated in the following to exemplify concepts for signal perception at the plant plasma membrane and its conversion into an intracellular response.

B.   Pattern Recognition Receptors

1. β-Glucan Receptor – Enzymatic Ligand
   Amplification and Optimization

Binding sites for β-glucans were described 20 years ago (Schmidt and Ebel 1987), but isolation, cloning and characterization of the receptors was only successful in recent years. Binding sites for 1,6-β-linked and 1,3-β-branched glucans of the soybean pathogen, *Phytophthora sojae*, were shown to exist in host membranes (Schmidt and Ebel 1987) and a structurally defined hepta-β-glucoside was found to possess the minimum requirements for elicitor activity and ligand specificity (Sharp et al. 1984; Cosio et al. 1990; Cheong et al. 1991; Cheong and Hahn 1991). Radiolabeling of the ligand allowed the identification in soybean of a low abundance 75-kDa β-glucan-binding protein (GBP) that was purified using affinity chromatography (Mithöfer et al. 1996; Umemoto et al. 1997). After cloning of the GBP cDNA from soybean and French bean, a protein sequence of 668 amino acid residues was deduced, which contained a single putative transmembrane helix, albeit in the absence of a membrane-targeting signal peptide (Umemoto et al. 1997; Mithöfer et al. 2000). Substantial amounts of the GBP were also detected in soluble protein fractions, where however it did not display β-glucan binding. Binding activity could at least to a certain extent be regained by reconstitution into lipid vesicles. This suggests that an as yet unknown mechanism or component is required for both membrane association and binding activity. Furthermore, GBP does not contain any recognizable functional domains indicating an involvement of the protein in signal transduction processes. This confirms the earlier assumption that GBP is part of a β-glucan receptor complex of 240 kDa (Mithöfer et al. 1996). Signaling may be accomplished by an as yet unidentified additional membrane protein, which mediates the observed early transient increase of cytosolic Ca^{2+} (Mithöfer et al. 1999), ion fluxes and membrane depolarization (Mithöfer et al. 2005) and the activation of a MAP kinase cascade (Daxberger et al. 2007).

In several species of the *Fabaceae*, high-affinity β-glucan binding correlates closely with the ability to respond with phytoalexin biosynthesis (Cosio et al. 1996), whereas neither binding sites nor elicitor activity can be demonstrated outside this plant family. However, recent data base searches revealed genes encoding GBP-related proteins in species from other plant taxa such as mosses, gymnosperms and mono- and dicotyledonous angiosperms (Fliegmann et al. 2004). Therefore, the ability to perceive and respond to β-glucans from *Phytophthora* spp. appears to be restricted to species of the *Fabaceae* family. Further evidence for this assumption comes

from transforming of tomato cells with the soybean GBP cDNA: a β-glucan binding site was generated, which failed however to mediate signal transduction, suggesting that the additional essential component is not conserved between different taxa (Mithöfer et al. 2000). Apparently, GBPs have been recruited to the defense system only in the *Fabaceae* family, where a hypothetical additional protein is needed for the response to β-glucan binding. In other species the function of these proteins remains unknown.

Detailed characterization of the GBP may shed further light on the function and evolution of β-glucan receptors. In *Saccharomyces cerevisiae*, two GBP-related proteins were shown to display a new type of endo-1,3-β-glucanase activity (Baladron et al. 2002) and a similar activity was also discovered in the soybean GBP. The catalytically active site is located in the carboxy terminal part of the protein and it is very unlikely to be identical with the β-glucan-binding site, as was shown by inhibitor studies (Fliegmann et al. 2004). This additional role of the GBP may explain the presence of the soluble protein. Its function may be the enzymatic degradation of *Phytophthora* cell walls and the concomitant release of soluble cell wall fragments, which ultimately may be enriched in units that are recognized by the elicitor-binding site of the GBP. Hence, GBP displays the ability to use the products of the intrinsic endoglucanase activity as ligands of a separate binding site localized in the same protein as part of a receptor complex. This would mean that a pathogen perception system amplifies and optimizes its own ligand.

2. Chitin Receptor – Heterodimerization of LysM RLP and RLK

Whereas the GBP appears to be associated with glucanases from fungi and plants, the chitin receptor resembles an entirely different class of proteins. As with the β-glucan receptor, it took a long time between identification of binding sites (Shibuya et al. 1993) and receptor identification (Kaku et al. 2006). Chitin is a major component of fungal cell walls and chitin fragments have been shown to induce defense reactions in mono- and dicotyledonous plants. A high-affinity binding protein, CEBiP, was purified from rice plasma membranes and the corresponding cDNA was cloned (Kaku et al. 2006). The mature protein consists of 328 amino acids and carries glycan chains. Its N-terminal signal peptide and C-terminal transmembrane region indicate a membrane-anchored extracellular localization. Since typical intracellular domains of membrane receptors are missing, CEBiP appears to require an additional protein for signal transduction. This is reminiscent of the plant CLAVATA system, where heterodimerization of the serine/threonine receptor kinase, CLV1, with CLV2 lacking the intracellular kinase domain, is thought to be required for regulation of meristem development (Diévart and Clark 2004). A similar principle has also been discussed for the function of the Cf resistance proteins in tomato (Joosten and de Wit 1999). In contrast to the latter proteins, which are characterized by extracellular leucine-rich repeat (LRR) domains, two LysM domains were identified as structural features in CEBiP. These short peptide domains were originally found in enzymes involved in bacterial cell wall degradation, in a chitinase from *Kluyveromyces lactis* and in a variety of peptidoglycan- and chitin-binding proteins (Butler et al. 1991), implying their direct involvement in oligosaccharide/chitin binding. LysM motifs are also found in the extracellular domains of legume serine/threonine receptor kinases (NRF1, NRF5 in *Lotus japonicus*; LYK3 in *Medicago truncatula*; Radutoiu et al. 2003). Since these kinases mediate the specific recognition of rhizobial lipochitooligosaccharide signals (nod factors), perception of chitin-related signals through LysM motifs represents a link between symbiosis formation and pathogen defense.

Recently, a chitin elicitor receptor kinase (CERK1) was identified in *Arabidopsis thaliana* (Miya et al. 2007). The *cerk1* mutant specifically lost the ability to respond to the chitin elicitor by activation of a mitogen-activated protein kinase (MAPK), generation of reactive oxygen species and defense gene expression. In addition, disease resistance in the incompatible interaction with *Alternaria brassicicola* was weakly affected, whereas the compatible interaction with *Colletotrichum higginsianum* was not altered. CERK1 is a plasma membrane protein with three extracellular LysM motifs and a functional intracellular serine/threonine kinase. If a similar receptor kinase were present in rice, this may be the missing additional protein that is recruited for signal perception. Hence, heterodimerization of a CERK1 analogon with CEBiP, possibly mediated by ligand binding to the LysM motifs, may be the mechanism of chitin perception and the ensuing signal transduction.

The defense system based on chitin recognition appears to be widely conserved among plant species, because CEBiP-like proteins are found in the plasma membranes from various plants that respond to chitin oligosaccharide elicitors such as barley, carrot, soybean and wheat (Stacey and Shibuya 1997; Day et al. 2001; Okada et al. 2002). In addition, BLAST searches detect proteins with high homology to CEBiP in many more plants, which have not yet been tested for elicitor responsiveness (data not shown). Interestingly, however, affinity-labeling experiments failed to detect a chitin-binding protein in membrane preparation from *Arabidopsis thaliana* (Miya et al. 2007). Therefore, more biochemical studies are required to fully unravel the chitin perception system of plants.

3. EIX Receptor – RLP-Mediated Endocytosis

Leucine-rich repeat (LRR) domains are found in a number of proteins with diverse functions and cellular locations and are usually involved in protein–protein interactions. Plant genomes are characterized by a high abundance of genes encoding proteins with extracellular LRR domains and single-pass plasma membrane-spanning transmembrane domains. If these receptor-like proteins (RLPs) contain a cytoplasmic serine/threonine protein kinase domain they are usually called receptor-like kinases (RLKs; Diévart and Clark 2004; Kruijt et al. 2005). The bacterial MAMPs, flagellin and EF-TU, are recognized by typical RLKs: FLS2 (Gomez-Gomez and Boller 2000) and EFR (Zipfel et al. 2006), respectively. Another member of this protein family, Xa21, is involved in the perception by rice of a specific effector protein, thus conferring resistance to the bacterial pathogen, *Xanthomonas oryzae* (Wang et al. 1998). In addition, RLKs involved in developmental regulation, such as the above-mentioned CLV proteins and the *Arabidopsis thaliana* brassinosteroid receptor, BRI1 (Li and Chory 1997; He et al. 2000), belong to this group of proteins. When the LRR and transmembrane domains of the BRI1 were fused to the serine/threonine kinase domain of Xa21, the chimeric receptor was able to activate defense response genes upon treatment with brassinosteroids (He et al. 2000). This demonstrates that the extracellular LRR domain is pivotal for signal perception and discrimination. Another RLK, the

BRI-associated receptor kinase, BAK1, associates with BRI1 (Li et al. 2002; Nam and Li 2002), FLS2 (Chinchilla et al. 2007) and possibly other PRRs (Kemmerling et al. 2007). This protein, therefore, appears to play a role in regulating receptors that are part of signaling pathways involved in such different processes as plant innate immunity and development.

RLPs lacking a kinase domain have been particularly studied in tomato, where they contain the large family of Cf resistance proteins (Kruijt et al. 2005). Another protein resembling these proteins acts as receptor of the fungal ethylene-inducing xylanase (EIX). In both tomato and tobacco, EIX recognition is controlled by the *LeEix* locus. Site-directed mutagenesis revealed that enzyme activity of EIX is not required for elicitor activity (Enkerli et al. 1999; Furman-Matarasso et al. 1999). The *LeEix* locus of tomato comprises three genes, two of which have been cloned (Ron and Avni 2004). The amino acid sequences of LeEix1 and LeEix2 show >81% identity with each other and ~30% identity (~48% similarity) with the tomato resistance protein, Cf-2. RNAi-mediated silencing of EIX-responsive tobacco using a sequence fragment from the *LeEix1* gene resulted in the suppression of EIX-induced cell death. Furthermore, fluorescein isothyocyanate-labeled EIX was found to interact only with wild-type cells but not with cells from silenced plants, indicating that EIX perception is mediated by one of the LeEix proteins. When complementation experiments were carried out using *LeEix1* and *LeEix2* cDNAs and EIX-nonresponding tobacco, it turned out that both, *LeEix1* and *LeEix2* can restore binding of EIX, but only the product of *LeEix2* is able to transmit the signal required for HR induction.

Both LeEix proteins contain an extracellular leucine zipper domain, indicating that dimerization may be required for receptor activation. Furthermore, both proteins contain the C-terminal endocytosis signal YXXΦ (where Φ represents an amino acid with hydrophobic side-chain), a motif that was also described to be present in EFR, the receptor involved in EF-Tu signaling (Zipfel et al. 2006). Site-directed mutagenesis of the motif in LeEix2 abolishes the capability to induce HR. This confirms the previous observation that after plasma membrane binding EIX is translocated into the plant cytoplasm (Hanania et al. 1999). Furthermore, a membrane-localized FLS2-GFP fusion protein was found to rapidly accumulate in intracellular vesicles upon addition of the ligand, flg22 (Robatzek et al.

2006). This suggests that ligand-induced receptor-mediated endocytosis plays a key role in signaling pathways leading to HR and MTI. The presence of the YXXΦ motif in the tomato resistance proteins Ve1, Ve2, Cf4 and Cf9 (Kawchuk et al. 2001) conferring resistance to races of *Verticillium dahliae*, *V. albo-atrum* and *Cladosporium fulvum*, respectively, suggests that endocytosis appears to not only be crucial in MTI but also in ETI.

As the consequence of receptor-mediated endocytosis, LeEix and/or EIX are able to interact with cytoplasmic host proteins, thus initiating intracellular signaling. One such protein, which may shed light on the downstream events that combine signal perception with intracellular signal transduction, was identified in a yeast two-hybrid system. EIX interacted with a tomato small ubiquitin-related modifier protein (T-SUMO; Hanania et al. 1999). EIX-induced ethylene biosynthesis was suppressed in transgenic plants expressing T-SUMO in the sense orientation, but induced when expressed in the antisense orientation. Although the mode of action of T-SUMO remains unknown, EIX may function through inhibiting or removing a repressor of plant defense reactions.

III. Signal Transduction

Activation of membrane-localized receptors through binding of their respective ligands is the first of a series of steps that finally lead to the expression of plant defense genes. Although many details have been unveiled to date, the molecular mechanisms linking signal perception with intracellular signaling events and signaling molecules with alterations in gene regulation still need to be unraveled. Nevertheless, changes in cytoplasmic Ca^{2+} levels and the production of reactive oxygen species (ROS) and nitric oxide (NO) usually occur as early events in plant–pathogen interactions. MAPK cascades are key players in the plant defense regulation. Finally, salicylic acid (SA), jasmonic acid (JA) and ethylene (ET) are signaling components that are part of networks organizing and integrating the plant defense response (Fig. 17.1; Chap. 18).

A. Calcium

Although present ubiquitously, calcium is well established as a second messenger in the response to

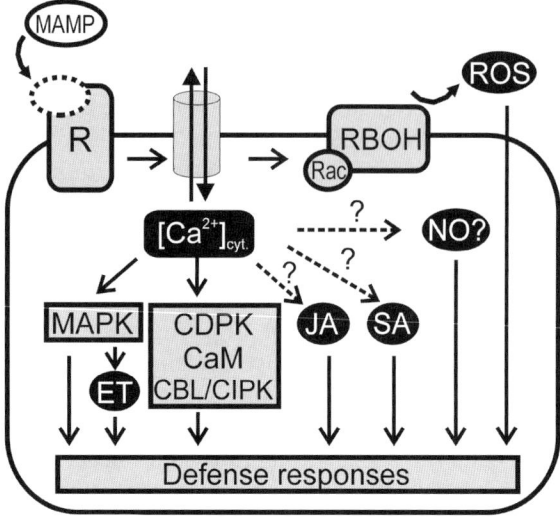

Fig. 17.1. Components of plant defense signaling. *CaM* Calmodulin, *CBL* calcineurin B-like protein, *CDPK* calcium-dependent protein kinase, *CIPK* CBL-interacting kinase, *ET* ethylene, *JA* jasmonic acid, *MAMP* microbe-associated molecular pattern, *MAPK* mitogen-activated protein kinase, *NO* nitric oxide, *R* receptor, *Rac* small GTP-binding protein, *RBOH* respiratory burst oxidase homologue, *ROS* reactive oxygen species, *SA* salicylic acid

various environmental, hormonal and pathogenic signals (Ward et al. 1995). The stimuli (often at the cell surface) are transduced to intracellular responses through a rise in free cytosolic Ca^{2+} concentration ($[Ca^{2+}]_{cyt}$; Sanders et al. 2002; Lecourieux et al. 2006).

1. Monitoring Changes in Ca^{2+} Levels During Pathogen Attack

Changes in Ca^{2+} levels can be addressed indirectly *via* patch clamp analysis to reveal the activities of membrane-localized channels and pumps (Zimmermann et al. 1997) or through direct measurement (Nürnberger et al. 1994; Lecourieux et al. 2002; Lecourieux et al. 2005; Garcia-Brugger et al. 2006; Lecourieux et al. 2006; Xiong et al. 2006). Older direct quantification of Ca^{2+} uptake using $^{45}Ca^{2+}$ tracers has gradually been replaced by optical methods such as microinjections of Ca^{2+}-sensitive dyes or stable transgenic plants expressing biolumi-nescence-based (aequorin) or fluorophore-based (cameleon) reporters (Allen et al. 1999; Rudd and Franklin-Tong 1999; Mithöfer and Mazars 2002).

One of the few case studies of Ca^{2+} response involving 'real' fungal pathogen is the correlation

of the $[Ca^{2+}]_{cyt}$ changes and the hypersensitive response after cowpea rust fungus infection (Xu and Heath 1998). Most other studies involve pathogen-derived molecules.

2. Ca^{2+} Transients by Fungal/Oomycete-Derived Elicitors

Table 17.1 summarizes a number of fungal or Oomycete-derived molecules that are reported to elicit Ca^{2+} transients in plants. Bacterial or viral elicitors are not included in this review. In most of these cases, pharmacological inhibitors (Nürnberger et al. 1994; Jabs et al. 1997) are used as a second line of evidence for the role of Ca^{2+} in defense signaling. The elicitors listed in Table 17.1 are predominantly of proteinaceous or polysaccharide nature. Additionally, lipid-based elicitors such as spingolipid or ergosterol (Umemura et al. 2002; Kasparovsky et al. 2004) also involve Ca^{2+} signaling, but these are solely inferred from inhibitor studies.

While inhibition of Ca^{2+} actions or Ca^{2+} chelators point to the requirement for $[Ca^{2+}]_{cyt}$ elevation in downstream signaling for almost all situations, the MAPK induction by BcPG1, a polygalacturonase elicitor from *Botrytis cinerea*, is strangely unaffected (Vandelle et al. 2006). This is reminiscent of the non-requirement of extracellular Ca^{2+} for defense gene activation by the bacterial harpin elicitor (Lee et al. 2001), suggesting that – while more likely an exception than the rule – Ca^{2+}-independent pathways do exist.

3. Calcium Signal Transduction: Sensors and Targets

Besides the biotic factors listed in Table 17.1, various abiotic stresses and hormones evoke Ca^{2+} signaling as well. In fact, one study proposed that abscisic acid (ABA) and elicitor (namely yeast elicitors/chitosan) stimuli converge on Ca^{2+} signaling in stomatal cells (Klüsener et al. 2002). This raises the question as to how a simple ion like Ca^{2+} can serve as second messenger in so many diverse pathways. One hypothesis suggests that it may alter protein conformation and therefore act as a chemical version of binary on-off switch (Plieth 2005). More widely accepted is the idea that well-defined spatiotemporal changes in $[Ca^{2+}]_{cyt}$ constitute a 'Ca^{2+} signature' that is further decoded

Table 17.1. Selected fungal/Oomycete elicitors (or effectors) that trigger Ca^{2+} signaling. *AM* Arbuscular mycorrhiza, '*AM signal+*' *Gigaspora margarita* culture filtrate containing putative component(s) that mediate symbiotic relationship, *Avr* avirulence proteins, *CBEL* cellulose-binding elicitor lectin, *NPP1* necrosis-inducing *Phytophthora* protein 1 (of the necrosis and ethylene-inducing1-like protein family), *Pep-13* a peptide of 13 residues from a transglutaminase from *P. sojae*, *PG* endopolygalacturonase

Elicitors/effectors	Microbe	Plant species investigated	Reference
β-Heptaglucan	*Phytophthora* spp.	*Glycine max*	Mithöfer et al. (2005)
Chitin	Various fungi	*G. max*	Ebel et al. (2001)
Elicitins	*Phytophthora* spp.	*Solanum lycopersicum/Nicotiana tabacum*	Lecourieux et al. (2002)
Avr2,4,9	*Cladosporium fulvum*	*S. lycopersicum* (race-specific)	de Wit et al. (2002)
NPP1 (NLP$_{pp}$)	*Phytophthora parasitica*	*Petroselinum crispum/Arabidopsis thaliana*	Fellbrich et al. (2002)
Pep-13	*P. sojae*	*P. crispum*	Blume et al. (2000)
CBEL	*P. parasitica nicotianae*	*N. tabacum*	Gaulin et al. (2006)
BcPG1	*Botrytis cinerea*	*Vitis vinifera*	Poinssot et al. (2003)
PG	*Sclerotinia sclerotiorum*	*G. max*	Zuppini et al. (2005)
Xylanase	*Trichoderma viride*	*N. tabacum*	Bailey et al. (1992)
'AM signal'	Mycorrhizal fungus	*G. max*	Navazio et al. (2007)
Laminarin[a]	*Laminaria digitata*	*V. vinifera*	Aziz et al. (2003)

[a] Laminarin is a β-1,3-glucan from the brown algae, *Laminaria digitata*, and hence *sensu stricto* not from a pathogen, but is used in agriculture as a defense-activating natural product.

by so-called 'Ca^{2+} sensors' (Sanders et al. 2002; Plieth 2005; Lecourieux et al. 2006).

In plants, the Ca^{2+} sensors are Ca^{2+}-binding proteins that typically contain multiple EF-hand domains. These include calmodulins (CaMs), calcineurin B-like proteins (CBLs) and calcium-dependent protein kinases (CDPKs), which are briefly discussed in the paragraphs below. Several proteins bind Ca^{2+} without EF-hands but *via* other domains, e.g. C2 domain (Kopka et al. 1998). This domain confers Ca^{2+}-dependent phospholipid binding, thus remobilizing the protein to another cellular location. Examples of plant C2-domain proteins include copines (Jambunathan et al. 2001), phospholipase-C (PLC) and phospholipase-D (Kopka et al. 1998; Laxalt and Munnik 2002), all of which can potentially regulate defense responses or further amplify the signal by generating other second messengers.

CaMs interact with a variety of downstream targets, such as Ca^{2+}-ATPase, nucleotide-gated ion channels or transcription factors, which presumably further transduce the Ca^{2+} signal or regulate the Ca^{2+} signal through Ca^{2+} homeostasis (Luan et al. 2002). A molecular target of CaM involved in plant defense is MLO, a seven-transmembrane receptor-like protein, that controls broad-spectrum resistance in barley. Recent microscopic analysis suggested an increase in fluorescence resonance energy transfer (FRET) signal – indicative of MLO/CaM interaction – around penetration sites that coincided with successful host cell entry (Bhat et al. 2005). This supports previous studies showing the interaction between MLO and CaM to modulate the defense response of barley to powdery mildew infection (Kim et al. 2002). The importance of CaM in pathogen response is also highlighted in earlier studies reporting the expression of two soybean CaM isoforms being induced by pathogen attack or fungal elicitors and that heterologous expression of these CaMs in tobacco led to enhanced resistance to a variety of pathogens (Heo et al. 1999).

Unlike CaMs, members from the second group of Ca^{2+} sensors, the CBLs, apparently interact with only one class of proteins that are referred to as CBL-interacting kinases (CIPKs; Batistic and Kudla 2004). Genome studies identified ten CBLs and 25–30 CIPKs, for *Arabidopsis*/rice, respectively (Batistic and Kudla 2004; Kolukisaoglu et al. 2004), which bring to light the potential pairs of combination for signaling processes. However, the few examples of CBL-CIPK pairs with known physiological function are implicated in abiotic stress (Batistic and Kudla 2004) and the involvement in pathogen defense has not yet been reported.

The third group of Ca^{2+} sensors, the CDPKs, comprises one of the largest families with 34 and 29 members predicted from the *Arabidopsis* and rice genomes, respectively (Hrabak et al. 2003; Asano et al. 2005). The first hint of involvement in pathogen response was based on elevated CDPK transcript levels after elicitor treatments (Yoon et al. 1999). The biochemical evidence for their involvement in pathogen response was shown by Romeis et al. (2001), where the Avr9 race-specific elicitor caused phosphorylation and activation of the tobacco NtCDPK2. It was, furthermore, shown that virus-induced gene-silencing of NtCDPK2 led to a delayed and reduced response to the race-specific elicitor. As in most plant kinases, the substrates of CDPKs await discovery, which will improve the understanding of how CDPKs regulate downstream responses. Two identified *in vitro* CDPK substrates, namely phenylalanine ammonia-lyase and serine acetyltransferase, point to modulation of metabolism

(Cheng et al. 2001; Liu et al. 2006). These may, respectively, deliver phenylpropanoid precursors or redox regulation components, such as glutathione, which are potentially important in response to pathogens. More recently, a potato CDPK was shown to phosphorylate the amino terminal region of a respiratory burst oxidase homologue (RBOH) and may possibly regulate production of reactive oxygen species (ROS; Kobayashi et al. 2007), thus further serving as an amplification in the signaling process.

4. Source of Calcium and Identity of Elicitor-Activated Channels/Pumps

While emphasis has been placed on the role of the extracellular pool as a Ca^{2+} source, the importance of internal stores, predominantly ER and vacuoles, is gaining appreciation. For instance, plant cells pre-treated with neomycin, a PLC inhibitor that interferes with phospholipid-mediated Ca^{2+} release, showed alterations in their elicitor-induced Ca^{2+} signature (Blume et al. 2000; Lecourieux et al. 2002).

A protein of the two-pore-channel family, TPC1, was purported to be the elicitor-responsive Ca^{2+} channel in plasma membranes of tobacco and rice cells (Kadota et al. 2004; Kurusu et al. 2005). However, Arabidopsis TPC1 was subsequently reported to be a slow vacuolar (SV) channel in the tonoplast (Peiter et al. 2005) and its role in Ca^{2+} signaling disproved (Ranf et al. 2008). Interestingly, the cyclic nucleotide-gated channel, AtCNGC2, is calcium-permeable (Ali et al. 2007) and responsible for the dnd1 (defense no death) phenotype that is characterized by constitutive expression of defense genes in the absence of HR-like cell death (Clough et al. 2000). Another member of the CNGC family, AtCNGC4, also controls HR in hlm1/dnd2 mutants (Balague et al. 2003; Jurkowski et al. 2004). The elevation of multiple defense markers in the cpr122 (constitutive PR gene expression) mutant was recently attributed to a genomic deletion that led to the expression of a chimeric CNGC11/12 protein (Yoshioka et al. 2006). On a whole, these findings suggested that members of CNGCs might be involved in ion transport, including Ca^{2+}, across the plasma membrane to control downstream defense signaling.

B. Reactive Oxygen Species

Reactive oxygen species (ROS), such as superoxide anion radical ($O_2^{\cdot-}$), hydrogen peroxide (H_2O_2), hydroxyl radical ($OH^{\cdot}$) and singlet oxygen, are recognized as important signal transduction elements in plants (Van Breusegem et al. 2008). Plants respond to infection by most pathogens with the rapid apoplastic generation of ROS, the so-called oxidative burst (Torres and Dangl 2005). Figure 17.2 shows ROS accumulation in an epidermal leaf cell of Arabidopsis thaliana upon attempted attack by Phytophthora infestans. While avirulent and non-adapted pathogens stimulate a long-lasting or biphasic oxidative burst, virulent pathogens usually only elicit a short burst of low intensity (Torres and Dangl 2005). Apoplastic ROS are generated by NADPH oxidases (respiratory burst oxidase homologs, RBOHs), extracellular peroxidases, type III peroxidases and polyamine oxidases (Torres and Dangl 2005; Bindschedler et al. 2006; Sagi and Fluhr 2006; Yoda et al. 2006; Choi et al. 2007). However, also cellular organelles such as chloroplasts, mitochondria and peroxisomes may contribute to ROS production during plant defense (Torres and Dangl 2005; Vidal et al. 2007). Recently, even the plant nucleus was described as a site of ROS generation in tobacco cells treated with the Oomycete elicitor, cryptogein (Ashtamker et al. 2007). In Arabidopsis thaliana, the NADPH oxidases, AtRBOHD and F, and in Nicotiana benthamiana, NbRBOHA and B, are primarily responsible for ROS production in response to infection (Torres et al. 2002; Yoshioka et al. 2003). While the Arabidopsis mutant, atrbohF, displayed reduced ROS accumulation but increased resistance against a weakly virulent isolate of the Oomycete, Hyaloperonospora parasitica (Torres et al. 2002),

Fig. 17.2. Accumulation of reactive oxygen species (ROS) at an epidermal leaf cell of Arabidopsis thaliana upon infection by Phytophthora infestans (ROS were stained with 3,3′-diaminobenzidine, as described by Halim et al. 2004)

virus-induced gene silencing of NbRBOHA and B in *Nicotiana benthamiana* reduced both, ROS production and basal defense against *Phytophthora infestans* (Yoshioka et al. 2003). However, silencing of NbRBOHB did not affect basal defense against *Colletotrichum orbiculare* (Asai et al. 2008). Most interestingly, ROS generated by AtRBOHD and F play a role in spatially limiting cell death to the sites of infection with a weakly pathogenic strain of *Botrytis cinerea* (Torres and Dangl 2005), thereby possibly limiting the spread of this necrotrophic fungus.

The activation of plant NADPH oxidases during pathogen defense is not well understood. Early studies suggested the involvement of Ca^{2+}, protein phosphorylation and small GTP-binding proteins (Jabs et al. 1997; Kawasaki et al. 1999; Lecourieux-Ouaked et al. 2000; Lecourieux et al. 2002). Plant NADPH oxidases contain an N-terminal extension that harbors two calcium-binding EF hands and an overlapping binding site for a small GTP-binding protein (Sagi and Fluhr 2006; Wong et al. 2007). In addition, phosphorylation of the N-terminal extension by a CDPK was found to be involved in activation of the enzyme (Kobayashi et al. 2007; Nühse et al. 2007). Finally, it was shown for rice that the small GTP-binding protein, OsRac1, directly interacts with the N-terminal extension of different NADPH oxidases and that Ca^{2+} regulates this interaction in a dynamic manner (Wong et al. 2007). The current model suggests that upon infection elevated cytosolic Ca^{2+} levels activate a CDPK, which phosphorylates the NADPH oxidase in its N-terminal extension. This initiates a conformational change that allows binding of Rac1 to the EF hand-containing domain resulting in enzyme activation and apoplastic ROS generation. ROS accumulation then stimulates another increase in cytosolic Ca^{2+} levels, which leads to occupation of the EF hands by Ca^{2+} followed by the release of Rac1 from its binding site and inactivation of the NADPH oxidase (Wong et al. 2007).

The involvement of ROS in plant cellular signaling has been extensively analyzed in suspension-cultured cells treated with fungal and Oomycete elicitors (Garcia-Brugger et al. 2006). In parsley cells, the Pep-13 elicitor, an oligopeptide derived from an extracellular transglutaminase of different *Phytophthora* species (Brunner et al. 2002), stimulates a strong long-lasting oxidative burst downstream of transient increases of cytosolic Ca^{2+} levels (Jabs et al. 1997; Blume et al. 2000). In these cells, ROS production is exclusively mediated by activation of an NADPH oxidase (Jabs et al. 1997), which furthermore requires activation of PLC and diacylglycerol kinase downstream of the calcium transient resulting in the generation of phosphatidic acid upstream of the oxidative burst (unpublished data). $O_2^{\cdot-}$ radicals produced during this burst are involved in activating a subset of defense-related genes, including those that encode phytoalexin biosynthetic enzymes (Jabs et al. 1997; Kroj et al. 2003). Pep-13 does not elicit programmed cell death in parsley indicating that the oxidative burst is not sufficient to stimulate this defense response (Jabs et al. 1997). In potato, however, where Pep-13 is recognized with specificity similar to that in parsley, the production of ROS and local programmed cell death are elicited, both downstream of elicitor-stimulated accumulation of salicylate (Halim et al. 2004). Such species-specific differences in embedding of ROS in defense signaling networks are also observed in tobacco and rice when triggered with other elicitors (Garcia-Brugger et al. 2006).

Oomycete-derived elicitins, such as cryptogein from *Phytophthora cryptogea* (Ricci et al. 1989), stimulate an oxidative burst in tobacco cells downstream of Ca^{2+} influx and transient increase of cytosolic Ca^{2+} levels (Lecourieux et al. 2002) by activating the NADPH oxidase, NtRBOHD (Allan and Fluhr 1997; Pugin et al. 1997; Simon-Plas et al. 2002). H_2O_2 accumulation in tobacco plants treated with cryptogein in the light results in lipid peroxidation, which together with H_2O_2 stimulates programmed cell death (Montillet et al. 2005). The *P. infestans* elicitin, INF1, stimulates ROS production *via* two alternative MAP kinase cascades (Asai et al. 2008).

Complex spatiotemporal patterns of ROS accumulation were found in barley infected with the powdery mildew fungus, *Blumeria graminis* f.sp. *hordei* (Hückelhoven and Kogel 2003). Despite the accumulation of H_2O_2 and $O_2^{\cdot-}$ at different phases of infection in the apoplast, ROS also accumulated in vesicles inside infected cells close to the infection site (Collins et al. 2003; Hückelhoven and Kogel 2003). In this interaction, accumulation of H_2O_2 correlates with programmed cell death, whereas $O_2^{\cdot-}$ appears to be involved in restriction of cell death (Hückelhoven and Kogel 2003).

In the interaction of plants with pathogenic fungi and Oomycetes, ROS are produced as components of early signal transduction processes in complex spatiotemporal patterns at the interface between plant and pathogen. They mediate multiple responses depending on the type of interaction.

C. Nitric Oxide

Nitric oxide (NO) appears to be generated upon pathogen attack concomitantly with ROS and is believed to be an important signal for local pro-grammed cell death during defense (Delledonne 2005; Garcia-Brugger et al. 2006; Wilson et al. 2008). Together with $O_2^{\cdot-}$ from the oxidative burst, NO can generate the highly reactive peroxinitrite radical, $OONO^{\cdot-}$ (Wilson et al. 2008). Although many studies have demonstrated NO generation upon attack of biotrophic bacterial pathogens or treatment with bacterial lipopolysaccharide elici-tor (Delledonne 2005), little data is available for fungi and Oomycetes.

Using 4,5-diaminofluorescein diacetate (DAF-2DA) for detection, NO production was visualized in *Medicago truncatula* leaves infected with an avirulent race of *Colletotrichum trifolii* (Ferrarini et al. 2008). Comparative gene expression analy-sis showed that many NO-responsive genes were activated upon infection with *C. trifolii*. Plant-derived NO was also detected in a compatible interaction of the necrotrophic fungus, *Botrytis elliptica*, with its host plant, lily (Van Baarlen et al. 2004). In response to treatment with the Oomycete elicitor, cryptogein, suspension-cultered tobacco cells and leaves were found to generate a monophasic burst of NO (Foissner et al. 2000; Lamotte et al. 2004). Similarly, Asai et al. (2008) detected NO production in *Nicotiana bentha-miana* expressing the *Phytophthora infestans* elicitin, INF1. In this case, INF1-stimulated NO production was downstream of a MAP kinase cascade involving an unknown MAPKKK, MEK2 and SIPK/NTF4.

Recently however, the specificity of DAF-2DA for NO detec-tion has been questioned, since physiological concentra-tions of dissolved NO, which were precisely quantifiable by chemiluminescence, did not result in DAF-2 fluorescence (Planchet and Kaiser 2006). In contrast, cryptogein-treated suspension-cultured tobacco cells produced DAF-2-respon-sive compounds, but those were not detectable by chemi-luminescence. Therefore, DAF-2 fluorescence appears not necessarily to be indicative for NO production.

In contrast to animals, the origin of NO in response to infec-tion of plants remains elusive (Wilson et al. 2008). While NO synthesis from *L*-arginine is catalyzed by NO synthases (NOS) in animals, this class of enzymes has not been iden-tified in plants (Zemojtel et al. 2006; Wilson et al. 2008). Nevertheless, inhibitors of animal NOS have been widely used in different plant systems (Wilson et al. 2008), which questions the conclusions drawn from those experiments. In addition, NO scavengers, such as 2-(4-carboxyphenyl)-4,4,5,5-tetramethylimidazoline-1-oxyl-3-oxide (cPTIO) apparently inhibit cryptogein-mediated programmed cell death not *via* NO scavenging (Planchet et al. 2006). There-fore, the origin and biological function of plant-derived NO in plant defense remains unclear.

Although plants can produce, perceive and respond to NO, details of the downstream sign-aling, in particular during plant defense against fungi and Oomycetes, is poorly understood (Wil-son et al. 2008). Several studies have analyzed plant responses to exogenously applied NO, such as accumulation of cyclic GMP, cyclic ADP-ribose, Ca^{2+} release from endogenous stores and activa-tion of defense-related genes (Durner et al. 1998; Polverari et al. 2003; Parani et al. 2004; Zago et al. 2006; Ferrarini et al. 2008). NO was also found to activate specific MAP kinases (MAPKs; Kumar and Klessig 2000; Pagnussat et al. 2004). Recently, the *S*-nitrosylation of proteins by NO has been described as possible downstream regulatory mechanism (Lindermayr et al. 2005; Lindermayr et al. 2006). However, the functional integration into the defense signaling network *via* NO, which is activated in response to fungi, Oomycetes or elicitors derived from these organisms, has not unequivocally been demonstrated.

D.   MAPK Cascades

Protein phosphorylation and especially the role of MAPK cascades has become an emerging theme in plant defense signaling. MAPK cascades con-sist of a hierarchical organization of three kinases: MAPK itself, the upstream MAPK kinase (MAPKK or MKK) and a MAPK kinase kinase (MAPKKK). The consecutive activation of the MAPK cascade components is instrumental in the signal trans-fer from extracellular signals into intracellular responses, which is often through the phospho-regulation of transcription factors or other cellu-lar targets (Gustin et al. 1998).

1.   Activation of MAPKs During Defense

Of these three classes of kinases, the MAPKs are the most amenable to biochemical analysis due to the utilization of so-called in-gel assays with artifi-cial substrates to follow their activities or through immunological methods that target a dual phos-phorylated motif upon activation. Furthermore, the development of specific antibodies for the MAPKs facilitated the coupling of immunoprecipitation techniques with *in vitro* kinase assays to distinguish individual MAPK activation profiles. Using these technologies, the activation of MAPKs by diverse elicitors (or effectors) from fungal/Oomycete pathogens has been reported for several plant species (see Table 17.2). Between one to three MAPKs, which are likely the orthologous proteins of the Arabidopsis MPK3, MPK6 and MPK4, were routinely found to be activated in these systems. It is unknown if this variation reflects true differences between plant species and/or treatments but, in all likelihood, all three might be activated. The discrepancy probably lies in technical details and sensitivity for the assays performed in different laboratories (e.g. poor renaturation of the MAPKs during in-gel assays).

While most studies merely correlate MAPK activation to plant innate immunity response (see references in Table 17.2), the importance of MPK3/MPK6 (and their orthologs) in positive control of MAMP-induced defense gene expression could be shown by introducing kinase-inactive MAPKs transiently into *Arabidopsis* and parsley protoplasts (Asai et al. 2002; Kroj et al. 2003). By contrast, MPK4 acts likely in negative regulation of the SA branch of defense gene activation but is required for full response to the JA/ET activation of defense genes in *Arabidopsis* (Petersen et al. 2000; Brodersen et al. 2006). Interestingly, Liu and Zhang (2004) identified the first MPK6 substrate as 1-aminocyclopropane-1-carboxylic acid (ACC) synthase, a key enzyme controlling ET biosynthesis, which suggests that ethylene signaling is activated downstream of MPK6 (Kim et al. 2003). Taken together, these data, albeit currently extrapolated from experiments performed in different plant species and systems, indicate that this complex network of pathogen-induced MAPKs and their substrates probably coordinates the fine-tuning of defense regulation.

2. Evidence for the Importance of MAPK Cascades in Disease Resistance

In addition to the monitoring of defense status using marker genes, infection assays show that certain MAPK mutants are indeed impaired or altered in disease resistance response. For instance, the *mpk4* mutant showed enhanced resistance to hemibiotrophs such as *Hyaloperonospora parasitica* and *Pseudomonas syringae* (Petersen et al. 2000), but was more susceptible to the necrotrophic fungus *Alternaria brassicicola* (Brodersen et al. 2006). Silencing of MPK6 compromised resistance to avirulent *H. parasitica* in *Arabidopsis* (Menke et al. 2004). In contrast, RNAi of the *OsMPK5* gene encoding a rice MPK3-like MAPK, led to increased *PR1* and *PR10* expression and elevated resistance to *Magnaporthe grisea* as well as *Burkholderia glumae*. These transgenic plants were, however, more sensitive to salt, drought and cold stresses (Xiong and Yang 2003), indicating that OsMPK5 is a negative regulator of innate immunity and inversely regulates stress responses to pathogens and abiotic factors.

The overexpression of selected MAPKs in heterologous plants increased resistance to *Alternaria alternata* or *Phytophthora parasitica* var

Table 17.2. MAPK activation by fungal/oomycete elicitors (or effectors)

Elicitors/effectors	Microbe	Plant material	Reference(s)
Avr4	*Cladosporium fulvum*	*Solanum lycopersicum*	Stulemeijer et al. (2007)
Avr9	*C. fulvum*	*S. lycopersicum/Nicotiana tabacum/N. benthamiana*	Romeis et al. (1999)
β-glucan	*Phytophthora sojae*	*Glycine max*	Daxberger et al. (2007)
BcPG1	*Botrytis cinerea*	*Vitis vinifera*	Poinssot et al. (2003)
CBEL	*P. parasitica nicotianae*	*N. tabacum*	Gaulin et al. (2006)
Chitin	Various fungi	*Arabidopsis thaliana*	Nühse et al. (2000), Wan et al. (2004)
Elicitins	*Phytophthora* spp.	*N. tabacum/N. benthamiana/S. lycopersicum*	Lebrun-Garcia et al. (1998, 2002), Zhang et al. (1998)
Hyphal Cell Wall	*P. infestans*	*Solanum b(tuberosum)*	Katou et al. (2005)
Laminarin[a]	*Laminaria digitata*	*V. vinifera*	Aziz et al. (2003)
NPP1 (NLP$_{pp}$)	*P. parasitica*	*Petroselinum crispum/A. thaliana*	Fellbrich et al. (2002)
Pep-13	*P. sojae*	*P. crispum*	Ligterink et al. (1997), Kroj et al. (2003)
Sphingolipid	*Magnaporthe grisea*	*Oryza sativa*	Lieberherr et al. (2005)
Xylanase	*Trichoderma viride*	*A. thaliana, S. lycopersicum, N. tabacum*	Suzuki et al. (1999), Nühse et al. (2000), Mayrose et al. (2004)

[a] Laminarin is a β-1,3-glucan from the brown alga, *Laminaria digitata*, and hence sensu stricto not from a pathogen, but is used in agriculture as a defense-activating natural product.

nicotianae, respectively (Cheong et al. 2003; Song et al. 2006). Resistance to *M. grisea* by overexpressing a pepper MAPK in rice (Lee et al. 2004) is also speculated, since these plants have enhanced *PR1/PR10* expression (cf. Xiong et al. 2003, where *PR1/PR10* expression correlated with resistance) and enhanced JA levels. Since MAPKs have to be activated by MKKs, enhancing resistance can be more elegantly achieved by activation of endogenous MAPKs through gain-of-function MKKs. An example of this is the introduction of constitutive active MKKs into *Arabidopsis*, which conferred resistance to both, bacterial and fungal pathogens (Asai et al. 2002). Similarly, an unbiased high-throughput screen with *Nicotiana benthamiana* also uncovered an MKK (NbMKK1) to mediate HR by the *Phytophthora infestans*-derived INF1 elicitin (Takahashi et al. 2007). Transgenic potato with a constitutive active MKK driven by a pathogen-responsive promoter conferring higher resistance to *Alternaria solani* and *P. infestans* was recently reported (Yamamizo et al. 2006) – indicating the possibility of using MKKs for enhancing pathogen resistance in crops of economic importance. To date, evidence points to the activation of ethylene (Kim et al. 2003) and ROS production (Ren et al. 2002; Yoshioka et al. 2003) by these active MKKs. The importance for ROS production in resistance was shown by Yoshioka et al. (2003). In this work, silencing of *RBOH* homologues in *N. benthamiana* caused a partial loss of resistance to *P. infestans* Race 0 (with a shift towards higher frequency of appressoria formation, penetration, secondary hyphae and sporangiophore emergence on the leaf underside). Nevertheless, before approaches with constitutive active MKKs can be implemented in agriculture, further understanding of the mechanistic mode of enhancing resistance is essential.

Compared to MAPKs and MKKs, much less is known about MAPKKKs. The first report about the involvement of MAPKKKs in plant innate immunity is the enhanced disease resistance 1 (*edr1*) mutant, which is more resistant to *Erysiphe cichoracearum via* an SA-dependent pathway, but is independent of JA/ET responses (Frye et al. 2001). Manipulation of the tobacco MAPKKK, NPK1, or the tomato MAPKKKα also affected plant innate immunity to bacterial and viral pathogens (Jin et al. 2002; del Pozo et al. 2004). The *Arabidopsis* MAPKKK, MEKK1, was initially reported to regulate the flg22 response leading to MKK4/MKK5 and MPK3/MPK6 activation (Asai et al. 2002),

but more recent data indicate that it controls the MKK1 and MPK4 pathway and is therefore likely to be involved in the negative regulation of defense responses (Ichimura et al. 2006; Meszaros et al. 2006; Suarez-Rodriguez et al. 2007). Since *mekk1* mutants are misregulated in cellular redox control and accumulate ROS, it is likely that MEKK1 can affect innate immunity through ROS homeostasis (Nakagami et al. 2006).

Virulence effectors of bacterial pathogens often target components of MAPK cascades. For instance, YopJ, a *Yersinia* effector, acetylates serine/threonine residues necessary for MKK activation and thus blocks the phosphorylation-based activation (Orth et al. 2000; Mukherjee et al. 2006, 2007). Recently, the *Shigella* type III effector OspF was identified to cleave the C-OP bond in the phosphothreonine residue of a MAPK (i.e. it has MAPK phosphothreonine-lyase activity), thus inactivating and, more importantly, preventing reactivation of MAPKs (Li et al. 2007). A homologous effector (HopAI1) from phytopathogenic *Pseudomonas syringae* bacteria had the same activity and can overcome PAMP-triggered immunity by inactivating MPK3 and MPK6 (Zhang et al. 2007). Thus, these findings of MAPK signaling interference by bacterial pathogen effectors highlight the importance of this pathway for pathogen defense.

E. Other Components in Signaling Systems

JA and SA are central signaling compounds in the plant's defense response (Fig. 17.1). Work with *Arabidopsis* has revealed that, despite exceptions, SA is generally involved in mediating defense against biotrophic pathogens while JA-activated responses are important for defense against pathogens with a necrotrophic life style (Glazebrook 2005).

1. Jasmonic Acid

JA is synthesized from α-linolenic acid (LnA) originating from chloroplast galactolipids (Fig. 17.3; Wasternack 2007). First, 13-hydroperoxylinolenic acid is produced from LnA by a plastid-localized 13-lipoxygenase (LOX) and subsequently converted by allene oxide synthase (AOS) to an unstable allene oxide (Laudert et al. 1996). This, in turn, is cyclized by allene oxide cyclase (AOC) to yield *cis*(+)-12-oxophytodienoic acid (OPDA), the first stereospecific isomer (Ziegler et al. 2000). After translocation from chloroplasts to peroxisomes, OPDA is reduced to 3-oxo-2-(2Z-pentenyl)cyclopentane-1-octanoic acid (OPC8) by an OPDA reductase (OPR3; Schaller et al. 2000). 4-Coumarate-CoA ligase-like proteins, such as OPCL1 of tomato, have been shown to activate the acyl group of OPDA and OPC8 (Schneider et al. 2005; Koo et al. 2006; Kienow et al. 2008). The first step in

the β-oxidation of the side chain of OPC8 is catalyzed by a peroxisomal acyl-CoA oxidase (i.e. ACX1 of tomato; Li et al. 2005). Subsequent shortening of the reaction product requires a multifunctional protein (MFP) with 2-trans-enoyl-CoA hydratase and 3-hydroxyacyl-CoA dehydrogenase activities, as well as a 3-ketoacyl-CoA thiolase (Cruz Castillo et al. 2004; Afitlhile et al. 2005; Li et al. 2005; Delker et al. 2007).

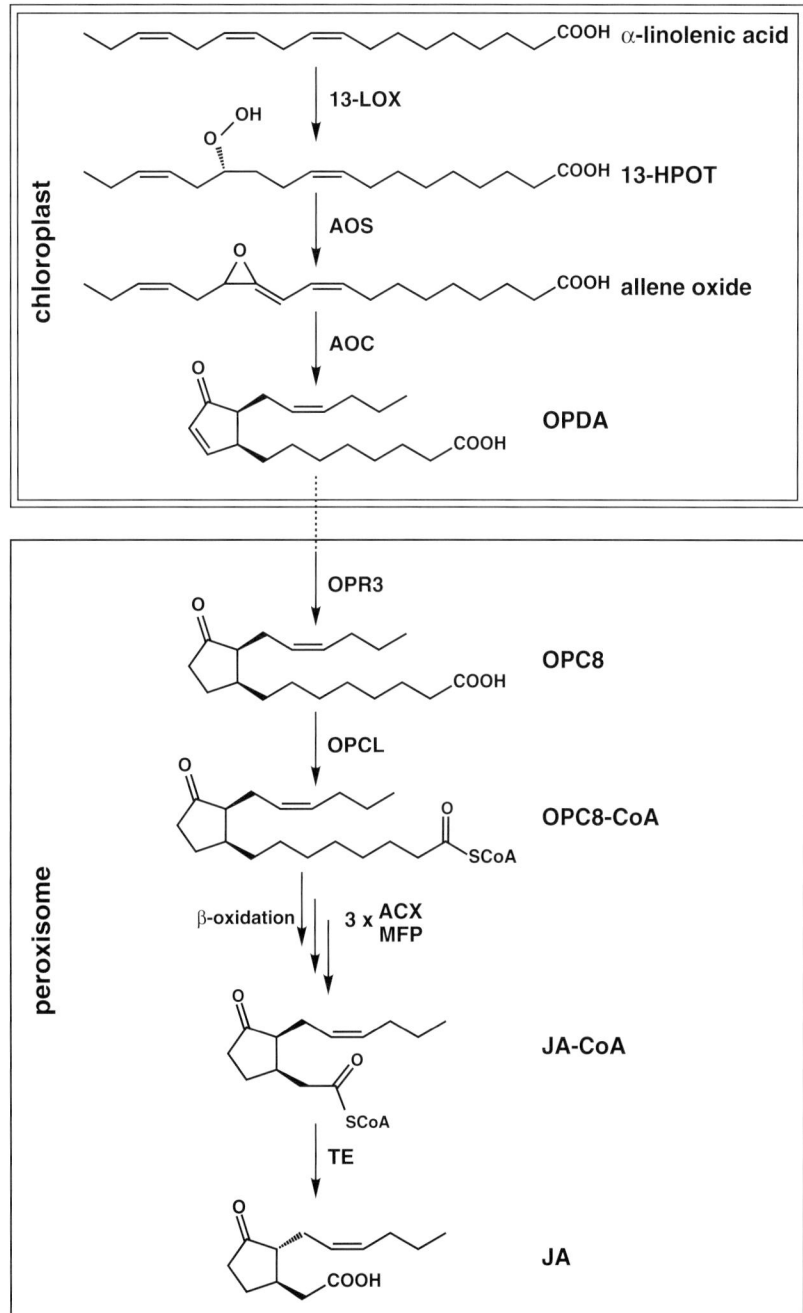

Fig.17.3. Biosynthetic pathway of jasmonic acid. *ACX* Acyl-CoA oxidase, *AOC* allene oxide cyclase, *AOS* allene oxide synthase, *13-HPOT* (13S)-hydroperoxyoctadecatrienoic acid, *JA* jasmonic acid, *13-LOX* 13-lipoxygenase, *MFP* multifunctional protein, *OPC8* 3-oxo-2-(2Z-pentenyl)cyclopentane-1-octanoic acid, *OPCL* OPC8-CoA ligase, *OPDA* cic(+)-12-oxophytodienoic acid, *OPR3* OPDA reductase, *TE* thioesterase

The F-box protein COI1, a central regulator of JA signaling, is part of a ubiquitin ligase complex, which is involved in the specific degradation of negative regulators of JA-induced responses (Xu et al. 2002). In the presence of the JA derivative jasmonoyl-isoleucine, COI1 interacts with the ZIM domain protein JAZ1, which is subsequently degraded by the proteasome (Thines et al. 2007). Since JAZ proteins repress transcription, in the case of JAI3 by interaction with the transcription factor AtMYC2, their removal affects JA-regulated genes (Chini et al. 2007). AtMYC2 negatively regulates pathogen response genes, while it activates transcription of wound response genes (Lorenzo and Solano 2005). JA signaling also involves the activation of MAPK cascades. Exogenous JA activates MPK6 *via* MKK3, which leads to a reduction in *AtMYC2* expression (Takahashi et al. 2007).

Consistent with the model that JA is important for efficient defense against necrotrophic fungi, *Arabidopsis* mutants defective in JA biosynthesis or signaling are more susceptible to pathogens such as *Botrytis cinerea*, *Alternaria brassicicola* or *Plectosphaerella cucumerina*. The triple mutant, *fad3-2fad7-2fad8*, which is defective in genes encoding fatty acid desaturases and thus compromised in the synthesis of trienoic acids including LnA, exhibits increased susceptibility to the soil Oomycete, *Pythium mastophorum*. Since exogenous application of methyljasmonate is able to restore wild-type responses, the mutant phenotype is apparently caused by the lack of JA (Vijayan et al. 1998).

Functional redundancy exists for LOX enzymes in *Arabidopsis*. Reduction of AtLOX2 levels in transgenic antisense plants results in the inability to accumulate JA after wounding (Bell et al. 1995). However, a knock-out mutation in *AtLOX2* is reported not to lead to decreased JA levels after wounding and alterations in defense against *Botrytis cinerea* (Dubugnon and Farmer 2007; www.tair.org). In contrast, an *Arabidopsis* mutant line with a knock-out of the single *AOS* gene shows enhanced disease symptoms in response to infection by *B. cinerea* (Raacke et al. 2006) and *Alternaria brassicicola* (Schilmiller et al. 2007). Interestingly, not JA, but its biosynthetic precursor OPDA appears to be required for resistance to *A. brassicicola*, since *opr3* mutants behave like wild-type plants upon fungal infection (Stintzi et al. 2001). *opr3* mutants also do not show significant differences to wild-type plants in response to infection with *B. cinerea* (Raacke et al. 2006). Whether the importance of OPDA for resistance against fungal pathogens can be extrapolated to other plants is not clear since the *acx1* mutant of tomato, which affects an enzymatic step downstream of OPDA, is at least still susceptible to insect attack (Li et al. 2005). Mutations in genes encoding peroxisomal β-oxidation enzymes were analyzed with respect to wound-induced JA

synthesis, but no data have yet been reported for alterations in resistance to filamentous pathogens. Thus, the *Arabidopsis acx1/5* double mutant is compromised in wound-induced JA accumulation, but does not differ from the wild type after infection with *Alternaria brassicicola*, suggesting that other ACXs are responsible for JA formation after pathogen attack (Schilmiller et al. 2007). The JA signaling mutant *coi1* is more susceptible to the necrotrophic fungi, *Botrytis cinerea* and *A. brassicicola*, but not to the obligate biotroph, *Hyaloperonospora parasitica*, (Thomma et al. 1998). *JAR1*, a gene encoding a JA-amino acid conjugase (Staswick et al. 2002), is required for resistance to the soil fungus, *Pythium irregulare* (Staswick et al. 1998). Mutations in the gene encoding the transcription factor, AtMYC2, which represses pathogen-response genes, results in increased expression of *PR* genes and enhanced resistance to the necrotrophic fungi, *B. cinerea* and *Plectosphaerella cucumerina* (Lorenzo et al. 2004).

Constitutive activation of JA responses correlates with enhanced resistance against fungi in mutants as well as in transgenic plants overexpressing JA biosynthetic genes. Release of LnA from chloroplast membranes requires the galactolipase, DGL, since *DGL* knock-down mutants show highly reduced JA levels at early time-points after wounding. Accordingly, a gain-of-function mutant overexpressing the *DGL* gene exhibits increased JA levels and enhanced resistance to *A. brassicicola* (Hyun et al. 2008). In rice, overexpression of *AOS* results in higher JA levels and increased resistance to *Magnaporthe grisea* (Mei et al. 2006). Moreover, overexpression of a JA methyl transferase in *Arabidopsis* leads to enhanced resistance to *B. cinerea* (Seo et al. 2001). The *cev* mutant of Arabidopsis, which shows constitutive expression of JA and ET responses, is more resistant to the powdery mildews, *Erysiphe cichoracearum*, *Golovinomyces orontii* and *Oidium lycopersicum* (Ellis and Turner 2001).

In accordance with the increased susceptibility of *Arabidopsis* JA mutants, exogenous application of JA protects *Arabidopsis* against fungal infection. The phytoalexin-deficient mutant *pad3*, which is susceptible to *Alternaria brassicicola*, is more resistant when treated with JA before inoculation with the fungus (Thomma et al. 1998). Similarly, Col-0 plants allowed less growth of *Plectosphaerella cucumerina* when pretreated with JA (Ton and Mauch-Mani 2004). Protection of potato against *Phytophthora infestans* can be achieved by exogenous application of JA (Cohen et al. 1993). In grapevine, JA treatment of leaf disks 24 h prior to infection with the obligate biotrophic Oomycete *Plasmopara viticola* results in reduced sporangia formation compared to water-treated control leaf disks. This effect was linked to the ability for callose formation, since application of callose inhibitors restored susceptibility (Hamiduzzaman et al. 2005). Similarly, exogenous application of JA led to the activation of defense gene expression and resistance against *Magnaporthe grisea* in rice (Mei et al. 2006).

However, instead of activating defense signaling, JA might act also as an inhibitor of fungal growth and development. A direct antimicrobial effect of JA on *Blumeria graminis* f.sp. *hordei* was postulated based on the strong inhibition of appressoria differentiation on JA-treated barley leaves (Schweizer et al. 1993). In contrast, there was no effect of JA on mycelial growth of *Phytophthora parasitica*, *Cladosporium herbarum* and *Botrytis cinerea* (Prost et al. 2005). On the other hand, OPDA, the biosynthetic precursor of JA, was the most active compound of 47 oxylipins tested for antimicrobial activity. In particular, OPDA inhibited spore germination of *B. cinerea*, *P. infestans* and *P. parasitica*. These differences were attributed to the structural features of the α,β-unsaturated carbonyl group present in OPDA (Prost et al. 2005), which, as an electrophile, is speculated to possess signaling functions by itself (Farmer and Davoine 2007).

Treatment of plants with pathogen elicitors can induce accumulation of signaling compounds suggesting that these are involved in the activation of defense responses. Functional analyses in the *coi1* mutant showed that JA is required for defense responses induced by the cellulose-binding elicitor lectin (CBEL) from *Phytophthora* species (Khatib et al. 2004). In contrast, silencing of *COI1* in *Nicotiana benthamiana* does not result in alterations of cell death induced by INF1, the elicitin from *P. infestans* (Kanneganti et al. 2006). The *Phytophthora*-specific MAMP, Pep-13, elicits accumulation of JA in potato (Halim et al. 2004). In transgenic potato plants with RNAi-mediated suppression of JA accumulation, Pep-13 is not able to induce defense responses to the same extent as on wild-type plants, suggesting that JA is required for Pep-13 signaling in potato. Conflicting data exist for the JA-dependence of Nep1-like proteins. Thus, PiNPP1.1 from *P. infestans* causes *COI1*-dependent cell death when expressed *via* potato virus X agroinfection in *Nicotiana benthamiana* (Kanneganti et al. 2006). However, cell death induced by the Nep1-like protein from *P. parasitica* (NLP_{pp}) was reported not to depend on JA in *Arabidopsis* (Qutob et al. 2006). Interestingly, NLP_{pp} does not induce the accumulation of transcripts of JA biosynthetic genes in *Arabidopsis* (Qutob et al. 2006).

2. Salicylic Acid

SA is an important signaling compound for both basal defense (MTI) and *R*-gene-mediated resistance (ETI), as well as for the establishment and/or maintenance of systemic acquired resistance (SAR) (Durrant and Dong 2004). SA levels increase in response to infection by fungal pathogens (Wildermuth et al. 2001; Govrin and Levine 2002). Synthesis of SA was reported to proceed *via* the phenylpropanoid pathway in a number of plant species (Ribnicky et al. 1998; Coquoz et al. 1998). However, in *Arabidopsis*, the majority of SA accumulating in response to pathogen attack is synthesized *via* the isochorismate pathway (Wildermuth et al. 2001). Mutants of the *sid2/eds16* gene coding for isochorismate synthase 1 (ICS1) show reduced SA accumulation in response to pathogen infection and are more susceptible to the biotrophic fungus, *Golovinomyces orontii* (Wildermuth et al. 2001).

SA signaling requires the ankyrin-repeat-containing protein NPR1, which was identified by mutant screens in *Arabidopsis*. NPR1 is present in the cytoplasm as an oligomer, and responds to pathogen-induced alterations in the cellular redox state by translocating to the nucleus in its monomeric form (Dong 2004). NPR1 interacts with distinct members of the TGA/OBF class of bZIP transcription factors in the nucleus, thus activating defense gene expression (Dong 2004). Overexpression of *NPR1* in crop plants, such as wheat, tomato and apple, resulted in resistance against fungal pathogens (Lin et al. 2004; Makandar et al. 2006; Malnoy et al. 2007). In contrast, loss of *NPR1* function leads to enhanced susceptibility. Thus, the *Arabidopsis* mutant, *npr1*, is more susceptible to infection with *Golovinomyces orontii* as are other mutants defective in SA signaling, such as *pad4* or *eds5* (Reuber et al. 1998).

In addition to mutants, transgenic plants unable to accumulate SA due to expression of the *NahG* gene have been instrumental in elucidating the role of SA for plant defense. *NahG* encodes a salicylate hydroxylase, which converts SA to catechol. The loss of SA accumulation in *NahG* plants correlates with increased susceptibility to biotrophic or hemibiotrophic filamentous pathogens, as exemplified by the increased susceptibility of *NahG Arabidopsis* plants to *G. orontii* (Reuber et al. 1998) and *Hyaloperonospora parasitica* (McDowell et al. 2005), as well as potato *NahG* plants to *Phytophthora infestans* (Halim et al. 2007).

Differences in pathogen defense responses in *Arabidopsis NahG* and *sid2* plants suggest a role for catechol or other degradation products of the NahG reaction in pathogen defense (van Wees and

Glazebrook 2003; Heck et al. 2003). Alternatively, *NahG* plants might reveal SA-dependent responses, which are independent of ICS1-catalyzed SA accumulation. Thus, *NahG Arabidopsis* plants show larger lesions after infection with *Botrytis cinerea* than *sid2* plants, which behave like wild type plants (Ferrari et al. 2003). Systemic symptoms, on the other hand, are similar in wild type, *NahG* and *sid2* plants. These observations imply that local resistance of *Arabidopsis* to *B. cinerea* requires SA, but not ICS1, and that SA (possibly synthesized *via* phenylalanine ammonia-lyase) induces cell death during lesion development, whereas systemic resistance requires SA produced *via* ICS1 (Ferrari et al. 2003; Wildermuth et al. 2001). In addition, *ICS1* gene expression is induced in *Arabidopsis* by infection with *G. orontii*, but not with the necrotrophic pathogen *B. cinerea* (Ferrari et al. 2003), suggesting that SA produced *via* ICS1 is part of a defense response induced by biotrophic pathogens.

Functional analyses using transgenic plants expressing the *NahG* gene indicate that SA is not required for CBEL-elicited HR in *Arabidopsis* but for the activation of a subset of defense genes (Khatib et al. 2004). Similarly, infiltration of tobacco leaves with the elicitin, cryptogein, resulted in an HR in *NahG* plants, which was equivalent to that in wild-type plants (Cordelier et al. 2003). In *NPR1*-silenced *Nicotiana benthamiana* plants, no alteration in cell death induced by INF1 is observed (Kanneganti et al. 2006). Thus, these effectors do not require SA for induction of cell death.

The NLP PiNPP1.1 from *P. infestans* causes *NPR1*-dependent cell death when expressed *via* potato virus X agroinfection in *N. benthamiana* (Kanneganti et al. 2006). However, SA is not required for cell death induced by the NLP from *P. parasitica* (NLP$_{PP}$) in *Arabidopsis* (Qutob et al. 2006). NLP$_{PP}$ induces the accumulation of transcripts of SA biosynthetic genes in *Arabidopsis* (Qutob et al. 2006), as well as *PR* gene expression in an SA-dependent manner, since no induction takes place in *NahG Arabidopsis* plants (Fellbrich et al. 2000). In potato, SA accumulation is required for the activation of defense responses by the *Phytophthora* MAMP, Pep-13, since they do not occur in *NahG* plants (Halim et al. 2004).

3. Cross-Talk

There is considerable cross-talk between SA- and JA-signaling pathways to optimize defense against infection by different pathogens. In *Arabidopsis*, mostly antagonistic interactions between SA and JA signaling have been described. SA can inhibit JA biosynthesis and has been shown to suppress JA-dependent defense responses (Pozo et al. 2004). Thus, in *NahG* plants, which do not accumulate SA, JA signaling is enhanced (Spoel et al. 2003). JA-defective mutants, in contrast, can exhibit increased SA-dependent defense as exemplified by the hyperactivation of SA responses in *coi1* (Kloek et al. 2001).

Increased JA responses in the *npr1* mutant, moreover, suggest a role of the central regulator of SA signaling in cross-talk (Spoel et al. 2003). Recently, a glutaredoxin that interacts with an SA-inducible TGA transcription factor was reported to inhibit JA-dependent gene expression, when ectopically expressed in transgenic plants, suggesting additional cross-talk at the level of redox regulation (Ndamukong et al. 2007).

The transcription factor, WRKY70, acts as an inducer of SA-responsive genes, but negatively regulates JA-dependent gene expression (Li et al. 2004). Overexpression of WRKY70 resulted in enhanced resistance to the biotrophic fungus, *Erysiphe cichoracearum*, and in reduced resistance to the necrotrophic fungus, *Alternaria brassicicola* (Li et al. 2006). Another point of convergence is the MAP kinase, MPK 4, which positively regulates JA responses, but acts as a negative factor for SA signaling (Brodersen et al. 2006).

Spatial considerations and the importance of pathogen type-specificity have been addressed by assessing the outcome of infections by multiple pathogens. Bacterial infections, which induce SA-mediated defense, resulted in suppression of the JA signaling pathway and enhanced susceptibility of *Arabidopsis* plants to subsequent infection by the necrotrophic pathogen, *Alternaria brassicicola* (Spoel et al. 2007). Interestingly, this effect was localized to the site of primary infection, whereas systemically no reduction in resistance occurred probably due to a gradient of SA signaling (Spoel et al. 2007).

IV. Conclusion

Plant signal transduction networks of defense responses against fungi and Oomycetes are highly complex. Some of the known elements of these networks and their interactions have been described above (Fig. 17.1). However, their position within individual signaling networks may differ depending on the specific plant-pathogen interaction, the receiver involved in MAMP recognition and the pathogenic strategy of a specific pathogen (e.g. biotrophic vs necrotrophic).

Many of the signaling mechanisms described above for plant defense against fungal and Oomycete pathogens are similarly involved in completely unrelated signaling networks, such as those activated upon wounding, by different abiotic stresses and during developmental processes. The mechanisms that maintain signal-response specificity are not at all understood, but are essential for the understanding of these processes and their possible application for the generation of plants better adapted to a changing environment.

References

Abramovitch RB, Anderson JC, Martin GB (2006) Bacterial elicitation and evasion of plant innate immunity. Nat Rev Mol Cell Biol 7:601–611

Afitlhile MM, Fukushige H, Nishimura M, Hildebrand DF (2005) A defect in glyoxysomal fatty acid β-oxidation reduces jasmonic acid accumulation in Arabidopsis. Plant Physiol Biochem 43:603–609

Ali R, Ma W, Lemtiri-Chlieh F, Tsaltas D, Leng Q, von Bodman S, Berkowitz GA (2007) Death don't have no mercy and neither does calcium: Arabidopsis CYCLIC NUCLEOTIDE GATED CHANNEL2 and innate immunity. Plant Cell 19:1081–1095

Allan AC, Fluhr R (1997) Two distinct sources of elicited reactive oxygen species in tobacco epidermal cells. Plant Cell 9:1559–1572

Allen GJ, Kwak JM, Chu SP, Llopis J, Tsien RY, Harper JF, Schroeder JI (1999) Cameleon calcium indicator reports cytoplasmic calcium dynamics in Arabidopsis guard cells. Plant J 19:735–747

Asai S, Ohta K, Yoshioka H (2008) MAPK signaling regulates nitric oxide and NADPH oxidase-dependent oxidative bursts in Nicotiana benthamiana. Plant Cell 20:1390–1406

Asai T, Tena G, Plotnikova J, Willmann MR, Chiu WL, Gomez-Gomez L, Boller T, Ausubel FM, Sheen J (2002) MAP kinase signalling cascade in Arabidopsis innate immunity. Nature 415:977–983

Asano T, Tanaka N, Yang G, Hayashi N, Komatsu S (2005) Genome-wide identification of the rice calcium-dependent protein kinase and its closely related kinase gene families: comprehensive analysis of the CDPKs gene family in rice. Plant Cell Physiol 46:356–366

Ashtamker C, Kiss V, Sagi M, Davydov O, Fluhr R (2007) Diverse subcellular locations of cryptogein-induced reactive oxygen species production in tobacco Bright Yellow-2 cells. Plant Physiol 143:1817–1826

Ayers AR, Ebel J, Finelli F, Berger N, Albersheim P (1976) Host–pathogen interactions. IX. Quantitative assays of elicitor activity and characterization of elicitor present in extracellular medium of cultures of Phytophthora megasperma var. sojae. Plant Physiol 57:751–759

Aziz A, Poinssot B, Daire X, Adrian M, Bezier A, Lambert B, Joubert JM, Pugin A (2003) Laminarin elicits defense responses in grapevine and induces protection against Botrytis cinerea and Plasmopara viticola. Mol Plant–Microbe Interact 16:1118–1128

Bailey BA (1995) Purification of a protein from culture filtrates of Fusarium oxysporum that induces ethylene and necrosis in leaves of Erythroxylum coca. Phytopathology 85:1250–1255

Bailey BA, Dean JFD, Anderson JD (1990) An ethylene biosynthesis-inducing endoxylanase elicits electrolyte leakage and necrosis in Nicotiana tabacum cv. Xanthi leaves. Plant Physiol 94:1849–1854

Bailey BA, Korcak RF, Anderson JD (1992) Alterations in Nicotiana tabacum L. cv Xanthi cell membrane function following treatment with an ethylene biosynthesis-inducing endoxylanase. Plant Physiol 100:749–755

Baladron V, Ufano S, Duenas E, Martin-Cuadrado AB, del Rey F, de Aldana CRV (2002) Eng1p, an endo-1,3-β-glucanase localized at the daughter side of the septum, is involved in cell separation in Saccharomyces cerevisiae. Eukaryot Cell 1:774–786

Balague C, Lin B, Alcon C, Flottes G, Malmstrom S, Kohler C, Neuhaus G, Pelletier G, Gaymard F, Roby D (2003) HLM1, an essential signaling component in the hypersensitive response, is a member of the cyclic nucleotide-gated channel ion channel family. Plant Cell 15:365–379

Batistic O, Kudla J (2004) Integration and channeling of calcium signaling through the CBL calcium sensor/CIPK protein kinase network. Planta 219:915–924

Bell E, Creelman RA, Mullet JE (1995) A chloroplast lipoxygenase is required for wound-induced jasmonic acid accumulation in Arabidopsis. Proc Natl Acad Sci USA 92:8675–8679

Bhat RA, Miklis M, Schmelzer E, Schulze-Lefert P, Panstruga R (2005) Recruitment and interaction dynamics of plant penetration resistance components in a plasma membrane microdomain. Proc Natl Acad Sci USA 102:3135–3140

Bindschedler LV, Dewdney J, Blee KA, Stone JM, Asai T, Plotnikov J, Denoux C, Hayes T, Gerrish C, Davies DR, Ausubel FM, Bolwell PG (2006) Peroxidase-dependent apoplastic oxidative burst in Arabidopsis required for pathogen resistance. Plant J 47:851–863

Blume B, Nürnberger T, Nass N, Scheel D (2000) Receptor-mediated rise in cytoplasmic free calcium required for activation of pathogen defense in parsley. Plant Cell 12:1425–1440

Brodersen P, Petersen M, Nielsen BH, Zhu S, Newman MA, Shokat KM, Rietz S, Parker J, Mundy J (2006) Arabidopsis MAP kinase 4 regulates salicylic acid- and jasmonic acid/ethylene-dependent responses via EDS1 and PAD4. Plant J 47:532–546

Brunner F, Rosahl S, Lee J, Rudd JJ, Geiler C, Kauppinen S, Rasmussen G, Scheel D, Nürnberger T (2002) Pep-13, a plant defense-inducing pathogen-associated pattern from Phytophthora transglutaminases. EMBO J 21:6681–6688

Butler AR, Odonnell RW, Martin VJ, Gooday GW, Stark MJR (1991) Kluyveromyces lactis toxin has an essential chitinase activity. Eur J Biochem 199:483–488

Cheng SH, Sheen J, Gerrish C, Bolwell GP (2001) Molecular identification of phenylalanine ammonia-lyase as a substrate of a specific constitutively active Arabidopsis CDPK expressed in maize protoplasts. FEBS Lett 503:185–188

Cheong JJ, Hahn MG (1991) A specific, high affinity binding site for the hepta-β-glucoside elicitor exists in soybean membranes. Plant Cell 3:137–147

Cheong JJ, Birberg W, Fuegedi P, Pilotti A, Garegg PJ, Hong N, Ogawa T, Hahn MG (1991) Structure-activity relationships of oligo-β-glucoside elicitors of phytoalexin accumulation in soybean. Plant Cell 3:127–136

Cheong YH, Moon BC, Kim JK, Kim CY, Kim MC, Kim IH, Park CY, Kim JC, Park BO, Koo SC, Yoon HW, Chung WS, Lim CO, Lee SY, Cho MJ (2003) BWMK1, a rice mitogen-activated protein kinase, locates in the nucleus and mediates pathogenesis-related gene expression by activation of a transcription factor. Plant Physiol 132:1961–1972

Chinchilla D, Zipfel C, Robatzek S, Kemmerling B, Nürnberger T, Jones JDG, Felix G, Boller T (2007) A flagellin-induced complex of the receptor FLS2 and BAK1 initiates plant defence. Nature 448:497–500

Chini A, Fonseca S, Fernandez G, Adie B, Chico JM, Lorenzo O, Garcia-Casado G, Lopez-Vidriero I, Lozano FM, Ponce MR, Micol JL, Solano R (2007) The JAZ family of repressors is the missing link in jasmonate signalling. Nature 448:666–671

Chisholm ST, Coaker G, Day B, Staskawicz B (2006) Host–microbe interactions: shaping the evolution of the plant immune response. Cell 124:803–814

Choi HW, Kim YJ, Lee SC, Hong JK, Hwang BK (2007) Hydrogen peroxide generation by the pepper extracellular peroxidase CaPO2 activates local and systemic cell death and defense response to bacterial pathogens. Plant Physiol 145:890–904

Clough SJ, Fengler KA, Yu IC, Lippok B, Smith RK Jr, Bent AF (2000) The Arabidopsis dnd1 "defense, no death" gene encodes a mutated cyclic nucleotide-gated ion channel. Proc Natl Acad Sci USA 97:9323–9328

Cohen Y, Gisi U, Niderman T (1993) Local and systemic protection against Phytophthora infestans induced in potato and tomato plants by jasmonic acid and jasmonic-methylester. Phytopathology 83:1054–1062

Collins NC, Thordal-Christensen H, Lipka V, Bau S, Kombrink E, Qiu JL, Hückelhoven R, Stein M, Freialdenhoven A, Somerville SC, Schulze-Lefert P (2003) SNARE-protein-mediated disease resistance at plant cell wall. Nature 425:973–977

Coquoz JL, Buchala A, Métraux JP (1998) The biosynthesis of salicylic acid in potato plants. Plant Physiol 117:1095–1101

Cordelier S, de Ruffray P, Fritig B, Kauffmann S (2003) Biological and molecular comparison between localized and systemic acquired resistance induced in tobacco by a Phytophthora megasperma glycoprotein elicitin. Plant Mol Biol 51:109–118

Cosio EG, Frey T, Verduyn R, Vanboom J, Ebel J (1990) High-affinity binding of a synthetic heptaglucoside and fungal glucan phytoalexin elicitors to soybean membranes. FEBS Lett 271:223–226

Cosio EG, Feger M, Miller CJ, Antelo L, Ebel J (1996) High-affinity binding of fungal β-glucan elicitors to cell membranes of species of the plant family Fabaceae. Planta 200:92–99

Cruz Castillo M, Martinez C, Buchala A, Métraux JP, Leon J (2004) Gene-specific involvement of β-oxidation in wound-activated responses in Arabidopsis. Plant Physiol 135:85–94

Daxberger A, Nemak A, Mithöfer A, Fliegmann J, Ligterink W, Hirt H, Ebel J (2007) Activation of members of a MAPK module in β-glucan elicitor-mediated non-host resistance of soybean. Planta 225:1559–1571

Day RB, Okada M, Ito Y, Tsukada K, Zaghouani H, Shibuya N, Stacey G (2001) Binding site for chitin oligosaccharides in the soybean plasma membrane. Plant Physiol 126:1162–1173

de Wit PJ, Brandwagt BF, van den Burg HA, Cai X, van der Hoorn RA, de Jong CF, van Klooster J, de Kock MJ, Kruijt M, Lindhout WH, Luderer R, Takken FL, Westerink N, Vervoort JJ, Joosten MH (2002) The molecular basis of co-evolution between Cladosporium fulvum and tomato. Antonie van Leeuwenhoek 81:409–412

del Pozo O, Pedley KF, Martin GB (2004) MAPKKKalpha is a positive regulator of cell death associated with both plant immunity and disease. EMBO J 23:3072–3082

Delker C, Zolman BK, Miersch O, Wasternack C (2007) Jasmonate biosynthesis in Arabidopsis thaliana requires peroxisomal β-oxidation enzymes – additional proof by properties of pex6 and aim1. Phytochemistry 68:1642–1650

Delledonne M (2005) NO news is good news for plants. Curr Opin Plant Biol 8:390–396

DeYoung BJ, Innes RW (2006) Plant NBS-LRR proteins in pathogen sensing and host defense. Nat Immunol 7:1243–1249

Diévart A, Clark SE (2004) LRR-containing receptors regulating plant development and defense. Development 131:251–261

Dong X (2004) NPR1, all things considered. Curr Opin Plant Biol 7:547–552

Durner J, Wendehenne D, Klessig DF (1998) Defense gene induction in tobacco by nitric oxide, cyclic GMP, and cyclic ADP-ribose. Proc Natl Acad Sci USA 95:10328–10333

Durrant WE, Dong X (2004) Systemic acquired resistance. Annu Rev Phytopathol 42:185–209

Ebel C, Gomez LG, Schmit AC, Neuhaus-Url G, Boller T (2001) Differential mRNA degradation of two beta-tubulin isoforms correlates with cytosolic Ca^{2+} changes in glucan-elicited soybean cells. Plant Physiol 126:87–96

Ellis C, Turner JG (2001) The Arabidopsis mutant cev1 has constitutively active jasmonate and ethylene signal pathways and enhanced resistance to pathogens. Plant Cell 13:1025–1033

Enkerli J, Felix G, Boller T (1999) The enzymatic activity of fungal xylanase is not necessary for its elicitor activity. Plant Physiol 121:391–397

Farmer EE, Davoine C (2007) Reactive electrophile species. Curr Opin Plant Biol 10:380–386

Felix G, Boller T (2003) Molecular sensing of bacteria in plants. The highly conserved RNA-binding motif

RNP-1 of bacterial cold shock proteins is recognized as an elicitor signal in tobacco. J Biol Chem 278:6201–6208

Fellbrich G, Blume B, Brunner F, Hirt H, Kroj T, Ligterink W, Romanski A, Nürnberger T (2000) *Phytophthora parasitica* elicitor-induced reactions in cells of *Petroselinum crispum*. Plant Cell Physiol 41:692–701

Fellbrich G, Romanski A, Varet A, Blume B, Brunner F, Engelhardt S, Felix G, Kemmerling B, Krzymowska M, Nürnberger T (2002) NPP1, a *Phytophthora*-associated trigger of plant defense in parsley and Arabidopsis. Plant J 32:375–390

Ferrari S, Plotnikova JM, De Lorenzo G, Ausubel FM (2003) Arabidopsis local resistance to *Botrytis cinerea* involves salicylic acid and camalexin and requires EDS4 and PAD2, but not SID2, EDS5 or PAD4. Plant J 35:193–205

Ferrarini A, De Stefano M, Baudouin E, Pucciariello C, Polverari A, Puppo A, Delledonne M (2008) Expression of *Medicago truncatula* genes responsive to nitric oxide in pathogenic and symbiotic conditions. Mol Plant–Microbe Interact 21:781–790

Fliegmann J, Mithöfer A, Wanner G, Ebel J (2004) An ancient enzyme domain hidden in the putative β-glucan elicitor receptor of soybean may play an active part in the perception of pathogen-associated molecular patterns during broad host resistance. J Biol Chem 279:1132–1140

Foissner I, Wendehenne D, Langebartels C, Durner J (2000) In vivo imaging of an elicitor-induced nitric oxide burst in tobacco. Plant J 23:817–824

Frye CA, Tang D, Innes RW (2001) Negative regulation of defense responses in plants by a conserved MAPKK kinase. Proc Natl Acad Sci USA 98:373–378

Furman-Matarasso N, Cohen E, Du QS, Chejanovsky N, Hanania U, Avni A (1999) A point mutation in the ethylene-inducing xylanase elicitor inhibits the β-1,4-endoxylanase activity but not the elicitation activity. Plant Physiol 121:345–351

Garcia-Brugger A, Lamotte O, Vandelle E, Bourque S, Lecourieux D, Poinssot B, Wendehenne D, Pugin A (2006) Early signaling events induced by elicitors of plant defenses. Mol Plant–Microbe Interact 19:711–724

Gaulin E, Drame N, Lafitte C, Torto-Alalibo T, Martinez Y, Ameline-Torregrosa C, Khatib M, Mazarguil H, Villalba-Mateos F, Kamoun S, Mazars C, Dumas B, Bottin A, Esquerre-Tugaye MT, Rickauer M (2006) Cellulose binding domains of a *Phytophthora* cell wall protein are novel pathogen-associated molecular patterns. Plant Cell 18:1766–1777

Glazebrook J (2005) Contrasting mechanisms of defense against biotrophic and necrotrophic pathogens. Annu Rev Phytopathol 43:205–227

Gomez-Gomez L, Boller T (2000) FLS2: An LRR receptor-like kinase involved in the perception of the bacterial elicitor flagellin in *Arabidopsis*. Mol Cell 5:1003–1011

Govrin EM, Levine A (2002) Infection of Arabidopsis with a necrotrophic pathogen, *Botrytis cinerea*, elicits various defense responses but does not induce systemic acquired resistance (SAR). Plant Mol Biol 48:267–276

Granado J, Felix G, Boller T (1995) Perception of fungal sterols in plants: subnanomolar concentrations of ergosterol elicit extracellular alkalinization in tomato cells. Plant Physiol 107:485–490

Gustin MC, Albertyn J, Alexander M, Davenport K (1998) MAP kinase pathways in the yeast *Saccharomyces cerevisiae*. Microbiol Mol Biol Rev 62:1264–1300

Hadwiger LA, Beckman JM (1980) Chitosan as a component of pea-*Fusarium solani* interactions. Plant Physiol 66:205–211

Halim VA, Hunger A, Macioszek V, Landgraf P, Nürnberger T, Scheel D, Rosahl S (2004) The oligopeptide elicitor Pep-13 induces salicylic acid-dependent and -independent defense reactions in potato. Physiol Mol Plant Pathol 64:311–318

Halim VA, Eschen-Lippold L, Altmann S, Birschwilks M, Scheel D, Rosahl S (2007) Salicylic acid is important for basal defense of *Solanum tuberosum* against *Phytophthora infestans*. Mol Plant–Microbe Interact 20:1346–1352

Hamiduzzaman MM, Jakab G, Barnavon L, Neuhaus JM, Mauch-Mani B (2005) β-Aminobutyric acid-induced resistance against downy mildew in grapevine acts through the potentiation of callose formation and jasmonic acid signaling. Mol Plant–Microbe Interact 18:819–829

Hanania U, Furman-Matarasso N, Ron M, Avni A (1999) Isolation of a novel SUMO protein from tomato that suppresses EIX-induced cell death. Plant J 19:533–541

He ZH, Wang ZY, Li JM, Zhu Q, Lamb C, Ronald P, Chory J (2000) Perception of brassinosteroids by the extracellular domain of the receptor kinase BRI1. Science 288:2360–2363

Heck S, Grau T, Buchala A, Metraux JP, Nawrath C (2003) Genetic evidence that expression of *NahG* modifies defence pathways independent of salicylic acid biosynthesis in the Arabidopsis-*Pseudomonas syringae* pv. *tomato* interaction. Plant J 36:342–352

Heo WD, Lee SH, Kim MC, Kim JC, Chung WS, Chun HJ, Lee KJ, Park CY, Park HC, Choi JY, Cho MJ (1999) Involvement of specific calmodulin isoforms in salicylic acid-independent activation of plant disease resistance responses. Proc Natl Acad Sci USA 96:766–771

Hrabak EM, Chan CW, Gribskov M, Harper JF, Choi JH, Halford N, Kudla J, Luan S, Nimmo HG, Sussman MR, Thomas M, Walker-Simmons K, Zhu JK, Harmon AC (2003) The Arabidopsis CDPK-SnRK superfamily of protein kinases. Plant Physiol 132:666–680

Hückelhoven R, Kogel KH (2003) Reactive oxygen intermediates in plant-microbe interactions: who is who in powdery mildew resistance? Planta 216:891–902

Hyun Y, Choi S, Hwang HJ, Yu J, Nam SJ, Ko J, Park JY, Seo YS, Kim EY, Ryu SB, Kim WT, Lee YH, Kang H, Lee I (2008) Cooperation and functional diversification of two closely related galactolipase genes for jasmonate biosynthesis. Dev Cell 14:183–192

Ichimura K, Casais C, Peck SC, Shinozaki K, Shirasu K (2006) MEKK1 is required for MPK4 activation and regulates tissue-specific and temperature-dependent cell death in Arabidopsis. J Biol Chem 281:36969–36976

Jabs T, Tschöpe M, Colling C, Hahlbrock K, Scheel D (1997) Elicitor-stimulated ion fluxes and O^{2-} from the oxidative burst are essential components in triggering defense gene activation and phytoalexin synthesis in parsley. Proc Natl Acad Sci USA 94:4800–4805

Jambunathan N, Siani JM, McNellis TW (2001) A humidity-sensitive Arabidopsis copine mutant exhibits precocious cell death and increased disease resistance. Plant Cell 13:2225–2240

Jin H, Axtell MJ, Dahlbeck D, Ekwenna O, Zhang S, Staskawicz B, Baker B (2002) NPK1, an MEKK1-like mitogen-activated protein kinase kinase kinase, regulates innate immunity and development in plants. Dev Cell 3:291–297

Jones JDG, Dangl JL (2006) The plant immune system. Nature 444:323–329

Joosten MHAJ, de Wit PJGM (1999) The tomato–Cladosporium fulvum interaction: a versatile experimental system to study plant-pathogen interactions. Annu Rev Phytopathol 37:335–367

Jurkowski GI, Smith RK, Jr., Yu IC, Ham JH, Sharma SB, Klessig DF, Fengler KA, Bent AF (2004) Arabidopsis DND2, a second cyclic nucleotide-gated ion channel gene for which mutation causes the "defense, no death" phenotype. Mol Plant–Microbe Interact 17:511–520

Kadota Y, Furuichi T, Ogasawara Y, Goh T, Higashi K, Muto S, Kuchitsu K (2004) Identification of putative voltage-dependent Ca^{2+}-permeable channels involved in cryptogein-induced Ca^{2+} transients and defense responses in tobacco BY-2 cells. Biochem Biophys Res Commun 317:823–830

Kaku H, Nishizawa Y, Ishii-Minami N, Akimoto-Tomiyama C, Dohmae N, Takio K, Minami E, Shibuya N (2006) Plant cells recognize chitin fragments for defense signaling through a plasma membrane receptor. Proc Natl Acad Sci USA 103:11086–11091

Kanneganti TD, Huitema E, Cakir C, Kamoun S (2006) Synergistic interactions of plant cell death pathways induced by Phytophthora infestans Nep1-like protein PiNPP1.1 and INF1 elicitin. Mol Plant–Microbe Interact 19:854–863

Kasparovsky T, Blein JP, Mikes V (2004) Ergosterol elicits oxidative burst in tobacco cells via phospholipase A2 and protein kinase C signal pathway. Plant Physiol Biochem 42:429–435

Katou S, Yoshioka H, Kawakita K, Rowland O, Jones JD, Mori H, Doke N (2005) Involvement of PPS3 phosphorylated by elicitor-responsive mitogen-activated protein kinases in the regulation of plant cell death. Plant Physiol 139:1914–1926

Kawasaki T, Henmi K, Ono E, Hatakeyama S, Iwano M, Satoh H, Shimamoto K (1999) The small GTP-binding protein rac is a regulator of cell death in plants. Proc Natl Acad Sci USA 96:10922–10926

Kawchuk LM, Hachey J, Lynch DR, Kulcsar F, van Rooijen G, Waterer DR, Robertson A, Kokko E, Byers R, Howard RJ, Fischer R, Prüfer D (2001) Tomato Ve disease resistance genes encode cell surface-like receptors. Proc Natl Acad Sci USA 98:6511–6515

Kemmerling B, Schwedt A, Rodriguez P, Mazzotta S, Frank M, Abu Qamar S, Mengiste T, Betsuyaku S, Parker JE, Mussig C, Thomma BPHJ, Albrecht C, de Vries SC, Hirt H, Nürnberger T (2007) The BRI1-associated kinase 1, BAK1, has a brassinolide-independent role in plant cell-death control. Curr Biol 17:1116–1122

Khatib M, Lafitte C, Esquerre-Tugaye M, Bottin A, Rickauer M (2004) The CBEL elicitor of Phytophthora parasitica var. nicotianae activates defence in Arabidopsis thaliana via three different signalling pathways. New Phytol 162:501–510

Kienow L, Schneider K, Bartsch M, Stuible HP, Weng H, Miersch O, Wasternack C, Kombrink E (2008) Jasmonates meet fatty acids: functional analysis of a new acyl-coenzyme A synthetase family from Arabidopsis thaliana. J Exp Bot 59:403–419

Kim MC, Panstruga R, Elliott C, Muller J, Devoto A, Yoon HW, Park HC, Cho MJ, Schulze-Lefert P (2002) Calmodulin interacts with MLO protein to regulate defence against mildew in barley. Nature 416: 447–451

Kim CY, Liu Y, Thorne ET, Yang H, Fukushige H, Gassmann W, Hildebrand D, Sharp RE, Zhang S (2003) Activation of a stress-responsive mitogen-activated protein kinase cascade induces the biosynthesis of ethylene in plants. Plant Cell 15:2707–2718

Kloek AP, Verbsky ML, Sharma SB, Schoelz JE, Vogel J, Klessig DF, Kunkel BN (2001) Resistance to Pseudomonas syringae conferred by an Arabidopsis thaliana coronatine-insensitive (coi1) mutation occurs through two distinct mechanisms. Plant J 26:509–522

Klüsener B, Young JJ, Murata Y, Allen GJ, Mori IC, Hugouvieux V, Schroeder JI (2002) Convergence of calcium signaling pathways of pathogenic elicitors and abscisic acid in Arabidopsis guard cells. Plant Physiol 130:2152–2163

Kobayashi M, Ohura I, Kawakita K, Yokota N, Fujiwara M, Shimamoto K, Doke N, Yoshioka H (2007) Calcium-dependent protein kinases regulate the production of reactive oxygen species by potato NADPH oxidase. Plant Cell 19:1065–1080

Koga J, Yamauchi T, Shimura M, Ogawa N, Oshima K, Umemura K, Kikuchi M, Ogasawara N (1998) Cerebrosides A and C, sphingolipid elicitors of hypersensitive cell death and phytoalexin accumulation in rice plants. J Biol Chem 273:31985–31991

Kolukisaoglu U, Weinl S, Blazevic D, Batistic O, Kudla J (2004) Calcium sensors and their interacting protein kinases: genomics of the Arabidopsis and rice CBL-CIPK signaling networks. Plant Physiol 134:43–58

Koo AJ, Chung HS, Kobayashi Y, Howe GA (2006) Identification of a peroxisomal acyl-activating enzyme involved in the biosynthesis of jasmonic acid in Arabidopsis. J Biol Chem 281:33511–33520

Kopka J, Pical C, Hetherington AM, Müller-Röber B (1998) Ca^{2+}/phospholipid-binding (C2) domain in multiple plant proteins: novel components of the calcium-sensing apparatus. Plant Mol Biol 36:627–637

Kroj T, Rudd JJ, Nürnberger T, Gäbler Y, Lee J, Scheel D (2003) Mitogen-activated protein kinases play an essential role in oxidative burst-independent expression of pathogenesis-related genes in parsley. J Biol Chem 278:2256–2264

Kruijt M, de Kock MJD, de Wit PJGM (2005) Receptor-like proteins involved in plant disease resistance. Mol Plant Pathol 6:85–97

Kumar D, Klessig DF (2000) Differential induction of tobacco MAP kinases by the defense signals nitric oxide, salicylic acid, ethylene, and jasmonic acid. Mol Plant–Microbe Interact 13:347–351

Kunze G, Zipfel C, Robatzek S, Niehaus K, Boller T, Felix G (2004) The N terminus of bacterial elongation factor Tu elicits innate immunity in Arabidopsis plants. Plant Cell 16:3496–3507

Kurusu T, Yagala T, Miyao A, Hirochika H, Kuchitsu K (2005) Identification of a putative voltage-gated Ca²⁺ channel as a key regulator of elicitor-induced hypersensitive cell death and mitogen-activated protein kinase activation in rice. Plant J 42:798–809

Lamotte O, Gould K, Lecourieux D, Sequeira-Legrand A, Lebrun-Garcia A, Durner J, Pugin A, Wendehenne D (2004) Analysis of nitric oxide signaling functions in tobacco cells challenged by the elicitor cryptogein. Plant Physiol 135:516–529

Laudert D, Pfannschmidt U, Lottspeich F, Holländer-Czytko H, Weiler EW (1996) Cloning, molecular and functional characterization of Arabidopsis thaliana allene oxide synthase (CYP 74), the first enzyme of the octadecanoid pathway to jasmonates. Plant Mol Biol 31:323–335

Laxalt AM, Munnik T (2002) Phospholipid signalling in plant defence. Curr Opin Plant Biol 5:332–338

Lebrun-Garcia A, Ouaked F, Chiltz A, Pugin A (1998) Activation of MAPK homologues by elicitors in tobacco cells. Plant J 15:773–781

Lebrun-Garcia A, Chiltz A, Gout E, Bligny R, Pugin A (2002) Questioning the role of salicylic acid and cytosolic acidification in mitogen-activated protein kinase activation induced by cryptogein in tobacco cells. Planta 214:792–797

Lecourieux-Ouaked F, Pugin A, Lebrun-Garcia A (2000) Phosphoproteins involved in the signal transduction of cryptogein, an elicitor of defense reactions in tobacco. Mol Plant–Microbe Interact 13:821–829

Lecourieux D, Mazars C, Pauly N, Ranjeva R, Pugin A (2002) Analysis and effects of cytosolic free calcium increases in response to elicitors in Nicotiana plumbaginifolia cells. Plant Cell 14:2627–2641

Lecourieux D, Lamotte O, Bourque S, Wendehenne D, Mazars C, Ranjeva R, Pugin A (2005) Proteinaceous and oligosaccharidic elicitors induce different calcium signatures in the nucleus of tobacco cells. Cell Calcium 38:527–538

Lecourieux D, Ranjeva R, Pugin A (2006) Calcium in plant defence-signalling pathways. New Phytol 171:249–269

Lee DE, Lee IJ, Han O, Baik MG, Han SS, Back K (2004) Pathogen resistance of transgenic rice plants expressing mitogen-activated protein kinase 1, MK1, from Capsicum annuum. Mol Cell 17:81–85

Lee J, Klessig DF, Nürnberger T (2001) A harpin binding site in tobacco plasma membranes mediates activation of the pathogenesis-related gene HIN1 independent of extracellular calcium but dependent on mitogen-activated protein kinase activity. Plant Cell 13:1079–1093

Li C, Schilmiller AL, Liu G, Lee GI, Jayanty S, Sageman C, Vrebalov J, Giovannoni JJ, Yagi K, Kobayashi Y, Howe GA (2005) Role of β-oxidation in jasmonate biosynthesis and systemic wound signaling in tomato. Plant Cell 17:971–986

Li H, Xu H, Zhou Y, Zhang J, Long C, Li S, Chen S, Zhou JM, Shao F (2007) The phosphothreonine lyase activity of a bacterial type III effector family. Science 315:1000–1003

Li J, Chory J (1997) A putative leucine-rich repeat receptor kinase involved in brassinosteroid signal transduction. Cell 90:929–938

Li J, Wen JQ, Lease KA, Doke JT, Tax FE, Walker JC (2002) BAK1, an Arabidopsis LRR receptor-like protein kinase, interacts with BRI1 and modulates brassinosteroid signaling. Cell 110:213–222

Li J, Brader G, Palva ET (2004) The WRKY70 transcription factor: a node of convergence for jasmonate-mediated and salicylate-mediated signals in plant defense. Plant Cell 16:319–331

Li J, Brader G, Kariola T, Palva ET (2006) WRKY70 modulates the selection of signaling pathways in plant defense. Plant J 46:477–491

Lieberherr D, Thao NP, Nakashima A, Umemura K, Kawasaki T, Shimamoto K (2005) A sphingolipid elicitor-inducible mitogen-activated protein kinase is regulated by the small GTPase OsRac1 and heterotrimeric G-protein in rice 1. Plant Physiol 138:1644–1652

Ligterink W, Kroj T, zur Nieden U, Hirt H, Scheel D (1997) Receptor-mediated activation of a MAP kinase in pathogen defense of plants. Science 276:2054–2057

Lin WC, Lu CF, Wu JW, Cheng ML, Lin YM, Yang NS, Black L, Green SK, Wang JF, Cheng CP (2004) Transgenic tomato plants expressing the Arabidopsis NPR1 gene display enhanced resistance to a spectrum of fungal and bacterial diseases. Transgen Res 13:567–581

Lindermayr C, Saalbach G, Durner J (2005) Proteomic identification of S-nitrosylated proteins in Arabidopsis. Plant Physiol 137:921–930

Lindermayr C, Saalbach G, Bahnweg G, Durner J (2006) Differential inhibition of Arabidopsis methionine adenosyltransferases by protein S-nitrosylation. J Biol Chem 281:4285–4291

Liu F, Yoo BC, Lee JY, Pan W, Harmon AC (2006) Calcium-regulated phosphorylation of soybean serine acetyltransferase in response to oxidative stress. J Biol Chem 281:27405–27415

Liu Y, Zhang S (2004) Phosphorylation of 1-aminocyclopropane-1-carboxylic acid synthase by MPK6, a stress-responsive mitogen-activated protein kinase, induces ethylene biosynthesis in Arabidopsis. Plant Cell 16:3386–3399

Lorenzo O, Solano R (2005) Molecular players regulating the jasmonate signalling network. Curr Opin Plant Biol 8:532–540

Lorenzo O, Chico JM, Sanchez-Serrano JJ, Solano R (2004) JASMONATE-INSENSITIVE1 encodes a MYC transcription factor essential to discriminate between different jasmonate-regulated defense responses in Arabidopsis. Plant Cell 16:1938–1950

Luan S, Kudla J, Rodriguez-Concepcion M, Yalovsky S, Gruissem W (2002) Calmodulins and calcineurin B-like proteins: calcium sensors for specific signal response coupling in plants. Plant Cell 14 [Suppl]:S389–S400

Mackey D, McFall AJ (2006) MAMPs and MIMPs: proposed classifications for inducers of innate immunity. Mol Microbiol 61:1365–1371

Makandar R, Essig JS, Schapaugh MA, Trick HN, Shah J (2006) Genetically engineered resistance to Fusarium head blight in wheat by expression of Arabidopsis NPR1. Mol Plant–Microbe Interact 19:123–129

Malnoy M, Jin Q, Borejsza-Wysocka EE, He SY, Aldwinckle HS (2007) Overexpression of the apple *MpNPR1* gene confers increased disease resistance in *Malus x domestica*. Mol Plant–Microbe Interact 20:1568–1580

Mayrose M, Bonshtien A, Sessa G (2004) LeMPK3 is a mitogen-activated protein kinase with dual specificity induced during tomato defense and wounding responses. J Biol Chem 279:14819–14827

McDowell JM, Williams SG, Funderburg NT, Eulgem T, Dangl JL (2005) Genetic analysis of developmentally regulated resistance to downy mildew (*Hyaloperonospora parasitica*) in *Arabidopsis thaliana*. Mol Plant–Microbe Interact 18:1226–1234

Medzhitov R, Janeway CA (1997) Innate immunity: the virtues of a nonclonal system of recognition. Cell 91:295–298

Mei C, Qi M, Sheng G, Yang Y (2006) Inducible overexpression of a rice allene oxide synthase gene increases the endogenous jasmonic acid level, PR gene expression, and host resistance to fungal infection. Mol Plant–Microbe Interact 19:1127–1137

Menke FL, van Pelt JA, Pieterse CM, Klessig DF (2004) Silencing of the mitogen-activated protein kinase MPK6 compromises disease resistance in Arabidopsis. Plant Cell 16:897–907

Mészáros T, Helfer A, Hatzimasoura E, Magyar Z, Serazetdinova L, Rios G, Bardóczy V, Teige M, Koncz C, Peck S, Bögre L (2006) The Arabidopsis MAP kinase kinase MKK1 participates in defence responses to the bacterial elicitor flagellin. Plant J 48:485–498

Meyers BC, Kozik A, Griego A, Kuang HH, Michelmore RW (2003) Genome-wide analysis of NBS-LRR-encoding genes in *Arabidopsis*. Plant Cell 15:809–834

Mithöfer A, Mazars C (2002) Aequorin-based measurements of intracellular Ca^{2+}-signatures in plant cells. Biol Proced Online 4:105–118

Mithöfer A, Lottspeich F, Ebel J (1996) One-step purification of the β-glucan elicitor-binding protein from soybean (*Glycine max* L) roots and characterization of an anti-peptide antiserum. FEBS Lett 381:203–207

Mithöfer A, Ebel J, Bhagwat AA, Boller T, Neuhaus-Url G (1999) Transgenic aequorin monitors cytosolic calcium transients in soybean cells challenged with β-glucan or chitin elicitors. Planta 207:566–574

Mithöfer A, Fliegmann J, Neuhaus-Url G, Schwarz H, Ebel J (2000) The hepta-β-glucoside elicitor-binding proteins from legumes represent a putative receptor family. Biol Chem 381:705–713

Mithöfer A, Ebel J, Felle HH (2005) Cation fluxes cause plasma membrane depolarization involved in beta-glucan elicitor-signaling in soybean roots. Mol Plant–Microbe Interact 18:983–990

Miya A, Albert P, Shinya T, Desaki Y, Ichimura K, Shirasu K, Narusaka M, Kawakami N, Kaku H, Shibuya N (2007) CERK1, a LysM receptor kinase, is essential for chitin elicitor signaling in *Arabidopsis*. Proc Natl Acad Sci USA 104:19613–19618

Montillet JL, Chamnongpol S, Rusterucci C, Dat J, van de Cotte B, Agnel JP, Battesti C, Inze D, Van Breusegem F, Triantaphylides C (2005) Fatty acid hydroperoxides and H_2O_2 in the execution of hypersensitive cell death in tobacco leaves. Plant Physiol 138:1516–1526

Mukherjee S, Keitany G, Li Y, Wang Y, Ball HL, Goldsmith EJ, Orth K (2006) *Yersinia* YopJ acetylates and inhibits kinase activation by blocking phosphorylation. Science 312:1211–1214

Mukherjee S, Hao YH, Orth K (2007) A newly discovered post-translational modification – the acetylation of serine and threonine residues. Trends Biochem Sci 32:210–216

Nakagami H, Soukupova H, Schikora A, Zarsky V, Hirt H (2006) A mitogen-activated protein kinase kinase kinase mediates reactive oxygen species homeostasis in Arabidopsis. J Biol Chem 281:38697–38704

Nam KH, Li JM (2002) BRI1/BAK1, a receptor kinase pair mediating brassinosteroid signaling. Cell 110:203–212

Navazio L, Moscatiello R, Genre A, Novero M, Baldan B, Bonfante P, Mariani P (2007) A diffusible signal from arbuscular mycorrhizal fungi elicits a transient cytosolic calcium elevation in host plant cells. Plant Physiol 144:673–681

Ndamukong I, Abdallat AA, Thurow C, Fode B, Zander M, Weigel R, Gatz C (2007) SA-inducible Arabidopsis glutaredoxin interacts with TGA factors and suppresses JA-responsive PDF1.2 transcription. Plant J 50:128–139

Nühse TS, Peck SC, Hirt H, Boller T (2000) Microbial elicitors induce activation and dual phosphorylation of the *Arabidopsis thaliana* MAPK 6. J Biol Chem 275:7521–7526

Nühse TS, Bottrill AR, Jones AM, Peck SC (2007) Quantitative phosphoproteomic analysis of plasma membrane proteins reveals regulatory mechanisms of plant innate immune responses. Plant J 51:931–940

Nürnberger T, Kemmerling B (2006) Receptor protein kinases – pattern recognition receptors in plant immunity. Trends Plant Sci 11:519–522

Nürnberger T, Nennstiel D, Jabs T, Sacks WR, Hahlbrock K, Scheel D (1994) High affinity binding of a fungal oligopeptide elicitor to parsley plasma membranes triggers multiple defense responses. Cell 78:449–460

Nürnberger T, Brunner F, Kemmerling B, Piater L (2004) Innate immunity in plants and animals: striking similarities and obvious differences. Immunol Rev 198:249–266

Okada M, Matsumura M, Ito Y, Shibuya N (2002) High-affinity binding proteins for N-acetylchitooligosaccharide elicitor in the plasma membranes from wheat, barley and carrot cells: conserved presence and correlation with the responsiveness to the elicitor. Plant Cell Physiol43:505–512

Orth K, Xu Z, Mudgett MB, Bao ZQ, Palmer LE, Bliska JB, Mangel WF, Staskawicz B, Dixon JE (2000) Disruption of signaling by *Yersinia* effector YopJ, a ubiquitin-like protein protease. Science 290:1594–1597

Pagnussat GC, Lanteri ML, Lombardo MC, Lamattina L (2004) Nitric oxide mediates the indole acetic acid induction activation of a mitogen-activated protein kinase cascade involved in adventitious root development. Plant Physiol 135:279–286

Parani M, Rudrabhatla S, Myers R, Weirich H, Smith B, Leaman DW, Goldman SL (2004) Microarray analysis of nitric oxide responsive transcripts in Arabidopsis. Plant Biotechnol J 2:359–366

Peiter E, Maathuis FJ, Mills LN, Knight H, Pelloux J, Hetherington AM, Sanders D (2005) The vacuolar Ca^{2+}-activated channel TPC1 regulates germination and stomatal movement. Nature 434:404–408

Pemberton CL, Salmond GPC (2004) The Nep1-like proteins – a growing family of microbial elicitors of plant necrosis. Mol Plant Pathol 5:353–359

Petersen M, Brodersen P, Naested H, Andreasson E, Lindhart U, Johansen B, Nielsen HB, Lacy M, Austin MJ, Parker JE, Sharma SB, Klessig DF, Martienssen R, Mattsson O, Jensen AB, Mundy J (2000) Arabidopsis map kinase 4 negatively regulates systemic acquired resistance. Cell 103:1111–1120

Planchet E, Kaiser WM (2006) Nitric oxide (NO) detection by DAF fluorescence and chemiluminescence: a comparison using abiotic and biotic NO sources. J Exp Bot 57:3043–3055

Planchet E, Sonoda M, Zeier J, Kaiser WM (2006) Nitric oxide (NO) as an intermediate in the cryptogein-induced hypersensitive response – a critical re-evaluation. Plant Cell Environ 29:59–69

Plieth C (2005) Calcium: just another regulator in the machinery of life? Ann Bot 96:1–8

Poinssot B, Vandelle E, Bentejac M, Adrian M, Levis C, Brygoo Y, Garin J, Sicilia F, Coutos-Thevenot P, Pugin A (2003) The endopolygalacturonase 1 from Botrytis cinerea activates grapevine defense reactions unrelated to its enzymatic activity. Mol Plant–Microbe Interact 16:553–564

Polverari A, Molesini B, Pezzotti M, Buonaurio R, Marte M, Delledonne M (2003) Nitric oxide-mediated transcriptional changes in Arabidopsis thaliana. Mol Plant–Microbe Interact 16:1094–1105

Pozo MJ, Van Loon LC, Pieterse CMJ (2004) Jasmonates – signals in plant-microbe interactions. J Plant Growth Regul 23:211–222

Prost I, Dhondt S, Rothe G, Vicente J, Rodriguez MJ, Kift N, Carbonne F, Griffiths G, Esquerre-Tugaye MT, Rosahl S, Castresana C, Hamberg M, Fournier J (2005) Evaluation of the antimicrobial activities of plant oxylipins supports their involvement in defense against pathogens. Plant Physiol 139:1902–1913

Pugin A, Frachisse JM, Tavernier E, Bligny R, Gout E, Douce R, Guern J (1997) Early events induced by the elicitor cryptogein in tobacco cells: involvement of a plasma membrane NADPH oxidase and activation of glycolysis and the pentose phosphate pathway. Plant Cell 9:2077–2091

Qutob D, Kemmerling B, Brunner F, Kufner I, Engelhardt S, Gust AA, Luberacki B, Seitz HU, Stahl D, Rauhut T, Glawischnig E, Schween G, Lacombe B, Watanabe N, Lam E, Schlichting R, Scheel D, Nau K, Dodt G, Hubert D, Gijzen M, Nürnberger T (2006) Phytotoxicity and innate immune responses induced by Nep1-like proteins. Plant Cell 18:3721–3744

Raacke IC, Mueller MJ, Berger S (2006) Defects in allene oxide synthase and 12-oxo-phytodienoic acid reductase alter the resistance to Pseudomonas syringae and Botrytis cinerea. J Phytopathol 154:740–744

Radutoiu S, Madsen LH, Madsen EB, Felle HH, Umehara Y, Gronlund M, Sato S, Nakamura Y, Tabata S, Sandal N, Stougaard J (2003) Plant recognition of symbiotic bacteria requires two LysM receptor-like kinases. Nature 425:585–592

Ranf S, Wünnenberg P, Lee J, Becker D, Dunkel M, Hedrich R, Scheel D, Dietrich P (2008) Loss of the vacuolar cation channel, AtTPC1, does not impair Ca^{2+} signals induced by abiotic and biotic stresses. Plant J 53:287–299

Ren D, Yang H, Zhang S (2002) Cell death mediated by MAPK is associated with hydrogen peroxide production in Arabidopsis. J Biol Chem 277:559–565

Reuber TL, Plotnikova JM, Dewdney J, Rogers EE, Wood W, Ausubel FM (1998) Correlation of defense gene induction defects with powdery mildew susceptibility in Arabidopsis enhanced disease susceptibility mutants. Plant J 16:473–485

Ribnicky DM, Shulaev VV, Raskin II (1998) Intermediates of salicylic acid biosynthesis in tobacco. Plant Physiol 118:565–572

Ricci P, Bonnet P, Huet JC, Sallantin M, Beauvais-Cante F, Bruneteau M, Billard V, Michel G, Pernollet JC (1989) Structure and activity of proteins from pathogenic fungi Phytophthora eliciting necrosis and acquired resistance in tobacco. Eur J Biochem/FEBS 183:555–563

Robatzek S, Chinchilla D, Boller T (2006) Ligand-induced endocytosis of the pattern recognition receptor FLS2 in Arabidopsis. Genes Dev 20:537–542

Romeis T, Piedras P, Zhang S, Klessig DF, Hirt H, Jones JD (1999) Rapid Avr9- and Cf-9-dependent activation of MAP kinases in tobacco cell cultures and leaves: convergence of resistance gene, elicitor, wound, and salicylate responses. Plant Cell 11:273–287

Romeis T, Ludwig AA, Martin R, Jones JD (2001) Calcium-dependent protein kinases play an essential role in a plant defence response. EMBO J 20:5556–5567

Ron M, Avni A (2004) The receptor for the fungal elicitor ethylene-inducing xylanase is a member of a resistance-like gene family in tomato. Plant Cell 16:1604–1615

Rudd JJ, Franklin-Tong VE (1999) Calcium signalling in plants. Cell Mol Life Sci 55:214–232

Sagi M, Fluhr R (2006) Production of reactive oxygen species by plant NADPH oxidases. Plant Physiol 141:336–340

Sanders D, Pelloux J, Brownlee C, Harper JF (2002) Calcium at the crossroads of signaling. Plant Cell 14 [Suppl]: S401–S417

Schaller F, Biesgen C, Mussig C, Altmann T, Weiler EW (2000) 12-Oxophytodienoate reductase 3 (OPR3) is the isoenzyme involved in jasmonate biosynthesis. Planta 210:979–984

Schilmiller AL, Koo AJ, Howe GA (2007) Functional diversification of acyl-coenzyme A oxidases in jasmonic acid biosynthesis and action. Plant Physiol 143:812–824

Schmidt WE, Ebel J (1987) Specific binding of a fungal glucan phytoalexin elicitor to membrane fractions from soybean Glycine max. Proc Natl Acad Sci USA 84:4117–4121

Schneider K, Kienow L, Schmelzer E, Colby T, Bartsch M, Miersch O, Wasternack C, Kombrink E, Stuible HP (2005) A new type of peroxisomal acyl-coenzyme A synthetase from *Arabidopsis thaliana* has the catalytic capacity to activate biosynthetic precursors of jasmonic acid. J Biol Chem 280:13962–13972

Schweizer P, Gees R, Mösinger E (1993) Effect of jasmonic acid on the interaction of barley (*Hordeum vulgare* L.) with the powdery mildew *Erysiphe graminis* f.sp. *hordei*. Plant Physiol 102:503–511

Seo HS, Song JT, Cheong JJ, Lee YH, Lee YW, Hwang I, Lee JS, Choi YD (2001) Jasmonic acid carboxyl methyltransferase: a key enzyme for jasmonate-regulated plant responses. Proc Natl Acad Sci USA 98:4788–4793

Sharp JK, McNeil M, Albersheim P (1984) The primary structure of one elicitor-active and seven elicitor-inactive hexa(β-D-glucopyranosyl)-D-glucitols isolated from the mycelial walls of *Phytophthora megasperma* f.sp. *glycinea*. J Biol Chem 259:11321–11336

Shibuya N, Kaku H, Kuchitsu K, Maliarik MJ (1993) Identification of a novel high-affinity binding site for *N*-acetylchitooligosaccharide elicitor in the membrane fraction from suspension-cultured rice cells. FEBS Lett 329:75–78

Simon-Plas F, Elmayan T, Blein JP (2002) The plasma membrane oxidase NtrbohD is responsible for AOS production in elicited tobacco cells. Plant J 31:137–147

Song D, Chen J, Song F, Zheng Z (2006) A novel rice MAPK gene, OsBIMK2, is involved in disease-resistance responses. Plant Biol 8:587–596

Spoel SH, Koornneef A, Claessens SM, Korzelius JP, Van Pelt JA, Mueller MJ, Buchala AJ, Metraux JP, Brown R, Kazan K, Van Loon LC, Dong X, Pieterse CM (2003) NPR1 modulates cross-talk between salicylate- and jasmonate-dependent defense pathways through a novel function in the cytosol. Plant Cell 15:760–770

Spoel SH, Johnson JS, Dong X (2007) Regulation of tradeoffs between plant defenses against pathogens with different lifestyles. Proc Natl Acad Sci USA 104:18842–18847

Stacey G, Shibuya N (1997) Chitin recognition in rice and legumes. Plant Soil 194:161–169

Staswick PE, Yuen GY, Lehman CC (1998) Jasmonate signaling mutants of Arabidopsis are susceptible to the soil fungus *Pythium irregulare*. Plant J 15:747–754

Staswick PE, Tiryaki I, Rowe ML (2002) Jasmonate response locus *JAR1* and several related Arabidopsis genes encode enzymes of the firefly luciferase superfamily that show activity on jasmonic, salicylic, and indole-3-acetic acids in an assay for adenylation. Plant Cell 14:1405–1415.

Stintzi A, Weber H, Reymond P, Browse J, Farmer EE (2001) Plant defense in the absence of jasmonic acid: the role of cyclopentenones. Proc Natl Acad Sci USA 98:12837–12842

Stulemeijer IJ, Stratmann JW, Joosten MH (2007) Tomato mitogen-activated protein kinases LeMPK1, LeMPK2, and LeMPK3 are activated during the Cf-4/Avr4-induced hypersensitive response and have distinct phosphorylation specificities. Plant Physiol 144:1481–1494

Suarez-Rodriguez MC, Adams-Phillips L, Liu Y, Wang H, Su SH, Jester PJ, Zhang S, Bent AF, Krysan PJ (2007) MEKK1 is required for flg22-induced MPK4 activation in Arabidopsis plants. Plant Physiol 143:661–669

Suzuki K, Yano A, Shinshi H (1999) Slow and prolonged activation of the p47 protein kinase during hypersensitive cell death in a culture of tobacco cells. Plant Physiol 119:1465–1472

Takahashi F, Yoshida R, Ichimura K, Mizoguchi T, Seo S, Yonezawa M, Maruyama K, Yamaguchi-Shinozaki K, Shinozaki K (2007a) The mitogen-activated protein kinase cascade MKK3-MPK6 is an important part of the jasmonate signal transduction pathway in Arabidopsis. Plant Cell 19:805–818

Takahashi Y, Nasir KH, Ito A, Kanzaki H, Matsumura H, Saitoh H, Fujisawa S, Kamoun S, Terauchi R (2007b) A high-throughput screen of cell-death-inducing factors in *Nicotiana benthamiana* identifies a novel MAPKK that mediates INF1-induced cell death signaling and non-host resistance to *Pseudomonas cichorii*. Plant J 49:1030–1040

Thines B, Katsir L, Melotto M, Niu Y, Mandaokar A, Liu G, Nomura K, He SY, Howe GA, Browse J (2007) JAZ repressor proteins are targets of the SCF(COI1) complex during jasmonate signalling. Nature 448:661–665

Thomma B, Eggermont K, Penninckx I, Mauch-Mani B, Vogelsang R, Cammue BPA, Broekaert WF (1998) Separate jasmonate-dependent and salicylate-dependent defense-response pathways in *Arabidopsis* are essential for resistance to distinct microbial pathogens. Proc Natl Acad Sci USA 95:15107–15111

Ton J, Mauch-Mani B (2004) Beta-amino-butyric acid-induced resistance against necrotrophic pathogens is based on ABA-dependent priming for callose. Plant J 38:119–130

Torres MA, Dangl JL (2005) Functions of the respiratory burst oxidase in biotic interactions, abiotic stress and development. Curr Opin Plant Biol 8:397–403

Torres MA, Dangl JL, Jones JD (2002) Arabidopsis gp[91phox] homologues AtrbohD and AtrbohF are required for accumulation of reactive oxygen intermediates in the plant defense response. Proc Natl Acad Sci USA 99:517–522

Umemoto N, Kakitani M, Iwamatsu A, Yoshikawa M, Yamaoka N, Ishida I (1997) The structure and function of a soybean β-glucan-elicitor-binding protein. Proc Natl Acad Sci USA 94:1029–1034

Umemura K, Ogawa N, Koga J, Iwata M, Usami H (2002) Elicitor activity of cerebroside, a sphingolipid elicitor, in cell suspension cultures of rice. Plant Cell Physiol 43:778–784

Van Baarlen P, Staats M, Van Kan J (2004) Induction of programmed cell death in lily by the fungal pathogen *Botrytis elliptica*. Mol Plant Pathol 5:559–574

Van Breusegem F, Bailey-Serres J, Mittler R (2008) Unraveling the tapestry of networks involving reactive oxygen species in plants. Plant Physiol 147:978–984

van der Biezen EA, Jones JDG (1998) Plant disease-resistance proteins and the gene-for-gene concept. Trends Biochem Sci 23:454–456

van Wees SC, Glazebrook J (2003) Loss of non-host resistance of Arabidopsis NahG to *Pseudomonas syringae* pv. *phaseolicola* is due to degradation products of salicylic acid. Plant J 33:733–742

Vandelle E, Poinssot B, Wendehenne D, Bentejac M, Pugin A (2006) Integrated signaling network involving calcium, nitric oxide, and active oxygen species but not mitogen-activated protein kinases in BcPG1-elicited grapevine defenses. Mol Plant–Microbe Interact 19:429–440

Vidal G, Ribas-Carbo M, Garmier M, Dubertret G, Rasmusson AG, Mathieu C, Foyer CH, De Paepe R (2007) Lack of respiratory chain complex I impairs alternative oxidase engagement and modulates redox signaling during elicitor-induced cell death in tobacco. Plant Cell 19:640–655

Vijayan P, Shockey J, Levesque CA, Cook RJ, Browse J (1998) A role for jasmonate in pathogen defense of *Arabidopsis*. Proc Natl Acad Sci USA 95:7209–7214

Wan J, Zhang S, Stacey G (2004) Activation of a mitogen-activated protein kinase pathway in Arabidopsis by chitin. Mol Plant Pathol l5:125–135

Wang GL, Ruan DL, Song WY, Sideris S, Chen L, Pi LY, Zhang S, Zhang Z, Fauquet C, Gaut BS, Whalen MC, Ronald PC (1998) *Xa21D* encodes a receptor-like molecule with a leucine-rich repeat domain that determines race-specific recognition and is subject to adaptive evolution. Plant Cell 10:765–779

Ward JM, Pei ZM, Schroeder JI (1995) Roles of ion channels in initiation of signal transduction in higher plants. Plant Cell 7:833–844

Wasternack C (2007) Jasmonates: an update on biosynthesis, signal transduction and action in plant stress response, growth and development. Ann Bot 100:681–697

Wildermuth MC, Dewdney J, Wu G, Ausubel FM (2001) Isochorismate synthase is required to synthesize salicylic acid for plant defence. Nature 414:562–565

Wilson ID, Neill SJ, Hancock JT (2008) Nitric oxide synthesis and signalling in plants. Plant Cell Environ 31:622–631

Wong HL, Pinontoan R, Hayashi K, Tabata R, Yaeno T, Hasegawa K, Kojima C, Yoshioka H, Iba K, Kawasaki T, Shimamoto K (2007) Regulation of rice NADPH oxidase by binding of rac GTPase to its N-terminal extension. Plant Cell 19:4022–4034

Xiong L, Yang Y (2003) Disease resistance and abiotic stress tolerance in rice are inversely modulated by an abscisic acid-inducible mitogen-activated protein kinase. Plant Cell 15:745–759

Xiong TC, Bourque S, Lecourieux D, Amelot N, Grat S, Briere C, Mazars C, Pugin A, Ranjeva R (2006) Calcium signaling in plant cell organelles delimited by a double membrane. Biochim Biophys Acta 1763:1209–1215

Xu H, Heath MC (1998) Role of calcium in signal transduction during the hypersensitive response caused by basidiospore-derived infection of the cowpea rust fungus. Plant Cell 10:585–598

Xu L, Liu F, Lechner E, Genschik P, Crosby WL, Ma H, Peng W, Huang D, Xie D (2002) The SCF(COI1) ubiquitin-ligase complexes are required for jasmonate response in Arabidopsis. Plant Cell 14:1919–1935

Yamamizo C, Kuchimura K, Kobayashi A, Katou S, Kawakita K, Jones JD, Doke N, Yoshioka H (2006) Rewiring mitogen-activated protein kinase cascade by positive feedback confers potato blight resistance. Plant Physiol 140:681–692

Yoda H, Hiroi Y, Sano H (2006) Polyamine oxidase is one of the key elements for oxidative burst to induce programmed cell death in tobacco cultured cells. Plant Physiol 142:193–206

Yoon GM, Cho HS, Ha HJ, Liu JR, Lee HS (1999) Characterization of NtCDPK1, a calcium-dependent protein kinase gene in *Nicotiana tabacum*, and the activity of its encoded protein. Plant Mol Biol 39:991–1001

Yoshioka H, Numata N, Nakajima K, Katou S, Kawakita K, Rowland O, Jones JD, Doke N (2003) *Nicotiana benthamiana* gp[91phox] homologs NbrbohA and NbrbohB participate in H_2O_2 accumulation and resistance to *Phytophthora infestans*. Plant Cell 15:706–718

Yoshioka K, Moeder W, Kang HG, Kachroo P, Masmoudi K, Berkowitz G, Klessig DF (2006) The chimeric Arabidopsis CYCLIC NUCLEOTIDE-GATED ION CHANNEL11/12 activates multiple pathogen resistance responses. Plant Cell 18:747–763

Zago E, Morsa S, Dat JF, Alard P, Ferrarini A, Inze D, Delledonne M, Van Breusegem F (2006) Nitric oxide- and hydrogen peroxide-responsive gene regulation during cell death induction in tobacco. Plant Physiol 141:404–411

Zemojtel T, Frohlich A, Palmieri MC, Kolanczyk M, Mikula I, Wyrwicz LS, Wanker EE, Mundlos S, Vingron M, Martasek P, Durner J (2006) Plant nitric oxide synthase: a never-ending story? Trends Plant Sci 11:524–525

Zhang J, Shao F, Li Y, Cui H, Chen L, Li H, Zou Y, Long C, Lan L, Chai J, Chen S, Tang X, Zhou J-M (2007) A *Pseudomonas syringae* effector inactivates MAPKs to suppress PAMP-induced immunity in plants. Cell Host Microbe 1:175–185

Zhang S, Du H, Klessig DF (1998) Activation of the tobacco SIP kinase by both a cell wall-derived carbohydrate elicitor and purified proteinaceous elicitins from *Phytophthora* spp. Plant Cell 10:435–450

Ziegler J, Stenzel I, Hause B, Maucher H, Hamberg M, Grimm R, Ganal M, Wasternack C (2000) Molecular cloning of allene oxide cyclase. The enzyme establishing the stereochemistry of octadecanoids and jasmonates. J Biol Chem 275:19132–19138

Zimmermann S, Nürnberger T, Frachisse JM, Wirtz W, Guern J, Hedrich R, Scheel D (1997) Receptor-mediated activation of a plant Ca^{2+}-permeable ion channel involved in pathogen defense. Proc Natl Acad Sci USA 94:2751–2755

Zipfel C, Felix G (2005) Plants and animals: a different taste for microbes? Curr Opin Plant Biol 8:353–360

Zipfel C, Kunze G, Chinchilla D, Caniard A, Jones JD, Boller T, Felix G (2006) Perception of the bacterial PAMP EF-Tu by the receptor EFR restricts *Agrobacterium*-mediated transformation. Cell 125:749–760

Zuppini A, Navazio L, Sella L, Castiglioni C, Favaron F, Mariani P (2005) An endopolygalacturonase from *Sclerotinia sclerotiorum* induces calcium-mediated signaling and programmed cell death in soybean cells. Mol Plant–Microbe Interact 18:849–855

18 Defence Responses in Plants

Chiara Consonni[1], Matt Humphry[1], Ralph Panstruga[1]

CONTENTS

I. Introduction............................ 363
II. Innate (PAMP-triggered) Immunity......... 364
  A. Plant Detection of
    Pathogens (Non-Self) 364
  B. Downstream Responses Following
    Pathogen Recognition................. 366
    1. Ion Fluxes 366
    2. Reactive Oxygen Species as Early
      Signalling Molecules 366
    3. Nitric Oxide as an Early
      Signalling Molecule................ 367
    4. MAP Kinase Signalling Cascades...... 368
    5. Signalling Molecule-Activated
      Defence Pathways................... 368
      a) The SA Signalling Pathway....... 368
      b) The JA/ET Signalling Pathway..... 369
      c) The ABA Signalling Pathway 370
      d) Cross-Talk Between
        Signalling Pathways.............. 370
  C. Execution of Defence Responses......... 371
    1. Subcellular Re-Arrangements 371
    2. Transcriptional Regulation
      of Defence Responses 371
    3. The Role of the Cell Wall.......... 372
    4. Biosynthesis and Accumulation
      of Secondary Metabolites 373
    5. Pathogenesis-Related Proteins........ 374
    6. Further Genetically Identified
      Components Involved in
      Defence Execution 374
III. Pathogen Effector-Triggered Defence 375
  A. Pathogen Effectors..................... 375
  B. Resistance Gene-Mediated Immunity
    to Plant Pathogens..................... 376
    1. Types of Plant Resistance Genes 376
    2. Direct or Indirect Recognition
      of Pathogen Effectors?.............. 377
IV. Resistance at the Whole
  Organism/Tissue Level 378
V. Conclusions 378
  References............................. 379

[1] Max-Planck Institute for Plant Breeding Research, Department of Plant–Microbe Interactions, Carl-von-Linné-Weg 10, 50829 Köln, Germany; e-mail: panstrug@mpiz-koeln.mpg.de

I. Introduction

In nature, plants are exposed to a wide range of adverse stimuli throughout their life cycle. These comprise abiotic stresses such as cold, salinity and drought, as well as biotic stresses including animal herbivory and microbial colonization. While some interactions between plants and micro-organisms can be beneficial for both partners (symbiosis), in most instances plants do not benefit from microbial invasion. Frequently, microbes, including fungi, attempt to invade and/or propagate in/on plants, which may ultimately result in disease, a condition which is obviously detrimental for the plant.

Most plants are, however, resistant to most micro-organisms that they encounter in nature. This widespread immunity relies in part on passive and pre-formed barriers, such as the waxy cuticle and cell wall components. Additionally, upon pathogen recognition, plants can activate multilayered active defence responses that co-operatively suffice to restrict most microbial attacks. Via these general defence mechanisms, which collectively are also referred to as "innate immunity", entire plant species are resistant to all genetic variants (e.g. races or isolates) of particular microbes. Consequently, in nature resistance is the rule rather than the exception. Owing to its multiple components and signalling pathways, which include both pre-formed and inducible responses, innate immunity to non-adapted (so-called "non-host" pathogens) is very robust and durable (Heath 2000).

Relatively few microbes are nevertheless adapted to the defence arsenal of particular plant species and are in principle able to colonize them. These pathogens evolved mechanisms to avoid or overcome the innate immune system and to exploit a given plant species as a host. In turn, affected plant species frequently gained a secondary recognition system, commonly based on cytoplasmic pattern recognition receptors termed resistance

Plant Relationships, 2nd Edition
The Mycota V
H. Deising (Ed.)
© Springer-Verlag Berlin Heidelberg 2009

(R) proteins that are able to detect virulent variants of a pathogen and activate a heightened defence state. This alternating success of either host or microbe is thought to be driven by an evolutionary "arms race", resulting in a still-ongoing "zigzag" evolution between plants and micro-organisms, a concept that was comprehensively highlighted in recent reviews (Maor and Shirasu 2005; Chisholm et al. 2006; da Cunha et al. 2006; Ferreira et al. 2006; Jones and Dangl 2006).

In this chapter we discuss the various molecular mechanisms that plants evolved to defend themselves against microbial pathogens, including fungi. Since the dicotyledonous reference species *Arabidopsis thaliana* currently is the best studied plant system at the molecular level, we primarily focus on this model species to illustrate fundamental principles of antimicrobial defence, ranging from pathogen recognition to defence signalling and the execution of antimicrobial downstream responses. Many of these principles, however, are shown or assumed to operate similarly in other plant species, including monocotyledonous plants. Likewise, while numerous seminal findings in *Arabidopsis* were obtained with the bacterial pathogen *Pseudomonas syringae*, many of the respective results are believed to have general validity for defence against other pathogens, too. A well studied *Arabidopsis*–fungus pathosystem is the interaction between *Arabidopsis* and powdery mildew fungi (Micali et al. 2008); we thus refer to this plant–microbe interaction in many instances.

In the last part we exemplify biotechnological and breeding aspects and how our current and future knowledge can be applied to aid crop productivity.

II. Innate (PAMP-triggered) Immunity

Constitutive, pre-formed defences are thought to comprise a major part and first step in resistance against pathogens. They include the plant cell wall, epidermal reinforcements such as the cuticle (composed of several wax layers) and leaf hairs (trichomes), as well as antimicrobial agents such as defensive peptides and toxic secondary metabolites (Heath 2000; Nürnberger and Lipka 2005). These pre-formed barriers are believed to delay the initial colonization by microbes, allowing the plant sufficient time to mount effective inducible defence responses (Ingle et al. 2006). Examples of preformed

antimicrobial metabolites acting as chemical barriers include the steroidal glycoalkaloid saponins α-tomatine (present in *Solanum lycopersicum*; Bouarab et al. 2002) and avenacin A-1 (produced by *Avena* sp.; Papadopoulou et al. 1999).

A. Plant Detection of Pathogens (Non-Self)

The activation of inducible defences requires that the plant is able to recognize the presence of a pathogen (a non-self organism). This is now generally thought to be mediated by cell surface pattern recognition receptors (PRRs), which can sense pathogen structures – so-called pathogen-associated molecular patterns or PAMPs (Nürnberger and Brunner 2002; Nürnberger and Lipka 2005; Robatzek 2007; Chap. 17). PAMPs are molecules that are essential for the microbial invader and are highly conserved within entire classes of micro-organisms, but are not present in the plant (Nürnberger and Lipka 2005; Robatzek 2007). These elicitors are ideally suited for the detection of intruders as owing to their essential functions it is very difficult or even impossible for microbes to eliminate or extensively modify these molecules during evolution. Thus, PAMP recognition is thought to constitute a general, sensitive, robust and durable way for discrimination of self from non-self.

PAMPs have been identified from a large range of pathogens and constitute a wide variety of molecules, including lipopolysaccharides (LPS), harpin and the motility factor flagellin from bacteria as well as β-glucans, xylanase and chitin from fungi (Nürnberger and Lipka 2005). Treatment of plants with PAMPs results in a series of stereotypical defence responses, including extracellular alkalinization, the production of reactive oxygen species and execution of cell death (Felix et al. 1999; Ron and Avni 2004), the release of nitric oxide (Zeidler et al. 2004) and the initiation of mitogen-activated protein (MAP) kinase cascades, ultimately leading to global changes in gene expression (Zipfel et al. 2004; Chap. 17). It is thought that these inducible defence responses suffice to prevent pathogenesis of the majority of non-adapted pathogens ("non-host resistance"; Nürnberger and Lipka 2005; Fig. 18.1) and at least partially limit propagation of adapted pathogens ("basal defence"; Zipfel et al. 2004, 2006). At the molecular level, basal defence and non-host resistance thus appear to have largely a common mechanistic base.

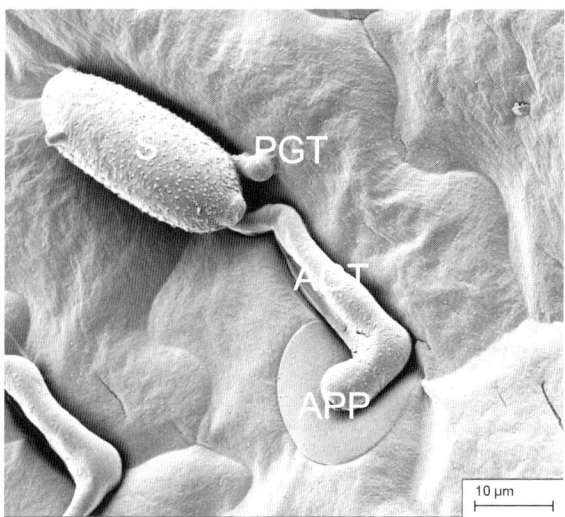

Fig. 18.1. *Arabidopsis* basal defences terminate the attempted attack by a non-adapted powdery mildew pathogen. The scanning electron micrograph depicts the failed penetration attempt of a sporeling of the barley powdery mildew pathogen, *Blumeria graminis* f.sp. *hordei*, into a leaf epidermal cell of *A. thaliana*. Please note the circular halo presumably originating from a fungal secretion that surrounds the appressorium. *AGT* Appressorial germ tube, *APP* appressorium, *PGT* primary germ tube, *S* spore. Micrograph courtesy of Elmon Schmelzer

To date, very few functional plant PRRs have been identified and experimentally shown to recognize pathogen-derived elicitors. Currently, the best characterized receptor of pathogen elicitors is the receptor-like kinase *flagellin sensing 2* (FLS2) of *Arabidopsis*, which senses and specifically interacts with an N-terminal 22-amino-acid oligopeptide (known as flg22) from the flagella of several Gram-negative bacteria (Felix et al. 1999; Gomez-Gomez and Boller 2000; Chinchilla et al. 2006). FLS2 consists of an extracellular leucine-rich repeat (LRR) domain that is involved in recognition of the bacterial elicitor and an internal serine/threonine kinase domain serving a role in cytoplasmic signalling (Gomez-Gomez and Boller 2000; Gomez-Gomez et al. 2001). Loss of FLS2 results in insensitivity to flg22, leading to an increased growth of the adapted bacterial pathogen *Pseudomonas syringae* pv. tomato DC3000 (*Pst*) in mutant plants, i.e. a loss of basal defence conditioned by PAMP recognition (Zipfel et al. 2004). Similarly, pre-treatment with flg22 primes *Arabidopsis* for defence against *Pst* infection, resulting in less growth and fewer disease symptoms of this pathogen in wild-type plants, but not in *fls2* mutants (Zipfel et al. 2004).

Recent results indicate that FLS2 is rapidly internalized into intracellular vesicles upon treatment with flg22, suggesting that ligand-induced endocytosis of FLS2 might be essential for its function (Robatzek et al. 2006). Additionally, FLS2 internalization is followed by degradation, potentially via a monubiquitination-dependent pathway (Robatzek et al. 2006; Robatzek 2007). This finding suggests that endocytosis might be important for signalling of cell surface receptors involved in plant immunity to activate downstream defence components.

A second receptor-like kinase PRR, *EF*-Tu *receptor* (EFR), recognizes an oligopeptide (elf18) from the bacterial elongation factor EF-Tu (Zipfel et al. 2006). Treatment with elf18 was found to trigger a comparable array of subcellular responses as application of flg22, including transcriptional activation of a similar set of genes (Zipfel et al. 2006). It thus appears that these two receptors activate a common set of signalling events, although they recognize entirely distinct elicitors. This result suggests that plants have developed a suite of receptors with seemingly overlapping activity, possibly to ensure that they are able to perceive pathogens that can evade one detection system.

Plant innate immune responses have been extensively compared with mammalian innate immunity (Nürnberger et al. 2004; Robatzek 2007). Indeed, both EFR and FLS2 show similarity to mammalian Toll-like receptors (TLRs) that are involved in immune responses. Recognition of non-self by this class of proteins triggers similar reactions in both kingdoms, including the release of nitric oxide (NO) and the activation of MAP kinase cascades, subsequently leading to defence responses (Nürnberger et al. 2004). For example, the receptor-like kinase TLR5 recognizes bacterial flagellin in mammals, resulting in similar downstream responses as in *Arabidopsis* (Hayashi et al. 2001). Similarly, PAMP recognition by the LPS receptor TLR4 leads to endocytosis and ubiquitination, which is thought to be important for its signalling function, as in case of FLS2 (Robatzek 2007). However, it is more likely that these similarities are the result of convergent rather than divergent evolution since both FLS2 and TLR5 recognize different flagellin epitopes and downstream signalling relies upon structurally different domains in either protein (Nürnberger et al. 2004).

B. Downstream Responses Following Pathogen Recognition

1. Ion Fluxes

One of the earliest responses of plant cells upon exposure to either PAMPs or microbes is the flux of ions, including Ca^{2+}, H^+ and K^+, across the plasma membrane which is followed by extracellular alkalinization and intracellular acidification (Atkinson et al. 1990). This was, for example, demonstrated in cultured parsley (*Petroselinum crispum*) cells, where stimulation by the fungal oligopeptide elicitor Pep-13 resulted in rapid influx of calcium ions and an increased extracellular pH concomitant with an efflux of potassium (Jabs et al. 1997). Similar results were obtained in a range of further experimental systems, e.g. tobacco cells treated with *Pseudomonas syringae* pv. *syringae* (Atkinson et al. 1990) or pea and cowpea treated with an elicitor from *Mycosphaerella pinodes* (Amano et al. 1997), indicating that ion fluxes are indeed a widely conserved biotic stress response.

In barley plants, rapid transient increases in apoplastic pH and calcium ion influx were observed upon inoculation with the powdery mildew fungus, *Blumeria graminis* f.sp. *hordei* (*Bgh*). However, long-lasting apoplastic alkalinization was only found in incompatible barley–*Bgh* interactions (Felle et al. 2004). These are typically characterized by occurrence of localized pathogen-triggered cell death, also known as the hypersensitive response (HR; Pennell and Lamb 1997), suggesting a link between ion fluxes and cell death (Felle et al. 2004). Indeed, in both of the above-mentioned studies with parsley and tobacco cell cultures, influx of calcium ions was also shown to be required for the production of reactive oxygen species (ROS) and HR induction (Atkinson et al. 1990; Jabs et al. 1997).

Given the assumed pivotal role of calcium ion fluxes in signalling for the activation of downstream defence responses (Lecourieux et al. 2006), cytoplasmic calcium sensors such as calcium-dependent protein kinases (CDPKs) and calmodulin represent bona fide candidates for sensing and transmitting intracellular changes in calcium levels. Genes encoding CDPKs have been shown to be transcriptionally induced upon treatment with elicitors as well as after pathogen attack in several species (Romeis et al. 2001; Ludwig et al. 2004), suggesting a possible role for these polypeptides during defence responses. Further evidence for a functional role for this gene family in early defence

reactions came from virus-induced gene silencing (VIGS) of a tobacco CDPK subfamily that resulted in reduced HR in response to race-specific fungal elicitation (Romeis et al. 2001). Besides the well supported function of CDPKs in plant defence, there is also increasing evidence for a role of distinct calmodulin isoforms in modulating plant defence responses (Heo et al. 1999; Kim et al. 2002).

2. Reactive Oxygen Species as Early Signalling Molecules

Recognition of incompatible pathogens by plants often results in a response characterized by localized cell death, known as the hypersensitive response (HR; Pennell and Lamb 1997; Torres et al. 2006; Chap. 12). The hypersensitive response is preceded by a biphasic apoplastic accumulation of reactive oxygen species (ROS) that are primarily composed of the superoxide anion radical ($O_2^{\cdot-}$) and hydrogen peroxide (H_2O_2), via the so-called oxidative burst (Hückelhoven and Kogel 2003; Torres et al. 2006). In addition, an increase in cytosolic calcium levels occurs in the cell, triggering a protein kinase-mediated cell death pathway that is similar to other programmed cell death pathways (Pennell and Lamb 1997). The production of ROS is typically fast and is thought to be both directly toxic to invading pathogens as well as aiding in reinforcing the cell wall to prevent pathogen ingress (Alvarez et al. 1998; Nimchuk et al. 2003). Additionally, ROS serve as signalling molecules that can trigger a plethora of downstream effects, such as gene transcription (Laloi et al. 2004). Interestingly, in interactions with compatible pathogens, only one transient phase of ROS accumulation occurs, suggesting that sustained ROS production is required for effective defence activation (Torres et al. 2006).

Studies on the interaction of barley with the barley powdery mildew pathogen demonstrated that the timing and spatial patterning of ROS accumulation plays a role in defence responses to this pathogen. In compatible interactions, the level of H_2O_2 accumulation in cell wall appositions under successful penetrations is much lower than under failed penetration attempts, highlighting a correlation between H_2O_2 production and defences (Hückelhoven and Kogel 2003). Additionally, during incompatible interactions in cells that undergo HR, H_2O_2 accumulates throughout whole cells.

Genetic evidence further indicates that the polarized delivery of H_2O_2-containing vesicles can influence defence responses to powdery mildews. In barley *mildew resistance locus o* (*mlo*) mutants that exhibit broad-spectrum powdery mildew resistance, a higher incidence of H_2O_2-containing vesicles was observed at fungal interaction sites, compared with *Mlo* wild-type plants, suggesting that *Mlo* may be acting to limit H_2O_2 accumulation in this interaction. This accumulation was significantly reduced in epistasis mutants with factors required for *mlo*-mediated resistance (i.e. *ror* mutants), suggesting a role for these proteins in H_2O_2 accumulation after fungal attack and strengthening the link between H_2O_2 accumulation and successful fungal entry (Piffanelli et al. 2002). In contrast to H_2O_2, superoxide accumulates only in cells that have undergone successful powdery mildew penetrations. This suggests that different reactive oxygen species may be involved in both inducing and restricting cell death, at least in the barley–powdery mildew interaction (Hückelhoven and Kogel 2003).

In *Arabidopsis*, two respiratory burst NADPH oxidases (RbohD, RbohF) were identified that are required for production of the majority of H_2O_2 accumulation and induction of cell death during interactions with both incompatible bacterial pathogens (Torres et al. 2002) and necrotrophic fungi (Torres et al. 2005). Interestingly, only one of these NADPH oxidases was required for H_2O_2 accumulation in response to an oomycete pathogen but it was also required for limiting hypersensitive cell death, suggesting that the link between the accumulation of specific reactive oxygen species and HR is not entirely causal, but may also rely on the relationship between H_2O_2 and superoxide (Torres et al. 2002). Genetic data also showed that these NADPH oxidases appeared to suppress cell death in areas surrounding sites of their activation, suggesting a role for ROS in negatively regulating HR after infection to limit defence responses to specific areas (Torres et al. 2005).

3. Nitric Oxide as an Early Signalling Molecule

Nitric oxide (NO) is a gaseous free radical of high chemical reactivity that has been shown to be an important signalling molecule in mammals, including humans. It can be biosynthesized from arginine and oxygen by nitric oxide synthase (NOS) enzymes and by reduction of inorganic nitrate. In animals NO acts at least in part by conferring post-translational modifications on target proteins, e.g. via S-nitrosylation of cysteine residues or by tyrosine nitration.

NO was initially identified in plants due to its involvement in multiple physiological processes, including control of root growth, closure of stomata and senescence (Neill et al. 2002; Wendehenne et al. 2004). Localized plant defences were also recently shown to trigger NO production and signalling (Nimchuk et al. 2003). The importance of NO in potentiating defence responses was first highlighted by Delledonne et al. (1998) who showed that ROS were only able to induce significant levels of hypersensitive cell death in soybean cell suspensions when treated with exogenous NO donors. The authors also demonstrated that treatment of the same cell suspensions with inhibitors of NO reduced the level of HR-like cell death induced by the bacterial pathogen *P. syringae* pv. *glycinea*, providing evidence for a role for endogenous NO in early defence responses. Similarly, Durner et al. (1998) found that NO synthase activity in tobacco plants was enhanced upon treatment with tobacco mosaic virus and that addition of exogenous NO donors resulted in the expression of several defence-related genes. Interestingly, treatment with NO donors also increased levels of cyclic GMP, which was shown to be involved in NO signalling in mammals (Durner and Klessig 1999). Moreover, the same defence genes that were transcriptionally upregulated upon application of NO donors were also induced by treatment with cyclic GMP (Durner and Klessig 1999). Together, these results suggest that at least some of the components of NO signalling are conserved between plants and animals.

Using inhibitors of NO synthesis, Delledonne et al. (1998) also demonstrated that NO was required for an early ROS-dependent HR in response to an incompatible interaction with *P. syringae* pv. *maculicola* and that loss of NO production resulted in increased pathogen growth. Likewise, pharmacological interference with its accumulation revealed a role for NO in the defence of barley against the fungal powdery mildew pathogen (Prats et al. 2005). An *Arabidopsis* mutant line with an insertion in a NO synthase was shown to produce significantly less NO in response to LPS treatment, compared with wild-type plants, and also exhibited enhanced disease susceptibility to

Pst (Zeidler et al. 2004). The latter finding provides genetic evidence for a role of NO in disease resistance. Though recent discoveries suggest that NO also contributes to post-translational modification of proteins in plants (Lindermayr et al. 2005), the exact role of NO in plant defence is still unclear.

4. MAP Kinase Signalling Cascades

In addition to CDPKs, mitogen activated protein kinases (MAPKs) are also activated soon after pathogen recognition, leading to a flood of downstream transcriptional changes (Romeis 2001; Nimchuk et al. 2003). MAP kinase cascades have been extensively studied in several plant species and are known to be composed of at least three components: a serine/threonine MAPKKK that activates a MAPKK via phosphorylation, which then in turn phosphorylates serine/threonine MAPKs, leading to the phosphorylation and activation of various transcription factors, protein kinases and other targets (Jonak et al. 2002). MAP kinase cascades are known to play essential roles in a diverse set of responses in both plants and animals, including the regulation of cell growth, differentiation and cell cycle control as well as responses to various abiotic and biotic stresses (Jonak et al. 2002).

In several plant species, MAP kinases have been identified that play a role in response to both abiotic and biotic stresses. The MAP kinase MPK4 from *Arabidopsis* was shown to be activated by a bacterial elicitor (Desikan et al. 2001) and reverse genetics indicated that its gene product negatively regulates defence responses that are effective against both bacterial and oomycete pathogens (Petersen et al. 2000). The orthologous MAP kinases MPK6 and SIPK from *Arabidopsis* and tobacco, respectively, are also activated by various pathogen signals (Jonak et al. 2002), but in contrast to *MPK4*, *Arabidopsis* lines silenced for *MPK6* showed loss of basal defences against oomycete and bacterial pathogens (Menke et al. 2004).

Recently, Asai et al. (2002) identified all components of a complete MAP kinase cascade that was activated upon perception of the bacterial elicitor flg22 by FLS2. MEKK1 was identified as the MAPKKK that specifically activated the functionally redundant MAPKK proteins MKK4 and MKK5, which in turn specifically phosphorylated MPK3 and MPK6, ultimately leading to defence gene activation. Interestingly and consistent with the loss-of-function *mpk6* phenotype, constitutively active forms of each MEKK1 and MKK4/MKK5 were able to confer increased resistance to

both bacterial and fungal pathogens (Asai et al. 2002; Menke et al. 2004). Similarly, expression of a constitutively active MAPKK (NtMEK2) in tobacco was shown to activate SIPK and another MAP kinase, WIPK, leading to defence gene activation and hypersensitive cell death (Yang et al. 2001). Kovtun et al. (2000) identified another *Arabidopsis* MAPKKK, ANP1, that was activated upon treatment with H_2O_2 and, similar to MEKK1, was able to initiate a kinase cascade involving MPK3 and MPK6. However, consistent with the idea that different stress responses activate distinct MAP kinase cascades, a constitutively active variant of ANP1 was only able to marginally activate MKK5, the immediate downstream kinase of MEKK1 (Kovtun et al. 2000).

5. Signalling Molecule-Activated Defence Pathways

The phenolic beta hydroxyl acid, salicylic acid (SA), cyclopentanone derivatives such as jasmonic acid (JA) and the gaseous alkene ethylene (ET) are phytohormones which serve key roles in defence signalling. Basal defences and resistance gene-mediated responses (described below III. B.) are usually associated with activation of SA-dependent signalling pathways that lead to the expression of pathogenesis-related (PR) proteins that are thought to contribute to resistance (Glazebrook 2005). Other plant defence responses are controlled by mechanisms dependent on ET and/or JA (described below). In many situations, induced downstream responses rely on a network of cross-communication between signalling pathways of which SA, JA and ET are the principal mediators (Broekaert et al. 2006). SA and JA are often mutually inhibitory for the expression of many defence genes, but they can also act in co-operation (Glazebrook 2005). This section discusses these three pathways, their interaction and the role of a further plant hormone, abscisic acid (ABA), in defence responses (see also Chap. 17).

a) The SA Signalling Pathway

Recognition of biotrophic pathogens belonging to kingdoms as different as fungi, oomycetes and bacteria commonly triggers high levels of SA accumulation. An *Arabidopsis* gene encoding for an isochorismate synthase (ICS/SID2) is a required component of SA biosynthesis. SA levels are significantly reduced in *sid2* mutant plants, indicating that the majority of SA in plants derives from isochorismate rather than from phenylalanine (Wildermuth et al. 2001). Furthermore, SID2 is required for downstream SA-dependent responses such as the expression op *pathogenesis-related (PR) genes*, the induction of systemic acquired resistance (SAR; see below IV.) and for basal defence against *Hyaloperonospora parasitica* (Wildermuth et al. 2001; Nawrath and Metraux 1999). In addition to SID2, EDS5 is also required for SA biosynthesis (Glazebrook et al. 2003). *EDS5* encodes a *m*ultidrug *a*nd *t*oxin *e*xtrusion (MATE) transporter thought to be involved in the transport of intermediates for SA biosynthesis. Mutations in *EDS5* have a similar effect as those in *SID2* (Nawrath and Metraux 1999; Nawrath et al. 2002).

Important players acting upstream of SA accumulation in the SA-mediated pathway include *phytoalexin deficient 4* (*PAD4*) and *enhanced disease susceptibility 1* (*EDS1*) which encode lipase-like proteins, though no lipase activity has been detected so far for either of them (Wiermer et al. 2005). EDS1 and PAD4 were found to physically interact, both in yeast two-hybrid assays and biochemically in planta (Feys et al. 2001). Both proteins co-operate in defence responses mediated by the so-called TIR-NB-LRR class of R genes, while the CC class of receptors trigger resistance independently of EDS1 and PAD4, suggesting that these two proteins might act at the cross-point between the TIR- and CC-class of R proteins (Wiermer et al. 2005; see below III. B. 1.). Additionally, EDS1 and PAD4 are both required for SA accumulation and for defence amplification involving ROS generation and HR at the infection site. The gene product of *senescence-associated gene 101* (*SAG101*) was identified as a further interactor of EDS1 (Feys et al. 2005). SAG101 shares a conserved *EDS1/PAD4* (EP) domain with EDS1 and PAD4 (Feys et al. 2005) and was reported to have acyl hydrolase activity in vitro (He and Gan 2002). It is thought that the three proteins are part of an SA-dependent signal feedback amplification loop which boosts basal defence responses.

Another crucial player in the SA-dependent defence pathway is encoded by *nonexpressor of pathogenesis-related genes 1* (*NPR1*), which acts downstream of SA accumulation. NPR1 is a regulatory protein that was identified in genetic screens for mutants unable to express PR proteins after pathogen challenge (Cao et al. 1994). *npr1* mutants are compromised in basal defence; they exhibit enhanced disease symptoms when infected with virulent pathogens (Dong 2004; Pieterse and van Loon 2004; Glazebrook 2005). In addition to its role in SA-dependent defences, NPR1 also plays a major role in regulating SAR (see below IV.).

b) The JA/ET Signalling Pathway

It is commonly believed that biotrophic pathogens predominantly trigger SA-mediated responses, while necrotrophic pathogens and herbivores activate defence through the signalling molecules JA and ET (Glazebrook 2005). Several studies suggest that the JA and ET signalling pathways often operate synergistically to induce defence responses (Glazebrook 2005; Lorenzo and Solano 2005; Beckers and Spoel 2006; Broekaert et al. 2006). Aside from its role in defence against pathogens and insects, the signalling molecule JA is involved in many processes as different as wound responses, root growth, fertility and senescence (Devoto and Turner 2005). ET is required, for instance, for cell elongation, seed germination and fruit ripening (Alonso and Stepanova 2004; Guo and Ecker 2004; Chen et al. 2005a).

Screens for *Arabidopsis* mutants that do not respond to JA or to the JA analogue coronatine led to the identification of *JA resistant* 1 (*jar1*) and *coronatine insensitive* 1 (*coi1*; Staswick et al. 1992; Feys et al. 1994). *JAR1* encodes a JA-amino synthetase that forms conjugates between JA and several amino acids, e.g. isoleucine. Mutants in this gene are unresponsive to JA and exhibit a strongly reduced expression of JA-responsive genes and impaired resistance to pathogens. These data suggest that this JA modification might be important for the role of JA in defence signalling (Lorenzo and Solano 2005; Beckers and Spoel 2006). *COI1* encodes a protein characterized by an F-box domain and 16 leucine-rich repeats. COI1 was found to interact with the proteins SKP1 and cullin1, which together form a SCF-ubiquitin complex involved in protein degradation (Devoto et al. 2002). *coi1* mutants were found to be more susceptible to necrotrophic fungal pathogens such as *Botrytis cinerea* and *Alternaria brassicicola*. Conversely, resistance to biotrophic pathogens such as *H. parasitica* and *Golovinomyces orontii* was not compromised in this mutant (Thomma et al. 1998; Bostock 2005).

In plant cells, the gaseous hormone ET is sensed by endoplasmic reticulum (ER)-localized receptors that are characterized by a hydrophobic domain and the presence of a copper cofactor (Alonso and Stepanova 2004; Chen et al. 2005a). The *Arabidopsis* genome encodes five ET receptors (ETR1, ETR2, EIN4, ERS1, ERS2), which exhibit a high degree of functional redundancy. Mutations in the hydrophobic domain of each of these receptors result in dominant ET insensitivity of the mutant plant. Additionally, loss of function mutants indicated that the receptors might act by repressing ET signalling (Alonso and Stepanova 2004). Another negative regulator of ET responses is the protein *constitutive triple response 1* (CTR1), which contains a Raf-like serine threonine kinase domain. CTR1 is located in the ER through its association with the ET receptor family. Association with ETR and kinase activity are required for the negative regulation ET-dependent downstream responses by CTR1 (Alonso and Stepanova 2004). Genetically located downstream of *CTR1*, *ET-in*sensitive 2 (EIN2) is a positive regulator of the ET-dependent pathway. *EIN2* encodes a plant-specific protein with a still-uncharacterized biochemical activity (Alonso and Stepanova 2004). *Arabidopsis ein2* mutants show increased susceptibility to the necrotrophic fungus *B. cinerea*, but not to *A. brassicicola*, suggesting that the ET-dependent pathway is not active during the response to the latter pathogen (Thomma et al. 1999a).

JA and ET can either co-operate or act as antagonists in the control of different stress responses and developmental processes. This regulation is mediated mainly by two transcription factors, ERF1 and MYC2. Overexpression of *ERF1* results in enhanced resistance against several necrotrophs. In addition, *ERF1* induction depends on both the JA and ET signalling pathways, as mutations impeding either of these signalling pathways block the induction of *ERF1* and its targets. Furthermore, when applied together, JA and ET synergistically activate *ERF1* transcription, suggesting that ERF1 plays a role in mediating the cross-talk between JA and ET (Lorenzo et al. 2003).

MYC2, also known as JIN1, differentially regulates two branches of the JA signalling pathway (Lorenzo et al. 2004). The negatively regulated branch controls the expression of pathogen defence genes, as supported by the fact that *myc2* mutants are more resistant to the necrotrophic pathogen *B. cinerea* (Nickstadt et al. 2004). The positively regulated branch is involved in wound responses. Interestingly, ERF1 also differentially regulates these two branches but in an opposing manner to MYC2. These results indicate how plants can share the same signalling molecules for different stress responses (Lorenzo et al. 2004).

c)   The ABA Signalling Pathway

The plant hormone abscisic acid (ABA) has long been known to regulate many aspects of plant development, such as stomatal aperture, adaptation to drought, low temperature and salinity (Christmann et al. 2006). Recent data provide evidence that ABA might also play a role in disease responses (Mauch-Mani and Mauch 2005). Either treatment with ABA or drought stress resulted in increased resistance to the bacterial pathogen *P. syringae*. In addition, the *Arabidopsis* ABA-deficient mutant *ABA-deficient 1* (*aba1*) displayed reduced susceptibility to the oomycete pathogen *H. parasitica*. However, an *Arabidopsis* mutant impaired in ABA signal transduction, *ABA insensitive 1* (*abi1*), showed no alteration in resistance to these two pathogens. These data suggest that the concentration of ABA rather than the presence of a functional ABA signalling pathway is important for the development of disease symptoms in *Arabidopsis* plants (Mohr and Cahill 2003).

Intriguingly, mutant screens for enhanced response to ABA led to the identification of a gene, *ERA3*, which was found to be allelic to the ET insensitive mutant *ein2*. This result suggests that the encoded protein might function as a point of convergence between the ET and ABA signalling pathways (Ghassemian et al. 2000). Further evidence for the interaction between the ET and ABA signalling pathways came from the ectopic application of ABA on *Arabidopsis* plants, which resulted in the suppression of defence genes triggered by JA and ET. Consistently, ABA-deficient mutants displayed an up-regulation of JA/ET-induced genes and an increased resistance to the necrotrophic fungus *Fusarium oxysporum*. These results indicate that an antagonistic interaction between ABA- and JA/ET-mediated pathways may modulate plant defence responses (Anderson et al. 2004).

There is, however, evidence that the ABA and JA signalling pathways do not always have opposing effects. Expression of the transcription factor BOS1 is induced through the JA signalling pathway after challenge with the necrotrophic fungi *B. cinerea* or *A. brassicicola*. Interestingly, loss of *BOS1* function results not only in enhanced susceptibility to necrotrophic pathogens, but also in impaired tolerance towards drought, salinity and oxidative stresses (Mengiste et al. 2003).

d)   Cross-Talk Between Signalling Pathways

Several studies indicate that the SA- and JA/ET-dependent pathways do not act independently, but rather interactively to reduce disease caused by pathogens. Fine tuning of the cross-talk between the SA- and JA/ET-mediated pathways appears to be regulated by key players of each pathway. The antagonistic interaction between JA and SA has been extensively studied and induction of SA is known to result in the repression of JA signalling (Glazebrook 2005; Lorenzo and Solano 2005; Ferreira et al. 2006).

The first gene identified as a regulator of the cross-talk between these two hormones was *MAP kinase 4* (*MPK4*). *mpk4* mutants exhibit constitutively high levels of SA and constitutive activation of *PR* gene expression and SAR. Thus, MPK4 activity is required for suppression of SAR. Furthermore, MPK4 is required for the induction of some JA-regulated defence genes such as *PDF1.2* and *THI2.1*. These results indicate that MPK4 can simultaneously repress SA biosynthesis while promoting responses to JA (Petersen et al. 2000).

NPR1, a key regulator of the SA pathway and SAR, appears to have an important role in regulation of cross-talk between SA and JA, as *npr1* plants are unable to suppress JA signalling upon exogenous application of SA. Upon SA activation, the monomeric form of NPR1 is translocated into the nucleus resulting in SAR activation. However, nuclear localization appears not to be required for the suppression of JA-mediated signalling by SA, suggesting that NPR1 might have a role in the SA-mediated repression of JA signalling, independently of its function in SAR (Spoel et al. 2003; Bostock 2005; Glazebrook 2005).

Enhanced disease resistance 1 (*EDR1*) encodes a CTR1-like kinase and was shown to act as a negative regulator of disease resistance (Frye et al. 2001). Mutations in *EDR1* resulted in enhanced stress responses and spontaneous necrotic lesions under drought conditions, suggesting that EDR1 is also involved in stress response signalling and cell death regulation. Double-mutant analysis revealed that these phenotypes are dependent on the SA, but not on the ET signalling pathway (Tang et al. 2005). Furthermore, *edr1* mutants present an enhanced senescence phenotype induced by ET treatment (Frye et al. 2001), which is suppressed by mutations that affect the ET pathway but not the SA-mediated pathway (Tang et al. 2005). These data suggest that EDR1 might function at a point of cross-talk between ethylene and salicylic acid signalling that impinges on senescence and cell death (Tang et al. 2005).

C. Execution of Defence Responses

1. Subcellular Re-Arrangements

It is well known that a rapid re-organization of plant cells occurs in response to microbial contact. For example, cytoplasmic aggregation at sites of pathogen assault has been observed in a wide range of non-adapted and adapted plant–pathogen interactions, suggesting that this is a general reaction upon pathogen detection (Takemoto et al. 2003; Koh et al. 2005; Koh and Somerville 2006). In many cases also organelles such as the nucleus move to the site of attempted attack (Schmelzer 2002; Takemoto and Hardham 2004). These movements usually coincide with the rapid rearrangement of the actin cytoskeleton towards sites of attempted invasion, suggesting that actin filaments and possibly

also microtubules provide the tracks for these translocations of organelles (Schmelzer 2002; Takemoto et al. 2003; Holub and Cooper 2004; Shan and Goodwin 2005; Shimada et al. 2006; Chap. 1).

In *Arabidopsis* and other plants, inhibition of actin cytoskeleton function by cytochalasin E treatment was demonstrated to allow a greater incidence of plant cell penetration by species of non-adapted *Colletotrichum* (Shimada et al. 2006) and powdery mildew (Kobayashi et al. 1997; Yun et al. 2003), indicating the importance of cytoskeletal rearrangement for effective defence responses. These pharmacological studies have been recently corroborated by enhanced powdery mildew host cell entry upon genetic interference with actin filaments upon *actin depolymerizing factor* (*ADF*) overexpression (Miklis et al. 2007). It is thought that, besides its role in organelle re-localization, the actin cytoskeleton facilitates the localized deposition of physical and chemical barriers at sites of attempted entry, thereby contributing to effective penetration resistance (Holub and Cooper 2004).

Novel cell biological tools recently allowed more detailed analysis of the dynamics of early cellular responses in plants upon attack by fungal and oomycete pathogens (Takemoto et al. 2003; Koh et al. 2005). Koh et al. (2005) used green fluorescent protein (GFP)-tagged marker proteins in transgenic *Arabidopsis* lines to visualize the live cell dynamics of different organelles after inoculation with the adapted powdery mildew pathogen, *Golovinomyces cichoracearum*. Consistent with previous observations using conventional techniques they observed aggregation of the cytoplasm and organelles towards penetration sites and an increase in the size of nuclei in infected cells. Interestingly, using marker lines for eight plasma membrane-resident proteins, they also found that all tested polypeptides were excluded from the extrahaustorial membrane (EHM). This finding suggests either that the EHM is synthesized de novo or that EHM formation involves differentiation of the plasma membrane in such a manner that the majority of plasma membrane-resident proteins are not incorporated into the newly formed structure.

2. Transcriptional Regulation of Defence Responses

In recent years the importance of transcriptional regulators in plant disease resistance has become increasingly evident. In particular, several members of the WRKY family have been shown to be involved in transcriptional reprogramming during plant immune responses (Ülker and Somssich 2004; Eulgem 2006). WRKY proteins are plant-specific transcription factors characterized by a DNA-binding motif referred to as the WRKY domain (Rushton et al. 1996). Plant WRKY families comprise a large number of members, e.g. 74 in *Arabidopsis* and 90 in rice (*Oryza sativa*; Ülker and Somssich 2004).

In *Arabidopsis*, WRKY70 was shown to contribute to basal resistance against the powdery mildew fungus *G. cichoracearum* and the bacterial pathogen *Erwinia carotovora*, as *wrky70* mutants displayed an increased susceptibility against both pathogens (Li et al. 2004, 2006). In addition, *WRKY70* expression was found to be up-regulated after *B. cinerea* infection and a T-DNA insertion in *WRKY70* resulted in an increased susceptibility against this necrotrophic fungus (AbuQamar et al. 2006). Moreover, mutations in *WRKY70* affect both basal defence against virulent *H. parasitica* and the *H. parasitica* strain carrying the *avr* gene *RPP4* (Knoth et al. 2007). In *Arabidopsis*, expression of the transcription factor *WRKY70* is induced by SA and repressed by JA. In addition, gene-silencing of *WRKY70* results in constitutive up-regulation of JA-responsive genes. These data indicate that WRKY70 might act as a positive regulator of the SA pathway and *PR* gene expression and as a negative regulator of the JA pathway (Li et al. 2004, 2006).

Three further members of the WRKY family, *WRKY18*, *WRKY40* and *WRKY60*, clustering in a phylogenetic clade distinct from *WRKY70*, were found to be up-regulated in response to virulent or avirulent strains of *P. syringae* and upon challenge with the necrotrophic fungus *B. cinerea* (Dong et al. 2003). Furthermore, WRKY18, 40 and 60 proteins were demonstrated to form both homo- and hetero-oligomeric complexes with each other (Xu et al. 2006). Mutations in *wrky18* resulted in slightly increased susceptibility against *Botrytis* and in partial resistance against a virulent strain of *P. syringae*. These phenotypes were exacerbated in the double mutant *wrky18/40* and *wrky18/60* and especially in the triple mutant *wrky18/40/60* (Xu et al. 2006). Recent data show that two potential barley orthologues of *Arabidopsis WRKY18, 40* and *60 – HvWRKY1* and *HvWRKY2* – are transcriptionally up-regulated after *Bgh* challenge. Furthermore, overexpression of *HvWRKY1* and *HvWRKY2* resulted in suppression of basal defence against a virulent strain of barley powdery mildew and silencing them resulted in enhanced basal defences (Shen et al. 2007). Consistently, *Arabidopsis* double and triple mutants of *wrky18/40* and *wrky18/40/60* are almost fully resistant against the *Arabidopsis* powdery mildew fungus, *G. orontii* (Shen et al. 2007).

WRKY factors have also been found to play a role in disease responses in other plant species. A WRKY transcription factor was identified from a grape berry library of *Vitis vinifera* (*VvWRKY1*) and found to be transcriptionally up-regulated after wounding and SA or H_2O_2 treatment. Ectopic expression in tobacco plants resulted in increased resistance against the oomycete *Pythium* spp. and against the ascomycete fungus, *G. cichoracearum* (Marchive et al. 2007).

In addition to WRKYs, further transcription factors have been found to play a role in disease resistance. Members of the plant-specific NAC-domain transcription factor family share a conserved N-terminal NAC domain, originally characterized from the NAM gene from petunia and the ATAF1, ATAF2 and CUC2 genes from *Arabidopsis*. Overexpression of the *Arabidopsis* ATAF2 transcription factor results in down-regulation of a number of pathogenesis-related genes, suggesting that ATAF2 functions as a repressor of pathogenesis-related genes in *Arabidopsis* (Delessert et al. 2005). *HvNAC6*, a barley NAC transcription factor, was recently found to positively regulate penetration resistance in barley upon inoculation with the virulent powdery mildew fungus, *Bgh* (Jensen et al. 2007). Transient gene-silencing of *HvNAC6* in barley epidermal cells resulted in reduced frequency of penetration resistance upon infection with *Bgh*. Consistently, transient overexpression of *HvNAC6* in barley epidermal cells led to increased penetration resistance against *Bgh*. These results suggest the involvement of NAC transcription factors in pre-haustorial defence to grass powdery mildew (Jensen et al. 2007).

3. The Role of the Cell Wall

The plant cell wall is a complex composite of cellulose, polysaccharides, proteins and aromatic substances. Cellulose is a polymer of β-1,4-glucose which is synthesized at the plasma membrane by a large protein complex formed by cellulose synthase (CESA) subunits. Generally, mutations in any of these subunits results in reduced levels of cellulose synthesis and modification of the composition and structure of the cell wall (Vorwerk et al. 2004). Mutations in *CESA3/CEV1*, which is involved in primary cell wall formation, additionally lead to constitutive activation of the JA/ET-dependent defence pathway and enhanced resistance to some pathogens (Ellis et al. 2002). Mutations in *COI1* and *ETR1* are able to suppress this phenotype, indicating that CEV1 might co-ordinate the synthesis of JA and ET in response to stress.

Disruption of *CESA8*, which is involved in secondary cell wall formation, causes an increase in ABA levels and an enhanced tolerance to drought and osmotic stress (Chen et al. 2005b). Recently, *cesa8* mutants were found to be more resistant to the necrotrophic fungi *B. cinerea* and *P. cucumerina* as well as to biotrophs such as powdery mildews (Hernandez-Blanco et al. 2007). Interestingly, mutations in *CESA8* result in increased endogenous levels of ABA (Hernandez-Blanco et al. 2007), suggesting that CESA8 might regulate the cross-talk between the ABA and JA/ET pathways. It appears that plants are able to sense perturbation of the cell wall by monitoring the integrity of its structure. Biotrophic fungi might manipulate this surveillance system for the establishment of biotrophy by subverting the interconnected plant defence signalling pathways or the underlying resistance mechanisms (Jones and Takemoto 2004; Schulze-Lefert 2004; Ferreira et al. 2006).

A common response to attack by fungal pathogens is the formation of local cell wall appositions named papillae. One of the main components of papillae is the β-1,3-D-glucan callose. In addition, papillae also contain phenolics, lignin, cellulose, pectin, chitin and lipids (Schmelzer 2002). It is believed that cell wall appositions are deposited beneath fungal penetration attempts to locally reinforce the cell wall. Owing to the correlation between papilla formation and resistance to infection, papillae (and by association callose) are assumed to represent an essential first barrier to prevent fungal ingress (Schmelzer 2002). However, results by Jacobs et al. (2003) and Nishimura et al. (2003) recently questioned the role of callose in papillae. Their results showed that loss of the pathogen-inducible callose synthase GSL5 (PMR4) increased resistance to adapted powdery mildew pathogens as well as the oomycete *H. parasitica*, but not to the bacterial pathogen *P. syringae*. Interestingly, resistance in mutant plants lacking this callose synthase was shown to be the result of a hyper-activated SA pathway, indicating that PMR4 might be a suppressor of the SA defence pathway (Nishimura et al. 2003). Surprisingly, all *pmr4* alleles identified in genetic screens are characterized by premature stop codons or by frame-shifts in the coding region, suggesting that the presence of the PMR4 protein per se rather than its biochemical activity or its reaction product (callose) is required

for susceptibility (Nishimura et al. 2003; O'Connell and Panstruga 2006).

4. Biosynthesis and Accumulation of Secondary Metabolites

Biosynthesis of secondary metabolites with antimicrobial activity is a common plant response to pathogen attack. These low molecular weight compounds, also referred to as phytoalexins, show high structural diversity and are often species-specific. The model plant *Arabidopsis thaliana* was found to accumulate the phytoalexin camalexin after treatment with bacterial pathogens (Glazebrook and Ausubel 1994), oomycetes (Glazebrook et al. 1997) or necrotrophic fungi (Thomma et al. 1999b). The camalexin precursor, tryptophan, is converted into the phytoalexin via indole-3-acetaldoxime (IAOx), a reaction catalysed by CYP79B2 and CYP79B3, two members of the large cytochrome P450 family (Glawischnig et al. 2004).

A genetic screen for camalexin deficient mutants led to the identification of *phytoalexin deficient 3* (*PAD3*) which was later found to encode cytochrome P450 CYP71B15 (Glazebrook and Ausubel 1994; Zhou et al. 1999). It was only recently demonstrated that PAD3 catalyses the final step in camalexin biosynthesis (Schuhegger et al. 2006). Although camalexin accumulation is triggered by pathogen challenge, *Arabidopsis pad3* mutants are not affected in resistance against avirulent *P. syringae* strains or the oomycete *H. parasitica*, suggesting that camalexin does not contribute to resistance against these pathogens (Glazebrook and Ausubel 1994; Glazebrook et al. 1997). In contrast, *pad3* mutants are more susceptible against the necrotrophic fungus *B. cinerea*, indicating a role for camalexin in resistance to necrotrophs or fungi (Thomma et al. 1999b; van Baarlen et al. 2007). The importance of phytoalexin production in defence against non-adapted pathogens was also demonstrated in the interaction between *Arabidopsis* and the non-adapted necrotrophic fungus *A. brassicicola* (Thomma et al. 1999b). After inoculation, *Arabidopsis pad3* mutants showed a marked increase in necrotic lesions compared with wild-type plants and in some cases were able to support production of blastoconidial spores. The authors also discovered that camalexin had a direct antimicrobial effect on both *A. brassicicola* and *Neurospora crassa*, further supporting its importance in plant defence.

Antimicrobial compounds have also been identified in other plant species. Tomato (*Solanum lycopersicum*) accumulates the steroidal glycoalkaloid saponin α-tomatine, which has potent antifungal activity. The tomato leaf spot fungus *Septoria lycopersici* produces tomatinase, an extracellular enzyme that degrades the α-tomatine to the less toxic compound β_2-tomatine (Bouarab et al. 2002). Pathogen strains impaired in this enzyme are still virulent; however infected tomato plants exhibit heightened defence responses to these strains (Martin-Hernandez et al. 2000). *Nicotiana benthamiana*, a normally susceptible host of *S. lycopersici*, is not known to produce α-tomatine. However, *N. benthamiana* plants are resistant to pathogen strains carrying a mutated tomatinase allele (Bouarab et al. 2002). In addition, pre-treatment of *N. benthamiana* with either purified tomatinase or β_2-tomatine resulted in loss of resistance to the *S. lycopersici* strain carrying a mutant tomatinase gene, whereas pre-treatment with α-tomatine led to full resistance to the pathogen (Bouarab et al. 2002). Taken together, these results suggest that the enzymatic hydrolysis products of α-tomatine are required for affecting plant resistance (Bouarab et al. 2002).

Phytoalexin accumulation upon infection has also been found to occur in monocotyledonous plants. For example, juvenile *Sorghum* plants synthesize the phytoalexins apigeninidin and luteolinidin in response to infection with the fungus *Colletotrichum graminicola* (Nicholson et al. 1987, 1988). These phytoalexins are synthesized in inclusion bodies within the cell under attack once a mature fungal appressorium has formed. The inclusions then move to the site of attempted penetration and release their content into the cytoplasm while additional phytoalexin production occurs in surrounding cells (Snyder and Nicholson 1990).

5. Pathogenesis-Related Proteins

In addition to subcellular rearrangements, local reinforcement of cell walls and the targeted accumulation of secondary metabolites, pathogen attack triggers the synthesis of pathogenesis-related (PR) proteins (van Loon et al. 2006; Ferreira et al 2007). PR proteins are generally considered as defence proteins, functioning in preventing or limiting pathogen invasion and spread. However, their mode of action and their contribution to resistance have not yet been elucidated in many cases. Plants synthesize PR proteins not only after pathogen attack, but also during normal development and in response to various environmental abiotic stresses, such as drought, salinity and heavy metals (van Loon et al. 2006; Ferreira et al 2007).

Based on sequence similarity or enzymatic or biological activity, PR proteins are grouped in 17 different families. Several PR proteins have been demonstrated to possess antimicrobial activity, such as members of the PR-1, thaumatin (PR-5), defensin (PR-12), thionin (PR-13) and lipid transfer protein (PR-14) families. The most prominent family is that of PR-1, whose members are strongly induced upon biotic stress and are often used as markers of defence activation in infected plants. Members of the PR-3, -4, -8 and -11 families possess chitinase activity and accumulate either in the apoplast or in the vacuole upon infection. Members of these families may limit pathogen invasion by hydrolysing fungal chitin into oligosaccharides that then act as secondary elicitors. Defensins (PR-12) and thionins (PR-13) are families of low molecular mass and cysteine-rich peptides possessing antifungal activity, possibly by inducing calcium uptake in the microbe (van Loon et al. 2006; Ferreira et al. 2007). Pathogen attack often triggers the synergistic induction of several PR proteins belonging to different families. Therefore, constitutively high levels of PR proteins with different modes of action and effective against different target pathogens may result in broad-spectrum, durable resistance to disease (Ferreira et al. 2007).

6. Further Genetically Identified Components Involved in Defence Execution

Recently, genetic screens for suppression of nonhost resistance to powdery mildew fungi (Fig. 18.1) identified three loci involved in plant defence responses to non-adapted pathogens. *Arabidopsis penetration 1, 2* and *3* (*pen1, pen2, pen3*) mutants exhibit increased penetration by the non-adapted powdery mildew *Bgh* which is associated with localized cell death and subsequent arrest of pathogen growth (Collins et al. 2003; Lipka et al. 2005; Stein et al. 2006). *Bgh* spores are therefore unable to complete their life cycle on *Arabidopsis pen* mutants, suggesting that a second layer of defence responses is responsible for preventing further epiphytic post-invasive growth of this pathogen.

pen1 individuals carry a mutation in the t-SNARE (syntaxin) *AtSYP121*. Syntaxins are members of the SNARE family of proteins, which are essential for vesicle trafficking in eukaryotic cells (Sanderfoot et al. 2000), suggesting a role for delivery and secretion of antimicrobial proteins or compounds in defence against non-adapted pathogens. The formation of papillae at sites of attempted penetration is delayed in *pen1* mutants, suggesting that a timely defence response cannot be mounted in these individuals, potentially due to an impaired secretory machinery (Assaad et al. 2004). Further evidence for SNARE-mediated defence responses to non-adapted powdery mildew pathogens comes from research in barley, where the SNAP25-like protein SNAP34 has been shown to be required for efficient non-host and *mlo*-mediated penetration resistance (Collins et al. 2003; Douchkov et al. 2005).

PEN2 encodes a family I glycosyl hydrolase that is localized to peroxisomes (Lipka et al. 2005), while *PEN3* encodes a plasma membrane-resident ATP binding cassette (ABC) transporter that is similar to yeast pleiotropic drug resistance (PDR) transporters (Stein et al. 2006). Fluorophore-tagged PEN1 and PEN3 polypeptides were shown to accumulate at attempted pathogen entry sites (Assaad et al. 2004; Bhat et al. 2005; Stein et al. 2006). Similarly, PEN2-containing peroxisomes were found to concentrate at sites of attempted fungal penetration (Lipka et al. 2005). It is thus thought that the accumulation of PEN2-containing peroxisomes may contribute to the activation of a metabolite that has broad-spectrum antimicrobial activity and which is possibly transported via PEN3 into the apoplastic space (Lipka et al. 2005; Stein et al. 2006).

Mutations in the lipase-like EDS1 protein and its sequence-related interaction partners PAD4 and SAG101 (Feys et al. 2005) were shown to support increased epiphytic fungal growth of powdery mildew non-host species. Interestingly, in triple mutants of *pen2*, *pad4* and *sag101* susceptibility to *Erysiphe pisi* was increased such that growth of this pathogen was virtually indistinguishable from macroscopic growth of *G. orontii* on wild-type plants (Lipka et al. 2005). These findings suggest that, besides defence responses which limit pathogen entry (such as the *PEN* genes in case of powdery mildews), a second layer of defence responses controls post-invasive pathogen spread. The organization of the plant defence machinery in multiple defence layers and presence of partial genetic redundancy within the

layers may ensure the evolutionary durability of basal defence against a broad spectrum of potentially pathogenic microbes.

Recently, *flavin mono-oxygenase 1* (*FMO1*) was recognized as a novel component of defence responses in *Arabidopsis* (Bartsch et al. 2006). *FMO1* was found in a microarray-based experiment designed to identify pathogen-induced genes whose expression was *EDS1*-dependent but SA-independent. The gene was shown to be involved in resistance conferred by the TIR-NBS-LRR-type of resistance proteins, but not in resistance mediated by CC-NBS-LRR immune receptors (Bartsch et al. 2006; for classification of resistance proteins see below III. B. 1.). Interestingly, it was found that a *FMO1* homologue is expressed at early time-points after powdery mildew *Bgh* challenge in barley, coincident with haustorium formation (Olszak et al. 2006). These data suggest a conserved function of FMO1-like proteins in disease resistance against different kind of pathogens in both monocots and dicots.

III. Pathogen Effector-Triggered Defence

A. Pathogen Effectors

Some plant pathogens are able to overcome the first barriers of plant defence and cause disease on particular plant species, generally by bypassing or actively suppressing PAMP-triggered defence responses (Robatzek 2007). This has been well studied for phytopathogenic bacteria which specifically interfere with host defence responses via the secretion of type III effector proteins into plant cells (Grant et al. 2006). Additionally, *Pst* was shown to produce the phytotoxin coronatine, a MeJA-mimicking compound, that suppresses SA-based defence responses through activation of JA signalling (Abramovitch and Martin 2004). There is also evidence that several plant-pathogenic bacteria evolved means to evade recognition of flagellin by FLS2 (Zipfel et al. 2006).

In fungi and oomycetes, less is known about the specific machinery involved in suppressing or evading host defences. However, few candidate effectors have been isolated from these pathogens which appear to have functions in suppressing defence responses (Chisholm et al. 2006). Most cloned fungal effectors are small peptides/proteins of unknown function, harbouring in most cases N-terminal secretion signals (Rivas and Thomas 2002;

Catanzariti et al. 2006; Chisholm et al. 2006). It is thought that fungal effectors might be delivered through their feeding structure (haustorium) into the plant apoplast (Chisholm et al. 2006).

B. Resistance Gene-Mediated Immunity to Plant Pathogens

In response to the evolution of pathogen effectors that are able to suppress PAMP-mediated defence responses, plants evolved a secondary surveillance system based on a second line of receptors, so-called resistance proteins, that are able to specifically recognize these effectors and mount a localized and pathogen-specific response, thereby potentiating PAMP-mediated defence (Chisholm et al. 2006). The specific genetic interaction between these resistance (*R*) genes and their cognate effector genes, for historical reasons also termed avirulence (*Avr*) genes, is termed "gene-for-gene" resistance (Flor 1971).

As opposed to PAMP-mediated immunity, which acts at the species level, gene-for-gene-mediated resistance is specific to particular genetic plant variants (e.g. cultivars) which carry a particular *R* gene and are thus able to recognize a subset of a pathogen species (strains or isolates) expressing distinct effectors. Microarray analysis indicates that incompatible *R* gene-mediated interactions result in the activation of a similar set of genes as compatible interactions (Tao et al. 2003). The largest difference between the two interactions appears to be quantitative, suggesting that *R* gene-mediated defence responses indeed potentiate basal defence responses.

1. Types of Plant Resistance Genes

The vast majority of plant *R* genes encode proteins containing a nucleotide-binding site (NBS) and a leucine-rich repeat (LRR) motif (Dangl and Jones 2001; Hammond-Kosack and Parker 2003). Similar to PRRs, the C-terminal leucine rich repeat motif in *R* genes appears to represent the main determinant involved in conferring the specificity of interactions, likely by directly binding pathogen ligands or their host targets. However, unlike PRRs, which recognize highly conserved structures, accumulating evidence indicates that the LRR motifs in R proteins are under diversifying selection to evolve

new specificities to counteract the evolution of pathogen effectors (Dangl and Jones 2001). The NBS is implicated in ATP or GTP binding (Saraste et al. 1990) and is part of a larger domain that shows homology to some eukaryotic cell death regulators (Dangl and Jones 2001). Unlike PRRs, the majority of R proteins are thought to be located in the cytoplasm, suggesting that they might recognize intracellular ligands (Dangl and Jones 2001). However, biochemical evidence indicates that at least some R proteins may be membrane-bound (e.g. Axtell and Staskawicz 2003).

In addition to their conserved domains, NBS-LRR class R proteins contain either a coiled coil (CC) or a TIR (for their sequence homology to *Drosophila* Toll and mammalian Interleukin 1 receptor) domain. It appears that this domain represents a major determinant specifying which particular defence signalling pathway is employed (Aarts et al. 1998). Despite few exceptions (Xiao et al. 2001; Chandra-Shekara et al. 2004), in *Arabidopsis* the majority of TIR class R proteins have been shown to require the lipase-like EDS1 protein (see above II. C. 6.), while those of the CC type entail the membrane-bound *n*onrace-specific *d*isease *r*esistance *1* (NDR1) protein (Aarts et al. 1998). Interestingly, a requirement for either of these components appears to be largely mutually exclusive, although some exceptions have been reported (Nimchuk et al. 2003).

Apart from canonical *R* genes that have been mainly characterized, resistance to plant pathogens can also be mediated by other types of atypical resistance genes. A number of loci mediating resistance to adapted powdery mildew pathogens were identified in various *Arabidopsis* ecotypes (Schiff et al. 2001; Wilson et al. 2001). Several of these loci were subsequently found to encode two naturally occurring dominant resistance genes residing at the same locus (*RPW8.1*, *RPW8.2*; Xiao et al. 2001), which differed from known resistance genes by encoding an N-terminal transmembrane domain and a C-terminal CC domain. Unlike the more characterized prototypical *R* genes, these genes conferred broad-spectrum (race non-specific) resistance to a range of powdery mildew pathogens, but interestingly did require EDS1, suggesting that they may employ a similar set of components of known signalling pathways (Xiao et al. 2001).

Other types of dominant resistance genes encode LRR kinase-like proteins that are predicted to be membrane-bound (Nimchuk et al. 2003). Examples include *Xa21* from rice that recognizes an effector from *Xanthomonas oryzae* pv. *oryzae* (Song et al. 1995) and Cf4/Cf9 (Thomas et al. 1997). Unlike the NBS-LRR type R proteins, the LRR-kinase class of proteins has also been assigned functions in plant development. It is thus thought that they may be derived from protein families with other developmental functions and recruited for plant defence (Nimchuk et al. 2003). It is also possible that these genes require an as yet unknown NBS-LRR protein for function, similar to the membrane-anchored tomato kinase *Pto* (Salmeron et al. 1996).

In addition to these atypical resistance genes, resistance to pathogens in plants can also be complex (or quantitative) in nature. Numerous examples exist of resistance to plant pathogens conferred by multiple loci (Young 1996). In contrast to immunity conferred by single *R* genes, resistance mediated by multiple loci is poorly understood. However, Quantitative Trait Locus (QTL) analysis of these resistance loci in multiple environments is shedding first light on the action of these genes (Young 1996). Additionally, marker-assisted breeding is a powerful tool that can be used to aid in providing resistant cultivars and determining interactions between general agronomic and pathology-based traits. Cloning of these loci, however, will be instrumental in comprehending the link between these resistance loci that are quantitative in nature and the dominant *R* genes which are inherited in a simple Mendelian manner.

2.   Direct or Indirect Recognition of Pathogen Effectors?

Plant *R* genes and pathogen *Avr* genes are thought to have evolved together in an evolutionary arms race where the pathogen develops new effectors or virulence factors to overcome plant defences and in turn the plant rapidly evolves new *R* gene specificities to counteract these pathogen effectors (Maor and Shirasu 2005). The first and simplest hypothesis for the specificity of the interaction between *R* gene products and *Avr* gene products is that they directly interact with each other. The so-called "guard hypothesis", however, suggests that R proteins do not directly identify pathogen effector molecules, but rather monitor the integrity of the host targets of those effectors (van der Biezen

and Jones 1998). In this way, R proteins would scan for changes in the target molecule and would activate only upon those changes, directing a specific and localized defence response. In either case, in the absence of a cognate R protein, the effector protein would be able to modify its host protein target, resulting in disease.

The guard hypothesis was originally put forward to explain why the protein kinase *Pto* from tomato (Martin et al. 1993) requires the NBS-LRR *Prf* (Salmeron et al. 1996) for activation of defence responses after recognition of *AvrPto* from *Pst*. Pto was found to directly interact with the avirulence product and is thought to activate a signal transduction cascade leading to a resistance response (Scofield et al. 1996). It was originally thought that the interaction between Pto and AvrPto leads to a specific resistance response, but now it is postulated that Pto represents a member of a non-specific defence signalling pathway and Prf confers specificity by detecting the interaction between Pto and AvrPto (Dangl and Jones 2001).

The interaction between the *Bgh* effector Avra10 and its cognate barley R protein Mla10 is another example of indirect recognition. The barley *mildew resistance locus A (Mla)* encodes for R proteins sharing 90% similarity among each other, but they recognize various *Bgh* isolate-specific effectors (Halterman et al. 2001). Recently, the *Bgh* effector Avra10, which is recognized by Mla10, was isolated and shown to belong to a large gene family (Ridout et al. 2006). Furthermore, Mla10 was unexpectedly shown to localize both in the cytoplasm and in the nucleus, and it is the nuclear pool that is essential for Mla10 activity (Shen et al. 2007). Using a yeast two-hybrid system, Shen and colleagues identified WRKY1 and WRKY2 as interactors of Mla10. Interestingly, this physical interaction could be confirmed in vivo only in the presence of the effector Avra10, indicating that the fungal effector stimulates the association between Mla10 and WRKY1 (Shen et al. 2007). This suggests that recognition of the *Bgh* effector Avra10 by Mla10 interferes with WRKY transcription factors that act as repressors of basal defence responses (see above II. C. 2.), thereby allowing enhanced localized defence responses to occur.

Further evidence for the guard hypothesis comes from the direct interaction between the *RPM1-in*teracting protein *4* (RIN4) and several *Pst* type III effector proteins. RIN4 acts as a negative regulator of basal defence responses (Mackey et al. 2002), but is phosphorylated by AvrB and AvrRpm1 via a direct interaction, potentially to enhance its action and facilitate growth of the pathogen. However, phosphorylation of RIN4 activates the *resistance to Pseudomonas syringae* pv. *maculicola* 1 protein (RPM1), which appears to monitor for changes in RIN4, rather than the presence of a particular pathogen effector. Similarly, RIN4 also directly interacts with the effector AvrRpt2, which then eliminates RIN4, leading to activation of another resistance gene, RPS2 (Axtell and Staskawicz 2003). This example suggests that effector-mediated suppression of basal defences may target a small set of host proteins and is also a good example of how the guard hypothesis can explain why plants are able to recognize a large range of effectors with a limited set of resistance genes.

Evidence for the direct interaction between plant *R* genes and fungal *Avr* genes also comes from several sources. Dodds et al. (2004) showed that the AvrL567 proteins expressed in the haustoria of the flax rust fungus (*Melampsora lini*) were recognized inside flax (*Linum usitatissimum*) cells. The authors then further showed (by yeast two-hybrid assays coupled with in planta data based on induction of a hypersensitive response) that recognition of AvrL567 proteins by their cognate R proteins L5, L6 and L7 was via direct interaction (Dodds et al. 2006). Additionally, direct interaction between R and Avr proteins was observed in the interactions between the tobacco N protein and the effector p50 from tobacco mosaic virus (Burch-Smith et al. 2007), the rice blast fungus (*Magnaporthe grisea*) resistance gene Pi-ta and AVR-Pita from *M. grisea* (Jia et al. 2000) as well as the broad-spectrum bacterial wilt resistance gene product RRS1-R and the *Ralstonia solanacearum* type III effector PopP2 (Deslandes et al. 2003).

In conclusion, it appears that both direct and indirect interactions between R and Avr proteins may be utilized by plants to identify the presence of virulent pathogens. Occurrence of one or the other in a given case may relate to the evolutionary pressures on both pathogen and plant or may depend on the nature of the interaction between the two.

IV. Resistance at the Whole Organism/Tissue Level

Pathogen challenge typically triggers local defence responses in order to restrict microbial growth. Consistently, most of the defence mechanisms discussed so far have been shown or are assumed to operate locally and in a cell-autonomous manner. However, there is also indication for defence mechanisms in plants which extend to neighbouring cells or even distantly located plant organs. Evidence for the former results from a phenomenon termed "induced inaccessibility". It refers to the increased defence responsiveness of cells next to a cell which successfully defeated an attempted microbial attack (Lyngkjaer and Carver 1999).

In case of necrotizing pathogens, local defence responses also activate a systemic signal throughout the plant, leading to a long-lasting immunity against subsequent infections by a broad spectrum of pathogens. This phenomenon is also known as systemic acquired resistance (SAR). SAR shares many features with local resistance, including the involvement of SA and the oxidative burst. It has been long thought that localized HR and production of ROS are decisive triggers of SAR (Alvarez et al. 1998; Durrant and Dong 2004). However, recent evidence indicated that recognition of PAMPs alone is sufficient to trigger SAR (Mishina and Zeier 2007) while the production of ROS and execution of the HR is dispensable.

The onset of SAR requires the accumulation of SA and the co-ordinate expression of *PR* genes (Dong 2004; Pieterse and van Loon 2004; Grant and Lamb 2006). In plants, the main pathway leading to de novo synthesis of SA is through chorismate via isochorismate synthase (ICS), whereas only a small proportion of SA is derived from phenylalanine. Therefore, SAR is strongly compromised in the isochorismate biosynthesis mutant, *sid2* (see above II. B. 5. a.). Not surprisingly, *SID2* expression is up-regulated in response to pathogens and in tissues exhibiting SAR (Wildermuth et al. 2001).

As already mentioned above, NPR1 is a key regulator of SA-mediated defence responses. However, it also plays a central role in controlling SAR. Recently, it was discovered that NPR1 needs to be re-localized after SA treatment to be able to activate SAR. At low SA levels (the uninduced state), NPR1 is present in the cytosol as a large oligomer formed through intermolecular disulfide bonds. In SA-treated samples (the induced state), the disulfide bonds of NPR1 oligomers are reduced and the NPR1 monomer moves to the nucleus where it binds to TGA transcription factors, resulting in the activation of *PR* genes (Mou et al. 2003; Dong 2004; Pieterse and van Loon 2004). An elegant study combining inducible *NPR1* expression with transcript profiling recently showed that NPR1 also controls the expression of genes encoding components of the secretory pathway. Mutations in some of these genes diminished the secretion of PR proteins, suggesting that the encoded proteins contribute to the delivery of antimicrobial cargo in the course of SAR (Wang et al. 2005).

V. Conclusions

One of the greatest issues facing plant pathology is that the resistance afforded by *R* genes is frequently and rapidly overcome by pathogens in the field. Typically, *R* genes have only a few years before they are no longer effective in agricultural settings (e.g. Brown et al. 1993). One technique for surmounting this problem is the pyramiding of *R* genes in crop species to provide multiple resistance specificities within one genotype (Kelly 1995; Young 1996; Huang et al. 1997). Additionally, multiple resistance specificities can be employed in an agricultural population rather than being bred into a single genotype (Dangl and Jones 2001). These strategies are thought to be more effective than employing a single resistance specificity as it is unlikely that a particular pathogen would be able to prevail over multiple R proteins by losing or modifying the corresponding avirulence products. This is because these pathogens would likely be less fit, considering that most avirulence proteins are likely to provide virulence in compatible interactions (Dangl and Jones 2001). Additionally, the latter approach would result in slower epidemics of pathogens that have overcome a given resistance specificity as fewer plants in a population would contain that resistance specificity.

The possibility of the generation of transgenic crop species with chimeric resistance genes expressing specificity to new avirulence products presents an interesting opportunity to defeat newly emerging pathogens in the field. Dodds et al. (2006) recently showed that a chimeric flax rust resistance gene containing regions of both the *L6* and *L11* resistance loci provided different recognition specificity to any of the known *L* alleles. In combination with advancements in knowledge of fungal effectors, development of chimeric genes could be further improved to generate resistance genes with multiple specificities to combat multiple pathogen isolates, much like the pyramiding of resistance specificities.

Alternatively, long-term resistance to plant pathogens may come from a greater understanding of the mechanisms of resistance and susceptibility in plants. Specifically, insights into the plant factors that are required for efficient resistance against non-adapted pathogens yielded new players of plant–pathogen interactions. Similarly, research on plant factors required for pathogen growth identified several proteins, such as MLO (Büschges et al. 1997), required for the growth of specific pathogens. These types of resistance are far more robust than resistance gene-mediated defences and would provide longer-term protection against pathogens. Once the pathways involving these host proteins are further unravelled, robust resistance to multiple pathogens may be engineered. Ultimately, greater understanding of the intimate interaction between plants and the pathogens that infect them will lead to the breeding and engineering of better defence. Potentially, combining knowledge about the mode of action of particular pathogens and the host factors required to prevent or allow their growth will aid in providing durable resistance to plant pathogens.

Acknowledgements Research in the lab of R.P. is supported by funds of the Max-Planck Society and the Deutsche Forschungsgemeinschaft (DFG; PA861/4).

References

Aarts N, Metz M, Holub E, Staskawicz BJ, Daniels MJ, Parker JE (1998) Different requirements for *EDS1* and *NDR1* by disease resistance genes define at least two irtalics gene-mediated signaling pathways in *Arabidopsis*. Proc Natl Acad Sci USA 95:10306–10311

Abramovitch RB, Martin GB (2004) Strategies used by bacterial pathogens to suppress plant defenses. Curr Opin Plant Biol 7:356–364

AbuQamar S, Chen X, Dhawan R, Bluhm B, Salmeron J, Lam S, Dietrich RA, Mengiste T (2006) Expression profiling and mutant analysis reveals complex regulatory networks involved in *Arabidopsis* response to *Botrytis* infection. Plant J 48:28–44

Alonso JM, Stepanova AN (2004) The ethylene signaling pathway. Science 306:1513–1515

Alvarez ME, Pennell RI, Meijer PJ, Ishikawa A, Dixon RA, Lamb C (1998) Reactive oxygen intermediates mediate a systemic signal network in the establishment of plant immunity. Cell 92:773–784

Amano M, Toyoda K, Ichinose Y, Yamada T, Shiraishi T (1997) Association between ion fluxes and defense responses in pea and cowpea tissues. Plant Cell Physiol 38:698–706

Anderson JP, Badruzsaufari E, Schenk PM, Manners JM, Desmond OJ, Ehlert C, Maclean DJ, Ebert PR, Kazan K (2004) Antagonistic interaction between abscisic acid and jasmonate-ethylene signaling pathways modulates defense gene expression and disease resistance in *Arabidopsis*. Plant Cell 16:3460–3479

Asai T, Tena G, Plotnikova J, Willmann MR, Chiu W-L, Gomez-Gomez L, Boller T, Ausubel FM, Sheen J (2002) MAP kinase signalling cascade in *Arabidopsis* innate immunity. Nature 415:977–983

Assaad FF, Qiu JL, Youngs H, Ehrhardt D, Zimmerli L, Kalde M, Wanner G, Peck SC, Edwards H, Ramonell K, Somerville CR, Thordal-Christensen H (2004) The PEN1 syntaxin defines a novel cellular compartment upon fungal attack and is required for the timely assembly of papillae. Mol Biol Cell 15:5118–5129

Atkinson MM, Keppler LD, Orlandi EW, Baker CJ, Mischke CF (1990) Involvement of plasma membrane calcium influx in bacterial induction of the K$^+$/H$^+$ and hypersensitive responses in tobacco. Plant Physiol 92:215–221

Axtell MJ, Staskawicz BJ (2003) Initiation of *RPS2*-specified disease resistance in *Arabidopsis* is coupled to the Avr-Rpt2-directed elimination of RIN4. Cell 112:369–377

Bartsch M, Gobbato E, Bednarek P, Debey S, Schultze JL, Bautor J, Parker JE (2006) Salicylic acid-independent ENHANCED DISEASE SUSCEPTIBILITY 1 signaling in *Arabidopsis* immunity and cell death is regulated by the monooxygenase *FMO1* and the Nudix hydrolase *NUDT7*. Plant Cell 18:1038–1051

Beckers GJM, Spoel SH (2006) Fine-tuning plant defence signalling: salicylate versus jasmonate. Plant Biol 8:1–10

Bhat RA, Miklis M, Schmelzer E, Schulze-Lefert P, Panstruga R (2005) Recruitment and interaction dynamics of plant penetration resistance components in a plasma membrane microdomain. Proc Natl Acad Sci USA 102:3135–3140

Bostock RM (2005) Signal crosstalk and induced resistance: straddling the line Between cost and benefit. Annu Rev Phytopathol 43:545–580

Bouarab K, Melton R, Peart J, Baulcombe D, Osbourn A (2002) A saponin-detoxifying enzyme mediates suppression of plant defences. Nature 418:889–892

Broekaert WF, Delaure SL, De Bolle MFC, Cammue BPA (2006) The role of ethylene in host–pathogen interactions. Annu Rev Phytopathol 44:393–416

Brown JKM, Simpson CG, Wolfe MS (1993) Adaptation of barley powdery mildew populations in England to varieties with 2 resistance genes. Plant Pathol 42:108–115

Burch-Smith TM, Schiff M, Caplan JL, Tsao J, Czymmek K, Dinesh-Kumar SP (2007) A novel role for the TIR domain in association with pathogen-derived elicitors. PLoS Biology 5:501–514

Büschges R, Hollricher K, Panstruga R, Simons G, Wolter M, Frijters A, van Daelen R, van der Lee T, Diergaarde P, Groenendijk J, Töpsch S, Vos P, Salamini F, Schulze-Lefert P (1997) The barley *Mlo* gene: A novel control element of plant pathogen resistance. Cell 88:695–705

Cao H, Bowling SA, Gordon AS, Dong XN (1994) Characterization of an *Arabidopsis* mutant that is nonresponsive to inducers of systemic acquired-resistance. Plant Cell 6:1583–1592

Catanzariti AM, Dodds PN, Lawrence GJ, Ayliffe MA, Ellis JG (2006) Haustorially expressed secreted proteins from flax rust are highly enriched for avirulence elicitors. Plant Cell 18:243–256

Chandra-Shekara AC, Navarre D, Kachroo A, Kang H-G, Klessig D, Kachroo P (2004) Signaling requirements and role of salicylic acid in *HRT*- and *rrt*-mediated resistance to turnip crinkle virus in *Arabidopsis*. Plant J 40:647–659

Chen Y-F, Etheridge N, Schaller GE (2005a) Ethylene signal transduction. Ann Bot 95:901–915

Chen Z, Hong X, Zhang H, Wang Y, Li X, Zhu J-K, Gong Z (2005b) Disruption of the cellulose synthase gene, *AtCesA8/IRX1*, enhances drought and osmotic stress tolerance in *Arabidopsis*. Plant J 43:273–283

Chinchilla D, Bauer Z, Regenass M, Boller T, Felix G (2006) The *Arabidopsis* receptor kinase FLS2 binds flg22 and determines the specificity of flagellin perception. Plant Cell 18:465–476

Chisholm ST, Coaker G, Day B, Staskawicz BJ (2006) Host-microbe interactions: shaping the evolution of the plant immune response. Cell 124:803–814

Christmann A, Moes D, Himmelbach A, Yang Y, Tang Y, Grill E (2006) Integration of abscisic acid signalling into plant responses. Plant Biol 314–325

Collins NC, Thordal-Christensen H, Lipka V, Bau S, Kombrink E, Qiu JL, Hückelhoven R, Stein M, Freialdenhoven A, Somerville SC, Schulze-Lefert P (2003) SNARE-protein-mediated disease resistance at the plant cell wall. Nature 425:973–977

da Cunha L, McFall AJ, Mackey D (2006) Innate immunity in plants: a continuum of layered defenses. Microbes Infect 8:1372–1381

Dangl JL, Jones JDG (2001) Plant pathogens and integrated defence responses to infection. Nature 411:826–833

Delessert C, Kazan K, Wilson IW, van der Straeten D, Manners J, Dennis ES, Dolferus R (2005) The transcription factor ATAF2 represses the expression of pathogenesis-related genes in *Arabidopsis*. Plant J 43:745–757

Delledonne M, Xia Y, Dixon RA, Lamb C (1998) Nitric oxide functions as a signal in plant disease resistance. Nature 394:585–588

Desikan R, Hancock JT, Ichimura K, Shinozaki K, Neill SJ (2001) Harpin induces activation of the *Arabidopsis* mitogen-activated protein kinases AtMPK4 and AtMPK6. Plant Physiol 126:1579–1587

Deslandes L, Olivier J, Peeters N, Feng DX, Khounlotham M, Boucher C, Somssich I, Genin S, Marco Y (2003) Physical interaction between RRS1-R, a protein conferring resistance to bacterial wilt, and PopP2, a type III effector targeted to the plant nucleus. Proc Natl Acad Sci USA 100:8024–8029

Devoto A, Turner JG (2005) Jasmonate-regulated *Arabidopsis* stress signalling network. Physiol Plant 123:161–172

Devoto A, Nieto-Rostro M, Xie DX, Ellis C, Harmston R, Patrick E, Davis J, Sherratt L, Coleman M, Turner JG (2002) COI1 links jasmonate signalling and fertility to the SCF ubiquitin-ligase complex in *Arabidopsis*. Plant J 32:457–466

Dodds PN, Lawrence GJ, Catanzariti A-M, Ayliffe MA, Ellis JG (2004) The *Melampsora lini AvrL567* avirulence genes are expressed in haustoria and their products are recognized inside plant cells. Plant Cell 16:755–768

Dodds PN, Lawrence GJ, Catanzariti AM, Teh T, Wang CIA, Ayliffe MA, Kobe B, Ellis JG (2006) Direct protein interaction underlies gene-for-gene specificity and coevolution of the flax resistance genes and flax rust avirulence genes. Proc Natl Acad Sci USA 103:8888–8893

Dong J, Chen C, Chen Z (2003) Expression profiles of the *Arabidopsis* WRKY gene superfamily during plant defense response. Plant Mol Biol 51:21–37

Dong X (2004) NPR1, all things considered. Curr Opin Plant Biol 7:547–552

Douchkov D, Nowara D, Zierold U, Schweizer P (2005) A high-throughput gene-silencing system for the functional assessment of defense-related genes in barley epidermal cells. Mol Plant-Microbe Interact 18:755–761

Durner J, Klessig DF (1999) Nitric oxide as a signal in plants. Curr Opin Plant Biol 2:369–374

Durner J, Wendehenne D, Klessig DF (1998) Defense gene induction in tobacco by nitric oxide, cyclic GMP, and cyclic ADP-ribose. Proc Natl Acad Sci USA 95:10328–10333

Durrant WE, Dong X (2004) Systemic acquired resistance. Annu Rev Phytopathol 42:185–209

Ellis C, Karafyllidis I, Wasternack C, Turner JG (2002) The *Arabidopsis* mutant *cev1* links cell wall signaling to jasmonate and ethylene responses. Plant Cell 14:1557–1566

Eulgem T (2006) Dissecting the WRKY web of plant defense regulators. Plos Pathogen 2:1028–1030

Felix G, Duran JD, Volko S, Boller T (1999) Plants have a sensitive perception system for the most conserved domain of bacterial flagellin. Plant J 18:265–276

Felle HH, Herrmann A, Hanstein S, Hückelhoven R, Kogel K-H (2004) Apoplastic pH signaling in barley leaves attacked by the powdery mildew fungus *Blumeria graminis* f. sp. *hordei*. Mol Plant–Microbe Interact 17:118–123

Ferreira RB, Monteiro S, Freitas R, Santos CN, Chen ZJ, Batista LM, Duarte J, Borges A, Teixeira AR (2006) Fungal pathogens: the battle for plant infection. Crit Rev Plant Sci 25:505–524

Ferreira RB, Monteiro S, Freitas R, Santos CN, Chen ZJ, Batista LM, Duarte J, Borges A, Teixeira AR (2007) The role of plant defence proteins in fungal pathogenesis. Mol Plant Pathol 8:677–700

Feys BJF, Benedetti CE, Penfold CN, Turner JG (1994) *Arabidopsis* mutants selected for resistance to the phytotoxin coronatine are male sterile, insensitive to methyl jasmonate, and resistant to a bacterial pathogen. Plant Cell 6:751–759

Feys BJ, Moisan LJ, Newman MA, Parker JE (2001) Direct interaction between the *Arabidopsis* disease resistance signaling proteins, EDS1 and PAD4. EMBO J 20:5400–5411

Feys BJ, Wiermer M, Bhat RA, Moisan LJ, Medina-Escobar N, Neu C, Cabral A, Parker JE (2005) *Arabidopsis* SENESCENCE-ASSOCIATED GENE101 stabilizes and signals within an ENHANCED DISEASE SUSCEPTIBILITY1 complex in plant innate immunity. Plant Cell 17:2601–2613

Flor HH (1971) Current status of gene-for-gene concept. Annu Rev Phytopathol 9:275–296

Frye CA, Tang D, Innes RW (2001) Negative regulation of defense responses in plants by a conserved MAPKK kinase. Proc Natl Acad Sci USA 98:373–378

Ghassemian M, Nambara E, Cutler S, Kawaide H, Kamiya Y, McCourt P (2000) Regulation of abscisic acid signaling by the ethylene response pathway in *Arabidopsis*. Plant Cell 12:1117–1126

Glawischnig E, Hansen BG, Olsen CE, Halkier BA (2004) Camalexin is synthesized from indole-3-acetaldoxime, a key branching point between primary and secondary metabolism in *Arabidopsis*. Proc Natl Acad Sci USA 101:8245–8250

Glazebrook J (2005) Contrasting mechanisms of defense against biotrophic and necrotrophic pathogens. Annu Rev Phytopathol 43:205–227

Glazebrook J, Ausubel FM (1994) Isolation of phytoalexin-deficient mutants of *Arabidopsis thaliana* and char-

acterization of their interactions with bacterial pathogens. Proc Natl Acad Sci USA 91:8955–8959

Glazebrook J, Zook M, Mert F, Kagan I, Rogers EE, Crute IR, Holub EB, Hammerschmidt R, Ausubel FM (1997) Phytoalexin-deficient mutants of *Arabidopsis* reveal that *PAD4* encodes a regulatory factor and that four *PAD* genes contribute to downy mildew resistance. Genetics 146: 381–392

Glazebrook J, Chen WJ, Estes B, Chang HS, Nawrath C, Metraux JP, Zhu T, Katagiri F (2003) Topology of the network integrating salicylate and jasmonate signal transduction derived from global expression phenotyping. Plant J 34:217–228

Gomez-Gomez L, Boller T (2000) FLS2: An LRR receptor-like kinase involved in the perception of the bacterial elicitor flagellin in *Arabidopsis*. Mol Cell 5:1003–1011

Gomez-Gomez L, Bauer Z, Boller T (2001) Both the extracellular leucine-rich repeat domain and the kinase activity of FLS2 are required for flagellin binding and signaling in *Arabidopsis*. Plant Cell 13:1155–1163

Grant M, Lamb C (2006) Systemic immunity. Curr Opin Plant Biol 9:414–420

Grant SR, Fisher EJ, Chang JH, Mole BM, Dangl JL (2006) Subterfuge and manipulation: type III effector proteins of phytopathogenic bacteria. Annu Rev Microbiol 60:425–449

Guo H, Ecker JR (2004) The ethylene signaling pathway: new insights. Curr Opin Plant Biol 7:40–49

Halterman D, Zhou F, Wei F, Wise RP, Schulze-Lefert P (2001) The MLA6 coiled-coil, NBS-LRR protein confers *AvrMla6*-dependent resistance specificity to *Blumeria graminis* f. sp. *hordei* in barley and wheat. Plant J 25: 335–48

Hammond-Kosack KE, Parker JE (2003) Deciphering plant-pathogen communication: fresh perspectives for molecular resistance breeding. Curr Opin Biotechnol 14:177–193

Hayashi F, Smith KD, Ozinsky A, Hawn TR, Yi EC, Goodlett DR, Eng JK, Akira S, Underhill DM, Aderem A (2001) The innate immune response to bacterial flagellin is mediated by Toll-like receptor 5. Nature 410:1099–1103

He YH, Gan SS (2002) A gene encoding an acyl hydrolase is involved in leaf senescence in *Arabidopsis*. Plant Cell 14:805–815

Heath MC (2000) Nonhost resistance and nonspecific plant defenses. Curr Opin Plant Biol 3:315–319

Heo WD, Lee SH, Kim MC, Kim JC, Chung WS, Chun HJ, Lee KJ, Park CY, Park HC, Choi JY, Cho MJ (1999) Involvement of specific calmodulin isoforms in salicylic acid-independent activation of plant disease resistance responses. Proc Natl Acad Sci USA 96:766–771

Hernandez-Blanco C, Feng DX, Hu J, Sanchez-Vallet A, Deslandes L, Llorente F, Berrocal-Lobo M, Keller H, Barlet X, Sanchez-Rodriguez C, Anderson LK, Somerville S, Marco Y, Molina A (2007) Impairment of cellulose synthases required for *Arabidopsis* secondary cell wall formation enhances disease resistance. Plant Cell 19:890–903

Holub EB, Cooper A (2004) Matrix, reinvention in plants: how genetics is unveiling secrets of non-host disease resistance. Trends Plant Sci 9:211–214

Huang N, Angeles ER, Domingo J, Magpantay G, Singh S, Zhang G, Kumaravadivel N, Bennett J, Khush GS

(1997) Pyramiding of bacterial blight resistance genes in rice: marker-assisted selection using RFLP and PCR. Theor Appl Genet 95:313–320

Hückelhoven R, Kogel K (2003) Reactive oxygen intermediates in plant-microbe interactions: who is who in powdery mildew resistance? Planta 216:891–902

Ingle RA, Carstens M, Denby KJ (2006) PAMP recognition and the plant-pathogen arms race. BioEssays 28:880–889

Jabs T, Tschope M, Colling C, Hahlbrock K, Scheel D (1997) Elicitor-stimulated ion fluxes and O_2^- from the oxidative burst are essential components in triggering defense gene activation and phytoalexin synthesis in parsley. Proc Natl Acad Sci USA 94:4800–4805

Jacobs AK, Lipka V, Burton RA, Panstruga R, Strizhov N, Schulze-Lefert P, Fincher GB (2003) An *Arabidopsis* callose synthase, GSL5, is required for wound and papillary callose formation. Plant Cell 15:2503–2513

Jensen M, Rung J, Gregersen P, Gjetting T, Fuglsang A, Hansen M, Joehnk N, Lyngkjaer M, Collinge D (2007) The *HvNAC6* transcription factor: a positive regulator of penetration resistance in barley and *Arabidopsis*. Plant Mol Biol 65:137–150

Jia Y, McAdams SA, Bryan GT, Hershey HP, Valent B (2000) Direct interaction of resistance gene and avirulence gene products confers rice blast resistance. EMBO J 19:4004–4014

Jonak C, Okresz L, Bogre L, Hirt H (2002) Complexity, cross talk and integration of plant MAP kinase signalling. Curr Opin Plant Biol 5:415–424

Jones DA, Takemoto D (2004) Plant innate immunity – direct and indirect recognition of general and specific pathogen-associated molecules. Curr Opin Immunol 16:48–62

Jones JDG, Dangl JL (2006) The plant immune system. Nature 444:323–329

Kelly JD (1995) Use of random amplified polymorphic DNA markers in breeding for major gene resistance to plant pathogens. Hortscience 30:461–465

Kim MC, Panstruga R, Elliott C, Müller J, Devoto A, Yoon HW, Park HC, Cho MJ, Schulze-Lefert P (2002) Calmodulin interacts with MLO protein to regulate defence against mildew in barley. Nature 416:447–450

Knoth C, Ringler J, Dangl JL, Eulgem T (2007) *Arabidopsis* WRKY70 is required for full *RPP4*-mediated disease resistance and basal defense against *Hyaloperonospora parasitica*. Mol Plant–Microbe Interact 20:120–128

Kobayashi Y, Kobayashi I, Funaki Y, Fujimoto S, Takemoto T, Kunoh H (1997) Dynamic reorganization of microfilaments and microtubules is necessary for the expression of non-host resistance in barley coleoptile cells. Plant J 11:525–537

Koh S, Somerville S (2006) Show and tell: cell biology of pathogen invasion. Curr Opin Plant Biol 9:406–413

Koh S, Andre A, Edwards H, Ehrhardt D, Somerville S (2005) *Arabidopsis thaliana* subcellular responses to compatible *Erysiphe cichoracearum* infections. Plant J 44:516–529

Kovtun Y, Chiu W-L, Tena G, Sheen J (2000) Functional analysis of oxidative stress-activated mitogen-activated protein kinase cascade in plants. Proc Natl Acad Sci USA 97:2940–2945

Laloi C, Apel K, Danon A (2004) Reactive oxygen signalling: the latest news. Curr Opin Plant Biol 7:323–328

Lecourieux D, Raneva R, Pugin A (2006) Calcium in plant defence-signalling pathways. New Phytol 171:249–269

Li J, Brader G, Palva ET (2004) The WRKY70 transcription factor: a node of convergence for jasmonate-mediated and salicylate-mediated signals in plant defense. Plant Cell 16:319–331

Li J, Brader G, Kariola T, Tapio Palva E (2006) WRKY70 modulates the selection of signaling pathways in plant defense. Plant J 46:477–491

Lindermayr C, Saalbach G, Durner J (2005) Proteomic identification of S-nitrosylated proteins in *Arabidopsis*. Plant Physiol 137:921–930

Lipka V, Dittgen J, Bednarek P, Bhat R, Wiermer M, Stein M, Landtag J, Brandt W, Rosahl S, Scheel D, Llorente F, Molina A, Parker J, Somerville S, Schulze-Lefert P (2005) Pre- and postinvasion defenses both contribute to nonhost resistance in *Arabidopsis*. Science 310:1180–1183

Lorenzo O, Solano R (2005) Molecular players regulating the jasmonate signalling network. Curr Opin Plant Biol 8:532–540

Lorenzo O, Piqueras R, Sanchez-Serrano JJ, Solano R (2003) ETHYLENE RESPONSE FACTOR1 integrates signals from ethylene and jasmonate pathways in plant defence. Plant Cell 15:165–178

Lorenzo O, Chico JM, Sanchez-Serrano JJ, Solano R (2004) *Jasmonate-insensitive 1* encodes a MYC transcription factor essential to discriminate between different jasmonate-regulated defense responses in *Arabidopsis*. Plant Cell 16:1938–1950

Ludwig AA, Romeis T, Jones JDG (2004) CDPK-mediated signalling pathways: specificity and cross-talk. J Exp Bot 55:181–188

Lyngkjaer MF, Carver TLW (1999) Induced accessibility and inaccessibility to *Blumeria graminis* f.sp. *hordei* in barley epidermal cells attacked by a compatible isolate. Physiol Mol Plant Pathol 55:151–162

Mackey D, Holt BF, Wiig A, Dangl JL (2002) RIN4 interacts with *Pseudomonas syringae* type III effector molecules and is required for RPM1-mediated resistance in *Arabidopsis*. Cell 108:743–754

Maor R, Shirasu K (2005) The arms race continues: battle strategies between plants and fungal pathogens. Curr Opin Microbiol 8:399–404

Marchive C, Mzid R, Deluc L, Barrieu F, Pirrello J, Gauthier A, Corio-Costet M-F, Regad F, Cailleteau B, Hamdi S, Lauvergeat V (2007) Isolation and characterization of a *Vitis vinifera* transcription factor, VvWRKY1, and its effect on responses to fungal pathogens in transgenic tobacco plants. J Exp Bot 58:1999–2010

Martin GB, Brommonschenkel SH, Chunwongse J, Frary A, Ganal MW, Spivey R, Wu T, Earle ED, Tanksley SD (1993) Map-based cloning of a protein kinase gene conferring disease resistance in tomato. Science 262:1432–1436

Martin-Hernandez AM, Dufresne M, Hugouvieux V, Melton R, Osbourn A (2000) Effects of targeted replacement of the tomatinase gene on the interaction of *Septoria lycopersici* with tomato plants. Mol Plant–Microbe Interact 13:1301–1311

Mauch-Mani B, Mauch F (2005) The role of abscisic acid in plant-pathogen interactions. Curr Opin Plant Biol 8:409–414

Mengiste T, Chen X, Salmeron J, Dietrich R (2003) The *BOTRYTIS SUSCEPTIBLE1* gene encodes an R2R3MYB transcription factor protein that is required for biotic and abiotic stress responses in *Arabidopsis*. Plant Cell 15: 2551–2565

Menke FLH, van Pelt JA, Pieterse CMJ, Klessig DF (2004) Silencing of the mitogen-activated protein kinase MPK6 compromises disease resistance in *Arabidopsis*. Plant Cell 16:897–907

Micali C, Göllner K, Humphry M, Consonni C, Panstruga R (2008) The powdery mildew disease of Arabidopsis: A paradigm for the interaction between plants and biotrophic fungi. The Arabidopsis Book Rockville, MD: American Society of Plant Biologists. doi: 10.1199/tab.0115, http://www.aspb.org/publications/arabidopsis/

Miklis M, Consonni C, Bhat RA, Lipka V, Schulze-Lefert P, Panstruga R (2007) Barley MLO modulates actin-dependent and actin-independent antifungal defense pathways at the cell periphery. Plant Physiol 144:1132–1143

Mishina TE, Zeier J (2007) Pathogen-associated molecular pattern recognition rather than development of tissue necrosis contributes to bacterial induction of systemic acquired resistance in *Arabidopsis*. Plant J 50:500–513

Mohr PG, Cahill DM (2003) Abscisic acid influences the susceptibility of *Arabidopsis thaliana* to *Pseudomonas syringae* pv. tomato and *Peronospora parasitica*. Funct Plant Biol 30:461–469

Mou Z, Fan WH, Dong XN (2003) Inducers of plant systemic acquired resistance regulate NPR1 function through redox changes. Cell 113:935–944

Nawrath C, Métraux JP (1999) Salicylic acid induction-deficient mutants of *Arabidopsis* express *PR-2* and *PR-5* and accumulate high levels of camalexin after pathogen inoculation. Plant Cell 11:1393–1404

Nawrath C, Heck S, Parinthawong N, Métraux JP (2002) EDS5, an essential component of salicylic acid-dependent signaling for disease resistance in *Arabidopsis*, is a member of the MATE transporter family. Plant Cell 14:275–286

Neill SJ, Desikan R, Clarke A, Hurst RD, Hancock JT (2002) Hydrogen peroxide and nitric oxide as signalling molecules in plants. J Exp Bot 53:1237–1247

Nicholson RL, Kollipara SS, Vincent JR, Lyons PC, Cadena-Gomez G (1987) Phytoalexin synthesis by the Sorghum mesocotyl in response to infection by pathogenic and nonpathogenic fungi. Proc Natl Acad Sci USA 84:5520–5524

Nicholson RL, Jamil FF, Snyder BA, Lue WL, Hipskind J (1988) Phytoalexin synthesis in the juvenile sorghum leaf. Physiol Mol Plant Pathol 33:271–278

Nickstadt A, Thomma BPHJ, Feussner I, Kangasjarvi J, Zeier J, Loeffler C, Scheel D, Berger S (2004) The jasmonate-insensitive mutant *jin1* shows increased resistance to biotrophic as well as necrotrophic pathogens. Mol Plant Pathol 5:425–434

Nimchuk Z, Eulgem T, Holt Iii BF, Dangl JL (2003) Recognition and response in the plant immune system. Annu Rev Genet 37:579–609

Nishimura MT, Stein M, Hou B-H, Vogel JP, Edwards H, Somerville SC (2003) Loss of a callose synthase results in salicylic acid-dependent disease resistance. Science 301:969–972

Nürnberger T, Brunner F (2002) Innate immunity in plants and animals: emerging parallels between the recognition of general elicitors and pathogen-associated molecular patterns. Curr Opin Plant Biology 5: 318–324

Nürnberger T, Lipka V (2005) Non-host resistance in plants: new insights into an old phenomenon. Mol Plant Pathol 6:335–345

Nürnberger T, Brunner F, Kemmerling B, Piater L (2004) Innate immunity in plants and animals: striking similarities and obvious differences. Immunol Rev 198:249–266

O'Connell RJ, Panstruga R (2006) Tête à tête inside a plant cell: establishing compatibility between plants and biotrophic fungi and oomycetes. New Phytol 171:699–718

Olszak B, Malinovsky FG, Brodersen P, Grell M, Giese H, Petersen M, Mundy J (2006) A putative flavin-containing mono-oxygenase as a marker for certain defense and cell death pathways. Plant Sci 170:614–623

Papadopoulou K, Melton RE, Legett M, Daniels MJ, Osbourn AE (1999) Compromised disease resistance in saponin-deficient plants. Proc Natl Acad Sci USA 96:12923–12928

Pennell RI, Lamb C (1997) Programmed cell death in plants. Plant Cell 9:1157–1168

Petersen M, Brodersen P, Naested H, Andreasson E, Lindhart U, Johansen B, Nielsen HB, Lacy M, Austin MJ, Parker JE, Sharma SB, Klessig DF, Martienssen R, Mattsson O, Jensen AB, Mundy J (2000) *Arabidopsis* MAP Kinase 4 negatively regulates systemic acquired resistance. Cell 103:1111–1120

Pieterse CMJ, van Loon LC (2004) NPR1: the spider in the web of induced resistance signaling pathways. Curr Opin Plant Biol 7:456–464

Piffanelli P, Zhou F, Casais C, Orme J, Jarosch B, Schaffrath U, Collins NC, Panstruga R, Schulze-Lefert P (2002) The barley MLO modulator of defense and cell death is responsive to biotic and abiotic stress stimuli. Plant Physiol 129:1076–1085

Prats E, Mur LAJ, Sanderson R, Carver TLW (2005) Nitric oxide contributes both to papilla-based resistance and the hypersensitive response in barley attacked by Blumeria graminis f. sp hordei. Mol Plant Pathol 6:65–78

Ridout CJ, Skamnioti P, Porritt O, Sacristan S, Jones JDG, Brown JKM (2006) Multiple avirulence paralogues in cereal powdery mildew fungi may contribute to parasite fitness and defeat of plant resistance. Plant Cell 18:2402–2414

Rivas S, Thomas CM (2002) Recent advances in the study of tomato *Cf* resistance genes. Mol Plant Pathol 3:277–282

Robatzek S (2007) Vesicle trafficking in plant immune responses. Cell Microbiol 9:1–8

Robatzek S, Chinchilla D, Boller T (2006) Ligand-induced endocytosis of the pattern recognition receptor FLS2 in *Arabidopsis*. Genes Dev 20:537–542

Romeis T (2001) Protein kinases in the plant defence response. Curr Opin Plant Biol 4:407–414

Romeis T, Ludwig AA, Martin R, Jones JDG (2001) Calcium-dependent protein kinases play an essential role in a plant defence response. EMBO J 20:5556–5567

Ron M, Avni A (2004) The receptor for the fungal elicitor ethylene-inducing xylanase is a member of a resistance-like gene family in tomato. Plant Cell 16:1604–1615

Rushton PJ, Torres JT, Parniske M, Wernert P, Hahlbrock K, Somssich IE (1996) Interaction of elicitor-induced DNA-binding proteins with elicitor response elements in the promoters of parsley PR1 genes. EMBO J 15:5690–5700

Salmeron JM, Oldroyd GED, Rommens CMT, Scofield SR, Kim H, Lavelle DT, Dahlbeck D, Staskawicz BJ (1996) Tomato Prf is a member of the leucine-rich repeat class of plant disease resistance genes and lies embedded within the Pto kinase gene cluster. Cell 86:123–133

Sanderfoot AA, Assaad FF, Raikhel NV (2000) The Arabidopsis genome. An abundance of soluble N-ethylmaleimide-sensitive factor adaptor protein receptors. Plant Physiol 124:1558–1569

Saraste M, Sibbald PR, Wittinghofer A (1990) The P-loop – a common motif in ATP- and GTP-binding proteins. Trends Biochem Sci 15:430–434

Schiff CL, Wilson IW, Somerville SC (2001) Polygenic powdery mildew disease resistance in Arabidopsis thaliana: quantitative trait analysis of the accession Warschau-1. Plant Pathol 50:690–701

Schmelzer E (2002) Cell polarization, a crucial process in fungal defence. Trends Plant Sci 7:411–415

Schuhegger R, Nafisi M, Mansourova M, Petersen BL, Olsen CE, Svatos A, Halkier BA, Glawischnig E (2006) CYP71B15 (PAD3) catalyzes the final step in camalexin biosynthesis. Plant Physiol 141:1248–1254

Schulze-Lefert P (2004) Knocking on heaven's wall: pathogenesis of and resistance to biotrophic fungi at the cell wall. Curr Opin Plant Biol 7:377–383

Scofield SR, Tobias CM, Rathjen JP, Chang JH, Lavelle DT, Michelmore RW, Staskawicz BJ (1996) Molecular basis of gene-for-gene specificity in bacterial speck disease of tomato. Science 274:2063–2065

Shan XC, Goodwin PH (2005) Reorganization of filamentous actin in Nicotiana benthamiana leaf epidermal cells inoculated with Colletotrichum destructivum and Colletotrichum graminicola. Int J Plant Sci 166:31–39

Shen QH, Saijo Y, Mauch S, Biskup C, Bieri S, Keller B, Seki H, Ulker B, Somssich IE, Schulze-Lefert P (2007) Nuclear activity of MLA immune receptors links isolate-specific and basal disease-resistance responses. Science 315:1098–1103

Shimada C, Lipka V, O'Connel R, Okuno T, Schulze-Lefert P, Takano Y (2006) Nonhost resistance in Arabidopsis–Colletotrichum interactions acts at the cell periphery and requires actin filament function. Mol Plant–Microbe Interact 19:270–279

Snyder BA, Nicholson RL (1990) Synthesis of phytoalexins in Sorghum as a site-specific response to fungal ingress. Science 248:1637–1639

Song WY, Wang GL, Chen LL, Kim HS, Pi LY, Holsten T, Gardner J, Wang B, Zhai WX, Zhu LH, Fauquet C, Ronald P (1995) A receptor kinase-like protein encoded by the rice disease resistance gene, Xa21. Science 270:1804–1806

Spoel SH, Koornneef A, Claessens SMC, Korzelius JP, van Pelt JA, Mueller MJ, Buchala AJ, Metraux JP, Brown R, Kazan K, van Loon LC, Dong XN, Pieterse CMJ (2003) NPR1 modulates cross-talk between salicylate- and jasmonate-dependent defense pathways through a novel function in the cytosol. Plant Cell 15:760–770

Staswick PE, Su WP, Howell SH (1992) Methyl jasmonate inhibition of root growth and induction of a leaf protein are decreased in an Arabidopsis thaliana mutant. Proc Natl Acad Sci USA 89:6837–6840

Stein M, Dittgen J, Sanchez-Rodriguez C, Hou BH, Molina A, Schulze-Lefert P, Lipka V, Somerville S (2006) Arabidopsis PEN3/PDR8, an ATP binding cassette transporter, contributes to nonhost resistance to inappropriate pathogens that enter by direct penetration. Plant Cell 18:731–746

Takemoto D, Hardham AR (2004) The cytoskeleton as a regulator and target of biotic interactions in plants. Plant Physiol 136:3864–3876

Takemoto D, Jones DA, Hardham AR (2003) GFP-tagging of cell components reveals the dynamics of subcellular re-organization in response to infection of Arabidopsis by oomycete pathogens. Plant J 33:775–792

Tang DZ, Christiansen KM, Innes RW (2005) Regulation of plant disease resistance, stress responses, cell death, and ethylene signaling in Arabidopsis by the EDR1 protein kinase. Plant Physiol 138:1018–1026

Tao Y, Xie ZY, Chen WQ, Glazebrook J, Chang HS, Han B, Zhu T, Zou GZ, Katagiri F (2003) Quantitative nature of Arabidopsis responses during compatible and incompatible interactions with the bacterial pathogen Pseudomonas syringae. Plant Cell 15:317–330

Thomas CM, Jones DA, Parniske M, Harrison K, Balint-Kurti PJ, Hatzixanthis K, Jones J (1997) Characterization of the tomato Cf-4 gene for resistance to Cladosporium fulvum identifies sequences that determine recognitional specificity in Cf-4 and Cf-9. Plant Cell 9:2209–2224

Thomma BPHJ, Eggermont K, Penninckx IAMA, Mauch-Mani B, Vogelsang R, Cammue BPA, Broekaert WF (1998) Separate jasmonate-dependent and salicylate-dependent defense-response pathways in Arabidopsis are essential for resistance to distinct microbial pathogens. Proc Natl Acad Sci USA 95:15107–15111

Thomma BPHJ, Eggermont K, Tierens KFMJ, Broekaert WF (1999a) Requirement of functional Ethylene-Insensitive 2 gene for efficient resistance of Arabidopsis to infection by Botrytis cinerea. Plant Physiol 121:1093–1101

Thomma BPHJ, Nelissen I, Eggermont K, Broekaert WF (1999b) Deficiency in phytoalexin production causes enhanced susceptibility of Arabidopsis thaliana to the fungus Alternaria brassicicola. Plant J 19:163–171

Torres MA, Dangl JL, Jones JDG (2002) Arabidopsis gp91(phox) homologues AtrbohD and AtrbohF are required for accumulation of reactive oxygen intermediates in the plant defense response. Proc Natl Acad Sci USA 99:517–522

Torres MA, Jones JDG, Dangl JL (2005) Pathogen-induced, NADPH oxidase-derived reactive oxygen intermediates suppress spread of cell death in *Arabidopsis thaliana*. Nat Genet 37:1130–1134

Torres MA, Jones JDG, Dangl JL (2006) Reactive oxygen species signaling in response to pathogens. Plant Physiol 141:373–378

Ülker B, Somssich IE (2004) WRKY transcription factors: from DNA binding towards biological function. Curr Opin Plant Biol 7:491–498

van Baarlen P, Woltering EJ, Staats M, van Kan JAL (2007) Histochemical and genetic analysis of host and non-host interactions of *Arabidopsis* with three *Botrytis* species: an important role for cell death control. Mol Plant Pathol 8:41–54

van der Biezen EA, Jones JDG (1998) Plant disease-resistance proteins and the gene-for-gene concept. Trends Biochem Sci 23:454–456

van Loon LC, Rep M, Pieterse CMJ (2006) Significance of inducible defense-related proteins in infected plants. Annu Rev Phytopathol 44:135–162

Vorwerk S, Somerville SC, Somerville CR (2004) The role of plant cell wall polysaccharide composition in disease resistance. Trends Plant Sci 9:203–209

Wang D, Weaver ND, Kesarwani M,, Dong XN (2005) Induction of protein secretory pathway is required for systemic acquired resistance. Science 308:1036–1040

Wendehenne D, Durner J, Klessig DF (2004) Nitric oxide: a new player in plant signalling and defence responses. Curr Opin Plant Biol 7:449–455

Wiermer M, Feys BJ, Parker JE (2005) Plant immunity: the EDS1 regulatory node. Curr Opin Plant Biol 8:383–389

Wildermuth MC, Dewdney J, Wu G, Ausubel FM (2001) Isochorismate synthase is required to synthesize salicylic acid for plant defence. Nature 414:562–565

Wilson IW, Schiff CL, Hughes DE, Somerville SC (2001) Quantitative trait loci analysis of powdery mildew disease resistance in the *Arabidopsis thaliana* accession Kashmir-1. Genetics 158:1301–1309

Xiao S, Ellwood S, Calis O, Patrick E, Li T, Coleman M, Turner JG (2001) Broad-spectrum mildew resistance in *Arabidopsis thaliana* mediated by *RPW8*. Science 291:118–120

Xu X, Chen C, Fan B, Chen Z (2006) Physical and functional interactions between pathogen-induced *Arabidopsis* WRKY18, WRKY40, and WRKY60 transcription factors. Plant Cell 18:1310–1326

Yang K-Y, Liu Y, Zhang S (2001) Activation of a mitogen-activated protein kinase pathway is involved in disease resistance in tobacco. Proc Natl Acad Sci USA 98:741–746

Young ND (1996) QTL mapping and quantitative disease resistance in plants. Annu Rev Phytopathol 34:479–501

Yun BW, Atkinson HA, Gaborit C, Greenland A, Read ND, Pallas JA, Loake GJ (2003) Loss of actin cytoskeletal function and EDS1 activity, in combination, severely compromises non-host resistance in *Arabidopsis* against wheat powdery mildew. Plant J 34:768–777

Zeidler D, Zahringer U, Gerber I, Dubery I, Hartung T, Bors W, Hutzler P, Durner J (2004) Innate immunity in *Arabidopsis thaliana*: lipopolysaccharides activate nitric oxide synthase (NOS) and induce defense genes. Proc Natl Acad Sci USA 101:15811–15816

Zhou N, Tootle TL, Glazebrook J (1999) *Arabidopsis PAD3*, a gene required for camalexin biosynthesis, encodes a putative cytochrome P450 monooxygenase. Plant Cell 11:2419–2428

Zipfel C, Robatzek S, Navarro L, Oakeley EJ, Jones JDG, Felix G, Boller T (2004) Bacterial disease resistance in *Arabidopsis* through flagellin perception. Nature 428:764–767

Zipfel C, Kunze G, Chinchilla D, Caniard A, Jones JDG, Boller T, Felix G (2006) Perception of the bacterial PAMP EF-Tu by the receptor EFR restricts *Agrobacterium*-mediated transformation. Cell 125:749–760

Biosystematic Index

A

Achlya, 124
    A. intricata, 125
Achnatherum, 281, 282, 285, 294
    A. eminens, 282
    A. inebrians, 282, 285
    A. robustum, 282, 285, 294
    A. sibiricum, 281
Acremonium. see Neotyphodium
Adenocarpus, 294
Aglaephyton, 260
Agropyron
    A. ciliare, 281
Agrostis
    A. hiemalis, 281, 294
    A. perennans, 281
    A. stolonifera, 281
    A. tenuis, 281
Albugo, 7, 124, 126
Alternaria, 340, 347, 348, 350, 352
    A. alternata, 184, 186, 202, 226
    A. brassicicola, 188, 190, 340, 347, 350, 352, 369, 370, 373
    A. citri, 184, 193
    A. eichorniae, 202
    A. solani, 348
Anthoxanthum, 281
    A. odoratum, 281
Aphanomyces, 124
    A. euteiches, 126
Arabidopsis, 16–18, 37, 45, 46, 224, 225, 227, 231, 262
    A. thaliana, 17, 80, 86, 88, 184, 247, 340, 343, 344, 347
Arrhenatherum
    A. elatius, 281
Arthrinium
    A. cupidatum, 124
Ascobolus, 121
    A. furfuraceus, 122
    A. immersus, 121–123
Aspergillus, 312, 324
    A. flavus, 183, 193
    A. fumigatus, 229, 230, 287
    A. nidulans, 139, 193, 230, 296, 299
    A. niger, 117, 190
    A. versicolor, 117
Athelia
    A. arachnoidea, 307
Atkinsonella, 275, 276

B

Bacidina
    B. scutellifera, 314

Balansia, 275
    B. hypoxylon, 276
Barley, 100, 102–108
Basidiobolus, 127
Blumeria
    B. graminis, 103
Blumeria graminis f. sp. *hordei,* 118, 119
Blumeria graminis f. sp. *hordei,* 227
Blumeria graminis f.sp. *hordei,* 345, 351
Blumeria graminis f.sp. *hordei (Bgh),* 365, 366, 372, 374, 375, 377
B. graminis f. sp. *tritici,* 227
Botanophila, 278
Botryosphaeria, 120
Botryotinia
    B. fuckeliana, 29
Botrytis, 120, 342, 343, 345, 347, 350–352, 372
    B. cinerea, 10, 29, 118, 138, 140, 183–186, 188, 190, 192, 193, 227, 231, 342, 343, 345, 347, 350–352, 369, 370, 372, 373
        alternative control strategies, 30
        apothecia, 31
        appressoria, 30, 38
        candidate gene approach, 38, 42
        cDNA libraries, 31, 38
        gene replacement, 31
        genetic variability, 31, 40
        genome sequence, 31, 38, 45, 46
        insertional mutagenesis, 31, 38
        knock-out mutants, 38, 44
        life- and disease cycle, 30
        quiescent stage, 30
        sclerotia, 31, 33, 34, 36, 41, 43
        secondary metabolites, 44, 46
        targeted gene inactivation, 31–37
        tetraspanin-like protein, 38–39
        transformation, 31, 38
    B. elliptica, 346
    B. squamosa, 31
Brachyelytrum
    B. erectum, 281
Brachypodium, 281
    B. pinnatum, 281
    B. sylvaticum, 281
Brassica
    B. napus, 183
Brassicaceae, 107
Brassicaceae, 338
Bromus
    B. auleticus, 283
    B. benekenii, 281
    B. erectus, 278, 281

Bromus (cont.)
    B. purgans, 281
    B. ramosus, 281
    B. setifolius, 279, 282, 283
Bulgaria
    B. inquinans, 202
Burkholderia, 109, 110
    B. glumae, 347

C
Caenorhabditis elegance, 226
Calamagrostis
    C. villosa, 281
Candida
    C. albicans, 230
Candidatus Glomeribacter gigasporarum, 109
Ceratobasidiales, 101
Cercospora, 201, 202, 204, 205, 207, 209, 210, 213–215
    C. beticola, 204, 206
    C. nicotianae, 205–209, 211, 214, 215
    C. petunia, 206
Cetraria
    C. islandica, 318
Cladosporium, 341, 343, 347, 351
    C. cladosporioides, 117, 202
    C. cucumerinum, 202
    C. fulvum, 13, 135–150, 174, 175, 176, 341, 343, 347
    C. herbarum, 202, 351
    C. phlei, 202
    C. sphaerospermum, 117
Claviceps
    C. fusiformis, 287, 291
    C. purpurea, 184–196, 193, 287, 290, 291
Clavicipitaceae, 275–284, 287, 292
Cochliobolus, 122
    C. carbonum, 38, 39, 182, 185, 186, 190, 191
    C. heterostrophus, 164, 165, 166, 168, 169, 190
Collema, 321
Colletotrichum, 120, 371
    C. coccodes, 226
    C. gloeosporioides, 184, 191, 193, 232
    C. graminicola, 374
    C. higginsianum, 340
    C. lindemuthianum, 183, 190
    C. magna, 105, 184
    C. orbiculare, 345
    C. trifolii, 230, 346
Colletotrichum species, 30, 69, 80
Conidiobolus, 124
Convolvulaceae, 276
Coprinopsis
    C. cinerea, 241–243, 245–248
Cryphonectria
    C. parasitica, 182, 183
Cryptococcus
    C. neoformans, 241–243, 245

D
Dactylis
    D. glomerata, 281
Dactylorhiza spp., 103, 107
Danthonia, 276

Deightoniella
    D. torulosa, 119–120
Dictyonema, 310
    D. glabratum, 317
Dipodascus, 121
Drechslera
    D. turcica, 118
Drosophila, 229

E
Echinopogon
    E. ovatus, 282, 294
Elsinoe, 202
Elymus
    E. canadensis, 281
    E. hystrix, 281
    E. villosus, 281
    E. virginicus, 281
Entomophthora, 124, 127
Epichloë
    E. amarillans, 279, 281, 283, 294, 297
    E. baconii, 279, 281, 283
    E. brachyelytri, 281
    E. bromicola, 278, 281, 283, 297
    E. clarkii, 281, 283
    E. elymi, 281, 283, 294, 297
    E. festucae, 105, 232, 277, 279, 281–284, 290–294,
        296–299
    E. glyceriae, 281
    E. sylvatica, 281, 283
    E. typhina, 278–283, 287, 291, 294, 296, 297
    E. yangzii, 282
Epicoccum
    E. nigrum, 124
Eremothecium, 121
Ericaceae, 100
Erwinia
    E. carotovora, 372
Erynia, 124
Erysiphe
    E. cichoracearum, 348, 350, 352
    E. pisi, 375
Erysiphe cichoracearum/Arabidopsis thaliana, 80
Eucalyptus
    E. globules, 244
Eurotium, 121
Evernia
    E. prunastri, 325

F
Fabaceae, 102, 294, 339
Festuca
    F. altissima, 283
    F. argentina, 283
    F. arizonica, 282, 283, 286
    F. arundinacea,
    F. hieronymi, 283
    F. longifolia, 281
    F. magellanica, 283
    F. obtusa, 282
    F. paradoxa, 283
    F. pratensis,

F. pulchella, 281
F. rubra, 86, 281
F. superba, 283
Fulvia fulva, 136
Furia, 124
Fusarium, 15, 120
    F. culmorum, 103
    F. graminearum, 103, 165, 166, 168, 187, 188, 190,
        192–194
    F. oxysporum, 140, 174, 176, 183–186, 190, 191, 194,
        232, 338, 370
    F. solani f. sp. pisi, 138, 184, 187–189, 193
    F. verticillioides, 186

G
Gaeumannomyces graminis var. graminis, 119
Gaultheria
    G. shallon, 101
Geosiphon
    G. pyriforme, 307, 309
Geosyphon
    G. pyriforme, 270
Gibberella, 121
    G. zeae, 122, 123
Gigaspora
    G. margarita, 262, 343
    G. rosea, 262
Glomerella
    G. cingulata, 184
Glomeromycota, 101, 109
Glomus
    G. intraradices, 71, 240, 259, 262, 270
    G. margarita, 109
    G. mosseae, 101, 107, 267, 270
Glyceria
    G. striata, 281
Glycine
    G. max, 343, 347
Golovinomyces
    G. cichoracearum, 371, 372
    G. orontii, 104, 108, 350–352, 369,
        372, 375
Graphis
    G. hematites, 202

H
Hakea
    H. actites, 247
Helminthosporium
    H. maydis, 117
    H. turcicum, 119, 185
Hemileia
    H. vastatrix, 70, 74
Holcus
    H. lanatus, 281
    H. mollis, 281
Hordelymus
    H. europaeus, 281, 283
Hordeum
    H. bogdanii, 283
    H. brevisubulatum, 281, 283
    H. chinense, 77

Hyaloperonospora
    H. parasitica, 14, 143, 174, 177, 178, 344, 347, 350–352,
        369, 370, 372, 373
Hymenomycetidae, 100
Hypocrella
    H. bambusae, 202
Hypomyces, 202

I
Ipomoea
    I. asarifolia, 276, 287
Italian ryegrass (Lolium multiflorum), 282, 286
Itersonilia
    I. perplexans, 128–130

K
Kluyveromyces
    K. lactis, 340
Koeleria
    K. cristata, 281

L
Labyrinthula, 124
Laccaria
    L. bicolor, 71, 179, 238, 240–248, 251, 253
        carbohydrate-degrading enzymes, 246, 250
        expansins, 247
        gene expression, 251
        genome, 238, 240–246, 252, 253
        Ras GTPases, 246
        rho GTPases, 246
        secretome, 243–244
        signal transduction pathways, 244
        small secreted proteins (SSP), 242, 244
        Transposable elements (TE), 240
Laccaria - Pinus, 252
Laccaria - Populus, 248, 250, 252
Laccaria - Pseudotsuga, 248, 252
Laccaria/Pseudotsuga, 250
Laminaria digitata, 343, 347
Lecanora
    L. vinetorum, 322
Leptosphaeria
    L. maculans, 174, 177
Letharia, 310, 322
Linum
    L. usitatissimum, 70, 378
Listronotus
    L. bonariensis, 286, 297
Lolium
    L. arundinaceum, 279, 282
    L. canariense, 282
    L. edwardii, 282
    L. multiflorum (Italian ryegrass), 282, 286
    L. perenne, 105, 232
    L. perenne (perennial ryegrass), 278, 281–283
    L. persicum, 282
    L. pratense, 283, 285
    L. remotum, 282
    L. rigidum, 282
    L. subulatum, 282
    L. temulentum, 277, 282, 294

Lolium arundinaceum (tall fescue), 279, 280, 284–287, 294, 296, 298, 299
Lolium multiflorum (annual ryegrass)
Lophodermium, 105
Lotus
    L. aponicus, 263, 264, 268, 340
Lycopersicon esculentum, 135, 138, 145

M
Magnaporthe
    M. grisea, 29, 78, 118, 123, 124, 143, 147, 157–161, 163, 165, 166, 168, 185, 187, 193, 224, 231, 232, 347, 350, 351, 378
    M. oryzae, 177, 244
    M. poae, 186
Malus floribunda, 144
Mastodia
    M. tessellata, 323
Medicago
    M. truncatula, 262–264, 266, 340, 346
Melampsora
    M. larici-populina, 71, 75, 83
    M. lini, 70, 71, 74, 84, 88, 140, 174, 176, 378
    M. medusae f. sp. *deltoidae*, 75
Melica
    M. ciliata, 282
    M. decumbens, 282
    M. stuckertii, 283
Metschnikowia, 121
Mucor
    M. racemosus, 229
Mycophycias
    M. ascophylli, 307
Mycosphaerella, 122, 187
    M. fijienis, 136
    M. graminicola, 136
    M. musicola, 205
    M. pinodes, 366
Myriogenospora, 275, 276

N
Nectria, 121
    N. haematococca, 184
Neotyphodium
    N. aotearoae, 282, 283, 294
    N. australiense, 282
    N. chisosum, 282
    N. coenophialum, 279, 280, 282, 284–287, 291, 294, 296–299
    N. funkii, 282, 294
    N. gansuense, 282, 283, 285
    N. guerinii, 282
    N. huerfanum, 282
    N. lolii, 278–280, 282–285, 287, 291–294, 296–299
    N. melicicola, 282
    N. occultans, 277, 282, 286, 294
    N. siegelii, 283, 294, 296
    N. tembladerae, 279, 283
    N. typhinum, 282
    N. uncinatum, 280, 283, 285, 294, 296
Neurospora
    N. crassa, 193, 373

Nicotiana, 5, 13
    N. attenuata, 107
    N. benthamiana, 13, 226, 344–348, 351, 352, 374
    N. sylvestris, 107
    N. tabacum, 343, 347
Nicotiana spp., 178
Nigrospora
    N. oryzae, 124
Nostoc, 307, 309, 314, 316, 321, 323

O
Ochromonas
    O. danica, 6
Oidium
    O. lycopersicum, 350
    O. neolycopersici, 118
Ophioparma
    O. ventosum, 321
Ophiostoma, 121
Orchidaceae, 103
Oryza
    O. sativa, 347

P
Passalora fulva, 135, 136
Plectosphaerella cucumerina, 373
Puccinia graminis f. sp. *avenae*, 72
P. graminis f. sp. *tritici*, 71–72, 74, 83, 87–89
P. recondita f. sp. *tritici*, 71
Pseudomonas syringae pv. *glycinea*, 367
P. syringae pv. *Maculicola*, 367
Puccinia syringea pv. *Tomato*, 226
Populus trichocarpa, 71
Puccinia triticina, 71, 82, 83, 87, 89, 89
Parepichloë, 275, 276
Paxillus
    P. involutus, 238, 250, 251
Paxillus/Betula, 248–250, 252
Peltigera, 314, 317, 322, 324
    P. aphthosa, 318, 321, 323
Penicillium, 116
    P. chrysogenum, 117
    P. melinii, 117
Perennial ryegrass *(Lolium perenne)*, 278–280, 283–290, 292, 294, 297–299
Peronospora
    P. tabacina, 118
Petroselinum
    P. crispum, 347, 366
Petroselinum crispum/Arabidopsis thaliana, 343
Pezizomycotina, 309, 312
Phakopsora, 70, 76, 89
    P. meibomiae, 89
    P. pachyrhizi, 70
Phanerochaete
    P. chrysosporium, 241–243, 245
Phialocephala
    P. ortinii, 102
Phleum
    P. commutatum, 279, 283
    P. pratense, 282
Phyllactinia, 121

Phytophthora, 124, 126, 244, 339, 344–348, 350, 351
    P. capsici, 11, 13, 177
    P. cinnamomi, 3, 6, 7, 9, 10, 17, 127
    P. cryptogea, 13, 18, 345
    P. infestans, 3, 4, 6, 7, 9–11, 13–16, 18–19, 29,
        174–178, 344–348, 350–352
    P. nicotianae, 3, 5–9, 11, 14–15, 19
    P. palmivora, 6, 11
    P. parasitica, 7, 14, 343, 347, 350–352
    P. parasitica nicotianae, 343, 347
    P. parasitica var, 347
    P. ramorum, 10, 13–15, 177
    P. sojae, 3, 4, 10–11, 13–19, 174–177, 339, 343, 347
    P. cinnamomi, 3, 6, 7, 9, 10, 17
    P. cryptogea, 13, 18
Pichia
    P. pastoris, 183
Pilobolus
    P. kleinii, 127
Pisolithus
    P. microcarpus, 239, 244
Pisolithus-Eucalyptus, 248, 252
Pisolithus-Populus, 252
Pisolithus/Eucalyptu, 249
Plasmodiophora
    P. brassicae, 124
Plasmodium, 177
    P. falciparum, 14, 178
Plasmopara, 7
    P. halstedii, 13
    P. viticola, 19, 350
Plectosphaerella
    P. cucumerina, 140, 350
Pleospora, 122
Poa
    P. ampla, 282
    P. autumnalis, 283, 294, 296
    P. huecu, 283
    P. pratensis, 282
    P. rigidifolia, 283
    P. silvicola, 282
    P. sylvestris, 282
    P. trivialis, 282
    P. nemoralis, 282
Poaceae, 107, 276
Podospora, 121
    P. anserina, 230–232
    P. dicipiens, 122
Pooideae, 278, 294
Populus
    P. trichocarpa, 71, 238, 252
Prosopis
    P. juliflora, 102
Pseudocercosporella
    P. herpotrichoides, 103, 104
Pseudomonas
    P. syringae, 19, 347, 348, 364–367, 370, 372,
        373, 377
    P. syringea, 225, 226
Pseudomonas syringae pv. *Syringae*, 366
Pseudomonas syringae pv. tomato DC3000 *(Pst)*, 365, 368,
    375, 377
Psychidae, 321

Puccinellia
    P. distans, 282
Puccinia, 72, 74, 86
    P. coronata, 84, 87
    P. graminis, 70–72, 74, 78, 81, 83, 84, 86–89,
        88–89
    P. hemerocallidinis, 80
    P. hordei, 77, 87
    Phakopsora pachyrrhizi, 71, 75, 77, 83, 88–89
    P. sorghi, 88
    P. thlaspeos, 88
    P. graminis f. sp. *tritici*, 117
Pueraria
    P. lobata, 89
Pyrenophora tritici-repentis, 143
Pythium, 7, 9, 13, 124, 126, 127, 372
    P. irregulare, 350
    P. mastophorum, 350
    P. middletonii, 126

R
Ralstonia
    R. solanacearum, 378
Rhamnaceae, 102
Rhizidiomyces, 127
Rhizobium, 262, 263, 265, 271
    R. radiobacter, 108
Rhizocarpon
    R. geographicum, 319
Rhizoctonia, 101, 102
    R. carotae, 307
Rhizopus
    R. microsporus, 109
Rhynchosporium secalis, 140
Roegneria
    R. kamoji, 282

S
Saccharomyces
    S. cerevisiae, 43, 45, 190, 194, 227, 228, 339
Saprolegnia, 125
    S. parasitica, 14
Schizoplasmodium
    S. cavostelioides, 130
Schizosaccharomyces
    S. pombe, 45
Sclerotinia, 121
    S. sclerotiorum, 31, 46, 183, 186, 191, 343
Scolecotrichum
    S. graminis, 202
Scytosiphon
    S. lomentaria, 6
Septoria
    S. lycopersici, 374
Shiraia
    S. bambusicola, 202
Solanaceae, 338
Solanum
    S. demissum, 15, 16
    S. lycopersicum, 364, 374
    S. tuberosum, 347
Sordaria, 121
    S. fimicola, 119, 122

Sorghum, 374

Sphaerobolus, 128

Sphenopholis
> *S. glomerata*
> *S. nitida,* 281
> *S. obtusata,* 281

Stachybotrys
> *S. chartarum,* 116, 117

Stagnospora
> *S. convolvuli,* 202

Stemphylium
> *S. botryosum,* 202

Sticta, 322
> *S. sylvatica,* 309

T

Tapesia
> *T. yallundae,* 103

Taphrina
> *T. deformans,* 121

Thamnolia, 310

Tilletia
> *T. caries,* 128

Trebouxia, 312–315, 317, 319, 321

Trichoderma
> *T. harzianum,* 107
> *T. reesei,* 190
> *T. viride,* 138, 144, 343, 347

Trifolium
> *T. subterraneum,* 102

Tuber
> *T. melanosporum,* 238, 240, 252, 253

Tuber-Tilia, 248, 252

Tuber/Tilia, 250

U

Uromyces, 71, 72, 81
> *U. appendiculatus,* 71, 77, 81, 83
> *U. fabae,* 70, 73, 174, 176–177

U. striatus, 77

U. viciae-fabae, 187, 188

U. vignae, 77, 80, 81, 84, 86, 88

Ustilago
> *U. maydis,* 157, 159, 161–164, 167–169, 184,
> 241–246

V

Venturia, 122
> *Venturia inaequalis,* 142, 144

Verticillium
> *V. albo-atrum,* 341
> *V. dahliae,* 138, 140, 142, 144, 341

Vicia
> *V. faba,* 70, 83, 84, 86, 88

Vitis
> *V. vinifera,* 343, 347, 372

W

Winfrenatia
> *W. reticulata,* 312

X

Xanthomonas
> *X. oryzae,* 340
> *X. campestris* pv. *zinniae* (XCZ), 215
> *X. oryzae* pv. *oryzae,* 377

Xanthoria, 317, 322, 323
> *X. elegans,* 319
> *X. fallax,* 321
> *X. parietina,* 310, 313, 319, 321

Xylosphaera
> *X. furcata,* 124

Z

Zea
> *Z. mays,* 17

Zizyphus
> *Z. nummularia,* 102

Subject Index

A

AAL, 226
    toxin, 224, 226
Abscisic acid (ABA), 368, 370, 373
Actin microfilaments, 11, 17, 18
Acyclic sugar alcohol, 314
Adhesion, 6, 7
Aeciospores, 124
Airflow and drying, 116–119
Air viscosity, 116, 123, 129, 130
Ammonia emissions, 323
Anemochory, 313
Antarctic lichens, 321, 322
Anthraquinones, 321, 322, 324
Antibiotic, 321, 324, 325
Antioxidant defense, 214
Antioxidants, 40
Apicomplexan, 3, 7, 14
Apoplastic continuum, 318, 324
Apothecia, 121
Apple
Appressorium, 7, 9, 10, 316, 317
α-proteobacterium, 108
Arbuscular mycorrhizal fungi, 101
Arbuscular mycorrhizas, 259–271
Arbuscules, 260, 261, 266, 267, 270
Arctic tundras, 313, 319, 323
Arg-Gly-Asp (RGD), 143
Arms race, 364, 377
Asbestos, 321
Ascomata, 121
Ascomycota, 121
Ascospores, 119, 121–123, 127, 129, 131
Aspartyl proteases, 186
Autofluorescence, 321
Avirulence (Avr) genes, 135–140, 142, 146, 148
Avirulence proteins, 12–14
Avr2, 137, 138, 140, 142, 143, 145–148
Avr4, 137–140, 142, 143, 145–150
Avr9, 136, 137, 139, 142, 143, 145–149
AVRAb, 61
Avr9/Cf-9 rapidly elicited (ACRE) gene, 148
Avr4E, 137, 139, 142, 143
Avr-Pita, 147
Avr responsive tomato (ART) gene, 148–149

B

Bagworm, 320
Balanced antagonism, 105
Ballistospore discharge, 128–131

Ballistospores, 116, 118, 128–131
Banana, 204, 205
Barley, 62, 64
Basidiomycete genomes
    gene duplications, 242
    gene families, 242
    genome size, 243
    protein families, 242, 243
    protein kinases, 244
    Ras-like GTPases, 245, 246
Basidiomycota, 118, 128, 130
Bax Inhibitor-1, 105, 107, 222–224, 227, 228, 232
Bcl-2 proteins
    Bak, 223
    Bcl-XL, 223
    Bid, 223
    Ced-9, 226
β-Glucanases, 185, 186, 191
BI-1, 227
Biosynthetic cluster, 208, 210, 213
Biotrophs, 4, 11, 16
Bir1p, 229
Blastidia, 313
Boundary layers, 116, 129
Buller's drop, 128–131

C

Ca^{2+}, 339, 342–346
Calcineurin B-like proteins (CBLs), 343
Calcium, 262–264
Calcium-dependent protein kinases (CDPKs),
    142, 149, 343
Callose, 373
Calmodulin (CaM), 343
Camouflage, 313, 321, 326
cAMP, 190
Carbon, 260, 270
Carbon catabolites, 190
Caspases, 222–226, 228–231
Catalase, 19
Cavendishioid mycorrhiza, 100–102
Cavitated fungal cells, 309
Cavitation, 115–120, 131
cDNA-AFLP analysis, 136, 148–149
Cell death 14, 100, 104–105, 107, 110
Cellobiohydrolase, 185, 186, 190, 194
Cellulases, 182, 185, 186, 188, 190, 194
Cell wall, 181–194, 363, 364, 366, 372–374
Cell wall appositions, 10, 16–18, 20
Cell wall degrading enzymes, 4, 6, 9, 10, 19

Cell wall degrading enzymes (CWDE), 29, 181–194
    cutinase, 38
    endo-polygalacturonase (endo-PG) genes, 39
    pectinases, 39
    pectin methylesterase, 38
Cell wall depolymerases, 182
Cell wall polysaccharides, 325
Cephalodia, 310, 312, 321, 323
Cercosporin, 201, 202, 204–215
Cercosporin facilitator protein (CFP), 205, 206, 210, 211, 214
Cf genes, 135
Cf-2, 135, 137, 138, 142–148
Cf-4, 137–139, 142, 143, 145–150
Cf-9, 137, 139, 142–149
Cf-4E, 137, 139, 143
Cf-Ecp, 137, 138, 140, 142, 143, 146
Cf-Ecp2, 146
Cf-Ecp3, 146
Cf-Ecp5, 146
Cf-Ecps, 140, 142
Cf-9 interacting thio-redoxin (CITRX), 142, 148
Chasmothecia, 121
Chelatinization, 322
Chemotaxis, 4, 6
Chernobyl accident, 323
Chitin, 323
Chitinases, 137–139, 147, 150
Chlorolichens, 312
Choke disease, 279
Chrysotyle fibres, 321
Chytridiomycota, 124, 127
CLAVATA, 340
Cleistothecia, 121
Co-evolution, 29, 45
Coffee, 204, 205
Comparative genomics, 31, 46, 194
Compatible interaction, 136–137
Conidia, 115–120, 123–124, 127
Conidiogenesis, 118
Contact dermatitis, 325
Contractile vacuole, 6, 7
Corn, 204, 205
COS cells, 147
Counterdefence, 11, 13, 18–19
Courgette (zucchini), plants, 52, 53
Cross-talk, 370, 371, 373
Cryptogein, 13, 14
Cuticle, 187
Cutin, 187, 189, 190
Cutinases, 182, 187, 193
Cyanobacterium, 270
Cyanolichens, 312
Cyphella, 309
Cyst germination, 7, 9
Cytochrome P450, 296
    CloA, 290, 291
Cytoplasmatic cavitation, 318
Cytoplasmic aggregation, 16, 18
Cytoskeleton, 264, 265

D
Dark septate endophytes (DSE), 102, 105
9DC, 143, 145

Defence strategies, 46
Desertification, 319, 325
Desiccation, 318, 319, 322–324
Detoxification, 212–215
Diacylglycerol (DAG), 142, 149
Diacylglycerol kinase (DGK), 142, 149
Dictyosomes, 11, 18
Differentiation, 41–44
Dihydrosphingosine (DHS), 226, 230
Dimethylallyltryptophan (DMATrp), 290
    *dma*W gene, 287, 288
Dothideomycetes, 201
Downy mildews
    oomycetes, 57
Drought stress, 309, 318, 319, 322

E
Ecological genomics, 252
Ectomycorrhiza, 100–103, 109, 110
Ectomycorrhizae-specific genes, 252
Ectomycorrhizal (ECM) fungi
    degradation of organic matter, 248
    extracellular fibrillar polymers, 243
    Extramatrical hyphae, 240
    extramatrical mycelium, 246, 248–251
    genotypic variation, 252
    Hartig net, 239, 249
    intra specific variation, 252
    Mantle, 239, 240, 244, 247, 248
    Nitrogen acquisition, 247
    proteases, 246–248, 253
    sheath, 237, 239, 240, 248
    transcript profiling, 248
Ectomycorrhizal symbiosis, 237–253
    The biotrophy-saprotrophism, 246
Ectomycorrhizal transcriptome, 248
Effectors, 4, 12–15, 18, 19, 136–150, 375–379
    apoplastic, 174, 175, 177
    cytoplasmic, 175, 176
    protein, 375, 377
Electrostatics, 116–119
Electrotaxis, 4, 6
Elicitins, 13, 15, 18
Elicitor(s), 4, 12, 13, 15, 18, 192
Endobacteria, 109, 110
Endo-β-1,3-glucanases, 13
Endocyanosis, 307
Endocytosis, 340–341
Endoglucanase, 186, 190, 194
    endoPG, 182–184, 190–192
    exoPG, 182, 183
Endophyte(s), 275–299, 310
Endophytism, 99, 100, 103, 105, 107, 110
Endoplasmic reticulum (ER), 264, 265
Endoxylanases, 185, 192
Epichloë (Epichloae), 275–299
Epidermis, 260, 261, 264, 266
Epiphyte, 275–276
Epoxy fatty acids, 187
Ergot alkaloids, 276, 284–286
    agroclavine, 288, 290
    chanoclavine, 287, 288, 290–291
    elymoclavine, 288–291

ergine, 291
ergonovine, 289–291
ergovaline, 287, 289–291
lysergic acid, 287–291
Ericoid mycorrhiza, 100, 101
Ethylene (ET), 107, 108, 338, 341, 343, 347, 348, 350, 368–373
Ethylene-inducing xylanase (EIX), 142, 144, 145
Extracellular space, 136

F
Faecal pellets, 313, 314
F-box protein, 350
Fixed nitrogen, 309, 323
Flagellin (Flagella), 3–6, 9, 12, 364, 365, 374
Flax plant, 59
Foliicolous lichens, 318–319, 325
Forest ecosystems, 237
Fossil lichens, 312
Functional redundancy, 188, 193, 194
Fungi, 173–178

G
Gene cluster, 206–208, 210, 211
Gene for gene, 70, 71, 72, 84, 87
Gene knockout, 187, 188, 193
Genes, 135–150
Genome sequences, 187, 193
Genomics, 193
Global warming, 325
Glomeromycota, 259, 260–261, 270
Glucanase inhibitor proteins, 13
Glucanases, 137, 150
Glucose, 185, 188, 190, 191, 314
Glutathione, 149, 150
Gooseberry plants, 52, 53
Grape plants, 52, 53, 55
Grey mould disease, 30
Groundsel plant, 60
Growth promotion, 103, 104, 107–109
Guard hypothesis, 147, 377, 378

H
Haustorium, 10–12, 29, 69, 70, 71, 76–82, 84, 85, 88, 175, 176
Hemibiotrophic fungi, 69
Hemibiotrophs, 4, 11, 69, 80
Heterotrimeric G protein, 190
High affinity binding site (HABS), 142, 147
Histone deacetylase, 191
Host cell death, 40
Host defence reactions, 45
Hybrids, 279–283, 298
Hydrogen peroxide (H_2O_2), 203–204
Hydrophobin(s), 137, 317, 318
Hydroxyl radical (OH·), 203, 204
Hypericin, 203
Hypersensitive cell death (HR), 221, 224–227, 231
Hypersensitive response (HR),103, 106, 137–140, 142, 144–149, 366, 367, 369, 378
Hyphal growth, 278, 283, 284
Hyphal polymorphism, 309, 315
Hyphopodium, 260–262, 265, 269

I
Ice nucleation, 322
Incompatible interaction, 137–143
Indole-diterpenes, 276, 284, 285, 292, 294
Induced systemic resistance, 108
Inhibition technique, 314–316
Inhibitor of apoptosis proteins (IAPs), 223–226, 228, 229
Innate immunity, 363, 365
Intragelatinous protrusions, 309, 317
Intraparietal haustoria, 317
Ion
    accumulation, 322
    exchange, 32
    uptake, 325
Irradiation damage, 324
Isidia, 313
Isotope trapping technique, 314

J
JA/ET, 369, 370, 372
Jasmonate, 108, 110
Jasmonic acid (JA), 341, 342, 347–352, 368–373, 375
Jungermannioid mycorrhiza, 100, 101

L
Lead deposition, 322
LeEix1, 142, 144
LeEix2, 142, 144
Lichen
    compounds, 324, 325
    mycobionts, 307, 311, 324
    photobionts, 312–314
Lichen-forming fungi, 307–326
Lichenicolous fungi, 310
Lichenin, 318
Lichenivorous invertebrates, 310, 313
Lichenization, 307, 311, 312, 318
Lichenized ancestors, 312, 324
Light, 203–208, 210–213
Light transmission, 315
Lipases, 182, 187, 188, 193
Lipid, 270
Lipid peroxidation, 204
LOL1, 225
Loline alkaloids, 278, 285–286, 297
    N-acetylloline, 294–295
    N-formylloline, 294–296
Lolitrems, 284–285, 292–294
Loss of lichenization, 312
Lsd1, 225
Lysergyl peptide synthetase (LPS)
    LPS1, 290–291
    LPS2, 290–291
    lpsA gene, 287, 290–291
    lpsB gene, 290–291
LysM, 339–340

M
Maize, 182, 185, 186, 188, 190, 191, 193
Major facilitator superfamily (MFS) transporter, 206, 207, 209–210, 214, 215
MAMPs, *See* Microbe associated molecular patterns

Mannitol, 137, 315
MAPK cascades, 346
    MAPK, 346, 347
    MAPKK, 346
Mastigonemes, 3, 5, 6
Medullary layer, 309, 316–318, 321
Melanin, 316
Melon plant, 60
Membrane, 260–264, 266–268
Membrane damage, 204
Metacaspases, 223, 229, 230, 231
Metal accumulation, 322
Metallophyte lichens, 322
Metalloprotease, 187
Microbe associated molecular patterns (MAMPs),
        12, 15, 337, 338
    BcPG1, 342, 343, 347
    β-glucan, 338, 339, 343, 347
    β-Heptaglucan, 343
    cellulosebinding elicitor lectin (CBEL), 338, 343,
        347, 351, 352
    Chitin, 338–340, 347
    cold-shock proteins, 338
    endopolygalacturonase, 338, 343
    ergosterol, 338, 342
    ethylene-inducing xylanase, 338, 341
    flagellin, 338, 340
    lipopolysaccharides, 338
    Nep1-like proteins, 338, 351
    Pep-13, 338, 343, 345, 347, 351, 352
    sphingolipids, 338
Microbe induced molecular patterns (MIMPs), 337
Microfilament, 264
Microtubule, 4, 11, 17, 18, 264
Mites, 313, 314
Mitochondrion-associated inducer of cell death (AMID),
        229, 231
Mitogen-activated protein kinases (MAPKs), 149, 190, 368
Mlo, 227
Mobile carbohydrates, 314–315
Mousse de chêne, 325
Mutualism, 99–101, 103, 105, 109, 110
Mycosphaerellaceae, 136

N
NADPH oxidase (NoxA), 36, 41, 149, 231, 283, 284, 344, 345
Namib desert, 321, 323
NB-LRRs, 143
Necrotrophic parasites, 69
Necrotrophs, 4, 11, 15, 69, 80
Newton's second law, 122
N-glycosylation sites (NxS/T), 144
NH_4^+, 268, 269
Nitric oxide (NO), 341, 342, 345–346, 365, 367
Nitrogen, 268–270
NO_3^-, 268
Non-expressor of PR genes, 1, 108
Nostoc, 307, 309, 314, 316, 321, 323
Nucleus, 260, 261, 265

O
Obligate biotrophy, 69, 70, 78, 81, 83, 84, 87, 89, 90
Oligogalacturonides, 182, 192

Oomycetes, 3, 7, 9, 12, 14–16, 19, 173–178
    pathogenicity, 173
Oomycota, 116, 126
Orchid mycorrhiza, 99–110
Organic acids, 40
Osmoregulation, 6
Osmotic pressure, 115, 122, 125, 126
Oxidative burst, 284, 344–346, 366, 378
Oxidative stress response, 40, 41, 43, 45

P
Papillae, 373, 375
Parasexuality, 279
Parasitism, 105
Pathogen-associated molecular patterns
    (PAMPs), 137, 139, 145, 337, 364–366, 375, 376, 378
Pathogenesis related genes, 103, 108
Pathogenesis-related (PR) proteins, 137, 142, 149,
    150, 374
Pathogenic fungi, 29, 42, 43, 45
    necrotroph/hemi-biotroph/biotroph, 29
Pathogenicity, 173
Pathogenicity determinants, 38–42
Pathogen recognition receptors (PRRs), 137
Pattern recognition receptors (PRRs), 337–339, 364,
    365, 376
    β-glucan receptor, 338–339
    chitin elicitor receptor kinase (CERK1), 340
    chitin receptor, 339–340
    EIX Receptor, 340–341
Pea plant, 60
Peanut, 204
Pectate lyases, 182, 184, 193, 194
Pectin, 181–184, 190, 192, 193
Pectinases, 182, 190
Pectin methylesterases, 182, 184
Peloton, 102, 103
Penetration resistance, 16, 18
Penguin benches, 323
Peramine, 284–285, 290, 297–298
Perfume, 325
Peripheral cortex, 315, 321, 322
Perithecia, 120–122
Peroxisomes, 17–19
Persimmon, 186
Perylenequinone, 201–215
Pezizomycotina, 309, 312
PGIPs, 192
pH, 191, 192
Phosphate, 266, 267, 269, 270
Phosphatidic acid (PA), 142, 149
Phosphingosine (PHS), 230
Photobiont diversity, 312–313
Photosensitizer, 203, 204, 210, 212–214
Photosymbiodeme, 319, 321
Phytoalexin, 274, 369, 373
Phytohormones, 40
    abscisic acid, 40
    ethylene, 40
Phytotoxic compounds, 29, 39–40
    botrydial, 40, 42, 44
Plant growth promoting bacteria (PGPR), 109
Plant innate immunity, 107

Poikilohydric water relations, 319
Polygalacturonase-inhibiting proteins, 192
Polygalacturonases, 10, 182
Polyketide pathway, 201, 205
Polyketide synthase (PKS), 205–209
Polyol, 314
Polyphosphate, 267
Post-invasion defence, 16
Potato virus X (PVX), 140, 178
Powdery mildews, fungi, 51, 53, 55, 57, 59, 62
Pre-penetration apparatus (PPA), 264–266
Pro-Glu-Ser-Thr (PEST), 144
Protease inhibitors, 13, 18
Proteases, 182, 186, 187, 190, 192–194
Proteomics, 193–194
Protists, 3, 6
Pyridoxine (vitamin B6) pathway, 214
Pyrrolizidines, 294

R
Radionuclides, 322, 323, 325
Raindrops and vibration, 119–120
Rcr3, 138, 146–148
Reactive oxygen intermediates (ROI), 221, 224–232
Reactive oxygen species (ROS), 40–42, 284, 340–342,
        344–345, 348, 366, 367, 369, 378
    H_2O_2, 40–41
    O_2, 41
Receptor-like kinases (RLKs), 143
    BAK1, 341
    BRI1, 340, 341
    CLV, 340
    EFR, 340, 341
    FLS2, 340, 341
    Xa21, 340
Receptor-like proteins (RLPs), 143, 144
Reductive detoxification, 213–214
Regulatory pathways, 210, 211
Reindeer, 311, 313, 323
    and caribou pastures, 323
    lichens, 311, 313
Relative humidity, 117
Remote sensing, 323, 325
Resistance genes, 15, 137, 368, 376–379
Restriction enzyme-mediated insertion (REMI), 206
Reynolds number, 116
Rhizocarpic acid, 321
Rhizomorphs, 239, 249, 252
Rhizosphere, 259–264
Rhododendron plants, 52, 53
RNAi, 268
Rodlet layer, 317, 318, 321
Root colonisation, 100, 102–108
Runway cell death (RCD), 225
Rusts fungi, 57

S
Salicylic acid (SA), 341, 342, 347, 348, 351, 352, 368–373,
        375, 378
Saprophytes, 40
Secondary metabolites, 313–318, 321, 322, 324, 325
Secretome, 173–178
Serine esterases, 187, 188

Serine proteases, 186, 193
Signaling, 262
Signalling factor, 33, 41
Signal transduction pathway
    Ca2+/Calmodulin-Dependent, 44
    cAMP-dependent, 42–43
    cell surface receptors, 44–45
    MAP kinase-controlled, 43
    small G-proteins, 43–44
    two-component histidine kinases, 45
Singlet oxygen ($^1O_2$), 203, 204, 210, 212–214
Small secreted proteins (SSP), 242, 250, 252
Small ubiquitin-related modifier protein (SUMO),
        144, 145
Soil crust
    communities, 319, 321, 325
    lichens, 321
    organisms, 319
Solute translocation, 318, 321
Soredia, 313, 321
Soybean, 204, 205
Space conditions, 319
Sporangiospores, 116
Sporangium, 3, 4, 6, 115–116, 118,
        124–127
Spore discharge, 115, 119–125, 127–131
Spore germination, 9
Spores, 115–131, 259, 260, 262, 263
Stramenopiles, 3, 5, 6, 14, 19
Strigolacton, 262, 266
Substrate induction, 188–190
Subtilisin, 186, 193, 194
Subtractive hybridization, 215
Sugar beet, 204, 205
Superoxide, 203, 204, 213
Surface tension, 115, 118, 120, 126, 128–131
Symbiosis, 103–105, 108, 110, 259, 260,
        262–265
    mutualistic, 275–278, 284
    pleiotropic, 276–277, 279, 299
Symbiotic propagules, 313–314, 319
Syntaxin, 142, 149
Systemic acquired resistance (SAR), 351,
        369–371, 378

T
Teliospores, 116
Terminal velocity, 120, 129
Tobacco, 204, 205, 213–215
Tobacco mosaic virus, 367, 378
Toll interleukin receptor (TIR), 143
Tomato, 183–186, 190–194
Toxin export, 213, 214
Toxins, 201–215
Transcriptome, 6, 7, 19
Transmission
    horizontal, 276–279
    vertical, 276–279, 286, 299
Transporters, 266–270
Triacylglycerols (TAG), 270
Tumor-inhibiting activities, 325
Turgor pressure, 119, 121–128, 131
Type three secretion system (TTSS), 140

U
Uredospores, 116, 117, 120
Usnic acid, 321, 325
UV light, 321, 324

V
Vagrant lichens, 319
Vesicle-associated protein 27 (VAP27),
            147, 148
Viability, 319
Virulence, 31–33, 36–45, 173–178
Virulence factor, 205
Virus-induced gene silencing (VIGS), 148, 149

W
Water expulsion vacuole, 5, 6
Weathering, 321, 322
Wheat plants, 53, 55, 59, 61–64
White rusts (oomycete), 57

X
Xanosporic acid, 213, 215
Xylan, 185, 190
Xylanase inhibitor protein (XIP), 192
Xylanases, 182, 185

Y
Yellow Sigatoka, 205

Z
Zinc finger transcription factor, 190, 191
Zn(II)Cys$_6$ transcriptional activator, 206, 208, 210
Zoochory, 313
Zoospores, 116, 118, 124–127
            encystment, 6, 7, 9
            motile, 3, 4, 6, 7, 9, 19
            motility, 5, 6, 9
            peripheral vesicles, 7, 8
Zygomycota, 116, 117, 127

Printing and Binding: Stürtz GmbH, Würzburg